Transportation Asset Management

Methodology and Applications

Transportation Asset Management

Methodology and Applications

Zongzhi Li

CRC Press
Taylor & Francis Group
Boca Raton London New York

CRC Press is an imprint of the
Taylor & Francis Group, an **informa** business

CRC Press
Taylor & Francis Group
6000 Broken Sound Parkway NW, Suite 300
Boca Raton, FL 33487-2742

First issued in paperback 2020

ISBN-13: 978-1-4822-1052-1 (hbk)
ISBN-13: 978-0-367-65708-6 (pbk)

Library of Congress Cataloging-in-Publication Data

Names: Li, Zongzhi, (Professor of transportation engineering), author.
Title: Transportation asset management : methodologies and applications / by Zongzhi Li.
Description: Boca Raton ; London : CRC Press, [2018] | Includes bibliographical references and index.
Identifiers: LCCN 2017030456| ISBN 9781482210521 (hardback : alk. paper) | ISBN 9781315117966 (ebook)
Subjects: LCSH: Transportation--Planning. | Transportation--Forecasting. | Transportation--United States--Management. | Infrastructure (Economics)--United States--Management. | Transportation--Management--Economic aspects--United States.
Classification: LCC HE147.5 .L5 2018 | DDC 388.068--dc23
LC record available at https://lccn.loc.gov/2017030456

Visit the Taylor & Francis Web site at
http://www.taylorandfrancis.com

and the CRC Press Web site at
http://www.crcpress.com

To my parents, uncles, and family who stimulated and encouraged my early work in transportation systems

Contents

List of symbols

$AAWDT$	Annual average weekday daily traffic
$a, b, c...$	Coefficients
$\hat{a}, \hat{b}...$	Coefficients estimation
a_k	Mode specific variable
AB	Agency benefits
AC	Average costs
B	Benefits
C	Costs
C_P^{const}	Total pavement construction cost
$C_{P,ESAL}^{const}$	Load-related portion of pavement construction cost
$CESALs$	Cumulative ESALs
CF	Crash frequency
c_{ij}	Size measure
c_l	Capacity of link l
Cov	Covariance
CR	Crash rate
CRF	Cost recovery factor
CS	Crash severity
CT	Crash type
D	Difference between ...
D	Pavement thickness
DHV	Directional hourly volume
DM	Dummy variable
DT	Delivery truck time
e	Superelevation
E_{ξ_L}	Mathematical expectation of the recourse function in stage L
E_P	Expected value of payoff
EUA	Equivalent uniform annual
$EUAAC$	Equivalent uniform annual agency costs
$EUAUC$	Equivalent uniform annual user costs
FG	Focus gain
FL	Focus loss
FTS	Fraction of time for shipping
$F_{w,link}$	Cross-section factor
G	Growth rate
GH	Generalized Hartley measure

I	Discount rate
IC	Inventory cost
K_l	Capital expenditure
L	Length of the relevant highway segment
$LCAC$	Life-cycle agency costs
$LCUC$	Life-cycle user costs
$L_{eq,t}$	Equivalent sound level
LOF	Layover time factor
MC	Marginal costs
ME	Maintenance expenditure
MEV	Million entering vehicles
MV	Maintenance volume
$MVMT$	Million vehicle miles of travel
N	Numbers of ...
N	Pavement design service life
OC	Operating costs
P	Pollutant type p
PCE	Passenger car equivalency
$PMPH$	Passenger miles per hour
PW	Present worth
$P(\theta_j\|A_i)$	Conditional probability
Q	Average annual ESALs
R	Radius
RD	Residential density
R_{ij}	Bayesian risk expectation
RM	Route-miles
R_{min}	Lowest total risk of expected project benefits
RR	Railroad rate
RT	Running time
RU	Risk and uncertainty
S	Crash severity level
S	Sample standard deviation
s^2	Sample variance
SB	Salvage benefits
SFG	Standardized focus gain
SFL	Standardized focus loss
t	Time
T	Component rehabilitation or replacement interval
TC	Travel time cost
TL	Tangent section length
TR	Tangent runout length
TR	Truck rate
TT	Travel time
v_l	Traffic volume on link l
$u(A_i, \theta_j)$	Corresponding utility value
UB	User benefits
UC	Unit cost, or User costs
$UTTC$	Unit travel time cost
$UVCC$	Unit vehicle crash cost

UVEC	Unit vehicle emission cost
UVOC	Unit vehicle operating cost
VC	Vertical curvature
VCC	Vehicle crash costs
VDC	Vehicle delay costs
VEC	Vehicle emission costs
VH	Vehicle hours
VHPD	Vehicle hours per day
VHPH	Vehicle hours per hour
VOC	Annual vehicle operating costs
VMPH	Vehicle miles per hour
VMT(t)	Total vehicle miles of travel for the pavement in service for *t* years from the base year 0
VMT(N)	Total vehicle miles of travel in *N*-year pavement service life-cycle
w	Local wage rate
W	Pavement width
WKT	Walking time
WSI	Weather severity index
WTT	Waiting time
WZ	Work zone
x_{ij}	Decision variable indicating whether the item *i* in class *j* is taken
$\bar{x}$	Sample mean
α	Value of travel time; Coefficients
β	Shadow price of early arrival
γ	Shadow price of late arrival
Δ	Differences
εi	The random error term
ξ	The vector of independent variables
ξ_L	Randomness associated with budgets in stage *L* and decision space
η	The vector of dependent variables in a SEM diagram
λ	Non-negative real Lagrange multipliers
$\tilde{\lambda}_l$	Context-dependent of normalized informational importance for performance criterion *l*
μ	Mean value
π	Pavement condition (PCI or IRI)
σ	Standard deviation
σ^2	Variance
φ	Passenger-car equivalent (PCE) index
ω	Odds ratio
ρ	Annualized factor of the capital expenditure *K*
Σ	Summation symbol

A	α	*alpha*
B	β	*beta*
Γ	γ	*gamma*
Δ	δ	*delta*
E	ε	*epsilon*
Z	ζ	*zêta*

Η	η	*êta*
Θ	θ	*thêta*
Ι	ι	*iota*
Κ	κ	*kappa*
Λ	λ	*lambda*
Μ	μ	*mu*
Ν	ν	*nu*
Ξ	ξ	*xi*
O	o	*omikron*
Π	π	*pi*
Ρ	ρ	*rho*
Σ	σ, ς	*sigma*
Τ	τ	*tau*
Υ	υ	*upsilon*
Φ	φ	*phi*
Χ	χ	*chi*
Ψ	ψ	*psi*
Ω	ω	*omega*

List of abbreviations

2D	Two-dimensional
3D	Three-dimensional
3R	Resurfacing, rehabilitation, and restoration
4R	Resurfacing, rehabilitation, restoration, and reconstruction
AAA	American Automobile Association
AADT	Annual average daily traffic
AASHO	American Association of State Highway Officials
AASHTO	American Association of State Highway and Transportation Officials
AC	Avoidance costs
ADA	Americans with Disabilities Act
ADT	Average daily traffic
AHP	Analytic hierarchy process
AI	Alignment index
AI	Artificial intelligence
ANN	Artificial neural networks
ANOVA	Analysis of variance
APC	Automatic passenger counters
APTA	American Public Transportation Association
AREMA	American Railway Engineering and Maintenance-of-Way Association
ARRB	Australian Road Research Board
ARS	Average rectified slope
ASTM	American Society for Testing and Materials
ATIS	Advanced traveler information system
AVC	Average variable cost
AVI	Automatic vehicle identification
B/C	Benefit-to-cost
BCA	Benefit-cost analysis
BCR	Benefit-to-cost ratio
BFS	Basic feasible solution
BHD	Bethlehem Highway Department
BOS	Bus-on-shoulder
BPA	Bonneville Power Administration
BPR	Bureau of Public Roads
BRM	Base routine maintenance
BRP	Binary recursive partitioning
BRT	Bus rapid transit
BSP	Bus signal priority
BSR	Bridge sufficiency rating

BTS	Bureau of Transportation Statistics
CAA	Clean Air Act
CALTRANS	California Department of Transportation
CART	Classification and Regression Trees (software)
CBD	Central business district
CBR	Case-based reasoning
CBR	California bearing ratio
CCD	Charge-coupled device
CD	Compact disc
CDF	Cumulative distribution function
CE	Certainty equivalent
CEB	Comite Euro-International du Beton (European Concrete Committee)
CEC	California Energy Commission
CFCs	Chlorofluorocarbons
CFF	Crash frequency factor
CFR	Code of Federal Regulations
CH_4	Methane
CIE	Commission Internationale de l'Eclairage (International Commission on Illumination)
CMEM	Comprehensive Modal Emission Model
CMF	Crash modification factor
CNG	Compressed natural gas
CO	Carbon monoxide
CO_2	Carbon dioxide
COPERT	COmputer Programme to calculate Emissions from Road Transport
CoRe	Commonly recognized
COV	Coefficient of variation
CP	Carpool
CPI	Consumer price index
CPR	Concrete pavement rehabilitation
CR	Commuter rail
CR	Condition rating
CR	Consistency ratio
CRC	Continuously reinforced concrete
CSF	Crash severity factor
CTP	Cumulative traffic passages
CV	Contingent valuation
CWR	Continuously welded rail
dB(A)	Decibels adjusted
DBMS	Database management system
DC	Delay costs
DC	Distribution center
DD	Driveway density
DEA	Data envelopment analysis
DF	Damage factor
DFA	Damage function approach
DGPS	Differential Global Positioning System
DH	Deck rehabilitation
DHDP	Deck rehabilitation with full depth patching
DMI	Distance measuring instrument

DMUs	Decision-making units
DOD	Department of Defense
DOT	Department of Transportation
DP	Deck replacement
DP	Dynamic programming
DPSH	Deck replacement and superstructure rehabilitation
DPSHW	Deck replacement, superstructure rehabilitation, and bridge widening
DSH	Deck and superstructure rehabilitation
DSHW	Deck and superstructure rehabilitation and bridge widening
DV	Detected values
DVD	Digital video disc
EB	East bound
EB	Empirical Bayesian
EF	Emission factor
EIS	Environmental impact statement
ELECTRE	ELimination Et Choix Traduisant la REalité (ELimination and Choice Expressing Reality)
ELECTRE-TRI	ELECTRE Tree
EOL	Expected value of opportunity loss
EOQ	Economic order quantity
EPA	U.S. Environmental Protection Agency
EPRI	Electric Power Research Institute
ER	Emission rate
E-R	Entity-relationship
ERFs	Exposure-response functions
ESALs	Equivalent single axle loads
EUAC	Equivalent uniform annual costs
EUACF	Equivalent uniform annual cost factor
EUDC	Equivalent uniform daily costs
EV	Entering vehicles
FAHA	Federal-Aid Highway Act
FHWA	Federal Highway Administration
FTA	Federal Transit Administration
FWD	Falling weight deflectometer
FYA	Flashing yellow arrow
GDF	Geoffrion–Dyer–Feinberg interactive algorithm
GDP	Gross domestic products
GI	Geometry index
GIS	Geographic information system
GPH	Gallon per hour
GPM	Gallon per mile
GPR	Ground penetrating radar
GPS	Global positioning system
GVW	Gross vehicle weight
GWP	Global warming potential
GWR	Geographically weighted regression
HA	Highly annoyed
HBO	Home-based-other
HBW	Home-based-work
HCM	Highway Capacity Manual

HERS-ST	Highway Economic Requirements System- State Version
HFC	Hydro fluorocarbon
HMA	Hot-mix asphalt
HOT	High-occupancy toll
HOV	High-occupancy vehicle
HP	Hedonic price
HPM	Hour per mile
HPMS	Highway Performance Monitoring System
HRB	Highway Research Board
HRT	Heavy rail transit
HSM	Highway Safety Manual
HT	Heavy truck
HVAC	Heating, ventilation, and air conditioning
ICC	Intra-class correlation
ID	Identification
IDAS	ITS Deployment Analysis System
IE	Impact-echo
INDOT	Indiana Department of Transportation
IP	Integer programming
IQR	Interquartile range
IR	Inventory rating
IRI	International roughness index
ISTEA	Intermodal Surface Transportation Efficiency Act
ITS	Intelligent transportation system
IVE	International Vehicle Emissions
IVTT	In-vehicle travel time
KKT	Karush–Kuhn–Tucker conditions
KP	Knapsack problem
KPM	Key performance measure
LED	Light-emitting diode
LEF	Load equivalency factor
LOS	Level of service
LP	Linear programming
LRI	Likelihood ratio index
LRMs	Linear referencing methods
LRS	Linear reference system
LRT	Light rail transit
LTL	Less than truckload
LVW	Loaded vehicle weight
LWR	Lighthill, Whitham, and Richards shockwave model
MAD	Mean absolute deviation
MAP-21	Moving Ahead for Progress in the 21st Century Act
MARR	Minimum attractive rate of return
MAUT	Multi-attribute utility theory
MAV	Multi-attribute value
MC	Marginal cost
MCDM	Multi-criteria decision-making
MCKP	Multi-choice knapsack problem
MCMDKP	Multi-choice multidimensional knapsack problem
MDKP	Multi-dimensional knapsack problem

ME	Margin of error
ME	Mean error
MFL	Magnetic flux leakage
MGT	Million gross tons
ML	Maximum likelihood
MLS	Maximum load sections
MOVES	Motor Vehicle Emission Simulator
MPG	Miles per gallon
MPO	Metropolitan Planning Organization
MRI	(Rail track) Maintenance/renewal interval
MSE	Mean-squared error
MT	Medium truck
MUTCD	Manual on Uniform Traffic Control Devices
N_2O	Nitrous oxide
NAAQS	National Ambient Air Quality Standards
NAC	Noise abatement criteria
NAL	New axle load
NBI	National Bridge Inventory
NBIS	National Bridge Inspection Standards
NCA	Noise Control Act
NCHRP	National Cooperative Highway Research Program
NDSI	Noise depreciation sensitivity index
NDT	Non-destructive testing
NEC	New England Council
NEPA	National Environmental Policy Act
NHB	Non-home-based
NHTSA	National Highway Traffic Safety Administration
NLP	Nonlinear programming
NMHC	Nonmethane hydrocarbon
NO_X	Nitrogen oxides
NP	Nondeterministic polynomial
NPSC	Nevada Public Service Commission
NPW	Net present worth
NTPEP	National Transportation Product Evaluation Program
NYDOT	New York State Department of Transportation
O_3	Ozone
O-D	Origin-destination
OLS	Ordinary least squares
OR	Odds ratio
OPC	Out-of-pocket cost
OPI	Overall pavement index
OS/OW	Oversize/overweight
OVTT	Out-of-vehicle travel time
P-A	Production-attraction
PCC	Portland cement concrete
PCI	Pavement condition index
PCR	Pavement condition rating
PDF	Probability density function
PDO	Property damage only
PM	Particle matters

PMF	Probability mass function
PMS	Pavement management system
PPP	Public-private partnership
PROMETHEE	Preference Ranking Organization Method for Enrichment Evaluations
PSD	Prevention of significant deterioration
PSI	Present serviceability index
PSI	Potential for safety improvement
PSR	Present serviceability rating
PV	Peak vehicles
PW	Present worth
PWF	Present worth factor
QA	Quality assurance
QUALIFLEX	Qualitative Flexible Multiple Criteria Method
RC	Reinforced concrete
RD	Rut depth
RE	Relative error
RFID	Radio-frequency identification
RH	Relative humidity
RHR	Roadside hazard rating
RHS	Right-hand side
RM	Retroreflectometer
RMEE	Reference mean energy emission
RMSE	Root mean square error
ROR	Rate of return
ROR	Run-off-road
ROW	Right-of-way
RPMs	Raised retroreflective pavement markers
RTRRMs	Response type road roughness meters
RWD	Rolling deflectometer
SA	Selective availability
SAFETEA-LU	Safe, Accountable, Flexible, Efficient Transportation Equity Act: A Legacy for Users
SCC	Standard cost categories
SCI	Sidewalk condition index
SD	Standard deviation
SE	Standard error
SEM	Structural equation model
SH	Safety hardware
SHD	Strategic highway designation
SHRP	Strategic Highway Research Program
SI	Safety index
SMARTS	Sign management and retroreflectivity tracking system
SN	(Pavement) Structure number
SO	System optimum
SO	System optimal
SO_2	Sulfur oxides
SP	Stochastic programming
SPF	Safety performance function
SPL	Sound pressure level
SPW	Superstructure replacement and bridge widening

SQL	Structured query language
SSCR	Sideway surface condition rating
SSE	Sum of squared errors
SSPE	Sum of squared pure errors
SSR	Sum of squared residuals
SST	Total sum of squares
SSW	Superstructure strengthening and bridge widening
ST	Structure type
STEAM	Surface Transportation Efficiency Analysis Model
STRAHNET	Strategic Highway Network
STURAA	Surface Transportation and Uniform Relocation Assistance Act
SUV	Sports utility vehicle
SV	Slope variance
SV	Superelevation variance
SWT	Surrogate worth trade-off method
TAZ	Traffic analysis zone
TC	Total cost
TCI	Track condition index
TDM	Travel demand management
TEA-21	Transportation Equity Act for the 21st Century
TEF	Traffic exposure factor
TGI	Track geometry index
TI	Twist index
TIFIA	Transportation Infrastructure Finance and Innovation Act
TL	Truckload
TM	Track-miles
TNM	Traffic noise model
TOD	Transit-oriented development
TQI	Track quality index
TRB	Transportation Research Board
TRLHP	Track road load horsepower
TSD	Total stopped delays
TSF	Traffic safety features
TSP	Total suspended particles
TTC	Travel time costs
TTI	Texas Transportation Institute
UE	User equilibrium
UI	Unevenness index
UIC	Union Internationale des Chemins de fer (International Union of Railways)
USACE	U.S. Army Corps of Engineers
UTM	Universal Transverse Mercator
UV	Ultraviolet
V/C	Volume-to-capacity
VIG	Visual interactive goal programming
VKT	Vehicle kilometers of travel
VKT	Vehicle kilometers travelled
VM	Vehicle miles
VMSs	Variable message signs
VMT	Vehicle miles of travel
VMT	Vehicle miles travelled

VNC	Vehicle noise cost
VOC	Vehicle operating costs
VP	Vanpool
VSP	Vehicle specific power
WATIB	Washington State Transportation Improvement Board
WB	West bound
WERD	Western European Road Directors
WHO	World Health Organization
WLS	Weighted least squares
WTA	Willingness to accept
WTP	Willingness to pay

Preface

Transportation enables safe and efficient movement of people and goods. In today's world, it is one of the primary reasons for the economic development of a region, and eventually the social and general growth of the region. The increasing population and ever-growing demand for safe and efficient passenger travel and the increased size and tonnage of freight shipments have significantly outpaced transportation development, leading to accelerated deterioration of physical facility conditions and degradation of system usage performance, such as mobility, safety, and environmental impacts. This leads to transportation agencies seeking effective means of transportation demand and supply management under limited budgets and resources to achieve system efficiency, effectiveness, and equity. More specifically, transportation agencies are tasked to provide users and nonusers with well-maintained facilities, good accessibility, and high mobility, along with safe, energy savings, and environmentally friendly transportation that will also support quality of life for the entire population. Typical solutions for achieving the above goals include travel demand management; transportation capacity expansion and multimodal integration concerned with integration of passenger and freight modes, auto and transit modes, and transit submodes (bus, bus rapid transit, fixed guideway, and shared transit modes) for passenger travel, as well as truck and rail modes for freight shipments; and efficient capacity utilization. Technological advancements in such fields as sensors, telecommunications, material science, computer, mathematical statistics, operations research and management science, and behavior science for collection, transmission, processing, storing, and analysis of data, as well as displaying information on transportation facilities, vehicles, and users/nonusers or freight have made it possible to adequately address this challenging task.

U.S. TRANSPORTATION DEVELOPMENT HISTORY

The vast land mass of the U.S. and its extensive coastlines are the factors for the usage of not only surface transportation but waterways as well. During the late 18th century, the primary modes of transportation in the country were horses for inland transportation and waterways as the majority of the population was concentrated along the coastlines and harbors, natural or manmade. Eventually, the inland surface transportation requirements and movements dictated expenditures of highways, but their cost intensive nature resulted in private turnpikes until the early 19th century, though these were not usable year-round. During the same time, economic expansion forced suppliers to transport goods faster, leading to the building of inland navigable waterways and canals throughout the country. However, the eventual advent of the railroads in the early 19th century decreased the use of waterways for both passenger and freight transportation, as the railroad was open to use year-round. By the end of the 19th century, automobiles coupled with industrialization led to a high demand for automobiles and, in turn, motorable highways. The development of the transportation system in the country was heralded by inventions and innovations in the field of automotive engineering, the industrial revolution, and two world wars, which is evident by its massive national highways, transcontinental railroads, canals, urban transit systems, major airports, and petroleum and natural gas pipelines. In the last 150 years, railroads, highways, transit lines, ports, and airports have helped to increase the range of cities and reduce the isolation of rural areas. They have brought the nation closer together.

IMPORTANT U.S. TRANSPORTATION LEGISLATION AND NEED FOR ASSET MANAGEMENT

The U.S. government aims to solve transportation problems systematically and progressively with the aid of legislation and enforcement by means of mandates in lieu of federal funding. In 1987, the Surface Transportation and Uniform Relocation Assistance Act (STURAA) was introduced that most notably allowed the speed limits on the rural Interstate highways to be increased to 65 mph. This was followed by the 1991 Intermodal Surface Transportation Efficiency Act (ISTEA), which initiated major changes in transportation planning and policy for the first time by U.S. legislation. The primary objective of the legislation was to develop an environmentally responsible, economically efficient, and intermodal transportation system for the movement of people and goods. It made sure that collaborative planning reached fruition by means of an intermodal approach to funding as well and gave additional powers to metropolitan planning organizations (MPOs) to develop multimodal transportation plans. One of the most important facets of this federal legislation was to have airbags in all passenger vehicles and light trucks manufactured post September 1998, thus to have airbags as standard equipment for drivers and the front seat passengers, which ensured safety of system users. Once the legislation ensured a holistic system perspective in terms of planning for the national transportation system, the next step was to fund the plans and ensure their fruition by means of enabling the organization with powers and funds.

The next piece of legislation, the Transportation Equity Act for the 21st Century (TEA-21), was enacted in 1998. It aimed at rebuilding the U.S. by means of streamlined planning in an environmentally responsible way, while improving user safety and creating jobs and opportunities. The legislation provided funding for investment in surface transportation while ensuring balanced investment among highways, transit, intermodal facilities, and Intelligent Transportation Systems (ITS) along with strong state and local flexibility in the use of the funds. The TEA-21 legislation was followed by the 2005 Safe, Accountable, Flexible, Efficient Transportation Equity Act: A legacy for Users (SAFETEA-LU), signed into public law. It extended and expanded the vision for its predecessor by providing 224.1 billion U.S. dollars of funding, the largest surface transportation investment in the nation's history. It not only addressed challenges of its time, such as safety improvements, congestion mitigation, intermodal connectivity, freight movement efficiency, and environmental impacts, but also laid the foundation for meeting future challenges.

With the budget share for spending on the U.S. transportation system over the years unable to close the gap of the deficit of system needs, it has caused increased deterioration of transportation facility conditions and system usage performance levels. The higher deterioration rates, in turn, increase the system deficit. The nation renewed its legislative outlook once again in 2012 by means of the Moving Ahead for Progress in the 21st Century Act (MAP-21) with an objective of decreasing the system deficit over the period 2012–2022. One of the major changes initiated by this piece of legislation, in line with the topics discussed in this book, is implementing performance-based management by all transportation agencies in the country to secure federal funding for pavements and bridges on the national highway system.

The traditional methods of working and institutional hierarchy of transportation agencies have seen a dramatic change over the past several decades to meet the eventual vision of performance-based planning, programming, design, construction, and operations management of the nation's transportation system that calls for inter- and intra-agency coordination and collaboration in a holistic manner. This is the essence of practicing transportation asset management.

TARGET AUDIENCE

This book will discuss various areas within transportation systems engineering and how to streamline transportation asset management concepts and methodology, as well as applications for estimating and evaluating the demand and supply required for the transportation system and monetizing or valuating them in support of efficient, effective, equitable, and accountable resource allocation decisions. This will help achieve the sustainable development of the transportation system.

The chapters focus on the systematic transportation asset management decision-making process, ranging from multimodal transportation asset management policy goals, objectives, performance measures, performance modeling, needs assessment, project evaluation, project selection and programming, and feedback, to institutional issues.

The use of this book as a textbook is to teach graduate students about the concepts and methodology and develop their skills by means of examples and computational studies. This book can be utilized by researchers to validate the proposed methodology and develop refined methodologies to advance transportation asset management. This book can also be adopted by practitioners to refine the existing and implement new asset management initiatives.

SPECIAL THANKS

The work presented in this book is a result of efforts undertaken by this author, his mentors, close colleagues, and students for the last two decades. In an effort to formalize this thinking, a large number of professors from Purdue University, West Lafayette, Indiana, are particularly noteworthy. Professors Kumares C. Sinha, Srinivas Peeta, and Patrick S. McCarthy (now at Georgia Institute of Technology) taught me transportation economics, demand forecasting, systems analysis and evaluation, and statistical and econometric methods. Professors Kumares C. Sinha, Jon Fricker, Fred L. Mannering (now at the University of South Florida), Tom F. Sparrow, and Thomas L. Morin, as well as Jennifer K. Ryan (now at the University of Nebraska) further taught me and enhanced my understanding of urban transportation planning, infrastructure management, transportation data analysis and performance modeling, discrete optimization, multicriteria decision-making, risk and uncertainty modeling, and supply chain management and inventory control. In parallel, Dr. Matthew Karlaftis (deceased), Dr. Samuel Labi, and Dr. Konstantinos Kepaptsoglou gave me valuable insights into research on multimodal travel demand forecasting, performance modeling, project evaluation, and optimization of resource allocation essential to transportation asset management.

In closing, I would like to thank my colleagues, professors David Arditi, Jotin Khisty, Jamshid Mohammadi, and P.S. Sriraj of Illinois Institute of Technology (IIT) for their help in the preparation of this manuscript. Special thanks also go to Dr. Hubert Ley of Argonne National Laboratory, Dr. Adrian Moore and Dr. Samuel Staley of the Reason Foundation, and some of my students, including Taqwa Alhadidi, Sang Hyuk Lee, Yongdoo Lee, Yi Liu, Xi Lu, Pu Lü, Mohammad Neishapouri, Harshingar Patel, Tung Truong, Lu Wang, Ji Zhang, and Bei Zhou for their contributions to the ideas in this manuscript. The long-term support of the Galvin Mobility Initiative, and Sustainable Transportation and Infrastructure Research Center and Transportation Engineering Laboratory at IIT is also gratefully acknowledged.

Zongzhi Li
Chicago, Illinois
June 2017

Author

Zongzhi Li, Ph.D. is a professor and director of Sustainable Transportation and Infrastructure Research (STAIR) Center and Transportation Engineering Laboratory, at Illinois Institute of Technology, Chicago, Illinois, USA.

Chapter 1

Introduction

1.1 OVERVIEW OF A MULTIMODAL TRANSPORTATION SYSTEM

A multimodal transportation system comprises three components: physical transportation facilities, vehicles, and users/nonusers corresponding to passenger travel and freight shipments. The mission of a transportation system is to ensure safe and efficient movements of people and goods, which are predominantly achieved by various types of vehicles. The use of vehicles is constrained by availability of physical transportation facilities that have limited usage capacities. Thus, the three components—facilities, vehicles, and users/nonusers—are interdependent on one another, and should, therefore, be addressed holistically.

1.1.1 Multimodal transportation facilities

A transportation system involves various types of physical facilities, such as roads, bridges, drainage structures, tunnels, traffic control devices and safety hardware (signs, lighting, signals, pavement markings, guardrails, and crash cushions), roadside furniture, transit stops and stations, rail tracks, navigable waterway channels and ports, and pipelines. Each facility plays a unique role in providing transportation services. For instance, roads, bridges, drainage structures, and tunnels carry traffic; traffic control devices and safety hardware foster smooth traffic flow and safety; and roadside furniture enhances convenience and aesthetics. For passenger travel or freight shipments, one type of transportation facility may be used by different types of vehicles or multiple types of transportation facilities may be used by one type of vehicle. Facilities used by vehicles for people or goods movements could be largely grouped according to travel modes.

1.1.2 Transportation vehicles

For passenger travel, typical travel modes include auto (private, taxi, rideshare), transit (bus and bus rapid transit, streetcar and light rail, subway and commuter rail, ferry), air, nonmotorized (bike, e-bike, scooter), and walking modes. Vehicle types may include automobiles, buses, streetcars, light rail vehicles, subway/metro trains, ferryboats, airplanes, bikes, e-bikes, and scooters. For freight shipments, key travel modes include trucking, freight rail, shipping, cargo air, and pipeline modes. Vehicle types generally consist of trucks, freight trains, ships, and cargo planes.

1.1.3 Passenger and freight movements

The service provided by a transportation system is generally related to passenger travel and freight shipments. For a given transportation system, the two categories of service occur simultaneously. Ensuring safe and efficient travel of people and shipping of goods from respective origins to destinations are at the core of transportation service.

1.2 TRANSPORTATION SYSTEM CHARACTERISTICS

1.2.1 Interdependent system components

The facility, vehicle, and user/nonuser components that make up a transportation system are interdependent on one another. People and goods are moved by different types of vehicles. The shared facilities used by vehicles have limited capacities. In temporal and spatial dimensions, the travel demand of people and goods for a certain area is assigned to traffic zones and travel modes, and then converted to traffic flow on routes. To provide acceptable levels of service, people travel and goods movements, travel mode, vehicles, and facilities need to be adequately matched. Traffic impacts could be analyzed to assess levels of service. Meanwhile, vehicle usage and facility performance could also be evaluated.

1.2.2 System component life cycle considerations

As part of dealing with the interdependent transportation system components (facilities, vehicles, users/nonusers) holistically, the useful service life cycles of system components need to be taken into consideration to ensure they are treated in an equitable manner. This is because different system components have different life cycles. Different types of facilities, vehicles or users/nonusers in the context of each system component also exhibit differences in their service lives. For a specific facility, its delivery process goes through planning, programming, design, construction, and in-service steps. The useful service life cycle is typically defined as the time interval between two consecutive construction interventions. During the service life cycle, repetitive maintenance and repair treatments are implemented in a coordinated manner to ensure the designed service life is achieved. Likewise, for a specific vehicle, the delivery process involves planning, design, manufacturing, and in-service steps. It also maintains a life span from production to retirement and will include repair and maintenance while in service. Although the entire life span of a person or a piece of goods may not be fully associated with the transportation system, the age of the person or the piece of goods does affect the performance of vehicles and, in turn, the performance of facilities. At any certain point in time, people and goods of various ages are using a transportation system. Similarly, a wide range of age distributions of people and goods may pass by one location of a transportation system over a certain period. Moreover, people or goods in the same age group may behave differently. Considering service life cycles of facility, vehicle, and user/nonuser components will help assess their temporal and spatial interdependency, interactions, and performance in a more rigorous manner.

1.2.3 Multidimensional impacts and multiple performance goals

With facility, vehicle, and user/nonuser components considered holistically over their respective service life spans, the impacts of the three system components in time and space dimensions are multidimensional in nature. The impacts can be classified into direct and indirect impacts. The direct impacts are directly relevant to the three components of a transportation system in time and space dimensions. The indirect impacts are related to those caused by the transportation system, such as economic growth triggered by transportation improvements. The overarching goals of managing the transportation system are to mitigate adverse impacts and promote positive impacts to achieve efficiency, effectiveness, and equity. For the facility component, specific performance management goals include preserving physical facility conditions at or above a desired level and minimizing transportation agency costs. For vehicle and user/nonuser components that are associated with system usage, the specific performance management goals are concerned with minimizing user/nonuser costs, ensuring certain levels of service for vehicle/user/nonuser mobility and safety, and reducing energy consumption and environmental impacts. The indirect performance management goals are geared toward job creation and freight shipment quantities in support of economic prosperity. The relative importance of various performance management goals among facility, vehicle, and user/nonuser components, and the goals dealing with different types of facilities, vehicles, and users/nonusers for each system component may be treated differently. Also, the relative importance may change over time. Tradeoffs among the goals need to be considered in the process of decision-making.

1.3 TRANSPORTATION ASSET MANAGEMENT PROCESS

Transportation asset management can be defined as a strategic, systematic process of constructing, preserving, expanding, and operating the transportation system to achieve a cost-efficient, effective, and equitable system performance. It focuses on engineering and business practices for predicting the travel demand that directly affects the performance of physical facilities and system usage. It may rely on innovative financing to maximize the budget and allocating the available funding to address the needs for performance improvements in a holistic and proactive manner (Li et al., 2002a,b; Li and Sinha, 2004). Figure 1.1 shows key components of the systematic transportation asset management process that generally accomplish the following functions:

- Establish transportation performance management goals and performance measures
- Carry out travel demand and traffic flow predictions

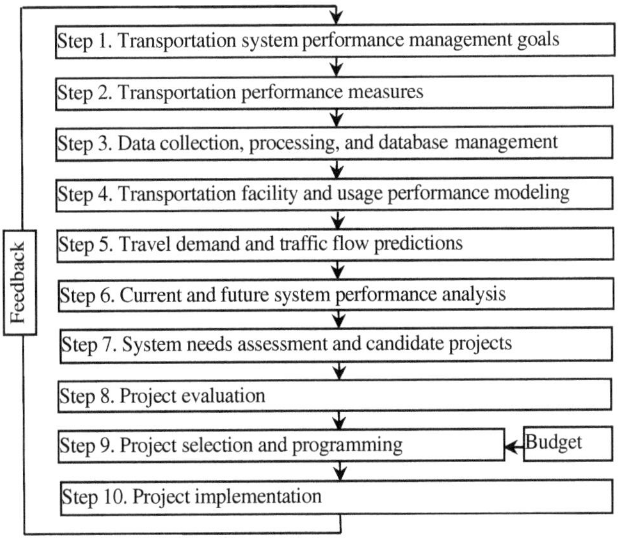

Figure 1.1 Framework of transportation asset management.

- Conduct transportation facility and system usage performance modeling
- Analyze performance trends according to predicted traffic and non-traffic conditions
- Assess needs for facility and system usage performance improvements, and recommend investment alternatives
- Evaluate the economic feasibility of investment alternatives
- Identify the best sub-collection of economically feasible investment alternatives under budget and other constraints
- Implement selected investment alternatives
- Provide feedback to refine the subsequent cycle decisions

1.3.1 Transportation goals, objectives, and performance measures

A policy goal is a general statement of a desired state or ideal function of a transportation system. A system management goal or an objective is a concrete step toward achieving a policy goal, stated in measurable terms. Goals and objectives reflect the perceptions of the transportation agency and the users of what the transportation system should achieve. The overarching policy goals are transportation efficiency, effectiveness, and equity, which can be stated in a more concrete form as transportation agency and user-related objectives, including facility preservation, agency and user costs, traffic mobility, safety, and environmental impacts, as well as the indirect objective of stimulating economic development. Performance measures are specific quantitative or qualitative indicators that directly or indirectly reflect the extent to which a transportation system stimulus realizes its agency and user objectives. They could provide information on the extent to which the expected facility and operation performance targets are met, the transportation users are satisfied with changes or improvements in the service, and the way in which available resources are transformed into performance improvements from resource allocation.

1.3.2 Data needs, collection, processing, and database management

The analyses associated with the key components of transportation asset management, ranging from performance modeling to project selection and programming, are intensively data-driven. Specifically, the historical and current data collected will help monitor the deterioration trend and the current condition of the existing physical facility and system usage performance. Such data can be employed to calibrate statistical models for predicting facility conditions and system usage performance levels at a future point or period in time. The prediction results will, in turn, assist in identifying the needs for improving the facility condition and system usage performance in the future, proposing investment alternatives for performance improvements, and estimating the expected benefits after the investments to assess the economic feasibility of the related investment alternatives. It could also support prioritization of economically feasible investment alternatives and tradeoff analysis of different investment strategies.

The data items that need to be collected depend on the types of physical facilities, available resources, and organizational units that will use the data. Typical data needs for a transportation asset management program can be related to transportation supply and demand, as well as the environment that directly affects transportation performance. The primary data categories can be classified as data on multimodal transportation network inventory; conditions of pavements, bridges, tunnels, transit fixed facilities and rolling stock, navigable inland waterways, freight facilities, and traffic and safety hardware; travel demand

and traffic; vehicle crashes; and climatic and environmental features. For each category of data, multiple data points should be collected in temporary and spatial domains through sampling. Different data sampling, collection techniques, and equipment used for data collection will inevitably lead to variations in data accuracy and precision. Measures of quality assurance/quality control need to be established for data collection. The raw data needs to be processed, compiled, and geo-coded in databases for storage and subsequent analysis.

1.3.3 Multimodal physical facility and system usage performance modeling

System performance refers to the way in which the physical facility condition and system usage performance in a transportation system deteriorate over time or after cumulative use. One of the purposes of this effort is to determine historical trends and to develop statistical models for forecasting the future facility condition and system usage performance. This information can be used to identify the needs for improving facility condition and system usage performance, help to recommend investment alternatives as countermeasures for performance improvements, and estimate the extent to which the performance improvements could be achieved after implementing the investment alternatives.

1.3.4 Travel demand and traffic flow predictions

Transportation investment decisions are based on the predicted facility condition and system usage performance, which are affected by traffic flows. Therefore, prediction of future traffic conditions becomes essential. Typically, the traffic flow predictions are obtained from traffic assignments using the travel demand estimates in the form of origin–destination (O–D) trip tables. In general, the travel demand estimates can be deterministic or time dependent and the traffic assignments can be carried out using deterministic, time dependent or dynamic approaches with added complexity. In this effort, the field traffic counts for the current period need to be employed first to calibrate and validate the traffic assignment models, which can then be utilized for traffic assignments to establish the future traffic flow through an iteratively computational process.

1.3.5 Transportation system performance trend analysis

The deterioration of the physical facility condition is a manifestation of the combined effects of traffic and non-traffic factors such as design standards, materials in use, construction quality, and climatic and environmental features. Also, system usage performance is greatly influenced by traffic and environmental factors. With the future traffic flows predicted from travel demand and traffic forecasting models, they could be utilized as inputs to assess the trends of facility deterioration and system usage performance levels.

1.3.6 Needs assessment and investment alternatives

Needs assessment aims to identify deficiencies of physical facility condition and system usage performance for a future point or period in time, typically predicted using performance models. It also focuses on developing countermeasures for mitigating or eliminating those identified problems in accordance with pre-specified minimum acceptable performance levels. As a result, a short list of investment alternatives that have been demonstrated to be cost-effective for each given set of circumstances can be compiled. Such a short list is preferably standardized for all units of a transportation agency. To facilitate cost estimation in the long run, it is necessary to estimate the resources needed to perform each of the identified investment alternatives. It is desirable to incorporate the standard resource requirements into the database to ensure automated resource needs and cost estimation for similar types of investment alternatives that may be proposed in the subsequent period.

1.3.7 Project evaluation

The implementation of an investment alternative will lead to changes in the values of performance measures under various system management objectives. The overall benefits of the investment alternative can be determined per multiple itemized benefits as captured by changes in related performance measure values. The benefits achieved from the investment alternative in terms of reductions in agency and user costs are conventionally estimated in dollar values. However, other benefits associated with traffic mobility, safety enhancements, and reductions in vehicle emissions and noise pollution are normally expressed in non-dollar values. Two approaches can be used to combine the individual benefit items measured in noncommensurable units. One approach is to convert the non-dollar-valued benefit items into dollar values and then express the overall benefits in dollar values. The other approach is to convert noncommensurable benefit items into utility values via weighting, scaling, and amalgamation according to the multicriteria decision theory. For each investment alternative, the difference in the utility values before and after the investment, triggered by changes in the values of performance measures, is considered as its overall non-dollar-valued benefits. One of the major issues encountered in project evaluation is interdependencies in the total benefits of multiple projects. That is, the total benefits of simultaneously implementing multiple projects may be equal to, greater than or smaller than the summation of the benefits of implementing the same set of investment alternatives individually. Based on either the dollar or non-dollar benefits and cost estimates, a subcollection of the investment alternatives will be screened out and the remaining investment alternatives are considered economically feasible or efficient enough to be further promoted for the subsequent project selection and programming analysis.

1.3.8 Project selection and programming

Due to budget limitations, even a smaller portion of the economically feasible or efficient investment alternatives can be selected for actual implementation. The objective is to ensure that the selected investment alternatives could yield the maximized overall benefits subject to budget and other constraints. Traditionally, ranking, prioritization, and optimization techniques with added complexity are used to conduct this analysis. Particularly, optimization models have been developed for solving real-world resource allocation problems due to their inherited mathematical rigor. As part of the optimization of investment alternatives constrained by budget and other factors, tradeoff analysis can be carried out to identify the best combination of investment alternatives by achieving maximized overall benefits. The criteria for tradeoff analysis may be consistent with dimensions for consideration in establishing transportation policy goals, objectives, and performance measures, such as tradeoffs among different system management objectives, between preserving facility condition against improving system usage performance, across transportation modes, and by flow entity concerning people and goods.

One of the key issues involved with optimization formulations is the uncertainty of budget available for resource allocation. This is because the project selection and programming decisions are based on an estimated budget many years ahead of the project programming period. The actual budget may be equal to, greater than or smaller than the expected budget.

As time passes and updated budget information becomes available, project selection and programming decisions must be updated accordingly to ensure realistic results.

1.3.9 Project implementation and feedback

The systematic asset management decision is an evolutionary process that is expected to be responsive to the needs of the transportation agency and the user. It is important that the analysis be flexible enough to keep abreast of the changing needs of transportation, yet robust enough to be applicable in a wide variety of areas related to transportation management. Feedback evaluation involves the routine collection and analysis of appropriate data, comparing the results with the previously established goals, objectives, and performance measures, and evaluating the performance of the strategies, policies, and operational procedures that comprise the program. The feedback component of the process for feedback evaluation allows practitioners to assess the effectiveness of their efforts, identify areas for improvements, justify these improvements, demonstrate the benefits realized, and support requests for additional resources.

Transportation systems are undergoing a time of great changes in travel demand, physical facility preservation, technology, and public expectations. These changes have had a corresponding impact on how these systems should be managed. There are institutional, strategic, measurement, integration, and analytic challenges that the transportation agency must overcome to be successful in implementing and managing the transportation system.

Chapter 2

Transportation goals, objectives, and performance measures

2.1 GENERAL

Transportation asset management goals may be classified into policy goals and system management goals. A policy goal is a general statement of a desired state or ideal function of a transportation system. An objective is a concrete step toward achieving a goal, stated in measurable terms. System management goals are related to system performance in that they reflect different perceptions of what the transportation system should achieve and are often developed through extensive public outreach efforts. Objectives are more concrete statements of system management goals. Thus, system management goals and objectives incorporate a broad range of agency and user perspectives on which elements of system performance are important.

Performance measures are specific quantitative or qualitative impact types that directly or indirectly reflect the extent to which a transportation system stimulus realizes system management goals or objectives. According to the U.S. National Cooperative Highway Research Program (NCHRP), a performance measure is implicitly defined as statistical evidence used to measure progress toward specific, defined organizational goals and objectives (NCHRP, 2009). As such, performance measures should reflect the satisfaction of the transportation service user as well as the concerns of the system owner or operator. In its National Performance Review, the U.S. Federal Highway Administration (FHWA) defines performance measurement as a process of assessing progress toward achieving predetermined goals, including information on the efficiency with which resources are transformed into goods and services (outputs), the quality of those outputs (how they are delivered to the clients and the extent to which clients are satisfied) and outcomes (the results of a program activity compared to its intended purpose), and the effectiveness of government operations in terms of their specific contribution to program objectives (FHWA, 1999). Performance measures are needed at various stages of the transportation program or project development process for purposes of decision-making at each stage and at various hierarchical levels of transportation management and administration.

2.2 TRANSPORTATION POLICY GOALS, SYSTEM MANAGEMENT GOALS, AND OBJECTIVES

Goals and objectives are related to system performance in that they reflect different perceptions of what the transportation system should achieve and are often developed through extensive public outreach efforts. As such, goals and objectives incorporate a broad user perspective in which elements of system performance are important. Understanding different goals and objectives is critical to identifying the different types of performance indicators

Table 2.1 Typical transportation policy goals, system management goals, and objectives

Policy goal	*System management goal*	*Objective*
Effectiveness	Facility preservation	Preserving transportation facility surface conditions Preserving transportation facility structural conditions Extending facility useful service lives
Efficiency	Delivery and operations efficiency	Reducing life cycle agency costs Enhancing agency's financial feasibility Reducing life cycle user costs Improving system operations efficiency Maintaining system levels of service
Effectiveness	Accessibility and mobility	Improving accessibility to destinations Integrating auto and transit modes Integrating transit sub-modes, including bus and bus rapid transit, streetcar and light rail, subway and commuter rail, and taxi and rideshare in facilities and operations Enhancing truck and freight rail intermodal connectivity Reducing O–D travel time Decreasing travel time buffer index for enhanced reliability Reducing traffic congestion and vehicle delays Reducing travel costs
Effectiveness and efficiency	Safety and security	Reducing crash frequencies Reducing crash impacts Reducing damage to travelers, freight, vehicles, and physical facilities from incidents and emergency events
Effectiveness	Environmental impacts	Reducing adverse impacts on ecology, water quality and quantity, air pollution, and noise Improving environmental aesthetics and general quality Avoiding damage to sites of cultural interest
Effectiveness	Economic development	Increasing employment Increasing business output, productivity, retainment, and new businesses Fostering mixed land-use for progressive community development Promoting transit-oriented development Promoting transit joint development
Effectiveness and equity	Quality of life and social impacts	Enhancing community cohesion Enhancing accessibility to social services Increasing recreational opportunities Providing transportation and opportunities for disadvantaged groups Providing transportation and opportunities for disabilities

that might be included in a highway asset management system. Table 2.1 lists some representative transportation policy goals, system management goals, and objectives.

2.3 TRANSPORTATION PERFORMANCE MEASURES

2.3.1 Desirable properties

The choice of performance measures should consider the following issues: (i) types of transportation networks, modes or facilities being dealt with; (ii) network- or project-level performance being handled; (iii) a policy intervention or a physical treatment being targeted; (iv) transportation development stage with evaluation performed; and (v) pre-versus post-implementation evaluation. Desirable properties of the selected performance measure include the following (Turner et al., 1996; Cambridge Systematics, 2000):

Pertinent: The performance measure should be pertinent to reflect one or more system management objectives that could provide the needed information to decision makers and a direct reflection of the effectiveness of the transportation action being evaluated.
Measurable: It should be possible and easy to measure the value of the performance measure in an objective manner with available tools and resources. Measurement results should be within an acceptable degree of accuracy and precision.
Dimensional: It should be able to capture the required levels of dimensions when evaluating the extent to which the corresponding objectives are achieved by transportation policy interventions or physical treatments. The dimensions may include transportation modes, facility types, temporal scopes, and spatial scopes over which they are measured, and administrative jurisdictions to which they are most relevant, by agency, operator or user/nonuser perspectives.
Realistic: It should help extract reliable information from the available data or analysis results without excessive efforts in terms of costs and time.
Defensible: It should be clear and concise so that the analysis results could be displayed in a simple and concise manner and be effectively communicated with transportation decision makers, users, and the general public.
Forecastable: It should facilitate predicting the performance levels at a future time using existing performance modeling methods, models, and tools.

2.3.2 Performance measures under policy goals

Any specific transportation program or project under evaluation is associated with a given level of each dimension shown above. For example, construction of a freeway to bypass a city may have the objective of congestion mitigation (a form of operational efficiency), be in the public good, be evaluated in terms of economic efficiency (reduction in travel time converted to dollars vis-à-vis project cost), and be evaluated for its contribution to freight transportation. Also, it is a single-mode (highway) project. The analyst may be interested in user impacts in the urban area (e.g., reduced noise and air pollution in the downtown area) in the medium term. The evaluation may be intended for policy making at management level. Thus, the analyst needs to identify the dimensions for the evaluation, after which he can decide the appropriate performance measures associated with each dimension. A given performance measure may be applicable to more than one dimension. The section below identifies the various performance measures that could be considered under each dimension.

Overall goals are the three Es (efficiency, effectiveness, and equity). Some performance measures involve an assessment of how much return can be "bought" for a given input—this is a concept of efficiency. Either the input or the returns or both are expressed in monetary terms. Examples include crashes saved per dollar of safety investment, average reduction in travel time per dollar of congestion mitigation investment, benefit/cost ratio, net present value, and so on. Other performance measures do not "selfishly" seek the returns per input, but focus solely on the returns, often from a nonmonetary perspective. Such performance measures describe the effectiveness of the transportation action and can be used to assess the degree to which operational goals are being attained. A relatively small and little-used set of performance measures relates the effectiveness of the transportation action to the distribution of the social fabric by ascertaining the extent to which the benefits or costs are being shared across various sectors of the affected population. Such equity-based performance measures are often backed by legislation and ensure that disadvantaged sections of the population do not incur relatively little benefit or higher cost compared to other, more privileged sectors of the population.

2.3.3 Performance measures under system management goals

Physical facility preservation: Facility preservation refers to the set of activities geared toward maintaining a minimum level of physical condition of transportation infrastructure and equipment and is generally considered a vital aspect of transportation management.

Delivery and operations efficiency: Delivery and operations efficiency measures efficient delivery of transportation projects and services to the customer, transportation system operations, and maintenance of service levels by focusing on the customer experience of using the system.

Accessibility and mobility: An important function of any transportation system is to provide accessibility for people between residential, employment, recreational or shopping locations, and goods and services between production and distribution points. Thus, any performance measure for accessibility should reflect the ease with which people and goods access services, use various transportation modes, and reach their various destinations. It is also desired that performance measures for accessibility capture the density of transportation service or land uses within a given area (TRB, 2016). Ensuring mobility of people and goods and services and factors of production is a basic role of any transportation system. Performance measures associated with mobility transcend several dimensions.

Safety and security: Transportation systems aim at enhancing accessibility and mobility while preserving the safety of those using the system (operators), those affected by system usage (pedestrians), and those involved in the system preservation and operations (agency field personnel). Security measures protection of travelers, freight, vehicles, and physical facilities from emergencies caused by noticeable and no-notice events such as severe weather conditions, terrorist attacks, and earthquakes.

Environmental impacts: Most transportation actions, while yielding agency benefits and operational efficiency, mobility, accessibility, and safety benefits, tend to have adverse impacts on the quality of the environment and quantity of resource consumption. One of the objectives of transportation provision is to simultaneously minimize such negative impacts. Performance measures for environmental impacts are typically expressed in terms of the amount of environmental damage, such as tons of various types of pollutants emitted. Performance measures for resource conservation are often expressed in terms of the amount of resources used per unit transportation effort.

Economic development: Most transportation improvements are geared toward enhancing operational efficiency and effectiveness, but the end goal may explicitly or implicitly be to provide the region with top-class transportation systems in order to retain existing businesses and attract new business establishments.

Quality of life and social impacts: Quality of life typically captures attributes such as overall well-being, community spirit, social equity, and concerns for the disadvantaged or disabled population groups. In contrast, social impacts measure the effects on broader society or on some specific population groups.

2.3.4 Network- and project-level performance measures

Performance measures may be applied at two levels:

Network level: Performance measures are used for management-level functions such as priority setting that focuses on determining the optimal use of limited funds for the entire network, estimating funding levels needed to achieve specified system-wide performance targets, such as average facility condition or average user's travel time delays, and estimating the system-wide impacts of alternative funding levels.

Project level: Performance measures are utilized to select an effective facility treatment strategy represented by coordinated timing and magnitude of construction, repair, and maintenance interventions in the service life cycle to ensure meeting a target performance level with the minimum possible life cycle agency and user costs. Project-level evaluation is typically more comprehensive than that at the network level, particularly on the data requirements and technical concerns.

Tables 2.2 through 2.10 present some typical network- and project-level performance measures (Poister, 1997; Li and Sinha, 2004; Cambridge Systematics et al., 2006).

Table 2.2 Network-level performance measures

System management goal	*Category*	*Performance measure*
Facility preservation	Pavements	Overall pavement condition index by functional classification Percent distribution of highway miles with pavement conditions rated excellent, good, fair, poor, and failed
	Bridges	Percent distribution of bridges with functional conditions rated excellent, good, fair, and obsolete Percent distribution of bridges with structural conditions rated excellent, good, fair, and deficient
	Traffic signs	Overall retroreflectivity level Age distribution of signs Percent of signs replaced every year Overall risk assessment scale
	Traffic signals	Age distribution of signals Overall conflict monitor grade scale
	Detection services	Percent of false detection rates
	Lighting	Overall illumination level Percent of lighting replaced every year
	Pavement markings	Overall retroreflectivity level Percent of paint loss per one-tenth of a mile Overall risk assessment scale
	Guide rails and crash cushions	Age distribution of guardrails Age distribution of crash cushions Percent of deformed linear length Percent of corroded surface area Percent of locations with structure height changed Overall risk assessment scale
	Multimodal facilities	Distribution of transit stops/stations in different condition ratings Distribution of transit routing facilities in different condition ratings Distribution of bikeways in different condition ratings Distribution of pedestrian walkways in different condition ratings

(*Continued*)

Table 2.2 (Continued) Network-level performance measures

System management goal	*Category*	*Performance measure*
Delivery and Operations efficiency	Construction, repair, maintenance	Cost per lane-mile of highway constructed Cost per lane-mile of highway repaired Cost per unit of maintenance work, labor cost per unit completed
	Multimodal cost effectiveness	Cost per vehicle mile of travel Cost per vehicle hour of travel Percent of cost recovery by farebox revenue
Accessibility	Highway	Percent of population within 10 min or 5 miles of public roads Percent of bridges with vertical over/under clearance restrictions Percent of bridges with weight restrictions
	Multimodal	Annual ridership Transit service coverage Transit hours of service Park and ride spacing Transit transfer condition Bikeway condition and use Pedestrian walkway condition
Mobility	Travel speed	Average speed versus peak-hour speed by highway classification Average speed versus peak-hour speed by transit mode Transit versus auto travel time ratio
	Delays	Average intersection delays per vehicle per cycle
	Congestion	Percent of congested vehicle miles of travel Percent of congested vehicle hours of travel Congestion index O–D travel time buffer index
Safety and security	Highway	Crash frequencies by severity level, type, and highway functional classification Percent changes in crash frequencies by severity level, type, and highway functional classification over time
	Multimodal	Transit related crash frequencies by severity level and type Bicycle related crash frequencies by severity level and type Pedestrian related crash frequencies by severity level and type
	Work zone	Work zone related crash frequencies by severity level and type
	Incidents and emergency events	Incident frequencies by travel mode, severity, and type Emergency event severity and type
	Incident and emergency prevention and preparedness	Percent of facilities with specific security features Level of incident response Level of emergency preparedness

(Continued)

Table 2.2 (Continued) Network-level performance measures

System management goal	*Category*	*Performance measure*
Environment impacts	Vehicle air emissions	Quantity of vehicle air pollutants emitted per vehicle mile of travel
	Vehicle noise pollution	Percent of vehicle miles of travel or vehicle hours of travel with noise levels exceeding the threshold levels
Economic development	Supported by transportation	Percent of wholesale and retail sales occurring in significant economic centers with quality transportation service
Quality of life and social impacts	Transportation accessibility and mobility	Percent of travelers satisfied with travel times for work and other trips
		Percent transit riders satisfied with travel times for work and other trips

Table 2.3 Project-level performance measures for system preservation goal

Category		*Performance measure*
Pavements		Pavement condition index Remaining service life
Bridges		Bridge deck condition rating Bridge superstructure condition rating Bridge substructure condition rating
Traffic control and safety hardware	Signs	Age of signs Retroreflectivity level Maintenance rating index Risk assessment scale
	Signals	Age of the signals Conflict monitor grade scale
	Detection devices	False detection rate
	Lighting	Illumination level
	Pavement markings	Retroreflectivity level Percent of paint loss Risk assessment scale
	Guardrail and crash cushions	Age of guardrails Age of crash cushions Risk assessment scale
Multimodal	Transit stops and stations	Stop/station condition rating
	Tracks	Track condition rating
	Bikeways	Bikeway condition rating
	Pedestrian walkways	Pedestrian walkway condition rating

Table 2.4 Project-level performance measures for delivery and operations efficiency goal

Category		*Performance measure*
Delivery	Construction efficiency	Average construction cost per lane-mile
	Repair and maintenance efficiency	Average resurfacing, rehabilitation, restoration, and reconstruction costs per lane-mile Average preventive maintenance costs per lane-mile Average routine maintenance costs per lane-mile
	Schedule and budget adherence	Percent of increase in cost overruns to complete contracts Percent of increase in days to complete contracts
Operations	System usage	Auto and transit mode shares for passenger travel Passenger transfer time between modes Truck and rail mode shares for freight shipment Freight tonnage transferred per hour
	Vehicle operating costs	Average vehicle operating costs per mile or hour Average fuel consumption per trip by type Annual fuel consumption per mile Transit vehicle operating cost per mile or hour
	Travel time, speed, delays	Traffic volumes Average travel speed Travel speed variations
User perception		Agency satisfaction with project delivery progress User satisfaction with completed projects Quality of transportation service in travel time, speed, and delays

Table 2.5 Project-level performance measures for accessibility goal

Category	*Performance measure*
Modal shares	Auto, transit, bicycle, walking modal shares for passenger travel Truck and rail modal shares for freight movements
Multimodal facility provisions	Transit lane-miles Bicycle compatible lane-miles Pedestrian walking compatible lane-miles Truck designated lane-miles Number of intermodal facilities Capacity of intermodal facilities
Travel time, distance	Average travel speed Access time to passenger or freight facilities Accessibility index Accessibility for persons with disabilities
User perception	Perceived accessibility deficiencies

Table 2.6 Project-level performance measures for mobility goal

Category	*Performance measure*
Amount of travel	Total passenger travel Total freight shipment
Travel time, speed	Average auto travel speed Average transit vehicle travel speed Average travel time Average transfer time
Delay, congestion	Volume-to-capacity ratio Vehicle miles or hours congested per day Vehicle miles or hours congested in peak hours Average and longest vehicle delay time in peak periods Average delays Average transfer delays Delay of trucks at facility per ton-mile
Reliability	Hourly, daily, weekly, monthly, and annual average travel speeds Variations in daily, weekly, monthly, and annual average travel speeds
User perception	Satisfaction with delay and congestion Perceived mobility deficiencies

Table 2.7 Project-level performance measures for safety and security goal

Category	*Performance measure*
Crash frequencies	Road segment related crash frequencies by crash severity level and type Intersection related crash frequencies by crash severity level and type Work zone related crash frequencies by crash severity level and type
Geometric design	Vertical curvature Horizontal curvature Number of lanes Lane width Paved shoulder width Access density and control Consistency of geometric design
Roadway condition	Pavement skid resistance
Traffic control and safety hardware	Traffic sign, pavement marking, guardrail, and crash cushion conditions
Roadside features	Roadside hazard ratings
Driver behavior	Actual driving speed compared with speed limit
Incident and emergency event impacts	Average response time for emergency services Percent of emergency road calls with responses
User perception	Satisfaction with safety and security

Table 2.8 Project-level performance measures for environmental impacts goal

Category	*Performance measure*
Vehicle air emissions	Vehicle air pollutant emission rates by vehicle type
Hazmat impacts	Amount of anti-icing or de-icing agents used per lane-mile Vehicle crashes involving hazardous waste
Noise impacts	Daily hours of noise levels at noise receptor sites above threshold
User perception	Perceived air quality, hazmat, noise deficiencies

Table 2.9 Project-level performance measures for economic development goal

Category	*Performance measure*
Job creation	Jobs supported or created
Support for freight movements	Road mileage upgraded to support truck traffic
Economic costs	Economic costs of pollution
User perception	Percent of businesses claiming transportation issues as a major factor in relocation, productivity, or expansion

Table 2.10 Project-level performance measures for quality of life and social impacts goal

Category	*Performance measure*
Accessibility, mobility	Average number of hours spent traveling Lost time due to congestion
Safety and security	Vehicle crash frequencies Incident frequencies Incident and emergency event response time
Environmental impacts	Vehicle air emission rates Daily hours of noise levels at noise receptor sites above threshold
User perception	Satisfaction with commute time Satisfaction with transit service Satisfaction with bikeways Satisfaction with pedestrian walkways Satisfaction with travel safety Satisfaction with incident and emergency responses Satisfaction with air quality

Chapter 3

Data collection, processing, and database management

The collection, processing, and management of data is key to the practice of asset management. Such data will help evaluate current performance levels of physical transportation facilities and system usage, develop performance models to predict performance trends over time to identify system needs, evaluate and prioritize countermeasures to address system needs, and examine the effectiveness of performance improvement measures after implementation. Ultimately, data is essential to transportation asset management decision making. First, agencies should be aware of the types of data required for analysis. Data collection is the next task in the process. In this step, the critical problem is the balance between the amount of data and the effort of data collection, which leads to the selection of data sampling techniques and data collection techniques. Once the data needed is collected, data processing should be conducted for current usage. A portion of data collected from the field may be used for traffic forecasting and performance prediction purposes, leading to newly generated data. Further, integrating the field measured and predicted data to the existing database is also necessary after the data processing step. Along with this, database management serves the purpose of storing and maintaining the data for future usage.

3.1 DIMENSIONS OF DATA NEEDS FOR TRANSPORTATION ASSET MANAGEMENT

Data needed for transportation asset management can be categorized into three dimensions, as illustrated in Figure 3.1. The first dimension is the transportation system, consisting of physical facilities and system usage (vehicles and users/nonusers). The second dimension is the transportation mode, including passenger and freight modes. Passenger modes can be split into auto, transit, bike, and so on. Freight modes can be classified into truck, waterway, and railway modes. The third dimension is travel and traffic demand and system supply. These three dimensions determine the broad category of the data. Under a certain data category, there are several data items, each of which also has its own sub-items known as detailed data items. Table 3.1 describes the primary data dimensions.

3.1.1 Data needs for pavement management

Within the framework of a pavement management system (PMS), the inventory of the pavement network is the foundation of PMS and must be capable of providing the support needed in the decision-making process for the PMS decision maker. Therefore, the inventory process needs to be developed for each data set to be collected. In this process, several questions need to be answered: (i) What data is needed? (ii) How will the data be used? (iii) How and when and even where should the data be collected? In other words, this process ensures

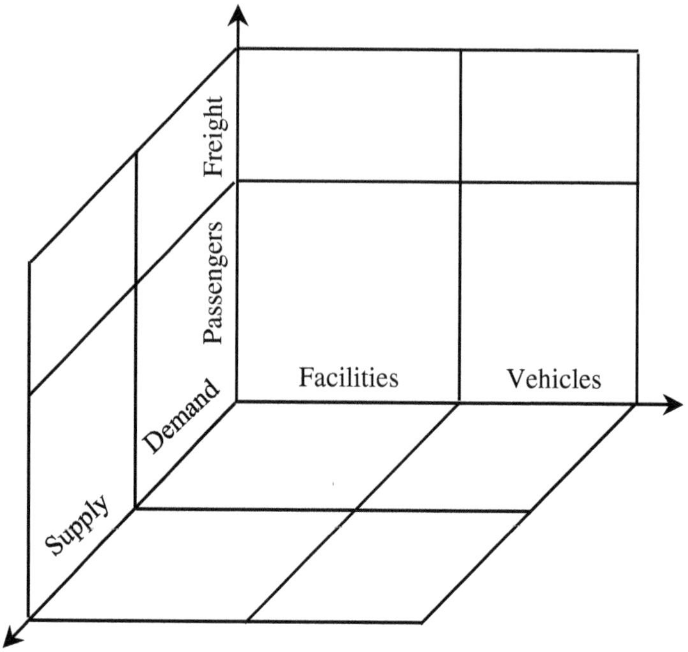

Figure 3.1 Dimensions of data needs for transportation asset management.

Table 3.1 Primary data categories

Mode			*Demand*	*Supply*	
				Facilities	*Vehicles*
Passengers	Auto		O–D-based trips	Roadway inventory Pavements Bridges Traffic control and safety hardware	Traffic count Speed Travel time Crashes
	Transit	Bus	Bus trips	Bus stops	
		BRT	BRT trips	BRT stops BRT travel ways	
		Light rail	Light rail trips	Light rail tracks Light rail stations	
		Heavy rail	Heavy rail trips	Heavy rail tracks Heavy rail stations	
		Cable/tram	Cable/tram trips	Cable/tram stations	
		Ferry	Ferry trips	Ferry terminals	
	Bike		Bike trips	Bike lanes	
	Pedestrian		Pedestrian trips	Sidewalks	
Freight	Truck		Truck trips	Terminals	
	Freight rail		Freight rail trips	Yards/terminals	

that the decision making can be supported by the collected data with the minimum data collection and management costs.

The Western European Road Directors (WERD, 2003) stressed the following criteria when selecting the kinds of data needed:

Relevance: Every data item collected and stored should support an explicitly defined decision need.
Appropriateness: The amount of collected and stored data and the frequency of their updating should be based on the needs and resources on hand.
Reliability: The data should exhibit the required accuracy, spatial coverage, completeness, and currency.
Affordability: The collected data is in accordance with the agency's financial and staff resources.

WERD (2003) further noted the following parameters that should be considered before data collection: (i) specification of the data to be collected; (ii) frequency of collection; (iii) accuracy and quality that the data should exhibit; and (iv) completeness and currency.

At the network level, data ought to be related to the transportation decision making including general planning, programming, and policy decisions. On the other hand, the project-level data should support the decision-making process on select optimal projects among all alternatives. Typically, data for network-level pavement management should contain the following types of data: (i) road inventory; (ii) traffic; (iii) crashes; (iv) climatic and environmental features; (v) productivity; and (vi) pavement costs and benefits (Johnson, 1983; FHWA, 1991; Haas et al., 1994).

Road inventory: This includes road classification data and pavement data. Road classification data items should contain the road functional classification, identification (ID) codes, location, history of construction, reconstruction, maintenance, and rehabilitation if any, and geoinformation including lane width, shoulder widths, number of lanes, and so forth. Pavement data items should contain type, thickness, and materials by layer, subgrade characteristics, drainage, and overlays.

Pavement structure and subgrade: Data items include pavement type, layer thickness, and material characteristics. Subgrade characteristics such as drainage coefficients, modulus of resilience, particle size distribution, and consistency limits and indices should also be taken into account. Data may also include groundwater levels and moisture content.

Pavement condition: Pavement condition may be represented by measures based on aggregate data (such as roughness) or disaggregate data (such as present serviceability index (PSI), which accounts for the occurrence frequencies and severities of various individual distresses). Other indicators of pavement condition include skid resistance, which is measured by a coefficient of friction and depends on the nature of and the amount of water on the pavement surface. The load-carrying capacity of an in-service pavement can be measured directly by full-scale load tests or indirectly by measuring material properties of each pavement layer, to be used subsequently in load-response calculation. Also, measurement of pavement material properties can be carried out by means of field tests, laboratory tests of cored samples or nondestructive tests using modern equipment and advanced technology such as deflection equipment.

Climatic and environmental features: Data on climatic conditions, such as precipitation and temperature-related factors (freeze index, freeze-thaw cycles, and depth of frost penetration) are useful in the development of pavement deterioration models. Also, relationships that enable the estimation of pavement temperature from air temperature are useful.

Maintenance and rehabilitation records: These include information on work activity type, types and levels of resources used (equipment, manpower, and materials), and cost of any routine or periodic pavement maintenance carried out in-house or by contract. Such data may be transformed to reflect work done per lane mile, and costs involved need to be brought to the constant dollar to address the changing values of money over time.

Economic evaluation: Alternatives are evaluated by comparing their costs and benefits. Cost data for maintenance, rehabilitation, and replacement actions are essential to estimate agency costs and budget requirements. For activities performed in-house, data on benefits can be combined with unit cost data to obtain estimates of costs of alternatives. Benefits include user and nonuser benefits. User benefits include reductions in vehicle operating costs, travel time, delay, accidents, and pollution. Data details regarding different components of vehicle operating costs, travel time, delay, crashes, and air emissions are used to estimate user benefits.

3.1.2 Data needs for bridge and tunnel management

Data needs for bridge and tunnel management can be generally grouped into five categories: bridge and tunnel inventory, traffic, component conditions, agency costs, and user costs (FHWA, 1987; Patidar et al., 2007).

Inventory: Detailed data on bridges and tunnels in the network. Bridge inventory data should include bridge location, route number, construction year, number and length of spans, and types and material use of substructure, superstructure, and deck components. Tunnel inventory data should cover items such as tunnel identification, age and level of service, classification, geometric data, inspection, load rating and postings, navigation, and structure type.

Traffic: Data mainly regarding annual average daily traffic (AADT), vehicle composition, truck traffic, and axle load distributions for different types of trucks.

Conditions: Data on bridge deck, superstructure, and substructure components provide information on the current conditions and predicting future conditions. Such data should include history of construction and rehabilitation, climatic and environmental conditions, traffic load, type, and condition inspection records.

Agency and user costs: Data on bridge and tunnel construction, repair, and maintenance records, along with the inventory, condition, and productivity data is needed for agency cost estimation. Similarly, data items on vehicle operating costs, average lengths of detours, bridge- or tunnel-related crash rates and costs, and toll fees if any are required.

3.1.3 Data needs for maintenance management

Generally, to perform maintenance management, the agency should have the following data: inventory data, activity definition and list, performance standards, work programs, performance budget, work calendar, and resource requirements (CALTRANS, 2006).

Inventory: Maintenance inventory data of a roadway are a file of maintenance-related road features, providing the physical foundation for estimating annual maintenance activity requirements. The inventory data should be organized to be consistent with the definition of maintenance activities. Besides the quality of items, the quantity of items requires maintenance by location and management unit. The inventory data provide support for estimating the amount of maintenance work. Inventories of physical assets should record their condition and functional obsolescence. Inventories of nonphysical assets should have the data items on the level of service being achieved.

Activity definition and list: This includes the definition of all work activities in a maintenance management system to facilitate planning, scheduling, and control requirements. Maintenance activities typically include maintenance of roadway and shoulder surfaces, maintenance of drainage facilities, roadside maintenance, winter maintenance, bridge maintenance, maintenance of traffic control hardware, emergency maintenance, and other maintenance, such as equipment repair and maintenance, materials handling and storage, and so forth. A bi-level maintenance management should be implemented, and all activities should accommodate the needs of both high- and low-level maintenance management. High-level management concerns policy, planning, programming, and budgeting. Low-level maintenance concerns more details of the day-to-day field operations to facilitate decentralized decision-making.

Performance standards: These typically include measures of activity accomplishment. Specifically, they should include accomplishment quantities per inventory unit for each road class, productivity measures, and resources' requirements for operation. In addition, bi-level classification is also recommended here. The high-level standards are designed for planning, programming, budgeting, and allocating resources to districts. The low-level standards are designed for the operations within the district based on local situations.

Work programs: These define physical targets as well as their allocation and limitations. A work program is developed for each work activity. The estimated work quantity is computed using productivity data, inventory data, and performance standards, constrained by resource limitations such as budget, time, and space.

Performance budgets: The estimated inputs and outputs corresponding to each activity unit and activity group, typically used to show the connection between invested resources and the potential outcomes of the investment. Therefore, the performance budget reflects a performance objective that forms the basis for allocating funds and resources.

Work calendar: An annual plan indicating how the amount of maintenance work is distributed over time seasonally. It provides guidelines for developing the detailed framework of schedules and evaluating the progress throughout the year.

Resource requirements: A month-by-month listing of resources needed, including more detailed resources. These data items should be included: labor by professions, equipment by classes, materials by types, and so on. The resource requirements guide the allocation of specific resources on hand.

3.1.4 Data needs for traffic control and safety hardware management

Traffic control and safety hardware includes traffic signals, vehicle detectors, data communication devices, and other types of traffic control-related devices. Roadway safety hardware includes signs, pavement markings, guardrails, crash cushions, and other types of safety devices, each of which plays an important role in the transportation safety enhancement. Every year, a large amount of resources is invested in installing, repairing, upgrading, and replacing both types of hardware to maintain the mobility and safety performance of the transportation system as intended (FHWA, 2005a,b).

Inventory: Hardware feature inventory data include data items such as type of hardware, identification codes, location, installation and repair history, as well as performance-related features.

Traffic: The necessity of traffic data and crash data depends on the type of hardware. For instance, a traffic sign can be safely assumed to be independent from traffic and crash data as long as it is still functional. However, the pavement markings and crash cushions are significantly influenced by traffic data and crash data.

Device and hardware age: Determining the age of each type of traffic control device and safety hardware in certain conditions is difficult but critical for deciding when to upgrade, repair or replace particular hardware, further predicting future budget needs.

Device and hardware condition: The condition measures vary by traffic control device and safety hardware type. For all types of devices and hardware, collecting data on maintenance rating index and risk assessment scale is essential. For signs and pavement markings, collecting data on retroreflectivity level is needed. For guardrails, data on corroded surface areas and deformation of structural heights is needed.

Crash data: Vehicle crash data can be useful for identifying the causes of crashes, to improve geometric and safety hardware designs, and to estimate crash cost impacts of different alternatives involving traffic control and safety hardware components.

Environmental factors: This includes temperature, lighting conditions, and weather which are of importance in the deterioration process of the transportation hardware. Therefore, these data items should be gathered to develop the life estimation model of transportation hardware.

3.1.5 Data needs for congestion management

Data collection for monitoring system performance is a basic element of congestion management process, which generally consumes the largest amount of resources. The data collection should be capable of providing the input of the functions to evaluate the congestion severity and performance measures. Furthermore, the data collection frequency should depend on the magnitude of congestion in a certain area (Raimond, 1997; FHWA, 2005c).

Transportation network: The network data include the links as well as their attributes, including number of lanes, functional class, capacity, free flow speeds, tolls, and parking restrictions by time of the day, which are always stored in the local geographic information system (GIS) database. Transportation network data is directly used to run the simulation and indirectly used as the basis of other analyses.

Traffic: Volume counts expressed by AADT or annual average weekday daily traffic (AAWDT) are basic data of congestion management. Specifically, the data set should include vehicle classification, land distribution, directional distribution, time-of-day counts, and turning movement counts in intersections. In network-level management, the raw traffic data can be aggregated into vehicle miles of travel (VMT) for further usage. Traffic volume counts can not only be obtained from traditional traffic volume data collection techniques including manual counts, test vehicle, and loop detector techniques, but can also be aggregated from trajectory data of each roadway user over time.

Vehicle trajectory: The related data provide the exact time and space distribution of all roadway users throughout a day. This is the most fundamental data to describe the traffic volume. Besides using it to compute the volume information, it provides the traffic demand between each O–D pair and each link, which can be used to develop effective means to conduct travel demand and supply management, and further dynamically improve the balance between demand and supply. Traffic trajectory data is always achieved by positioning technologies such as global positioning system (GPS) and hybrid positioning system.

Speed and travel time: Such data samples directly indicate the magnitude of congestion. Speed data can be collected using a probe vehicle method or extracted from traffic trajectory data.

Transit: Transit data is supplementary to the traffic data. The traffic demand in passenger trips can be estimated based on traffic volume data and transit data since the total passenger travel demand will be split into different modes: mainly into auto and transit. Knowing the transit data can help us get a better understanding and calibration of total travel demand. Transit data could be collected from local transit agencies, including route, ridership, and transfer data, as well as frequency of usage, schedule, mode of access, transit vehicle capacity, and so forth.

Truck traffic: Truck traffic can be a major component of the total traffic on some highways. The truck flow is always assumed to be the freight flow. Knowing freight movement demand and passenger trip demand can help us enhance the travel demand model, get a better vision of the total travel demand in the future, and implement demand management and congestion management.

3.1.6 Data needs for safety management

In safety management, besides the aforementioned roadway inventory, traffic, condition, and crash data, as well as some other data such as local driving regulations and design preference (e.g., geometric design, safety hardware design) might also be included.

3.1.7 Data needs for transit performance management

In transit performance management, a comprehensive cross-sectional set of data is required to describe the entire transit operation. Specifically, the following transit-related data is significant in transit performance management (Fielding, 1987, 1992):

Physical inventories of lines and facilities: This can include, for example, locations and characteristics of bus stops, rail transit rights-of-way, track layout, signals, stations, workshop, and so on.
Vehicle data: This should include vehicle maintenance history, mileage, age, current condition, and so on.
Operating conditions on lines: Operating conditions refer to traffic regulations, their speeds, standing time, service reliability, speed limit, and so on.
Usage of services: Data items under this category should be able to indicate the usage of services, such as passenger boarding, alighting volume, volume along the lines, and so on.
Miscellaneous information on events in operations, fares, and passenger attitudes: This data category should include transit-related crashes, security incidents, fare collection methods and fare types, passenger riding experience regarding fares, and seats and schedules, as well as other service characteristics.

3.1.8 Network-wide forecasted traffic data

In the context of transportation asset management, there are many scenarios for which future traffic estimates are required. The forecasted traffic is a part of the input to forecast the physical facility condition or system usage performance over time, which forms the basis for assessing the needs of the transportation network, evaluating the impacts and economic merits of transportation interventions or treatments, and prioritizing the effective alternatives under budget constraints.

Two general approaches are used in practice. The first approach is to use regression techniques based on historical data. However, this approach could be questionable when the regional land use or transportation supply system changed significantly and consequently traffic demand gets redistributed. The lack of historical data may also result in the loss of reliability of the first approach.

The second approach is the travel demand forecasting process to predict traffic conditions that could capture traffic dynamics in a show time window or conduct traffic predictions at a certain point in time or for a certain period in the future. The typical travel demand forecasting process covers trip generation, trip distribution, modal split, and traffic assignment steps to establish network-wide traffic forecasts (Wilson, 1974; Ben-Akiva and Lerman, 1985; Oppenheim, 1995; Meyer and Miller, 2000).

Trip generation: Trip generation is conducted from either the trip production or attraction side. Trip production and attraction are estimated for home-based work (HBW), home-based other (HBO), and non-home-based (NHB) trip purposes, respectively. Data needed for trip generation are mainly related to the socioeconomic characteristics of the transportation network.

Socioeconomic data includes household data (dwelling units, income levels, and auto ownership) for estimating trip productions, and employment data for estimating trip attractions, and is generally expressed in geographic units called traffic analysis zones (TAZs). Although some activity-based travel forecasting models are based on a more disaggregate level than the traffic zone, they are still used extensively by various travel demand forecasting models. The zonal boundaries are usually major roadways, jurisdictional borders, and geographic boundaries that form polygons around homogeneous land uses. This polygon layer is useful in organizing socioeconomic data and serves as a guide when placing centroids and centroid connectors. Ultimately, zones are represented by centroids in the roadway network where the trips are either produced from or attracted to them.

Transportation network data is basically associated with the transportation system supply side. A complete transportation network typically consists of networks accommodating passenger travel by auto, transit, non-motorized modes, and freight movements by truck, rail, and waterway modes.

The socioeconomic and transportation data is applied to methods of trip rate, cross-classification, and regression analysis to estimate and balance the total of trip productions and attractions for individual traffic zones by trip purpose. Further, the transportation network is overlapped with the zonal centroids for individual traffic zones of the study area. This will help to determine the zonal centroid connectors necessary to allow inter- and intra-zonal trips gaining access to the transportation network and completing the trips.

Trip distribution: For each traffic zone, the properly balanced HBW, HBO, and NHB trips estimated by production and attraction can be classified into external–internal, internal–external, internal–internal, and external–external types of trips. For the first three types of trips, the gravity model is used for trip distribution, whereas in the last type of trips, the Fratar model is typically utilized for trip distribution. The application of the gravity model requires the use of friction factors which are determined separately by trip purpose as inversely nonlinear functions of zonal travel times. The inter-zonal travel time is referred to as the total travel time between zonal centroids, comprising the travel times of centroid connectors on both ends and travel time of using the transportation network estimated according to the peak-hour average travel speed. The intra-zonal travel time is roughly estimated as one-half of the travel time to traverse through the zonal boundaries.

The friction factor measures the impedance between zones that should be calibrated for the study area. This factor may be obtained approximately from the friction factor functions for different trip purposes for similar cities and transportation systems. In some cases, trip monetary costs and distances are significant enough to be incorporated into the friction factor to provide a more realistic measure of the travel impedance.

Trip distribution will help determine inter- and intra-zonal production–attraction (P–A) trips by trip purpose, which need to be converted to O–D trips by trip purpose. Typically, the O–D trips by HBW or HBO trip purpose are established as the weighted sum of P–A trips and the transpose of P–A trips. The O–D trips by NHB trip purpose are consistent with the P–A trips. The O–D trips for HBW, HBO, and NHB trip purposes are synthesized to arrive at the overall O–D trips for all traffic zone pairs, commonly termed as the O–D trip matrix. The O–D trip matrix is used as the basis of modal split and traffic assignment analyses.

Modal split: The O–D trip matrix that shows the total person trips for all zonal pairs is split across competing travel modes. This is accomplished by applying the binomial or multinomial logit model where a utility function for each travel mode is established as a function of several attributes, including O–D in-vehicle and out-vehicle travel times, out-of-pocket costs, and so forth. Finally, modal split will help establish mode-specific O–D person trips for all zonal pairs.

Traffic assignment: The O–D person trips by auto are converted to O–D vehicle trips for assigning auto trips to various O–D paths. The user equilibrium (UE) and system optimum (SO) principles are largely followed for auto (and truck) traffic assignment, with the UE principle being more practical. In this context, both analytical and simulation-based methods, models, and tools have been developed. Analytical models could effectively generate close-formed assignment results, but are limited by some oversimplified or unrealistic assumptions. This renders them to be less practical for real-world applications. Conversely, simulation-based models could incorporate various parameters to capture real-world traffic conditions, but need extensive computational capability to produce modeled results. Moreover, significant efforts are required for model calibration and validation using field data prior to utilizing the simulation-based models for traffic predictions. Similarly, the transit-specific O–D person trips are assigned to transit lines that are feasible to accommodate the O–D trips.

Auto, truck, and transit traffic assignments will help establish vehicular traffic details in terms of traffic volumes, speeds, and travel times associated with a transportation network. The modeled traffic details could be traffic predictions at a certain time point or for a specific time period in the future as a result of travel demand changes or traffic predictions after implementing one or more transportation interventions or treatments causing significant traffic redistributions in the study area. With traffic predictions in place, it could help assess the impact of changes in travel demand and transportation supply on asset management goals of facility preservation, agency and user costs, mobility, safety, and air emissions. That is, the traffic predictions become essential to the practice of asset management.

3.2 DATA SAMPLING METHODS

Asset management is data-intensive and the decision making is data-driven. However, where the data comes from, how it is gathered, how to ensure reliability and representability are issues to be considered in the first place.

Various methods exist in practice to gather or obtain data for transportation analysis, among which are three most popular methods: sample surveys, observational studies, and experiments (Savage, 1972; Siegel and Castellan, 1988; Box and Tiao, 1992; Box et al., 2005; Montgomery and Runger, 2006). A sample survey is a study that acquires data from a subset of the target population to infer some information of the population. An experiment is a controlled study conducted to seek the cause-and-effect relationships, in which there is group assignment of subjects and the treatment of each group is controlled. Similarly, an observational study attempts to reveal cause-and-effect relationships. But the group assignment and the corresponding treatment of an observational study cannot be controlled.

Sample surveys are widely used in transportation practice for reasons of cost and practicality. In any case, the sampled population and the target population should be mutually similar. Ideally, sampling methods could be adopted to achieve data accurately and cost effectively. Sampling methods are categorized into two types: probability sampling and non-probability sampling. Probability sampling is the method in which each element's probability of being chosen in the target population is known and is nonzero. With nonprobability sampling methods, each element's probability of being chosen in the target population is unknown. Nonprobability sampling methods have two potential advantages: convenience and cost. But the magnitude of the difference between sample statistics and population parameters cannot be estimated. Only probability sampling methods can help overcome this disadvantage.

3.2.1 Simple random sampling method

Simple random sampling is the purest and simplest form of probability sampling. Every sample in the entire target population has an equal and known chance of being selected. There are many ways to conduct simple random samplings. One way is the lottery method. Each of N population elements is assigned a unique number known as element ID. These numbers are placed in a container and completely mixed. Then, a blind selection of n numbers will be performed. Those elements with selected numbers compose the final sample.

The advantage is that the method will simplify analysis of the results and avoid bias. In particular, the variance between individual results within the sample is a good indicator of variance in the overall population, which makes it relatively easy to assess the precision of estimates, while the disadvantages are that there might be some practical constraints in terms of time or resources available, and the samples will lose representativeness if the population covers large distinct areas and the sample size is small. Therefore, this method can be used at the project level when survey objects have similar characteristics.

3.2.1.1 *Sample size*

The proper sample size for a particular study needs to provide sufficient statistical power, depending on many interacting factors, which include (i) budget for data collection; (ii) administrative consideration (complexity of the operation, duration of the process, etc.); (iii) minimum acceptable precision level; (iv) confidence level; (v) variability within the population or the subset of population; and (vi) sampling method.

For the simple random sampling method, the following inputs help determine the proper sample size:

- The accuracy and precision level, measured by the maximum expected difference between the true population parameter and the corresponding sample estimate expressed by a probability statement (e.g., confidence level), also known as the margin of error (ME). For example, 85.5% of the local highway segments have PSI over 2. The ME is +5% with a confidence level of 90%. That means the probability of the event that the true percentage falls into 85.5% + 5% is 90%
- Significance level α. For an estimation problem, α is 1 − confidence level
- Critical standard z-score. For an estimation problem, the critical standard score is the value for which the cumulative probability is $1 - \alpha/2$
- The population size N
- The population variance σ^2

When the mean of the population is of interest, the preliminary sample size can be determined by using the following formulas:

If the sampling process is without replacement

$$n = \frac{z^2\sigma^2 \dfrac{N}{N-1}}{ME^2 + \dfrac{z^2\sigma^2}{N-1}} \tag{3.1}$$

If the sampling process is with replacement

$$n = \frac{z^2\sigma^2}{ME^2} \tag{3.2}$$

When the proportion is of interest, the following formulas are useful to determine the sample size:

If the sampling process is without replacement

$$n = \frac{z^2 p(1-p) + ME^2}{ME^2 + \dfrac{z^2 p(1-p)}{N}} \tag{3.3}$$

If the sampling process is with replacement

$$n = \frac{z^2 p(1-p) + ME^2}{ME^2} \tag{3.4}$$

In Equations 3.3 and 3.4, p is the preliminary estimator of the population proportion. If this value is unknown, use $p = 0.5$ to be conservative. Equations 3.2 and 3.3 are also applicable to large or unknown populations.

3.2.1.2 Data sampling analysis

With the following notation, the precision of a sample can be quantified:

σ = The known standard deviation (SD) of the population
σ^2 = The known variance of the population
N = The number of elements in the population
n = The number of elements in the sample
$\overline{x}$ = The sample mean, also the estimate of the population mean
S = The sample SD, also the estimate of the population SD
s^2 = The sample variance, also the estimate of the population variance
p = The proportion of elements possessing certain attributes (successes) in the sample
P = The true population proportion of elements possessing certain attributes
SD = The standard deviation of the sampling distribution
SE = The standard error of the sampling distribution
Σ = Summation symbol (e.g., $\Sigma x_i = x_1 + x_2 + \cdots + x_n$)

The precision of a sample is inversely related to the variability of the sample statistics, which is always measured by the SD of sampling distribution or standard error (SE). Equations 3.5 and 3.6 show how to calculate the variability of a mean score in four cases, respectively. In the first two cases, the SD can be directly computed based on the known population variance. In the second two cases, the SE is calculated to estimate the SD since the population variance is unknown.

$$SD = \begin{cases} \sqrt{\dfrac{\sigma^2}{n}}, & \text{if sampling is with replacement and } \sigma^2 \text{ is known} \\ \sqrt{\dfrac{N-n}{N-1}\dfrac{\sigma^2}{n}}, & \text{if sampling is without replacement and } \sigma^2 \text{ is known} \end{cases} \tag{3.5}$$

$$SE = \begin{cases} \sqrt{\dfrac{s^2}{n}}, & \text{if sampling is with replacement and } \sigma^2 \text{ is unknown} \\ \sqrt{\dfrac{N-n}{N-1}\dfrac{s^2}{n}}, & \text{if sampling is without replacement and } \sigma^2 \text{ is unknown} \end{cases} \tag{3.6}$$

Similarly, Equations 3.7 and 3.8 show how to calculate variability of a proportion in four cases, respectively. In the first two cases, the SD can be directly computed based on the known population proportion. In the second two cases, the SE can be calculated to estimate the SD since the population proportion is unknown.

$$SD = \begin{cases} \sqrt{\dfrac{P(1-P)}{n}}, & \text{if sampling is with replacement and } P \text{ is known} \\ \sqrt{\dfrac{N-n}{N-1}\dfrac{P(1-P)}{n}}, & \text{if sampling is without replacement and } P \text{ is known} \end{cases} \tag{3.7}$$

$$SE = \begin{cases} \sqrt{\dfrac{p(1-p)}{n}}, & \text{if sampling is with replacement and } p \text{ is unknown} \\ \sqrt{\dfrac{N-n}{N-1}\dfrac{p(1-p)}{n}}, & \text{if sampling is without replacement and } p \text{ is unknown} \end{cases} \tag{3.8}$$

EXAMPLE 3.1

A state transportation agency wants to evaluate the pavement conditions of the state-maintained highways. A simple random sampling method will be performed in the following steps: (1) dividing all highway segments of the same pavement type into equal lengths. These pavement segments form the population and each segment is an element of the population; and (2) randomly selecting pavement segments from the population for condition evaluation. Suppose that the total mileage of the state-maintained highways with flexible pavement is 72596 miles. Roughly 1452 pavement segments with 50 miles of equal length can be established. The agency is interested in the proportion of pavement segments with a PSI greater than 4 and the overall average PSI value. The required margin error for proportion and average PSI are 0.05 and 0.1, respectively, with 95% confidence level ($z = 1.96$). The budget limits the maximum sample size to be controlled at 450.

Solution

Step 1: Determine the sample size

The experience indicates the preliminary proportion to be 450/1452 = 38%. Since $N = 1452$ is a relatively large population size, Equation 3.4 is used. The sample size is estimated by

$$n_1 = \frac{z^2 p(1-p) + ME^2}{ME^2} = \frac{1.96^2 \times 38\%(1-38\%) + 5\%^2}{5\%^2} = 363.03$$

Given that the sample variance obtained from the previous round of condition survey is 0.842, it is then used as the estimate of population variance. If the current variance increases over time, a factor of 1.2 is considered to create a conservative sample size. Equation 3.2 can be used to determine the sample size as

$$n_2 = \frac{z^2\sigma^2}{ME^2} = \frac{1.96^2 \times 1.2 \times 0.842}{0.1^2} = 388.15$$

Therefore, the minimum sample size is

$$n = Min\{450, Max\{n_1, n_2\}\} = Min\{450, Max\{363.03, 388.15\}\} = 388.15$$

Hence, the sample size is to select 389 pavement segments for condition survey.

Step 2: Simple random sampling and field data collection

Assign a unique ID to each pavement segment from 1 to 1452. Use a computer-based random number generator to generate a number within the interval from 1 to 1452 for 389 times. This process can be viewed as a simple random sampling process with replacement. Therefore, some IDs may appear more than one time in the list. The sample consists of segments whose IDs are in the list. Then field survey can be conducted to obtain PSI values.

Step 3: Data sampling analysis

If the sample mean and sample variance of the PSI values from the field survey of the 389 pavement segments are 3.24 and 0.903, and the proportion of pavement segments with PSI values over 4 is 33.9%, the following calculations can be made:

For the mean PSI value, the SE can be computed using Equation 3.6

$$SE = \sqrt{\frac{s^2}{n}} = \sqrt{\frac{0.903}{389}} = 0.048$$

For the proportion of pavement segments with PSI values greater than 4, the SE can be calculated according to Equation 3.8

$$SE = \sqrt{\frac{p(1-p)}{n}} = \sqrt{\frac{33.9\%(1-33.9\%)}{389}} = 2.4\%$$

As seen in the above, both SEs are remarkably smaller than their corresponding estimators. Therefore, the sampled data is reasonably precise. Moreover, the sample size is determined by fixing the required margin errors to ensure accuracy. The two sample statistics are reasonably accurate as well.

3.2.2 Systematic random sampling method

Systematic random sampling method is also called an Nth name selection technique. After the required sample size is calculated, every N^{th} element is selected from the randomly ranked population, where N is computed by dividing the number of elements in the population by the required sample size. Compared with the random sampling method, its main advantage is its simplicity. However, the disadvantage is that the sample will not be representative if the list contains any hidden order. Because the systematic random sampling method is a varied version of simple random, the analysis can be performed in the same way.

EXAMPLE 3.2

For another state-wide pavement condition survey, given that the computed sample size required is 200 pavement segments and the total number of pavement segments in the population is 4000, then the value of N can be computed as $4000/200 = 20$. Randomly pick one number between 1 and 20, noted as r. Then the final sample will consist of sample rank number r, $r + 20$, $r + 40$, until $r + 3980$ out of the randomly ranked list of population elements.

3.2.3 Stratified random sampling method

"Stratified" means "in layers." The stratified random sampling method first identifies different types of elements that make up the entire population and the corresponding proportions. Next, the proportion and the total required sample size are used to compute the required sample size in each group of the same type of elements. The last step involves random sampling within each group.

The advantage is that the sample should be highly representative of the population, especially when the population contains multiple independent strata. Generally, a stratified sample could offer higher precision than a simple random sample with the same sample size. Therefore, with the same precision required, the stratified random sampling method requires a smaller sample size compared with the simple random sampling method. The major disadvantage is that it may need more administrative effort to achieve the preliminary information than the simple random sampling method.

3.2.3.1 *Sample size*

In the stratified random sampling method, two approaches are always used to assign a sample to strata. One approach is proportionate stratification. With proportionate stratification, the proportion of each stratum in the sample is consistent with the proportion of the same stratum in the population. Equation 3.9 can express this property.

$$n_h = n\frac{N_h}{N} \tag{3.9}$$

where

N = The number of elements in the population
N_h = The number of elements in stratum h of the population
n = The number of elements in the sample
n_h = The number of elements in stratum h of the sample

The second approach is disproportionate stratification, which may have a lot of potential advantages, such as lower cost and more precision if it is properly used. The ideal allocation plan should provide the highest precision under the minimum expense. A linear cost function is assumed, meaning the cost of data collection is computed as

$$C = c_0 + \sum_{h=1}^{H} n_h c_h \tag{3.10}$$

where

C = The total data collection cost
c_0 = The "overhead" cost
H = The number of strata in the population
c_h = The cost per sampling unit in the stratum h

Equation 3.11 indicates the optimal sampling fraction that minimizes the cost for a fixed variance or minimizes the variance for a fixed cost

$$n_h/n = \frac{N_h\sigma_h/\sqrt{c_h}}{\sum_{h=1}^{H} N_h\sigma_h/\sqrt{c_h}} \tag{3.11}$$

where

σ_h = The known SD in stratum h of the population

The variance of the mean estimator is computed by

$$V = \frac{1}{N^2}\sum_{h=1}^{H} N_h^2\left(1 - \frac{n_h}{N_h}\right)\frac{\sigma_h^2}{n_h} \tag{3.12}$$

The total sample size n is computed by

$$n = \begin{cases} \dfrac{\left(\sum_{h=1}^{H}\dfrac{N_h\sigma_h}{\sqrt{c_h}}\right)\left(\sum_{h=1}^{H} N_h\sigma_h\sqrt{c_h}\right)}{N^2V + \sum_{h=1}^{H} N_h{\sigma_h}^2}, & \text{if cost } C \text{ is minimized for a given variance } V \\[2ex] \dfrac{(C - c_0)\sum_{h=1}^{H} N_h\sigma_h \,/\, \sqrt{c_h}}{\sum_{h=1}^{H} N_h\sigma_h\sqrt{c_h}}, & \text{if variance } V \text{ is minimized for a given cost } C \end{cases} \tag{3.13}$$

The total sample size n calculated by Equation 3.13 can be combined with Equation 3.11 to determine the optimal sample size for each stratum.

3.2.3.2 *Data sampling analysis*

With the following notation, the analysis based on the sample can be conducted:

σ = The known SD of the population
σ^2 = The known variance of the population
$\bar{x}$ = The sample mean, also the estimate of the population mean
$\bar{x}_h$ = The sample mean in stratum h
s_h = The sample SD in stratum h
s_h^2 = The sample variance in stratum h
p_h = The proportion of elements possessing certain attributes in stratum h of the sample
P_h = The true population proportion of elements possessing certain attributes in stratum h
SD = The standard deviation of the sampling distribution
SE = The standard error, also the estimate of the SD of the sampling distribution

In the stratified sampling method, the formulas are more complicated since more strata are involved. Equations 3.14 and 3.15 show how to calculate the variability of a mean score in four cases. In the first two cases, we can directly compute the SD base on the known population variance. In the second two cases, we compute the SE to estimate the SD since the population variance is unknown.

$$SD = \begin{cases} \dfrac{1}{N}\sqrt{\sum \dfrac{N_h^2{\sigma_h}^2}{n_h}}, & \text{if sampling is with replacement and } \sigma^2 \text{ is known} \\[2ex] \dfrac{1}{N}\sqrt{\sum \dfrac{N_h^3}{N_h - 1}\dfrac{1 - n_h}{N_h}\dfrac{{\sigma_h}^2}{n_h}}, & \text{if sampling is without replacement and } \sigma^2 \text{ is known} \end{cases} \tag{3.14}$$

$$SE = \begin{cases} \frac{1}{N}\sqrt{\sum \frac{N_h^2 s_h^2}{n_h}}, & \text{if sampling is with replacement and } \sigma^2 \text{ is unknown} \\ \frac{1}{N}\sqrt{\sum \frac{1-n_h}{N_h}\frac{N_h^2 s_h^2}{n_h}}, & \text{if sampling is without replacement and } \sigma^2 \text{ is unknown} \end{cases} \quad (3.15)$$

Similarly, Equations 3.16 and 3.17 show how to calculate the variability of a proportion in four cases. In the first two cases, we can directly compute the SD base on the known population proportion. In the second two cases, we compute the SE to estimate the SD since the population proportion is unknown.

$$SD = \begin{cases} \frac{1}{N}\sqrt{\sum \frac{N_h^2 P_h(1-P_h)}{n_h}}, & \text{if sampling is with replacement and } P \text{ is known} \\ \frac{1}{N}\sqrt{\sum \frac{N_h^3}{N_h-1}\frac{1-n_h}{N_h}\frac{P_h(1-P_h)}{n_h}}, & \text{if sampling is without replacement and } P \text{ is known} \end{cases} \quad (3.16)$$

$$SE = \begin{cases} \frac{1}{N}\sqrt{\sum \frac{N_h^2 P_h(1-P_h)}{n_h-1}}, & \text{if sampling is with replacement and } P \text{ is unknown} \\ \frac{1}{N}\sqrt{\sum \frac{1-n_h}{N_h}\frac{N_h^2 P_h(1-P_h)}{n_h-1}}, & \text{if sampling is without replacement and } P \text{ is unknown} \end{cases} \quad (3.17)$$

In addition, the central tendency is measured by mean score or proportion as follows:

$$\text{Mean score} = \sum \frac{N_h}{N}\bar{x}_h \quad (3.18)$$

$$\text{Mean proportion} = \sum \frac{N_h}{N} p_h \quad (3.19)$$

EXAMPLE 3.3

For the above state-maintained highways that contain 1452 pavement segments, each of which is 50 miles in length, suppose that the proportions of flexible, rigid, and composite pavement are 40%, 20%, and 40%, respectively. If the state transportation agency is interested in the overall average PSI, determine the data sampling details using the stratified sampling method. Other information is summarized in Table 3.2. The "overhead" cost of the survey is $50000.

Table 3.2 Information for determining the stratified sample size

Pavement	*Population size*	*Estimated population SD of PSI*	*Cost per unit (Dollars)*
Flexible	653	1.23	740
Rigid	290	0.78	850
Semirigid	508	1.07	640

Solution

Step 1: Determine the sample size

If the agency wants to minimize the cost with a given SD of 0.048 for the estimator of population mean PSI, the total sample size can be calculated by Equation 3.13

$$n = \frac{\left(\sum_{h=1}^{H} \frac{N_h \sigma_h}{\sqrt{c_h}}\right)\left(\sum_{h=1}^{H} N_h \sigma_h \sqrt{c_h}\right)}{N^2 V + \sum_{h=1}^{H} N_h \sigma_h^2}$$

$$= \frac{\left(\frac{290 \times 0.78}{\sqrt{850}} + \frac{508 \times 1.07}{\sqrt{640}} + \frac{653 \times 1.23}{\sqrt{740}}\right)\left(\begin{array}{c}290 \times 0.78 \times \sqrt{850} + 508 \times 1.07 \times \sqrt{640} \\ + 653 \times 1.23 \times \sqrt{740}\end{array}\right)}{1452^2 \times 0.048^2 + 290 \times 0.78^2 + 508 \times 1.08^2 + 653 \times 1.23^2}$$

$$= 375.81$$

Therefore, the total sample size $n = 376$.

With a sample size of 376 determined, it could provide information for the required statistical power and simultaneously minimize the data collection cost.

Using Equation 3.11, the proportion of each stratum in the sample can be calculated as

$$\frac{n_1}{n} = \frac{\frac{N_1 \sigma_1}{\sqrt{c_1}}}{\sum_{h=1}^{H} \frac{N_h \sigma_h}{\sqrt{c_h}}} = \frac{\frac{653 \times 1.23}{\sqrt{740}}}{\frac{290 \times 0.78}{\sqrt{850}} + \frac{508 \times 1.07}{\sqrt{640}} + \frac{653 \times 1.23}{\sqrt{740}}} = 50.22\%$$

$$\frac{n_2}{n} = \frac{\frac{N_2 \sigma_2}{\sqrt{c_2}}}{\sum_{h=1}^{H} \frac{N_h \sigma_h}{\sqrt{c_h}}} = \frac{\frac{290 \times 0.78}{\sqrt{850}}}{\frac{290 \times 0.78}{\sqrt{850}} + \frac{508 \times 1.07}{\sqrt{640}} + \frac{653 \times 1.23}{\sqrt{740}}} = 13.24\%$$

$$\frac{n_3}{n} = \frac{\frac{N_3 \sigma_3}{\sqrt{c_3}}}{\sum_{h=1}^{H} \frac{N_h \sigma_h}{\sqrt{c_h}}} = \frac{\frac{508 \times 1.07}{\sqrt{640}}}{\frac{290 \times 0.78}{\sqrt{850}} + \frac{508 \times 1.07}{\sqrt{640}} + \frac{653 \times 1.23}{\sqrt{740}}} = 36.54\%$$

Of the total of 376 pavement segments sampled, they consist of 189 flexible, 50 rigid, and 137 composite pavement segments.

Step 2: Stratified random sampling and field data collection

In each of the three strata of flexible, rigid, and composite pavement segments, the simple sampling method could be used to select 189 out of the 653 flexible pavement segments, 50 out of the 290 rigid pavement segments, and 137 segments out of the 508 composite pavement segments.

Step 3: Data sampling analysis

Furthermore, given the following information from the field survey, the overall sample mean and sample SD of PSI are 3.15 and 0.692; the sample mean and sample SD of PSI for flexible pavement stratum are 3.128 and 0.651; the sample mean and sample SD of PSI for rigid pavement stratum are 3.162 and 0.787; and the overall sample mean and sample SD of PSI for composite pavement stratum are 3.146 and 0.714. The variance of the mean PSI estimator can be computed using Equation 3.12 as

$$V = \frac{1}{N^2}\sum_{h=1}^{H} N_h^2\left(1 - \frac{n_h}{N_h}\right)\frac{\sigma_h^2}{n_h} = \frac{1}{1452^2}\sum_{h=1}^{H} N_h^2\left(1 - \frac{n_h}{N_h}\right)\frac{\sigma_h^2}{n_h} = 0.0479$$

This value meets the requirement in the sampling design process.

3.2.4 Cluster sampling method

The cluster sampling method will divide the total population into clusters, and simple random sampling is performed within each cluster. There are generally two types of cluster sampling methods: one-stage sampling and two-stage sampling. One-stage sampling refers to the sampling method in which all the elements in the selected clusters are included in the sample. Two-stage sampling refers to the sampling method in which only a subset of elements in selected clusters is randomly chosen for inclusion in the sample. Multistage sampling is an extension of two-stage sampling and is conducted in the same way.

Stratified sampling and cluster sampling are complementary of each other. In the stratified sampling method, the population is divided into groups (known as strata) that are in some meaningful way mutually distinct. Thus, we would doubt the satisfaction of the sample if one group were to be underrepresented in the sample. Whereas in the cluster sampling method, the population is divided into groups (known as the clusters) that are all basically similar. Unlike the stratified sampling method, in cluster sampling it does not matter if some of these groups are missed.

The advantage of the cluster sampling method is the ability to reduce the time and resources consumed in the sampling process. But the quality of the sample highly depends on how the clusters differ among themselves. Generally, cluster sampling provides less precision than either simple random sampling or stratified sampling when the sample size is assumed to be constant in this comparison. This is the major disadvantage of cluster sampling. Therefore, cluster sampling is highly recommended if saved costs can overcome losses in precision.

3.2.4.1 *Sample size*

Ideally, the selected clusters are representative of the population. However, the elements from the same cluster are somewhat mutually similar. As a result, adding one more element from the same cluster in cluster sampling is less informative than adding one more independent element would be. So, with the same actual sample size, cluster sampling has a less effective sample size when compared with simple random sampling. This is known as the design effect. The design effect is basically the ratio of the actual variance under the sampling method actually used to the variance computed under the assumption of simple random sampling. For example, a design effect of 3 means that the variance of the estimator from using a certain sampling method is three times larger than it would be if the sampling method were simple random sampling with the same sample size. Another interpretation is that only one-third as many sample elements are needed to measure the same statistic if a simple random sample were used instead of a cluster sample.

To compute the design effect, the cluster sample size and the intra-class correlation (ICC) are needed. The design effect is calculated as follows. It is evident that the design effect increases as the cluster size increases and as ICC increases.

$$DEFF = 1 + \rho(m - 1) \tag{3.20}$$

where

$DEFF$ = The design effect
m = The average number of observations in each cluster
ρ = The intraclass correlation

The ICC is an index for comparing the variability within a cluster to the variability across clusters, representing the likelihood that two elements in the same cluster have the same value as one chosen completely at random in the population. An ICC of 0.01 indicates that the elements in the same cluster are 0.01 more likely to have the same value than two elements selected randomly from the entire population. The smaller the value is, the more homogeneity the population would have, and consequently the more reliable the sample statistic is. There are several ways to measure variability, which leads to several ways to compute ICC. The most common formula used for ICC is expressed as follows:

$$\rho = \frac{MSB - MSW}{MSB + (n-1)MSW} \tag{3.21}$$

where

MSW = The mean square error within clusters
MSB = The mean square error between clusters

Sometimes the design effect can be found in published reports (e.g., demographics). It can be estimated using the formula provided above. With the sample size determined using the simple random sampling method and design effect, the sample size required for cluster sampling is computed as follows:

$$m_{cls} = DEFF \times m_{srs} \tag{3.22}$$

where

m_{cls} = The sample size required for cluster sampling
m_{srs} = The sample size required for simple random sampling with certain requirements

The next step is to determine the number of elements in each cluster.

3.2.4.2 Data sampling analysis

With the following notation, the analysis based on the sample can be carried out:

N = The number of clusters in the population
M_i = The number of elements in cluster i of the population
$\bar{X}_i$ = The population mean of cluster i
M = The total number of elements in the population
P = The true population proportion of elements possessing certain attributes
P_i = The true population proportion of elements possessing certain attributes in cluster i
n = The number of clusters in the sample
m_i = The number of elements in cluster i of the sample
x_{ij} = The score of the element j from cluster i
$\bar{x}_i$ = The sample mean of cluster i

$\bar{x}$ = The overall sample mean, also the estimator of population mean
$\bar{p}$ = The overall sample mean of proportion of elements possessing certain attributes
p = The proportion of elements possessing certain attributes in population
p_i = The proportion of elements possessing certain attributes in cluster i of the sample

Equations 3.23 and 3.24 show how to calculate central tendency values in terms of the estimated population mean $\bar{x}$ and proportion $\bar{p}$ for one- and two-stage samplings, respectively.

$$\bar{x} = \begin{cases} \dfrac{\sum_{i=1}^{n} M_i \bar{X}_i}{\sum_{i=1}^{n} M_i}, & \text{one-stage sampling} \\ \dfrac{\sum_{i=1}^{n} m_i \bar{x}_i}{\sum_{i=1}^{n} m_i}, & \text{two-stage sampling} \end{cases} \tag{3.23}$$

$$\bar{p} = \begin{cases} \dfrac{\sum_{i=1}^{n} M_i P_i}{\sum_{i=1}^{n} M_i}, & \text{one-stage sampling} \\ \dfrac{\sum_{i=1}^{n} m_i p_i}{\sum_{i=1}^{n} m_i}, & \text{two-stage sampling} \end{cases} \tag{3.24}$$

The precision is measured by SEs for the estimated population mean $\bar{x}$ and proportion $\bar{p}$, $SE_{\bar{x}}$ and $SE_{\bar{p}}$, as calculated by Equations 3.25 and 3.26. These equations only apply for the case where all select clusters have equal size ($m_i = m$). Unequal cluster size will lead to much messier equations. For this reason, the sampling strategy with unequal cluster size is not recommended.

$$SE_{\bar{x}} = \begin{cases} \sqrt{\dfrac{(N-n)}{Nn}\sum_{i=1}^{n}\dfrac{(\bar{X}_i - \bar{x})^2}{(n-1)} + \dfrac{1}{NM}\sum_{i=1}^{n}\sum_{j=1}^{M}\dfrac{(x_{ij} - \bar{X}_i)^2}{(M-1)}}, & \text{one-stage sampling} \\ \sqrt{\dfrac{(N-n)}{Nn}\sum_{i=1}^{n}\dfrac{(\bar{x}_i - \bar{x})^2}{(n-1)} + \dfrac{1}{NM}\sum_{i=1}^{n}\sum_{j=1}^{m}\dfrac{(x_{ij} - \bar{x}_i)^2}{(m-1)}}, & \text{two-stage sampling} \end{cases} \tag{3.25}$$

$$SE_{\bar{p}} = \begin{cases} \sqrt{\dfrac{(N-n)}{Nn}\sum_{i=1}^{n}\dfrac{(P_i - \bar{p})^2}{(n-1)} + \dfrac{1}{N}\sum_{i=1}^{n}\dfrac{P_i(1-P_i)}{(M-1)}}, & \text{one-stage sampling} \\ \sqrt{\dfrac{(N-n)}{Nn}\sum_{i=1}^{n}\dfrac{(p_i - \bar{p})^2}{(n-1)} + \dfrac{1}{N}\sum_{i=1}^{n}\dfrac{p_i(1-p_i)}{(m-1)}}, & \text{two-stage sampling} \end{cases} \tag{3.26}$$

EXAMPLE 3.4

Since selecting 200 sample segments out of 4000 highway segments across the state in the previous pavement condition survey effort is both time- and resource-consuming, the cluster sampling method will be performed. First, the population of pavement segments can be divided into clusters based on eight administrative divisions within the state transportation agency. Historical data indicate that the design effect is 1.4. Suppose the goal of the survey is to achieve the proportion of highway segments with PSI greater than 3.75.

Solution

Step 1: Determine the sample size

$$m_{cls} = DEFF \times m_{srs} = 1.4 \times 200 = 280$$

This shows that it requires a total of 280 samples.

Step 2: Cluster random sampling and field data collection

The first stage of cluster sampling is to randomly select several divisions to administer the pavement condition survey. Considering the cost and time constraints, the agency decides to select four out of the eight divisions for survey. For each division, 280/4 = 70 pavement segments are allocated.

Step 3: Data sampling analysis

Given that historical records show that $p_i = 0.68, 0.74, 0.57, 0.71$ for the four divisions, the population mean estimator and its SE become

$$p = \frac{40 \times 0.68 + 40 \times 0.74 + 40 \times 0.57 + 40 \times 0.71}{40 + 40 + 40 + 40} = 0.675$$

$$SE = \sqrt{\frac{(8-4)}{8 \times 4} \sum_{i=1}^{n} \frac{(p_i - 0.675)^2}{(4-1)} + \frac{1}{N} \sum_{i=1}^{n} \frac{p_i(1-p_i)}{(40-1)}} = \sqrt{0.000688 + 0.00276} = 0.059$$

3.2.5 Combined sampling

Data is gathered by using combinations of various sampling techniques, which depend on the characteristics of the data needed as well as the advantages and disadvantages of each sampling technique. Agencies need to know the target data to be gathered and resources on hand so that they can collect the information as intended at relatively small costs.

EXAMPLE 3.5

A planning organization wants to collect data on travel demand data in a metropolitan area. The population residing in the area could be split into multiple clusters by TAZ divisions (cluster sampling), among which 15 clusters are taken randomly (simple random sampling) after the sample size is determined. For any of the 15 clusters, identify the strata and corresponding percentage based on the occupation attributes, such as professional, clerical, blue collar, and other.

To begin the data collection effort, we need to first determine the strata for each of the 15 clusters. Then simple random samples can be drawn from each stratum. After this process is done for all 15 clusters, the complete sample is created.

3.3 DATA COLLECTION TECHNIQUES

Data collection is generally the most resource-consuming part of transportation asset management. Data collection methods that are more expensive are also usually more accurate, more precise, and have the greatest resolution. Thus, transportation agencies face a trade-off between data collection cost and data quality in terms of accuracy and precision. They need to be aware to what extent the data needs to be accurate and detailed. It is not necessary to collect all data at each level. Some measures, such as structural evaluation, may only be collected at the project level. Other measures, such as pavement surface friction, may only be used when a specific problem has been identified.

3.3.1 Passenger demand data collection

Obtaining data on traffic predictions via travel demand forecasting typically involves trip generation, trip distribution, modal split, and traffic assignment steps (Wilson, 1974; Ben-Akiva and Lerman, 1985; Oppenheim, 1995; Meyer and Miller, 2000). For the execution of a simulation-based travel demand forecasting model, calibration and validation efforts are needed at each step before moving to the subsequent step. In the end, traffic assignment results generated by the model need to be compared with field traffic data using quantitative methods, coupled with realistic checks and reasonableness evaluation. The simulation-based model could be employed for traffic forecasts only after model validation is successfully accomplished. One unique focus of executing the simulation-based travel demand forecasting model is creation of an O–D trip matrix for the study area. This section mainly introduces various methods used for survey data collection to help develop the O–D trip matrix.

Household travel surveys or travel diaries: In this method, a sample of households within the study is selected to participate in a survey. Members of participating households are asked to record their travel activity in a travel diary or through a recall interview for a certain time. The travel activity records include the trip start time, travel time, trip length, origin, destination, mode, trip purpose, and vehicle occupancy of each trip. This information pertaining to other similar households is aggregated to establish average trip rates and trip lengths for that type of household. This method is expanded to all types of households in the study area. By combining this information with trip production and attraction analysis, an O–D matrix can be defined. The household travel survey indicates general travel patterns in a study area, but fails to provide detailed information on any roadway within the study area.

Roadside station surveys: In this type of O–D study, drivers are directly recruited for interviews or vehicles are monitored, typically via license plates matching techniques at a set of roadside stations to obtain their travel characteristics in the study area. This survey method is generally used to determine the number of trips on a roadway segment with both trip ends external to the study area. This type of study is useful and practical. The roadside station survey is often used to supplement the household travel survey. Particularly, the roadside station survey helps in creating and validating travel demand models by collecting data at critical internal cordons and screen-lines.

Employer and special generator travel surveys: This type of survey is always used to collect information on establishments with great trip attractions. These establishments include airports, train stations, and tourist points of interest. This type of survey is often conducted because the establishments with high attraction rates are critical in the study area. If the survey is being conducted for a small number of establishments, the survey can be conducted as an intercept survey as people enter and exit a building, similar to a roadside station survey. Otherwise, to gather information on a large number of establishments, the survey may be conducted within a sample of the population at each establishment, similar to a household travel survey.

Commercial vehicle travel surveys: This survey type is used to obtain O–D data for special vehicles such as trucks and other commercial vehicles. This information can be utilized to determine and refine trip rates, commodity flows, and air quality modeling within an area.

On-board transit surveys: This type of survey is widely used by modelers or transit agencies. Characteristics of the transit users and their travel patterns are typically collected. The transit O–D matrix can be used in travel mode split models or by transit agencies to identify current problems and improve the performance of the transit system.

Hotel and visitor surveys: This type of survey tends to collect travel data on visitors who stay in hotels and who are not included in the household survey. Typically, these groups of people have much more complex travel patterns than the residents of the same city.

Parking surveys: Parking surveys are used to obtain detailed parking information for the purpose of evaluating parking supply, costs, and subsidies on travel decisions. This can be performed as an intercept survey as people enter and exit parking facilities or through a questionnaire placed on the windshield of the vehicle that is voluntarily completed and returned.

Because of technological advancements, methods relying on license plate matching, GPS, and Bluetooth technologies can collect O–D trip data directly. These are getting increasingly popular due to the unique advantages of low cost and being nonintrusive to travelers.

3.3.2 Truck demand data collection

Several options exist for collecting truck demand data from different perspectives, such as driver/carrier, producer/recipient, vehicle/trip, and consignment. Some key freight data collection methods come from this typology, which include trip diary, administrative by-product, automated counts/identification, roadside intercepts, and enterprise surveys (Turner et al., 1998).

Trip diaries: In freight demand studies, a trip diary is a commonly used tool to collect trip and vehicle information in the form of a survey filled by a truck driver. This method could easily collect freight movement information, especially when there exists a special requirement for origins and destinations. Although drivers may not be willing or able to reveal some important or sensitive characteristics of their trips, such as economic sector, a trip diary is still a great way to collect the freight data.

Administrative by-product: This type of data source includes consignment notes, vehicle log books, positioning information on freight, vehicle trajectory record, and others. This type of data source is kept by administrative offices or operators. But it is hard to know whether they will be willing to divulge this information. They may not be willing to provide the economic and employment details about trips but may provide some general information on the freight trips.

Automated counts/identification: With advancements in technology, there are more and more automatic methods available to collect freight trip data in the market. They can collect the number of vehicles passing a certain intercept point during a certain time interval. The type and size of each passing vehicle can also be recognized and recorded in the system. In this way, truck traffic can be separated from the background traffic stream. The most commonly used methods include license plate recognition and radio-frequency identification (RFID). The data collected from each location will be uploaded to the center system to be further analyzed to extrapolate the rough trajectory of each vehicle.

Roadside intercepts: Similar to the roadside station survey in the passenger mode, this method involves intercepting trucks during a trip and conducting a short interview or

Table 3.3 Key freight data collection methods

Method	Sampling unit	Location of data collection		
		Workplace	Roadside	Electronic
Trip diary	Driver	√	√	√
Administrative by-product	Carrier/consignment	√	√	√
Automated counts/identification	Vehicle/trip		√	√
Roadside intercepts	Driver/vehicle/trip	√	√	√
Enterprise surveys	Producer/recipient	√		√

questionnaire. The drivers are asked to record the nature of that trip and commodity. Traffic counts are usually conducted at the same time and in the same place in order to gather sample information that can best reveal the characteristics of the population in this area or corridor. Similar to a trip diary, drivers may not know commodity and origin/destination details.

Enterprise surveys: Enterprise surveys consist of brief interviews of businesses, accompanied by aggregate data collected from other sources, such as published annual reports and historical data, and are combined with statistical inference to obtain commodity flow information. Commodity flows can to some extent help find trips by applying a vehicle loading model. A more thorough enterprise survey might contain a personal interview and site visit covering a wide range of variables pertaining to commodity movement characteristics and vehicle movements into and out of the facilities. Every trip and freight flow has a producer and a recipient. Thus, it is possible to collect or infer origin, destination, and commodity information from the producers and recipients of freight flows. Surveying at the enterprise level also provides accurate information on the type of land use at that location (Table 3.3).

3.3.3 Pavement data collection

3.3.3.1 Pavement segment delineation

Pavement data is collected and stored in discrete units typically referred to as segments. For data to be compatible, it is necessary for such segments to be well defined. Two approaches for defining pavement segments are generally used: the equal length method and the uniform characteristics method (AASHTO, 1990; FHWA, 1991; Haas et al., 1994; Li et al., 2002a).

Equal length pavement segments: The use of equal length pavement segments is convenient both for data collection purposes and for representation in the database. However, it is possible that some segments have unequal lengths. For network-level analyses, the characteristics of equal length segments are uniform enough within each segment to obtain results of sufficient accuracy. For project-level analyses, more accuracy is required. Therefore, shorter segment lengths or segments with uniform characteristics should be used.

Pavement segments with uniform characteristics: Pavement segments with uniform characteristics may be identified using classification-based, response-based, and cumulative difference approaches. With the classification-based approach, pavement characteristics are chosen and segments are identified so that the pavement characteristics are uniform within each segment. Usually these are characteristics that influence the deterioration of the segment and the type of rehabilitation action to be applied to the segment. Such pavement characteristics are pavement type, material type, layer thickness, subgrade type, highway classification, and traffic loading. With the response-based approach, a number of pavement response variables are chosen and segments are identified so that the response variables remain fairly constant within each segment. Pavement response variables that can be chosen include roughness, rut depth, skid resistance, some surface distresses, and structural characteristics such as deflection. These variables are chosen according to their importance for predicting pavement deterioration and for determining the type of rehabilitation action to be implemented. The cumulative difference approach identifies the boundary between adjacent segments as the point where the cumulative pavement characteristic versus distance function changes slope. The boundaries and their stability over time will depend on the characteristic chosen for segment delineation.

3.3.3.2 Surface condition data collection

Pavement skid resistance: Skid resistance is the force developed when a tire that is prevented from rotating slides along the pavement surface. Skid-resistance measurements of highways, roads, and streets are generally for safety analysis in locations where crashes are hypothesized to be caused by deficiencies in surface skid resistance. Skid resistance is generally quantified in terms of friction factor or skid number, computed by dividing the frictional resistance to motion in the plane of interface by the load perpendicular to the interface times 100. The friction factor of a certain pavement segment is not a constant value; it is influenced by pavement humidity, vehicle running speed, temperature, and tire type. Therefore, skid resistance needs to be tested under standard environmental conditions using standard tires.

The locked wheel tester is the most commonly used method in the U.S. for skid-resistance testing. During testing, the data is recorded in terms of friction force on a locked wheel as it is dragged over the test pavement surface under constant speed (typically at 64 km/h, or 40 mph). Water is spread in front of the test wheels. The outputs are proposed to estimate a pavement surface skid number. In addition, several portable field devices have been developed to measure skid resistance, including the keystone tester, California skid tester, and pendulum skid tester.

Pavement roughness: Pavement roughness measurements indicate whether irregularities in the pavement surface adversely affect the ride quality of a vehicle. Roughness is an important characteristic because it can also affect the vehicle delay costs, fuel consumption, and deterioration rate of vehicles. Roughness currently is quantified in terms of either present serviceability rating (PSR) or international roughness index (IRI).

The PSR was defined by the American Association of State Highway Officials (AASHO) Road Test to represent the pavement serviceability based on individual rating for the rideability. PSR is defined as "the judgment of an observer as to the current ability of a pavement to serve the traffic it is meant to serve" (HRB, 1972; FHWA, 2001a). To generate the original AASHO Road Test PSR scores, observers rode on the test pavement surface and rated their ride using the quantitative scale. This subjective scale ranges from 5 (excellent) to 0 (essentially impassable). PSR is based on passenger interpretations of ride quality, which is generally determined by road roughness. Therefore, PSR can reflect the road roughness.

The IRI was developed by the World Bank in the 1980s to define a characteristic of the longitudinal profile. The commonly recommended units are meters per kilometer (m/km) or millimeters per meter (mm/m). The IRI is based on the average rectified slope (ARS), which is a filtered ratio of a standard vehicle's accumulated suspension motion (in mm, inches, etc.) divided by the distance traveled by the vehicle during the measurement (km, mi, etc.). IRI is then equal to ARS multiplied by 1000. For example, if the accumulated suspension motion of a standard vehicle is 60 mm and the distance traveled is 1 km, the ARS is 0.06 mm/m and the IRI is 60. The IRI was measured by using response type devices, which can be grouped into four categories: (i) rod and level survey and the dipstick profiler; (ii) profilograph; (iii) response-type road roughness meters (RTRRMs); and (iv) profiling devices.

The rod and level survey can be used to determine the accurate profile of a road at any desired spacing. However, this requires normal rod and level measurements that are extremely time-consuming and has to occupy the road during the survey. A dipstick profiler is always used as the first step of the rod and level survey. It consists of an inclinometer in a case supported by two legs separated by 12 inches. Two digital displays are on the top of the case, right above the two legs. Each display indicates the elevation of the respective end relative to the elevation of the other leg. The operator then "walks" the dipstick to the next predetermined pavement section by alternatively using the legs as the pivot. The microcomputer application will analyze the data measured and provide a profile with the error controlled within 0.005 inch.

The truss-type California profilograph is constructed with a metal truss frame with about 25 ft, or 7.62 m, between the front and rear wheel assembly supports. Each wheel assembly consists of six averaging rubber-tired wheels arranged so that the support wheels can reflect the mean deviations of the center wheel. Because its front and rear support wheels are in contact with the pavement surface, the profilograph cannot measure the pavement profile accurately.

RTRRMs, also known as "road meters," have a long history of collecting pavement roughness data. RTRRMs measure the vertical motions of the rear axle of an automobile or the axle of a trailer relative to the entire vehicle body as the vehicle travels over a pavement. The transducer records small vertical distance changes between the axle and the vehicle body. The two major limitations are that (i) RTRRMs cannot measure pavement profile. They record the dynamic response of the mechanical system traveling over a pavement at a constant speed. The characteristics of the mechanical system and the traveling speed affect the data; and (ii) RTRRM measures made by one system can hardly be reproducible by another, which leads to a high cost if accurate, consistent, and repeatable data is required. Therefore, this method is not recommended.

Profiling devices provide accurate, scaled, and complete reproductions of the pavement profile within a certain range. They eliminate the disadvantages of RTRRMs—for example, they are time-consuming and labor intensive in the data calibration process—and can also be used to calibrate RTRRMs. The devices measure, compute, and store the profile through the creation of an inertial reference by using one or two accelerometers on the body of the vehicle to measure the body vertical motion in one or both wheel paths. The accelerometer located on top of the height sensor records the vertical acceleration of the vehicle over time. The relative displacement between the accelerometer and the pavement surface is measured by an acoustic, optical, or laser height sensor. The acceleration is mathematically integrated to vertical displacement. The distance measuring system keeps tracking the distance from the reference starting point over time. Data from the height sensor and the accelerometer are combined to compute the profile of the pavement using Equation 3.27

$$E(t) = \iint_{0}^{t} A(\tau)d\tau - H(t) \tag{3.27}$$

where

$E(t)$ = The relative elevation of the vehicle with respect to the reference starting point at time t

$A(\tau)$ = The vertical acceleration rate measured by the accelerometer at time τ

$H(t)$ = The relative displacement between the accelerometer and the pavement surface measured by the height sensor

With data on travel distance collected over time, the elevation over time can be converted to a profile.

Pavement rut depth: Rut depth is another important pavement performance indicator. It is computed as the difference between the height of the hump in the center and average height of left and right wheel tracks. Rut depth data is collected every 2 ft, and averaged and recorded every 10 ft. A few states used a 5-sensor system to achieve more accurate rut depth. Several devices are available to measure rut depth in the market and most of them have combined rut depth measurement and profile measurement into one system so that only one trip is needed to measure both rut depth and transverse profile if the accuracy is acceptable. The major difference among them is the types of sensors and algorithms designed to eliminate noises.

3.3.3.3 Pavement structural capacity data collection

The function of the pavement structure is to effectively carry traffic and transfer wheel loads to the roadbed soils. Structural testing is the evaluation of the load-carrying capacity of the existing pavement subsoils. Structural data is not routinely collected for pavement monitoring by most agencies. Surface deflection data is mainly used for selecting and designing specific rehabilitation strategies for pavement sections under consideration.

Exact location and frequency of structural testing within specified road sections should be carefully determined prior to seeking testing services. The tests should be limited to locations where distress and roughness surveys indicate structural problems and areas where overlays are anticipated. The results of these tests reflect the degree of structural adequacy that exists in the pavement structure. Although expensive, structural testing can considerably reduce maintenance and rehabilitation costs.

Many agencies use minimum or standard thickness for overlays. Thus, if a 2-inch, or 50-mm, overlay is the standard design and structural testing indicates that a 1.5-inch, or 40-mm, overlay will provide adequate strength, a savings of approximately 20% is realized. Structural testing can also determine the need for varying overlay thickness within a single project, thereby realizing considerable savings. For even a small project, reduced material costs easily justify the cost of structural testing.

On the other hand, an inadequate standard design can be even more costly. Nothing undermines support of a highway agency more quickly than a pavement that fails soon after construction. The same considerations apply to aggregate surfaces. The cost of maintaining aggregate surfaces in rural areas can be a significant portion of the total maintenance funds. Proper design results in one-time construction and eliminates the costly addition of aggregate at regular intervals.

3.3.3.3.1 Destructive testing

Structural evaluation includes both destructive and nondestructive testing (NDT). Destructive testing involves coring and removing surface, base, and subsoil samples for laboratory testing to determine the load-carrying capacity of the roadway. Another destructive procedure involves the excavation of pits for tests such as on-site plate bearing or field California bearing ratio (CBR) test. Samples of pavement layers and supporting soils are retrieved and tested in the laboratory to determine layer properties. The strength of the materials and types of damage present in each layer are used to determine the load-carrying capacity, the damaged layers, and the cause of structural failure. This information can then be used in a design and analysis procedure to determine whether the pavement is structurally adequate for current and projected traffic loadings.

3.3.3.3.2 Nondestructive testing

The NDT can also be used to evaluate the structural adequacy and load-carrying capacity of an existing pavement. NDT provides measurements of the overall pavement response to an external force or load without disturbing or destroying the pavement components. NDT has many advantages over the destructive testing methods in that it provides *in situ* properties of the pavement conditions, does not damage the pavement, minimizes laboratory tests, and is fast.

The application of loads on a pavement surface includes strains (ε) in the underlying layers causing stresses in all layers. The summation of all vertical strains in the pavement structure and in the underlying subgrade represents the surface deflection (δ) of the pavement. The deflection value is considered an excellent indicator of pavement strength; in other words, once deflection exceeds a certain limit, the pavement is certain to show some kind of structural weakness. Thus, a weaker pavement will deflect much more than a stronger pavement at a given load. A number of NDT devices have been developed in recent years and are being used in the pavement structural evaluation analysis. All of the NDT devices provide some measure of surface deflection of in-service pavements in response to an external load.

The NDT devices that are available in the U.S. to evaluate the *in situ* properties of pavements are (i) Benkelman beam; (ii) Dynaflect; (iii) Road Rater; (iv) falling weight deflectometer (FWD); (v) rolling deflectometer; and (vi) ground-penetrating radar (GPR). The first five devices operate by measuring the pavement response to an imposed force. The response is generally in terms of surface deflections at one or more points on the pavement. Major differences between these devices include the load levels, the way the load is applied to the pavement, and the number of points at which deflections are measured. A device that applies a static or slowly moving load is the Benkelman beam. The common devices that apply a vibratory steady state load to the pavement surface are the Dynaflect and the Road Rater. The devices that uses an impulse loading are the FWDs. The rolling deflectometer is still under development in the U.S.

The two primary NDT methods are vibratory and falling weight. Although both devices produce useful analyses of low-volume pavement structures, the FWD more closely approximates a heavy moving wheel load. FWDs induce a heavy enough load to yield meaningful results in rigid pavements. NDT analysis requires knowledge of the existing pavement structure in terms of layer types and thickness. Coring of pavements may be necessary to support the analysis of NDT data. An NDT analysis will typically result in an evaluation of remaining service life of a pavement in terms of 18-kip equivalent single axle loads (ESALs), and an overlay thickness design.

3.3.4 Bridge condition data collection

Bridge inspection methods can also be categorized into destructive methods and nondestructive methods (FHWA, 1987). Destructive methods including coring and chipping, and nondestructive methods are those that test the properties of a material, component or system without causing any damage, including hammer sounding or chain dragging. This section will introduce some commonly used NDT methods.

Hammer sounding: The most inexpensive nondestructive method is the hammer sounding or chain dragging. This method has chain dragged or a hammer moved over the deck surface of the bridge; the worker focuses on listening to the change in sound (tone or pitch) made and marks the spots considered suspect. The method may not work well on bridges with asphalt overlays. Moreover, people have different degrees of hearing ability, which makes the result subjective to some extent. In addition, environmental noise such as traffic can affect the hearing result. All these factors could reduce the reliability of this method. Note that this method is usually used to detect delamination conditions where rebar is significantly corroded. However, this method will not detect an area where corrosion is well advanced but not yet at the delamination stage.

Visual inspection: Visual inspection is a low-tech method for identifying obvious cracks, spalling, and potholes. It only pinpoints apparent surface damage and it certainly cannot infer condition of the concrete's interior. In this way, it tells you only the obvious.

Half-cell potential: The half-cell potential technique is a more sophisticated method. It measures the voltage between the rebar inside the concrete and a reference electrode on the surface of the concrete. Because of the high sensitivity to rebar corrosion, this approach could detect corrosion before delamination has happened. However, this approach can only work on bare concrete, so it is not applicable for bridges with asphalt overlays. The implementation of this method is time-consuming and requires closing the bridge. The half-cell potential method is best for knowing whether a deck needs repair, without knowing the exact location and type of repairs.

Infrared: The infrared method uses the signal received to detect corrosion. More specifically, it indicates corrosion and cracks according to variations in infrared radiation emitted from the surface of the concrete. This work can be easily done by driving a vehicle or boat with the equipment mounted, capturing the infrared image of the bridge. However, the infrared technique has to be done when there is a big difference between the temperature of the bridge surface and the ambient air. Also, it cannot be done on bridges with asphalt overlays. This method could be considered a more advanced version of visual inspection, belonging to image processing methods. Similar methods use high-definition video and digital camera, to name only two.

Ground-penetrating radar: The most technically complicated nondestructive method uses GPR. This method uses radar pulses to generate an image of the subsurface. Generally, GPR can be widely used in soil, rock, ice, fresh water, pavements, and other structures. It is similar to an x-ray: electromagnetic radiation in the microwave band is collected when bouncing back from the detected structures. The reflected signals indicate changes in objects, material, voids, and cracks in different patterns. Therefore, GPR can be used for testing asphalt pavement, concrete pavement, and bridge decks.

Ultrasonic: The ultrasonic method is a relatively new method to test welds, bolts, and rivets on steel members. The ultrasonic method also allows the visualization of perpendicularly arranged reinforcement bars or tendon ducts. This method relies on measuring the trip time of ultrasonic pulses propagating through an object to be tested. The test system consists of three main components. The ultrasonic pulse echo transducer produces the ultrasonic pulse traveling through the structural component to be tested. A computerized data collection system gathers the data of the ultrasonic pulse signal. Last of all, a spatial control system tracks the movement of the transducer while testing and it consequently could correspond the signal of discontinuities to the location on or inside the component.

Eddy current: The eddy current method could test welds and detect residual stresses in steel components of any shape. The eddy current method works by placing an energized probe on the surface of the steel component to be tested. A current will be induced on the surface of the test component with a certain magnitude and phase if the energized probe is well calibrated with a correct frequency. The eddy currents could reflect the conductivity of the steel, and hence indirectly reflect the interior structure of the steel component. When the eddy current travels through a discontinuity (e.g., crack) in the weld or steel component, the current will have a disruption as a result, which could be instantly visualized on a device with the size and location mapped.

Impact-echo: The impact-echo (IE) method is one of the most commonly used methods of testing concrete and masonry structures. The IE method relies on the use of impact-generated stress (sound) waves that travel through concrete and masonry. The internal flaws and external surfaces with size and location details will be reflected consequently. Such flaws include cracks, delamination, voids, honeycombing, and debonding in plain, reinforced, and post-tensioned concrete structures. The IE method is capable of testing plates (e.g., slabs, walls, decks), layered plates (e.g., concrete with asphalt surface), columns and beams, and hollow cylinders (e.g., pipes, tunnels, mine shaft liners), which could be used to map voids in the grouted tendon ducts of many types of post-tensioned structures. It also has a relatively high accuracy performance in measuring the thickness of concrete slabs, and it can locate voids beneath slabs and pavements in the subgrade. The method is commonly used to locate cracks, voids, and other flaws in masonry structures which use mortar to bond the brick or block units.

Magnetic flux leakage: Magnetic flux leakage (MFL) testing is a widely used method for detecting corrosion and pitting in steel structures. MFL is often used for assessing the integrity of pipelines and storage tanks, but the principle can also be applied to transportation industries. The elementary principle of MFL involves magnetizing a ferrous metal object to saturation level with a powerful magnetic field. The magnetic flux will stay undisturbed if the test object is flawless. However, if there exist internal or external metal defects (such as cracks or corrosion), the magnetic flux leaks from the object.

3.3.5 Traffic control and safety hardware data collection

3.3.5.1 Traffic sign and pavement marking data collection

Various kinds of reflective sheeting are used on road signs and markers to enhance the readability and perception of information displayed during low light and night time conditions. Unfortunately, the effectiveness of these reflective materials tends to deteriorate over time. Retroreflectivity (defined as the ability of a material to reflect incident light back toward its source), specified in candelas per lux per square meter (cd/lux/m^2), is an important characteristic utilized by transportation agencies to assess the night time visibility of road signs.

RetroView: This is an automated system that provides a safe, cost-effective method to measure the retroreflectivity of the signs and markers. Accurate reflectivity data can be collected from a moving vehicle, called a Digilog VX data collection system, traveling at posted road speeds. Luminance values are determined on the fly using an active infrared light sensor triggered at periodic intervals along a route with a high-intensity black and white digital image.

Digital imaging: The camera is attached to an adjustable mount inside the vehicle on the front windshield for easy access and protection from outside elements. The camera is positioned at the driver's eye level with a right-of-way view. This field of view includes the lane of travel, street signs, guide signs, mile markers, pavement markings, and overhead signs. When the system has a second camera installed, this camera is also mounted at the driver's eye level, but angled to the right approximately 55°. The dual camera extended field of view includes guide signs, roadside features, billboards, vegetation, and terrain. The camera positions are fully adjustable and can be locked once the desired position is acquired. The data collection software is designed to handle multiple cameras simultaneously.

RetroChecker: This is a convenient rechargeable measuring instrument for the determination of the coefficient of retroreflection of retroreflective materials (night time visibility). Additional equipment like data storage, GPS, barcode reader, and corresponding software are embedded with the RetroChecker RC-2000 to build up a comprehensive traffic sign inventory. To carry out a measurement, the RC-2000 is positioned directly on the traffic sign under examination and it immediately delivers results with laboratory precision. GPS data for localization and bar codes for identification of traffic signs are recorded and stored automatically and can be transferred to a personal computer via a serial interface. The RC-2000 can be used for all types and colors of retroreflective sheeting.

Line-inspector: The line-inspector (Vonderohe et al., 1993) combines precision measurement of the night time visibility (R_L) and daytime visibility (Q_d) in one compact instrument. It is suitable for all types of road markings—smooth, textured, profiled colored or white, with or without aggregates and reflective beads, for wet and dry measurement.

RoadVista: The RoadVista is the retroreflectometer designed for use in field measurements of raised retroreflective pavement markers (RPMs). It is designed to measure the retroreflection (R_I) of all RPMs. The internal memory will store up to 10000 measurements before downloading to a computer. The 1200F connects to any standard computer using an RS-232C interface. Its mapping software allows the user to plot the data on a clickable and printable map to quickly identify problematic areas that need attention.

StripeMaster II: This measures retroreflectivity of pavement markings accurately on wet or dry surfaces and it has an RS-232C interface that allows data to be downloaded into any traffic inventory system. The StripeMaster II is designed for quick and easy measurements with the touch of a button. An internal GPS and printer come standard, allowing the field user to quickly verify the data. The internal memory will store up to 10000 measurements before downloading to a computer.

Retroreflectometer: Retroreflectometer (RM) is a sophisticated mobile highway retroreflectometer system with proven optical head technology, and the system objectively measures the retroreflectivity of pavement markings using a scanning laser source. The system ensures that measurements are collected quickly and accurately by incorporating new high-speed data acquisition electronics and software, and precise angular settings of illumination and angle of observation. The system provides real-time pavement marking reflectivity and can be used either day or night at variable traffic speeds. Some notable products include Model 930C Laboratory/Field Retroreflectometer and Laserlux® CEN 30 mobile retroreflectometer.

Impulse RM for signs: This is a hand-held device manually directed toward, or precisely at, a target object and then manually fired. Once fired, the hand-held device bounces a laser beam off the target object and measures the reflected laser energy that is then used to determine a retroreflectivity value. Sign retroreflectivity (R_A) is usually measured with an impulse RM that is pointed at the face of a sign at a distance of about 100 ft. The resulting R_A value correlates roughly to a geometry of 0.2/−4 (observation angle and entrance angle), which has become a generally accepted standard geometry in the industry for signs. The hand-held device can measure a single R_A value for a single sign, and can only measure either foreground or background R_A with a single measurement.

Sign Management and Retroreflectivity Tracking System (SMARTS): Another technique for determining the night time visibility of signs has been introduced by the Federal Highway Administration (FHWA, 2001a). The SMARTS is a vehicle that contains one high-intensity flash source, one color camera, two black and white cameras, one range-sensing device, and a GPS-positioning system. The SMARTS vehicle requires two people—a driver and a system operator—for proper operation. As the SMARTS vehicle travels down the road, the system operator locks on to a sign up ahead by rotating the camera and light assembly to point at the sign. At a distance of 60 m, the system triggers the flash source to illuminate the sign surface, an image of which is captured by one of the black and white cameras. A histogram is produced of the sign's legend and background that is then used to calculate retroreflectivity. A GPS system stores the location of the vehicle along with the calculated retroreflectivity in a computer database.

3.3.5.2 Traffic lighting illumination level data collection

The quantity of light or the light output and light levels are measured in lumens, lux, and foot-candles. Initial lumens/foot-candles reflect the amount of light produced by a lamp when it is installed. Supply voltage variations, the lamp's interaction with the ballast, and dirt build-up reduce the produced amount of light. Light level standards are affected by light quantity, the desired quality of fixture efficiency, and other applicable factors. The following sections describe some of the technologies applicable to measure the level of illumination emitted from a light source.

HISLAT: The HISLAT® unit is a very portable, laptop computer-based unit for mobile street light assessment. It is designed to be used at normal driving speeds (up to 62 mph), thus eliminating the need for road closures or other expensive traffic management plans. The data is collected automatically from a moving survey vehicle by using specialist sensors and equipment that can be easily installed on that vehicle. Although the survey equipment has been designed for simple operation and automatic data collection, two operatives are required to carry out the survey. This allows the driver to concentrate purely on safe driving. The transducer is the only item of equipment that is permanently installed in the survey vehicle. The transducer sends a series of pulses to the displacement meter, which is programmed to display to the nearest meter. The displacement meter then sends a suitable signal to the data sampler. An initial displacement calibration is required for each survey vehicle. The photodiode is located on the roof of the car in the appropriate location. The photodiode is powered from the light adaptor inside the car. The luminance figures are sent to the data logger on a continuous basis. The photodiode needs to be checked for calibration on a regular basis. The data logger then sends both figures to the laptop for recording in a database. This allows the light level to be recorded for each displacement pulse along the road. The data sampler software allows recording and displaying of the luminance values in the graph or spreadsheet format and transfer into Microsoft Excel. The results can be represented graphically as a lux level/displacement plot. The key performance measure (KPM) is defined as the average lux per subgroup function that is the calculated area beneath the plot line divided by the length (m) over which the survey is completed.

Spectroradiometer: This is able to measure the spectrum of different light sources from 250 to 800 nm or from 200 to 400 nm depending on the calibration and luminous intensity. The system saves the measured values in an ASCII-file on a laptop computer from where it is easy to calculate various lighting quantities directly after the measurement or at a later stage.

Luminancephotometer: This is used for luminance measurements and analysis of indoor and outdoor lighting. The photometer consists of a Peltier-cooled charge-coupled device (CCD)-based still-frame camera and a computer. Simultaneous luminance values of the whole scene are captured in a few seconds. The image consists of 250000 pixels. The results can be saved as numerical values for later analysis. The luminance range that can be measured with the ProMetric 1400 is from 0.005 to 1010 cd/m^2. For road lighting measurements, a program has been developed at the Lighting Laboratory which calculates the International Commission on Illumination (CIE) road lighting parameters from the ProMetric measurement results.

3.3.6 Travel time data collection

3.3.6.1 Conventional travel time data collection

Test vehicle techniques: The test vehicle technique has been used for travel time data collection since the late 1920s. Traditionally, this technique has involved the use of a data collection vehicle within which an observer records cumulative travel time at predefined checkpoints along a travel route. This information is then converted to travel time, speed, and delay for each segment along the survey route. There are several different methods for performing this type of data collection, depending upon the instrumentation used in the vehicle and the driving instructions given to the driver: (i) manual; (ii) distance measuring instrument (DMI); and (iii) GPS.

Historically, the manual method has been the most commonly used travel time data collection technique. This method requires a driver and a passenger to be in the test vehicle. The driver operates the test vehicle while the passenger records travel time at predefined checkpoints using pen and paper, a tape recorder or a portable computer.

Technology has automated the manual method with the use of an electronic DMI. It is connected to a portable computer in the test vehicle and receives pulses at given intervals from the transmission of the vehicle. Distance and speed records are then determined from these pulses. The DMI unit is able to send the data to a portable computer for storage. Specialized software can be used to record the electronic information, eliminating the data entry and errors associated with the older models. Notes can also be added to the end of the file to describe incidents or other relevant information about the travel time run. The technique is used for a variety of applications, such as route numbering, emergency 911 addressing, and acreage and volume calculations, as well as general linear distance measuring for pavement markings. These instruments are very accurate once calibrated, at plus or minus 0.19 m/km (1 foot per mile). However, there were some problems with this technology. Wheel sensors would fall off or not read properly and sometimes unbalance the wheel. Data media was paper format, either circular graphs or adding machine tape, which were difficult to read and required large amounts of data entry.

The GPS was originally developed by the U.S. Department of Defense (DOD) for tracking military ships, aircraft, and ground vehicles. Signals are sent from the 24 satellites orbiting the earth at 20120 km. These signals can be utilized to monitor location, direction, and speed anywhere in the world. A consumer market has quickly developed for many civil, commercial, and research applications of GPS technology including recreational (e.g., backpacking, boating), maritime shipping, international air traffic management, and vehicle navigation. The vehicle location and navigation advantages of GPS have found many uses in the transportation profession. Due to the level of accuracy that GPS technology provides, the DOD has altered the accuracy of the signal for civilian use. This is called selective availability (SA), and when it is activated precision can be degraded to about 300 ft, or 91 m. In the absence of SA activation, accuracy can be within 60 ft, or 18 m. However, with the use of the differential global positioning system (DGPS), accuracy can be improved. It utilizes a receiver placed at a known location to determine and correct the signal that is being provided when SA is activated.

License plate matching techniques: License plate matching techniques consist of collecting vehicle license plate numbers and arrival times at various checkpoints, matching the license plates between consecutive checkpoints, and computing travel times from the difference in arrival times. Four basic methods of collecting and processing license plates are commonly used: (i) manual; (ii) portable computers; (iii) video with manual transcription; and (iv) video with character recognition.

For manual license plate recognition, pen and paper or audio tape recorders are used to manually enter license plates and arrival times into a computer. When portable computers are used in the field for license plate recognition, the arrival time stamp for each vehicle is automatically provided. The video with manual transcription technique collects license plates in the field using video cameras or camcorders, which are manually transcribed using human observers. The video with character recognition technique collects license plates in the field using video, and then automatically transcribes license plates and arrival times into a computer using computerized license plate character recognition.

3.3.6.2 Real-time traffic data collection

Probe vehicle techniques: The probe vehicle techniques are unique in that they are typically intelligent transportation system (ITS) applications designed primarily for collecting traffic data in real time. Their primary application is for a specific purpose other than travel time data collection, such as real-time traffic operations monitoring, incident detection, and route guidance applications. However, these techniques can be used for the collection of travel time data. Since the probe vehicles are used for travel time data collection but are already in the traffic stream for a different purpose, they are sometimes referred to as "passive" probe vehicles. Coordination is often necessary between the agency responsible for the system operation and the agency that would like to utilize the system for travel time data collection. This distinction removes these probe vehicle techniques from other techniques, such as the test vehicle and license plate matching techniques.

Probe vehicle techniques generally include (i) automatic vehicle identification (AVI); (ii) ground-based radio navigation; (iii) cellular geo-location; and (iv) GPS-equipped probe vehicle technique. For the AVI technique, probe vehicles are equipped with electronic tags that communicate with roadside transceivers to identify unique vehicles and collect travel times between transceivers. The ground-based radio navigation technique is often used for transit or commercial fleet management. This system is similar to the GPS technique with data collected by communication between probe vehicles and a radio tower infrastructure. The cellular geo-location technique can collect travel time data by discretely tracking cellular telephone call transmissions. For the GPS-equipped probe vehicle technique, probe vehicles are equipped with GPS receivers and two-way communication to receive signals from earth-orbiting satellites. The positional data determined from the GPS signals are transmitted to a control center to display real-time positions of the probe vehicles. Travel time information can be extracted from the data.

Nonintrusive detection technologies: Nonintrusive detection techniques use devices that cause minimal disruption to normal traffic operations and can be deployed more safely than conventional detection methods. These devices do not need to be installed in or on pavements, but can be mounted overhead, to the side or beneath pavements by inserting the devices from the shoulder. The nonintrusive detection techniques mainly include (i) infrared detection; (ii) magnetic detection; (iii) Doppler microwave devices; (iv) acoustic detection; (v) ultrasonic detection; and (vi) video detection.

Infrared detection could be passive or active. Passive infrared devices detect the presence of vehicles by comparing the infrared energy naturally emanating from the road surface with the change in energy caused by the presence of a vehicle. Since the roadway may generate either more or less radiation than a vehicle depending on the season, the contrast in heat energy is what is detected. Active infrared devices detect the presence of vehicles by emitting a low-energy laser beam(s) at the road surface and measuring the time for the reflected signal to return to the device. The presence of a vehicle is measured by the corresponding reduction in time for the signal return.

Magnetic detection could be passive or active. Passive magnetic devices measure the change in the earth's magnetic flux created when a vehicle passes through a detection zone. Active magnetic devices, such as inductive loops, apply a small electric current to a coil of wires and detect the change in inductance caused by the passage of a vehicle.

Doppler microwave devices transmit low-energy microwave radiation at a target area on the pavement and then analyze the signal reflected back to the detector. According to the Doppler principle, the motion of a vehicle in the detection zone causes a shift in the frequency of the reflected signal. This can be used to detect moving vehicles and to determine their speed. Radar devices use a pulsed, frequency- or phase-modulated signal to determine the time delay of the return signal, thereby calculating the distance to the detected vehicle. Radar devices have the additional ability to sense the presence of stationary vehicles and to sense multiple zones through their range-finding ability. A third type of microwave detector, passive millimeter, operates at a shorter wavelength than other microwave devices. It detects the electromagnetic energy in the millimeter radiation frequencies from all objects in the target area.

Passive acoustic devices consist of an array of microphones aimed at the traffic stream. The devices are passive in that they are listening for the sound energy of passing vehicles.

Ultrasonic detection could be based on pulse or Doppler principles. Pulse devices emit pulses of ultrasonic sound energy and measure the time for the signal to return to the device. Doppler devices emit a continuous ultrasonic signal and utilize the Doppler principle to measure the shift in the reflected signal.

Video devices use a microprocessor to analyze the video image input from a video camera. Two basic analysis techniques are used: tripline and tracking. Tripline techniques monitor specific zones on the video image to detect the presence of a vehicle. Video-tracking techniques employ algorithms to identify and track vehicles as they pass through the field of view. The video devices use one or both of these techniques.

3.3.7 Transit performance data collection

Many of the transit performance data, especially supply data, could be found from transit planning and operation records, such as transit right-of-way, location, equipment on bus stops, and schedules. Data on transit usage must be obtained through field observations (Vuchic, 2005).

3.3.7.1 Speed-and-delay survey

The objective of a speed-and-delay survey on transit lines is to investigate the time profile over the service. Specifically, the speed-and-delay survey should be able to present the time a transit unit spent on running, accelerating, decelerating, waiting at stops, and other types of delay. This measured time profile is important in operation speed computation and service reliability evaluation.

A manual method is performed by asking an observer to ride the transit vehicle, recording the time profile as well as the reason for each slowdown and stop. This process should be conducted several times for morning peaks, afternoon peaks, off-peaks, and weekends. The method is simple and the equipment required is just a field sheet, clipboard, pen, watch, and stopwatch. However, the labor cost and human factor are two of largest disadvantages of this method. The observer is trained by detailed instructions on how to define the component of travel time (e.g., stop time, dwell time) and the reason that causes the delay. Therefore, this reason-identification process could be more or less subjective.

Speed-and-delay data could also be measured by using some automated methods. For example, it could be measured by a system that could detect drivers' maneuvers, and communicate the vehicle's power supply and door opening and closing system as well as the vehicle positioning system. This system could record the spatial and temporal distribution of the transit vehicle along with the doors' opening-or-closing status. The data generated from this system could be directly loaded into an analysis program or database after a minor reformatting. The major advantages of this automated method over a manual method are low long-term costs, accuracy, and less subjectivity.

3.3.7.2 Passenger volume and load count

The objective of passenger volume count is to investigate the passenger volume on transit vehicles over a transit line. The result should be the passenger load counts at several sections along the transit line. The result can identify the sections with heavy passenger loads, which is critical to locate the maximum load sections (MLS).

The manual method will require an observer to count the passenger load at each counting station. The number of observers required depends on the demands of the transit line. If the route is heavily loaded or multiple vehicles might be loading simultaneously, more observers may be required. Similarly, the observer needs to have several skills. For example, the observer should be able to estimate the number of passengers within a short time interval because the standing time is usually short. The equipment needed consists of a field sheet, clipboard, pen, watch, or a specially designed electronic recorder.

3.3.7.3 *Passenger boarding and alighting counts*

The detailed passenger boarding and alighting counts, as indicated by the name, provide the number of boarding and alighting passengers at each station along a transit route. These counts could provide a lot of other information, including the aforementioned passenger volume and load counts, passenger miles of travel, and trip production and attraction of each station during a certain time interval.

A manual method would require one or more observers, depending on the size of the transit vehicle. For a regular bus with two doors on a low-demand transit route, one observer can cover each bus. For a light rail train with multiple doors, more observers should be set as assistants, especially at busy stations. The observer should also have the fast-estimation skill required in passenger volume and load count. The equipment required, similarly, consists of a field sheet, clipboard, pen, watch, or a specially designed electronic recorder.

The invention of automatic passenger counters (APC) makes it possible to record the boarding/alighting passenger numbers by scanning passenger movements or detecting passenger steps near doors. The advantages of using an APC to conduct passenger count include the labor cost savings in the long run, achieving more accurate and reliable data, and simplifying the follow-up work load, such as typing data into a database. With electronic collection becoming increasingly popular, the fare collection system could also help conduct boarding/alighting passenger counts. One extreme case is the fully controlled bus rapid transit (BRT) system, in which the passenger needs to slide or insert the "BRT pass" (e.g., a prepaid transit ticket, cellphone, credit card, etc.) when boarding/alighting the bus. In this case, the system could achieve not only the number of boarding/alighting passengers, but also the complete trips table along this route. This system could be expanded to the entire transit network, meaning that the transfer data could also be possibly collected.

3.4 DATA COLLECTION FREQUENCY

The data details on transportation facility and system usage need to be upgraded at regular intervals to assist the asset management decision-making process. Data collection is cost- and time-intensive and consumes many resources. Depending on the needs of demand versus supply, passenger versus freight, and facility versus system usage, as well as geographic coverage, the sampled data can be collected annually, semiannually, quarterly, monthly, weekly, daily, hourly, every 15 minutes, or in real time.

3.5 DATA QUALITY ASSURANCE

Quality assurance (QA) for data collection is an ongoing process that could be achieved using system design, statistical methods, training, and auditing to maximize various attributes of data. Quality is not free, but the expense of QA does tend to pay for itself later. For example, increased quality of bridge inspection saves the expense of sending crews to a bridge site for repairs that turn out not to be needed or setting up emergency repairs on a bridge where an existing problem had not been detected. Several important attributes are essential to data QA (Fekpe et al., 2003; FHWA, 2013):

Accuracy: Accuracy is a measure of rightness, indicating whether the estimation achieves the correct value. It refers to how close a sample statistic is to a population parameter. Thus, if a sample mean of the collected data is 99 and the true population mean is 100, the sample accuracy of data is assured.

Precision: Precision is the measure of exactness, indicating the size of estimation range of the parameter. It refers to how close estimates from different samples are to each other. For instance, the SE is a measure precision. A smaller SE indicates higher precision and vice versa, as illustrated in Figure 3.2.

Coverage: The extent of coverage of a data set is a key design decision and a key limitation on its usefulness. An inventory of state highways, for example, is of no help to project-level needs identification for local roads.

Timeliness: Timeliness refers to the age of data at the time they are used. Timeliness must balance several competing requirements: for example, the appropriate point or season of the year in which to conduct inspection surveys; the need to process data for use in management system(s); and the need for the resulting information by one or more organizational units in assessing current condition, comparing actual to planned accomplishments, identifying work needs, developing a program budget, and so on.

Detail: An appropriate level of detail is an application-specific requirement that is often a matter of definition. PMS, BMS, and MMS need unit cost data at a level of detail that matches the definitions of treatments that their models can analyze.

Accessibility: This attribute of data quality refers to the ease with which the data can be put to use. Weigh-in-motion data, for example, are an excellent resource for truck weights, but are often useless unless processed to the needed level of detail.

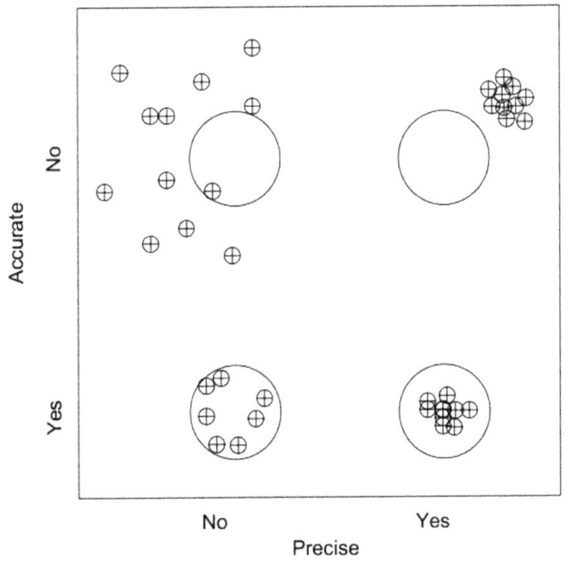

Figure 3.2 Illustration of data precision and accuracy.

Assumptions and definitions: Data sources may have definitional differences or inherent assumptions that make them more or less useful for asset management applications. For instance, definition and interpretation of pavement condition data may differ between pavement management and maintenance management.

The 2010 Highway Capacity Manual (TRB, 2010) suggests that, in most cases, the estimates can only be expected to be accurate to within 5%–10% of the true value, as represented by the circles in Figure 3.2. The precision of such an estimate is the range that would be acceptable from an analyst's perspective in providing an accurate estimate, as represented by the deviation of the estimate group. Since the error of the estimates or observations will be cumulated in further computations, the computations performed with these inputs cannot be expected to be extremely accurate, and the final results could be considered as accurate and precise only if the inputs are under some constraints.

The level of data accuracy and precision required varies from place to place depending on the usage. For example, traffic volume data is widely used for most cases. AADT with seasonal adjustment factors are precise enough for long-term transportation planning purposes. However, in the case of traffic control operations, such as adaptive signal timing control, the volume data collected is summarized for each 5 min to represent the variation.

3.5.1 Quality standards

When two or more information systems share the same data source, it is important to have a formal, documented quality standard, describing minimum and maximum requirements along all the dimensions noted above, that meet the needs of the stakeholders using the systems. This serves as a multilateral agreement among the end users, data collectors, and system developers, an agreement that should not be modified without involving all of these stakeholders. Upper managers do not have to be involved in developing these standards, but they do need to insist that the standards are developed. Data quality standards are an essential management tool: they are directly connected to budgetary requirements for data collection, and they provide a streamlined way for upper management to ensure that conflicts regarding data quality are resolved. With this tool, a manager responsible for a data collection budget can express the impacts of budgetary increases or decreases in terms of changes to the data quality standard, and their effect on specific end users.

3.5.2 Quality assurance

The quality assurance (QA) processes require a context of documented standards, and they are the mechanism by which adherence to standards is measured. Senior managers are not typically involved in QA personally, but the existence of QA processes, and periodic effectiveness measures, are what provide managers the needed control and assurance. The first point of QA is the training of data collectors and equipment operators.

QA with respect to the use of fully automated data collection equipment includes defining standards for measurement, planned equipment testing and certification, applying calibration procedures prior to surveys, and verifying calibration following surveys. After data is collected, a number of methods are available to measure adherence to the quality standards. These include reinspection strategies, consistency checks, stakeholder surveys, and formal data audits.

Reinspection: This is a standard procedure in any sizeable data collection process to devote a portion of the resource, often 5%–20% depending on the consequences of error, to recollect a sample of data using similar or better equipment and/or personnel. For example, after a section of road is completed with a pavement survey vehicle, an agency might use an alternative vehicle, a different crew or even profile measurements made by land surveying equipment to double check the initial data. Locations for the recheck are typically chosen by random sampling, and statistical methods are available for deciding how many locations to check. The results of these checks are tracked over time as a performance measure. Sometimes crews compete and are rewarded according to the results of the process.

Consistency checks: Often data sets have built-in redundancy. For example, a roadway inventory may include the number of lanes, lane and shoulder widths, and traveled way width. An automated process could easily identify discrepancies needing evaluation. A well-designed information system should be able to perform these checks automatically and flag potential problems for later resolution. The ability to resolve such problems at a later time is important, since it may have to wait until the next data collection cycle or until someone can be dispatched to visit the facility. After resolving the issue, it should be possible to record an explanation and turn off the flag even if no correction is warranted. The number of such errors in newly collected data, and their resolution status, should be tracked as a quality measure.

Stakeholder surveys: For certain attributes of quality, it is efficient to ask stakeholders to report the level of quality they perceive in the information they receive, including their level of satisfaction. Although stakeholders generally cannot easily measure accuracy (except anecdotally) or precision, they can often uncover problems with coverage, timeliness, detail, accessibility, and definitions.

Data audits: Occasionally, it is useful to employ an outside agency or consultant for an independent review of data quality, especially if the consequences of incorrect information are dire. In bridge inspection, for example, it is common for districts within a state to swap inspectors periodically to give a fresh perspective.

It is very important for senior managers to recognize that data quality for asset management is relatively easy to define using the approach described here, and is highly measurable at reasonable cost. For each data item (or group of items) in an asset management database, it is reasonable to identify, along with the source of the data, the quality control process that ensures that the data will be sufficiently accurate for its intended use, according to all relevant quality dimensions. Doing this in an organized way is less expensive and more effective than an ad hoc approach, and certainly less expensive than the consequences of poor decisions that could result from incorrect or insufficient data.

3.6 DATA INTEGRATION

Transportation agency-related information management activities refer to tools and methodologies for data collection, storage, processing, analysis, and representation to support management analyses. The information achieved from these information management activities facilitates analysts in creating a systematic decision-making framework. Data integration and sharing can familiarize agencies with their existing system, and consequently support transportation decision-making with the limited resources and other constraints considered.

3.6.1 Data integration process

For transportation asset management, a data integration team contains decision makers, asset managers, database users, information technology and database management experts,

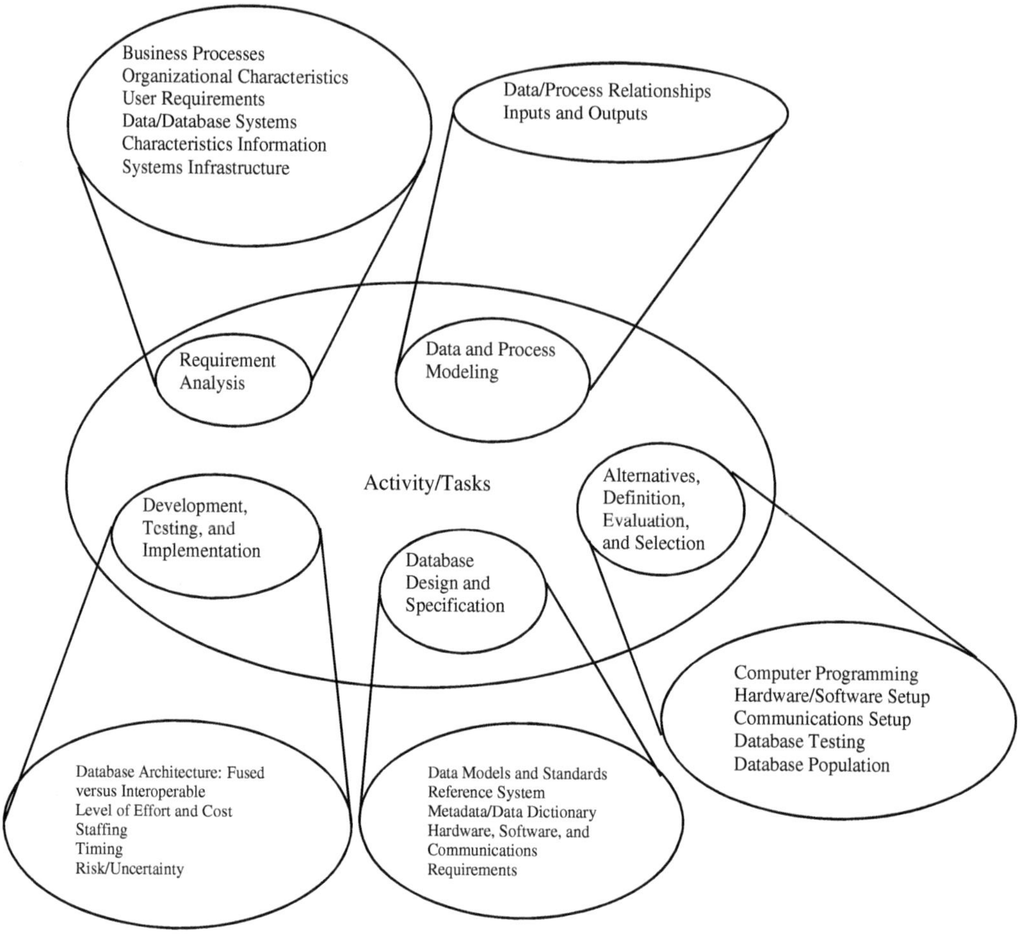

Figure 3.3 Illustration of data integration process.

and other key stakeholders (FHWA, 2001b). An external group of data integration professionals or consultants may be needed by the agency at any phase of the process.

Figure 3.3 (FHWA, 2001b) shows a generic framework along with the corresponding elements for the key activities involved in data integration. The first step in the process is to conduct a requirements analysis, including identification and analysis of the target data to be integrated. Then data and process flow modeling creates diagrams presenting the work flow and data utilization based on the information obtained from the previous step. The comprehensive information obtained can help identify and evaluate alternative data integration strategies in this process. A strategy can be selected, and its explicit database design specifications and plans developed. The last step is the integrated data strategy development, testing, and implementation.

Requirements analysis: Conducting requirements analysis is the most essential step in the data integration process. In this step, the requirements of the data integration system will be identified. These requirements include the business processes to be supported, the data to be shared, the goals to be achieved, and the expected risks or constraints that might be encountered in the process. The analysis could be very complicated and time-consuming depending on the extent of integration and data size.

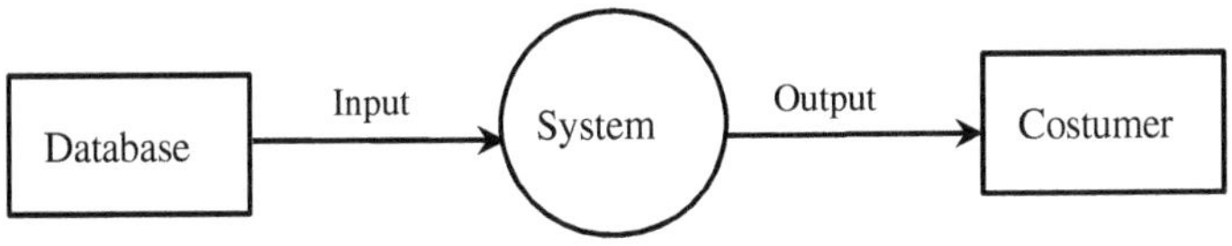

Figure 3.4 An example of a data flow diagram.

Data and process flow modeling: Following the requirements analysis step, data and process flow modeling uses the information obtained from the first step to develop diagrams illustrating the flow of data within and crossing the business processes. The purpose of this step is to create a map depicting the connections between the data and the business functions supported by data. Database technicians and analysts will benefit from these diagrams when determining the design specifications. Flow diagrams use arrows to connect data and business processes. Arrows indicate relationships where the direction indicates information flow and the interdependence between business processes. Figure 3.4 presents a simple example of the data flow diagram.

Alternatives definition, evaluation, and selection: Both the information and flow diagrams obtained from the previous two steps serve as the basis for identifying feasible data integration alternatives. Generally, fused databases and interoperable databases are two of the most commonly used alternative approaches. Data fusion, also known as data warehousing, refers to the process that combines information originated from multiple sources within a one-time integration. When data is moved to the warehouse, these data source may be removed or be kept to serve specific business processes. Interoperable database systems, also known as federated or distributed systems, comprise self-communicable databases where communications use multi-database queries. Figure 3.5 illustrates these two database options.

Database design and specifications: In this step, the detailed plans and methods are generated to guide data integration strategy implementation, as well as the overall approach to the database development effort. In the integrated database design, no matter which

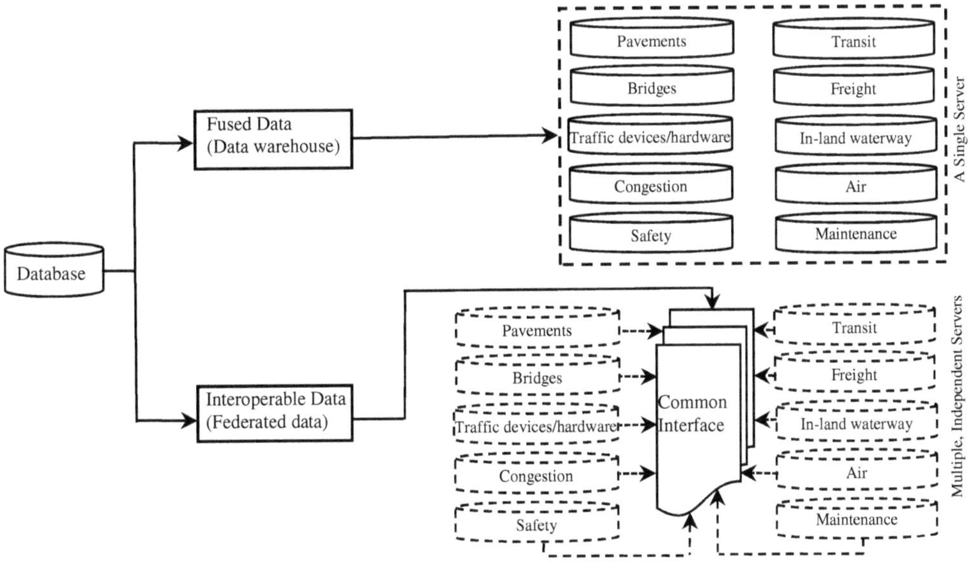

Figure 3.5 Transportation asset management database alternatives.

Table 3.4 Comparison of fused and interoperable databases

Characteristic	*Data warehouse*	*Interoperable database*
Data servers	Centralized	Distributed
Location of data server(s)	Single site	Multiple sites
Data replication	Yes	No
Advantages	Easy to manage and control the databases Maximum data processing power (quick access to the database) Able to handle large amounts of data and processing requests Provides data security	Can keep data in independent locations and file servers (autonomy of sites) No reliance on a single site that can become a point of failure. Changes made to data at one location can propagate quickly to become visible at other locations Unified description of all data; no need to know database models Allows access to resources in the computer network
Disadvantages	Requires considerable time and resources to implement Data is generally in read-only format and cannot be updated online Storage requirements can become a major problem	Hard to support and maintain integrated (global) data model Need to rebuild the database system every time data export protocols change Requires rigorous procedures for database access and updates

type of database environment is applied, the following elements are included: data models, standards, and reference systems; metadata and data dictionary; computer communication requirements, software, hardware, staffing, and data management requirements. Table 3.4 gives a thorough comparison between a fused database and an interoperable database.

Development, testing, and implementation: Software development and system implementation, serving as the last step in the data integration process, include prototyping and use case applications development, computer systems and network communications setup, and populating the database with data. Specifically, development activities refer to testing activities, evaluation activities, database model(s) modification, data management applications, and communications' interfaces. In order to easily accommodate any possible future changes to any element of the system, the development approach is recommended to be modular and incremental so that the integrated environment is flexible and robust.

3.6.2 Data transformations

Transformation of data is a necessary and essential step in making various datasets interoperable. The advancement of data collection and communication techniques enables transportation agencies to collect a large amount of variable data at a lower and lower cost. But in this process, the data format changes over time, resulting in data heterogeneity. To enhance the value of the information, newly collected data in different formats should keep being added to legacy databases. Therefore, data integration often needs to standardize and transform the data format. For example, the geoinformation coded in different spatial referencing systems should be transformed into a required spatial referencing system.

In practice, transportation agencies use a lot of location referencing methods to satisfy the data needs required by the asset management system. Transformations between various linear location referencing methods (LRMs) are always needed to support their business work flows (Baker and Blessing, 1974; White and Griffin, 1985; O'Neill and Harper, 1997;

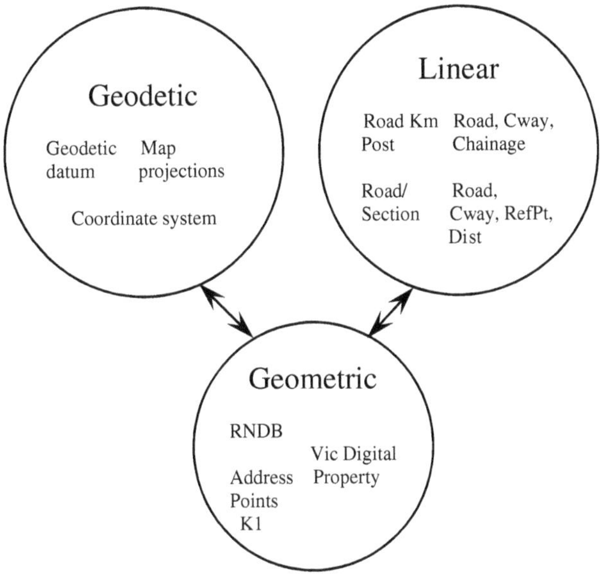

Figure 3.6 LRM groups and their relationships.

Vonderohe et al., 1997; Ries, 2000; Adams et al., 2001; ESRI, 2003). LRMs in the context of transportation data management, in general, can be organized into three groups: geodetic, geometric, and linear. Figure 3.6 illustrates these three groups. Locations are described on the earth's surface in geodetic LRMs. Discrete features on the earth's surface are represented in geometric LRMs. Locations are described along discrete features in linear LRMs.

Two approaches can be used to transform the data between two interoperable data sets: direct approach and indirect approach. The direct transformation approach transforms the data format from the current one to other required formats without requiring a neutral location reference and affecting the legacy system. On the contrary, the indirect transformation approach requires a neutral location reference to perform transformations. Figure 3.7 explains these two approaches.

The indirect approach has several advantages. First, transformations to be maintained in the indirect approach are significantly fewer than those of the direct approach. Second, if a new LRM is introduced, only two transformations (transformations between the new LRM and the existing neutral LRM) are required. Third, it makes the system interoperable. The main advantage of the direct approach is that no neutral LRM and the consequent intermediated transformations are required.

Categories of transformations: Transformation of data sets can be split into several categories. *The first category* refers to purely spatial transformation with setting a constant time variable. Transformations among linear methods, from one linear method to a two-dimensional (2D) or three-dimensional (3D) location referencing method and transformation among 2D or 3D location referencing methods fall into this category. *The second category* of transformation refers to transformations from an LRM address at a specific point of time to the equivalent LRM address at a different time. This category of transformation enables historical data analysis because the spatial and temporal features are considered. *The third category* of transformation refers to transferring a linear (distance) LRM to a linear (temporal) LRM.

Nature of transformations: Linear transformation refers to transformation among linear LRMs. Nonlinear transformation refers to the transformation from a LRM to a geometric

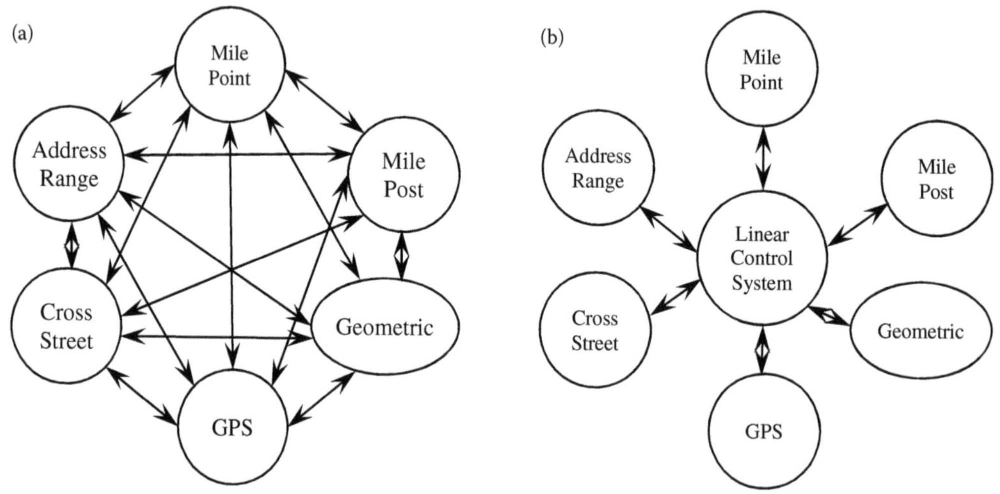

Figure 3.7 Data transformation approaches. (a) Direct approach. (b) Indirect approach.

or geodetic LRM, as well as the transformation between geometric and geodetic LRMs. Temporal transformation refers to the transformation among temporal references between temporal reference and location reference.

3.6.3 Transformation between linear reference systems

The geographic information of an object in a transportation field is always recorded and represented in a certain location referencing system. Each location referencing system records a location in a particular way, with a particular metric. WGS84 uses distance offsets from each of the X, Y, and Z axes. NAD83 uses angular measures of latitude (I) and longitude (O). The linear reference system (LRS) is a system where geographic features are localized by a measure along a linear element.

The LRS is highly suitable for the data with linear features like roads, railways, rivers, and pipelines. An event is defined in LRS by a unique route ID and a measure. The route is a basic linear element on the network, which has more geographic features in the base map. A measure is a distance measured along the linear object. For example, in the following address, 254 W 31st St., Chicago, "31st St." is the route ID and "254 W" is the measure. Figure 3.8 gives an example of recording a linear event (e.g., highway segment under reconstruction).

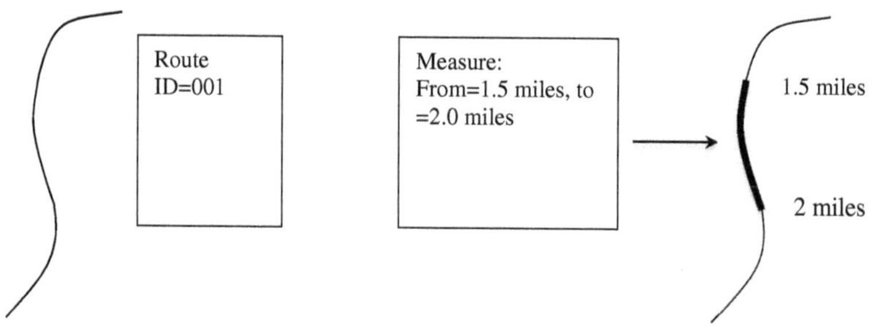

Figure 3.8 An example of a linear reference (LR) system.

The main types of LRSs-related transformations for transportation applications are as follows (Fekpe et al., 2003):

Type 1: Transformation of a 2D/3D location to a linear referenced location. For example, the conversion of collected GPS data to positions in a given LRS.
Type 2: Transformation of a linear referenced location expression to a 2D/3D expression. For example, the locations of signal lights coded in an LRS need to be converted into GPS data for some analytical purpose.
Type 3: Transformation among linear location references. For example, the standard linear location reference is updated and the old location reference needs to be updated accordingly.

3.6.4 Transformation between GPS and LRS

As illustrated in Figure 3.9, GPS captures transportation spatial features in terms of coordinates. Further, these coordinates are transformed to Universal Transverse Mercator (UTM) or state plane. The following steps present details of the transformation process (Fekpe et al., 2003):

- Computing the distance of the road centerline in 3D, starting at the beginning of the road
- Calculating the mileposts of all intersections serves as log points. This road centerline and the distance references of the log points compose the network
- For each roadway asset inventoried feature, finding the closest point on the centerline and its corresponding milepost, as well as the feature's offset from the centerline

Distance and offset are the final outputs of this transformation process. Linear features can also be positioned using this method if its endpoints can be measured. Figure 3.10 illustrates the methodology.

3.6.5 Data integration techniques

Areal interpolation: Areal interpolation means the reorganization of data from one geographic set to another (e.g., polygon). For example, transportation planning sometimes needs to downscale or upscale the study area units of their data. If population was obtained at the census block level, a transportation engineer may need the population data at the traffic analysis zonal level. In the case of large-scale redistricting, population data may be predicted for a new set of polygons. A widely used approach is the Arial weighting method

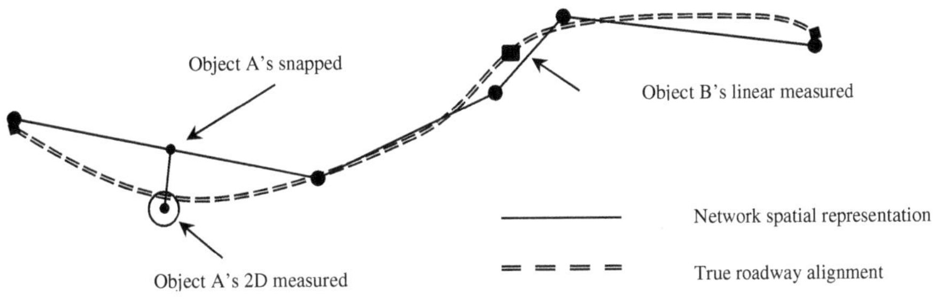

Figure 3.9 Transformation of 2D-measured position to LR position.

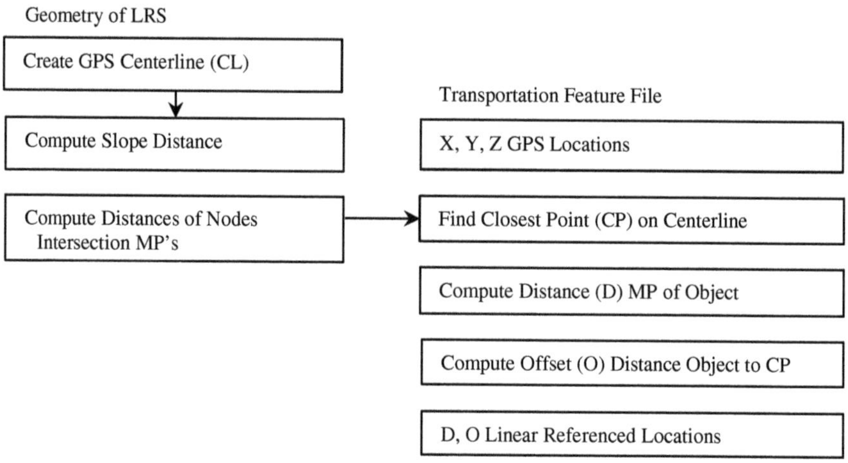

Figure 3.10 Steps in transformation of data from GPS to LRS.

that estimates an attribute value for a target zone based on degree of overlap with the source zone. Figure 3.11 illustrates this method.

Network conflation: Conflation is the process that matches GIS network layers and transfers the geographic attributes of objects from one layer to another. The goal of conflation is to merge the best-quality elements of both data sets to create a composite dataset that is better than either of them. The target layer should be preselected to better represent the geographic attributes required for decision making. Network conflation involves two major steps: (i) match links in the geometric layer with those in the attribute layer; and (ii) transfer link attribute data from the attribute layer to the corresponding links in the geometric layer. Figure 3.12 illustrates network conflation.

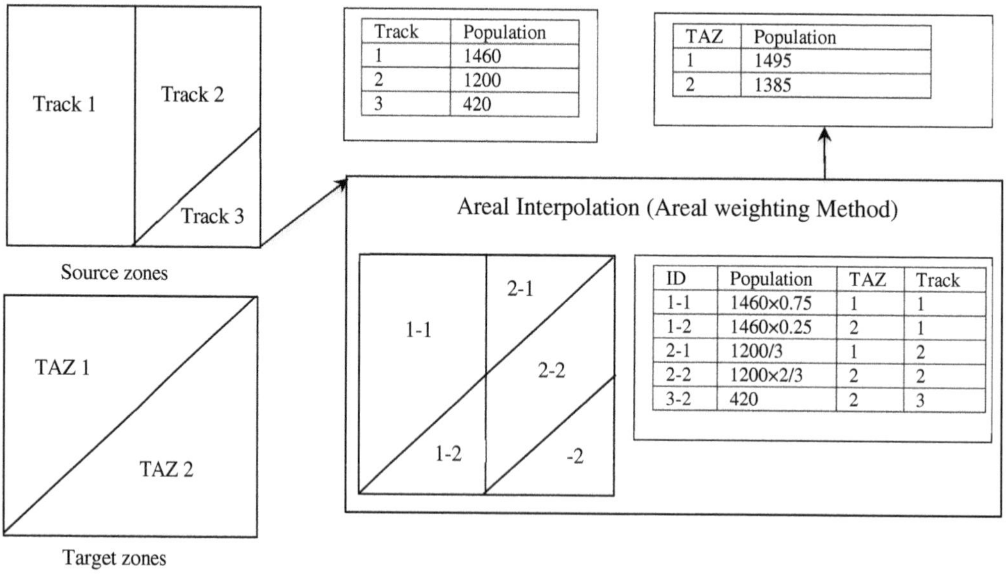

Figure 3.11 An example of the Arial weighting method.

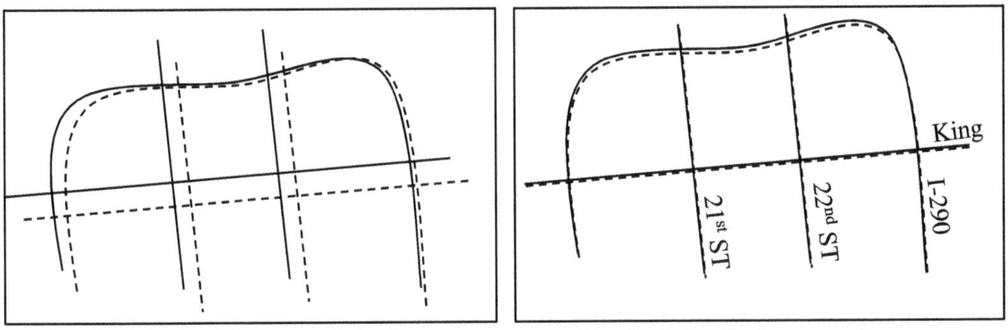

Before Conflation (Dash lines: network data from source one; Solid lines: network data from source two)

After Conflation (Road names are from data source two)

Figure 3.12 An example of network conflation.

Dynamic segmentation: Dynamic segmentation is the process of measuring the map location of an event according to the attribute table. Dynamic segmentation is developed on an LRS. Multiple sets of attributes are allowed to be jointly related with any position in a linear object. It uses two data structures: route dataset and event table dataset. In fact, many geographic locations are stored as events along the route. For instance, locations with a relatively higher crash frequency are given along with detailed geographic information. The relative position points displayed in the route base map provide information for best visualization. Figure 3.13 presents an example.

Geocoding: Geocoding is the computational process of transforming location information such as a postal address to a location in terms of geographic coordinates. This is quite useful for integrating address-based data with other types of spatial data.

Structured query language relationships: Structured query language (SQL) is the standard language for relational database management systems. SQL statements are applied to conduct data manipulation tasks, such as update data on a database or retrieve data from a database. The standard SQL commands such as "Select," "Insert," "Update," "Delete," "Create," and "Drop" can be used to accomplish almost every task needed.

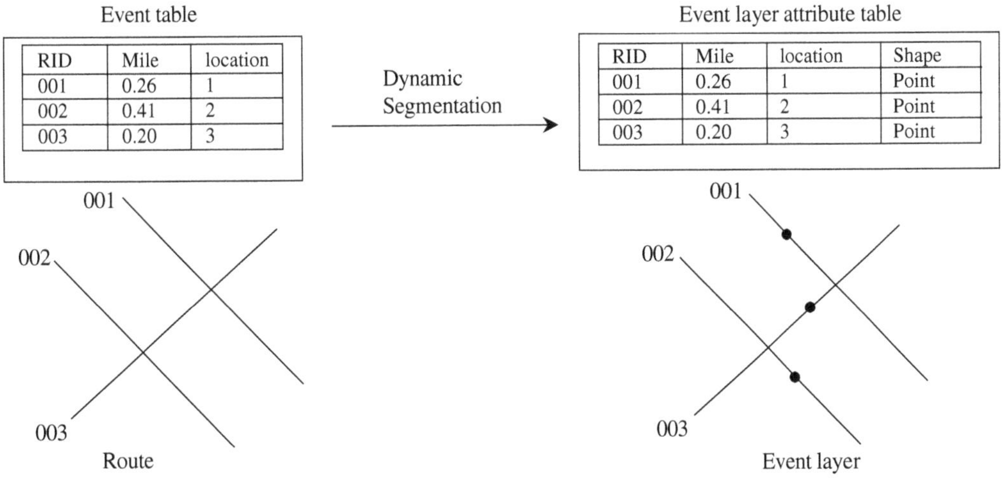

Figure 3.13 An example of dynamic segmentation.

Challenges of data integration: The following lists some challenges that may be encountered in data integration: (i) heterogeneous data; (ii) bad data; (iii) lack of storage capacity; (iv) unanticipated cost; (v) inadequate cooperation from staff; and (vi) lack of data management expertise.

3.7 DATABASE DEVELOPMENT AND MANAGEMENT

3.7.1 Database planning, design, and construction

Database planning is concerned with the data life cycle—identification of data in the needs assessment, inclusion of the data in the data model, creation of the metadata, collection and entry into the database, updating and maintenance, and data retention based on the appropriate record retention schedule. It also taps into the conceptual design of the GIS database, including identification of functions of hardware and GIS software, estimates of usage, and scoping the size of the GIS database system. The conceptual design needs to address the issue of how the existing data processing environments will interface with the database.

Database design is usually divided into three major activities: conceptual data modeling, logical database design, and physical database design. Conceptual data modeling aims to describe what the GIS database needs to do, what data contents are required conceptually, and how the logical and physical processes will be handled. Logical database design is translation of the conceptual database model into the data model of a specific software system. Physical database design represents the logical data model in the schema of the software.

3.7.1.1 Data standards

Data standards give acceptable rules by which data is represented, accessed, manipulated, transferred, and reported. These standards or rules would be identified or established in the stage of requirements analysis. Evaluation on both the data elements and the processes that draw information for specific applications is required when developing new data standards.

There are three common data standards. The first type of data standard gives rules by which a specific data is stored in the database (e.g., format). The second type of data standards pertains to data access and manipulation (e.g., database communications protocol). The third type of data standards pertains to data transfer and reporting (e.g., some applications can export data from the database in an Excel file format).

3.7.1.2 Metadata and data dictionary

Beside defining the data models, reference systems, and data standards, a crucial component of database design is the creation of metadata and data dictionaries, which are essentially detailed descriptions of the data. "Metadata" is data about the data. Metadata describes the data's meaning in the real world (e.g., its formal names, definitions, integrity, and accuracy). Metadata also indicates the data's physical nature (e.g., the way it is stored), the data types, structure (e.g., relational, object-oriented), and other properties that may assist the database user to understand and manage the data.

The data dictionary is a subset of the metadata containing an organized catalog of the data files pertaining to the definition, type, structure, and other information of the data. The use of a data dictionary is crucial in the data integration process by ensuring data definition and usage consistency in the databases and by clearly differentiating various data items.

3.7.1.3 Requirements

Computer communication requirements: The process by which integrated data will be accessed by various users or clients from their computer terminals and workstations is also included in the database design and specifications. Access to the data by end users and application programs is normally carried out through a computer interface. Depending on whether a fused or interoperable database is selected, the user may access data directly from a warehouse in which the database resides or dial up to a computer network to access the data from other computers or database servers and use these data on a local machine.

A computer network is a system where multiple computers share the resources of all computers connected to the network via a high-speed communications link. The computer network allows the databases stored in one location to be accessed by various users working on separate computers in different places. These users can communicate with one another or exchange information via the network. The communications' requirements for integrated data that will be identified in the database design include dial up and communications procedures and other components, such as software, needed by the computer network.

Software and hardware requirements: Software and hardware requirements for the integrated database depend upon the database design and specifications described above. These include software and hardware choices and requirements for database servers, network communications, data mapping, user interfaces, computer operating systems, and programming environments. Agencies have the option of building the database management system from scratch or adopting a commercial software package, which may be customized.

Database server: Choosing the server software and hardware is vital. Important factors to be considered are the possible highest demand measured by the expected maximum number of users trying to access the database at the same time, the extent of uptime needed, the programs that will communicate with the database, the hardware and operating system the server will use, and so forth. Ideally, the server will have sufficient speed bandwidth and data storage capacity to deal with large and complex data processing tasks. Recent software market trends indicate that implementing integrated databases is getting easier and less costly.

Software: Spatial reference and mapping software, which is used to display and analyze location data, is collectively referred to as GIS software. The spatial nature of most transportation data makes GIS a powerful tool for asset management. GIS software is used for constructing spatial databases of transportation networks and features, conducting various types of analyses and applications on the spatial data, and integrating many management and decision-making information and processes.

Some GIS products have external database integrators that enable them to coexist and be integrated with an organization's GIS infrastructure. This functionality provides GIS users with the ability to access and use data from a number of relational database management systems. Existing transportation spatial databases or warehouses developed by a number of highway agencies use GIS software and modules to link databases or perform specific database functions such as querying or reporting. GIS software options include several commercial products or suites of products. Each software product offers various data management, analytical, and reporting capabilities. Some products are designed for web-based mapping and analysis applications. The software runs in different computer operating systems and network environments.

Staffing and data management requirements: Finally, in the stage of database design and specifications, administrative and management responsibilities for the integrated databases are established by identifying the people who will be managing database programming, prototyping, and testing (database development), software and hardware purchases (procurement), computer network setup (systems administration), and database management, maintenance, and upkeep.

3.7.1.4 Testing and implementation

Software development and system implementation is the last step in data integration. The tasks involved include prototyping and use-case applications development activities, computer systems and network communications setup activities, and populating the database with data. Development activities include testing, evaluation, and modification of database models, data management applications, and communications interfaces.

Prototypes and use-case applications are usually created by developers to implement the integrated database models. These two tasks are ideally conducted in tandem in order to concentrate on the data users and their use of the system. The very first version of the prototype will comprise major program modules written to move data between the screens, the database, reports, and the inputs and outputs used to communicate with other data systems. These prototypes may do little preliminary data processing in the beginning. As the prototyping continues, updated programs that perform full-scaled data processing will replace the original versions.

3.7.1.5 Database evaluation

Database evaluation gives feedback on the selected set of data formats, standards, and procedures and helps agencies to unceasingly update the database management system.

A database management system is a tool to facilitate all activities involved with transportation decision making. Among commercially available database management systems, such as ORACLE database management system (DBMS) and MySQL, GIS is proven to be a very handy tool for managing the data in an effective way (Poling et al., 1994; Harvey and Shaw, 2001). In general, GIS consists of attribute data and spatial data that are seamlessly integrated to allow users to apply spatial analysis to the attribute data. Attribute data consist of descriptions attached to an object. For example, a highway segment may contain attribute data, such as geometric design details, traffic, and repair records. Spatial data consist of geocoded objects in terms of nodes, polylines, and polygons that have an orientation and relationship in a 2D or 3D space. These objects have precise definitions and are interrelated with the topological rules.

3.7.2 Data modeling

Data modeling is a process of extracting information from various data stored in the database and making such information into relations. Objects of data modeling are the physical-geographical sphere and socioeconomic components. Figure 3.14 illustrates the data modeling process. It begins with the physical environment. Geodetic, cartographic, and photogrammetric methods help create the subjected environment model with a certain structure, content, and accuracy. The environmental model can recreate its functional data model. Aided by computer technologies, the data model could use the database for a wide range of analyses and applications.

The general tool for describing transportation elements is the entity—attributes—relational concept. It is based on classification of transportation entities and their classes, as well as on definitions of attributes and one-to-one, many-to-one, and many-to-many relations among them. The layering-relational concept of modeling is based on mapping and

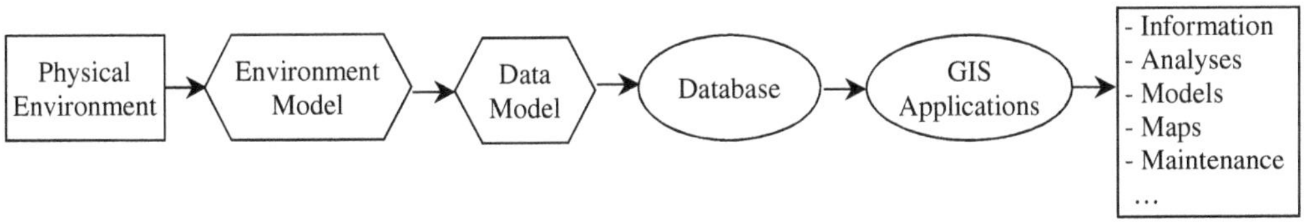

Figure 3.14 Procedure of data modeling.

relational database technologies. Transportation entities are situated in the layers of digital maps with attributes processed based on organized tables.

Data model types: A data model is a formal definition of the data required in a GIS database that typically takes the form of a structured list and an entity–relationship diagram. It ensures that the identified data is described and defined in a completely rigorous and unambiguous manner. It is then the formal specification for the entities, their attributes, and all relationships between the entities. For a subjected area, its photogrammetric or satellite image is transformed and digitized in a defined coordination system suitable for technical purposes. The contents of a digital map are usually divided into levels with a standard cell library and geometrically graphical elements including points, lines, areas, and text. With the digital map in place, different data models are executed. First, a geometric model can be created and used for graphical presentation and documentation. Next, an object model can be developed to contain complex graphical elements with closed and exactly defined lines and areas representing real-world entities. Then a subjected area model can be created by attaching data on user attributes to geometric descriptions of graphical objects. In such a model, manipulations and queries can be made and supported by graphical functions. Further, a following model can be established by defining mutual relations among objects' topology. Finally, a topologic model of the environment is developed to create geographic or spatial data. In this way, geospatial analysis can be effectively carried out using the spatial data that maintains a unique definition for a physical or abstract transportation entity (e.g., a highway segment) and a definition of a corresponding spatial entity (i.e., a polygon to represent its footprint).

Entity–relationship data model: An entity-relationship (E–R) model is one of the most popular object-based models studied in database literature. The E–R model is easy to understand and also powerful enough to model complex scenarios. An E–R model uses three components to describe data: entities, relationships between entities, and attributes of entities or relationships. An entity is an object that exists uniquely. It could be an event or a location. Entities in an organization cannot stay isolated. The relationship is defined to describe the association among entities. The normal relationships included in an E–R model are those of (i) belonging to; (ii) set and subset relationships; (iii) parent–child relationships; and (iv) component parts of an object. An attribute is a property of an entity type that is of interest for some purpose. An entity is thoroughly described through adding a set of attributes to be associated with it.

Topological data model: Topology can be applied to the modeling of basic relations among real-world entities. By topology, they can be exactly defined by planar graphs, their nodes and polygons, and borders of shapes. This tool can be employed to express continuity, polygonization, and adjacency.

Node-arc model: A node-arc model represents interactions or movements between point locations of a network. *Nodes* are point locations where flow originates, terminates or relays. Arcs connect nodes serving as physical conducts (such as highway segments) and logical relationships (like airline services) for flow between nodes. Arcs are directed or undirected, and each has a weight that represents the cost incurred by one unit of flow when traversing the arc. The node-arc model can be utilized to represent a transportation network that only consists of directed arcs. The node-arc model has some limitations, which include enforcement of planar topological consistency and integrity, homogeneity of arc properties, and difficulty in supporting one-to-many relationships among transportation entities.

Relational data model: A relational data model groups data using attributes associated with the data set. As we can see in most of the GIS software, geographic entities are grouped by layers, and a layer is a group of geographic entities with one or more required attributes. In this process, the interconnection between geographic entities and attribute tables are established.

Object-based data model: The object-based data model stores the spatial entities as objects. These entities are simplified into graphical representations such as polygons and lines. For example, a territory is viewed as a continuous leveled surface where altitude information is stored as an attribute.

Object-oriented data model: An object-oriented data model organizes objects by their attributes and behaviors. Any entity of interest is uniformly modeled as an object with a unique identifier. Every object has a state and a behavior. The state is a set of attribute values associated with the object. The behavior is a set of methods that could change the state of the object. For example, a roadway segment is an object, the pavement surface condition measured by PSI is an attribute, and resurfacing is a behavior that could change the value of the PSI.

Navigable data model: A navigable data model is a spatial database that can facilitate travel navigation. In order to realize this function, this model should be able to transfer a street address to the corresponding coordinates and vice versa. This model should also support map matching and vehicle routing. As we can find now, the navigable data model-based products in the market have advanced greatly in the past 10 years, such as GPS navigator and cell phone-based rideshare applications.

3D data model: A shift from 2D to 3D GIS can eliminate the fundamental flaws of implementing the LRS data models. By capturing the third spatial dimension, a 3D object-oriented data model can help deal with the planar topological consistency issue associated with complex roadway design features encountered by the node-arc model. It can also effectively support route guidance, vision enhancement, and automated piloting.

3.7.3 New requirements for database management

3.7.3.1 Storage capacity

The data acquired are stored in various formats and storage media. Formats includes paper, electronic databases, and geo-referenced database systems. The storage media includes paper forms, hard disks, magnetic tapes, compact discs (CDs), digital video discs (DVDs), and combinations of the aforementioned media. Electronically documented data is most preferred because they are the easiest to share and can exist and be presented in multiple forms such as text, graphics, figures, and videos. The storage can be as flexible as either in standalone files or in structured database files (relational, object-oriented).

As a result of advancements in technology, the data collected could be more detailed and fine-grained. More accurate and detailed data enables transportation professionals to perform data analysis in multiple ways and from different points of view. However, the rapid growth of data size demands more capacity of storage media, although a well-constructed database can minimize data storage. The exponentially increasing need for storage may cause the cost to exceed the benefit of data integration, as agencies have limited resources to be allocated to storage media expansion. Directly merging such large amounts of data can push the hardware and software to the limits. This may put excessive loads on the existing system and hence trigger unanticipated negative impacts.

Another challenge resulting from the modernization of data collection technology is to convert the format of the legacy data to the format used currently, which involves synchronizing a huge amount of variable, heterogeneous data resulting from internal legacy systems that vary in data format. It is extremely time- and labor-consuming if the previous data storage media is nonelectronic, such as paper. The situation will become slightly better when it comes to CDs, DVDs, and other electronically recorded data, but that process still needs a huge amount of effort, along with which other challenges may pop up during the data format conversion, such as heterogeneity issues and data quality issues.

3.7.3.2 Heterogeneous data

As data has been collected at different times by different equipment through different approaches stored in various formats and media, data integration is necessarily needed to organize the raw data. Data integration is critical to convert the data into information to support decision making in multiple ways. Transportation agencies must standardize the available data into proper forms depending upon certain applications in the decision-making process.

Data standards are a set of rules designed for representing, accessing, manipulating, transferring, and reporting information. Current data standards put a lot of emphasis on data format standardization, but can hardly handle the data heterogeneity issue. For example, the traffic count of a roadway segment has been monitored by loop detectors for the first five years, and by surveillance cameras for the following 5 years. Supposing we have the daily traffic counts for both data collection approaches, the problem is how to integrate these two data sets considering the different accuracy and precision levels from using different data sources before standardizing the format. There are more variations of this problem, such as how to improve the data quality when two techniques are employed at the same time. In fact, the effort spent on data homogenization becomes a major task in data integration, which may not necessarily improve the quality of data.

3.7.3.3 Bad data and missing data

Data quality is certainly a top consideration in any kind of data integration strategy. In the earlier stage, the data collection techniques are not able to provide clean data, which brings noise from the legacy data. Missing data is another issue resulting from data collection device failure. Noise in the legacy data must be cleaned up prior to conversion and integration, otherwise an agency will almost certainly face serious data problems later. Legacy data impurities have a compounding effect: by nature, they tend to concentrate around high-volume data users. If this information is corrupted, so too will be the decisions made from it. It is not unusual for undiscovered data quality problems to emerge in the process of cleaning information for use by the integrated system. The issue of bad data leads to procedures for regularly auditing the quality of information used. But who holds the ultimate responsibility for this job is not always clear.

PROBLEMS

3.1 One state transportation agency plans to install an automatic vehicle counting station near an on-ramp of a limited access highway. Three options are proposed: (a) Inductive burial loop detection; (b) pneumatic pressure detection; and

(c) non-directional wireless vehicle counters. As shown in Table P3.1, thirteen experiments are conducted on the same highway segment by temporarily installing all three types of counting equipment during different time periods of a day. As a benchmark to evaluate the quality of vehicle counts from these experiments, manual vehicle counts are treated as true values.

Table P3.1 Experiment results of all three techniques

True values		346	357	602	354	744	546	465	597	591	624	523	511	476
a	DV	338	320	552	330	635	486	463	526	530	507	523	477	443
	RE (%)	2.31												
b	DV	341	352	596	343	678	506	424	566	522	591	512	481	471
	RE (%)													
c	DV	302	356	555	351	692	492	407	548	546	594	523	482	450
	RE (%)													

where DV = Detected Values; and RE = Relative Error $= \frac{|\text{Detected Value} - \text{True Value}|}{\text{True Value}} \times 100\%$.

a. Compute the RE of each experiment using each type of the counting equipment. Round your answers to three decimal places;

   Example: $RE_{11} = \frac{|338 - 346|}{346} \times 100\% = 2.31\%$

b. Compute the sample mean and sample variance of RE of each type of the counting equipment;

c. Comment on type of the counting equipment in terms of accuracy and precision; Which type of the counting equipment should be selected? Explain why.

3.2 What data should be included in a network-level pavement management inventory? Explain the use of each type of data.

3.3 What data should be included in a network-level bridge management inventory? Explain the use of each type of data.

3.4 Explain the importance of vehicle trajectory data complementary to the vehicle count data.

3.5 Explain how the GPS technology could help with transportation data collection. Use transit vehicle location tracking as an example.

3.6 Some emerging techniques such as smartphone devices and social media could assist in collection of travel demand, travel counting, and facility condition data. Explain how they could support transportation data collection.

3.7 A local transportation agency needs to conduct an O–D survey for a region of 90000 residents living in 50 TAZs. It is necessary to collect data on household income characteristics in each traffic analysis zone. If a stratified sampling method is applied, how can one determine the sample size in each zone? Describe additional datasets needed in this process.

3.8 A local public transit authority wants to assess the current condition of the buses that have been in service for 3 years. They put together a list of all these buses. Then, the analyst randomly selects a bus from the first three buses on the list. Starting with that bus, the analyst selects every third bus for the assessment. For example, if bus number 2 were the first bus selected, the sample would consist of bus numbers 2, 5, 8, 11, 14, etc. Is this an example of simple random sampling?

3.9 Describe the difference between stratified sampling and cluster sampling methods.
3.10 Fill in the blanks in the following figure using the Arial weighting method.

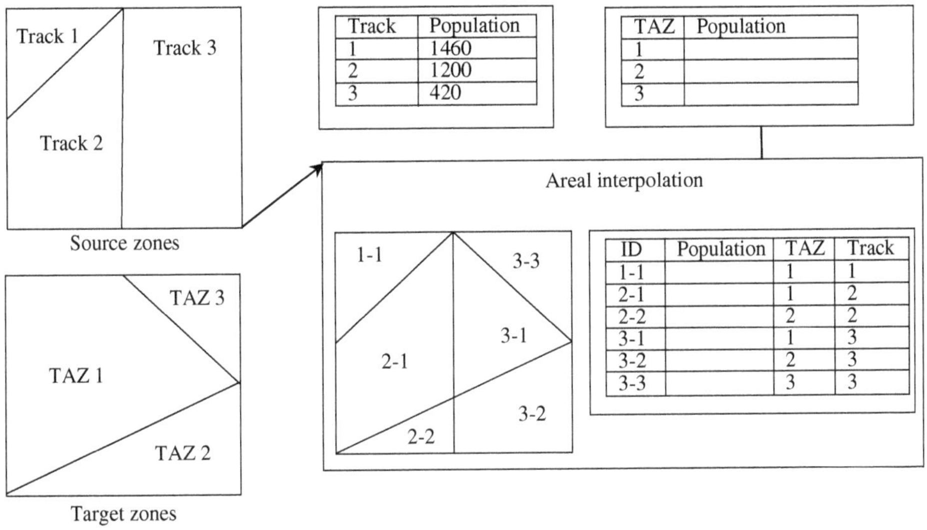

Track	Population
1	1460
2	1200
3	420

TAZ	Population
1	
2	
3	

ID	Population	TAZ	Track
1-1		1	1
2-1		1	2
2-2		2	2
3-1		1	3
3-2		2	3
3-3		3	3

3.11 Describe the major challenges and opportunities of data integration.
3.12 Describe main steps of GIS database development.

Chapter 4

Transportation facility performance modeling

4.1 GENERAL

4.1.1 Characteristics of facility performance measures and models

For a multimodal surface transportation system, the primary types of physical facilities include highway pavements, bridges, traffic control devices, safety hardware, transit installations, rail tracks, bikeways, sidewalks, and so forth. Each type of facility has a limited useful service lifespan that terminates when the facility condition deteriorates to an unacceptable level due to material aging and cumulative use over time. Performance measures are typically utilized to assess the current condition of a facility and develop performance models for predicting its conditions in the future. With a minimum acceptable condition level pre-specified, the remaining service life of the facility can be estimated. In some cases, performance measures are employed to directly estimate or predict the remaining service life of a facility at a given point in time. Depending upon the nature of the condition or service life measurements for a facility, performance measures could be either continuous or discrete. When using a performance measure as the response variable to calibrate a facility performance model, different modeling techniques such as statistical and econometric methods and artificial intelligence approaches could be utilized. If a statistical or an econometric model is calibrated, regardless of having a linear/nonlinear, static/dynamic, or continuous/discrete model form, it is stochastic in nature since randomness of the performance measure as the response variable of the calibrated model presents, and variable states are described not by unique values, but rather by probability distributions.

4.1.2 Facility condition deterioration and service life expectancy

As explained in the above, performance assessment of a facility can be carried out according to condition deterioration or service life expectancy (or remaining service life) over time or its cumulative use. For instance, the concept of condition index was introduced to correlate the subjective rating of riding quality of a pavement segment or a bridge deck component with distress data objectively measured to help analyze the performance of the pavement segment or bridge deck. Such a condition index may deal with a single type of distresses of a pavement segment, such as roughness, rutting, and cracking; distresses of a bridge component; or a composite condition index synthesized from condition indices of multiple types of distresses of a pavement segment or multiple bridge components. The condition deterioration of a pavement segment or bridge component is a manifestation of traffic loading and non-load factors, including material types, design standards, construction quality, environmental conditions, and weather and climatic features.

The facility service life expectancy is defined as the length of time between a given point when a facility starts the service and a later time when it must be reconstructed, replaced or removed from service. The starting point can usually be determined by (i) the manufacturing date; (ii) the date when the facility is placed into service; (iii) the present date from which remaining life is measured; or (iv) the date of some future action or decision. However, determination of the ending point often must be carried out with due circumspection. The ending point can be decided according to (i) completion of the design service life; (ii) the date of reaching cumulative use; (iii) the date of reaching the threshold condition level; (iv) the date of functional obsolescence or structure deficiency, whichever takes place first; and (v) the date of failure (Thompson et al., 2011). Thus, the service life expectancy could be estimated using a single or multiple performance measures associated with facility condition deterioration, a single or multiple performance measures directly related to facility service life, or multiple performance measures relevant to both facility condition deterioration and service life. Similar to facility condition deterioration, the service life expectancy of a facility depends on traffic loading and non-load factors, including material types, design standards, construction quality, environmental conditions, and weather and climatic features. In this respect, different types of facilities or different facilities of the same type could potentially have different condition deterioration rates and service life expectancies. The use of condition deterioration and service life expectancy for facility performance evaluation will assist in facility deterioration predictions, maintenance and repair treatment recommendations, and life cycle agency and user cost calculations.

4.2 PERFORMANCE MODEL TYPES

4.2.1 Condition deterioration model types

With data on condition levels collected for a specific type of facility or a group of facilities in a transportation network over time, it could be utilized to develop facility condition deterioration models for facility performance analysis. Depending on characteristics of performance measures, data availability, and modeling techniques, different types of facility condition deterioration models could be developed (FHWA, 1985, 1987, 1991; Lytton, 1988; Jackson and Mahoney, 1990).

4.2.1.1 Regression models

Regression methods are statistical tools to judge how well an equation fits the actual data or the predictive power of the model, including coefficient of determination, root mean square error, and hypothesis tests on model coefficients. An advantage of regression analysis is that it provides statistical methods to analyze facility condition data and develop condition prediction models that can be updated using future analysis results and engineering judgment. However, regression analysis requires an accurate, precise, and abundant set of continuous condition data in a time series and needs to consider all significant variables affecting facility deterioration. Equation 4.1 exhibits the general form of a linear regression model

$$Y_i = \beta X_i + e_i \tag{4.1}$$

where

Y_i = The dependent variable

X_i = The vector of independent variables, $X = \{X_1, X_2, \dots, X_{K-1}\}$

β = The vector of estimated coefficients of independent variables, $\beta = \{\beta_0, \beta_1, \beta_2, \dots, \beta_{k-1}\}$ with β_0 being the estimated constant
ε_i = The random error term
i = 1, 2, ... , N for the number of observations in the dataset

The linear regression models can be calibrated using the ordinary least squares (OLS) and maximum likelihood (ML) estimation techniques (Pindyck and Rubinfeld, 2000).

4.2.1.2 Econometric models

Econometrics may be defined as the field of economics in which mathematical statistics and statistical inference tools are applied to the empirical analysis of economic phenomena (Goldberger, 1964). In recent decades, there has been a trend toward the use of econometric modeling techniques to explain the behavior of facility condition deterioration (Ben-Akiva et al., 1991; Gopinath et al., 1996; Mohammed et al., 1997; Ramaswamy and Ben-Akiva, 1997). Compared with the regression models, econometric methods are equipped with appropriate tools available to help avoid selectivity, simultaneity, and endogeneity biases that would compromise the model predictivity.

Latent variable models: The latent variable approach considers the facility condition as a set of unobservable or latent variables that depend on other variables, such as previous maintenance, environmental features, and traffic loading. The observed characteristics, such as the condition measurements, in turn simultaneously depend on the underlying latent variables. Because variables in the models are measured with a large degree of error, the observed variables can be modeled as functions of the true values as well as stochastic measurement errors. The model can also be enhanced by using lagged variables and by simultaneously modeling deterioration and maintenance and repair. This treatment is especially important in that deterioration tends to increase with decreasing maintenance and repair when all other factors are held constant. However, maintenance tends to increase with increasing deterioration. As seen in Bentler and Weeks (1980), Hoyle (1995), and Arminger et al. (1995), Equation 4.2 specifies a general framework of structural equation models (SEMs)

$$\eta = \beta\eta + \gamma\xi + \varepsilon \tag{4.2}$$

where
η = The vector of dependent variables in an SEM diagram with one-way arrows pointing to them
ξ = The vector of independent variables
β = The vector of estimated coefficients of dependent variables
γ = The vector of estimated coefficients of independent variables
ε = The vector of random error terms

Probabilistic models with qualitative dependent variables: The most commonly used is the linear probability function. The dependent variable can only take two values: 0 (if the event does not occur) and 1 (if the event occurs). Then the estimators are generated by treating the dichotomous dependent variable problem as an ordinary linear regression problem. The drawback of this approach is that the assumption of homoscedasticity of the error term is untenable. The absence of negative values of the qualitative dependent variable also tends to keep the regression line above the axis over the relevant range of independent variables. This will result in an overestimation of the value at the lower end and an underestimation of the value at the higher end. Furthermore, the linear probability function allows the expected values of the dependent variable to fall outside the interval between 0 and 1. Alternatively, the probit or logit analysis model can keep the expected values of the binary dependent variable within the unit interval from the start by using a critical value with standard normal distribution. The dependent variable will be 1.0 only if the predictor exceeds the threshold value; it will otherwise take the value of zero. Equations 4.3 and 4.4 present a binomial probit model specification

$$T_{1n} = \beta_1 X_{1n} + \varepsilon_{1n} \tag{4.3}$$

$$T_{2n} = \beta_2 X_{2n} + \varepsilon_{2n} \tag{4.4}$$

where

T_{1n} = A linear function that determines discrete outcome 1 for observation n
T_{2n} = A linear function that determines discrete outcome 2 for observation n
X_{1n} = The vector of observable characteristics that determine discrete outcome 1 for observation n
X_{2n} = The vector of observable characteristics that determine discrete outcome 2 for observation n
β_1 = The vector of estimated parameters for discrete outcome 1
β_2 = The vector of estimated parameters for discrete outcome 2
ε_{1n} = The random error term of discrete outcome 1 for observation n
ε_{2n} = The random error term of discrete outcome 2 for observation n

Probabilistic models with limited dependent variables: The Tobit (Tobin's probit or censored regression) model first introduced by James Tobin (1958) is one of the typical cases. The basic idea of the Tobit model is to create a threshold value for the occurrence of an event, such as choosing a facility for maintenance or repair treatment. All facilities are then grouped into two categories: those involving treatments and those without treatments. The joint ML function for the two categories of facilities is formulated to obtain the point estimators. Tobit models are appropriate in cases where the distribution of a dependent variable is censored by a threshold value using a data sample with a mixture of discrete and continuous distributions. Equation 4.5 exhibits the Tobit model

$$Y_i^* = \beta X_i + e_i \tag{4.5}$$

with

$$Y_i = \begin{cases} Y_i^* & \text{if } Y_i^* > 0 \\ 0, & \text{if } Y_i^* \leq 0 \end{cases}$$

where

Y_i^* = An implicit stochastic index or latent variable observed only when positive from the observations in the dataset
Y_i = The dependent variable
X_i = The vector of independent variables, $X = \{X_1, X_2, \dots, X_{K-1}\}$
β = The vector of estimated coefficients of independent variables, $\beta = \{\beta_0, \beta_1, \beta_2, \dots, \beta_{k-1}\}$ with β_0 being the estimated constant
ε_i = The random error term
i = 1, 2, ... , N for the number of observations in the dataset

4.2.1.3 Markov models

If conditions of a facility or a facility component are classified into discrete states, the condition deterioration process that undergoes transitions from a state at one stage to a state at the next can be modeled as a Markov chain. The state of each facility or facility component, the proportion of facilities or the portion of facility components in each state can be measured during field inspection. An underlying assumption of Markov chains is that given the present state of the process, the future states are independent of the past. To make better use of Markov chains for condition prediction, the states of a facility or facility component would have to be defined as per its current condition and the factors significantly influencing its deterioration.

To make provision for changing transition probabilities over a facility or facility component's age, different transition matrices can be used for the facility or facility component of different ages. One approach is to estimate regression or econometric models having the state as a dependent variable, assume a probability distribution for the random error term, and then convert interval probabilities to transition probabilities. To use a Markov chain, individual states must be defined as intervals on a continuum. Alternative models like multinomial logit models can also be used. Another approach is to use subjective judgment of bridge maintenance experts to obtain estimates of transition probabilities in Markov chains and to update and improve the initial estimates based on regular inspections using the Bayesian estimation technique (FHWA, 1987, 1991).

4.2.1.4 Reliability-based models

The Markov models focus on the probability of transition from one discrete condition state to another. One drawback is that there is limited capability to relate the qualitative measurement of condition state of a facility such as a bridge deck component to its data on quantitative parameters such as material properties, stress conditions, and structural behavior. To overcome the drawbacks, Estes and Frangopol (1999) proposed a reliability-based performance modeling that focuses on relevant modes of bridge failures, as opposed to considering the condition states and transitions. In the reliability model, bridges are modeled as systems of completely independent or correlated components with upper and lower bounds of probability of a structural failure. A series–parallel combination of individual failure modes is established to represent the system reliability of bridges over time. Despite the accuracy of these models for deterioration prediction, they can be inefficient and impractical to apply to bridge components of a highway network that contains many bridges where diverse consequences of bridge failures and failure modes need to be considered (Morcous and Lounis, 2007).

4.2.1.5 Ordered probability models

The ordered probability models (both probit and logit) have been widely used for more than four decades (McKelvey and Zavoina, 1975). Their applications have also spanned to the modeling of transportation facility condition deterioration. Similar to Markov models for condition state predictions, these models could quantify the level of uncertainty and provide sensitivity to age, cumulative utilization, and other independent variables for every condition state. The model calibration is relatively complex, requiring time series condition inspection data and knowledge of past facility maintenance and repair records. Equation 4.6 specifies the ordered probability model

$$z = \beta X + \varepsilon \tag{4.6}$$

where

z = The unobserved variable
X = The vector of independent variables determining the discrete ordering for observation n
β = The vector of estimated coefficients of independent variables
ε = The random error term

4.2.1.6 Bayesian models

The Bayesian method allows for combination of subjective data from opinions and objective data to develop predictive models. In traditional regression analysis, the unknown regression coefficients are based on the observed data and are assumed to have unique values. In Bayesian regression analysis, the regression parameters are assumed to be random variables with associated probability distributions analogous to a mean and an SD, which can be used for updating the estimated probabilities of future facility conditions (Attoh-Okine and Bowers, 2006). It is particularly well suited for updating the estimates of transition probabilities in Markov chain analysis as additional data become available with inspection. Under suitable assumptions, the updated estimate (posterior mean) equals a weighted average of the previous estimate (prior mean) and the mean of the new data. The weights represent the value attached to the data, from which the prior mean is estimated relative to the new data. Usually, the relative numbers of observations are used as weights. When the estimates are later updated, the posterior values become the prior values for the new estimates. In this way, the effects of initial estimates are reduced as new data becomes available.

4.2.1.7 Nonparametric binary recursive partitioning method

Binary recursive partitioning (BRP) is a nonparametric statistical method commonly used to tackle the classification and decision problem (Gibbons, 1976; Breiman et al., 1984; Therneau and Atkinson, 1997). It has been applied to predict bridge component condition deterioration. In this respect, the term "binary" implies that each group of bridges represented by a "node" in a decision tree can only be split into two groups. Thus, each node can be split into two child nodes, in which case the original node is called a parent node. The term "recursive" refers to the fact that the process can be repeated iteratively by executing the following steps: (i) selecting the explanatory variables to obtain maximum reduction in the heterogeneity of deck condition as the response variable; and (ii) determining the value of the selected explanatory variable that could result in the maximum reduction in the heterogeneity of the response variable. This process can be applied repetitively until a desirable convergence condition is met. Thus, each parent node can give rise to two child nodes and, in turn, each of these child nodes may themselves be split, forming additional children. The term "partitioning" refers to the fact that the dataset is split into sections or partitioned. The entire analytical process consists of four basic steps: tree building, stopping tree building, tree pruning, and optimal tree selection.

4.2.1.8 Artificial intelligence models

Artificial intelligence (AI) models rely on computer techniques to analyze facility deterioration behavior, which can be classified into expert systems, artificial neural networks (ANN), and case-based reasoning (CBR), among others. The most commonly used ANN models for predicting transportation facility condition deterioration are ANN-based backward prediction models (Miwa and Simon, 1993; Morcous et al., 2002).

In general, facility condition deterioration has been viewed as a manifestation of combined effects of traffic loading and non-load factors. Taking bridges, for example, when the non-load factors such as environmental conditions and climatic features are primary causes of bridge component condition deterioration, the change in condition rating is a long progressive process over time, which would not be large for a short period in time. As such, extensive historical data on bridge component condition ratings with high quality is required if more traditional methods are used for calibrating the facility condition deterioration model to ensure its prediction power. Conversely, the ANN-based backward prediction model can generate artificial historical data on bridge component condition ratings. It could establish correlations between existing condition ratings and corresponding years' non-bridge factors such as environmental conditions and climatic features. When historical data on non-bridge factors is made available, it can be applied to the formed correlations to derive the historical condition ratings of the bridge component in its service life cycle. The data on derived historical and current condition ratings can be utilized to predict the future conditions of the bridge component. Applications of ANN models require complementary tools to utilize the generated data for condition predictions and validation of prediction results.

4.2.2 Service life expectancy model types

Similar to facility condition deterioration model development, statistical methods and AI techniques could be used to develop a facility service life expectancy model (Kutner et al., 2004; Box et al., 2005; Montgomery and Runger, 2006; Wooldridge, 2015; Greene, 2017).

4.2.2.1 *Regression models*

For facility service life expectancy predictions, a regression model can be developed by using facility age at reconstruction or replacement as the dependent variable. Therefore, data on facility reconstruction or replacement records is needed to ensure model predictability.

4.2.2.2 *Parametric survival models*

Parametric survival models aim to link the remaining service life of a physical facility such as a pavement segment or a bridge component to one or more influencing factors using a specified probability distribution. Notables include exponential, Weibull, gamma, generalized gamma, log-normal, log-logistic, and generalized F distributions (Wooldridge, 2015; Greene, 2017). The hazard function of the Weibull regression model in proportional hazards form is

$$\ln[h(t)/[p \cdot \lambda^p \cdot t^{p-1}]] = \beta X + \varepsilon \quad (4.7)$$

where

$h(t)$ = The expected hazard or failure rate at time t
X = The vector of independent variables
β = The vector of estimated coefficients of independent variables
p = The shaping parameter that controls facility condition deterioration rate
λ = The scaling parameter
ε = The random error term
ln[.] is the natural logarithm

4.2.2.3 *Semi-parametric survival models*

One of the most popular semi-parametric survival analysis models is the Cox proportional hazards model (Cox, 1972). There are no assumptions about the shape of the baseline hazard function, which makes it a semi-parametric model. Particular to facility service life expectancy modeling, the Cox model could rely on a hazard or failure rate to assess the risk of service life termination for a facility that is in-service up to a specific time in association with one or more independent variables. The model is of the following specification

$$\ln[h(t)/h_0(t)] = \beta X + \varepsilon \tag{4.8}$$

where

$h(t)/h_0(t)$ = The hazard ratio as the relative risk of facility service life termination
$h(t)$ = The expected hazard at time t
$h_0(t)$ = The baseline hazard at time t, representing the hazard when all independent variables are at baseline levels
X = The vector of independent variables
β = The vector of estimated coefficients of independent variables
ε = The random error term
ln[.] is the natural logarithm

4.2.2.4 *Neural network models*

Essentially, neural network models that are adaptive could predict conditions based on pattern identification from past data. Statistically, an ANN is a nonlinear form of 3-stage least squares regression where variables used to represent relationships between other variables are estimated with input data to predict future events. To facilitate neural network learning, this approach updates estimates as posterior means by applying weighted averages based on previous estimates as prior means. This treatment is consistent with the Bayesian modeling concept. The weights can be established based on the number of observations. Activation functions within the network generally include hyperbolic tangent, log-sigmoid, and bipolar-sigmoid functions. Applications of neural network models require developing learning algorithms or customizing commercial software tools to the prediction needs.

4.3 MODEL ESTIMATION TECHNIQUES

4.3.1 Preliminary data analysis

Prior to model estimation, preliminary data analysis needs to be carried out using descriptive measures to gain some insights into the nature of the sampled data from which population inferences are made. Measures of data-descriptive statistics mainly include (i) percentile ranking for relative standing; (ii) central tendency; (iii) variability; (iv) skewness and kurtosis; and (v) association (Washington et al., 2011).

The data can be further inspected to determine relationships and trends, identify outliers and influential observations, and derive summary information on the dataset, which will help refine the model data and identify well-suited modeling techniques to ensure prediction power of the calibrated model. Data inspection methods generally cover (i) histograms; (ii) ogives; (iii) box plots; (iv) scatter plots; (v) bar charts; and (vi) line charts (Washington et al., 2011).

Tables 4.1 and 4.2 briefly describe descriptive statistical measures and inspection methods for preliminary data analysis. For illustration purposes, two random variables,

Table 4.1 Descriptive statistical measures for preliminary data analysis

Measure		*Description*
1. Ranking for relative standing	Percentile	Defined as that value where p% of values in the remaining sample lies below it. For a sufficiently large sample of n observations, the position of the pth percentile is given by $p\text{th percentile} = \frac{[(n+1)\cdot p]}{100}$
	Quartile	Percentage points separating the data into quarters: First quarter (Q_1) corresponds to the 25th percentile, below which lies 25% of the data. Second and third quarters (Q_2, Q_3) correspond to 50th percentile (i.e., median) and 75th percentile, respectively. The first and third quarters are often referred to as lower and upper quartiles
2. Central tendency	Sample mean	The average of a sample of observations $\bar{X} = \frac{\left(\sum_{i=1}^{n} x_i\right)}{n}$
3. Variability	Interquartile range (IQR)	The numerical difference between the first and third quarters that measures the overall data dispersion $IQR = x_{Q3} - x_{Q1}$
	Range	The difference between the largest and smallest values in a sample of observations $Range = x_{max} - x_{min}$
	Average sample deviation	A more detailed measure of precision of a sample of observation values
	Sample variance	The average squared deviation of individual observations from the mean, also termed as the second moment around the sample mean $s^2 = \frac{\left[\sum_{i=1}^{n}(x_i - \bar{X})^2\right]}{(n-1)}$
	Sample SD	The square root of sample variance as an absolute measure of dispersion $s = \sqrt{\frac{\left[\sum_{i=1}^{n}(x_i - \bar{X})^2\right]}{(n-1)}}$
	Sample coefficient of variation (COV)	Relative measure dispersion as a proportion of the sample mean $COV = s/\bar{X} = \sqrt{\frac{\left[\sum_{i=1}^{n}(x_i - \bar{X})^2\right]}{(n-1)}} \bigg/ \frac{\left[\left(\sum_{i=1}^{n} x_i\right)\right]}{n}$
4. Degree of asymmetry	Skewness	The third moment around the sample mean as a measure of the degree of asymmetry of a frequency distribution $m_3 = \frac{\left[\sum_{i=1}^{n}(x_i - \bar{X})^3\right]}{n}$
5. Flatness	Kurtosis	The fourth moment around the sample mean as a measure of the degree of "flatness" of a frequency distribution $m_3 = \frac{\left[\sum_{i=1}^{n}(x_i - \bar{X})^4\right]}{n}$

(Continued)

Table 4.1 (Continued) Descriptive statistical measures for preliminary data analysis

Measure		*Description*
6. Association	Sample covariance	Sample covariance of two random variables X and Y is the expected value of the product of the deviation of X and Y from their means $$\text{COV}_s(X,Y) = \frac{\sum_{i=1}^{n}(x_i - \bar{X}) \cdot (y_i - \bar{Y})}{(n-1)}$$
	Sample correlation coefficient	Standardized information about the strength of the linear relationship between two random variables $r = \text{COV}_s(X,Y)/(s_X \cdot s_Y)$ $$= \frac{\left[\left(\sum_{i=1}^{n}(x_i - \bar{X}) \cdot (y_i - \bar{Y})\right)/(n-1)\right]}{\left[\sqrt{\left[\sum_{i=1}^{n}(x_i - \bar{X})^2\right]/(n-1)} \cdot \sqrt{\left[\sum_{i=1}^{n}(y_i - \bar{Y})^2\right]/(n-1)}\right]}$$

Table 4.2 Data Inspection methods for preliminary data analysis

Method	*Description*
1. Histogram	A chart that consists of bars of varying heights with the height of each bar in proportion to the frequency of values of a random variable, which can help identify skewness and kurtosis of the sample of observations
2. Ogive	A graph of the cumulative relative frequency of values of a random variable
3. Box plot	Illustrates the center point and dispersion level in the values of observations of a data sample for a random variable, which can help identify central tendency, spread, and skewness of data using information on median, IQR, whisker that is 1.5 times of the IQR, and smallest and largest observation values within the whiskers, and diagnose outliers that are observation values outside of the range between the smallest and largest values
4. Scatter plot	Typically used to uncover underlying relationships between two continuous variables, which can be further examined in greater depth with more quantitative statistical methods
5. Pie chart	Graphical examination of nominal data
6. Bar chart	Graphical representation of the (relative) frequency of data by category as a bar with the height of each bar in proportion to the (relative) frequency
7. Line chart	Graphical display of the frequency of data by category and joining the points representing the frequencies with straight lines to uncover trends of a variable over time

X and Y, are used. A sample of n observations x_i or y_i, $i = 1, 2, \ldots, n$, is assumed for each random variable.

4.3.2 Classical regression model assumptions

Table 4.3 lists six assumptions of regression models (Kutner et al., 2004; Wooldridge, 2015).

4.3.3 OLS estimation

For regression model development, the OLS estimation is one of the popular methods to derive "good" estimators of the regression parameters $\beta_0, \beta_1, \ldots$, and β_{p-1}. Without loss

Table 4.3 General assumptions of classical regression models

Statistical assumption	*Math expression*	*Description*
1. Linearity of the functional form	$Y_i = \beta X_i + \varepsilon_i$	Requires that the dependent variable and independent variables are linearly correlated after adopting necessary transformations. Often, a suitable linear relationship is found
2. Nonmulticollinearity of regressors and error terms	$Cov[X_i, \varepsilon_j] = 0$, for all i and j	Implies that values of regressors are determined by influences independent of the model where the dependent variable does not directly affect the value of an exogenous regressor, or exogenous regressors are not correlated with error terms
3. Nonautocorrelation of error terms	$Cov[\varepsilon_i, \varepsilon_j] = 0$, if $i \neq j$	Specifies independency of error terms across observations
4. Normality of error terms	$\varepsilon_i \sim N(0, \sigma^2)$	Demands that the error terms be approximately normally distributed to help make meaningful inferences about the model coefficients according to the central limit theorem. Combined with the assumption of nonautocorrelation of error terms, it leads to error terms following an independent and identical normal distribution
5. Zero mean of error terms	$E[\varepsilon_i] = 0$	States that on an average model, overpredictions are equal to model underpredictions, making model errors sum to zero
6. Homoscedasticity of error terms	$Var[\varepsilon_i] = \sigma^2$	Shows that the variance of error term is independent across observations, which implies that the impacts on model uncertainty by unobserved variables omitted from the model, measurement errors in the dependent variable or the imprecision in measuring the dependent variable, and true random variation inherent in the underlying data-generating process are not systematic across observations, instead such impacts are random across observations and covariates

of generality, considering the ith observation $(X_{i1}, X_{i2}, \ldots, X_{i,p-1}, Y_i, i = 1, 2, \ldots, n)$ among the n observations, the OLS method computes the deviation of Y_i from its expected value

$$Y_i - (\beta_0 + \beta_1 X_{i1} + \cdots + \beta_{p-1} X_{i,p-1}) \tag{4.9}$$

In addition, the method calculates the sum of squared deviations, denoted Q

$$Q = \sum_{i=1}^{n} (Y_i - \hat{Y}_i)^2 = \sum_{i=1}^{n} (Y_i - \beta_0 X_{i0} - \beta_1 X_{i1} - \cdots - \beta_{p-1} X_{i,p-1})^2 \tag{4.10}$$

where $X_{i0} \equiv 1$.

The estimators of $\beta_0, \beta_1, \ldots$, and β_{p-1} obtained should minimize Q. To achieve this, the first-order partial derivatives of Q with respect to $\beta_0, \beta_1, \ldots$, and β_{p-1} are separately calculated and set to zero as follows:

$$\frac{\partial Q}{\partial \beta_j} = -2 \sum_{i=1}^{n} X_{ij} (Y_i - \beta_0 X_{i0} - \beta_1 X_{i1} - \cdots - \beta_{p-1} X_{i,p-1}) = 0 \tag{4.11}$$

Equation 4.11 can be expressed in a matrix form as

$$X'(Y - X\beta) = 0 \tag{4.12}$$

where

$$X = \begin{bmatrix} X_{10} & X_{11} & \cdots & X_{1,p-1} \\ X_{20} & X_{21} & \cdots & X_{2,p-1} \\ \vdots & \vdots & \ddots & \vdots \\ X_{n0} & X_{n1} & \cdots & X_{n,p-1} \end{bmatrix} Y = \begin{bmatrix} Y_1 \\ Y_2 \\ \vdots \\ Y_n \end{bmatrix} \beta = \begin{bmatrix} \beta_0 \\ \beta_1 \\ \vdots \\ \beta_{p-1} \end{bmatrix}$$

Setting $B = [\hat{\beta}_0, \hat{\beta}_1, \ldots, \hat{\beta}_{p-1}]'$ as the solution to $\beta = [\beta_0, \beta_1, \ldots, \beta_{p-1}]'$, the OLS estimators as the solution can be obtained by solving Equation 4.12

$$B = (X'X)^{-1}X'Y \tag{4.13}$$

4.3.4 ML estimation

ML estimation is another popular method to calibrate regression model coefficients. This method uses the product of the densities or probabilities as the measure of consistency of the parameter value with the sample data. The product is called likelihood value of parameter value θ, denoted by $L(\theta)$. Equation 4.14 presents the computation of the likelihood value using sample data or observations $(x_1, x_2, \ldots, x_n)$. The basic idea of ML estimation is that, if the value of θ is consistent with the data, the densities will be relatively large and so will be the likelihood.

$$L(\theta) = f(x_1, x_2, \ldots, x_n, \theta) = \prod_{i=1}^{n} f(x_i \mid \theta) \tag{4.14}$$

Therefore, the ML estimator is obtained by solving the maximization problem of Equation 4.14 with respect to the parameter θ. If the likelihood is continuously differentiable (which it is in most of the cases), the estimator of θ is obtained by solving Equation 4.15

$$\frac{\partial L(\theta)}{\partial \theta} = 0 \tag{4.15}$$

For a linear multiple regression model, normality of sample observations is assumed

$$Y_i \sim N(\beta_0 + \beta_1 X_{i1} + \beta_2 X_{i2} + \cdots + \beta_{p-1} X_{i,p-1}, \sigma^2) \tag{4.16}$$

The likelihood function for the model is of the following specification:

$$L(\beta, \sigma^2) = \left[\frac{1}{(2\pi\sigma^2)^{n/2}}\right] \exp\left[-\frac{1}{2\sigma^2}\sum_{i=1}^{n}(Y_i - \beta_0 - \beta_1 X_{i1} - \cdots - \beta_{p-1} X_{i,p-1})^2\right] \tag{4.17}$$

Maximizing this likelihood function with respect to $\beta_0, \beta_1, \ldots,$ and β_{p-1} will help derive the ML estimators of the regression parameters $\beta_0, \beta_1, \ldots, \beta_{p-1}$.

4.3.5 Model validation

In the process of developing a regression model, model validation is one of the most important tasks. Only a successfully validated model can be used for predictions. Model validation aims to confirm (i) goodness of fit reflecting the model prediction power; (ii) statistically significant relation between the dependent variable and independent variables; and (iii) no violation to each of the remaining five assumptions made to the classical regression models (Pindyck and Rubinfeld, 2000; Kutner et al., 2004; Box et al., 2005; Wooldridge, 2015; Greene, 2017). Table 4.4 shows the model validation methods.

4.3.6 Model predictability evaluation

The main purpose of model development is to utilize the calibrated models for predictions. Table 4.5 list measures for evaluating the accuracy of model predictability (Washington et al., 2011).

4.4 PAVEMENT PERFORMANCE MODELING

4.4.1 Pavement types

Highway pavements are generally classified into flexible, rigid, and composite types (HRB, 1961, 1962; Yoder, 1985; Haas et al., 1994; AASHTO, 1998; Huang, 2004), which are briefly described below.

Flexible pavements: Flexible pavements are composed of a bituminous material surface course and underlying base and subbase courses. Specifically, a typical flexible pavement has a surface layer that is fully made of an asphalt and aggregate mix laid over a granular treated or untreated base layer, and sometimes an untreated natural gravel subbase layer. In general, an asphalt pavement is one where all layers (surface, base, and subbase) contain an asphalt binder in varying proportions and aggregate gradations and quality.

Rigid pavements: Rigid pavements are constructed from cement concrete or reinforced concrete slabs. They can be generally classified into three types: jointed plain concrete pavements, jointed reinforced concrete pavements, and continuously reinforced concrete (CRC) pavements. All three rigid pavement types are typically constructed on a layer of granular subbase layer. In some cases, an additional but lower-quality natural gravel or crushed rock layer is used to separate the granular layer from the subgrade.

Composite pavements: Composite pavements usually utilize both asphalt and concrete. They are mainly constructed from existing flexible or rigid pavements that are resurfaced with asphalt or concrete overlays after many years of service. The detailed classification of various pavement types is illustrated in Figure 4.1.

4.4.2 Pavement distresses

Pavement distresses can be generally grouped into six categories: (i) cracking; (ii) patching and potholes; (iii) joint deficiencies; (iv) surface deformation; (v) surface defects; and (vi) miscellaneous. Within each category of distresses, some distresses are common to different types of pavements. The extent and severity of a distress are key to assigning a value

Table 4.4 Methods for regression model validation

Focus	*Plot analysis and statistical test*	*Remedying measure*
1. Goodness of fit	• OLS estimation: Adjusted R^2 $R^2_{adj} = 1 - \left(\frac{n-1}{n-K}\right)\frac{SSE}{SST}$ • ML estimation: Likelihood ratio test $LRI_{McFadden} = 1 - \frac{\ln(\hat{L}_U)}{n(\hat{L}_R)}$	Possibly acquire a better dataset
2. Linearity relation between the dependent variable and independent variables	H_0: $\beta_k = 0$ $(k = 0, 1, 2, \ldots, K-1)$ H_a: $\beta_k \neq 0$ • OLS estimation 1. *t*-test If $\lvert t^* = b_k/s\{b_k\}\rvert > t_{(1-\alpha/2,\ n-K)}$, reject H_0; Otherwise, reject *Ha* 2. *F*-test for lack of fit If $F^* = \frac{(SSE_R - SSE_F)/(p_F - p_R)}{SSPE/(n-p_F)} \le F_{1-\alpha, p_F-p_R, n-p_F}$, fail to reject H_0; Otherwise, reject H_0 p_F: Twice the no. of variables in time period models p_R: No. of variables in the all-data model 3. *F*-test for lack of fit If $F^* = \frac{(SSE - SSPE)/(c-2)}{SSPE/(n-c)} \le F_{1-\alpha, c-2, n-c}$, fail to reject H_0; Otherwise, reject H_0 *c*: No. of levels considered • ML estimation: Likelihood ratio test If $\chi^2 = -2\ln(\hat{L}_R/\hat{L}_U) \ge \chi^2_{1-\alpha,K-1}$, reject H_0; Otherwise, reject H_a	Use parabolic and hyperbolic transformations, as well as exponential and inverse exponential functions for *X*, *Y*, or both
3. Nonmulticollinearity	• OLS estimation: Variance inflation factor H_0: No multicollinearity of regressors and error terms H_a: There is multicollinearity of regressors and error terms $\overline{VIF} = \frac{\sum_{k=1}^{K-1}\left(1/\left(1-R_k^2\right)\right)}{K-1}$ If $\overline{VIF}$ value is considerably larger than 1, reject H_0; Otherwise, reject H_a	Use ridge regression
4. Nonautocorrelation	• OLS Estimation: Durbin–Watson Test H_0: There is no autocorrelation of error terms H_a: There is autocorrelation of error terms $D = \frac{\sum_{i=2}^{n}\left(e_i - e_{i-1}\right)^2}{\sum_{i=2}^{n} e_i^2}$ • If $D > d_U$, fail to reject H_0; if $D < d_L$; fail to reject H_a; if $d_L \le D \le d_U$, the test is inconclusive. *D* should be approximately equal to 2 when error terms are independent	• Add independent variables • Use transformations based on Cochrane–Orcutt, Hildreth–Lu, and first difference procedures

(*Continued*)

Table 4.4 (Continued) Methods for regression model validation

Focus	*Plot analysis and statistical test*	*Remedying measure*
5. Normality	Plot analysis • Normal probability plot of preparing model fitted values on the X axis versus residuals on the Y axis, a plot that is nearly linear suggests agreement with normality • OLS estimation: Looney–Gulledge correlation test H_0: The error term is normally distributed H_a: The error term is not normally distributed If $\lvert r \rvert = \sqrt{1 - \frac{SSE}{SST}} \geq r_{1-\alpha,n}$, fail to reject H_0; otherwise, reject H_0	• Use transformations for OLS regression • Use weighted least squares (WLS) regression • Use ridge regression
6. Zero mean	When normality and homoscedasticity assumptions are satisfied, this assumption will generally be satisfied	
7. Homoscedasticity	Plot analysis • Plot model fitted values on the X axis versus residuals on the Y axis to identify heteroscedasticity • Plot independent variables on the X axis versus residuals on the Y axis to identify culprit Heteroscedasticity test • H_0: The error term is homoscedastic • H_a: The error term is heteroscedastic OLS estimation 1. Modified Levene test If $t_L^* = \frac{\bar{d}_1 - \bar{d}_2}{s\sqrt{(1/n_1) + (1/n_2)}} \leq t_{1-\alpha/2,n-2}$, fail to reject H_0; otherwise, reject H_0 2. Breusch–Pagan Test If $\chi^2_{BP} = \frac{SSR^*/2}{(SSE/n)^2} \leq \chi^2_{1-\alpha,1}$, fail to reject H_0; otherwise, reject H_0 • ML estimation: Chi-square test If $\chi^2 = -2\ln\left(\frac{\hat{L}_{original}}{\hat{L}_{expanded}}\right) < \chi^2_{1-\alpha,K}$, fail to reject H_0; otherwise, reject H_0	

Note: SSE-Sum of squared errors; SSPE-Sum of squared pure errors; SSR-Sum of squared residuals; and SST-Total sum of squares.

Table 4.5 Measures for evaluating model predictability

Measure	*Mathematical expression*
Mean error	$ME = \left(\frac{\sum_{i=1}^{n} e_i}{n}\right)$
Mean absolute deviation	$MAD = \frac{\left(\sum_{i=1}^{n} \lvert e_i \rvert\right)}{n}$
Sum-of-squared errors	$SSE = \sum_{i=1}^{n} e_i^2$
Mean-squared error	$MSE = \frac{\left(\sum_{i=1}^{n} e_i^2\right)}{n}$
Root-mean-squared error	$RMSE = \sqrt{\frac{\left(\sum_{i=1}^{n} e_i^2\right)}{n}}$
Standard deviation of errors	$SDE = \sqrt{\frac{\left(\sum_{i=1}^{n} e_i^2\right)}{(n-1)}}$
Percentage error	$PE_i = \left[\frac{(x_i - \hat{x}_i)}{x_i}\right] \cdot 100\%$
Mean percentage error	$MPE = \frac{\left(\sum_{i=1}^{n} PE_i\right)}{n}$
Mean absolute percentage error	$MAPE = \left(\sum_{i=1}^{n} \lvert PE_i \rvert\right) / n$

Note: Error e_i is the difference between the observed value x_i and predicted value $\hat{x}_i$, that is, $e_i = x_i - \hat{x}_i$; and n is the number of observations.

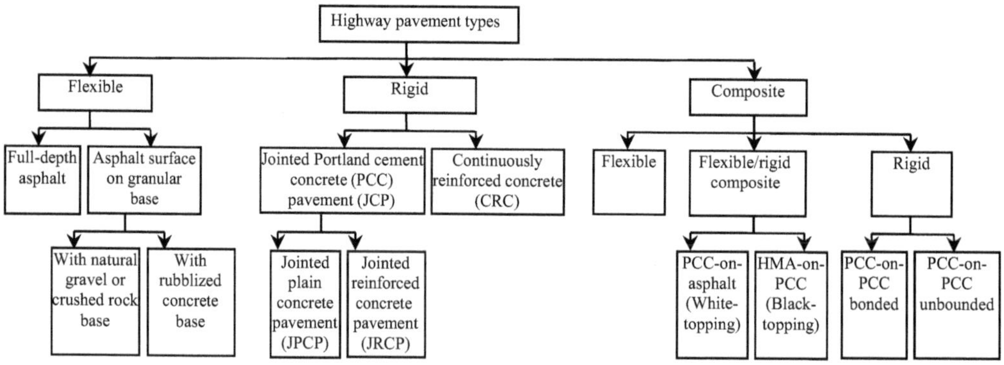

Figure 4.1 Categorization of typical highway pavement types.

to the pavement condition index (PCI). Table 4.6 presents surface distress types for flexible, jointed PCC (Portland Cement Concrete), and CRC pavements (FHWA, 1991; Li et al., 2002a).

4.4.3 Pavement condition indices

4.4.3.1 *Index scales*

The index scale is the scale used for assigning the PCI. There are an infinite number of index scales to choose from. For instance, the scale for the roughness measure in terms of international roughness index (IRI) are open-ended or "unbounded," and can theoretically go to infinity. Others are bound by minimum and maximum values, such as 0 (worst) to 100 (best), 1 (worst) to 10 (best), and 0 (worst) to 5 (best). The first step in developing individual indices is to decide on a scale. In order to compare one distress index with another, it is important that the same scale be used for each index.

4.4.3.2 *Typical index types*

Present serviceability index: One of the earliest pavement condition indices is the present serviceability rating (PSR) developed at the AASHO Road Test in the 1960s. The PSR data was collected from panels riding in an automobile to assign a pavement condition value for each test section using a 0–5 scale that indicated the ride quality rating of the pavement. It was found that highly consistent ratings were given to the same segments by different panels.

Because the efforts of panel ratings are both time consuming and expensive, the subjective PSR is substituted by Present Serviceability Index (PSI) that could be derived by correlating subjective PSR values with objectively measured field data on pavement longitudinal roughness, patch work, rutting, and cracking using regression analysis. Therefore, a PSI value of a pavement is defined to be an algebraic function of subjective PSRs correlating with actual measurements of distresses concerning roughness, rutting, and cracking. Federal Highway Administration (FHWA) developed guidelines for pavement condition data collection, as summarized in Table 4.7. The American Association of State Highway and Transportation Officials (AASHTO) and state transportation agencies in the U.S. have developed guidelines and methods for design and analysis of flexible, rigid, and composite pavements (HRB, 1961, 1962; Yoder, 1985; FHWA, 1991; AASHTO, 1998; Huang, 2004).

Pavement condition index: Subsequent to the AASHO road test, the U.S. Army Corps of Engineers (USACE) Construction Engineering Research Laboratory developed a very complete condition rating system for pavement management (Shahin et al., 1976). This included pavement condition survey procedures and a detailed method for calculating a PCI that is still used today by many agencies. The PCI reflects the composite effects of pavement condition rating (PCR) values assigned to different types of distresses, severity, and extent upon the overall condition of the pavement based upon visual inspections. Table 4.8 shows the PCR scale.

The model for computing PCR is based upon the sum of deducted points for each type of observable distress. Deduct values are a function of distress type, severity, and extent. Deduction for each distress type is calculated by multiplying distress weight times the weights for severity and extent of the distress. Distress weight is the maximum number of deductible points for each different distress type. The PCR computation is as follows:

$$PCR = 100 - \sum_{i=1}^{n} w_i \times w_{s,i} \times w_{e,i} \tag{4.18}$$

Table 4.6 Surface distress types for flexible and rigid pavements

	Pavement type		
Distress	*Flexible*	*Rigid—JCP*	*Rigid—CRC*
Cracking	Fatigue cracking		
	Block cracking		
	Edge cracking		
		Corner breaks	
		D-cracking	D-cracking
	Longitudinal cracking	Longitudinal cracking	Longitudinal cracking
	Transverse cracking	Transverse cracking	Transverse cracking
	Longitudinal reflection cracking		
	Transverse reflection cracking		
Patching and potholes	Patch/patch deterioration		
	Potholes		
Joint deficiencies	Longitudinal joint seal damage		
	Transverse joint seal damages		
	Spalling of longitudinal joints		
	Spalling of transverse joints		
Surface deformation	Rutting		
	Shoving		
Surface defects	Bleeding		
	Polished aggregate	Polished aggregate	Polished aggregate
	Raveling		
		Map cracking	Map cracking
		Scaling	Scaling
		Pop outs	Pop outs
Miscellaneous distress	Lane-to-shoulder drop-off	Lane-to-shoulder drop-off	Lane-to-shoulder drop-off
	Water bleeding and pumping	Water bleeding and pumping	Water bleeding and pumping
		Blowups	Blowups
		Faulting of transverse joints and cracks	
		Lane-to-shoulder separation	Lane-to-shoulder separation
		Patch/patch deterioration	Patch/patch deterioration
			Transverse construction joint deterioration
			Punch-outs
			Spalling of longitudinal joints
			Longitudinal joint seal damage

Table 4.7 FHWA guidelines for PCR data collection

PCR	*Pavement condition*	*Description*
4–5	Very good	Only new, nearly new, or superior pavements are likely to be smooth enough and distress-free to qualify for this category. Most pavements constructed or resurfaced during the data year would normally be rated very good.
3–4	Good	Pavements in this category, although not quite as smooth as those described above, give a first-class ride and exhibit few, if any, visible signs of surface deterioration. Flexible pavements may be beginning to show evidence of rutting and fine random cracks. Rigid pavements may be beginning to show evidence of slight surface deterioration, such as minor cracks and spalling.
2–3	Fair	The riding qualities of pavements in this category are noticeably inferior to those of new pavements and may be barely tolerable for high-speed traffic. Surface defects of flexible pavements may include rutting, map cracking, and extensive patching. Rigid pavements in this category may have a few joint failures, faulting and cracking, and some pumping.
1–2	Poor	Pavements in this category have deteriorated to such an extent that they affect the speed of free-flow traffic. Flexible pavements may have large potholes and deep cracks. Distresses include raveling, cracking, rutting, and occurs over 50% or more of the surface. Distresses of rigid pavements include joint spalling, faulting, patching, cracking, scaling, and may include pumping and faulting.
0–1	Very poor	Pavements in this category are in an extremely deteriorated condition. The facility is passable only at reduced speeds and with considerable ride discomfort. Large potholes and deep cracks exist. Distresses occur over 75% or more of the surface.

where

w_i = Weight for distress type i
$w_{s,i}$ = Weight for the severity level of distress type i
$w_{e,i}$ = Weight for the extent of distress type i
n = Number of observable distresses

For each type of distress, three severity levels (low, medium, and high) and three extent levels (occasional, frequent, and extensive) are defined. The weight of a severity level or an extent level ranges between 0 and 1, with a higher weight assigned to a higher severity level or extent level. The weights associated with the severity level and extent of a specific distress may be established based on expert opinion and engineering or mathematical approaches. Mathematically, relative weights could be established using the straight line, semi-log, and log–log approaches for varying extents of a specific type of distress at low, medium, and high severity levels, respectively.

For instance, if a jointed PCC pavement segment has exposed two types of distresses—faulting (low severity and frequent extent) and surface deterioration (medium severity

Table 4.8 PCR scale developed by USACE

PCR	*Pavement Condition*
85–100	Good
70–85	Satisfactory
55–70	Fair
40–55	Poor
25–40	Very poor
10–25	Serious
<10	Failed

and extensive extent)—and the weight for each type of distress is 10, weights for low and medium severity levels are 0.4 and 0.7, weights for frequent and extensive extents are 0.8 and 1.0, the PCR is calculated as PCR $= 100 - (10 \times 0.4 \times 0.8 + 10 \times 0.7 \times 1.0) = 89.8$. Therefore, the pavement condition is good.

Composite condition index: As pavement analysis evolves more complex systems, the form and utility of the condition indices have changed as well. Composite indices provide an effective means of representing the general condition of pavements. They indicate when a pavement treatment is needed but may not be sufficient to help identify what treatment is most appropriate so that pavement condition is well preserved while keeping the life cycle costs to a minimum. More distress-specific indices are used to provide more information for the required analysis. Detailed distresses are considered within one type of distress to refine the condition index computation.

PCR System: The PCRs and roughness in terms of IRI can be collectively used to establish a PCR system with ratings ranging from very good to very poor, as in Table 4.9. The condition rating system is particularly suited for network-level pavement condition assessment.

4.4.4 Pavement performance models

4.4.4.1 AASHTO pavement condition deterioration model

The AASHTO pavement condition deterioration model correlates PSI with roughness or slope variance, rut depth, cracking, and patching measurements as follows (HRB, 1961, 1962):

$$PSI = 5.03 - \log_{10}(1 + SV) - 1.38(RD^2) - 0.01\sqrt{C + P} \tag{4.19}$$

where

PSI = Present serviceability index of a pavement segment, a statistical estimate of the mean of present serviceability ratings on a scale of 1–5 given by panels
SV = Slope variance of a pavement segment as an early roughness measurement, inches/mi
RD = Mean rut depth, inches
C = Cracking, ft/1000 ft^2
P = Patching, ft^2/1000 ft^2

With field data on pavement roughness, rut depth, cracking, and patching measured over time, the condition deterioration model could be used to establish pavement serviceability-time history. Since the concept of pavement performance incorporates cumulative amount of traffic being served, the serviceability-time history of a pavement is necessary but not sufficient to determine pavement performance. The primary performance index used to

Table 4.9 PCR range

	Pavement condition measure		
PCR system	*PCR by FHWA*	*PCR by USACE*	*IRI*
Very good	4–5	85–100	≤60
Good	3–4	75–85	61–95
Fair	2–3	55–75	96–120
Poor	1–2	40–55	121–170
Very poor	0–1	<40	>170

analyze AASHO Road Test data was the common logarithm of cumulative axle loads of a given weight that the pavement carried at a serviceability level above the threshold value.

4.4.4.2 George pavement condition deterioration models

George et al. (1989) developed a model to predict pavement deterioration considering the effective factors: traffic load, pavement ages, pavement strength, and pavement condition. These models are expressed by the following:

Flexible pavements

$$PCI_{Flexible}(t) = 90 - 0.6349 \cdot [e^{(Age^{0.4203})} - 1] \cdot \log_{10}\left[\frac{ESALs}{((SNC)^{2.7062})}\right] \tag{4.20}$$

with

$$SNC = \sum_{i=1}^{n}(a_i \cdot h_i) + SN_g$$

$$SN_g = 3.51 \cdot \log_{10}(CBR) - 0.85 \cdot [\log_{10}(CBR)]^2 - 1.43$$

where

$PCI_{Flexible}(t)$ = Pavement condition index at time *t*
Age = Period during which the pavement has been in service
$ESALs$ = Annual equivalent single-axle loads
SNC = Modified strength and condition of pavement structure
a_i = Material layer coefficients
h_i = Layer thicknesses, inches
SN_g = Subgrade contribution of modified pavement strength
CBR = *In-situ* California bearing ratio (CBR) of subgrade, %

Composite pavements

$$PCI_{Composite}(t) = 90 - 1.7661 \cdot [e^{((Age^{0.2826})/T)} - 1] \cdot [\log_{10}(ESALs)] \tag{4.21}$$

where

T = Thickness of last overlay

EXAMPLE 4.1

A newly constructed flexible pavement accommodates 3.6 million ESALS per year. The layer thicknesses for the wearing surface, base, and subbase are 8, 7, and 11.5 inches. The corresponding layer coefficients are 0.42, 0.14, and 0.08. The drainage coefficients for the base and subbase are both 1.0. The CBR of the subgrade is 4.5. Given the initial PCI of 90, estimate the percent of condition deterioration after 4 years of service.

Solution

$$\begin{aligned} SN_g &= 3.51 \cdot \log_{10}(CBR) - 0.85 \cdot [\log_{10}(CBR)]^2 - 1.43 \\ &= 3.51 \cdot \log_{10}(4.5) - 0.85 \cdot [\log_{10}(4.5)]^2 - 1.43 \\ &= 0.862 \end{aligned}$$

$$
\begin{aligned}
SNC &= \sum_{i=1}^{n} (a_i \cdot h_i) + SN_g \\
&= (0.42 \times 8 + 0.14 \times 7 \times 1.0 + 0.08 \times 11.5 \times 1.0) + 0.862 \\
&= 14.942
\end{aligned}
$$

$$
\begin{aligned}
PCI_{Flexible}(t) &= 90 - 0.6349 \cdot [\ e^{(Age^{0.4203})} - 1] \cdot \log_{10} \left[\frac{ESALs}{((SNC)^{2.7062})} \right] \\
&= 90 - 0.6349 \cdot [\ e^{(4^{0.4203})} - 1] \cdot \log_{10} \left[\frac{(3.6 \times 10^6)}{((14.942)^{2.7062})} \right] \\
&= 79.28
\end{aligned}
$$

Percent condition deterioration: $(79.28 - 90)/90 = -11.9\%$

4.4.4.3 SHRP distress-based deterioration models

Using data from 244 General Pavement Studies (GPS) of in-service flexible pavement test sections across the U.S., the Strategic Highway Research Program (SHRP) conducted a study to evaluate the AASHTO design equations (SHRP, 1994). It revealed that the design equations tended to overestimate the level of ESALs needed to cause a measured loss of PSI, relative to observed values. The study also indicated that the use of composite PSI presents some limitations in the use of AASHTO equations. With composite indices of this type, where all distresses are lumped together, it is difficult to identify one single distress type or combination of multiple distress types responsible for a reduction in performance. Moreover, by lumping all the structural properties together, the contribution each specific layer makes to the performance of the pavement structure is also masked. For instance, one inch of asphalt will not always be equivalent to 3.1 inches of granular base as the structural number concept suggests. This relationship will naturally vary, depending on the structural properties of the other layers incorporated in the pavement, the environmental conditions in which the pavement is situated, and numerous other factors. Thus, equations for individual distress, including alligator cracking, rutting, transverse or thermal cracking, increases in roughness, and loss of surface friction, were developed in the SHRP study. The distress-based models are of the general form shown below:

$$D = CESALs^{B} \cdot 10^{C} \tag{4.22}$$

where

D = Distress in appropriate units, for example, inches of rutting or inch/mi of roughness

$CESALs$ = Cumulative ESALs, in 1000s

B $= b_0 + b_1 \cdot X_1 + b_2 \cdot X_2 + \cdots + b_{n-1} \cdot X_{n-1}$

C $= c_0 + c_1 \cdot X_1 + c_2 \cdot X_2 + \cdots + c_{n-1} \cdot X_{n-1}$

$X_0, X_1, X_2, \ldots, X_{n-1}$ = Pavement design and construction, and climatic feature parameters

$b_0, b_1, b_2, \ldots, b_{n-1}, c_0, c_1, c_2, \ldots$, and c_{n-1} are model coefficients

4.4.4.4 ARRB pavement condition deterioration model

Australian Road Research Board (ARRB) (Martin, 1994) developed a pavement condition deterioration model to correlate pavement roughness with a series of independent variables, including cumulative traffic loading, pavement and subgrade strength, environmental effects, pavement maintenance (containment and restoration), and rehabilitation (improvement) practices. The model involves a simple addition of load-related roughness changes (heavy vehicle axle loads) and non-load-related roughness changes (environmental factors). The model is of the following specification:

$$R(t) = R_0 + R_0 \cdot A \cdot \left[\frac{I+100}{SNC}\right]^a \cdot t^b \cdot \left[\frac{(B \cdot L^c)}{(ME+G)^d} + \frac{1}{(ME+G)^d}\right]$$
$$-R(t)^{rehab}_{before} \cdot g - T \cdot h - K \quad (4.23)$$

where

$R(t)$ = Pavement roughness measured at time t
$R(t)^{rehab}_{before}$ = Pavement roughness measured at time t before rehabilitation treatment
R_0 = Pavement initial roughness, $R_0 = R(t=0)$
SNC = Modified structural number
L = Average annual ESALs, in millions
ME = Annual average routine maintenance expenditure, dollars/lane-km
G = Constant for maintenance expenditure for each arterial road group
T = Asphalt pavement overlay thickness, mm
a, b, c, d, g, h, K, A, and B are model coefficients

4.4.4.5 Pavement routine maintenance and rehabilitation interval tradeoff model

Tradeoff relationships between pavement routine maintenance and rehabilitation interval have been extensively studied using field data on Indiana state highways by Labi (2001). Tradeoff models by highway functional classification for different types of pavements are of the following specification:

$$T_P^{rehab} = a + b \cdot c^{\left[C_P^{maint}/(ESALs \cdot WSI)\right]} \quad (4.24)$$

where

T_P^{rehab} = Pavement rehabilitation interval, years
C_P^{maint} = Annual pavement routine maintenance expenditure, dollars/lane-mi/year
$ESALs$ = Average annual ESALs, in millions/lane-mi
WSI = A weighted index of precipitation (millimeter), freeze index (degree-day), and freeze–thaw cycles (number of days) with weights being 0.3, 0.35, and 0.35, respectively
a, b, and c are model coefficients, as shown in Table 4.10

Table 4.10 Pavement routine maintenance and rehabilitation interval tradeoff model coefficients

	Highway class				
	Interstate		Non-interstate		
Model coefficient	Rigid	Composite	Flexible	Rigid	Composite
a	34.0793	37.5909	11.9249	39.0608	61.8719
b	−19.8293	−23.3918	−8.2775	−27.0552	−50.7040
c	0.2410	0.5394	0.7678	0.9281	0.9824

Table 4.11 Pavement preventive and corrective maintenance tradeoff model coefficients

Model coefficient	Pavement remaining service life	
	>50%	<50%
a	3.8866	39.0885
b	0.5582	0.7601
c	0.7632	0.1978

4.4.4.6 Pavement preventive and corrective maintenance tradeoff models

Tradeoff relationships between preventive and corrective maintenance over a short-term period of 3 years have also been studied using field data on Indiana state highways by Labi (2001). Tradeoff models are of the Gompertz growth curve form as follows:

$$C_P^{3YCM} = c \cdot a^{b^{C_P^{3YPM}}} \tag{4.25}$$

where

C_P^{3YCM} = Three-year pavement corrective maintenance expenditure, dollars/lane-mi
C_P^{3YPM} = Three-year pavement preventive maintenance expenditure, dollars/lane-mi
a, b, and c are model coefficients, as shown in Table 4.11

4.5 BRIDGE PERFORMANCE MODELING

4.5.1 Bridge types

A bridge is composed of substructure, superstructure, and deck components. Two types of substructure (solid stem and pile type) and five types of superstructure (reinforced concrete slab or box-beam, concrete I-beam, steel beam, and steel girder) are commonly encountered in practice. Bridge decks commonly use reinforced concrete and composite materials. The thorough classification of highway bridge types is illustrated in Figure 4.2 (FHWA, 1987, 1995).

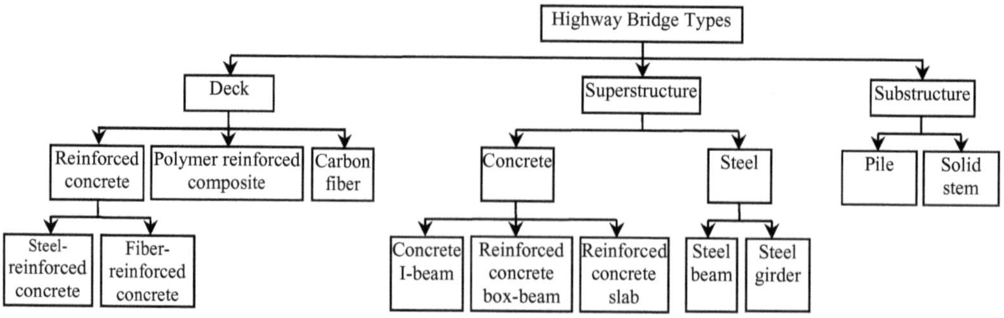

Figure 4.2 Categorization of typical highway bridge types.

4.5.2 Bridge element condition ratings

Bridge element condition ratings: Typically, a bridge may last 35–100 years or longer, and each of the substructure, superstructure, and deck components has its own useful service life. Over a bridge's long lifespan, certain bridge components or elements of a component need to be replaced at intervals that are relatively frequent, allowing the larger and more expensive components to last longer. Since 1978, all federally supported bridges in the U.S. have been inspected every two years according to two sets of standards: (i) Federal National Bridge Inspection Standards (NBIS) created in the early 1970s based on a Congressional mandate to ensure functionality and safety (FHWA, 1987, 1995); and (ii) AASHTO Guide for Commonly-Recognized (CoRe) Structural Elements created in 1992 for maintenance management (AASHTO 1997, 2002, 2010a). The national bridge inventory (NBI) database maintained by FHWA is updated annually (Table 4.12).

Bridge sufficiency rating: In bridge performance assessment, condition ratings assigned to different elements associated with the substructure, superstructure, and deck components of a bridge are utilized to evaluate their functional adequacy. Meanwhile, the loading capacity rating designated to the bridge reflects its structural adequacy to carry more loads or to safely accommodate the current loading in good riding condition. The two sets of ratings can assist in performance assessment at both project and network levels. Practically, the bridge sufficiency rating (BSR) can be used as a composite index to evaluate bridge condition based on functional obsolescence, structural adequacy, level of service, and efficiency for public use as below (FHWA, 1995):

$$BSR = S_1 + S_2 + S_3 - S_4 \tag{4.26}$$

with

$$S_1 = 55\% - (A_1 + B_1)$$
$$S_2 = 30\% - [J_2 + (G_2 + H_2) + I_2]$$

Table 4.12 NBI deck, superstructure, and substructure condition rating

BCR	*Bridge element condition*	*Description*
9	Excellent	As new
8	Very good	No problems noted
7	Good	Some minor problems
6	Satisfactory	Structural elements show some minor deterioration
5	Fair	All primary structural elements are sound, but may have minor section loss, cracking, spalling, or scour
4	Poor	Advanced section loss, deterioration, spalling, or scour
3	Serious	Loss of section, deterioration, spalling, or scour has seriously affected primary structural components; local failures are possible; and fatigue cracks in steel, or shear cracks in concrete may be present
2	Critical	Advanced deterioration of primary structural elements; fatigue cracks in steel or shear cracks in concrete may be present or scour may have removed substructure support; unless closely monitored it may be necessary to close the bridge until corrective action is taken
1	"Imminent" failure	Major deterioration, or section loss present in critical structural components or obvious vertical or horizontal movements affecting structure stability; bridge is closed to traffic, but corrective action may resume the bridge in light service
0	Failed	Out of service that is beyond corrective action

$S_3 = 15\% - (A_3 + B_3)$
$S_4 = (A_4 + B_4 + C_4)$

where
BSR = Bridge sufficient rating
S_1 = Bridge structural adequacy
S_2 = Bridge serviceability and functional obsolescence
$J_2 = A_2 + B_2 + C_2 + D_2 + E_2 + F_2$
S_3 = Efficiency for public use
S_4 = Special reductions, 0%–13%
A_1, B_1, A_2, B_2, C_2, D_2, E_2, F_2, A_3, B_3, A_4, B_4, and C_4 are BSR factors as shown in Table 4.13.

EXAMPLE 4.2

A 4-lane bridge with a total width of 12 m accommodates 12800 vehicles daily. Compute the bridge sufficiency rating using the following: superstructure or substructure condition rating 6, loading capacity 20 tons, deck condition rating 6, structural evaluation 5, deck geometry rating 3, vertical clearance under 4 and over deck 5, water adequacy 4, approach road alignment 6, approach road width 14 m, Strategic Highway Network (STRAHNET) designation greater than zero, no detour length, 2nd and 3rd digits of traffic safety feature 12, and 2nd, 3rd, and 4th digits of main structural type all non-zero.

Solution

The following table summarizes the calculated bridge sufficiency rating:

BSR		*Calculation*	
S_1 Structural adequacy and safety (55% max)	a.	Condition rating reduction (55% max)	
	A_1	Superstructure, substructure	$A_1 = 0\%$ (CR=6)
	b.	Loading capacity reduction (55% max)	
	B_1	Load IR	$B_1 = 0.3254(32.4-20)^{1.5} = 14.2\%$
S_2 Serviceability and functional obsolescence (30% max)	a.	Condition rating reductions (13% max)	
	A_2	Deck condition	$A_2 = 0\%$ (CR = 6)
	B_2	Structural evaluation	$B_2 = 1\%$ (CR = 5)
	C_2	Deck geometry	$C_2 = 4\%$ (CR ≤ 3)
	D_2	Vertical clearances—U	$D_2 = 2\%$ (CR = 4)
	E_2	Waterway adequacy	$E_2 = 2\%$ (CR = 4)
	F_2	Approach road alignment	$F_2 = 0\%$
	b.	Highway insufficiency width: X = 12800/4 = 3200 veh/day; Y = 12/4 = 3 m (15% max)	
	G_2	N	$G_2 = 5\%$, if 12 + 0.6 m < 14
	H_2	ADT W_B W_A ST	$N = 4 > 2$, $H_2 = 15\%$ (X=3, 200 > 1350, Y = 3 < 4.6)
	c.	Vertical clearance insufficiency (2% max)	
	I_2	VC_O SHD	$I_2 = 0\%$ (SHD > 0, $VC_O = 5 > 4.87$)

Continued

BSR				Calculation
S_3	Efficiency for public use	A_3	L_D ADT	$A_3 = 15[(12800 \times 0)/(320000((S_1 + S_2)/85))] = 0\%$ A_3 (15% max)
		B_3	SHD	$B_3 = 0\%$ ($SHD > 0$)
S_4	Special reductions	A_4	L_D	$A_4 = 0 \times (7.9 \times 10^{-9}) = 0\%$ A_4 (5% max)
		B_4	TSF	$B_4 = 5\%$ (2nd and 3rd digits of $TSF = 12$)
		C_4	ST	$C_4 = 0\%$ (2nd, 3rd, and 4th digit of $ST \neq 0$)

$$S_1 = 55\% - (A_1 + B_1) = 55\% - (0\% + 14.2\%) = 40.8\%$$
$$S_2 = 30\% - [J_2 + (G_2 + H_2) + I_2] = 30\% - [9\% + (5\% + 15\%) + 0\%] = 1\%$$
$$S_3 = 15\% - (A_3 + B_3) = 15\% - (0\% + 0\%) = 15\%$$
$$S_4 = (A_4 + B_4 + C_4) = 0\% + 5\% + 0\% = 5\%$$
$$BSR = S_1 + S_2 + S_3 - S_4 = 40.8\% + 1\% + 15\% - 5\% = 51.8\%, \text{ or } 51.8.$$

4.5.3 Bridge performance models

4.5.3.1 Markov bridge element condition transition models

Markovian techniques are the most commonly used techniques for condition deterioration modeling of bridge elements (Jiang et al., 1988; Jiang and Sinha, 1989; Cesare et al., 1992; DeStefano and Grivas, 1998). These models rely on the Markov decision process to define states of condition transition of a bridge component from one state to another during one transition period. In the FHWA bridge element condition rating system on a scale of 0–9, the lowest acceptable condition rating before the bridge element is repaired or replaced is 3. Thus, seven bridge element condition ratings from 9 to 3 are typically considered for analyzing the transition probability. Without loss of generality, the condition ratings from 9 to 3 are defined as seven states with each condition rating corresponding to one of the states. The condition rating 9 is defined as state 1, rating 8 as state 2, and so on. Unless rehabilitation or repair is applied, the bridge structure deteriorates gradually so that the bridge condition ratings are either unchanged or change to a lower number in the two-year rating period. Hence, the condition ratings of the bridge element should monotonically decrease with increase in bridge age and there is a probability of condition transition from state i to another state j during a one-year period, which is denoted by p_{ij}. Naturally, $p_{ij} = 0$ for $i > j$, indicating that probability of the condition rating changing from a state of worse condition to a state of better condition is zero without rehabilitation or repair treatment.

In order to establish transition probability p_{ij}, two kinds of predictions need to be performed: (i) percentage of bridge elements with a particular condition rating at any given time t; and (ii) bridge element conditions for different bridge age groups. First, the number of bridge element condition transitions from one state to another can be obtained from historical data on bridge element condition ratings. This will help establish the percentage of bridge elements with a particular condition rating at any given time t. Next, bridge element inspection data can be divided into different bridge age groups, that is, 0–10, 10–20, 20–30 years, and a fourth group to meet the homogeneity requirements of Markov chain. Within each group, Markov chain is assumed to be homogeneous. Therefore, the transition matrix for each bridge element is established by bridge age group. Unlike the above percentage estimation, a one-year transition period is considered to develop the transition probability. As a result, transition probability p_{ij} from state i to state j in a one-year period is estimated for each

Table 4.13 Bridge sufficiency rating calculation

Bridge sufficiency rating		*Calculation*		
S_1	Structural adequacy and safety (55% max)	a.	Condition rating reduction (55% max)	
		A_1	Superstructure, substructure or culvert	$A_1 = 55\%$ (CR ≤ 2), 40% (CR = 3), 25% (CR = 4), 10% (CR = 5)
		b.	Loading capacity reduction (55% max)	
		B_1	Load inventory rating (IR)	$B_1 = 0.3254(32.4 - IR)^{1.5}$, if $IR \leq 32.4$; 0, o/w
S_2	Serviceability and functional obsolescence (30% max)	a.	Condition rating reductions (13% max)	
		A_2	Deck condition	$A_2 = 5\%$ (CR ≤ 3), 3% (CR = 4), 1% (CR = 5)
		B_2	Structural evaluation	$B_2 = 4\%$ (CR ≤ 3), 2% (CR = 4), 1% (CR = 5)
		C_2	Deck geometry	$C_2 = 4\%$ (CR ≤ 3), 2% (CR = 4), 1% (CR = 5)
		D_2	Vertical clearances—U	$D_2 = 4\%$ (CR ≤ 3), 2% (CR = 4), 1% (CR = 5)
		E_2	Waterway adequacy	$E_2 = 4\%$ (CR ≤ 3), 2% (CR = 4), 1% (CR = 5)
		F_2	Approach road alignment	$F_2 = 4\%$ (CR ≤ 3), 2% (CR = 4), 1% (CR = 5)
		b.	Highway insufficiency width: $X = ADT/lane$; $Y = W_B/lane$, meters (15% max)	
		G_2	Lanes on structure (*N*)	$G_2 = 5\%$, if $W_B + 0.6\ m < W_A$
		H_2	Average daily traffic (ADT) Bridge roadway width (W_B) Approach road width (W_A) Main structure type	• If $N = 1, H_2 = 15\%$ ($Y < 4.3$), $H_2 = 15[(5.5 - Y)/1.2]\%$ ($4.3 \leq Y < 5.5$), $H_2 = 0\%$ ($Y \geq 5.5$) • If $N \geq 2, H_2 = 0\%$ ($N = 2, Y \geq 4.9; N = 3, Y \geq 4.6; N = 4, Y \geq 4.3;$ *or* $N = 5, Y \geq 3.7$) • If $N \geq 2, H_2 = 15\%$ ($X > 50, Y < 2.7$), $H_2 = 7.5\%$ ($X \leq 50, Y < 2.7$), $H_2 = 0\%$ ($X \leq 50, Y \geq 2.7$) • If $N \geq 2, H_2 = 15\%$ ($50 < X \leq 125, Y < 3.0$), $H_2 = 15(4 - Y)\%$ ($50 < X \leq 125, 3.0 \leq Y < 4.0$), $H_2 = 0\%$ ($50 < X \leq 125, Y \geq 4.0$) • If $N \geq 2, H_2 = 15\%$ ($125 < X \leq 375, Y < 3.4$), $H_2 = 15(4.3 - Y)\%$ ($125 < X \leq 375, 3.4 \leq Y < 4.3$), $H_2 = 0\%$ ($125 < X \leq 375, Y \geq 4.3$) • If $N \geq 2, H_2 = 15\%$ ($375 < X \leq 1350, Y < 3.7$), $H_2 = 15[(4.9 - Y)/1.2]\%$ ($375 < X \leq 1350, 3.7 \leq Y < 4.9$), $H_2 = 0\%$ ($375 < X \leq 1350, Y \geq 4.9$) • If $N \geq 2, H_2 = 15\%$ ($X > 1350, Y < 4.6$), $H_2 = 15[(4.9 - Y)/1.2]\%$ ($X > 1350, 4.6 \leq Y < 4.9$), $H_2 = 0\%$ ($X > 1350, Y \geq 4.9$)
		c.	Vertical clearance insufficiency (2% max)	
		I_2	Vertical clearances—O (VC_O) Strategic Highway Network (STRAHNET) highway designation (SHD)	$I_2 = 0\%$ (SHD > 0, $VC_O \geq 4.87$), 2% (SHD > 0, $VC_O < 4.87$) $I_2 = 0\%$ (SHD = 0, $VC_O \geq 4.26$), 2% (SHD = 0, $VC_O < 4.26$)

(*Continued*)

Table 4.13 (Continued) Bridge sufficiency rating calculation

Bridge sufficiency rating				*Calculation*
S_3	Efficiency for public use (15% max)	A_3	Detour length (L_D) ADT	$A_3 = 15[(\text{ADT} \cdot L_D)/(320000 \cdot ((S_1 + S_2)/85))]$ A_3 (15% max)
		B_3	STRAHNET highway designation (SHD)	$B_3 = 0\%$ (SHD > 0), 2% (SHD $= 0$)
S_4	Special reductions (13% max)	A_4	Detour length	$A_4 = L_D \cdot (7.9 \times 10^{-9})$ A_4 (5% max)
		B_4	Traffic safety features (TSF)	$B_4 = 5\%$ (if 2nd and 3rd digits of TSF = 10, 12, 13, 14, 15, 16, or 17)
		C_4	Main structure type (*ST*)	$C_4 = 1\%$ (if 2nd digit of $ST = 0$), 2% (if 3rd digit of $ST = 0$), 3% (if 4th digit of $ST = 0$)

bridge element by bridge age group. To simplify the initial computations, the bridge condition rating is assumed to potentially drop by no more than one state in a single year. That is, the bridge element condition rating would either stay at the current state or drop to the immediate next lower state in the one-year period. The transition matrix P is then defined as

$$P_g = \begin{bmatrix} p_1 & q_1 & 0 & 0 & 0 & 0 & 0 \\ 0 & p_2 & q_2 & 0 & 0 & 0 & 0 \\ 0 & 0 & p_3 & q_3 & 0 & 0 & 0 \\ 0 & 0 & 0 & p_4 & q_4 & 0 & 0 \\ 0 & 0 & 0 & 0 & p_5 & q_5 & 0 \\ 0 & 0 & 0 & 0 & 0 & p_6 & q_6 \\ 0 & 0 & 0 & 0 & 0 & 0 & 1 \end{bmatrix} \tag{4.27}$$

where, $q_i = 1 - p_i$; p_i corresponds to p_{ii} and q_i corresponds to $p_{i,i+1}$. Hence, p_1 is the probability of transition from condition rating 9 (state 1) to condition rating 9 (state 1) and q_1 is the probability of transition from condition rating 9 (state 1) to condition rating 8 (state 2). It is noticed that the lowest condition before a bridge is repaired is condition rating 3 (state 7). Therefore, the corresponding transition probability q_7 is 1.

The transition matrix for each bridge element for bridge age group P_g, $g = 1, 2, \ldots, G$ in Equation 4.27 can be calculated by formulating the nonlinear programming objective function as

$$\text{Min} \sum_{t=1}^{T_g} |R_{pt,g} - S_g(t)| \tag{4.28}$$

$$\text{Subject to} \quad 0 \le p_i \le 1, \quad i = 1, 2, \ldots, n-1$$

where

$R_{pt,g}$ = Estimated condition rating of the bridge element by Markov chain for age group g at time t

$S_g(t)$ = Average of condition ratings of the bridge element calculated for time period t, by applying a polynomial time regression model calibrated using historical condition rating data collected over time

t = 1, 2, …, T_g years of analysis for age group g

i = 1, 2, …, 6 for number of unknown probabilities for the seven condition states

A new bridge element is almost always given a condition rating of 9, which represents a near-perfect condition. In other words, a bridge element at bridge age 0 has a condition rating 9. Thus, the initial state vector Q_0 for the deck, superstructure, or substructure of a new bridge is always [1 0 0 0 0 0 0], where the numbers are the probabilities of having condition ratings of 9, 8, 7, 6, 5, 4, and 3 at age 0, respectively. The state vector of bridge elements in age group 1 for any age from 1 to 10 can be solved by multiplication of the initial state vector Q_0 and the transition matrix for age group 1, $P_{g=1}$, as $P^t = Q_0 \cdot P^t_{g=1}$ $(t = 1, 2, \ldots, 10)$. Initial value for age group 2 is taken as the last state vector of age group 1. In general, group $g + 1$ takes the last-state vector of group g as its initial state vector. By following this analysis for all age groups $g = 1, 2, \ldots, G$ condition rating versus age relationship can be obtained for the bridge element using the Markov chain approach. After obtaining the transition matrix for a specific bridge element by bridge age group, the future condition state of the bridge element can be predicted by multiplying the initial condition state and the transition probability matrix.

Markov models are found to have several limitations by assuming discrete transition time intervals, constant bridge population, and stationary transition probabilities (Collins, 1972). To facilitate developing the transition matrix, they are also limited by assuming that the future condition of a bridge element depends on its current condition only, the condition of a bridge element can either stay the same or deteriorate, and no interactive effects in deterioration of different bridge elements (Madanat et al., 1995, 1997; Sianipar and Adams, 1997; DeStefano and Grivas, 1998).

4.5.3.2 BRP bridge element condition deterioration models

The BRP method is a nonparametric statistical method that can be employed for bridge element condition predictions. Denote x_i to be the ith observation of the M-explanatory variable vector, $x_i = \{x_{1i}, x_{2i}, \ldots, x_{pi}\}$ $(p = 1, 2, \ldots, M;\ i = 1, 2, \ldots, N)$, y_i the ith observation of the response variable of such a bridge deck element condition rating taking values in a prior class j $(j = 1, 2, \ldots, C)$, π_j the prior probability of class j, $L(j, k)$ the loss matrix for incorrectly classifying a class j as k, A is the node of the tree, $\tau(x_i)$ the true class of the ith observation of the vector x_i, $\tau(A)$ the class assigned to node A if it is the final node, n_j the number of observations in class j, n_A the number of observations in node A, n_{jA} the number of observations in class j and node A, $P(A)$ the probability of node A for future samples, $P(A) = \sum_{j=1}^{C}[\pi_j \cdot P(x_i \in A \mid \tau(x_i) = j)] \approx \sum_{j=1}^{C}[\pi_j \cdot (n_{jA}/n_j)]p(j \mid A)$, will be the proportion of class j in node A for future samples, $p(j \mid A) = P(\tau(x_i) = j \mid x_i \in A) \approx [\pi_j \cdot (n_{jA}/n_j)]/\sum_{j=1}^{C}[\pi_j \cdot (n_{jA}/n_j)]$, $R(A)$ is the risk of node A, $R(A) = \sum_{j=1}^{C}[p(j \mid A) \cdot L(j, \tau(A))]$, where $\tau(A)$ is chosen to minimize this risk, and $R(t)$ is the risk of a decision tree T, $R(T) = \sum_{k=1}^{C}[p(A_k) \cdot R(A_k)]$. If $L(j, k) = 1$ for all $j \neq k$, and set the prior probability π_j equal to the observed class frequency in the sample observations, then $p(j \mid A) = n_{jA}/n_A$ and $R(T)$ is proportionately misclassified (Pittou et al., 2009). The detailed calculations include four steps: (i) tree building; (ii) stop tree building; (iii) tree pruning; and (iv) optimal tree selection.

Tree building: This is intended to generate a classification tree for which each node is class-wise purer than its parent node. It begins at the root node, which includes all observations in the learning dataset, and then seeks the best possible explanatory variable affecting the response variable x_p to split the node into two child nodes. For this purpose, all possible splitting variables (called splitters) along with their possible values are examined. More formally, let f be some impurity function and define the impurity of a node A as

$$I(A) = \sum_{j=1}^{C} f[p(j \mid A)] \tag{4.29}$$

Since $I(A) = 0$ when node A is pure, the impurity function f must be concave with $f(0) = f(1) = 0$. The two candidate functional forms for impurity function f are the information index $f(p) = -p.\log(p)$ and the Gini index $f(p) = p(1 - p)$ (Breiman et al., 1984). The measures for the two indices differ only slightly. Empirical evidence shows that they nearly always choose the same split point (Therneau and Atkinson, 1997). Equation 4.29 implies that a "pure" node, whose impurity index equals zero, is one where every individual bridge in the node makes the same mode selection. Equivalently, the least "pure" node is one where individual bridges are equally split between modes.

A partition (split) S of node A into A_L and A_R results in a proportion $p(A_L)$ of cases in node A going to A_L and the remaining proportion $p(A_R)$ going to A_R. The impurity reduction for split S is given by

$$\Delta I(S, A) = I(A) - p(A_L) \cdot I(A_L) - p(A_R) \cdot I(A_R) \quad (4.30)$$

When an unclassified bridge is "passed-down" a decision tree from the root node to a leaf node along the path, it is assigned to the class that is the most frequent among those bridges present in the leaf node. As a result, node A is partitioned into two descendant nodes, A_L and A_R, with respect to the response variable x_p. We will then use the split out of all explanatory variables $x_1, x_2, \ldots, x_p$, that yields maximum reduction in class heterogeneity or equivalently yields the largest maximization of class-purity. That is, given a class of s possible splits, the optimal split S^* is defined by maximizing Equation 4.30, namely, $\Delta I(S^*, A) = \max_{S \in s} \Delta I(S, A)$. When the primary splitting variable is missing for an observation, a surrogate splitting variable whose pattern within the dataset resembles the primary splitter is sought. This partitioning is recursively applied to each leaf node.

Stopping tree building: The tree building process is stopped when (i) there is only one observation in each of the child nodes; (ii) all observations within each child node have the identical distribution of predictor variables, making splitting impossible; or (iii) an external limit on the number of levels in the maximal tree has been set by the user ("depth" option). The "maximal" tree that is created is generally much overfit. In other words, the maximal tree follows every idiosyncrasy in the learning dataset, many of which are unlikely to occur in a future independent group of bridges. The later splits in the tree are more likely to represent overfitting than the earlier splits, although one part of the tree may need only one or two levels, while a different branch of the tree may need many levels to fit the true information in the dataset.

Tree pruning: The complete tree built could possibly be quite large and/or complex and a sequence of simpler and simpler trees must be created through the cutting off of increasingly important nodes. Let $T_1, T_2, \ldots, T_k$ be the terminal nodes of a complete tree T. Define $|T|$ to be number of terminal nodes and risk of T to be $I(A) = \sum_{j=1}^{C}[P(T_j) \cdot R(T_j)]$. Let α be a complexity parameter between 0 and ∞, which measures the cost of adding another variable to the complete tree T. Let $R(T_0)$ be the risk for the zero-split tree. Define $R_\alpha(T) = R(T) + \alpha|T|$ to be the cost for the tree and denote $T\alpha$ to be that subtree of the full model that has minimal cost. Obviously, T_0 represents the full model and T_∞ for the model with no splits. The following results are shown in Breiman et al. (1984): (i) If T_1 and T_2 are subtrees of T with $R_\alpha(T_1) = R_\alpha(T_2)$, then either $|T_1| < |T_2|$ or $|T_2| < |T_1|$; (ii) If $\alpha > \beta$, then either $T_\alpha = T_\beta$ or T_α is a strict subtree of T_β; and (iii) given some set of numbers $\alpha_1, \alpha_2, \ldots, \alpha_m$; both $T_{\alpha 1}, T_{\alpha 2}, \ldots, T_{\alpha m}$ and $R(T_{\alpha 1}), R(T_{\alpha 2}), \ldots, R(T_{\alpha m})$ can be computed efficiently. Using the result in (i), we can uniquely define T_α as the smallest tree T for which $R_\alpha(T)$ is minimized. The result in (ii) implies that all possible values of α can be grouped into m intervals ($m \leq |T|$), I1 = $[0, \alpha_1]$, $I_2 = (\alpha_1, \alpha_2]$, $\ldots$, $I_m = (\alpha_{m-1}, \infty]$, where all $\alpha \in I_i$ share the same minimizing subtree.

Optimal tree selection: The goal in selecting the optimal tree, defined with respect to expected performance on an independent set of data, is to find the correct complexity parameter α so that the information in the learning dataset is fit but not overfit. As the number of nodes increases, the decision cost decreases monotonically for the learning data. This corresponds to the fact that the maximal tree will always give the best fit to the learning dataset. In contrast, the expected cost for an independent dataset reaches a minimum, and then increases as the complexity increases. This reflects the fact that an overfitted and overly complex tree will not perform well on a new set of data.

The best value for complexity parameter α can be determined by the following steps (Therneau and Atkinson, 1997): (i) fit the full model on the dataset, compute I_1, I_2, … , I_m. Set $\beta_1 = 0$, $\beta_2 = \sqrt{\alpha_1\alpha_2}, \ldots, \beta_{m-1} = \sqrt{\alpha_{m-2}\alpha_{m-1}}$, $\beta_m = \infty$; (ii) divide the dataset into s groups g_1, g_2, … , g_s each of size s/n, and for each group separately fit a full model on the data set for everyone except "g_i" and determine $T_{\beta_1}, T_{\beta_2}, \ldots, T_{\beta_m}$ for this reduced data set, compute the predicted class for each observation in g_i, under each of the models $T_{\beta j}$ for $1 \leq j \leq m$, and compute the risk for each subject; (iii) sum over the g_i to get an estimate of risk $R(T_{\beta_1}), R(T_{\beta_2}), \ldots, R(T_{\beta_m})$; and (iv) for the β_j with smallest risk compute $T_{\beta j}$ for the full dataset, this is chosen as the best pruned tree.

EXAMPLE 4.3

A U.S. state bridge inventory database contains 40-year data details for approximately 5500 bridges that can be used to assess the quality of the bridge network. Particular to the quality of structural safety, data on condition ratings of deck, superstructure, and substructure components, as well as inventory load ratings can be used. The deck, superstructure, and substructure ratings are given by a rating scale of 0–9 from the poorest condition to near-perfect condition. Condition ratings less than or equal to 3 are identified as "serious" and greater than or equal to 7 are indicative of "good." Condition ratings 4, 5, and 6 represent "poor," "fair," and "satisfactory," respectively. Data on inventory load ratings indicates the maximum weight in tons that could withstand the passage of a vehicle without causing structural damage. Apply the BRP method for bridge deck condition prediction.

Solution

The CART6.0 ProEx (Classification and Regression Trees) software (Salford Systems, 2003) is utilized for the BRP method application. The response variable considered is DROP, which represents units of deck condition deterioration since last inspection. Table 4.14 lists the explanatory variables employed for bridge deck condition prediction.

To successfully predict bridge deck deterioration, different trees are created and compared. All explanatory variables are used as predictors in the hierarchical tree-based regression analysis. The data is partitioned into relatively homogeneous (low SD) terminal nodes and the mean value observed in each node is taken as its predicted value; exclusion of some variables is done based on Chi-square tests. For calculating variable importance scores, the CART software looks at the improvement measure attributable to each variable in its role as a surrogate to the primary split. The values of these improvements are summed over each node and aggregated and are then scaled relative to the "best" performing variable. The "best" performance variable as most influential is scored 100 and the remaining variables will have lower scores ranging downwards toward zero.

Table 4.15 shows the significance level of the explanatory variables. The most important variable is deck condition rating at the time of last inspection. The next most significant explanatory variable is the deck age. The remaining variables that appear to be significant are highway class, daily traffic, deck width, and wearing surface protection systems. The deck structure type, region, number of directional travel lanes, and bridge skew effects have negligible significance in affecting the bridge deck condition deterioration.

Table 4.14 Explanatory variables used for BRP method application

Variable name	*Variable description*
AGE	Deck age at beginning of time-in-state or observation period, in years
AVGADT	Average of daily traffic for all bridge inspection records
CONSTRTR	Year of construction
DECKCOND	Deck operational condition
DECKWID	Deck width, in tenths of foot
HWCLASS	Highway functional class
LANEPDIR	Number of traffic lanes per direction
NUMSPANS	Number of main bridge spans
REGION	Dummy variable: North = 1, South = 0
SKEW	Bridge skew effect
SPANLEN	Length of the bridge main span
TIS	Time in condition state
TYPE	Deck structural type: simple concrete
WEARSURF	Wearing surface protective systems: concrete, no protective system

Table 4.15 Significance level of explanatory variables used for bridge deck condition prediction

<table>
<tr><th>Explanatory variable</th><th>Score</th><th>Illustration of significance</th></tr>
<tr><td>DECKCOND</td><td>100.00</td><td>||||||||||||||||||||||||||||||||||||||||||</td></tr>
<tr><td>AGE</td><td>49.61</td><td>||||||||||||||||||||</td></tr>
<tr><td>HWCLASS</td><td>21.35</td><td>||||||||</td></tr>
<tr><td>AVGADT</td><td>15.60</td><td>||||||</td></tr>
<tr><td>DECKWID</td><td>14.89</td><td>|||||</td></tr>
<tr><td>WEARSURF</td><td>14.43</td><td>|||||</td></tr>
<tr><td>NUMSPANS</td><td>8.98</td><td>|||</td></tr>
<tr><td>SPANLEN</td><td>6.80</td><td>||</td></tr>
<tr><td>TYPE</td><td>6.80</td><td>||</td></tr>
<tr><td>REGION</td><td>4.38</td><td>|</td></tr>
<tr><td>LANEPDIR</td><td>1.59</td><td></td></tr>
<tr><td>SKEW</td><td>0.33</td><td></td></tr>
</table>

To evaluate predictions yielded by the BRP method, the test sample cross-validation algorithm is used for the computation from the learning sample. The predictive accuracy is tested by applying it to predict class membership in the test sample. The two sample datasets are created by equally splitting the initial dataset using the simple random sampling method.

4.6 TRAFFIC CONTROL AND SAFETY HARDWARE PERFORMANCE MODELING

4.6.1 Traffic control and safety hardware categories, components, and performance

Traffic control and safety hardware, including traffic signs, lighting, signals, pavement markings, and guardrails and crash cushions, collectively contribute to safety performance of travelers and users (FHWA, 2005a,b, 2007). In the U.S., the Manual on Uniform Traffic

Control Devices (MUTCD) serves as minimum standards for deploying traffic control and safety hardware (FHWA, 2009a). Table 4.16 lists the main components of different categories of traffic control and safety hardware.

Traffic signs: For traffic signs, the useful service life of sign sheeting (10–15 years) is generally less than that of sign posts, and much less than that of sign structures (30–50 years). As such, the timing for replacing different sign components would vary. Practically, the replacement of sign components is largely due to changes in sign installation standards, required accuracy of information display, traffic or severe weather damages, age-related deterioration in physical conditions, and degradation of retroreflectivity.

Highway lighting: Highway lighting provides safety and comfort operations of vehicles, especially at night when the natural lighting illumination intensity is insufficient. For electrical components and luminaires of the light system, their useful service lives are terminated when no economical repair options exist to restore the deteriorated conditions.

Traffic signals: Traffic signal systems provide traffic control and communication among physical facilities, vehicles, and drivers. The signal systems consist of signal heads, flashers, detectors, controllers, support structures, enclosures, communications equipment, and other electronic components. Signal heads and flashers contain conventional lamps that are regularly replaced in 1–1.5 years and light-emitting diode (LED) lamps that last for 5 years or more. Traffic signal components are replaced based on their condition reaching the minimum acceptable level, remaining service life close to termination or technological improvements.

Pavement markings: Pavement markings include longitudinal lane, shoulder, and center lines; raised markers; and various symbols and guidance and warning messages on pavement surfaces. They have a fast pace of deterioration due to frequency of contact with vehicle tires; rain and snow falls; accumulation of salt, dirt, and debris; and sunlight exposure. Replacement of pavement markings is attributable to condition deterioration, and changes in highway geometric designs or standards. The most commonly used performance measure for pavement markings is the retroreflectivity representing their ability to reflect light from the headlight of a vehicle back to the driver's eyes.

Concrete barriers: Concrete barriers can provide positive separation for directional traffic movements. They could also effectively split the construction vehicles, machinery, and workers from the traveling public in the presence of work zones. The shrinkage strain of concrete barriers can be used for condition assessment.

Table 4.16 Main components of traffic control and safety hardware

Category	*Component*	*Material example*
Highway signs	Sign sheeting Sign posts Sign bridges and supports	American Society for Testing and Materials (ASTM) type or grade, color, and product designation
Lighting	Lamps Structural supports Other	Incandescent, mercury vapor, sodium, metal halide, and fluorescent
Traffic signals	Signal displays	
	Controller systems	Tubular: steel and aluminum
	Structural supports	Pole: wood, concrete, and steel
	Sensors and detectors	Loop, infrared, microwave, ultrasonic, video camera, weigh-in-motion scale, and radar
Pavement markings	Lane and edge strips Pavement markers Other	Paint: epoxy and non-epoxy Thermoplastic: plastic, polyester, and tape
Guardrails, barriers, and crash cushions	Guardrails, barriers, and crash cushions	Steel, concrete, and plastic

Table 4.17 FHWA traffic sign retroreflectivity deterioration models

Sign type		*Deterioration model*
Red	Engineering	$R_{SIGN} = 21.47 - 1.269(Age) - 0.0004(DegDays) + 0.124(Precip) + 0.003(Elv)$
Age ≤ 5	High intensity	$R_{SIGN} = 38.97 - 3.574(Age) - 0.0001(DegDays) + 0.240(Precip) + 0.001(Elv)$
Age > 5		$R_{SIGN} = 19.765 + 2.496(Age) - 0.0003(DegDays) + 0.067(Precip) + 0.0001(Elv)$
Yellow	Engineering	$R_{SIGN} = 78.794 - 3.906(Age) - 0.002(DegDays) + 0.115(Precip) + 0.002(Elv)$
	High intensity	$R_{SIGN} = 247.85 - 4.578(Age) - 0.001(DegDays) + 0.174(Precip) + 0.002(Elv)$
White	Engineering	$R_{SIGN} = 103.09 - 5.451(Age) + 0.002(DegDays) + 0.178(Precip) + 0.002(Elv)$
	High intensity	$R_{SIGN} = 304.09 - 4.815(Age) + 0.002(DegDays) + 0.06(Precip) + 0.001(Elv)$
Green	Engineering	$R_{SIGN} = 15.990 - 0.637(Age) + 0.0003(DegDays) - 0.036(Precip) + 0.0001(Elv)$
	High intensity	$R_{SIGN} = 53.386 - 1.345(Age) - 0.002(DegDays) + 0.337(Precip) + 0.003(Elv)$

Note:

R_{SIGN} = Predictive sign retroreflectivity
Age = Age category of sign sheeting, in years
DegDays = Annual heating degree-days
Precip = Annual precipitation, in inches
Elv = Average ground elevation, in ft

4.6.2 Traffic sign retroreflectivity deterioration models

FHWA model: Black et al. (1991) collected retroreflectivity measurements of over 6000 signs in the U.S. to develop models for predicting sign sheeting retroreflectivity performance. Signs are classified by color, type, age, and geographic location. Factors considered include sheeting types, colors, annual heating degree-days, annual precipitation, and ground elevations. Table 4.17 presents sign retroreflectivity deterioration models for different sheeting types (engineering grade and high-intensity grade) and colors (red, green, white, and yellow). It is observed that the retroreflectivity of traffic signs decreases over time except red high-intensity sheeting where retroreflectivity value decreases in the first 5 years and then starts to increase, probably due to fading of red ink.

Louisiana models: Wolshon et al. (2002) and Wolshon (2003) collected data on 3646 traffic signs in Louisiana to calibrate sign sheeting deterioration models. Traffic signs were divided into two types: engineering grade and high-intensity grade. Table 4.18 shows estimated coefficients

Table 4.18 Louisiana traffic sign retroreflectivity deterioration models

Sheeting grade	*Condition*	*Color*	*Deterioration model*
Engineering	Unwiped	Green	$R_{SIGN} = 1.49507 - 0.00531(Age) + 0.00039(D_{PEDGE}) + K_{ORIENT}$
		White	$R_{SIGN} = 1.33813 - 0.00763(Age) + 0.00039(D_{PEDGE}) + K_{ORIENT}$
		Yellow	$R_{SIGN} = 1.29848 - 0.00455(Age) + 0.00039(D_{PEDGE}) + K_{ORIENT}$
	Wiped	Green	$R_{SIGN} = 1.80944 - 0.00573(Age) + 0.00011(D_{PEDGE}) + K_{ORIENT}$
		White	$R_{SIGN} = 1.49909 - 0.00726(Age) + 0.00011(D_{PEDGE}) + K_{ORIENT}$
		Yellow	$R_{SIGN} = 1.48635 - 0.00432(Age) + 0.00011(D_{PEDGE}) + K_{ORIENT}$
High intensity	Unwiped	Green	$R_{SIGN} = 0.89382 - 0.00077(Age) + 0.00934(D_{PEDGE}) + K_{ORIENT}$
		White	$R_{SIGN} = 1.07122 - 0.00226(Age) + 0.00934(D_{PEDGE}) + K_{ORIENT}$
		Yellow	$R_{SIGN} = 1.33811 - 0.00481(Age) + 0.00934(D_{PEDGE}) + K_{ORIENT}$
	Wiped	Green	$R_{SIGN} = 1.24338 - 0.00138(Age) + 0.00927(D_{PEDGE}) + K_{ORIENT}$
		White	$R_{SIGN} = 1.19448 - 0.00122(Age) + 0.00927(D_{PEDGE}) + K_{ORIENT}$
		Yellow	$R_{SIGN} = 1.57808 - 0.00493(Age) + 0.00927(D_{PEDGE}) + K_{ORIENT}$

Table 4.19 K_{ORIENT} values for traffic sign retroreflectivity deterioration models

Sheeting grade	*Condition*	*East*	*North*	*South*	*West*
Engineering	Unwiped	−0.1269	−0.10394	−0.04057	0.00000
	Wiped	−0.1408	−0.04285	0.00173	0.00000
High intensity	Unwiped	0.0083	0.09113	0.06191	0.00000
	Wiped	−0.0614	0.05652	−0.00145	0.00000

Note:

R_{SIGN} = Adjusted coefficient of retroreflectivity
Age = Age category of sign sheeting, in years
EOPD = Distance from the edge of pavement to signs
K_{ORIENT} = Constant for sign orientation adjustment

of factors to traffic sign retroreflectivity, which include two conditions, unwiped and wiped, and three colors, green, white, and yellow. From the results, it is observed that high-intensity grade sheeting performs better than engineering grade sheeting during the service life.

Table 4.19 lists K_{ORIENT} values to capture impacts of exposure of signs to sun over the service life cycle that depend on sheeting grades of signs, as well as conditions and directions of sign installations.

4.6.3 Lighting illuminance deterioration model

LED lamps are commonly used in traffic lighting due to energy saving, low environmental impacts, and long service lives. Song et al. (2010) introduced a methodology to determine the service life of the LED lamp. The light output model was based on the junction temperature inside the light and light output deterioration. The light output in lumen of the LED lamp at time t is formulated as

$$I_{LED} = I_0 \cdot e^{[-\alpha \cdot (T_j) \cdot t]} \tag{4.31}$$

with

$$T_j = T_j^0 + \left[\frac{\log_{10}(m \cdot t + 1)}{\log_{10}(m \cdot t_{life} + 1)}\right] \cdot \left(T_j^f - T_j^0\right),$$

$$T_j = T_j^0 + \left(\frac{t}{t_{life}}\right) \cdot \left(T_j^f - T_j^0\right),$$

or

$$T_j = T_j^0 + \left[\frac{\lambda^{n \cdot t} - 1}{(\lambda^{n \cdot t_{life}} - 1)}\right] \cdot \left(T_j^f - T_j^0\right)$$

where

I_{LED} = Light output at time t
I_0 = Initial light output and
T_j^0 = Initial junction temperature
T_j^f = Final junction temperature
T_j = Junction temperature at time t
α = Light output's deterioration rate, α = 10%, 20%, 30%, 40%, or 50%
m, n = Increase rates of junction temperature

λ = Coefficient of junction temperature increase function
t = Operation time, hours
t_{life} = Service life operation time, hours

4.6.4 Traffic signal head deterioration model

Data on light intensity were collected from 372 LED signals in Missouri to evaluate the service life of LED traffic signal heads based on the realities of manufacturer, indicated life, color, and directional view (Long et al., 2011). Light intensity deterioration models for the LED signal heads were developed, as shown in Table 4.20.

4.6.5 Pavement marking retroreflectivity deterioration models

Alabama models: The Laserlux mobile retroreflectometer was used to collect retroreflectivity data from pavement markings installed in Alabama at different times for a period of 36 months. Two types of pavement markings were involved, including thermoplastic and profiled pavement markings. The dataset was used to develop retroreflectivity deterioration models for pavement markings (Lindly and Wijesundera, 2003). The primary independent variable considered is a composite index termed as cumulative traffic passages (CTP) defined by

$$CTP = \frac{(AADT \cdot Age)}{(Lanes)} \tag{4.32}$$

where
CTP = Cumulative traffic passages
$AADT$ = Annual average daily traffic, veh/day
Age = Age of pavement marking, days
$Lanes$ = Number of travel lanes

The pavement marking retroreflectivity deterioration models are as below

$$\text{Linear form: } R_{PM} = \begin{cases} 310 - 31.1 \cdot CTP, \text{ flat thermoplastic} \\ 239 - 28.9 \cdot CTP, \text{ profiled} \end{cases} \tag{4.33}$$

Table 4.20 Missouri light intensity deterioration models for LED traffic signal heads

Signal head type	*Manufacturer*	*Regression equation*	*Threshold*	*Service life (Years)*	
Green	Circular	General Electric (GE)	$I_{SIGNAL} = 386.60 - 28.139(Age)$	257	5
	Circular	Dial	$I_{SIGNAL} = 531.07 - 32.415(Age)$	257	8
	Arrow	GE	$I_{SIGNAL} = 116.46 - 9.8846(Age)$	41	7
	Arrow	Dial	$I_{SIGNAL} = 154.61 - 12.681(Age)$	41	9
Yellow	Circular	Dial	$I_{SIGNAL} = 530.28 - 8.185(Age)$	491	5
	Arrow	GE	$I_{SIGNAL} = 274.37 - 33.366(Age)$	79	6
	Arrow	Dial	$I_{SIGNAL} = 115.56 - 5.9974(Age)$	79	6
Red	Circular	GE	$I_{SIGNAL} = 473.85 - 30.77(Age)$	197	9
	Circular	Dial	$I_{SIGNAL} = 282.74 - 16.406(Age)$	197	5

Note:

I_{SIGNAL} = light intensity of LED signal head
Age = LED signal head age, years

$$\text{Exponential form: } R_{PM} = \begin{cases} 329[\lambda^{-0.16(CTP)}], \text{ flat thermoplastic} \\ 244[\lambda^{-0.16(CTP)}], \text{ profiled} \end{cases} \tag{4.34}$$

NCHRP models: Apart from the cumulative passages of traffic, performance of pavement markings is also dependent upon materials used such as waterborne paint, thermoplastic, epoxy, solvent paint, permanent tape, and methyl methacrylate. Bahar et al. (2006) utilized data on pavement markings from National Transportation Product Evaluation Program (NTPEP) established in Minnesota, Wisconsin, Pennsylvania, and California to develop retroreflectivity deterioration models for pavement markings using different types of materials categorized into two age groups. The model specifications are as follows:

$$R_{PM} = \begin{cases} 1/(\beta_0 + \beta_1 \cdot \text{Age} + \beta_2 \cdot \text{Age}^2), \text{ Age} \leq 25 \text{ months} \\ -\lambda \cdot (\text{Age} - 25) + \delta, \text{ Age} > 25 \text{ months} \end{cases} \tag{4.35}$$

Tables 4.21 and 4.22 list model coefficients.

4.6.6 Concrete barrier deterioration models

The effect of material properties, ambient temperature, relative humidity, and age would cause shrinkage strain on concrete barriers. The Comite Euro-International du Beton (CEB)-FIP 90 and BaZant and Baweja B3 models are typical ones developed for concrete barrier deterioration prediction.

Table 4.21 Model coefficients for pavement markings younger than 25 months

Pavement material	*Snow removal impacts*		$\beta_0 \cdot 10^{-3}$	$\beta_1 \cdot 10^{-5}$	$\beta_2 \cdot 10^{-7}$
Epoxy	White	All	2.95	6.96	−6.01
	Yellow	All	4.22	13.3	−8.40
Methyl methacrylate	White	All	2.65	8.81	3.00
	Yellow	All	5.56	8.76	65.7
Permanent tape	White	All	1.82	19.6	7.40
	Yellow	All	2.59	40.1	−1.10
Solvent	White	All	5.49	33.5	125
	Yellow	All	7.82	43.0	140
Thermoplastic	White	None	2.42	13.2	−11.8
	Yellow	None	4.89	18.5	−0.80
	White	Low to medium	2.37	12.2	−21.6
	Yellow	Low to medium	4.75	15.3	−1.60
	White	High	2.63	−1.29	392
	Yellow	High	4.72	−8.33	637
Waterborne	White	None	2.81	21.9	−41.4
	Yellow	None	4.20	38.3	−17.5
	White	Low to medium	2.92	14.4	−12.2
	Yellow	Low to medium	4.45	30.1	−0.0476
	White	High	3.03	−8.32	0.068
	Yellow	High	4.78	−9.95	0.00103

Table 4.22 Model coefficients for pavement markings older than 25 months

Pavement material	*Snow removal impacts*		λ	δ
Epoxy	White	All	2.17	231.51
	Yellow	All	1.89	142.49
Methyl methacrylate	White	All	4.13	198.46
	Yellow	All	3.02	84.34
Permanent tape	White	All	4.65	139.21
	Yellow	All	2.60	79.78
Solvent	White	All	2.11	46.13
	Yellow	All	1.56	36.61
Thermoplastic	White	None	3.03	200.51
	Yellow	None	2.06	105.61
	White	Low to medium	1.30	243.10
	Yellow	Low to medium	2.06	117.99
	White	High	2.86	37.30
	Yellow	High	1.82	23.56
Waterborne	White	None	0.50	175.52
	Yellow	None	1.89	78.75
	White	Low to medium	2.58	173.77
	Yellow	Low to medium	2.15	83.51
	White	High	1.81	22.33
	Yellow	High	1.20	14.99

CEB-FIP 90 model: This model was developed by European Concrete Committee and it considered the time concrete started drying and the ultimate shrinkage as follows (CEB, 1993):

$$\hat{\varepsilon}_s(t) = \varepsilon_{s_{max}} \cdot \beta_s \cdot (t - t_c) \tag{4.36}$$

with

$$\varepsilon_{s_{max}} = \beta^{RH}_{s_{max}} \cdot \left[160 + 10 \cdot \beta_{CT} \cdot \left(\frac{9 - f'_c}{1450}\right)\right] \cdot 10^{-6}$$

$$\beta_s = \sqrt{\frac{1}{[350 \cdot (((2 \cdot (A_s/A))/3.937)^2) \cdot ((t - t_c)) + (t - t_c)^2]}}$$

$$\beta^{RH}_{s_{max}} = \begin{cases} -1.55 \cdot \left[1 - (RH/100)^3\right], 40\% \le RH \le 99\% \\ 0.25, RH \ge 99\% \end{cases}$$

where

$\hat{\varepsilon}_s(t)$ = The predicted shrinkage
$\varepsilon_{s_{max}}$ = The ultimate shrinkage
β_s = Coefficient of shrinkage development over time
$\beta^{RH}_{s_{max}}$ = Coefficient of ultimate shrinkage incorporating the effect of relative humidity
RH = Relative humidity
β_{CT} = Cement type factor according to the European standard, $\beta_{CT} = 4$ for low heat development cement and $\beta_{CT} = 5$ for rapid heat development cement

f'_c = The 28-day standard concrete strength
A_s = The surface area of shrinkage
A = The surface area of concrete barrier
t_c = The age curing or the age at which concrete started drying
t = The age of concrete

BaZant and Baweja model: BaZant and Baweja (1995) developed a model for concrete barrier shrinkage strain prediction that explicitly considered cement type, curing, water content, and 28-day standard concrete's strength. The model is of the following specification:

$$\hat{\varepsilon}_s(t) = \varepsilon_{s_{max}} k_H \cdot \tanh \sqrt{\frac{(t-t_c)}{\left[190.8 \cdot \left(t_c^{-0.08}\right) \cdot \left((f'_c)^{-0.25}\right) \cdot \left((k_s \cdot (A_s/A))^2\right)\right]}} \tag{4.37}$$

with

$$\varepsilon_{s_{max}} = \alpha_1 \cdot \alpha_2 \cdot [26 \cdot (w^{2.1}) \cdot ((f'_c)^{-0.28}) + 270] \cdot 10^{-6}$$

$$k_H = \begin{cases} 1 - RH^3, RH \leq 98\% \\ 0.25, 98\% \leq RH \leq 100\% \end{cases}$$

where

$\hat{\varepsilon}_s(t)$ = The predicted shrinkage
$\varepsilon_{s_{max}}$ = The ultimate shrinkage
k_H = Coefficient of the humidity
β_s = Coefficient of shrinkage development over time
$\beta_{s_{max}}^{RH}$ = Coefficient of ultimate shrinkage incorporating the effect of relative humidity
RH = Relative humidity
β_{CT} = Cement type factor according to the European standard, $\beta_{CT} = 4$ for low heat development cement and $\beta_{CT} = 5$ for rapid heat development cement
f'_c = The 28-day standard concrete strength
k_s = Shrinkage shape factor
A_s = The surface area of shrinkage
A = The surface area of concrete barrier
w = Water content
t_c = The age curing or the age at which concrete started drying
t = The age of concrete
α_1 = Model coefficient, 1.0, 0.85, and 1.1 for types I, II, and II cements
α_2 = Model coefficient, 0.75 and 1.0 for steam and water curing conditions

4.7 MULTIMODAL FACILITY PERFORMANCE ASSESSMENT

4.7.1 Transit condition rating

Transit condition rating may be concerned with transit facility condition, fleet condition or performance of the transit system that consists of facilities, fleet, and users. Table 4.23 presents primary transit facility components and element definitions. Table 4.24 shows the transit facility element condition rating scale introduced by the Federal Transit Administration (FTA) (FTA, 1994).

Table 4.23 Transit facility components and elements

Transit component	*Transit element*
A. Substructure	Foundations: Walls, columns, pilings other structural components Basement: Materials, insulation, slab, floor underpinnings
B. Shell	Superstructure/structural frame: Columns, pillars, wall Roof: Roof surface, gutters, eaves, skylights, chimney surrounds Exterior: Windows, doors, and all finishes (paint, masonry) Shell appurtenances: Balconies, fire escapes, gutters, downspouts
C. Interiors	Partitions: Walls, interior doors, fittings such as signage Stairs: Interior stairs and landings Finishes: Materials used on walls, floors, and ceilings
D. Conveyance	Elevators, escalators, and lifts
E. Plumbing	Fixtures, water distribution, sanitary waste, and rain water drainage
F. Heating, ventilation, and air conditioning (HVAC)	Energy supply Heat generation and distribution systems Cooling generation and distribution systems Testing, balancing, controls, and instrumentation Chimneys and vents
G. Fire protection	Sprinklers, standpipes, and hydrants and other fire protection specialties
H. Electrical	Electrical service and distribution Lighting and branch wiring (interior and exterior) Communications and security Other pieces such as lightning protection, generators, and emergency lighting
I. Equipment	Facility-related equipment including maintenance and vehicle service valued over $10000
J. Fare collection	Turnstiles, ticket machines, and other major equipment with capital for replacement
K. Site	Roadways, driveways, parking lots and associated signs, markings, and equipment Pedestrian areas and associated signage, markings, and equipment Site development such as fences, walls, and miscellaneous structures Landscaping and irrigation Site utilities

Table 4.24 Primary transit facility element condition rating

Transit element rating	*Transit element condition*	*Description*
5	Excellent	No visible defects, new or near new condition, may still be under warranty if applicable
4	Good	Good condition, but no longer new, may have some slightly defective or deteriorated component(s), but is overall functional
3	Adequate	Moderately deteriorated or defective components; but has not exceeded useful life
2	Marginal	Defective or deteriorated component(s) in need of replacement, exceeded useful life
1	Poor	Critically damaged component(s) or in need of immediate repair, well past useful life

4.7.2 Rail track condition rating and modeling

Track condition index: Rail track performance is assessed in terms of track durability and riding comfort. Both quantitative and qualitative measures are used for condition evaluation. The first track condition index (TCI) was developed by the International Union of Railways (UIC), which used data on 1000 m rail sections to calculate the SD as the TCI (UIC, 1981). Ebers Öhn and Conrad (1998) introduced track roughness index (R^2) to assess railway conditions in the

U.S., followed by *J* index introduced by Madejski and Grabczyk (1999) and track geometry index (TGI) (Mundrey, 2003). More recently, the five-parameter track defectiveness index (W_5) (Madejski and Grabczyk, 2002) and track quality index (TQI) (Zhang et al., 2004) were developed in various countries to measure track conditions. Table 4.25 presents their specifications.

With the introduction of various track performance indices to track performance assessment, field data can be collected to compute the respective index values. Having established the minimum acceptable track performance levels for individual indices, the actual condition index values can be compared with the minimum acceptable levels to identify the current needs for track maintenance and renewal on a project by project basis (Burrow et al., 2009). In addition, historical data on track performance indices can be used for trend analysis that could predict the point in time when the track performance index values fall below the minimum acceptable level. This will help identify the future needs for track maintenance and renewal at the project level.

Track condition rating system: Based on the values of track performance indices that fall in different ranges, a track condition rating system could be introduced to rate track conditions from excellent, good, fair, to poor condition levels. As shown in Table 4.26, Sadeghi and Askarinejad (2007, 2008) proposed such a track condition rating system that uses four track performance indices, and each index is specified over a value range for excellent, good, fair, and poor condition ratings. The condition rating system is well suited for identifying needs for track maintenance and renewal at the network level. First, the distribution of all tracks in a rail network with

Table 4.25 Typical track condition indices

Track condition index	*Specification*
Standard deviation (SD)	$SD = \sqrt{\frac{\sum_{i=1}^{N}(x_i - \bar{x})^2}{N-1}}$
Track roughness index (R^2)	$R^2 = \frac{\sum_{i=1}^{N}(x_i - \bar{x})^2}{N}$
J index	$J = (S_z + S_y + S_W + 0.5\ S_l)/3.5$
Track geometry index (*TGI*)	$TGI = (2UI + TI + 6AI + GI)/10$
Five-parameter track defectiveness index (W_5)	$W_5 = 1 - (1 - w_e)(1 - w_g)(1 - w_w)(1 - w_x)(1 - w_y)$
Track quality index (TQI)	$TQI = (l_s/l_o - 1) \times 10^6$

Note:

x_i = The ith data value of mid-chord profile and alignment or cross level and gauge measurement
$\bar{x}_i$ = The average value of mid-chord profile and alignment or cross level and gauge measurement
S_z = SD for track gauge
S_y = SD for horizontal irregularities
S_w = SD for twist
S_l = SD for vertical irregularities
UI = Unevenness index
TI = Twist index
AI = Alignment index
GI = Geometry index
W_5 = Five parameter index
W_e = Track gauge deficiency
W_g = Superelevation deficiency
W_w = Twist deficiency
W_x = Horizontal irregularities
W_y = Vertical irregularities
l_s = Traced length of space curve
l_0 = The fixed length of track segment
N = Number of data points in the data sample

Table 4.26 Rail track condition rating system

Track condition rating system	*Track condition index*			
	SD	R^2	*TGI*	W_5
Excellent	SD ≤ 1	$R^2 \leq 1$	TGI ≥ 85	$W_5 \leq 0.1$
Good	1 < SD ≤ 2	$1 < R^2 \leq 4$	50 ≤TGI < 85	$0.1 < W_5 \leq 0.2$
Fair	2 < SD ≤ 4	$4 < R^2 \leq 16$	36 ≤TGI < 50	$0.2 < W_5 \leq 0.6$
Poor	SD > 4	$R^2 > 16$	TGI < 36	$W_5 > 0.6$

condition ratings being excellent, good, fair, and poor rating levels can be determined by applying the condition rating system. Next, the rail authority can establish a policy of only allowing the percentage of tracks in fair or poor rating level being controlled within an upper limit. For a given year or multiyear period, if the percentage of fair or poor condition rating level exceeds the upper limit, track maintenance and renewal treatments need to be implemented for track condition improvements to ensure that the percentages of fair and poor condition rating levels get reduced to meet the respective upper limits. Further, trend analysis using historical data can be carried out to predict track performance index values and then predict excellent, good, fair, and poor condition rating distributions in the future. Accordingly, track maintenance and renewal treatments can be proposed for track preservation in a proactive manner.

EXAMPLE 4.4

A rail track is selected for condition inspection. Forty samples are taken along the track alignment with roughness measures obtained as follows: 20, 18, 16, 15, 18, 19, 15, 20, 17, 18, 16, 19, 20, 16, 19, 15, 19, 17, 15, 16, 17, 20, 18, 19, 19, 18, 15, 16, 16, 16, 18, 22, 18, 15, 16, 16, 20, 19, 21, and 16 mm. Evaluate the track condition rating using the track roughness index.

Solution

$$R^2 = \frac{\sum_{n=1}^{40} (x_i - \bar{x})^2}{40}$$

$$= \frac{\begin{bmatrix} [20-(20+18+\cdots+16)/40]^2 + [18-(20+18+\cdots+16)/40]^2 \\ +\cdots+[16-(20+18+\cdots+16)/40]^2 \end{bmatrix}}{40}$$

$$= 3.54, \text{ track condition rating is good}$$

Network-level TCI: The network TCI can be evaluated based on (i) network maintenance efficiency, TCI_1; and (ii) network ballast geometric condition, TCI_2. For maintenance efficiency index, a higher value of TCI_1 reflects a larger amount of maintenance cost to maintain or restore the poor track condition. For ballast geometric condition, TCI_2 may be regarded as a measure of passenger comfort or ride quality. In contrast with the TCI_1, a higher value of TCI_2 represents a better network condition. The two track condition indices are of the following specifications:

$$TCI_1 = \frac{\sum_{j=1}^{n} \sum_{i=1}^{m} MV_{ij} \times MC_i}{L} \tag{4.38}$$

$$TCI_2 = \frac{\sum_{j=1}^{n} p_j \times d_j}{L} \tag{4.39}$$

where

MV_{ij} = Maintenance volume of type i implemented to network j
MC_i = Maintenance cost of type i
p_j = Percentage of network j that has fair or better condition ratings
d_j = Length of network j
m = Number of maintenance treatments
n = Number of subnetworks
L = Total network length

EXAMPLE 4.5

An 85-mile long railway network consists of six subnetworks. The respective network lengths are 12, 16, 18, 15, 13, and 12 miles. The corresponding maintenance volumes are 6, 8, 9, 7, 7, and 6 miles; maintenance costs equal to 0.2, 0.4, 0.3, 0.6, 0.1, and 0.5 million dollars; and the percentages of subnetworks in fair or better conditions are 20%, 60%, 67%, 50%, 69%, and 82%. Determine the network track condition indices for TCI_1 and TCI_2.

Solution

$$TCI_1 = \frac{\sum_{j=1}^{n}\sum_{i=1}^{m} MV_{ij} \times MC_i}{L}$$

$$= \frac{6\times0.2+8\times0.4+9\times0.3+7\times0.6+7\times0.1+6\times0.5}{85}$$

$$= 0.18 \text{ million dollars}$$

$$TCI_2 = \frac{\sum_{j=1}^{n} p_j \times d_j}{L}$$

$$= \frac{12\times0.2+16\times0.6+18\times0.67+15\times0.5+13\times0.69+12\times0.85}{85} = 0.6$$

Track maintenance/renewal interval: Rail track condition can be assessed based on surface defects and internal defects. Surface defects are the defects that are mainly caused by the rolling contact fatigue, corrosion, wheel burns, wear, and indentures, whereas the internal defects are caused by cracks on the web, foot, or head of the rail. The defects can be measured based on either visual inspection or using nondestructive testing. The interval of rail track maintenance or renewal is determined by cumulative usage based on traffic density as below (Schoech, 2007; Zarembski, 2015):

$$MRI_{\text{Axle Load}} = \left(\frac{NAL^{\text{DFE}}}{33}\right) \times BRM \times MGT \tag{4.40}$$

where

$MRI_{\text{Axle Load}}$ = Rail track maintenance/renewal interval determined by axle loads per year, in hours
33 = 263000 lbs freight car axle load
NAL = New axle load
DFE = Damage factor, computed by $[p/p_o]^n$
p = New axle load
p_o = Old axle load
n = Damage exponent
BRM = Base routine maintenance, hours/million gross tons (MGT)/mi
MGT = Traffic density measured by MGT/mi/year

Table 4.27 Levels of service for lock facilities by service frequency

Lock service frequency rating	*Lock service frequency condition*	*Description*
1	Full service	More than 1000 commercial lockages per year
2	Reduced service	500–1000 commercial lockages per year
3	Limited service	Less than 500 commercial lockages per year or greater than 1000 recreational lockages per year
4	Scheduled service	Limited commercial and/or substantial recreational traffic, with a more consistent pattern of lockage
5	Service on weekends and holidays	Little to no commercial lockages with significant recreational lockages of 500 or more per year with no consistent pattern
6	Service by appointment	Limited commercial traffic with no consistent pattern of lockage

Table 4.28 Levels of service for lock facilities by lock processing time

Lock service time rating	*Lock service time condition*	*Description*
A	Minimal delays with no unplanned outages	Delays and outages will be in line with best service levels historically in presence of minor queuing
B	Moderate delays with no unplanned outages	Queues, delays, and outages are expected, but the average is kept within a certain variance from historical best conditions
C	Significant delays with possible unplanned outages	Delays and outages are unpredictable and windows of service may be constrained
D	Severe delays with high potential for unplanned outages	Delays are expected to be lengthy and windows of service will be constrained, with an imminent risk of failure

4.7.3 Inland waterway condition rating

The performance of inland waterway transportation is typically concerned with levels of service in terms of how often the waterway facilities perform a service (USACE, 2015). It could also be measured by lock processing time or lost time for a vessel to traverse through it (TRB, 2016). Tables 4.27 and 4.28 present definition of levels of service for lock facilities by service frequency and according to lock processing time, respectively.

4.7.4 Bikeway condition rating

The key measure for bike lane condition evaluation is concerned with bike lane pavement surface condition rating. Table 4.29 illustrates the condition rating used by the FHWA highway performance monitoring system (HPMS) (FHWA, 2014).

4.7.5 Pedestrian walkway condition rating

Pedestrian walkways or sidewalks support pedestrian walking, especially in densely populated urban areas. The pedestrian walkway system also serves as feeders for transit riders to gain access to transit stops or stations and from there to reach the destinations. The following rating schemes are currently used for sidewalk condition assessment.

Sidewalk condition index: The sidewalk condition index (SCI) is calculated by (FSTINC, 2014)

$$SCI = 100[1 - (A_D / A_T)] \tag{4.41}$$

where

SCI = Sidewalk condition index, ranging from 0 to 100
A_D = Damaged sidewalk area
A_T = Total sidewalk area

Table 4.29 Bikeway condition rating

Bikeway condition rating	*Bikeway condition*	*Description*
5	Very good	Only new or nearly new pavements are likely to be smooth enough and free of cracks and patches to qualify for this category
4	Good	Pavement, although not as smooth as described above, gives a first-class ride and exhibits signs of surface deterioration
3	Fair	Riding qualities are noticeably inferior to those above; may be barely tolerable for high-speed traffic. Defects may include rutting, map cracking, and extensive patching
2	Poor	Pavements have deteriorated to such an extent that they affect the speed of free-flow traffic. Flexible pavement has distress over 50% or more of the surface. Rigid pavement distress includes joint spalling, patching, etc.
1	Very poor	Pavements that are in an extremely deteriorated condition. Distress occurs over 75% or more of the surface

Table 4.30 Pedestrian walkway condition rating

Pedestrian walkway condition rating	*Pedestrian walkway condition*	*Description*
9	Excellent	Pavement is new
8		Less than 1/8 inch vertical edge, little or no depressed or raised areas; no more than 2 pieces of cracked squares for cement concrete, no spalling of concrete surface or raveling of bituminous surface, no horizontal separation, or debris/vegetation
7	Good	Same as 6, but in slightly better condition, needs routine maintenance of spot patch repairs/crack filling
6		Vertical edge between 1/8 inch and 1/4 inch, 0–1 inch raised/depressed, no more 3 cracked squares of cement concrete, less than 25% spalled concrete surface or bituminous surface, less than an inch of horizontal separation, less than 25% covered by debris/vegetation
5	Fair	Same as 4, but in slightly better condition, needs preservative treatment of fractural sealcoat/joint grinding
4		Vertical edge between 1/4 inch and 1/2 inch, 1–2 inch raised/depressed, no more than 4 cracked squares of cement concrete, 25%–50% spalled concrete surface or raveled bituminous surface, less 1–1½ inch of horizontal separation, 25%–50% covered by debris/vegetation
3	Poor	Same as 2, but in slightly better condition, needs milling and/or structural overlay/patching or replacement of sections full width
2		Significant aging, vertical edge between 1/2 inch and 3/4 inch, 2–3 inch raised/depressed, no more than 5 cracked squares of cement concrete, 50%–75% spalled concrete surface or bituminous surface, less 1½–2 inch of horizontal separation, 50%–75% covered by debris/vegetation
1	Failed	Vertical edge between 3/4 inch and 1 inch, greater than 3 inch raised/depressed, more than 5 cracked squares of cement concrete, more than 75% spalled concrete surface or bituminous surface, more than 2 inch of horizontal separation, more than 75% covered by debris/vegetation

Table 4.31 Pedestrian walkway system evaluation criteria

Criteria (weight)	Measure	
	Category (weight)	*Item*
Pedestrian safety (55)	Existing conditions (≤30)	Posted speed
		Pedestrian walk route:
		In travel lane, on shoulder, on sidewalk
		Existing sidewalk condition
		Existing Americans with Disabilities Act (ADA) barriers
	Crash history (≤25)	
	Existing hazards (≤15)	Sight distance
		Deep ditches
		Traffic volume
		Truck volume
		Obstructions
		Existing lighting
		Drainage/snow issues
		Posted school zone
Pedestrian connectivity (30)	Pedestrian destinations (≤30)	Central business district
		Commercial development
		Industrial area
		Schools
		Public facilities
		Recreational facilities
		Medical facilities
		Senior centers
		Signed transit stop
	Sidewalk connectivity (≤5)	Completes gap
		Extends existing sidewalk
Local support (5)	Local match (≤5)	
Sustainability (10)		Adopted greenhouse gas emissions policy
		Sidewalk width greater than 3 ft
		Sidewalk network development
		Low-energy street lighting or signal
		Recycled material usage
		Low-impact drainage practice

The work activities are generally categorized as (i) reconstruction for $SCI = 0–49$; (ii) localized repairs for $SCI = 50–79$; and (iii) do nothing, otherwise.

Sideway surface condition rating: Table 4.30 lists pedestrian sidewalk surface condition rating (SSCR) used in current practice (BHD, 2010).

Table 4.31 presents exemplary criteria for pedestrian walkway system evaluation (WATIB, 2017).

PROBLEMS

4.1 Why do we need performance modeling in transportation asset management?

4.2 Define the difference between pavement serviceability and performance.

4.3 What are the adverse impacts of having an incorrect performance model?

4.4 List various categories of performance models used in transportation asset management. How would you go about validation of a performance model?

4.5 What are the common distresses of a flexible pavement and a rigid pavement?
4.6 What primary types of data are needed for highway bridge deck, superstructure, and substructure condition assessment?
4.7 What are the primary performance measures for bus rapid transit performance analysis?
4.8 The yearly transition probabilities of various condition states of deck component of bridges on urban Interstate highways in a U.S. state are given in Table P4.1.
In 2018, there are 1000, 2000, 1000, and 500 bridges in the Excellent, Good, Fair, and Poor condition states, respectively. Find the number of bridges that are expected to be in the Excellent, Good, Fair, and Poor conditions in 2022. Assume that no major repairs were done during the period, so the original transition matrix stays unchanged.

Table P4.1 Transition probabilities of bridge conditions

	Excellent	*Good*	*Fair*	*Poor*	*Very poor*
Excellent	0.6	0.3	0.1	0	0
Good	0	0.7	0.2	0.1	0
Fair	0	0	0.6	0.3	0.1
Poor	0	0	0	0.7	0.3
Very poor	0	0	0	0	1.0

4.9 Consider the following dataset regarding the bike lane condition rating in a specific area. Plot a histogram and a box plot for the dataset and comment on the distribution. (Data: 7.4, 8.1, 9.2, 6.7, 8.5, 9.1, 7.6, 6.8, 5.9, 8.2, 7.9, 7.5, 8.2, 9.3, 6.3, 6.3, 9.8, 7.8, 6.3, 8.5, 7.9, 9.5)
4.10 Use the ML estimation technique to estimate the mean and variance of a normal distribution with the given data: 7.4, 8.1, 9.2, 6.7, 8.5, 9.1, 7.6, 6.8, 5.9, 8.2, 7.9, 7.5, 8.2, 9.3, 6.3, 6.3, 9.8, 7.8, 6.3, 8.5, 7.9, 9.5.
4.11 A pool of rail track segments is randomly selected for condition inspection. The following roughness indices are collected along the track alignment: 19, 18, 16, 15, 18, 19, 15, 16, 12, 12, 11, 19, 16, 18, 13, 14, 16, 18, 19, 16, 19, 17, 18, 19, 15, 14, 15, 16, 15, 15, 18, 21, 18, 15, 17, 16, 21, 14, 21, and 16 mm. Evaluate the track condition rating using the track roughness index.

Chapter 5

Transportation impacts modeling

5.1 TRANSPORTATION AGENCY COST MODELING

5.1.1 General

5.1.1.1 Short- and long-run costs

The common description of technology, processing multiple inputs into outputs, can be defined using a transformation function

$$F(q, x; \theta) = 0 \tag{5.1}$$

where

q = The vector of outputs
x = The vector of inputs
θ = A vector of parameters, which could include service-quality indication

The cost function for a certain producer gives the minimum expenditure C of producing output vector q, according to the supply relations for inputs as well as the transformation function. Commonly, these supply relations are presumed to be a constant price vector w. Thus, the problem then becomes minimizing input costs $w'x$. The optimal solution gives an input vector x^* if unique. The resulting minimum input cost, $w'x^*$, varies with q, w, and θ, so the cost function can be expressed as a function form, $C(q, w; \theta)$. In the cases where input prices w are dependent on the producer's input choices and can, therefore, be varied parameters, the vector w used in the cost function must be redefined to reflect these complex relations.

If inputs vector x includes elements that only vary in a long run, denoted as $\tilde{C}$ when the short and long run are required to be treated distinctly, if there exist inputs being treated as constant during the input cost minimization, the achieved cost is known as a short-run cost function. Normally, the fixed input is denoted as x_n, and its fixed value $\bar{x}_n$ becomes an additional argument of the derived short-run cost function, expressed as $C(q, w; \theta, \bar{x}_n)$. Then the long- and short-run cost could be represented as

$$\tilde{C}(q, w; \theta) = C(q, w; \theta, \bar{x}_n) \tag{5.2}$$

When pushing q to 0, the cost function, including both short- and long-run cost, may converge to a positive constant C_0. C_0, if it exists, is known as the fixed cost, and the remaining portion, $C - C_0$, is called the variable cost. The fixed cost always exists in the short-run cost because fixed capital expenditures are contained. The rest of the short-run cost is called operating cost due to the reflection of operations. Note that the operating cost could also contain fixed elements independent of q. Fixed cost is also different from the sunk cost, a cost that has already been incurred and cannot be recovered. Regardless of what money is spent on, sunk costs are dollars already spent and permanently lost. For example, the rent and market analysis belong to sunk cost.

5.1.1.2 Marginal cost and economies of scale

Another important concept is the marginal cost, defined as the partial derivative of cost against one specific output, q_i. That is, $mc_i = \partial C / \partial q_i$. The degree of scale economies, extended from the marginal cost, is defined as the ratio of average cost over marginal cost. The degree of scale economies, s, indicates how fast costs rise against output(s).

$$s = \left(\frac{C}{q_i} \right) \Bigg/ \left(\frac{\partial C}{\partial q_i} \right) \tag{5.3}$$

If $s < 0$, it is called diseconomies of scale. If $s > 0$, it is called economies of scale. If $s = 0$, it is called neutral scale economies. A short-run cost function is more likely to have scale economics since it usually has a larger proportion of fixed cost than a long-run cost function.

5.1.1.3 Methods of measurements

Cost functions can be empirically measured using at least three general approaches. The accounting approach examines budgetary accounts, combined with opportunity cost, and then attributes specific accounts to corresponding outputs. The engineering approach establishes a detailed analysis of cost behavior based on a detailed engineering evaluation of the inputs required by production process and of the prices of those inputs. The statistical approach infers the relationship between cost and levels of output as well as other variables based on the observation of the costs historically incurred in different situations. There is significant intersection among these approaches, and more than one of them should be used in any given study.

5.1.1.4 External and total social costs

Some costs of producing an item such as emission, noise, underground-water contamination, and wildlife disruption are borne by neither the producer nor the individual users. These costs are called external costs because they are borne by third parties. An example of external costs in the transportation sector could be more severe congestion and longer travel times for other drivers caused by someone driving a car in the network.

Total social costs are the full cost society pays for the production or for taking actions in the economy, including any external costs. The concept extended is the marginal social cost. In the transportation sector, the marginal social cost of a particular trip represented by vehicle miles of travel (VMT) is the derivative of social cost against that of travel. This concept is the basis of congestion cost modeling.

5.1.2 Highway agency costs

Figure 5.1 presents primary categories of physical highway facilities. The facility service life cycle is defined as the time interval between two successive construction interventions. Agency costs are mainly composed of capital costs related to facility construction and the

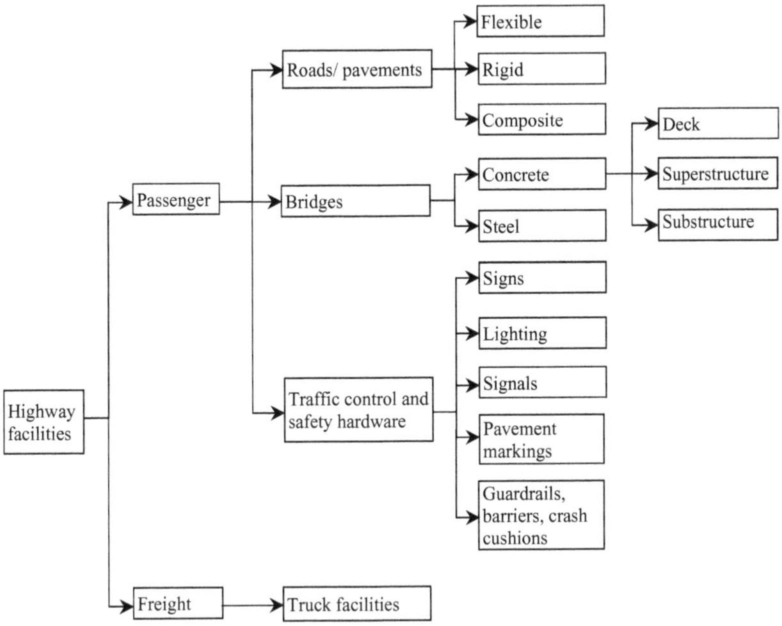

Figure 5.1 Physical highway facilities.

discounted future costs of repair and maintenance. Construction cost covers cost items of pre-investigation, preliminary planning, right-of-way acquisition, final design, and so forth. Repair treatments may be related to rehabilitation, restoration, and reconstruction (3Rs), and sometimes include resurfacing (4Rs). Maintenance may be classified as routine, including regular maintenance and special maintenance in snow and cold weather.

5.1.2.1 Highway agency cost estimation methods

Disaggregate method: For estimating the total cost of a transportation project, the disaggregate method relies on bid prices that separately list the cost of each payment related to labor, material, and supervision. The itemized costs are summed to arrive at the total costs, as shown in Equation 5.4

$$TC = \sum_{i=1}^{n} C_i \tag{5.4}$$

where

TC = Total cost

C_i = The cost of a specific work item i

Aggregate method: Usually a large transportation project contains extensive work items. Practically, several work activities of a similar nature can be grouped together as one large work activity, and an average unit rate from historical bids may be utilized. Next, the average unit rate is multiplied by the quantity for each of the grouped work activities and the amounts for all grouped work activities are further aggregated to establish the total project cost estimate. The total cost is computed as Equation 5.5

$$TC = \sum_{i=1}^{n} R_i \cdot Q_i \tag{5.5}$$

where

TC = Total cost
R_i = Cost rate of specific work activity group i
Q_i = The quantity of specific work activity group i

Statistical method: The statistical method considers both the historical cost data and the actual cost data in statistically estimating the unit cost of a work activity. Regression methods are the most popular approaches to develop statistical estimation equations. Note that the price of labor and materials may vary from place to place, so there is no general standard equation applicable to all the states and cities.

5.1.2.2 Highway agency cost models

Total highway construction cost: A statistical model can be used to estimate total highway construction, which is of the following specification:

$$C_{const} = \beta_0 + \beta_1 \cdot N + \beta_2 \cdot P + \beta_3 \cdot R + \beta_4 \cdot T \tag{5.6}$$

where

C_{const} = Total highway construction cost, million dollars/mi
N = Number of lanes
P = Annual property tax revenue density, thousand dollars/mi^2
R = Gross residential density, persons/mi^2
T = Terrain relief measure, dimensionless
β_0, β_1, β_2, β_3, and β_4 = Corresponding coefficients

Major highway project costs: In general, the costs of major highway projects can be approximately estimated by the following:

$$C_i = UC_i \cdot N \cdot L \tag{5.7}$$

where

C_i = Cost of major highway project type i
UC_i = Unit cost of major highway project type i, dollars/lane-mi
N = Number of lanes
L = Length of road segment, mi

Cost rates of highway projects tend to vary significantly by time and location. The amounts derived from historical data are valuable but need adjustments to reflect site-specific conditions. Moreover, the cost rate estimates need to consider price fluctuations of materials and labor, inflation, interest rates, and other factors. Table 5.1 lists cost rates of different types of highway projects in the U.S. (FHWA, 2016) (Table 5.1).

Pavement rehabilitation cost models: Regression models developed by Li et al. (2001) can be utilized for estimating in-service pavement rehabilitation costs for flexible, jointed concrete pavement (JCP), and composite pavements, which are presented in the following:

Table 5.1 Cost rates of highway work activities in the United States (1000 dollars/lane-mile, 2014 values)

	Rural		*Urban*	
Functional classification	*Freeways and expressways*	*Principal arterials*	*Freeways and expressways*	*Principal arterials*
New construction	3225.55	3743.48	4838.87	9337.19
Reconstruction-added capacity	1048.42	1535.97	8788.43	2211.96
Reconstruction-no added capacity	542.15	790.46	1527.20	1395.24
Restoration and rehabilitation	251.60	245.99	578.92	663.85
Resurfacing	140.26	197.04	963.99	451.35

$$C_{flexible}^{rehab} = \alpha_0 + \alpha_1 \cdot (CESALs/THICK) + \alpha_3 \cdot (P200 \cdot MOIST \cdot FZI) + \alpha_5 \cdot DRAINCO + \alpha_9 \cdot [(AGE - 10) \cdot DM_{AGE}] \quad (5.8)$$

$$C_{rigid}^{rehab} = \alpha_0 + \alpha_1 \cdot CESALs + \alpha_2 \cdot SLABTH + \alpha_6 \cdot (DAYS > 32°F) + \alpha_8 \cdot AGE \quad (5.9)$$

$$C_{composite}^{rehab} = \alpha_0 + \alpha_1 \cdot (CESALs/THICK) + \alpha_4 \cdot (MOIST \cdot FZI) + \alpha_7 \cdot MINTEM + \alpha_9 \cdot [(AGE - 10) \cdot DM_{AGE}] \quad (5.10)$$

where

$C_{flexible}^{rehab}, C_{rigid}^{rehab}, C_{composite}^{rehab}$ = Flexible, rigid, and composite pavement rehabilitation costs, 1997 dollars/lane-mi
$CESALs$ = Cumulative equivalent single axle loads (ESALs) applied to the pavement in the rehabilitation interval
$THICK$ = Total pavement thickness, in
$SLABTH$ = Slab thickness of rigid pavements, in
$P200$ = Subgrade material percent passing number 200 sieve, %
$MOIST$ = Subgrade moisture content, %
FZI = Average freeze index during one life-cycle, degree-days
$DRAINCO$ = Drainage coefficient of the subgrade
$DAYS > 32°F$ = Average annual number of days >32°F during rehabilitation interval
$MINTEM$ = Average annual minimum temperature during one life-cycle, °F
AGE = Pavement age at time of rehabilitation, years
DM_{AGE} = Dummy variable, 1 if age >10 years and 0, otherwise

Table 5.2 lists model coefficients α_0, α_1, α_2, α_3, α_4, α_5, α_6, α_7, α_8, and α_9.

Pavement annual routine maintenance cost models: The Tobit model initiated by Tobin (1958) and popularized by Maddala (1983) and Long (1997) is well suited to estimate pavement routine maintenance costs for its ability to recognize the lagged relationship between routine maintenance and pavement condition change caused by traffic loading and non-load factors. The following Tobit models developed by Li et al. (2002b) can be employed to estimate routine maintenance costs for flexible, JCP, CRC, and composite pavements:

$$\begin{cases} \log_{10}\left(C_{flexible_{t+1}}^{maint}\right) = \alpha_0 + \alpha_1 \cdot \log_{10}(1 + \Delta IRI_t/IRI_t) + \alpha_2 \cdot DM_{IS} + \alpha_3 \cdot DM_{ST} \\ \log_{10}(1 + \Delta IRI_t/IRI_t) = \beta_1 \cdot \log_{10}(ESALs_t/THICK) + \beta_2 \cdot FZTHAW_t + \beta_3 \cdot DAYS32°FSF_t \end{cases} \quad (5.11)$$

Table 5.2 Pavement rehabilitation cost model coefficients

Independent variable	*Coefficient*	*Flexible*	*Rigid*	*Composite*
Constant	α_0	172431	−628986	540918
CESALs/THICK	α_1	0.3113	0.0380	0.2409
SLABTH	α_2	–	−40021	–
P200·MOIST·FZI	α_3	301	–	–
MOIST·FZI	α_4	–	–	71
DRAINCO	α_5	−145378	–	–
DAYS >32°F	α_6	–	13274	–
MINTEM	α_7	–	–	−10393
AGE	α_8	–	34760	–
(AGE−10)·DM$_{AGE}$	α_9	701	–	10593

$$\begin{cases} \log_{10}\left(C^{maint}_{JCP_{t+1}}\right) = \alpha_0 + \alpha_1 \cdot \log_{10}\left(1 + \Delta IRI_t / IRI_t\right) + \alpha_2 \cdot DM_{IS} \\ \log_{10}(1 + \Delta IRI_t / IRI_t) = \beta_1 \cdot \log_{10} ESALs_t + \beta_3 \cdot DAYS32°FSF_t + \beta_6 \cdot DM_N + \beta_7 \cdot DM_C \end{cases} \tag{5.12}$$

$$\log_{10}\left(C^{maint}_{CRC_{t+1}}\right) = \beta_1 \cdot \log_{10}(ESALs_t / THICK) + \beta_8 \cdot (AGE_t - 8) \cdot DM_{AGE} \tag{5.13}$$

$$\begin{cases} \log_{10}\left(C^{maint}_{composite_{t+1}}\right) = \alpha_0 + \alpha_1 \cdot \log_{10}(1 + \Delta IRI_t / IRI_t) + \alpha_2 \cdot DM_{IS} \\ \log_{10}(1 + \Delta IRI_t / IRI_t) = \beta_1 \cdot \log_{10}(ESALs_t / THICK) + \beta_4 \cdot MAXTEM_t \\ \qquad + \beta_5 \cdot MINTEM_t + \beta_8 \cdot (AGE_t - 8) \cdot DM_{AGE} \end{cases} \tag{5.14}$$

where

$C^{maint}_{flexible_{t+1}}, C^{maint}_{JCP_{t+1}}, C^{maint}_{CRC_{t+1}}, C^{maint}_{composite_{t+1}}$ = Annual routine maintenance cost for flexible, JCP, CRC, and composite pavements in year $t+1$, 1998 dollars/lane-mi/year
IRI_t = Pavement international roughness index (IRI), in/mi
$(1 + \Delta IRI_t / IRI_t)$ = Ratio of roughness values between year $t+1$ and t
$ESALs_t$ = Annual ESALs applied to the pavement in year t, ESALs/lane-mi
$THICK$ = Pavement total thickness, in
$FZTHAW_t$ = Number of annual freeze-thaw cycles in year t, days
$DAYS32°FSF_t$ = Number of days of first min. of 32°F between last spring and current fall
$MAXTEM_t$ = Average annual maximum temperature in year t, °F
$MINTEM_t$ = Average annual minimum temperature in year t, °F
AGE_t = Pavement age in year t, years
DM_{IS} = Dummy variable: 1 for interstate highways and 0, otherwise
DM_{ST} = Dummy variable: 1 for state highway and 0, otherwise
DM_N = Dummy variable: 1 for northern Indiana and 0, otherwise
DM_C = Dummy variable: 1 for central Indiana and 0, otherwise
DM_{AGE} = Dummy variable: 1 if pavement age is greater than 8 years and 0, otherwise

Table 5.3 lists model coefficients α_0, α_1, α_2, α_3, β_0, β_1, β_2, β_3, β_4, β_5, β_6, β_7, and β_8.

Bridge replacement costs: The total bridge replacement cost is the sum of four cost elements, including superstructure replacement cost, substructure replacement cost, approach cost, and

Table 5.3 Pavement annual routine maintenance cost model coefficients

Independent variable	*Coefficient*	*Flexible*	*Rigid-JCP*	*Rigid-CRC*	*Composite*
Constant	α_0	0.3770	0.7924	–	−114.48
$(1+\Delta IRI_t/IRI_t)$	α_1	46.02	37.73	–	8061.98
$DM_{IS}DM_{IS}$	α_2	0.8864	0.2627	–	373.64
DM_{ST}	α_3	0.2537	–	–	–
Constant	β_0	–	0.0756	–	–
$ESALs_t/THICK$ or $ESALs_t$	β_1	0.0052	0.0081	0.4449	0.0062
$FZTHAW_t$	β_2	0.0003	–	–	–
$DAYS32°FSF_t$	β_3	−0.0002	−0.0006	–	–
$MAXTEM_t$	β_4	–	–	–	0.00035
$MINTEM_t$	β_5	–	–	–	−0.0007
DM_N	β_6	–	−0.0087	–	–
DM_C	β_7	–	−0.0116	–	–
$(AGE_t-8)\cdot DM_{AGE}$	β_8	–	–	0.2567	0.0012

other costs. A previous study suggested that the Cobb–Douglas function could appropriately represent the relationship between bridge characteristics and replacement cost (Labi, 2001).

$$C_B^{rep} = A\cdot X_1^{\alpha}\cdot X_2^{\beta}\cdot\cdots\cdot X_N^{\gamma} \tag{5.15}$$

where

C_B^{rep} = Cost of bridge replacement
A = Constant
$X_1^{\alpha}, X_2^{\beta} \ldots X_N^{\gamma}$ = Variables represent bridge physical characteristics
$\alpha, \beta \ldots \gamma$ = Coefficient variables

Table 5.4 presents cost rates of bridge activities in the U.S. (FHWA, 2016).

Tunnel construction costs: Over many decades, highway tunnels have been constructed worldwide as the tunnel construction method is getting more advanced and this allows tunnels that were impossible to build to be built now to support more direct, less obtrusive traffic movements. A highway tunnel primarily consists of the tunnel structure and portals, lighting system, ventilation system, traffic operations control system, and safety and security system. In general, highway tunnels are classified by the cross section shape, lining, invert type, and construction method (Cascadia Center, 2008; Li et al., 2008). Table 5.5 shows costs rates of some of the longest highway tunnels constructed worldwide (Table 5.5).

Table 5.4 Cost rates of bridge activities in the United States (1000 dollars/bridge, 2014 values)

	Rural		*Urban*	
Functional classification	*Freeways and expressways*	*Principal arterials*	*Freeways and expressways*	*Principal arterials*
New bridge	2953.00	5484.25	5282.14	55512.00
Bridge replacement	2337.52	3644.66	4325.26	6407.41
Major bridge rehabilitation	868.67	332.46	1818.01	1948.32
Minor bridge rehabilitation	290.52	520.88	1545.17	2293.78

Table 5.5 Typical highway tunnel construction cost rates (completion year dollar values)

Tunnel	*Year completed*	*Bores*	*Alignment length (miles)*	*Tunnel length (miles)*	*Cost (million dollars)*	*Cost rate (million dollars/mile)*
Wesertunnel, Germany	2001	Twin	1	2	358	180
Lefortovo tunnel, Russia	2005	Single	1.4	1.4	600	439
Wuhan, China	2008	Twin	1.7	3.4	288	85
M-30 Madrid tunnel, Spain	2008	Twin	2.2	4.3	570	131
4th Tube of the Elbe, Germany	2002	Single	2.6	2.6	775	303
Dublin Port Tunnel, Ireland	2006	Twin	2.8	5.6	530	94
Westerschelde tunnel, Netherlands	2002	Twin	4.1	8.2	490	60
Shanghai River Crossing, China	2008	Twin	4.6	9.3	245	27
SMART tunnel, Malaysia	2007	Single	6	6	515	85
Folgefonn tunnel, Norway	2001	Single	6.9	6.9	89	13
Gudvanga tunnel, Norway	1991	Single	7.1	7.1	69	10
Mont-Blanc road tunnel, France	1965	Single	7.2	7.2	58	8
Fréjus road tunnel, France/Italy	1980	Single	8.0	8.0	443	55
Hsuehshan, Taiwan	2006	Twin	8.1	16.1	580	36
Arlberg road tunnel, Austria	1978	Single	8.7	8.7	21.2	2
St. Gotthard road tunnel, Switzerland	1980	Single	10.5	10.5	626	60
Zhongnanshan, China	2007	Twin	11.2	22.4	410	18
Laerdal road tunnel, Norway	2000	Single	15.2	15.2	114	7

Traffic control and safety hardware costs: Traffic control and safety hardware mainly include signals, traffic signs, lighting, pavement markings, crash cushions, and guardrails. They are an essential part of physical highway facilities ensuring highway safety performance. Table 5.6 gives the references of related traffic control and safety hardware costs (Markow, 2007). Since these kinds of facilities are quantifiable, the total cost is simply expressed as the product of quantity and unit price.

5.1.3 Transit agency costs

5.1.3.1 Transit accounting costs

Accounting cost establishes the relation between costs and intermediate outputs for transit agencies. The underlying assumption of these cost estimates is a linearity between cost and a few measures of intermediate outputs. The following equation reflects the general form of cost function:

$$C_{transit} = c_1 \cdot RM + c_2 \cdot PV + c_3 \cdot VH + c_4 \cdot VM \tag{5.16}$$

where

$C_{transit}$ = Total transit cost
RM = Route miles
PV = Peak vehicles in service

Table 5.6 Traffic control and safety hardware costs (2005 U.S. dollars)

Designation	*Traffic control and safety hardware type*		*Amount*
Traffic sign	Sign panel	Regulatory/warning/marker	15–18/sq.ft
		Large guide signs	20–25/sq.ft
		Electronic variable message sign	50000–150000 each
	Sign post	U-channel	125–200 each
		Square tube (telespar)	10–15 per foot
		Large steel breakaway posts	15–30 per foot
		Cantilever sign	15000–20000 each
		Sign bridge	30000–60000 each
	Supporting structure	Square tube	150–250 each
		Breakaway post	250–750 each
		Cantilever/bridge	6000–7000 each
Lighting		In-pavement lighting	17620–18250 total
		Street lighting	3600–4880 each
Traffic signal		Signal controller	Approximately 10000 each
		Signal update	2500–3100 per signal
		Technician maintenance	Approximately 56000/year
Pavement markings		Advance stop/yield line	320–380 each
		Advance stop/yield line	10/ft^2
		Island marking	1.49–1.94/ft^2
		Painted curb/sidewalk	1.21–3.40/ft^2
		Painted curb/sidewalk	2.57–3.06/linear ft
		Pedestrian crossing	310–360 each
		Shared lane/bicycle marking	160–180 each
		School crossing	470–520 each

VH = Vehicle-hours of service
PV = Peak vehicles in service
VM = Vehicle-miles of service

In this approach, all cost items are assigned to one of the outputs exclusively, meaning that there are no fixed costs and no joint costs. The cost function could be further refined to treat the costs per vehicle-hours during peak hours and off-peak hours separately as

$$C_{transit} = c_1 \cdot RM + c_2 \cdot PV + c_b \cdot VH_b + c_p \cdot VH_p + c_4 \cdot VM \tag{5.17}$$

where
VH_b = Vehicle-hours belonging to base service
VH_p = Vehicle-hours belonging to peak service

With historical data made available, statistical methods could be employed to calibrate the cost functions. As an example, Wunsch (1996) estimated the transit labor cost function in the following form:

$$C_{transit,i}^{labor} = \left[c_{1a,i} \cdot TM_i + c_{1b}^3 \cdot Station_i + c_3 \cdot VH_i + (c_{4a,i} + c_{4b,i} \cdot n_i) \cdot VM_i\right] \cdot (w/\bar{w}) \tag{5.18}$$

where

i = Transit sub-mode index, $i = 1$ for buses, 2 for streetcars or bus rapid transit (BRT), and 3 for light rail
TM_i = Number of track-miles for streetcars or light rail, 0 for buses
$Station_i$ = Number of light rail stations for transit mode i, 0 for other transit sub-modes
n = Capacity of a convoy in persons measured as square meters of floor space divided by four
w = Local wage rate
$\bar{w}$ = Average wage rate over the sample

5.1.3.2 Transit engineering costs

Engineering cost establishes the cost function with respect to detailed engineering specifications. Nowadays, most of the handbooks for transit cost estimation use this approach. Figure 5.2 shows primary physical transit facilities.

BRT: BRT is a bus-based public transport system with larger capacity and operating speed. Typically, an exclusive bus lane is established for BRT buses. Bus signal priority (BSP) system is a useful tool to improve the efficiency of BRT. BRT costs include costs of purchasing, repair, and maintenance of BRT buses, as well as construction and preservation of stops and stations. Costs may vary in different locations and projects.

Light rail: Light rail transit (LRT) is one type of urban public transport, which is similar to tram but with higher capacity. LRT is known in some places as fast tram or rapid streetcar. LRT systems generally operate in an exclusive right-of-way. The running speed is usually between 15 and 40 mph.

Guideway construction cost is considered to take 16%–38% of total cost of light rail construction (Black, 1995). An equation was proposed to estimate the guideway construction cost.

$$C_{LR}^{guideway} = e^{(\beta_0 + \beta_1 \cdot L^{\alpha_1} + \beta_2 \cdot S^{\alpha_2})} \tag{5.19}$$

where

$C_{LR}^{guideway}$ = Light rail guideway construction cost, million dollars
L = Guideway length, mi
S = Number of stations
$\alpha_1, \alpha_2, \beta_0, \beta_1, \beta_2$ = Model coefficients

The following model coefficients are proposed by Black (1995):

$$C_{LR}^{guideway} = e^{(-1997.92 + 1448.22 \cdot L^{0.0005} + 553.55 \cdot S^{0.0005})} \tag{5.20}$$

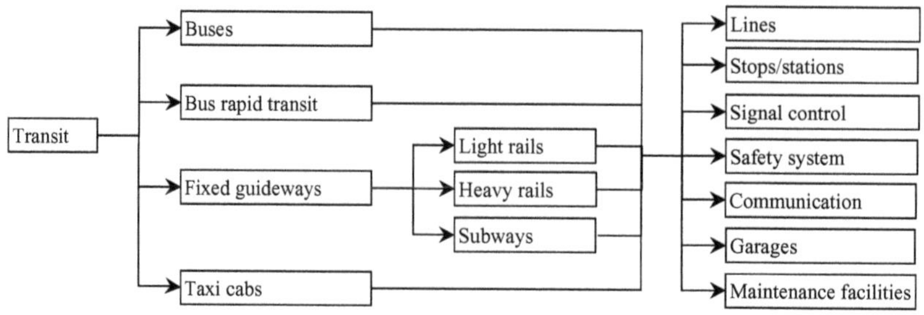

Figure 5.2 Physical transit facilities.

Heavy rail: The American Public Transportation Association defines a heavy rail as a railway system with the capacity to handle a heavy volume of traffic. Typically, heavy rail lines are operated on exclusively owned right-of-way and grade-separated from general traffic. The following model specification can be used to estimate the heavy rail guideway construction cost:

$$C_{HR}^{guideway} = \alpha \cdot L^{(1+\beta)} \cdot U^{\gamma} \cdot ST^{\delta} \tag{5.21}$$

where

$C_{HR}^{guideway}$ = Heavy rail guideway construction cost, million dollars
L = Number of lane-miles
U = Underground faction of the system
ST = Number of stations
$\alpha, \beta, \gamma, \delta$ = Model coefficients

The Federal Transit Administration (FTA) implemented the Standard Cost Categories (SCC) in 2005 to develop a consistent format for the reporting, estimating, and managing of capital costs for new transits projects. Information collected from major projects across the country has been integrated into a database called the Capital Cost Database, which serves a cost-estimation tool helpful to FTA and the transit industry. Across the entire life cycle of a project, tracking, evaluating, and controlling cost changes benefits from the consistent format of SCC worksheets, which could be downloaded from the FTA website. Table 5.7 provides definitions of the FTA standard transit cost categories and Table 5.8 lists transit vehicle purchasing cost rates (FTA, 2016).

5.1.3.3 Transit operating costs with user inputs

With intermediate outputs provided by transit agencies, travelers could also provide some inputs, typically in the form of their time or money, to produce final outputs such as travel movements. For transit travel, the main cost to users is time, including waiting time for vehicles and riding time in vehicles. Mohring (1972) introduced an aggregated model considering transit waiting time cost as below:

$$C_{transit}^{VO} = UC_V \cdot V \tag{5.22}$$

$$C_{transit}^{WT} = \frac{UC_W \cdot q}{2V} \tag{5.23}$$

where

$C_{transit}^{VO}$ = Transit vehicle operating costs, dollars/hr
$C_{transit}^{VO}$ = Transit user waiting time costs
V = Vehicles passing a given transit stop or station per peak hour
UC_V = The unit cost of V, dollars/veh
q = Number of passengers per peak hour on the route
W = The user-supplied input defined as the aggregate waiting time per peak hour
UC_W = The unit cost of W

The underlying assumption of using these two equations is that the average waiting time per passenger, w/q, is equal to half the headway, $1/V$. Find the minimum sum of costs by manipulating V, subject to a transit vehicle capacity constraint.

$$q \leq NV; N = nL/d \tag{5.24}$$

Table 5.7 FTA standard transit cost categories

Category	*Description*
Guideway and track elements	Include guideway and track costs for all transit modes (heavy rail, light rail, commuter rail, BRT, rapid bus, bus, monorail, cable car, etc.) The unit of measure is route miles of guideway, regardless of width. As associated with the guideway, include costs for rough grading, excavation, and concrete base for guideway where applicable. Include all construction materials and labor regardless of who is performing the work
Stations, stops, terminals, and intermodal	Include costs for rough grading, excavation, station structures, enclosures, finishes, equipment; mechanical and electrical components including HVAC, ventilation shafts and equipment, station power, lighting, public address/customer information system, safety systems such as fire detection and prevention, security surveillance, access control, life safety systems, and so on. Include all construction materials and labor regardless of who is performing the work
Support facilities, yards, shops, and administration buildings	Include costs for rough grading, excavation, support structures, enclosures, finishes, equipment; mechanical and electrical components including HVAC, ventilation shafts and equipment, facility power, lighting, public address system, safety systems such as fire detection and prevention, security surveillance, access control, life safety systems, and so on. Include fueling stations. Include all construction materials and labor regardless of who is performing the work
Sitework and special conditions	Include all construction materials and labor regardless of who is performing the work
Systems	Include all construction materials and labor regardless of who is performing the work
Right of work, land, and existing improvements	Include professional services associated with the real estate component of the project. These costs may include agency staff oversight and administration, real estate and relocation consultants, legal counsel, court expenses, insurance, and so on.
Vehicles	Include professional services associated with the vehicle component of the project. These costs may include agency staff oversight and administration, vehicle consultants, design and manufacturing contractors, legal counsel, warranty and insurance costs, and so on.
Professional services	Include all professional, technical, and management services related to the design and construction of fixed infrastructure
Unallocated contingency	Include unallocated contingency, project reserves. Document allocated contingencies for individual line items on the main worksheets
Finance charges	Include finance charges expected to be paid by the project sponsor/grantee prior to either the completion of the project or the fulfillment of the New Starts funding commitment, whichever occurs later in time

Table 5.8 Transit vehicle purchase cost rates in the United States (dollars/vehicle, 2015 values)

	Type of fuel					
Type of vehicle	*Diesel*	*Gasoline*	*Compressed natural gas (CNG)*	*Biodiesel*	*Electric*	*Hybrid electric*
≤30 ft bus	117485	70464	150726	61653	424000	500000
≤40 ft bus	337773	132957	363084	316623	890538	343884
Articulated bus	450701	Not applicable (NA)	740330	1065212	827838	747363
Bus commuter/ suburban	149379	99360	352925	461333	NA	NA
Bus trolley	378014	122667	261329	NA	883856	NA
Sedan/station wagon	23120	29018	NA	NA	NA	18229
Vans	82064	39340	171921	NA	NA	NA

where

N = Number of passengers a transit vehicle can pick up and drop off as it travels the entire route
n = The physical capacity of a single transit vehicle
d = The average passenger's trip length
L = The length of the route

The solution of this optimization problem is

$$V^* = \sqrt{\frac{UC_W \cdot q}{2 \cdot UC_V}}$$

$$W^* = \frac{q}{2V^*} = \sqrt{\frac{UC_V \cdot q}{2 \cdot UC_W}}$$

$$C_{transit}^{VO\ *} = UC_V \cdot V^* = \sqrt{\frac{UC_V \cdot UC_W \cdot q}{2}}$$

$$C_{transit}^{WT\ *} = UC_W \cdot W^* = \sqrt{\frac{UC_V \cdot UC_W \cdot q}{2}}$$

Two features of this solution are worth noting. First, the optimal bus frequency V^* is proportional to the square root of the passenger travel demand q, which is known as the square root rule for operating policy. Second, the cost function is also proportional to which it gives economies of scale ($s = 2$). This indicates that a generalized price set equal to marginal cost will involve a fare that could not cover its average cost. Also, this cost model is more realistic. We could treat peak and off-peak travel as separate outputs. We could consider the effect of additional passengers boarding or leaving the bus on average speed, and a peak hour broaden effect.

5.1.4 Non-motorized transportation agency costs

Non-motorized transportation refers to human-powered transportation modes, including biking and walking. These modes offer excellent commuting options that support our overall mission to reduce traffic congestion and improve air quality. Communities that are pedestrian- and bicycle-friendly are also "livable," providing residents with opportunities for recreation and community-enhancing economic development (Bushell et al., 2013).

5.1.4.1 Bike facility costs

Bicycle parking: Bicycle parking has several forms of facilities, including bicycle racks, bicycle lockers, and bicycle stations. Bicycle racks are fixed structures which provide a space for bikes to be securely locked at a fixed place. Bicycle locker is a closed space serving the purpose of storing a single bike. Bicycle station is a more systematic bike parking facility providing additional services such as maintenance. The cost of building a bicycle station varies from site conditions, choice of contractor, and other factors. Table 5.9 gives the cost for bicycle parking facilities.

Table 5.9 Bicycle parking facility costs (dollars)

Category	*Median*	*Average*	*Unit*
Bicycle locker	2140	2090	Each
Bicycle rack	540	660	Each

Table 5.10 Bikeway and path costs (dollars)

Category	*Median*	*Average*	*Unit*
Bicycle lane	89470	133170	Mile (with 5 ft width)
Signed bicycle route	27240	25070	Mile (with 5 ft width)
Signed bicycle route with improvement	241230	239440	Mile (with 5 ft width)
Multiuse trail path-paved	261000	481140	Mile (with 8 ft width)
Multiuse trail path-unpaved	83870	121390	Mile (with 8 ft width)
Boardwalk	1957040	2219470	Mile (with 8 ft width)

Bikeway: Bikeway is a roadway used by bicycles exclusively or shared by cyclists and other users. It includes bicycle paths, bicycle lanes, and signed bicycle routes. A bicycle path is usually a misnomer for a multi-use pathway, while bicycle lanes are marked lanes adjacent to roads and highways, including protected paths and shared lanes. Signed bicycle route is a series of signs showing the safest way from two locations for common commuting across a city (Table 5.10).

5.1.4.2 Pedestrian facility costs

In the planning stage of transportation improvements, the agency shall consider the travel demand of all transportation modes including bicycle and walking modes. If sufficient demand is found in the demand study, the improvement project should consider providing the appropriate accommodations. This section introduces the cost of predestrian accommodation treatments.

Bollards: A bollard is a sturdy, short, vertical post embedded in the ground. It can protect pedestrians by reducing vehicular speed or limiting vehicle access to keep pedestrians away from vehicle streams. Bollards could be categorized into permanent bollards (including fixed bollard, rising bollard, security bollard, removable bollard, telescopic bollard, lay-flat bollard, and bell bollard) and temporary bollards (including flexible bollard, Qwick Kurb, breakaway bollard, and planters) based on the usage. The cost of each bollard is approximately \$650–\$730.

Curb ramps: Curb ramps provide access between the sidewalk and the street surface for wheelchair users. Curb ramps are widely used at intersections, bus stops, midblock crossings, and other places that have certain needs. Curb ramps can be categorized by their structural characteristics (how the components are assembled) and their relative position on the street in practice (e.g., perpendicular curb ramp, diagonal curb ramp, etc.). The cost of each curb ramp is \$740–\$810 or a per square foot cost of \$12.

Fence: Fence helps separate the pedestrian and cyclists from the roadways, creating a shield for bike/pedestrian paths. The fence could be made from steel, wood, plastic, and other materials. The cost of fence is \$120–\$130 per linear foot.

Overpass/underpass: Pedestrian overpasses and underpasses enable pedestrians to have exclusive right-of-way, separated from regular vehicle traffic and other impassable barriers. Overpasses and underpasses are usually expensive, visually intrusive, and sometimes could be poorly used if pedestrians have an option to move on the street level. The cost of overpass/underpass depends on the structure, material, size, required right-of-way, and site characteristics. Pedestrian overpass/underpass can be very costly. A bridge over an arterial street may cost over $1500000. The general cost ranges from $500000 to $4 million (Table 5.11).

Railing: Pedestrian railing is installed on the outside edge of a bridge sidewalk/walkway for pedestrian safety. Standard pedestrian railing has a four-rail, with the top rail 3'-7" above the sidewalk/deck. The pedestrian rail costs about $95–$100 per linear foot.

Street furniture: Well-designed walking environments can scientifically improve pedestrians' walking experience, attracting more auto-active mode shifts. Street furniture, including benches, bus shelters, water fountains, and so on can enhance a walking environment and increase walkability (Martincigh et al., 2010) (Table 5.12).

Crosswalks: Marked crosswalks indicate a legal crosswalk for pedestrians. In practice, marked crosswalks are used to provide safer and easier crossings for pedestrians at complex locations, supplemented with other treatments (e.g., raised crossings, traffic signals, roadway narrowing, enhanced overhead lighting, traffic-calming measures, etc.) (Table 5.13).

Sidewalks: Sidewalks are the most fundamental pedestrian facility providing a safe path to walk along that is separated from the motorized traffic for pedestrian travel. Sidewalks provide many benefits including safety, mobility, and healthier communities (Table 5.14).

Table 5.11 Overpass/underpass costs (dollars)

Category	*Median*	*Average*	*Unit*
Wooden bridge	122610	124670	Each
Prefab steel bridge	191400	206290	Each

Table 5.12 Street furniture costs (dollars)

Category	*Median*	*Average*	*Unit*
Street trees	460	430	Each
Bench	1660	1550	Each
Bus shelter	11490	11560	Each
Trash/recycling	1330	1420	Each

Table 5.13 Crosswalk costs (dollars)

Category	*Median*	*Average*	*Unit*
High visibility crosswalk	3070	2540	Each
Marked crosswalk	340	770	Each
Marked crosswalk	5.87	8.51	Linear foot
Marked crosswalk	6.32	7.38	Square foot

Table 5.14 Sidewalk costs (dollars)

Category	*Median*	*Average*	*Unit*
Asphalt paved shoulder	5.81	5.56	Square foot
Asphalt sidewalk	16	35	Linear foot
Brick sidewalk	60	60	Linear foot
Concrete paved shoulder	6.10	6.64	Square foot
Concrete sidewalk	27	32	Linear foot
Concrete sidewalk – patterned	38	36	Linear foot
Concrete sidewalk – stamped	45	45	Linear foot
Concrete sidewalk + curb	170	150	Linear foot
Sidewalk unspecified	34	45	Linear foot
Sidewalk pavers	70	80	Linear foot

5.1.4.3 Traffic calming facility costs

Traffic calming measures are engineering and traffic management approaches used to reduce vehicle speed and enhance the safety of motorists, pedestrians, and cyclists. Common traffic calming measures can be categorized into four types: vertical deflections (e.g., speed hump, speed table, and raised intersection), horizontal shifts (e.g., chicane, neighborhood traffic circle), closures, and roadway narrowing. This section will introduce the cost of chicanes, curb extensions (neckdowns/bulb-outs), closure, crossing islands, raised intersections, and so on.

Chicanes: Chicanes are a series of curb extensions that alternate from one side of the street to the other forming S-shape curves. The cost of each chicane is about $8050–$9960.

Curb extensions: Curb extension, also known as a choker, extends the curb at midblock or intersection corners that narrow a street by extending the sidewalk or widening the planting strip. They are also called parallel chokers, angled chokers, twisted chockers, angle points, pinch points, or midblock narrowings if designed at midblock. If they are designed at intersections, they are also called neckdowns, bulb-outs, knuckles, or corner bulges. Sometimes, the curb extensions are combined with crosswalks, which are called safe crosses. The cost of a curb extension (choker/bulb-out) is approximately $10150–$13000.

Closure: Closure is an island built at an intersection that prevents certain vehicular movements (Table 5.15). There are four types of closures, including diagonal diverters (with barriers placed diagonally across an intersection), half closures (with barriers installed to block one direction of travel lane for a short distance), full-street closures (with barriers placed across a street to completely prevent through movement for both directions), and median barriers (with islands raised in the centerline of a street and extended across an intersection).

Crossing islands: Crossing islands are raised medians that are placed at the center line of the street or at the intersection. It can help protect pedestrians crossing the street by enabling pedestrians to deal with one direction of traffic at a time (Table 5.16).

Table 5.15 Closure costs (dollars)

Category	*Median*	*Average*	*Unit*
Diagonal diverters	22790	26040	Each
Half closures	15000	15060	Each
Median barriers	6.00	7.26	Square foot

Table 5.16 Crossing island costs (dollars)

Category	*Median*	*Average*	*Unit*
Crossing islands	10460	13520	Each
Crossing islands	9.80	10	Square foot

Table 5.17 Raised intersection costs (dollars)

Category	*Median*	*Average*	*Unit*
Raised crosswalk	7110	8170	Each
Raised intersections	59160	50540	Each

Table 5.18 Traffic circle and roundabout costs (dollars)

Category	*Median*	*Average*	*Unit*
Roundabout/traffic circle	27190	85370	Each

Table 5.19 Speed hump/table costs (dollars)

Category	*Median*	*Average*	*Unit*
Speed hump	2130	2640	Each
Speed table	2090	2400	Each

Raised intersections: A raised intersection is a flat raised area covering the entire intersection, with ramps on all approaches. They are also called raised junctions, intersection humps, or plateaus (Table 5.17).

Neighborhood traffic circles: Traffic circles are small raised islands placed in intersections that channelize traffic around to circulate. Traffic circles can also reduce vehicle speed by adding extra horizontal curves. Traffic circles are generally used at intersections of local or collector streets. Traffic circles are different from roundabouts. A study shows traffic circles can reduce middle block speed by 10% and reduce intersection collisions by 70% on average.

Roundabouts: Similarly, roundabouts could also reduce vehicle speeds and help organize traffic flow at intersections at a relatively lower cost than signalized intersections. The cost of a roundabout varies from place to place depending on the size of the intersection, site condition, and other factors (Table 5.18).

Speed humps: Speed humps are rounded raised areas of pavement with a general height from 3 to 4 inches, always placed in a series with a certain distance (e.g., 300–600 feet). Speed humps are usually placed on residential streets. Observations indicate that the speed is reduced by 20%–25% on average due to speed humps.

Speed tables: Speed tables are long (typically 6 feet long) raised speed humps with a flat section in the middle. They are always used on local and collector streets (Table 5.19).

5.2 TRANSPORTATION USER COST MODELING

User costs are those costs that are borne by the users, not the transportation agency, including vehicle operating costs (VOC), travel time costs, and safety costs, as well as environmental impacts costs. These used to be called external costs. Each of the cost components

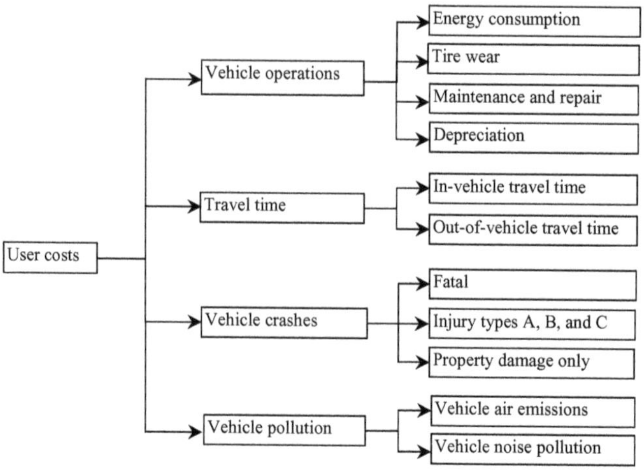

Figure 5.3 Transportation user cost components.

is estimated for normal traffic operating conditions and excessive costs occurring in work-zone conditions (Figure 5.3).

5.2.1 Vehicle operating costs

VOCs refer to costs that vary with vehicle usage, such as fuel, tires, maintenance, repairs, and mileage-dependent depreciation costs (Booz Allen and Hamilton et al., 1999). These could be categorized as three components: fluids (fuel, engine oil, brake fluid, power-steering fluid, etc.), light consumable parts (tires, brake pads, wiper blades, etc.), and heavy consumable parts (battery, radiator, etc.). Each of them is affected by vehicle type, vehicle speed, acceleration/deceleration, vertical and horizontal alignment, road surface condition, and experience of drivers. Another kind of factor is called vehicle ownership costs, which refers to the costs not directly affected by vehicle mileage, including vehicle depreciation, insurance cost, registration fees, and parking fees. Land use or city planning changes may alter the mode of traveling choice. They will affect the VOC indirectly.

5.2.1.1 Vehicle energy consumption

The industrial revolution brought great changes to the world, including the invention of machines and the revolution in energy. The steam engine combined these parallel revolutions in machinery and energy. Fossil fuels, including petroleum, coal, and lately natural gas, have been widely used ever since. In the transportation industry, the rapid construction of infrastructure and growth of the vehicle industry jointly resulted in a sharp increase of energy consumption. By the 1970s, a large majority of families owned a vehicle in the U.S., and now the entire world has become dependent upon transportation that requires an incredible amount of energy—particularly petroleum—to operate.

In the U.S., transportation consumed about 28% of all the energy resources each of the past five years, and this trend is likely to continue over the next few years. The major source of energy for transportation is petroleum products. Commonly, petroleum is refined into various types of petroleum products, such as gasoline, diesel, jet fuel, residual fuel oil, and other fractional distillation products. Recently, new fuel-type vehicles have become popular in research. Biofuel (ethanol and biodiesel), natural gas, and electric vehicles have been released or are under testing.

This section will introduce the vehicle energy consumption models. The following factors are considered in estimation of vehicle energy consumption (AASHTO, 2003).

Vehicle model: Typically, the fuel consumption of vehicles is measured by the unit of miles per gallon (MPG). U.S. Department of Energy website provides average MPG for almost all available vehicle manufacturers and models in the country with consideration of vehicle age. If detailed information of vehicle maker or model is unknown, the Bureau of Transportation Statistics (BTS) suggests an average MPG for all types of vehicles on road as 17.5 mpg (BTS, 2014).

Fuel price: To a certain extent, the fuel price affects the willingness of people to travel. In transportation terms, fuel price (supply) may influence the travel demand, which includes trip generation and mode split. Specifically, the increase of fuel price will reduce the number of unnecessary trips. As for the necessary trips, more travelers tend to use transit service instead of driving private automobiles as fuel prices increase. Consequently, the overall fuel consumption tends to decline.

Vehicle speed: It is well acknowledged that the optimal speed for achieving highest MPG exists. For most vehicles, the optimal speed lies in the interval from 55 to 60 mph. The fuel efficiency will be reduced with speed increase or decrease from the optimal speed. Table 5.20 gives the general fuel economy loss by speed.

Geometric design: The term road design here mainly refers to geometric features of the roadway, including grade of vertical curves and alignments of horizontal curves. As for vertical curves, vehicles need more energy to crawl the vertical slope. As for horizontal curves, extra energy is used to overcome centrifugal forces. The fuel consumption model is given below

$$Q_{fuel} = \begin{cases} a_{1,0} + \dfrac{a_{1,1}}{V_f}, \; V_f < 35 \text{ mph} \\ a_{2,0} + a_{2,1} \cdot v_{ff} + a_{2,2} \cdot V_f^2, \; V_f \geq 35 \text{ mph} \end{cases} \tag{5.25}$$

where

Q_{fuel} = Fuel consumption rate, gallon per mile (gpm)
V_f = Average vehicle speed, mph
$a_{1,0}$ = Fuel used to overcome resistance to reach operating speed, gpm
$a_{1,1}$ = Fuel used to maintain operating speed, gallon per hour (gph)
$a_{2,0}$ = Calibration parameter
$a_{2,1}$ = Parameter proportional to vehicle idling fuel flow rate
$a_{2,2}$ = Parameter related to vehicle and kinetic energy changes (Table 5.21)

Table 5.20 Fuel economy by speed

Speed (mph)	60	65	70	75	80
Fuel economy loss	3%	8%	17%	23%	28%

Table 5.21 Model coefficients of fuel consumption affected by vehicle speed

Vehicle type	a_1	a_2
Passenger cars	0.0362	0.0746
Tractor trailers	0.17	2.43

Traffic flow condition: The general equation is used to solve problems without considering traffic flow and stops' impacts. However, it rarely happens during the peak hour or driving on local streets. Fuel consumption in the process of deceleration, acceleration, or idling needs to be considered. The following model is used to calculate extra fuel consumption due to delays (FHWA, 2000):

$$\Delta Q_{fuel} = a_{3,0} + a_{3,1} \cdot \Delta T_{congestion} \tag{5.26}$$

where

ΔQ_{fuel} = Rate of change in fuel consumption, gpm
$a_{3,0}$ = Coefficient for congested distance, gpm
$a_{3,1}$ = Coefficient for congested time, gph
$\Delta T_{congestion}$ = Added travel time per mile due to congestion, hour per mile (hpm) (Table 5.22)

Due to traffic congestion or signalized control, vehicles may experience a full stop or a decelerate–accelerate process. Figure 5.4 shows the full stop process. Vehicles first travel at the desired speed and then decelerate until fully stopped. After a while, they accelerate to the desired speed again. The decelerating, idling, and accelerating times are $t_2 - t_1$, $t_3 - t_2$, and $t_4 - t_3$, respectively. The extra time delay is calculated as

Decelerating process: $(t_2 - t_1) - \dfrac{d_2 - d_1}{v_{ff}}$

Idling process: $t_3 - t_2$

Table 5.22 Model coefficients of fuel consumption affected by traffic flow condition

Vehicle type	a_M	a_D
Automobiles	0.04	0.42
Heavy trucks	0.16	0.87
Buses	0.25	–

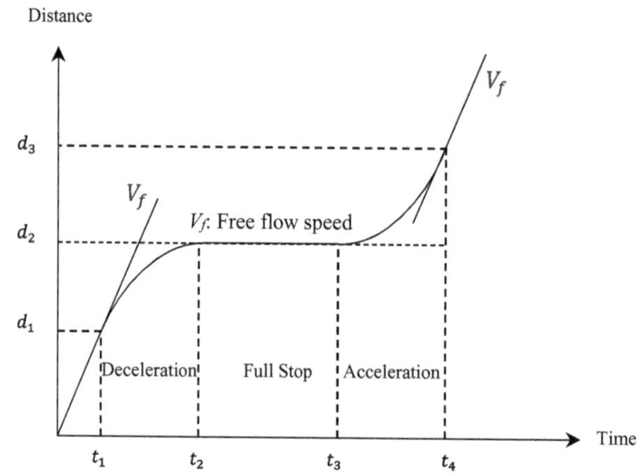

Figure 5.4 Vehicle full stop process at an intersection.

Accelerating process: $(t_4 - t_3) - \frac{d_4 - d_3}{v_{ff}}$

$$\Delta T_{delays} = (t_2 - t_1) - \frac{d_2 - d_1}{V_f} + t_3 - t_2 + (t_4 - t_3) - \frac{d_4 - d_3}{V_f} \quad (5.27)$$

where

V_f = Desired free flow speed
ΔT_{delays} = Total travel delays

The extra miles are calculated as follows:

$$\Delta D = (t_4 - t_1) \cdot V_f - (d_3 - d_1) \quad (5.28)$$

5.2.1.2 *Tire cost*

Tire cost mainly depends on the retail price of a set of tires. To estimate the tire cost of a vehicle intuitively, use the total price of a set of tires divided by the life cycle in miles. The tire cost will be measured in dollars per mile.

$$C_{tire} = \frac{\text{Price of a set of tires}}{\text{Average life in miles}} \quad (5.29)$$

Note that this equation does not consider the impact of pavement roughness or extreme weather, which may lead to an early replacement of tires. Also, as we discussed in the section on fuel cost, economic inflation is usually not considered. Therefore, this measure is not perfect.

5.2.1.3 *Vehicle maintenance and repair*

Maintenance cost mainly depends on the manufacturer's maintenance schedule recommendation and the corresponding labor costs. Maintenance service schedule and labor cost vary with the make and model of vehicle. Additional costs may be required if vehicle parts need to be replaced.

To estimate future maintenance cost, inflation and price fluctuation need to be considered. However, inflation and the consumer price index (CPI) are relatively stable in the U.S., so they may not significantly affect the cost estimation. Another factor that affects the maintenance cost is road condition. Poor road conditions will accelerate the deterioration rate of the vehicle, which can lead to an increase in vehicle maintenance and repair cost. Other factors such as driving skill and climate conditions will also influence the cost of maintenance and repair.

5.2.1.4 *Vehicle depreciation value*

Depreciation cost cannot strictly be categorized into fixed costs or variable costs, because some of the depreciation costs are independent of usage and some are not. The non-usage costs are caused by market fluctuation and inflation. The usage-dependent costs are mainly based on the usage as indicated by the name, which takes a large share of the total depreciation cost. This is understandable through the reality is that the mileage of travel is still a key attribute when evaluating the price of a used car.

The depreciation cost could be assessed by several methods, including (i) straight-line method; (ii) sum-of-the-years digits method; (iii) declining-balance method; and (iv) units-of-production method. The straight-line method is a basic method to estimate the vehicle depreciation, which spreads the cost of the fixed asset uniformly over its lifespan. The sum-of-the-years digits method computes depreciation cost by summing all years of the fixed asset's expected useful life and factoring in present worth in the base year, as compared to the total number of years. The declining balance method is an accelerated method of depreciation; it results in higher depreciation expense in the earlier years of ownership. The units-of-production method charges the cost according to the actual usage of the vehicle. Equations 5.30 through 5.32 present the various depreciation methods.

$$D_{SL} = \frac{\text{Vehicle purchase cost-salvage}}{\text{Useful life}} \tag{5.30}$$

$$D_{SYD} = \left(\frac{\text{Vehicle purchase cost-salvage}}{\text{Sum of remaining useful life}}\right) \cdot \text{Remaining useful life} \tag{5.31}$$

$$D_{DB} = \text{Undepreciated balance} \cdot \text{Accelerator} \cdot \text{Straight line rate} \tag{5.32}$$

EXAMPLE 5.1

A vehicle has a \$15000 acquisition cost, a three-year useful life, and an estimated salvage value at the end of its useful life of \$3000. Compute the annual depreciation amounts.

Solution

As summarized in Table 5.23.

In practice, the most commonly used depreciation methods will set a standard mileage for a given year and adjust the depreciation value based on the comparison of standard and actual mileage. The excessive mileage will lead to an additional depreciation value. Usually, according to the price, vehicles are categorized into four classes to perform the depreciation analysis. Economy cars are marked as Class I and luxury cars as Class IV. Expensive vehicles usually have a higher depreciation rate. The depreciation rate in the first five years is higher than in the following years.

Table 5.23 Depreciation value

	Straight-line	*Sum-of-years digits*	*Declining balance (with accelerator = 2)*
Year	*(15000–3000)/3*	*(15000–3000)× (remaining life/(3 + 2 + 1))*	*Undepreciated balance× (2 × (100%/3))*
1	12000/3 = 4000	12000 × (3/6) = 6000	15000 × 66.7% = 10005
2	12000/3 = 4000	12000 × (2/6) = 4000	Min{(15000−10005) × 66.7% (15000−3000−10005)} = 1995
3	12000/3 = 4000	12000 × (1/6) = 2000	0

5.2.2 Vehicle travel time

Travel time costs refer to the value of time spent in travel and include costs to businesses of time by their employees, vehicles, and goods, and costs to consumers of personal unpaid time spent on travel, including time spent parking and walking to and from a vehicle. Travel time mainly consists of two major parts: in-vehicle travel time (IVTT) and out-of-vehicle travel time (OVTT) reflecting the time spent in and out of a transportation facility. For example, time used to walk from home to vehicle and waiting time for transit are OVTT. IVTT is much more dependent on traffic conditions, where segment congestions will surely result in a longer IVTT. OVTT is related to the parking lot or road network design, as well as the transit schedule. In the following sections, travel time prediction procedure will be covered.

Travel time cost is an essential factor to evaluate the mobility of a transportation system. It is estimated as the product of travel time and the unit travel time value. Actually, time is not a merchandise that can be traded or valued in terms of money. However, it is a limited resource to perform any kind of activity, including value creation. Therefore, value of travel time is required to quantitatively measure the importance of this resource. Moreover, saving travel time is the ultimate goal of providing transportation mobility.

5.2.2.1 *Travel time estimation using instantaneous speed–density–flow static models*

Intuitively, IVTT can be indirectly calculated as the travel distance divided by the corresponding travel speed

$$\text{Travel time} = \frac{\text{Travel distance}}{\text{Average speed}} \tag{5.33}$$

Therefore, the essence of travel time estimation determines travel speed. Historically, various instantaneous speed–density–flow relationships have been developed for various classes of highways, which include the following:

a. Greenshields model for a two-lane, two-way road (Greenshields, 1935)

$$V = V_f \cdot (1 - D/D_j) \tag{5.34}$$

where
- V = Space mean speed of the traffic stream, mph
- V_f = Free flow space mean speed, mph
- D = Traffic density, passenger car per mile per lane (pcpmpl)
- D_j = Jam traffic density, pcpmpl

b. Boardman and Lave model for a four-lane road (Boardman and Lave, 1977)

$$q = 2490 - 0.523 \cdot (V - 35.34)^2 \tag{5.35}$$

where
- q = Flow rate of the traffic stream, passenger car per hour per lane (pcphpl)
- V = Space mean speed of the traffic stream, mph

c. Inman model handling traffic congestion (Inman, 1978)

$$q^{-2.95} = 3.351 \times 10^{9} - 231.4 \cdot (V - 7.2)^{-4.06} \tag{5.36}$$

where
q = Flow rate of the traffic stream, pcphpl
V = Space mean speed of the traffic stream, mph

5.2.2.2 Travel time estimation using space-averaged static models

Some space-averaged speed–density–flow models are listed below:

a. Smeed model for urban streets (Smeed, 1968)

$$q/W = 68 - 0.13 \cdot V^{2} \tag{5.37}$$

where
q = Flow rate of the traffic stream, pcphpl
W = Lane width, m
V = Space mean speed, km/hr

b. Keeler and Small model for expressways (Keeler and Small, 1977)

$$q/c = 0.8603 - 0.001923 \cdot (V - 45.68)^{2} \tag{5.38}$$

where
q = Flow rate of the traffic stream, pcphpl
c = Highway capacity, pcphpl
V = Space mean speed of the traffic stream, mph

c. Ardekani and Herman model for urban central area (Ardekani and Herman, 1987)

$$V = 18.38 \cdot \left[1 - (0.01D)^{1.239}\right]^{2.58} \tag{5.39}$$

where
V = Space mean speed of the traffic stream, mph
D = Traffic density, pcpmpl

d. *Highway Capacity Manual (HCM) method*: The HCM 2010 (TRB, 2010; Garber and Hoel, 2014) provides the speed flow curves and for basic highway segments by highway functional classification. Some of the speed calculation equations are shown in the following:
- For basic freeway segments

$$V_f = V_f^{base} - f_{LW} - f_{LC} - f_{N} - f_{ID} \tag{5.40}$$

- For multilane roadway segments

$$V_f = V_f^{base} - F_{M} - F_{LW} - F_{LC} - F_{A} \tag{5.41}$$

where

V_f = Estimated free flow speed, mph
V_f^{base} = Base free flow speed, mph
f_{LW} = Speed adjustment for lane width, ft
f_{LC} = Speed adjustment for right-shoulder lateral clearance
f_N = Speed adjustment for number of lanes per direction
f_{ID} = Speed adjustment for interchange density
F_M = Speed adjustment for median type
F_A = Speed adjustment for access points

5.2.2.3 *Travel time estimation using time-averaged static models*

The space-averaged traffic stream relationships could not handle cases where the volume exceeds the capacity. In such situations, speed depends not only on contemporaneous flow but also on past flows. In this respect, time-averaged relationships are sought where travel time is directly estimated.

a. *Bureau of public roads (BPR) function*: The BPR function introduced by Bureau of Public Roads (BPR)—the predecessor of Federal Highway Administration (FHWA), in 1964 gives an alternative idea to state the relationship between resistance and traffic volume of each link. This method is developed based on the Frank–Wolfe algorithm to deal with the traffic equilibrium problem. The function is presented as follows:

$$T_l(v_l) = T_{f,l} \cdot \left[1 + a \cdot \left(\frac{v_l}{c \cdot c_l}\right)^b\right] \tag{5.42}$$

where

$T_l(v_l)$ = Average travel time on link l, in hr
$T_{f,l}$ = Travel time of traversing link l at a free flow speed, in hr
v_l = Traffic volume on link l
c_l = Capacity of link l
a, b, c = Model coefficients, the most commonly used values are $a = 0.15, b = 4.0$, and $c = 1$

b. *Small's uniform arrival piecewise linear duration-dependent model*: One of the drawbacks of the BPR functions lies in the fact that it does not account for the time duration of demand exceeding capacity. The Small's duration-dependent model could mitigate this limitation by assuming a uniform inflow arrival rate (Small, 1982; Small and Verhoef, 2007). The model is of the following specification:

$$T_l(v_l) = \begin{cases} T_{f,l}, & \text{if } v_l \le c_l \\ T_{f,l} + 0.5 \cdot p \cdot (v_l/c_l - 1), & \text{otherwise} \end{cases} \tag{5.43}$$

where

$T_l(v_l)$ = Average travel time on link l
$T_{f,l}$ = Travel time of traversing link l at a free flow speed
p = Peak period of fixed duration
v_l = Traffic volume on link l
c_l = Capacity of link l

c. *Akcelik's random arrival model*: Akcelik (1991) developed a smooth travel time function by incorporating a delay parameter with random arrivals of the inflow rate as given below:

$$T_l(v_l) = T_{f,l} + \left[(v_l/c_l - 1) + \sqrt{(v_l/c_l - 1)^2 + (8 \cdot J_l \cdot (v_l/c_l))/(c_l \cdot P)}\right] \quad (5.44)$$

where

$T_l(v_l)$ = The average travel time on link l
$T_{f,l}$ = Travel time of traversing link l at a free flow speed
J_l = Delay parameter, the Small's model is a special case when $J_l = 0.26$
P = Peak period of fixed duration
v_l = Traffic volume on link l, assumed to follow a uniform arrival rate
c_l = Capacity of link l

5.2.2.4 Travel time estimation using dynamic models

Compared with the above static models, dynamic models are advantageous in dealing with extreme time-varying traffic congestion. The following briefly describes four types of dynamic models: (a) queueing model; (b) shockwave model; (c) car-following model; and (d) schedule-varying dynamic model.

a. *Queueing model*: Dynamic modeling of queueing at a bottleneck can deal with extreme traffic congestion (Newell, 1971). The following introduces a queueing model.

Denote:

c_k = Capacity of a bottleneck location k
$v_a(t)$ = Volume of traffic arriving at a bottleneck point at time t
$v_b(t)$ = Volume of traffic departing from the bottleneck at time t
$N(t)$ = Number of vehicles stored in the queue, where the queue is assumed to be vertical rather than horizontal
$\dot{N}(t)$ = Time derivative of $N(t)$
t_q = Time when $v_a(t)$ is first equal to capacity c_k
$t_{q'}$ = Time when the queue is fully dissipated
$T_D(t)$ = Travel time delay
$\bar{T}_D(t)$ = Average travel time delay

The following relationship holds for a vehicle entering and exiting the bottleneck:

$$v_b(t) = \begin{cases} v_a(t), & \text{if } v_a(t) \leq c_k \text{ and } N(t) = 0 \\ c_k, & \text{otherwise} \end{cases} \quad (5.45)$$

$$\dot{N}(t) = v_a(t) - v_b(t) \quad (5.46)$$

$$T_D(t) = N(t)/c_k = \int_{t_q}^{T} [v_a(\tau)/c_k - 1] \cdot d\tau, t_q \leq \tau \leq t_{q'} \quad (5.47)$$

In the above, $[t_q, t_{q'}]$ are denoted as the queue beginning and ending times.

Considering a special case of a fixed peak period volume of $v_a(t)$ arriving inside the bottleneck during the time interval of $[t_p, t_{p'}]$, the queueing delay for $t \in [t_p, t_{p'}]$ can be computed using Equation 5.45 as

$$T_D(t) = \begin{cases} 0, & \text{if } v_a(t) \le c_k \\ [v_a(t)/c_a - 1] \cdot (t - t_p), & \text{otherwise} \end{cases} \tag{5.48}$$

The average travel delay can be calculated as

$$\begin{aligned} \bar{T}_D(t) &= \left[\int_{t_p}^{t_{p'}} T_D(t) \cdot dt \right] / (t_{p'} - t_p) \\ &= \begin{cases} 0, & \text{if } v_a(t) \le c_k \\ 0.5 \cdot (t_{p'} - t_p) \cdot [v_a(t)/c_k - 1], & \text{otherwise} \end{cases} \end{aligned} \tag{5.49}$$

Equation 5.49 for $v_a(t) \ge c_k$ with $p = t_{p'} - t_p$, is then expressed as $0.5 \cdot p \cdot [v_a(t)/c_k - 1]$, which is consistent with the average travel time delay expression in Equation 5.43.

b. *Shockwave model*: The shockwave model developed by Lighthill and Whitham (1955) and Richards (1956) is commonly called the LWR model (Daganzo, 1997). This model considers that the space mean speed V, density D, and flow rate q of a traffic stream are continuous functions of location x and time t that satisfy the follow conditions: (i) existence of a space mean speed–density relationship for both uncongested and congested traffic, $V(x, t) = f[D(x, t)]$; (ii) flow rate is always the product of space mean speed and density, $q(x, t) = V(x, t) \cdot D(x, t)$; and (iii) conservation of flow, $\partial q(x, t)/\partial x + \partial D(x, t)/\partial t = 0$.

   The presence of a traffic disturbance creates a shockwave as a moving boundary that splits the traffic stream into upstream steady state, V_U, D_U, and q_U, and downstream steady state, V_D, D_D, and q_D. The wave speed V_W is computed by

$$\begin{aligned} V_W &= (q_U - q_D)/(D_U - D_D) \\ &= (V_U \cdot D_U - V_D \cdot D_D)/(D_U - D_D) \end{aligned} \tag{5.50}$$

   If the direction of the wave speed is opposite to the direction of traffic stream, it shows a backward-moving shockwave. Otherwise, a forward-moving shockwave is formed. Also, a forward-moving shockwave can never travel faster than the traffic that carries it for a concave flow–density function that crosses the origin.

c. *Car-following model*: Similar to the LWR model, the car-following model is another type of dynamic model that allows for continuous space and time traffic dynamics (May, 1990). However, vehicles are treated as discrete entities that could incorporate specific behavior in terms of acceleration and deceleration relying on spacing and speed difference between the leading and lagging vehicles. The generalized car-following model proposed by General Motors is of the following specification:

$$\ddot{x}_{n+1}(t + \delta) = \frac{\alpha \cdot [\dot{x}_{n+1}(t + \delta)]^m}{[x_n(t) - x_{n+1}(t)]^l} \cdot [\dot{x}_n(t) - \dot{x}_{n+1}(t)] \tag{5.51}$$

where

$x_n(t)$ = Location of the nth vehicle at time t
$x_{n+1}(t)$ = Location of the $(n+1)$th vehicle at time t
$\dot{x}_n(t)$ = Speed of the nth vehicle at time t
$\dot{x}_{n+1}(t)$ = Speed of the $(n+1)$th vehicle at time t
$\ddot{x}_{n+1}(t+\delta)$ = Acceleration of the $(n+1)$th vehicle at time $(t+\delta)$
δ = Reaction time of the $(n+1)$th vehicle
α, l, m = Nonnegative model parameters

When $m = 0$ and $l = 0$, the above expression reduces to the linear car-following model as

$$\ddot{x}_{n+1}(t+\delta) = \alpha \cdot [\dot{x}_n(t) - \dot{x}_{n+1}(t)] \tag{5.52}$$

When $m = 0$ and $l = 1$, the above expression reduces to the inverse-space car-following model as

$$\ddot{x}_{n+1}(t+\delta) = \frac{\alpha}{[x_n(t) - x_{n+1}(t)]} \cdot [\dot{x}_n(t) - \dot{x}_{n+1}(t)] \tag{5.53}$$

For Equation 5.51, taking the integration over t yields $\dot{x}_{n+1} = \alpha \cdot \ln[e^{(C_0 \cdot \alpha)}/(1/(x_n - x_{n+1}))]$, or $V = \alpha \cdot \text{In}(D_j/D)$, where $V = \dot{x}_{n+1}$, $D_j = e^{(C_0 \cdot \alpha)}$, and $D = 1/(x_n - x_{n+1})$. This is consistent with the Greenberg model for traffic stream analysis (Greenberg, 1959).

d. *Schedule-varying dynamic model*: The schedule-varying dynamic model assumes that travelers intend to apply an optimal trip schedule by trading off travel time cost against schedule-delay or schedule-early cost. The weighted total travel time for a traveler departing at time t with travel delay and schedule-varying tradeoffs is

$$T'(t) = T(t) + T_S(t) \tag{5.54}$$

with

$$T(t) = T_f + T_D(t)$$

$$T_D(t) = \begin{cases} 0, & \text{if } v_d(t) \le c_k \\ \left[v_d(t)/c_k - 1\right] \cdot (t - t_p), & \text{otherwise} \end{cases}$$

$$T_S(t) = \begin{cases} (\beta/\alpha) \cdot [(t_d - t) - T(t)], & \text{if } T(t) \le t_d - t \\ (\gamma/\alpha) \cdot [T(t) - (t_d - t)], & \text{if } T(t) > t_d - t \end{cases}$$

where

$T'(t)$ = Weighted total travel time incurred when departing at t incorporating trade-offs between actual and desirable arrival times
$T(t)$ = Travel time incurred when departing at t
T_f = Free flow travel time
$T_D(t)$ = Travel delay from queueing at a bottleneck during peak period when departing at t

$N_D(t)$ = Number of travelers in the queue at a bottleneck during peak period when departing at t
$T_S(t)$ = Schedule-varying trade-off time when departing at t
t = Departure time
t_d = Desired arrival time
t' = Time at which a traveler exits from the queue at the bottleneck
c_k = Capacity of a bottleneck location k
$v_d(t)$ = Departure rate at time t
$v_a(t)$ = Queue entry rate at time t
$v_b(t)$ = Queue exit rate at time t
$v_d(t')$ = Departure rate for travelers exiting from the queue at time t'
v_d = Constant departure rate during the peak interval $[t_p, t_{p'}]$
α = Value of travel time
β = Shadow price of early arrival, where $\beta < \alpha$
γ = Shadow price of late arrival
t_p = Peak interval beginning time
$t_{p'}$ = Peak interval ending time
p = Peak period fixed duration, $p = t_{p'} - t_p$
t_q = Queue beginning time
$t_{q'}$ = Queue dissipation time

Assumptions: Without loss of generality, vehicle occupancy is assumed to be one—that is, one vehicle per traveler. The free flow travel time T_f before and after the bottleneck is set to zero when there is no route choice. Then, travel time $T(t)$ is equal to travel delay $T_D(T)$, traveler's queue exit time t' is equivalent to the trip arrival time $t + T(t)$ or $t + T_D(t)$. If there is no travel delay, that is, $T_D(t) = 0$, the traveler's departure time t is equal to queue entry time as well as queue exit time t'. The departure rate of travelers exiting from the queue $v_d(t')$ would be the desired trip arrival rate. Therefore, the analysis is centered on estimation of travel delays for individual travelers. For the queueing analysis, travel delays are assumed to occur solely through vertical queueing behind the bottleneck.

If $v_d(t') \leq c_k$, no travel delay or schedule delay exists for all t'. The queue entry rate $v_a(t)$ and exit rate $v_b(t)$ are both equal to the desired departure rate $v_d(t')$. Conversely, queue entry rate $v_a(t)$ needs to be adjusted by considering the trade-offs between the travel delay $T_D(t)$ and schedule-varying tradeoff time $T_S(t)$.

For the given peak interval $p = t_{p'} - t_p$, suppose that the departure rate $v_d(t)$ is constant at v_d and zero outside the interval. The total number of travelers departing during the interval becomes $v_d \cdot p$. If $v_d > c_k$, the desired queue exit rate $v_d(t')$ cannot be reached and both travel delay $T_D(t)$ and schedule-varying trade-off time $T_S(t)$ will occur. Therefore, the queue entry rate $v_a(t)$ for a traveler exiting the queue before or after the desired arrival time t_d must be adjusted.

Second-best travel time for queue entry time equilibrium: For a traveler exiting the queue before the desired arrival time t_d, equilibrium requires that the queue entry time minimizes the weighted total travel time $T_D(t) + (\beta/\alpha) \cdot [(t_d - t) - T_D(t)]$ or $T_D(t) \cdot [1 - (\beta/\alpha)] - (\beta/\alpha){\cdot}t + (\beta/\alpha){\cdot}t_d$. Setting $T_D(t) \cdot [1 - (\beta/\alpha)] - (\beta/\alpha) \cdot t = 0$, the rate of change in $T_D(t)$ against departure time t is $T_D(t)/t = [\beta/(\alpha/\beta)]$. Similarly, for a traveler exiting the queue after the desired arrival time t_d, equilibrium requires that the queue entry time minimizes the weighted total travel time $T_D(t) + (\gamma/\alpha) \cdot [T(t) - (t_d - t)]$ or $T_D(t) \cdot [1 + (\gamma/\alpha)] + (\gamma/\alpha) \cdot t - (\gamma/\alpha) \cdot t_d$. Setting $T_D(t) \cdot [1 + (\gamma/\alpha)] + (\gamma/\alpha) \cdot t = 0$, the rate of change in $T_D(t)$ against departure time t is $T_D(t)/t = [-\gamma/(\alpha + \gamma)]$.

For the constant queue entry rate v_a, $T_D(t)/t = (v_a/c_k - 1)$. The queue entry rate for travelers arriving before the desired arrival time t_d, $(v_a^{early}/c_k - 1) = \beta/(\alpha - \beta)$, this gives

$$v_a^{early} = [\alpha/(\alpha - \beta)] \cdot c_k \tag{5.55}$$

The queue entry rate for travelers arriving after the desired arrival time t_d, $(v_a^{late}/c_k - 1) = -\gamma/(\alpha + \gamma)$, which yields

$$v_a^{late} = [\alpha/(\alpha + \gamma)] \cdot c_k \tag{5.56}$$

As seen in Figure 5.5, the proportion of travelers arriving before the desired arrival time t_d will maintain a queue entry rate $v_a^{early} = [\alpha/(\alpha - \beta)] \cdot c_k > c_k$, the proportion of travelers arriving after the desired arrival time t_d will keep a queue entry rate $v_a^{late} = [\alpha/(\alpha + \gamma)] \cdot c_k < c_k$, the queue discharge rate is c_k. Because of existence of travel delays, the duration of first traveler entering the queue to the last traveler exiting the queue will be longer than the peak interval $p = [t_p, t_{p'}]$, which is $[(v_d/c_k) \cdot p] = [t_q, t_{q'}]$. This means that delay begins at time t_q prior to the earliest desired queue exit and ends at $t_{q'}$ that lasts beyond the latest desired queue exit.

Further, define t_m^{entry} and t_m^{exit} as the queue entry and exit times for the traveler experiencing the longest travel delay $T_{D,m}$. The highest number of travelers in the queue is at time t_m^{exit}, denoted as $N_{D,m}$. Intuitively, the last traveler arriving before the desired arrival time t_d that with a queue entry rate v_a^{early} will encounter the longest travel delay $T_{D,m}$. Specifically, for duration $[t_q, t_m^{entry}]$, the early arrival travelers will follow queue entry rate v_a^{early} with the last traveler arriving at t_m^{entry}, experiencing the longest delay $T_{D,m}$, and exiting at $t_m^{exit} = t_m^{entry} + T_{D,m}$. For duration $[t_m^{exit}, t_{q'}]$, the late arrival travelers will follow queue entry rate v_a^{late} with the first traveler arriving at t_m^{exit} and the last traveler exiting at $t_{q'}$.

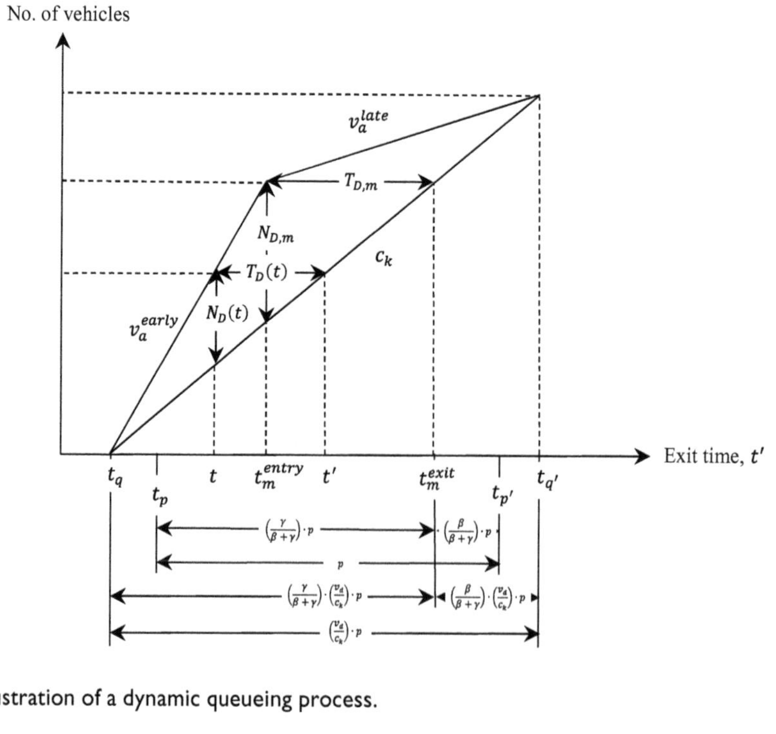

Figure 5.5 Illustration of a dynamic queueing process.

The portion of travelers who exit the queue before t_m^{exit}, σ, can be computed as

$$\sigma = (t_m^{exit} - t_p)/p = \left[(v_a^{early}/c_k)\cdot(t_m^{entry} - t_q)\right]/p \tag{5.57}$$

The portion of travelers who enter the queue after t_m^{entry} at queue entry rate v_a^{late}, $1-\sigma$, can be computed as

$$1-\sigma = (t_{p'} - t_m^{exit})/p = \left[(v_a^{late}/c_k)\cdot(t_{q'} - t_m^{entry})\right]/p \tag{5.58}$$

Solving for σ, it yields

$$\sigma = \gamma/(\beta+\gamma) \tag{5.59}$$

with

$$t_q = t_p - [\gamma/(\beta+\gamma)]\cdot[(v_d/c_k)-1]\cdot p$$

$$t_{q'} = t_{p'} + [\beta/(\beta+\gamma)]\cdot[(v_d/c_k)-1]\cdot p$$

$$t_m^{entry} = t_p + [\gamma/(\beta+\gamma)]\cdot p - T_{D,m}$$

$$t_m^{exit} = t_p + [\gamma/(\beta+\gamma)]\cdot p$$

$$T_{D,m} = [((\beta\cdot\gamma)/\alpha)/(\beta+\gamma)]\cdot(v_d/c_k)\cdot p$$

At this point, travel delay $T_D(t')$, schedule-varying trade-off time $T_S(t')$, and weighted total travel time $T'(t')$ as functions of queue exit time t' can be derived

$$T_D(t') = \begin{cases} (\beta/\alpha)\cdot(t'-t_q), & \text{if } t_q \le t' \le t_m^{exit} \\ (\gamma/\alpha)\cdot\left[\dfrac{\beta}{(\beta+\gamma)}\cdot\left(\dfrac{v_d}{c_k}\right)\cdot p - \left(t' - t_m^{exit}\right)\right], & \text{if } t_m^{exit} \le t' \le t_{q'} \end{cases} \tag{5.60}$$

$$T_S(t') = \begin{cases} (\beta/\alpha)\cdot\left[\dfrac{\gamma}{(\beta+\gamma)}\cdot\left(\dfrac{v_d}{c_k}-1\right)\cdot p - \left(1-\dfrac{c_k}{v_d}\right)\cdot(t'-t_q)\right], & \text{if } t_q \le t' \le t_m^{exit} \\ (\gamma/\alpha)\cdot\left[\left(1-\dfrac{c_k}{v_d}\right)\cdot\left(t' - t_m^{exit}\right)\right], & \text{if } t_m^{exit} \le t' \le t_{q'} \end{cases} \tag{5.61}$$

$$T'(t') = \begin{cases} (\beta/\alpha)\cdot\left[\dfrac{\gamma}{(\beta+\gamma)}\cdot\left(\dfrac{v_d}{c_k}-1\right)\cdot p + \left(\dfrac{c_k}{v_d}\right)\cdot(t'-t_q)\right], & \text{if } t_q \le t' \le t_m^{exit} \\ (\gamma/\alpha)\cdot\left[\dfrac{\beta}{(\beta+\gamma)}\cdot\left(\dfrac{v_d}{c_k}\right)\cdot p - \left(\dfrac{c_k}{v_d}\right)\cdot\left(t' - t_m^{exit}\right)\right], & \text{if } t_m^{exit} \le t' \le t_{q'} \end{cases} \tag{5.62}$$

Figure 5.6 illustrates changes in the travel delay $T_D(t')$, schedule-varying trade-off time $T_S(t')$, and weighted total travel time $T'(t')$ with respect to the queue exit time t'.

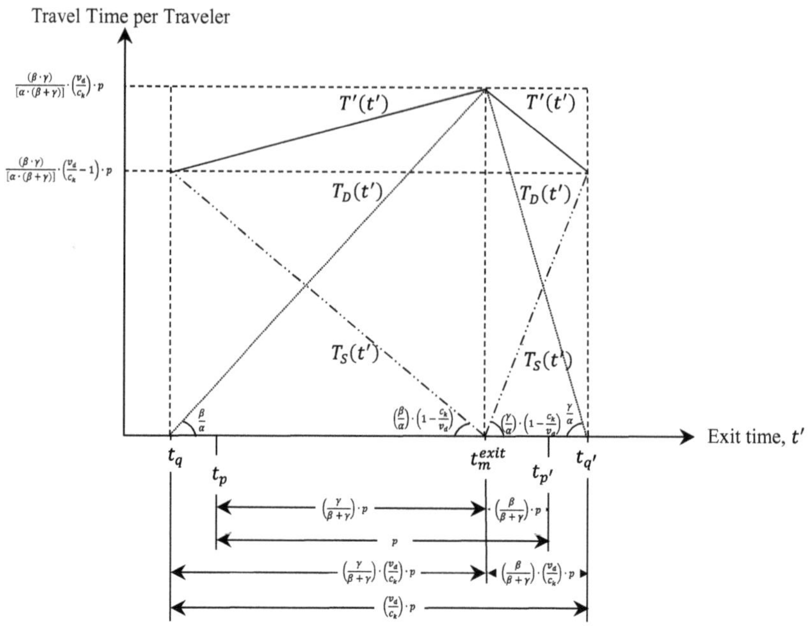

Figure 5.6 Queue entry time equilibrium travel delay, schedule-varying tradeoff time, and weighted total travel time.

Travel delay per traveler $T_D(t')$ increases linearly from zero to $T_{D,m} = [((\beta \cdot \gamma)/\alpha)/(\beta + \gamma)] \cdot (v_d/c_k) \cdot p$ for $t' \in \left[t_q, t_m^{exit}\right]$ and decreases linearly from $T_{D,m} = [((\beta \cdot \gamma)/\alpha)/(\beta + \gamma)] \cdot (v_d/c_k) \cdot p$ to zero for $t' \in \left[t_m^{exit}, t_{q'}\right]$. Meanwhile, schedule-varying trade-off time decreases linearly from the maximum of $[((\beta \cdot \gamma)/\alpha)/(\beta + \gamma)] \cdot (v_d/c_k - 1) \cdot p$ to zero for $t' \in \left[t_q, t_m^{exit}\right]$ and increases linearly from zero to the maximum of $[((\beta \cdot \gamma)/\alpha)/(\beta + \gamma)] \cdot (v_d/c_k - 1) \cdot p$ for $t' \in \left[t_m^{exit}, t_{q'}\right]$. The weighted total travel time $T'(t')$ increases linearly from $[((\beta \cdot \gamma)/\alpha)/(\beta + \gamma)] \cdot (v_d/c_k - 1) \cdot p$ to $[((\beta \cdot \gamma)/\alpha)/(\beta + \gamma)] \cdot (v_d/c_k) \cdot p$ for $t' \in \left[t_q, t_m^{exit}\right]$ and decreases linearly from $[((\beta \cdot \gamma)/\alpha)/(\beta + \gamma)] \cdot (v_d/c_k) \cdot p$ to $[((\beta \cdot \gamma)/\alpha)/(\beta + \gamma)] \cdot (v_d/c_k - 1) \cdot p$ for $t' \in \left[t_m^{exit}, t_{q'}\right]$.

At queue entry time equilibrium, the weighted total travel time is not held constant during the period $[t_q, t_{q'}]$. This is attributable to the differences in selecting the desired arrival time t_d by individual travelers. The time-averaged travel delay, schedule-varying trade-off time, and weighted total travel time are

$$\bar{T}_D = 0.5 \cdot T_{D,m} = 0.5 \cdot [((\beta \cdot \gamma)/\alpha)/(\beta + \gamma)] \cdot (v_d/c_k) \cdot p \tag{5.63}$$

$$\bar{T}_S = 0.5 \cdot [((\beta \cdot \gamma)/\alpha)/(\beta + \gamma)] \cdot (v_d/c_k - 1) \cdot p \tag{5.64}$$

$$\bar{T}' = \begin{cases} 0, & \text{if } v_d \leq c_k \\ \bar{T}_D + \bar{T}_S = \left[((\beta \cdot \gamma)/\alpha)/(\beta + \gamma)\right] \cdot (v_d/c_k - 0.5) \cdot p, & \text{otherwise} \end{cases} \tag{5.65}$$

First-best travel time for unpriced equilibrium: When the peak duration interval $p = t_{p'} - t_p$ approaches zero, the interval beginning time t_p and ending time $t_{p'}$ will be identical to the time of first traveler entering the queue t_q and to the last traveler exiting the queue $t_{q'}$, respectively. The queue entry rates v_a^{early} and v_a^{late} for travelers arriving before and after the desired arrival time t_d will ensure that the average weighed total travel time $\bar{T}'^{,0}$

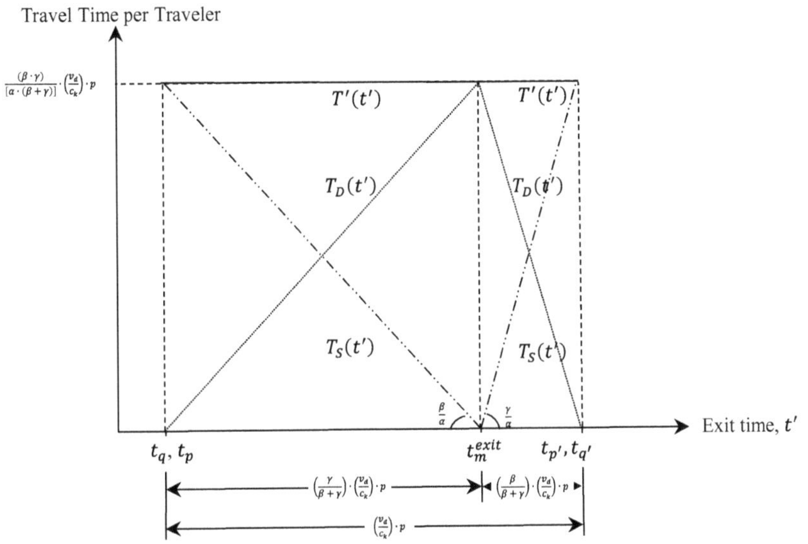

Figure 5.7 Unpriced equilibrium travel delay, schedule-varying tradeoff time, and weighted total travel time.

remains constant over all queue-entry times. As seen in Figure 5.7, both the average travel delay $\bar{T}_D^0$ and average schedule-varying trade-off time $\bar{T}_S^0$ are equal to one-half of the average weighted total travel time $\bar{T}'$ for the unpriced equilibrium

$$\bar{T}_D^0 = \bar{T}_S^0 = 0.5 \cdot [((\beta \cdot \gamma)/\alpha)/(\beta + \gamma)] \cdot (v_d / c_k) \cdot p \tag{5.66}$$

$$\bar{T}'^{,0} = [((\beta \cdot \gamma)/\alpha)/(\beta + \gamma)] \cdot (v_d / c_k) \cdot p \tag{5.67}$$

The total unpriced equilibrium travel time and marginal travel time is equal to

$$\bar{T}'^{,0}_{total} = [((\beta \cdot \gamma)/\alpha)/(\beta + \gamma)] \cdot [(v_d \cdot p)^2 / c_k] \tag{5.68}$$

$$\bar{T}'^{,0}_{marginal} = 2 \cdot [((\beta \cdot \gamma)/\alpha)/(\beta + \gamma)] \cdot (v_d / c_k) \cdot p = 2 \cdot \bar{T}'^{,0} \tag{5.69}$$

5.2.2.5 Additional travel time due to intersection delays

Considering a travel path that consists of roadway segments and intersections, vehicle delays could be experienced at intersections because of traffic control. The 2010 HCM method could be utilized for intersection delay estimation (TRB, 2010), which is of the following specification:

$$D_{int,i} = D_{1,i} + D_{2,i} + D_{3,i} \tag{5.70}$$

where

$D_{int,i}$ = Total vehicle delays at intersection i, sec/veh
$D_{1,i}$ = Uniform delay at intersection i, sec/veh
$D_{2,i}$ = Incremental delay at intersection i, sec/veh
$D_{3,i}$ = Initial queue delay at intersection i, sec/veh

The uniform delay is an estimate of delay under random uniform arrivals, stable flow, and no initial queue condition, and can be calculated as

$$D_{1,i} = \frac{0.5 \cdot c \cdot (1 - g/c)^2}{1 - \min\{1, v_i/c_i\} \cdot (g/c)} \tag{5.71}$$

where
c = Cycle length of traffic signal, sec
g = Effective green time, sec
v_i = Traffic volume entering intersection i, veh/hr
c_i = Transverse capacity of intersection i, veh/hr

The incremental delay aims to estimate the extra delay due to non-uniform arrivals and temporary cycle failures (random delay), as well as delay caused by sustained periods of oversaturation. It can be calculated as

$$D_{2,i} = 900 \cdot T \cdot \left[(v_i/c_i - 1) + \sqrt{(v_i/c_i - 1)^2 + (8 \cdot K \cdot I \cdot v_i)/(T \cdot c_i^2)}\right] \tag{5.72}$$

where
T = The duration of analysis period, sec
K = The incremental delay factor, a value of 0.5 is recommended for pre-timed signals
I = The upstream filtering/metering adjustment factor, a value of 1.0 is suggested for the analysis of an isolated intersection

The initial queue delay is to account for the additional delay caused by an existing initial queue, which is the result of unmet demand in the previous period, and can be estimated as

$$D_{3,i} = \frac{3600}{v_i \cdot T} \cdot \left[\frac{t \cdot (Q_{b,i} + Q_{e,i} - Q_{eo,i})}{2} + \frac{(Q_{e,i}^2 - Q_{eo,i}^2)}{2 \cdot c_i} - \frac{Q_{b,i}^2}{2 \cdot c_i}\right] \tag{5.73}$$

with

$$Q_{b,i}(k) = Q_{b,i}(k-1) + T \cdot [v_i(k) - c_i]$$

$$Q_{e,i} = Q_{b,i} + t \cdot (v_i - c_i)$$

$$Q_{eo,i} = \begin{cases} T \cdot (v_i - c_i) \text{ and } t = T, & \text{if } v_i > c_i \\ 0 \text{ and } t = Q_{b,i}/(c_i - v_i), & \text{otherwise} \end{cases}$$

where
Q_b = Initial queue at the start of the analysis period, veh
Q_e = Queue at the end of the analysis period, veh
Q_{eo} = Queue at the end of analysis period when traffic volume exceeds capacity and there is no initial queue, veh
k = Period of calculation
t = Period of time when the approach is working on dissipating a queue, hr

5.2.2.6 *Additional travel time due to parking search*

In many urban areas, the excessive demand for parking facilities makes drivers waste a large amount of time in searching for available parking slots, an activity known as cruising. Parking is linked to congestion because vehicles engaged in entering, exiting, or looking for parking spaces slow down other vehicles. Conversely, additional travel time due to parking search is negligible when the parking supply is in abundance. In the total travel time estimation process, the added travel time from parking search from excessive parking demand needs to be included.

5.2.2.7 *Short-run travel time cost*

With travel time prediction models and value of travel time made available, the travel time cost functions can be established. The traffic flow that is used for the cost calculation could be static flow, time-averaged flow or desired arrival rate.

Stationary-state flow on a homogeneous roadway segment: Stationary-state flow is defined as a situation where traffic flow is constant over time and space, and the flow is equal to the rates at which trips are originated and terminated. This situation is an ideal case because the traffic condition always changes over time and space rapidly, which is less practical. Nonetheless, the travel time cost function is built as follows:

$$c_{TT}(v_l) = \alpha \cdot T_l(v_l) = \alpha \cdot L/V(v_l) \tag{5.74}$$

where

$c_{TT}(v_l)$ = The average travel time cost of link l, dollars
v_l = Traffic volume on link l, pcphpl
α = Value of travel time, dollars/hr
$T_l(v_l)$ = Average travel time on link l, hr or min
L = Segment length, mi
$V(v_l)$ = Space mean speed on link l, mph

Time-averaged static flow consideration on a roadway segment: Time-averaged functions establish the relation between average travel time and traffic flow rate. The cost model using time-average models can accommodate the case that the inflow exceeds capacity in a short period. The time-average model irons out the time dependency of traffic conditions by averaging the variation during a certain period to mitigate the issue. With the Akcelik's random arrival model, the travel time cost function can be specified as

$$c_{TT}(v_l) = \alpha \cdot \left\{ T_{f,l} + \left[(v_l/c_l - 1) + \sqrt{(v_l/c_l - 1)^2 + (8 \cdot J_l \cdot (v_l/c_l))/(c_l \cdot P)} \right] \right\} \tag{5.75}$$

where

$c_{TT}(v_l)$ = The average travel time cost of link l, dollars
α = Value of travel time, dollars/hr
$T_{f,l}$ = Travel time of traversing link l at a free flow speed, hr or min
J_l = Delay parameter, the Small's model is a special case when $J_l = 0.26$
P = Peak period of fixed duration, hr or min
v_l = Traffic volume on link l, assumed to follow a uniform arrival rate, pcphpl
c_l = Capacity of link l, pcphpl

Dynamic schedule-varying trade-off flow: With the dynamic model with endogenous scheduling, the travel time cost function can be created as below:

$$c_{TT}(v_l) = \alpha \cdot T(t) + \begin{cases} \beta \cdot [(t_d - t) - T(t)], & \text{if } T(t) \leq t_d - t \\ \gamma \cdot [T(t) - (t_d - t)], & \text{if } T(t) > t_d - t \end{cases} \tag{5.76}$$

where

$c_{TT}(v_l)$ = The average travel time cost of link l, dollars
α = Value of travel time, dollars/hr
β = Cost of early arrival, dollars/min
γ = Cost of late arrival, dollars/min
t = Departure time
$T(t)$ = Travel time incurred when departing at t
t_d = Desired arrival time

5.2.2.8 Long-run travel time cost

The short-run travel time cost functions could rely on the time duration of a 24-h daily period. The calculated travel time cost could be extrapolated to an annual travel time cost. Considering the annual vehicle volume growth rate, the total travel time cost associated with the highway facility service life cycle could be computed.

5.2.3 Vehicle crashes

5.2.3.1 Safety performance model types

Poisson models: Poisson models are a well-acceptedmethod of modeling discrete rare events such as highway vehicle crashes (AASHTO, 2010b; FHWA, 2010). It is assumed that occurrences of vehicle crashes at a highway site (segment or intersection) are independent of one another, and that the average number of crashes per unit time (i.e., the mean) is characteristic of the given site and of other sites with similar properties. The mean itself is assumed to depend on highway variables and must be greater than or equal to zero. The safety performance function (SPF) calibrated using the Poisson modeling technique has a generalized linear form given by

$$\ln(\mu_i) = \beta_0 + \beta_1 \cdot X_{1i} + \beta_2 \cdot X_{2i} + \cdots + \beta_{p-1} \cdot X_{p-1,i} \tag{5.77}$$

where

μ_i = Average number of vehicle crashes per unit time at site i, $\mu_i = E[k_i]$
$X_{p-1,i}$ = The $(p-1)$th highway variables that contributes to crashes at site i
β_{p-1} = Coefficient of variable $X_{p-1,i}$ to be estimated by the modeling
i = Number of highway sites with data observations, $i = 1, 2, \ldots, n$
ln (.) = The natural logarithm

In Poisson distribution, the variance of number of crashes at site i, σ_i^2, is equal to the mean μ_i. The Poisson model takes the form

$$P(k_i) = \frac{\mu_i^{k_i} \cdot e^{-\mu_i}}{k_i!} \tag{5.78}$$

where

k_i = Expected number of crashes at site i
μ_i = Average number of vehicle crashes per unit time at site i
$P(k_i)$ = Probability of having k_i crashes at site i
i = Number of sites with data observations, $i = 1, 2, \ldots, n$

The coefficients $\beta_0, \beta_1, \beta_2, \ldots, \beta_{p-1}$ are estimated by maximizing the log-likelihood function $L(\beta_0, \beta_1, \beta_2, \ldots, \beta_{p-1})$ for the Poisson distribution

$$L(\beta_0, \beta_1, \beta_2, \ldots, \beta_{p-1}) = \sum_{i=1}^{n} \left[k_i \cdot \ln(\mu_i) - \mu_i - \ln(k_i!)\right]$$

Negative binomial models: A limitation of the Poisson distribution is that the mean equals the variance. The actual field data may not always support this assumption. Suppose that the variance of the data exceeds the estimated mean of the vehicle crash data distribution. It is then said to be overdispersed, and the underlying assumption of variance being equal to the mean for the Poisson distribution is violated. The negative binomial distribution, which is a discrete distribution, provides an alternative model to deal with overdispersion in vehicle crash count data. The negative binomial model takes the form

$$P(k_i) = \frac{\Gamma\left(\frac{1}{K} + k_i\right)}{k_i!\,\Gamma\left(\frac{1}{K}\right)} \left(\frac{K \cdot \mu_i}{1 + K \cdot \mu_i}\right)^{k_i} \left(\frac{1}{1 + K \cdot \mu_i}\right)^{\frac{1}{K}} \tag{5.79}$$

where

k_i = Expected number of crashes at site i
μ_i = Average number of vehicle crashes per unit time at site i
$P(k_i)$ = Probability of having k_i crashes at site i
K = Overdispersion parameter
i = Number of sites with data observations, $i = 1, 2, \ldots, n$

The variance of negative binomial distribution is $\mu_i + K.\mu_i^2$. If K equals 0, the negative binomial model reduces to the Poisson model. The model coefficients $\beta_0, \beta_1, \beta_2, \ldots, \beta_{p-1}$ and parameter K are estimated by maximizing the log-likelihood function $L(\beta_0, \beta_1, \beta_2, \ldots, \beta_{p-1})$ for the negative binomial distribution

$$L(\beta_0, \beta_1, \beta_2, \ldots, \beta_{p-1}) = \sum_{i=1}^{n} \left[\sum_{j=0}^{k_i} \ln(1 + K \cdot j) - \ln(1 + K \cdot k_i)k_i + k_i \cdot \ln(\mu_i) - (k_i + 1/K) \cdot \ln(1 + K \cdot \mu_i) - \ln(k_i!)\right].$$

Table 5.24 lists tests to evaluate the hypothesis of a Poisson model against the alternative negative binomial model that is with overdispersion (Vogt, 1999). In addition to a plausible basis for the underlying distribution assumptions, an acceptable model needs to maintain the following characteristics: (i) the estimated model coefficient for each explanatory variable should be statistically significant; (ii) the sign and rough magnitude of each estimated coefficient should be confirmed by engineering and intuitive judgments; and (iii) goodness-of-fit

Table 5.24 Tests for determining Poisson model against negative binomial model

Test	*Test statistic*	*Conclusion*
Pearson Chi-square statistic	$\chi^2 = \sum_{i=1}^{n} \left[\frac{(k_i - \hat{k}_i)^2}{\hat{k}_i} \right]$	If $\chi^2/(n-p)$ is significantly larger than 1, Poisson model is rejected and a negative binomial model needs to be considered
Dean and Lawless standard normal statistic	$T_1 = \frac{\sum_{i=1}^{n} \left[(k_i - \hat{k}_i)^2 - k_i \right]}{\sqrt{2 \cdot \sum_{i=1}^{n} (\hat{k}_i)^2}}$	If T_1 is large positive, Poisson model is rejected and a negative binomial model needs to be considered

Note: Where $\hat{k}_i$ is estimated crashes at site i using the calibrated model, n is total number of sites, and p is number of explanatory variables included in the calibrated model.

Table 5.25 Tests for evaluating goodness-of-fit

Ordinary R^2	*Weighted R^2*	*Explanation*
$R^2 = 1 - \frac{\sum_{i=1}^{n} (k_i - \hat{k}_i)^2}{\sum_{i=1}^{n} (k_i - \bar{k})^2}$	$R_w^2 = 1 - \frac{\sum_{i=1}^{n} [(k_i - \hat{k}_i)^2 / \hat{k}_i]}{\sum_{i=1}^{n} [(k_i - \bar{k})^2 / \hat{k}_i]}$	Variation of vehicle crashes explained by the model
$P^2 = 1 - \frac{\sum_{i=1}^{n} \hat{k}_i}{\sum_{i=1}^{n} (k_i - \bar{k})^2}$	$P_w^2 = 1 - \frac{n}{\sum_{i=1}^{n} [(k_i - \bar{k})^2 / \hat{k}_i]}$	Explainable random variation to be expected when independent crashes of mean frequency $\hat{k}_i$ occur
$R_p^2 = \frac{R^2}{P^2}$	$R_{pw}^2 = \frac{R_w^2}{P_w^2}$	Proportion of potentially explainable systematic variation that can be explained from the causal factors considered

Note: Where $\hat{k}_i$ is estimated crashes at site i using the calibrated model, n is total number of sites, and $\bar{k}$ is average number of vehicle crashes from n site observations.

measures should indicate that the variables do have explanatory and predictive power. Table 5.25 presents commonly used tests to evaluate goodness-of-fit (Vogt, 1999).

For Poisson and negative binomial models, R_P^2, being unnormalized, will make observations with large predicted means more influential, while R_{PW}^2 tends to exaggerate the estimation errors associated with small predicted means.

Zero inflation considerations: The crash data set may exhibit a large number of observations with no crash occurrences in a given time period. This case can be handled by zero-inflated Poisson model or zero-inflated negative binomial model. In the model calibration process, both model forms with or without zero crash inflation can be tested.

Lognormal models: Lognormal regression models are based on the assumption that the natural logarithm of vehicle crashes k_i follows a normal distribution with mean μ_i, that is, $E[\ln(k_i)]$, and variance $V[\ln(k_i)]$. This model is a reasonable choice whenever the data are inherently nonnegative and maintain large mean value as well as positive skewness. In this case, the relationship between the expected number of vehicle crashes μ_i at site i and explanatory variables for modeling calibration can be written as

$$P(\ln(k_i)) = \frac{1}{\sqrt{2\pi}\sigma_i} e^{-\frac{(ln(k_i)-\mu_i)^2}{2\sigma_i^2}} \tag{5.80}$$

where

k_i = Expected number of crashes at site i
μ_i = Average number of vehicle crashes per unit time at site i, $\mu_i = E[\ln(k_i)]$
σ_i^2 = Variance of vehicle crashes per unit time at site i, $\sigma_i^2 = V[\ln(k_i)]$
$P(\ln(k_i))$ = Probability of having $\ln(k_i)$ crashes at site i
i = Number of sites with data observations, $i = 1, 2, \ldots, n$
ln (.) = The natural logarithm

Geographically weighted regression (GWR) models: Regardless of Poisson, negative binomial, zero-inflated Poisson, zero-inflated negative binomial, or lognormal regression techniques for calibrating SPFs, model coefficients should capture variability of localized factors that contribute to a vehicle crash occurring to improve model prediction power. GWR is a relatively new technique that permits the parameter estimates to vary locally rather than globally (Fotheringham et al., 2003), which is well suited to predicting vehicle crashes. The SPF calibrated using the GWR modeling technique has the following general form:

$$\ln(k_i) = \beta_0(x,y) + \beta_1(x,y)\cdot X_{1i} + \beta_2(x,y)\cdot X_{2i} + \cdots + \beta_{p-1}(x,y)\cdot X_{p-1,i} \tag{5.81}$$

where

μ_i = Mean number of vehicle crashes per unit time at site i, $\mu_i = E[k_i]$
$X_{p-1,i}$ = The $(p-1)$th highway variables that contributes to crashes at site i
$\beta_{p-1}(x, y)$ = Spatially varied regression coefficient of variable $X_{p-1,i}$
i = Number of sites with data observations, $i = 1, 2, \ldots, n$
ln (.) = The natural logarithm

The GWR model coefficient estimates have spatial coordinates and can therefore be mapped to show how the relationship between vehicle crashes and highway variables varies over space. A regression model is fitted at each data point i, weighting all observations by a function of distance from that data point. Therefore, observations sampled near to the observation where the regression is centered have more influence on the resulting regression model coefficients at that data point than observations further away. This then produces a set of model coefficient estimates for a regression model at each point in space.

The GWR technique uses a weighting function that deals with spatial position of a site in terms of its latitude and longitude coordinates to derive model coefficients. A higher weight is assigned to the proximal data points, while the weight diminishes according to the increase of distance between data points. Two types of statistical tests are performed for model selection: (i) analysis of variance (ANOVA) to determine improved predictability of GWR models over conventional global regression models; and (ii) Monte Carlo tests to verify localized nonstationary effect of individual model coefficients in the GWR models. Once the nonstationary effect of any model coefficient is rejected, a mixed GWR model is calibrated.

5.2.3.2 SPFs for highway segments

Highway segments can be generally classified into rural and urban interstate, multilane (divided median and undivided median), and two-lane segments. The predicted annual crash frequency is generally calculated as given below:

$$\hat{k} = \hat{k}_{SPF} \cdot (CMF_1 \cdot CMF_2 \cdots CMF_N) \cdot C \tag{5.82}$$

where

$\hat{k}$ = Vehicle crash frequency predicted for a highway segment, crashes/seg/year
$\hat{k}_{SPF}$ = Vehicle crash frequency predicted for the highway segment by SPF
CMF_i = Crash modification factor i
C = Calibration factor for the highway segment
i = 1, 2, …, N

According to AASHTO, SPFs are used to estimate the expected crash frequency for a highway facility with specified base conditions. Crash modification factors (*CMFs*) are the ratio of the effectiveness of a site-specific condition in comparison with the base condition and are multiplied with the crash frequency predicted by the SPF to account for the difference between site-specific condition and the base condition. The calibration factor is multiplied with the crash frequency predicted by the SPF to account for differences between jurisdictions and time periods for which the SPFs were developed and applied for crash predictions (AASHTO, 2010b).

SPFs for highway segments: In general, an SPF for a highway segment is a function of daily traffic and segment length as follows:

$$\hat{k}_{SPF} = e^{[a + b \cdot \ln(AADT) + \ln(L)]} \tag{5.83}$$

where

$\hat{k}_{SPF}$ = The predicted base case crash frequency for a highway segment, crashes/seg/year
$AADT$ = Annual average daily traffic for the highway segment, veh/day
L = The length of the highway segment, mi
a, b = Model coefficients

Table 5.26 lists values of model coefficients a, b for different classes of highway segments. The base roadway conditions mainly include the following:

- Lane width: 12 ft
- Shoulder width: 6 ft
- Horizontal curvature: none
- Vertical curvature: none

Table 5.26 SPF coefficients for different classes of highway segments

	Urban			*Rural*		
Coefficient value	*Multilane divided*	*Multilane undivided*	*Two-lane*	*Multilane divided*	*Multilane undivided*	*Two-lane*
a	−12.34	−11.63	−15.22	−8.643	−8.902	−9.586
b	1.36	1.33	1.68	0.913	1.002	1.301

- Grade: 0%
- Driveway density: five driveways per mile
- Passing lane for two-lane road segments: none
- Median width for divided segments without barriers: 15 ft for urban and 30 ft for rural
- Side slope for undivided segments: 1:7 or flatter
- Roadside hazard rating for rural segments: 3
- Lighting: none
- Automated speed enforcement for urban segments: none
- On-street parking for urban streets: none
- Roadside fixed objects for urban streets: none

CMFs for segments: A *CMF* is used to modify the crash frequency predicted for a highway segment in base conditions to reflect the relative change in crash frequency due to a change in one specific localized condition when all other conditions and site characteristics remain constant. As such, a *CMF* is the ratio of the crash frequency of a site under the localized and base conditions. Therefore, *CMFs* of base conditions are equal to 1.

$$CMF = N^b / N^a \tag{5.84}$$

where

CMF = The crash modification factor for a highway segment

N^a, N^b = Expected crash frequency with conditions "a" and "b"

Table 5.27 lists the crash modification factors for different functional classes of highway segments.

a. *Lane width*: Table 5.28 presents *CMF* values for different lane widths on highway segments. The base condition for lane width is 12 ft, and the *CMFs* are separately determined for different highway function classes by daily traffic.

Table 5.27 Crash modification factors for roadway segments

	Urban			*Rural*		
Factor	*Multilane divided*	*Multilane undivided*	*Two-lane*	*Multilane divided*	*Multilane undivided*	*Two-lane*
a. Lane width	X	X	X	X	X	X
b. Shoulder width and type	X	X	X	X	X	X
c. Horizontal alignment	X	X	X	X	X	X
d. Grade	X	X	X	X	X	X
e. Driveway density	X	X	X	X	X	X
f. Passing lane			X			X
g. Median width	X			X		
h. Side slope		X			X	
i. Roadside hazard rating				X	X	X
j. Lighting	X	X	X	X	X	X
k. Speed enforcement	X	X	X			
l. On-street parking	X	X	X			
m. Roadside fixed object	X	X	X			

Note: "X" means this factor is considered for adjusting highway segment crash predictions.

Table 5.28 *CMF* values for different lane widths

Segment type	*Lane width*	*AADT (veh/day)* <400	400–2000	>2000
Urban and rural multilane divided	9 ft or less	1.03	1.03 + 0.000138(*AADT*- 400)	1.25
	10 ft	1.01	1.01 + 0.0000875(*AADT*- 400)	1.15
	11 ft	1.01	1.01 + 0.0000125(*AADT*- 400)	1.03
	12 ft or more	1.00	1.00	1.00
Urban and rural multilane undivided	9 ft or less	1.04	1.04 + 0.000213(*AADT*- 400)	1.38
	10 ft	1.02	1.02 + 0.000131(*AADT*- 400)	1.23
	11 ft	1.01	1.01 + 0.0000188(*AADT*- 400)	1.04
	12 ft or more	1.00	1.00	1.00
Urban and rural two-lane	9 ft or less	1.05	1.05 + 0.000281(*AADT*- 400)	1.50
	10 ft	1.02	1.02 + 0.000175(*AADT*- 400)	1.30
	11 ft	1.01	1.05 + 0.000025(*AADT*- 400)	1.05
	12 ft or more	1.00	1.00	1.00

Table 5.29 *CMF* values for different shoulder widths

Shoulder width	*AADT(veh/day)* <400	400–2000	>2000
0 ft	1.10	1.10 + 0.000250(*AADT*- 400)	1.50
2 ft	1.07	1.07 + 0.000143(*AADT*- 400)	1.30
4 ft	1.02	1.02 + 0.00008125(*AADT*- 400)	1.15
6 ft	1.00	1.00	1.00
8 ft or more	0.98	0.98 + 0.00006875(*AADT*- 400)	0.87

Table 5.30 *CMF* values for different shoulder types

Shoulder type	*Shoulder width (ft)* 0	1	2	3	4	6	8	10
Paved	1.00	1.00	1.00	1.00	1.00	1.00	1.00	1.00
Gravel	1.00	1.00	1.01	1.01	1.01	1.02	1.02	1.03
Composite	1.00	1.01	1.02	1.02	1.03	1.04	1.06	1.07
Turf	1.00	1.01	1.03	1.04	1.05	1.08	1.11	1.14

Table 5.31 Proportion of total crashes attributable to the shoulder factor

Value	*Urban/Rural multilane undivided*	*Urban/Rural multilane divided*	*Urban/Rural two-lane*
p	0.27	0.5	0.574

b. *Shoulder width and type*: The *CMF* for shoulders comprises *CMFs* for shoulder width and type synthesized by Equation 5.85. The base condition of shoulder width and type is a 6-ft paved shoulder, which is assigned a *CMF* value of 1.00. Tables 5.29 through 5.31 list *CMF* values for shoulders.

$$CMF_{shoulder} = (CMF_w \cdot CMF_t - 1) \cdot p + 1 \quad (5.85)$$

where

$CMF_{shoulder}$ = The CMF for shoulders of a highway segment
CMF_w = The CMF for paved shoulder width
CMF_t = The CMF for shoulder type
p = Proportion of total crashes attributable to the shoulder factor

c-1. *Horizontal curves: length, radius, and spiral transition*: The base condition for horizontal alignment is a tangent roadway segment. A CMF has been developed to represent the manner in which crash experience on curve alignments differs from that of tangents. The CMF for length, radius, and presence or absence of spiral transitions on horizontal curves is determined as below:

$$CMF_{HC}^{LRS} = \frac{(1.55L_c) + (80.2/R) - (0.012S)}{(1.55L_c)} \tag{5.86}$$

where

CMF_{HC} = CMF for horizontal curve of a highway segment
L_c = Length of horizontal curve, which includes spiral transitions if exist, mi
R = Radius of horizontal curve, ft
S = 1 for presence of spiral transition curves on both sides, 0.5 for presence of spiral transition in one side, and 0, otherwise

c-2. *Horizontal curves: superelevation*: The base condition of superelevation of a horizontal curve is the amount of superelevation recommended by the AASHTO Policy on Geometric Design of Highways and Streets (AASHTO, 2011). The CMF for superelevation is based on the superelevation variance (SV) that is the difference between the actual superelevation and the superelevation recommended by AASHTO policy as follows:

$$CMF_{HC}^{SE} = \begin{cases} 1.00, & \text{if } SV < 0.01 \\ 1.00 + 6(SV - 0.01), & \text{if } 0.01 < SV < 0.02 \\ 1.06 + 3(SV - 0.02), & \text{if } SV \geq 0.02 \end{cases} \tag{5.87}$$

c. *Grades*: The base condition for grade is a level roadway with 0% grade. Table 5.32 shows CMF values for different grades.

d. *Driveway density*: The base condition for driveway density is five driveways per mile. The CMF for driveway density is determined by

$$CMF_{DD} = \frac{0.322 + DD \cdot [0.05 - 0.005 \cdot \ln(AADT)]}{0.322 + 5[0.05 - 0.005 \cdot \ln(AADT)]} \tag{5.88}$$

where

CMF_{DD} = CMF for driveway density of a highway segment
DD = Driveway density considering driveways on both sides of the highway, driveways/mi
$AADT$ = Annual average daily traffic, veh/day

Table 5.32 CMF values for different grades

Grade (0%)	*0*	*2*	*4*	*6*	*8*
CMF	1.00	1.04	1.08	1.12	1.16

Table 5.33 *CMF* values for different median widths

Median width (ft)	*CMF value*	
	Urban	*Rural*
10	1.01	1.04
15	1.00	1.03
20	0.99	1.02
30	0.98	1.00
40	0.97	0.99
50	0.96	0.97
60	0.95	0.96
70	0.94	0.96
80	0.93	0.95
90	0.93	0.94
100	0.92	0.94

e. *Passing lanes*: Passing lanes are installed only on urban and rural two-lane highway segments. The base condition for passing lanes is in the absence of a passing lane. Passing lanes bring safety benefits where *CMF* becomes 0.75 with passing lanes and 1.00 otherwise.

f. *Median width*: Median width factor is considered for divided highway segments without barriers. The base condition is a median width of 30 ft for rural segments and 15 ft for urban segments. Table 5.33 shows *CMF* values for different median widths.

g. *Side slopes*: Side slopes factor is considered for undivided segments only. The base condition is for a side slope of 1:7 or flatter. Table 5.34 lists *CMFs* for different side slopes of undivided highway segments.

h. *Roadside hazard rating*: Roadside hazard rating represents the level of safety concerns of roadside design on a scale of 1–7 only applied for rural segments. The base value of roadside hazard rating for a roadway segment is 3. The *CMF* of roadside hazard rating is calculated as

$$CMF_{RHR} = \frac{e^{(-0.6869+0.0668 \cdot RHR)}}{e^{(-0.4865)}} \tag{5.89}$$

where

CMF_{RHR} = *CMF* for roadside hazard rating of a rural highway segment
RHR = A rating scale of 1–7

i. *Lighting*: The base condition for lighting is the absence of roadway segment lighting. Table 5.35 presents *CMF* values for lighted roadway segments by functional class.

j. *Automated speed enforcement*: Automated speed enforcement systems use video or photographic identification in conjunction with radar or laser devices to detect speeding drivers. These systems automatically record data on vehicle identification without the need for the presence of police. The base condition for automated speed enforcement is

Table 5.34 *CMF* values for different side slopes of undivided highway segments

Side slope	1:2 or steeper	1:4	1:5	1:6	1:7 or flatter
CMF value	1.18	1.12	1.09	1.05	1.00

Table 5.35 CMF values for highway lighting

	Urban			Rural		
Highway class	Multilane divided	Multilane undivided	Two-lane	Multilane divided	Multilane undivided	Two-lane
CMF value	0.91	0.92	0.93	0.91	0.95	0.97

the absence of the system. With the presence of automated speed enforcement, drivers tend to be more careful in driving which could potentially reduce crash occurrences. The value of *CMF* is 0.95 with the presence of automated speed enforcement for all types of urban segments.

k. *On-street parking*: The on-street parking factor is considered for urban segments only. The base condition is the absence of on-street parking. If presence, the *CMF* is determined as

$$CMF_{pk}^{on-street} = 1 - p_{pk} \cdot (f_{pk} - 1) \tag{5.90}$$

with

$$p_{pk} = 0.5 \cdot (L_{pk}/L)$$

where

$CMF_{pk}^{on-street}$ = *CMF* for on-street parking
p_{pk} = Proportion of curb length with on-street parking
L_{pk} = Total curb length with on-street parking for both sides of the highway segment
L = Length of the highway segment
f_{pk} = Factor of on-street parking, as seen in Table 5.36

l. *Roadside fixed objects*: Roadside fixed-objects factor is considered for urban segments only. The base condition is the absence of roadside fixed objects. If presence, the *CMF* is determined by

$$CMF_{FO} = f_{offset} \cdot D_{FO} \cdot p_{FO} + (1 - p_{FO}) \tag{5.91}$$

where

CMF_{F_O} = *CMF* for roadside fixed objects
f_{offset} = Fixed-object offset factor, as seen in Table 5.37
D_{FO} = Fixed-object density for both sides of the highway segment
p_{FO} = Fixed-object crashes as a proportion of total crashes, as listed in Table 5.38

Table 5.36 f_{pk} for determining *CMF* values for on-street parking

	Parallel parking		Angle parking	
Urban segment type	Residential	Commercial or industrial/institutional	Residential	Commercial or industrial/institutional
Multilane divided	1.100	1.709	2.574	3.999
Multilane undivided	1.100	1.709	2.574	3.999
Two-lane	1.465	2.074	3.428	4.853

Table 5.37 Fixed-object offset factor

Offset to fixed-object (ft)	f_{offset}
2	0.232
5	0.133
10	0.087
15	0.068
20	0.057
25	0.049
30	0.044

Table 5.38 Proportion of fixed-object crashes

Urban segment type	p_{FO}
Multilane divided	0.036
Multilane undivided	0.037
Two-lane	0.059

5.2.3.3 SPFs for highway intersections

The prediction of crash frequencies for highway intersections is based on SPFs, *CMFs*, and calibration factor C. The types of intersections considered mainly include rural and urban signalized and stop-controlled 3-leg and 4-leg intersection.

SPFs for intersections: The SPF for a highway intersection is a function of $AADT$ values for major and minor intersections approaches, which is of the following specification:

$$\hat{k}_{SPF}^{int} = e^{\left[a+b\cdot\ln(AADT_{major})+c\cdot\ln(AADT_{minor})\right]} \tag{5.92}$$

where

$\hat{k}_{SPF}^{int}$ = The predicted base case crash frequency for an intersection, crashes/int/year

$AADT_{major}$ = $AADT$ for the intersection major approach, veh/day

$AADT_{major}$ = $AADT$ for the intersection minor approach, veh/day

a, b, c = Model coefficients

Table 5.39 lists values of model coefficients a, b, c for different types of intersections. The base intersection conditions include the following:

- Skew angle for rural stop-sign intersections: zero degree
- Left-turn lanes: none
- Left-turn signal phasing for urban signalized intersections: none

Table 5.39 Model coefficients for different types of intersections

	Urban				Rural		
Coefficient	*3-leg, stop sign*	*3-leg, signalized*	*4-leg, stop sign*	*4-leg, signalized*	*3-leg, stop sign*	*4-leg, stop sign*	*4-leg, signalized*
a.	−13.36	−12.13	−8.9	−10.99	−11.193	−9.284	−6.156
b.	1.11	1.11	0.82	1.07	0.997	0.724	0.661
c.	0.41	0.26	0.25	0.23	0.49	0.529	0.2685

Table 5.40 List of crash contributing factors for intersections

	Urban				*Rural*		
Factor	*3-leg, stop sign*	*3-leg, signalized*	*4-leg, stop sign*	*4-leg, signalized*	*3-leg, stop sign*	*4-leg, stop sign*	*4-leg, signalized*
a. Skew angle					X	X	
b. Left-turn lanes	X	X	X	X	X	X	X
c. Left-turn signal phasing		X		X			
d. Right-turn lanes	X	X	X	X	X	X	X
e. Right-turn on red		X		X			
f. Lighting	X	X	X	X	X	X	X
g. Red light enforcement		X		X			
h. Roadside facilities	X	X	X	X			

Note: "X" means this factor is considered for adjusting intersection crash predictions.

- Right-turn lanes: none
- Right-turn on red for urban signalized intersections: permitting
- Lighting: none
- Red light enforcement for urban signalized intersections: none
- Facilities on roadside for urban intersections: none

CMFs for intersections: The definition and the calculation of *CMFs* for intersections are consistent with those for highway segments. Table 5.40 lists the crash contributing factors for all types of intersections (AASHTO, 2011).

a. *Intersection skew angle*: This factor is considered for rural stop-sign controlled intersections only. The base condition for intersection skew angle is zero degree of skewness where the skew angle is defined as the absolute value of the deviation from an intersection angle of 90°. The *CMFs* for intersection skew angle at rural stop-sign 3-leg and rural stop-sign 4-leg are determined by

$$CMF_{skew}^{3-leg} = e^{(0.004SKEW)} \tag{5.93}$$

$$CMF_{skew}^{4-leg} = e^{(0.0054SKEW)} \tag{5.94}$$

where

$CMF_{skew}^{3-leg}, CMF_{skew}^{4-leg}$ = The *CMF* for a rural 3-leg, or 4-leg stop-controlled intersection

$SKEW$ = The skew angle of the intersection, which is the absolute value of the deviation from an intersection angle of 90°

b. *Intersection left-turn lanes*: The base condition for intersection left-turn lanes is the absence of left-turn lanes on the intersection approach. Table 5.41 shows *CMF* values for the presence of left-turn lanes. These *CMFs* apply to installation of left-turn lanes on any approach to a signalized intersection, but only on uncontrolled major-road approaches to a stop-controlled intersection.
c. *Intersection left-turn signal phasing*: This factor is considered for urban signalized intersections only. The base condition for intersection left-turn signal phasing is the absence of left-turn signal phasing on the signalized intersection approaches. Table 5.42 lists *CMFs* for the presence of left-turn signal phasing. Types of left-turn

Table 5.41 *CMF* values for installation of left-turn lanes

Intersection type		*Traffic control*	*Number of approaches with left-turn lanes*			
			One	*Two*	*Three*	*Four*
Rural	3-leg	Stop sign	0.56	0.31	–	–
	4-leg	Stop sign	0.72	0.52	–	–
		Signalized	0.82	0.67	0.55	0.45
Urban	3-leg	Stop sign	0.67	0.45	–	–
		Signalized	0.93	0.86	0.8	–
	4-leg	Stop sign	0.73	0.53	–	–
		Signalized	0.90	0.81	0.73	0.66

Table 5.42 *CMF* values for different types of left-turn signal phasing

Type of left-turn signal phasing	*Permissive*	*Protected/Permissive*	*Protected*
CMF	1.00	0.99	0.94

Table 5.43 *CMF* values for installation of right-turn lanes

Intersection type		*Traffic Control*	*Number of approaches with right-turn lanes*			
			One	*Two*	*Three*	*Four*
Rural	3-leg	Stop sign	0.86	0.74	–	–
	4-leg	Stop sign	0.86	0.74	–	–
		Signalized	0.96	0.92	0.88	0.85
Urban	3-leg	Stop sign	0.86	0.74	–	–
		Signalized	0.96	0.92	–	–
	4-leg	Stop sign	0.86	0.74	–	–
		Signalized	0.96	0.92	0.88	0.85

signal phasing considered include permissive, protected, protected/permissive, and permissive/protected.

d. *Intersection right-turn lanes*: The base condition for intersection right-turn lane factor is the absence of right-turn lanes on the intersection approaches. Table 5.43 lists *CMF* values for the presence of right-turn lanes. These *CMFs* apply to installation of right-turn lanes on any approach to a signalized intersection, but only on uncontrolled major-road approaches to stop-controlled intersections.

e. *Intersection right turn on red*: This factor is considered for urban signalized intersection only. The base condition for this factor is permitting a right turn on red at all approaches of a signalized intersection. The *CMF* is determined by

$$CMF_{RTOR} = 0.98^{n} \tag{5.95}$$

where

CMF_{RTOR} = The *CMF* for intersection right turn on red

n = Number of signalized intersection approaches for which right turn on red is prohibited

f. *Lighting*: The base condition for lighting is the absence of intersection lighting. Table 5.44 presents *CMF* values for different types of intersections with lighting.

Table 5.44 *CMF* values for intersection with lighting

	Urban				*Rural*		
Value	*3-leg, stop sign*	*3-leg, signalized*	*4-leg, stop sign*	*4-leg, signalized*	*3-leg, stop sign*	*4-leg, stop sign*	*4-leg, signalized*
CMF	0.91	0.91	0.91	0.891	0.90	0.91	0.89

Table 5.45 *CMF* values for roadside facilities

	Number of bus stops			*Presence of school*		*Number of alcohol stores*		
Value	*0*	*1 or 2*	*3 or more*	*No*	*Yes*	*0*	*1–8*	*9 or more*
CMF	1.00	2.78	4.15	1.00	1.35	1.00	1.12	1.56

g. *Red light enforcement*: This factor is considered for urban signalized intersections only. The base condition for red light enforcement is the absence of enforcement. The *CMF* for installation of red light photo enforcement at a signalized intersection is calculated by

$$CMF_{RLPE} = 1 - 0.26 \cdot p_{RA} - 0.82 \cdot p_{RE} \tag{5.96}$$

where
CMF_{RLPE} = *CMF* for red light photo enforcement
p_{RA} = Proportion of multiple vehicle, right-angle crashes
p_{RE} = Proportion of multiple vehicle, rear-end crashes

h. *Roadside facilities*: This factor is considered for urban intersections only. The base condition for the factor of roadside facilities is the absence of bus stops, schools, and alcohol sales stores within 1000 ft of the center of an intersection. Table 5.45 lists *CMFs* for roadside facilities.

5.2.3.4 Vehicle crash prediction

Figure 5.8 depicts the method for vehicle crash prediction that generally consists of 11 steps. Through the detailed steps, the expected crash frequency can be estimated.

Step 1. Identify facility types: Vehicle crash prediction procedure can be focused on a highway network that comprises highway segments and intersections. It can be related to an existing highway segment or a new segment that is a design alternative. Further, crash prediction could be limited to only one specific site, a group of sites or a wide range of highway networks for safety improvement screening.

Step 2. Determine analysis period: In general, the year of interest is determined by the availability of data on vehicle crashes, traffic exposure, and geometric design details.

Step 3. Collect data on traffic volumes: Crash prediction requires data on traffic volumes as inputs. Typically, historical traffic data are recorded that can be used to determine the traffic volume of the analysis year. Table 5.46 presents rules for determining traffic volumes.

Step 4. Collect data on geometric conditions: In order to avoid unnecessary efforts of data collection, it is necessary to understand the base conditions of applying SPFs and the *CMFs* for the selected sites. Emphasis needs to be given to collecting detailed data on geometrics and traffic control features that vary from the base conditions.

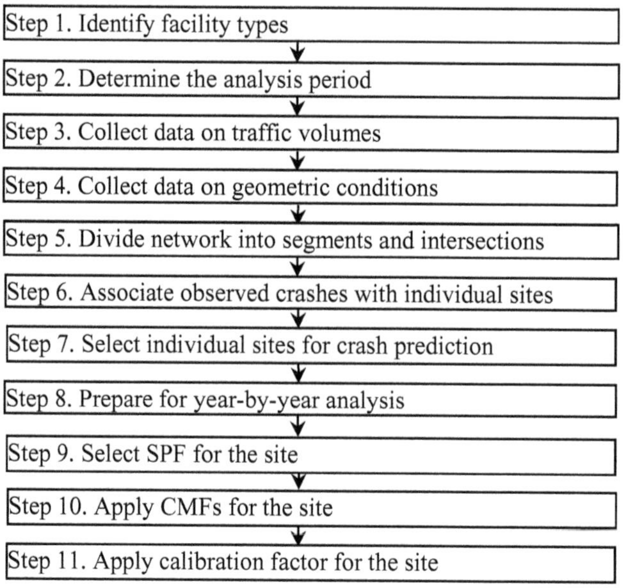
Step 1. Identify facility types
Step 2. Determine the analysis period
Step 3. Collect data on traffic volumes
Step 4. Collect data on geometric conditions
Step 5. Divide network into segments and intersections
Step 6. Associate observed crashes with individual sites
Step 7. Select individual sites for crash prediction
Step 8. Prepare for year-by-year analysis
Step 9. Select SPF for the site
Step 10. Apply CMFs for the site
Step 11. Apply calibration factor for the site

Figure 5.8 Framework of vehicle crash prediction.

Table 5.46 Rules for determining traffic volumes for analysis

Traffic volume availability	*Default rule*
Only a single year	Same value is assumed to apply to all years of the before period
Two or more years	The traffic volumes for intervening years are computed by interpolation
Years before the first year	The available data are assumed to be equal to the volume of the first year
Years after the last year	The available data are assumed to be equal to the last year

Step 5. Divide highway network into roadway segments and intersections: The highway network is divided into individual sites, consisting of homogenous highway segments and intersections by using the information from Steps 1 and 4. The length of a homogeneous highway segment should not be less than 0.1 mile. Also, a buffer zone should be considered to associate crashes with an intersection. For instance, a crash is treated as intersection-related if it occurred within 250 ft from the center of any intersection approach.

Step 6. Associate observed crashes with individual sites: This is an optional step if the predicted crashes are used in conjunction with field observed crash data for empirical Bayesian (EB) before–after comparison group analysis to assess effectiveness of safety improvements.

Step 7. Select individual sites for crash prediction: Once the highway segments and intersections are determined, a single, some or all highway segments and intersection sites of a highway network can be selected for crash predictions.

Step 8. Prepare for year-by-year analysis: Since crash frequencies are affected by changes in time-varying traffic exposure, crash predictions are desirable to be carried out for individual sites for each year of the study period.

Step 9. Select SPF for the site: Based on the site-specific facility type identified, geometric and traffic control conditions, and traffic exposure, an SPF applicable to the selected site can be chosen for crash predictions by crash severity level that represents expected crash frequencies for the base conditions.

Step 10. Apply CMFs for the site: The predicted crash frequencies for the base conditions need to be adjusted to incorporate site-specific conditions using *CMFs*. The adjustments should explicitly address the interdependency issue of multiple crash contributing factors in need for applying *CMFs*.

Step 11. Apply calibration factor for the site: Although applicable to the identified site, the adopted SPFs for crash predictions for the base conditions might have used data that exhibit differences in jurisdictions and time periods. A calibration factor needs to be introduced to handle such differences.

EXAMPLE 5.2

Given data on geometrics, traffic, and observed crashes for the following highway segment, predict the annual crash frequency.

Solution

Steps 1–8. Data collection

Geometrics and traffic data:

- Category of segment: urban four-lane divided arterial
- Length of segment: 0.53 mile
- Number of lanes: 4
- Lane width: 12 ft
- Shoulder width and type: 9 ft, paved
- Alignment: straight
- Speed limit: 50 mph
- Grade: 0%
- Driveway density: five driveways per mile
- Centerline rumble strips: none
- Roadside hazard rating: 5
- Segment lighting: present
- Automated speed enforcement: none
- Local calibration factor (C_r): 2.63 as given
- $AADT = 27700$ veh/day

Crash data: The observed vehicle crashes in 2010–2016 are 275, 136, 112, 116, 124, 157, and 109.

Step 9. SPF selection

To calculate the expected vehicle crash frequency for the segment, the following equation can be used:

$$\hat{k}_{SPF,i} = e^{\alpha_0} \cdot AADT_i^{\alpha_1} \cdot L_i$$

where

$\hat{k}_{SPF,i}$ = The expected number of crashes for highway segment i, crashes/seg/year
$AADT_i$ = $AADT$ for segment i, veh/day
L_i = The length of segment i, mi
α_0, α_1 = Model coefficients, $\alpha_0 = -3.53$ and $\alpha_1 = 0.60$

$$\hat{k}_{SPF,i} = e^{-3.53} \cdot (27700)^{0.60} \cdot (0.53) = 7.19 \text{ crashes}$$

Step 10. CMF applications

- *CMF* for lane width: lane width is 12 ft and *AADT* is over 2000 veh/day, $CMF_{LW} = 1.00$
- *CMF* for shoulder width and type: shoulder width is 9 ft, *AADT* is over 2000 veh/day, paved shoulder for a *CMF* of 1.00, and $P_{ra} = 0.103$ for this segment

$$CMF_{shoulder} = (CMF_{wra} \times CMF_{ra} - 1.0) \times p_{ra} + 1.0$$

where

CMF_{2r} = *CMF* for shoulder width and type on total crashes
CMF_{wra} = *CMF* for paved shoulder width
CMF_{ra} = *CMF* for shoulder type
p_{ra} = Proportion of total crashes constituted by related crashes

$$CMF_{shoulder} = (1.00 \times 1.00 - 1.0) \times 0.103 + 1.0 = 1.103$$

- *CMF* for horizontal curve: segment is straight, $CMF_{HC}^{LRS} = 1.00$ and $CMF_{HC}^{SE} = 1.00$
- *CMF* for grades: target segment is a level roadway, so $CMF_{grade} = 1.00$
- *CMF* for driveway density: there are less than five driveways per mile on the segment, $CMF_{DD} = 1.00$
- There are no passing lanes and centerline rumble strips, $CMF_{passing}$ and CMF_{median} are 1.00
- *CMF* for roadside design: on the segment, fixed objects and lighting are installed very near to the roadway, the roadside hazardous rating is 5

$$CMF_{RHR} = \frac{e^{(-0.6869 + 0.0668 \times RHR)}}{e^{(-0.4865)}}$$

where

CMF_{RHR} = *CMF* for the effect of roadside design
RSH = Roadside hazard rating

$$CMF_{RHR} = \frac{e^{(-0.6869 + 0.0668 \times 5)}}{e^{(-0.4865)}} = 1.143$$

- *CMF* for lighting: for the four-lane divided roadway segment, *pnr*, *pinr*,and p_{pnr} are 0.41, 0.364, and 0.636, respectively,

$$CMF_{lighting} = 1.0 - [p_{nr} \cdot (1.0 - 0.72 \cdot p_{inr} - 0.83 \cdot p_{pnr})]$$

where

$CMF_{lighting}$ = *CMF* for the effect of roadway segment lighting on total crashes
p_{inr} = Proportion of total night-time crashes for unlighted roadway segments that involve a fatality or injury
p_{pnr} = Proportion of total night-time crashes for unlighted roadway segments that involve property damage only
p_{nr} = Proportion of total crashes for unlighted roadway segments that occur at night

$$CMF_{lighting} = 1.0 - [0.41 \times (1.0 - 0.72 \times 0.364 - 0.83 \times 0.636)] = 0.914$$

- *CMF* for speed enforcement: no enforcement system is deployed on target segment, $CMF_{SE} = 1.00$.

Step 11. Calibration factor (C_r) *application*

For a rural two-lane, two-way roadway segment, the calibration factor is 1.00, $N_{predicted}$ is computed by

$$\hat{k}_i = \hat{k}_{SPF,i} \times C_r \times (CMF_{1r} \times CMF_{2r} \times \ldots)$$

$$\hat{k}_i = 7.19 \times 2.63 \times (1.103 \times 1.143 \times 0.914) = 21.79 \text{ crashes/seg/year.}$$

EXAMPLE 5.3

Given data on geometrics, traffic, and observed crashes for the following highway intersection, predict the annual crash frequency.

Steps 1–8. Data collection

Geometric and traffic data:

- Three-legs intersection
- Major road: one-way
- Minor road: stop control
- No right-turn lanes
- No left-turn lanes
- Skew angle: 25°
- Intersection lighting: presence
- Local calibration factor (C_r): 0.87 as given
- ADT of major road: 8536 veh/day
- ADT of minor road: 4509 veh/day

Crash data: the observed vehicle crashes in 2010–2016 are 10, 12, 6, 6, 6, 9, and 8.

Step 9. SPF selection

To calculate the expected vehicle crash frequency for the 3-leg stop-controlled intersection, the following equation can be used:

$$\hat{k}_{SPF,i} = e^{\alpha_0} \cdot (AADT_{Major,i})^{\alpha_1} \cdot (AADT_{Minor,i})^{\alpha_2}$$

where

$\hat{k}_{SPF,i}$ = The expected crash frequency for intersection i
$AADT_{Major,i}$ = $AADT$ of major approach of intersection i
$AADT_{Minor,i}$ = $AADT$ of minor approach of intersection i
$\alpha_0, \alpha_1, \alpha_2$ = Model coefficients, $\alpha_0 = -5.35$, $\alpha_1 = 0.34$, and $\alpha_2 = 0.28$.

$$\hat{k}_{SPF,i} = e^{-5.35} \cdot (8536)^{0.34} \cdot (4509)^{0.28} = 1.09 \text{ crashes.}$$

Step 10. CMF applications

- Intersection skew angle CMF adjustment:

  Intersection skew angle = 25°

$$CMF_{SKEW} = \frac{0.016 \times SKEW}{(0.98 + 0.16 \times SKEW)} + 1.0 = \frac{0.016 \times 25}{(0.98 + 0.16 \times 25)} + 1.0 = 1.0803$$

- Intersection left-turn lane and right-turn lane CMF adjustments:

  $CMF_{LT} = 1.00$ for no left-turn lanes are present
  $CMF_{RT} = 1.00$ for no right-turn lanes are present

- Intersection lighting *CMF* adjustment:

$$CMF_{lighting} = 1 - 0.38 \times p_{ni}$$

where

$CMF_{lighting}$ = *CMF* for the effect of lighting on total crashes
p_{ni} = Proportion of total crashes for unlighted intersections that occur at night

Night time crash proportions for unlighted intersections in 3-leg intersection is 0.276, $CMF_{lighting} = 1 - 0.38 \times 0.276 = 0.895$.

Step 11. Calibration factor (C_r) application

For the 3-leg stop-controlled intersection, $C_r = 0.87$. $\mu_{predicted}$ is calculated by

$$\hat{k}_i = \hat{k}_{SPF,i} \cdot C_r \cdot (CMF_{SKEW} \cdot CMF_{LT} \cdot CMF_{RT} \cdot CMF_{lighting})$$

$$\hat{k}_i = 1.09 \times 0.87 \times (1.08 \times 1.00 \times 1.00 \times 0.895) = 0.917 \text{ crashes/int/year.}$$

5.2.4 Vehicle air emissions

5.2.4.1 Vehicle air pollutant types

Air pollutants are particulates, biological molecules, gas, liquid, or other harmful materials that pose a threat to flora, fauna, climates, and human beings if discharged to the air without strict control. Various activities and factors contribute to pollutant emissions to the atmosphere. With running an internal combustion engine, different gases and particles such as nonmethane hydrocarbon (NMHC), carbon monoxide (CO), carbon dioxide (CO_2), oxides of nitrogen (NO_X), and sulfur dioxide (SO_2) , and particle matters are generated and released into the atmosphere. The pollutants are harmful to the organism of human bodies, resulting in diseases, including cancers. In addition, CO_2 pollutants also contribute to the greenhouse effect and NO_X and SO_2 are primary pollutants in the formation of acid rain. Table 5.47 lists different types and impacts of air pollutants related to vehicle use.

5.2.4.2 Vehicle emission factors

Vehicle emissions of pollutants depend on vehicle characteristics, fuel characteristics, operating conditions, and fleet characteristics, with details listed in Table 5.48.

5.2.4.3 Vehicle air emission models

Vehicle specific power: Vehicle specific power (VSP) that represents the tractive power exerted by a vehicle to move forward is computed by vehicle speed, acceleration rate, and mass.

$$VSP_{V,t} = \frac{A \cdot v_t + B \cdot v_t^2 + C \cdot v_t^3 + m \cdot v_t \cdot a_t}{m} \tag{5.97}$$

where

$VSP_{V,t}$ = Vehicle specific power, kw/ton
m = Mass, kg
v_t = Speed at time t, m/s
a_t = Acceleration at t, m/s^2
A, B, C = Coefficients of rolling resistance, rotational resistance, and aerodynamic drag, kW-s/m

Table 5.47 Vehicle air emitted pollutants and impacts

Air pollutant	*Description*	*Impact*
NMHC	React with nitrogen oxides with sunlight to form ground-level ozone	A primary ingredient in smog causing coughing, choking, and reduced lung capacity
CO	Odorless, colorless, and poisonous gas	Blocks oxygen from the brain, heart, and other vital organs
CO_2	Combustion of fossil fuels in which transportation accounts for around 30% of total CO_2 emissions	Results in climate change
Chlorofluorocarbons (CFCs)	Colorless and nonpoison gases or liquids stable, nonflammable, and nontoxic components	Results in climate change
SO_2	A toxic gas with a pungent, irritating smell	Reacts in the atmosphere to form fine particles and poses the largest health risk to young children and asthmatics
NO_X	A binary compound of oxygen and nitrogen contributes to ozone formation	Causes lung irritation and weakens the body's defenses against respiratory infections, assists in the formation of ground level ozone and particulate matter
Total suspended particles (TSP)	Less than one-tenth the diameter of a human hair	Threat to human health as they can penetrate deep into lungs

Table 5.48 Factors affecting emission of vehicle pollutants

Designation	*Factor*	
Vehicle characteristic	Engine type	Two-stroke, four-stroke; diesel, Otto, Wankel, other engines
	Engine mechanical condition	
	Vehicle appurtenances	Air conditioning, trailer towing, and others
	Maintenance condition	
	Age and mileage	
Fuel characteristic	Fuel properties and quality	Gasoline, diesel
	Alternative fuels	Ethanol, electricity
Operating condition	Altitude, temperature, humidity	
	Traffic congestion level, highway capacity, pavement condition, and traffic control hardware quality	
	Demand management programs	
	Vehicle use patterns	Hot start, cold start, hot stabilized
Fleet characteristic	Vehicle mix	Number and type of vehicles in use
	Vehicle utilization	Annual vehicle mileage by vehicle type
	Adequacy of fleet maintenance	

For different types of vehicles, different sets of coefficients, A, B, and C are utilized. For example, for light duty vehicles, the coefficients are estimated in accordance with "track road load power (TRLP) at 50 mph" as follows:

$$A = PF_A \times \left(\frac{TRLHP \times c_1}{v_{50} \times c_2} \right) \tag{5.98}$$

$$B = PF_B \times \left(\frac{TRLHP \times c_1}{(v_{50} \times c_2)^2} \right) \tag{5.99}$$

$$C = PF_C \times \left(\frac{TRLHP \times c_1}{(v_{50} \times c_2)^3} \right) \tag{5.100}$$

where

PF_A = Default power fraction for coefficient A at 50 mph, $PF_A = 0.35$
PF_B = Default power fraction for coefficient B at 50 mph, $PF_B = 0.10$
PF_C = Default power fraction for coefficient C at 50 mph, $PF_C = 0.55$
c_1 = Constant, converting TRLP from hp to kW, 0.74570 kW/hp
c_2 = Constant, converting mph to m/s, 0.447 m × h/m × s
v_{50} = Constant, vehicle speed at 50 mph

Vehicle emission factors for highway segments: Vehicle emission factors on highway segments are correlated with vehicle volumes, age distribution, composition, speed, speed change cycle, and seasonal effects. The vehicle air emission factor model in the MOVES tool is of the following specification (EPA, 2011):

$$EF_{k,l}^{p} = e^{\beta_0 + \omega_1 \times \gamma_l + \omega_2 \times \gamma_l^2 + \omega_3 \times \gamma_l^3 + \omega_4 \times \gamma_l \times S_l + \beta_1 \times \mu + \beta_2 \times S_l + \delta} \tag{5.101}$$

where

$EF_{k,l}^{p}$ = Emission factor for pollutant p for vehicles of type k using highway segment l, g/mi
γ_l = Vehicle specific power, kw/ton
μ = Vehicle age distribution
$\beta_0, \beta_1, \beta_2$ = Model coefficients
$\omega_1, \omega_2, \omega_3$ = Coefficients of γ_l
S_l = Average vehicle running speed of highway segment l

Vehicle air emissions for highway segments: Air emissions of different types of vehicles are estimated using emission factors multiplying VMT. According to the U.S. Environmental Protection Agency (EPA), an emission factor is defined as a representative value that attempts to relate the quantity of a pollutant released to the atmosphere with an activity associated with the release of that pollutant (EPA, 2011). Vehicle emission quantity of a specific pollutant is estimated by

$$E_l^p = \sum_k E_{k,l}^p \tag{5.102}$$

$$E_{k,l}^p = (EF_{k,l}^p \cdot (1 - E_r/100)) \cdot VMT_{k,l} \tag{5.103}$$

where

E_l^p = Estimated emission quantity of pollutant p for all vehicles using highway segment l, g

$E_{k,l}^p$ = Estimated emission quantity of pollutant p for vehicles of type k using highway segment l, g

$EF_{k,l}^p$ = Emission factor for pollutant p for vehicles of type k using highway segment l, g/mi

E_r = Overall emission reduction efficiency, %

$VMT_{k,l}$ = Vehicle miles of travel by vehicles of type k using highway segment l

Vehicle air emissions at intersections: Additional vehicle air emissions will be produced due to deceleration, acceleration, and idling of vehicles at intersections in response to signalized or unsignalized traffic control. The additional emissions caused by the above vehicle maneuvers are correlated with vehicle delays at intersections. At a given time interval, a portion of the vehicles entering an intersection will experience the process of deceleration, stop, and acceleration to cope with the traffic control. First, vehicles will travel on an intersection approach at free flow speed prior to reaching the stop bar. If a red interval or a stop sign presents, the vehicles will decelerate until they come to full stops. Then, vehicles will be in the idling condition while waiting for the signals to change to green or regaining the right-of-way to traverse through the intersection. Finally, vehicles will accelerate in the beginning of the green interval or right after obtaining the right-of-way until reaching the saturation flow speed. As depicted in Figure 5.4, the total vehicle delays are the summation of deceleration, idling, and acceleration delays with respective durations of $t_{dec} = (t_2 - t_1) - (d_2 - d_1)/V_f$, $t_{idling} = t_3 - t_2$, and $t_{acc} = (t_4 - t_3) - (d_4 - d_3)/V_f$. Therefore, the additional vehicle air emissions due to intersection delays are estimated as

$$E_i^{int,p} = \left(ER_{dec,i}^{int,p} \cdot t_{dec,i} + ER_{idling,i}^{int,p} \cdot t_{idling,i} + ER_{acc,i}^{int,p} \cdot t_{acc,i} - ER_{cruise,i}^{int,p} \cdot t_{cruise,i}\right) \cdot EV_i \qquad (5.104)$$

where

$E_i^{int,p}$ = Estimated additional emission quantity of pollutant p for vehicles experiencing delays at intersection i, g

$ER_{dec,i}^{int,p}, ER_{idling,i}^{int,p}, ER_{acc,i}^{int,p}, ER_{cruise,i}^{int,p}$ = Emission rate of pollutant p for a vehicle decelerating, idling, accelerating, and cruising at intersection i, mg/sec/veh

EV_i = Total of number of entering vehicles at intersection i

Table 5.49 presents emission rates of various air-emitted pollutants by vehicle operating mode.

Table 5.50 shows EPA's vehicle air-emission standards for light duty vehicles and trucks (EPA, 2017).

Table 5.49 Emission rates of various air pollutants by vehicle operating mode (mg/s/vehicle)

	On-board emission measurement			*Correction with distribution of fleet*		
Mode	*NMHC*	*CO*	*NO_X*	*NMHC*	*CO*	*NO_X*
Deceleration	0.4	7.5	0.5	3.5	10.8	3.2
Idling	0.25	1.5	0.1	4.7	4.3	0.7
Acceleration	1.1	22.5	1.5	9.5	32.3	9.5
Cruise	0.6	10	1.25	4.9	24.8	8.1

Table 5.50 U.S. EPA's emission standards for light duty vehicles and trucks (g/mile)

Vehicle type	*Emission category*	*Useful life*	*Test weight (lbs)*	*NMHC*	*Formaldehyde (CH_2O)*	*CO*	*NO_X*	*Particle matters (PM)*
Light duty vehicle	Transitional low	Intermediate	All	0.125	0.015	3.4	0.4	–
	Low			0.075	0.015	3.4	0.2	–
	Ultra-low			0.04	0.008	1.7	0.2	–
	Transitional low	Full		0.156	0.018	4.2	0.6	0.08
	Low			0.09	0.018	4.2	0.3	0.08
	Ultra-low			0.055	0.011	2.1	0.3	0.04
Light duty truck	Transitional low	Intermediate	Loaded vehicle weight (LVW)	0.0125	0.015	3.4	0.4	–
	Low			0.075	0.015	3.4	0.2	–
	Ultra-low		0–3750	0.04	0.008	1.7	0.2	–
	Transitional low		LVW	0.16	0.018	4.4	0.7	–
	Low		3751–5750	0.1	0.018	4.4	0.4	–
	Ultra-low			0.05	0.009	2.2	0.4	–
	Transitional low	Full	LVW	0.15	0.018	4.2	0.6	0.08
	Low		0–3750	0.09	0.0188	4.2	0.3	0.08
	Ultra-low			0.055	0.011	2.1	0.3	0.04
	Transitional low		LVW	0.2	0.023	5.5	0.9	0.08
	Low		3751–5750	0.13	0.023	5.5	0.5	0.08
	Ultra-low			0.07	0.013	2.8	0.5	0.04
Heavy light duty truck	Low	Intermediate	LVW 0–3750	0.125	0.015	3.4	0.4	–
	Ultra-low		Adjusted LVW	0.075	0.008	1.7	0.2	–
	Low		LVW 3751–5750	0.16	0.018	4.4	0.7	–
	Ultra-low		Adjusted LVW	0.1	0.009	2.2	0.4	–
	Low		LVW >5750	0.195	0.022	5	1.1	–
	Ultra-low		Adjusted LVW	0.117	0.011	2.5	0.6	–
	Low	Full	LVW 0–3750	0.018	0.022	5	0.6	0.08
	Ultra-low		Adjusted LVW	0.107	0.012	2.5	0.3	0.04
	Low		LVW 3751–5750	0.23	0.027	6.4	1	0.1
	Ultra-low		Adjusted LVW	0.143	0.013	3.2	0.5	0.05
	Low		LVW >5750	0.28	0.032	7.3	1.5	0.12
	Ultra-low		Adjusted LVW	0.157	0.016	3.7	0.8	0.06

5.2.5 Vehicle noise emissions

According to World Health Organization (WHO), noise has become the second greatest threat to the health of human beings after only air pollution (Berglund et al., 1999). Excessive noise may lead to hearing damage, heart disease, learning impairment in children, and sleep disturbance. Unlike other pollutants, noise pollution is spontaneous in that the noise generated at a particular time is neither affected by previous activity nor does it affect future activities. It is, therefore, important to understand the nature of noise impacts and effectively assess the level of expected noise due to transportation system operations (Cohen and McVoy, 1982).

5.2.5.1 Noise level measurements

Noise level: The noise level also refers to the sound pressure level (SPL), which is measured by decibels (dB). Figure 5.9 illustrates the threshold of human hearing. The sound ranges to be heard at different frequencies is quite different. For example, 60 dB to be heard requires at least about 30 Hz. The standard threshold of hearing at 1000 Hz is usually set to be 0 dB, but is actually measured to be about 4 dB.

Unit of noise (decibels): The basic unit of noise is decibels. It is a logarithmic unit used to express the ratio of two values.

$$SPL = 10\log(P/P_0)dB \tag{5.105}$$

where

SPL = Sound pressure level, decibels
P_0 = Reference pressure, 2×10^{-5} newton/m^2
P = The concerning sound pressure

This equation transfers the sound pressure from newton/m^2 to dB which is expressed as SPL. P_0 is equal to 2×10^{-5} newton/m^2, which is the minimum pressure that is audible by humans. Note that a 10 dB increase means double the noise for human hearing.

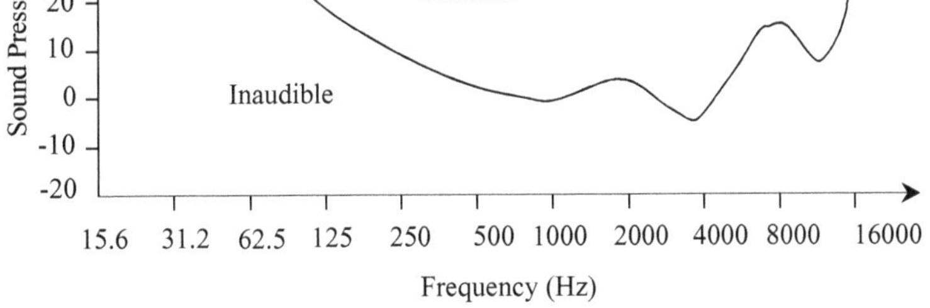

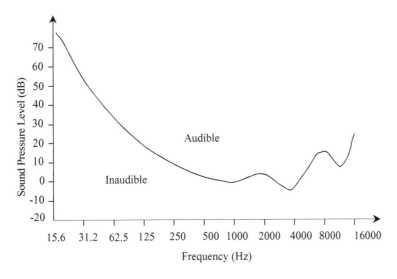

Figure 5.9 Threshold of human hearing.

Table 5.51 Examples of common sound levels

dB(A)	*Outdoor*	*Indoor*
120		
110	Jet flyover 1000 ft Horn noise-train at 100 ft	
100	Gas lawnmower at 3 ft	Inside subway train
90	Diesel truck at 50 ft General freight train at 100 ft Noisy urban daytime	Food blender at 3 ft
80	Gas lawnmower at 100 ft	Garbage disposal at 3 ft Very loud speed at 3 ft
70	Commercial area	Vacuum cleaner at 10 ft Normal speed at 3 ft
60	Heavy traffic at 300 ft	Large business office Quite speed at 3 ft Dishwasher in the next room
50	Quiet urban night time	Small theater Large conference room
40		Library
30	Quiet suburban night time	Bedroom at night Concert hall
20	Quiet rural night time	Broadcast and recording studio

Weighted sound level: Since human ears are not sensitive to all audio frequencies, a weighted sound level is introduced as a sound level measurement to consider ear sensitivity perceived by human ears. The loudness of a sound that a human receives not only depends on the SPL, but also the frequency. For example, human beings cannot differentiate a sound source with 67 dB and 100 Hz versus another of 60 dB and 1000 Hz because they have exactly the same feeling to us. A weighted sound level is expressed in a decibels adjusted dB(A) value. Table 5.51 shows exemplary weighted sound levels for outdoor and indoor environments.

5.2.5.2 *Traffic noise sources*

Usually, the level of traffic noise increases along with increases in traffic flow and speed. Different types of pavement materials will affect the noise level. Concrete pavements are more likely to produce a higher noise level than that of asphalt pavements. Deteriorated pavement surface conditions tend to exacerbate the noise level. Several factors that influence the traffic noise level are listed as follows.

Vehicle types: Noise could be made from engines, horns, brakes, or other mechanical parts of vehicles. Vehicles with larger horsepower engines are inclined to make more noise. The relative contributions of noise coming from the engines and the exhaust systems vary by vehicle type.

Tire–road interactions: Tire–road friction is the fundamental force that drives a vehicle. Conversely, it is also the major source of noise. The noise due to the contact between the tires and pavement surface becomes dominant especially at high speeds.

Pavement surface condition: Intuitively, smoother pavement surface produces less noise than rougher pavement surface. For a rougher road surface, noise may not only be from friction, but also from the impact of tires and pavement distresses such as cracking, rutting, or shoving.

5.2.5.3 *Traffic noise models*

Equivalent noise level for one type of vehicle in the traffic stream: One of the commonly used models for traffic noise analysis is the FHWA's Traffic Noise Model (TNM), which is capable of predicting sound levels in the vicinity of highways and serves as the basis to design highway noise barriers (FHWA, 1998a). Like other models, the TNM estimates the noise level by conducting a series of adjustments to a reference sound level called vehicle noise emission level as the maximum sound level radiated by a vehicle at a reference distance of 50 ft (15 m). Major adjustments to the vehicle noise emission level include traffic flow, distance, finite length of the highway segment, and shielding as given below:

$$L_{eq,k,t} = L_{0,k} + L_{flow,k,t} + L_d + L_{fl} + L_s \tag{5.106}$$

where

$L_{eq,k,t}$ = Hourly equivalent sound level of vehicles of type k in hour t, dB(A)
$L_{0,k,t}L_{0,k,t}$ = Reference energy mean emission level of vehicles of type k in hour t, dB(A)
$L_{flow,k,t}$ = The adjustment for traffic flow, vehicle volume, and speed for vehicle type k, dB(A)
L_d = The adjustment for distance between the highway and receiver, dB(A)
L_{fl} = The adjustment for finite length of the highway segment, dB(A)
L_s = The adjustment for shielding and ground effects between the highway and receiver, dB(A)

a. *Reference energy mean emission (REME) level*: In the U.S., the REME level was determined by experiments conducted in 1994 and 1995. The noise levels of approximately 6000 vehicles of different types such as automobiles, light, medium and heavy trucks, buses, and motorcycles running on different types of highway segments were tested. Data details of vehicle status, decelerating, idling, accelerating, and cruising features on grades were recorded to derive the reference noise levels for different facility, vehicle, and system usage conditions. The REME model is given as follows:

$$L_{0,A,t} = 38.1 \times \log_{10}(S_A) - 2.4 \tag{5.107}$$

$$L_{0,MT,t} = 33.9 \times \log_{10}(S_{MT}) + 16.4 \tag{5.108}$$

$$L_{0,HT,t} = 38.1 \times \log_{10}(S_{HT}) + 38.5 \tag{5.109}$$

where

$L_{0,A,t}, L_{0,MT,t}, L_{0,HT,t}$ = Reference energy mean emission level vehicles for automobiles, medium trucks, and heavy trucks in hour t
$S_{A,t}, S_{MT,t}, S_{HT,t}$ = Vehicle running speeds for automobiles, medium trucks, and heavy trucks in hour t

b. *Traffic flow adjustment factor*: Traffic flow adjustment factor, $L_{flow,k,t}$, is a function of traffic flow and vehicle speed computed for each type of vehicle as follows:

$$L_{flow,k,t} = 10 \times \log_{10}\left(\frac{q_{k,t} \cdot D_0}{S_{k,t}}\right) - 25 \tag{5.110}$$

where

$q_{k,t}$ = Traffic flow of vehicle type k in hour t, veh/hr

D_0 = Reference distance at which the emission levels are measured, $D_0 = 15$ m as the default value

$S_{k,t}$ = Vehicle speed of vehicle type k in hour t, km/hr

c. *Distance adjustment factor*: The adjustment factor L_d is a function of distance from the highway segment to the receiver and for the length of the segment, which is computed by

$$L_d = 10 \times \log_{10}\left[\left(\frac{D_0}{D}\right)^{1+\alpha}\right] \tag{5.111}$$

where

D_0 = Reference distance at which the emission levels are measured, $D_0 = 15$ m as the default value

D = Perpendicular distance from the receiver point to the highway segment

α = Site condition parameter, $\alpha = 0$ for hard surface and $\alpha = 0.5$ for soft surface

d. *Finite highway length adjustment factor*: The adjustment factor L_{fl} is a function of distance from the highway segment to the receiver and for the length of the segment, which is computed by

$$L_{fl} = 10 \times \log_{10}\left[\left(\frac{\psi(\Delta\varphi)}{180^\circ}\right)\right] \tag{5.112}$$

where

$\psi(\Delta\varphi)$ = Function of adjustment for finite length highway segment

$\Delta\varphi$ = Subtended angle from the receiver point connecting both ends of the highway segment, degrees

e. *Shielding and ground effects*: Unlike the distance adjustment that considers effects along the plane of the ground surface, shielding and ground effects take the third dimension into account. Specifically, shielding effects consider the obstructions located between the noise source and receiver. These obstructions could be trees and buildings. Ground effects are those of the surface of the ground as a reflector/absorber in the path of the sound waves. Therefore, ground effects vary from one terrain to another. The two factors are interconnected and need to be jointly considered in noise-level adjustment.

Equivalent noise level for all vehicles in the traffic stream: Once the equivalent noise level for a given type of vehicle in the traffic stream in hour t is established, the noise levels for all types of vehicles in the traffic stream in hour t could be calculated:

$$L^d_{eq,t} = 10 \times \log_{10}\left[10^{\left(L^d_{eq,A,t}/10\right)} + 10^{\left(L^d_{eq,MT,t}/10\right)} + 10^{\left(L^d_{eq,HT,t}/10\right)}\right] \tag{5.113}$$

$$L_{eq,t} = 10 \times \log_{10}\left[10^{\left(L^{d=1}_{eq,t}/10\right)} + 10^{\left(L^{d=2}_{eq,t}/10\right)}\right] \tag{5.114}$$

where

$L_{eq,t}$ = Hourly equivalent sound level of all vehicles in hour t, dB(A)
$L_{eq,t}^{d}$ = Hourly equivalent sound level of directional vehicles in hour t, dB(A)
$L_{eq,A,t}^{d}$ = Hourly equivalent sound level of directional automobiles in hour t, dB(A)
$L_{eq,MT,t}^{d}$ = Hourly equivalent sound level of directional medium trucks in hour t, dB(A)
$L_{eq,HT,t}^{d}$ = Hourly equivalent sound level of directional heavy trucks in hour t, dB(A)

EXAMPLE 5.4

An engineer is tasked to assess the total equivalent sound level of the East-West U.S. route 64 connecting his work place in Chicago, IL, to his home in the west suburb of Chicagoland. The East bound hourly volume for automobile, medium truck, and heavy truck for this route is 500, 60, and 40 vehicles, and the hourly volume for the West bound is 400, 20, and 30 vehicles, respectively. The average speed for automobile is 85 km/h, for medium and heavy trucks is 60 km/h. The lane width of the road is 3.6 m, distance to East bound centerline is 50 m, and the sight angle to both angles is 90°. The surface of the route is made by reflective, with no obstruction materials. Help the engineer estimate the total equivalent sound level of the route using the FHWA traffic noise models.

Solution

From the given information, the specifications of the route can be summarized in Table 5.52 as follows.

The total equivalent sound level can be calculated top-down as presented in Table 5.53.

5.2.5.4 *Traffic noise costs*

The FHWA's TNM could help determine the hourly equivalent sound level $L_{eq,t}$ caused by vehicles using a highway segment in hour t of a day. Next, the social costs of noise can be estimated by calculating the depreciation in the value of residential units alongside highways. The closer a house is to a highway, the higher the social costs are. With the noise depreciation sensitivity index (NDSI) introduced by Nelson (1982), the percentage reduction in the house value caused by the net increase in the equivalent noise level from the maximum acceptable noise level at 50 dB(A) can be computed. Specifically, the house value depreciation function in hour t is defined as follows:

$$VNC_t = N_{HH} \cdot \left(\frac{i \cdot W_{avg}}{365 \times 24} \right) \cdot D \cdot (L_{eq,t} - L_{max}) \tag{5.115}$$

Table 5.52 Route specifications for noise level calculation

Item	*Unit*	*Automobile*	*Medium truck*	*Heavy truck*
East bound (EB) volume ($N_{i,E}$)	Veh/h	500	60	40
West bound (WB) volume ($N_{i,W}$)	Veh/h	400	20	30
Speed (S_i)	Km/h	85	60	60
Lane width (L)	m		3.6	
Distance to EB centerline (D_E)	m		50	
Sight angle to both angels ($\Delta\varnothing$)	Degree		90	
Surface type			Reflective, no obstruction	

Table 5.53 Total equivalent sounds level calculation steps

Item		*Automobile*	*Medium truck*	*Heavy truck*
REME level	$L(o)_{E,i,E}$ (*EB*), $L(o)_{E,i,W}$ (*WB*)	$38.1 \times \log_{10}(S_i) - 2.4$ $= 38.1 \times \log_{10}(85) - 2.4 = 71.1$	$33.9 \times \log_{10}(S_i) + 16.4$ $= 33.9 \times \log_{10}10(60) + 16.4 = 76.7$	$24.6 \times \log_{10}(S_i) + 38.5$ $= 24.6 \times \log_{10}10(60) + 38.5$ $= 82.2$
Traffic flow adjustment	(EB) (WB)	$10 \times \log_{10}(N_i \times D_0 / S_i) - 25 =$ $10 \times \log_{10}(400 \times 15/85) - 25 = -6.5$ $10 \times \log_{10}(N_i \times D_0 / S_i) - 25 =$ $10 \times \log_{10}(350 \times 15/85) - 25 = -7.1$	$10 \times \log_{10}(30 \times 15/60) - 25$ $= -16.3$ $10 \times \log_{10}(20 \times 15/60) - 25$ $= -18.0$	$10 \times \log_{10}(40 \times 15/60) - 25$ $= -15.0$ $10 \times \log_{10}(30 \times 15/60) - 25$ $= -16.3$
Distance adjustment	(EB) (WB)	$10 \times \log_{10}(D_0 / D_E) =$ $10 \times \log_{10}(15/50) = -5.2$ $10 \times \log_{10}(D_0 / D_W) =$ $10 \times \log_{10}(15/(50+3.6)) = -5.5$	$10 \times \log_{10}(15/50) = -5.2$ $10 \times \log_{10}(15/(50+3.6)) = -5.5$	$10 \times \log_{10}(15/50) = -5.2$ $10 \times \log_{10}(15/(50+3.6)) = -5.5$
Adjustment for finite length	(EB), (WB)	−3 (Monograph)	−3	−3
Shielding adjustment	(EB), (WB)	0 (No object located between observer and road)	0	0
Hourly ESL For each vehicle type	$L_{eq}(h)_{i,E}$ (EB) $L_{eq}(h)_{i,W}$ (WB)	$71.1 - 6.5 - 5.2 - 3 - 0 = 56.4$ $71.1 - 7.1 - 5.5 - 3 - 0 = 55.5$	$76.7 - 16.3 - 5.2 - 3 - 0 = 52.2$ $76.7 - 18.0 - 5.5 - 3 - 0 = 50.2$	$82.2 - 15.0 - 5.2 - 3 - 0 = 59.0$ $82.2 - 16.3 - 5.5 - 3 - 0 = 57.4$
Hourly ESL	$L_{eq}(h)_E$ (EB)	$L_{eq}(h)_E = 10 \times \log_{10}[10L_{eq}(h)_{A,E}/10 + 10L_{eq}(h)_{MT,E}/10 + 10L_{eq}(h)_{HT,E}/10]$ $= 10 \times \log_{10}[1056.4/10 + 1052.2/10 + 1059.0/10] = 61.5$		
For each vehicle type	$L_{eq}(h)_W$ (WB)	$L_{eq}(h)_W = 10 \times \log_{10}[10L_{eq}(h)_{A,W}/10 + 10L_{eq}(h)_{MT,W}/10 + 10L_{eq}(h)_{HT,W}/10$ $= 10 \times \log_{10}[1055.5/10 + 1050.2/10 + 1057.4/10 = 60.04$		
Total ESL	$L_{eq}(h)$	$L_{eq}(h) = 10 \times \log_{10}[10L_{eq}(h)_E/10 + 10L_{eq}(h)_W/10]$ $= 10 \times \log_{10}[1061.5/10 + 1060.04/10] = 63.84$		

with

$$N_{HH} = [(RD) \cdot D \cdot (2 \times L)]/5280$$

where

VNC_t = Vehicle noise cost in hour t, dollars/hr
$L_{eq,t}$ = Equivalent sound level, dB(A)
D = Distance to the highway, ft
N_h = Number of houses affected per square mile
RD = Average residential density around a highway, houses/mi²
L = Length of the relevant highway segment, mi
L_{max} = Maximum acceptable noise level, $L_{max} = 50$ dB(A)
d = Percentage discount in value per dB(A) increase in the ambient noise level, $d = 0.4\%$
W_{avg} = Average house value, dollars/house
i = Discount rate

5.3 TRANSPORTATION INDIRECT IMPACTS MODELING

5.3.1 Land use impacts

Transportation and land use are mutually affected. Transportation networks designed by planners aim to shape the land and improve the accessibility and mobility. However, developed transportation networks will attract more people and stimulate more travel to the accessible locations. This phenomenon is an example of the interaction between supply and demand.

Land use refers to a given area of land, which is developed to satisfy desired purposes. Transportation investments will significantly affect land use directly and/or indirectly. For direct impacts, new transportation investments will determine the land use, such as a highway, parking lot or other transportation facilities. For indirect impacts, the land value in the area may be changed due to changes in accessibility and mobility. Both direct and indirect impacts will further affect transportation behavior and social economy. Figure 5.10 illustrates the connection between land use and transportation.

5.3.1.1 Land use impacts on transportation

Transportation projects typically increase the accessibility and mobility of land area, and further alter the spatial distribution of service facilities in the area. Accessibility is a measurement assessing the number of travel opportunities of a location to be reached or reach other places within a particular travel radius, usually measured in distance or travel time. Mobility is the ability and level of easiness to move people, goods, and services. Improved accessibility and mobility generally lead to a growth in travel demand. As a result, increasing travel demand requires expansion or optimization of the transportation network.

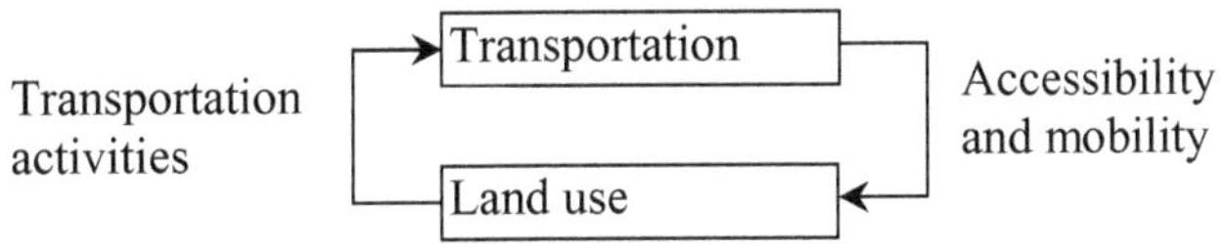

Figure 5.10 Relationship of transportation and land use.

Table 5.54 Elasticity of travel with land use changes

Influencing factor	*Vehicle trips*	*Vehicle miles of travel*
Local density	−0.05	−0.05
Local diversity	−0.03	−0.05
Local design	−0.05	−0.03
Regional accessibility	–	−0.20

As mentioned earlier, reshaping of land use will affect the travel demand. The classic four-step travel demand forecasting method provides an acceptable estimation based on three general trip purposes: home-based work, home-based others, and non-home-based trips. The change in land use will alter the number of residents, customers, or employees; then the trip generation is directly affected. The change in number of residents will change the average population density. The travel pattern is more likely to be influenced by travel cost, rather than population density. Ewing and Cervero (2001) proposed elasticity of travel caused by land-use changes in local density, diversity, design, and regional accessibility, as shown in Table 5.54. Local density refers to number of people per unit of area. Local diversity is a job-population balance measurement. Local design is the combination of sidewalk completeness, route directness, and street network density. Regional accessibility is an accessibility index derived using a gravity model.

5.3.1.2 Transportation impacts on land use

Transportation interventions will directly or indirectly affect land use. Table 5.55 summarizes some impacts of transportation activities and policy on land use.

5.3.1.3 Monetization of land use impacts

Bein (1997) proposed a generic cost structure in 1997 to estimate the cost of converting land from current use to new use. A positive value means benefits and a negative value means costs. For example, for converting an area of wetland to highway pavements, the environmental external benefits are −100000/hectare/year. Inversely, convert highway pavement to an area of wetland, the environmental external benefits are 100000/hectare/year. Indirect impacts, such as air and noise pollution within 500 m of a highway is considered to impose one-half of the costs.

Table 5.55 Exemplary land use impacts by transportation activities

Transportation activity	*Land use impacts*
New facilities	Redistribute the growth of people and economics to new facilities area Increase the land value around new facilities Decentralize the population
Added intersections	Same as above but with lesser degree
Vehicle/fuel costs	Encourage to change transportation mode to transit
Parking pricing	Encourage to change transportation mode to transit Possible shift of population and employment to exurban areas

5.3.2 Economic development impacts

Transportation activities may have various types of economic impacts on a specific area, such as employment, business activities, property values, and government revenues at local, regional or national level. Transportation investments will improve the accessibility and mobility of the transportation network, making it much easier and faster to transport cargo from one place to another. Commodity circulation speeds up or productivity increases. On the other hand, accessibility and mobility improvements could reduce travel time of a trip. With more trips attracted, traffic mobility may deteriorate. Both positive and negative impacts should be considered in economic impact analysis.

5.3.2.1 Economic impacts dimensions

From accessibility and mobility viewpoints, transportation investments can lead to accelerated circulation of commodities and make economic and social activities more frequent. The economic impacts caused by transportation investments can be categorized as follows (Weisbrod and Reno, 2009):

a. Impacts on transportation supply
   - Fare and wage income from transportation operations
   - Access to wider distribution markets and niches
b. Impacts on transportation demand
   - Improved accessibility
   - Travel time and cost savings
   - Productivity gains
   - Division of labor
   - Access to wider range of consumers
   - Economies of scale
c. Impacts on microeconomy
   - Rent income
   - Lower price of commodities
   - Higher supply of commodities
d. Impacts on macroeconomy
   - Formation of distribution networks
   - Attraction and accumulation of economic activities
   - Increased competitiveness
   - Consumption growth
   - Mobility needs fulfillment

5.3.2.2 Economic impact types

Direct impacts: Direct impacts are directly caused by transportation investments. Typically, direct impacts will alter the traffic mode and traffic choice by improving the mobility and accessibility of a road network. Meanwhile, more traffic travels are produced and traffic costs are reduced.

Indirect impacts: Different from direct impacts, indirect impacts refer to the impacts that are not caused by transportation projects directly but provide support to direct impacts. For example, new rail transit projects may increase property values and spark new construction along the line—a direct impact. For the new buildings, the demand for facilities, furniture or maintenance services grows as well—an indirect impact.

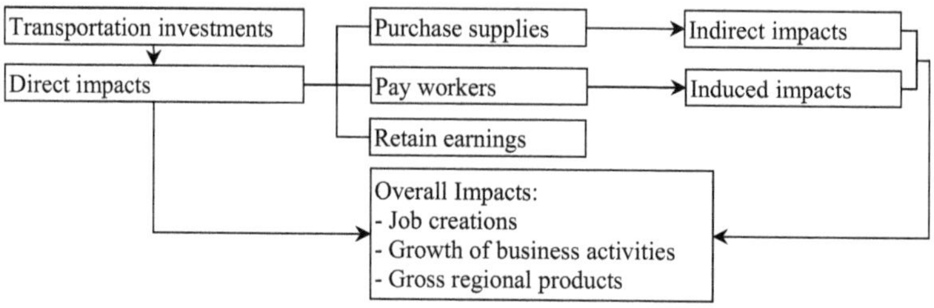

Figure 5.11 Economic impacts of transportation investments.

Induced impacts: The primary reason for induced impacts is increasing household income in the area impacted by the transportation project—in other words, consumption stimulation. More travel into the area means more expenditure or other economic activities. Even in the future, the structure of industries and residents will change. The outcomes of induced impacts are typically price increases for necessaries, goods, and services (Figure 5.11).

The FHWA periodically estimates impacts of highway capital expenditures on creating direct, indirect, and induced job opportunities (NEC, 2008; FHWA, 2009b, 2017). Direct jobs are known as construction-oriented employment, including all jobs that work directly on construction, repair, and maintenance of transportation facilities. Indirect jobs are referred to as supporting industry employment, which are jobs that provide support to transportation facility construction, repair, and maintenance work activities. Induced jobs are the induced employment that includes all jobs caused by consumer expenditures from salaries and wages of employees earned from transportation work. It should be cautioned that investments in transportation need to be carefully thought through, as they may support growth and help aggregate people and capital around it or trigger dispersion. Also, investments in transit should help achieve urban planning goals—quality of life, walkability, local shopping and jobs, complete neighborhoods—only then does it really generate economic growth.

PROBLEMS

5.1 Describe total cost, average cost, marginal cost, and economies of scale.

5.2 For a typical transportation project, what are the agency cost items? What are the user cost items?

5.3 List possible costs that may be involved in the transportation construction process.

5.4 The city engineers have proposed to construct a new highway through existing farmland to bypass the city for use by through traffic. List all major positive and negative impacts that this project could have on the downtown and bypass area.

5.5 Suppose a transportation facility has an initial value of 100000 and the estimated salvage value is 10000 after a 10-year service life, what is the corresponding depreciation values by using straight-line, sum-of-year-digit, and double decline methods?

5.6 What are the essential elements that need to be considered when determining the value of travel time?

5.7 What are the possible effects of vehicle air emissions when a traffic management measure is applied, such as decreasing the speed limit or adding stop signs or signals? How could you measure the costs and benefits of the traffic management actions?

5.8 How will economic factors affect transportation project delivery and how will transportation investments stimulate economic development?

5.9 Explain the internal relationship and impacts of land use and transportation project investments.

5.10 A reconstruction project of a 10-mile 4-lane concrete highway is proposed for next year. The initial cost per lane-mile is estimated to be \$250000 with zero salvage value at the end of its service life of 30 years. The annual maintenance cost is expected to be \$2500 per lane-mile. For a discount rate of 5%, compute the equivalent uniform annual cost.

5.11 At a specific instant in time, 2 vans, 2 trucks, and 1 car pass near a residential area. These noise sources generate noise levels of 45, 55, 80, 100, and 40 dB, respectively, at the area. Determine the combined loudness that is experienced by the residential area when they all pass at the same time.
[Hint: Combined noise level (SPL_{total}), in decibels, from n noise sources is given by:

$$SPL_{total} = 10\log_{10}\sum_{i=1}^{n} 10^{(SPL_i)/10}$$

where SPL_i is the sound pressure level (decibels) experienced by receptor due to each individual source $k(k = i,n)$.]

Chapter 6

Transportation needs assessment

6.1 GENERAL

A transportation needs assessment aims to identify performance levels of physical transportation facilities and system usage that do not meet the minimum acceptable standards. In the temporal dimension, the needs for performance improvements could focus on addressing those performance issues that are currently in existence or are to be expected in the future. In the spatial dimension, the needs for performance improvements could address performance issues at a specific location or many locations of a transportation network concerning a single transportation mode or multiple modes. The needs assessment for performance improvements can be carried out in accordance with static, time-dependent or dynamic traffic with added complexity.

6.2 PHYSICAL TRANSPORTATION FACILITY NEEDS ASSESSMENT

For a multimodal transportation system, the primary types of physical facilities include pavements, bridges, traffic control and safety hardware, and multimodal facilities related to transit components, rail tracks, bike ways, and pedestrian walkways. The performance of physical facilities can be analyzed according to facility condition deterioration over time, damage by cumulative traffic passage, or remaining service life. Once the remaining service life of a facility or facility element has been reached, it needs to be replaced. With useful service life of a facility or facility element still remaining, a treatment is required to slow down the condition deterioration rate to achieve the lowest possible life-cycle agency and user costs; without such treatment the condition is expected to reach the threshold level sooner than desired, leading to higher agency and user costs in the long run. Therefore, the essence of physical facility needs assessment is establishing threshold condition levels for different types of physical facilities.

6.2.1 Threshold condition levels for pavement treatments

The performance measures for pavement condition assessment mainly include subjective measures of PSR and PCR and objective measurements of roughness in terms of IRI, along with rut depth, cracking, and patching. The subjective ratings can be correlated with load- and non-load-related factors that collectively lead to pavement condition deterioration in order to develop pavement condition deterioration models. These models can then be utilized to estimate objective condition assessments in terms of PSI and PCI, respectively. With PSI, PCI, and IRI measures in place, they could be utilized individually or collectively to establish threshold condition levels for pavement treatments. In practice, pavement treatments are not

Table 6.1 Threshold condition levels for pavement treatments

Threshold condition	*Pavement performance measure*		
	PSR	*PCR*	*IRI*
Good	3–4	75–85	61–95
Fair	2–3	55–75	96–120

needed if the pavement condition is at "very good" level, and it will not be cost-efficient if the treatment is deferred too long until the pavement condition has deteriorated to "poor" or "very poor" condition level. Therefore, the minimum acceptable pavement condition can be set at "fair" condition level where pavement treatments are required. Also, some preventive maintenance treatments may be implemented even when the pavement condition is at "good" condition level to potentially reduce the frequency, timing, and costs of repair treatments in pavement service life cycle. This may result in lower life-cycle agency costs. Since higher classes of highways such as interstates and major arterial roads typically carry more traffic and truck traffic, the overall pavement condition needs to be maintained at a better condition level to ensure riding quality. In this respect, the minimum acceptable pavement condition for higher classes of highways may be set at both "good" and "fair" condition levels.

Table 6.1 shows threshold levels of "good" and "fair" conditions for pavement treatments. The threshold condition levels for PSI, PCR, and IRI measures can be used separately or collectively. If they are used collectively, a pavement treatment is required if the current or expected condition level of any one of the three measures reaches the threshold level. The separate or collective use of the condition threshold levels can be focused on a single pavement segment or on all pavement segments in a highway network to help identify current or future pavement treatment needs.

6.2.2 Threshold condition levels for bridge treatments

A bridge contains deck, superstructure, and substructure components. Each component consists of many CoRe structural elements. In the U.S., bridge inspection is largely carried out at two-year intervals where a condition rating is assigned to each bridge element during each inspection. The condition ratings focus on two aspects: structural deficiency and functional obsolescence on a 0–9 rating scale. Functional obsolescence implies that the approach alignment, deck geometry, and capacity of loading areas of a bridge fail to meet the current standards. These are reflected by low load-carrying capacity, insufficient clearances for horizontal and vertical (over and under the bridge) alignments, poor roadway alignment, narrow deck width, and low waterway adequacy. Structure deficiency refers to the poor condition of bridge load carrying elements due to damage or deterioration.

According to the NBI bridge element inspection procedure, condition ratings of 2, 3, and 4 for bridge elements represent "critical," "serious," and "poor" condition levels. In practice, functional obsolescence ratings target deck geometry, horizontal and vertical clearance, bridge approach alignment, structural evaluation, and waterway adequacy. Structural deficiency ratings concentrate on deck, superstructure, and substructure, structural evaluation, and waterway adequacy. Table 6.2 lists proposed minimum acceptable condition ratings for bridge functional obsolescence and structural deficiency treatments. Once the current or expected condition rating of any of the performance measures falls below the threshold condition level, a bridge component/element treatment is required. Since a bridge component often contains multiple bridge elements, the repair treatment of a bridge component may contain multiple treatments of constituent bridge elements.

Table 6.2 Threshold condition levels for bridge component/element treatments

Threshold condition	*Bridge component/element performance measure*				
Functional obsolescence rating	Deck geometry	Insufficient clearance	Approach alignment	Structural evaluation	Waterway adequacy
	≥3	≥3	≥3	≥3	≥3
Structural deficiency rating	Deck	Superstructure	Substructure	Structural evaluation	Waterway adequacy
	≥4	≥4	≥4	≥3	≥3

Apart from establishing threshold condition levels for different bridge elements, threshold condition levels could also be created based on the overall BSR that simultaneously addresses functional obsolescence, structural adequacy, efficiency for public use, and special conditions such as detour length, safety features, and structural types. For BSR on a 0–100 scale, a minimum acceptable BSR can be determined. A bridge can be identified as a candidate for treatment if the current or expected BSR deteriorates below the threshold level. In addition, the use of threshold condition levels for bridge elements or BSR dealing with an overall bridge can be focused on individual elements of a single bridge, one type of bridge elements or components of all bridges in a highway network or all types of bridge elements or components of all bridges in a highway network.

6.2.3 Threshold condition levels for traffic control and safety hardware treatments

Traffic control and safety hardware, including traffic signs, lighting, signals, pavement markings, and guardrails and crash cushions, coupled with geometric design, consistency of geometric design, and traffic exposure, affect highway safety performance. Compared with pavements and bridges, all types of traffic control and safety hardware maintain shorter useful service lives, as seen in Table 6.3 (Lindly and Wijesundera, 2003; Markow, 2007; Long et al., 2011).

For traffic signs and pavement markings, retroreflectivity is the most important performance measure used for condition assessment. For the traffic lighting system, the light output is the primary performance measure. For concrete barriers, the shrinkage strain is the key performance measure for condition assessment. The retroreflectivity, light output, and shrinkage strain are correlated with the age. Therefore, the threshold condition levels for traffic control and safety hardware can be determined based on the minimum acceptable measurements of the above performance measures. Tables 6.4 and 6.5 illustrate the minimum acceptable retroreflectivity levels for traffic signs and pavement markings for replacement for sign sheeting components (FHWA, 2007) and pavement marking strips (Debaillon et al., 2007).

For traffic lighting, the need for replacing light bulbs, poles, and supporting structures can be made according to their respective service time spans after installation. For concrete barriers, replacement can follow the time intervals of pavement or bridge reconstruction.

Table 6.3 Typical useful service lives of traffic control and safety hardware

Traffic control and safety hardware	*Signs*	*Lighting*	*Signals*	*Pavement markings*	*Concrete barriers*
Typical service life (years)	9–20	20	5–9	3–10	25–40

Table 6.4 FHWA minimum retroreflectivity standards for sign sheeting (cd/lx/m²)

		Sheeting type (ASTM D4956-04)[a]			
		Beaded sheeting			Prismatic sheeting
Sign color	*Additional criteria*	*I*	*II*	*III*	*III–X*
White on green	Overhead	W*, G ≥ 7	W*, G ≥ 15	W*, G ≥ 25	W ≥ 250, G ≥ 25
	Ground-mounted	W*, G ≥ 7	W ≥ 120, G ≥ 15		
Black on yellow[b]		Y*, O*	Y ≥ 50, O ≥ 50		
Black on orange[c]		Y*, O*	Y ≥ 75, O ≥ 75		
White on red[d]		W ≥ 35, R ≥ 7			
Black on white		W ≥ 50			

Note:

[a] The retroreflectivity levels are measured at an observation angle of 0.2° and an entrance angle of −4.0°
[b] For text and fine symbol signs measuring at least 1.2 m and for all sizes of bold symbol signs
[c] For text and fine symbol signs measuring less than 1.2 m
[d] Minimum sign retroreflectivity contrast ratio ≥3:1
* This sheeting type should not be used for this color for this application

Table 6.5 Minimum retroreflectivity standards for pavement markings (mcd/sq.m/lux)

	Without raised reflective pavement markings (RRPMs)			
Pavement marking configuration	*≤50 mph*	*55–65 mph*	*≥70 mph*	*With RRPMs*
Fully-marked highways	40	60	90	40
Centerlines only	90	250	575	50

For the cases of barriers being damaged by vehicle crashes, they could be handled on a case-by-case basis.

6.2.4 Threshold condition levels for multimodal facility treatments

Threshold condition levels for transit facility treatments: Physical transit facilities cover components of substructure, shell, interiors, conveyance, plumbing, HVAC, fire protection, electrical, equipment, fare collection, and site. The service lives of individual components vary significantly. The replacement of elements under each component can be handled according to the service lives stipulated by technical specifications or whenever repairs are needed on a case-by-case basis.

Threshold condition levels for rail track treatments: The performance measures for rail track condition assessment generally include SD, R^2, J index, TGI, W_5, and TQI. The above measures could be utilized individually for track condition assessment. Practically, the SD, R^2, W_5, and TGI measures can be used collectively as the track condition rating system to inspect the track condition. If the track condition is at "very good" level, it is not safe and economically feasible if the treatment is deferred too long until the track condition deteriorates to "poor" condition level. Therefore, the minimum acceptable track condition can be set at "fair" condition level where treatments are required. Also, some preventive maintenance treatments may be applied even when the track condition is at "good" condition level to potentially reduce the frequency, timing, and costs of repair treatments in track service life. This may help reduce the life-cycle track construction, repair, and maintenance costs. Table 6.6 shows threshold levels

Table 6.6 Threshold condition levels for rail track treatments

	Rail track condition measure			
Threshold condition	*SD*	R^2	*TGI*	W_5
Good	$1 < SD \leq 2$	$1 < R^2 \leq 4$	$50 \leq TGI < 85$	$0.1 < W_5 \leq 0.2$
Fair	$2 < SD \leq 4$	$4 < R^2 \leq 16$	$36 \leq TGI < 50$	$0.2 < W_5 \leq 0.6$

of "good" and "fair" conditions for rail track treatments. When threshold condition levels for SD, R^2, W_5, and TGI measures are used collectively, a track treatment is required if the current or expected condition level of any one of the four measures reaches the threshold level. The use of the condition threshold levels can be focused on a single-track segment or all track segments in a railway network to help identify current or future track treatment needs.

Threshold condition levels for inland waterway lock facility treatments: The performance of lock facilities greatly affects performance of the overall inland waterway transportation. Typically, the replacement or reconstruction of lock facilities is made once the design useful service lives are reached.

Threshold condition levels for bikeway treatments: Bikeways may be installed adjacent to the outermost vehicular travel lanes (or even converted from them), on shoulders or shared with regular travelways. The treatments of bikeways can be handled at the same time when conducting travel lane and/or shoulder repair or maintenance treatments. Also, bike traffic control signs, signals, and pavement markings can be treated as an integrated part of the treatments of traffic control and safety hardware for the general traffic.

Threshold condition levels for pedestrian sidewalk treatments: The condition of pedestrian sidewalks can be assessed by the SCI and SSCR. The SCI is an index on a 0–100 scale with points deduced proportional to the area of sidewalk cracked against the total area. Once the SCI drops below 49 out of 100, the sidewalk needs to be reconstructed and for the SCI value between 50 and 79, repair treatments are required. Therefore, the threshold SCI level is 79.

Another sidewalk condition assessment measure is the SSCR, which generally utilizes a rating scale similar to the deck condition ratings on a 0–9 scale. An SSCR of 3 or lower indicates that the sidewalk is in "poor" or "failed" condition, which can be adopted as the minimum acceptable condition level. However, for sidewalks accommodating extensive pedestrian traffic, a higher threshold condition level of 4 or 5 representing "fair" condition is preferred.

6.3 TRANSPORTATION USAGE NEEDS ASSESSMENT

For the use of transportation system by various types of vehicles, the direct impacts are concerned with vehicle operations, mobility, safety, and environmental impacts. These impacts will carry on during transportation facility service life cycles, leading to life-cycle user costs of vehicle operations, travel time, crashes, and air emissions, respectively. At any given point in time, user impacts mitigation measures need to be implemented if such impacts exceed the predefined threshold levels. Hence, the focus of transportation usage needs assessment is to determine the threshold system usage performance levels corresponding to vehicle operations, mobility, safety, and environmental impacts.

6.3.1 Threshold performance levels for vehicle operating costs

Normalized vehicle operating cost estimates: VOCs contain cost components of fuel consumption, tire wear, maintenance and repair, and mileage-dependent depreciation. The costs of fuel consumption are affected by vehicle type, age, fuel price, vehicle speed, highway geometric design, and traffic flow condition. Given the geometric design standard, traffic operations may deal with uninterrupted or interrupted flow. If the highway facility accommodates uninterrupted flow, vehicles would be running on consecutive highway segments. Prior to the traffic volume reaching capacity of the highway segment, free-flow speed is maintained. Once the traffic volume exceeds capacity, traffic congestion will present, leading to rapid vehicle speed reductions and delays accompanied by additional fuel consumption. If the highway facility accommodates interrupted flow, vehicles would be running on highway segments and utilizing intersections. The flow condition associated with a highway segment could be uncongested or congested like that of a highway segment handling uninterrupted flow. For vehicles traversing through an intersection, vehicles in perpendicular directions will take turns. This will naturally cause vehicle delays and idling with additional fuel consumption. Consequently, the added fuel consumption caused by reductions in vehicle running speeds and increases in vehicle speed change cycles will prompt a higher total amount of VOCs. To facilitate performance-based analysis, the costs can be normalized to VOCs per VMT for the highway segment and per entering vehicle (EV) for the intersection.

VOC reduction needs assessment: When dealing with a highway network, the normalized VOCs can be separately calculated for different classes of highway segments and different types of intersections. A threshold value can be predefined for each class of highway segment or each type of intersection. Taking freeway segments for instance, the highest fuel efficiency is generally achieved at 55 mph. The corresponding value of normalized VOCs could be calculated. Further, a certain extent of deviation from the computed value could be defined. The range of normalized VOCs could be used as the threshold performance levels of vehicle operations. If the current or expected normalized VOCs associated with a highway segment or an intersection fall outside of the threshold levels, measures need to be recommended to improve the current or future traffic operational condition by reducing the normalized VOCs.

6.3.2 Threshold performance levels for mobility

People/freight versus traffic mobility: Mobility performance may be assessed by people/freight mobility or vehicular traffic mobility that aims to maximize throughput of people/freight movements or throughput of vehicles, which could be for the current or future time horizon. For assessment of mobility performance in the current time, field data on people/freight movements and vehicle volumes by transportation mode and facility type within each mode need to be collected.

Current or future people/freight or traffic flow details: For assessment of mobility performance at a certain point or for a certain period in the future, details are needed for predicted person trips, freight shipments, and traffic volumes by transportation mode and facility type within each mode. This is typically accomplished by developing and executing a multimodal travel demand forecasting model. The issue of multimodal integration needs to be addressed. For passenger travel, the integration of multiple transportation modes contains two layers of efforts. The first layer of integration is between auto and transit modes and the second layer of integration is among transit sub-modes, including bus and BRT, streetcar and light rail, heavy rail and commuter rail, and taxi and rideshare, as well as bike and pedestrian walking modes. For freight transportation, multimodal integration is concerned with intermodal integration between trucking and freight rail modes. Multimodal integration could help utilize the available capacity of the multimodal transportation system to accommodate the demand for passenger travel and freight shipments in an efficient manner. As a result, each transportation mode will receive a certain share of passenger travel or freight shipments. The O–D person trips or tonnage of freight shipments will be further assigned to the multimodal transportation network with travel paths determined. Ultimately, the auto-based O–D person trips will be converted to vehicle volumes on individual highway segments or intersection approaches of the highway network. Transit O–D trips will be converted to person and transit vehicle volumes along individual bus, rail, and rideshare routes of the transit network. Bike and pedestrian walking O–D trips will be converted to number of bikes riding on individual bikeways and number of people walking along individual pedestrian walkways, respectively. For freight transportation, the O–D truck shipments will be converted to truck volumes on individual truck route segments of the highway network. The O–D freight rail shipments will be converted to number of freight trains running on different track segments of the freight rail network.

For either passenger travel or freight shipments in the current or future period, the traffic details by transportation mode and facility type within each mode can be utilized for assessing mobility improvement needs. For the current period assessment, fined-grained data on person trips, freight shipments, and vehicle volumes by transportation mode and facility type within each mode are readily available from field measurements. For future period assessment, the high-fidelity travel demand forecasting model could derive related data with accuracy and precision. For instance, auto traffic details on individual highway segments or intersection approaches can be obtained or predicted by travel lane on a second-by-second basis, making it possible to assess highway mobility performance incorporating traffic dynamics in time and space domains (Roshandeh et al., 2013).

Vehicular traffic mobility needs assessment: Without loss of generality, vehicular traffic mobility is hereby utilized to illustrate the basic concept of mobility needs assessment. With vehicular traffic details on a highway segment or an intersection approach in terms of vehicle volume, vehicle composition, speed, and occupancy in a short time interval (say 15-second time interval) made available, different mobility performance measures could be employed for mobility assessment (Table 6.7).

(a) *Levels of service analysis*: A highway network can be partitioned into highway segments and intersections. The mobility performance assessment through levels of service (LOS) analysis can be carried out for highway segments and intersections, respectively. For segment-related mobility assessment, highway segments could be classified by highway function class, such as rural/urban freeways, multilane highways with divided or undivided medians, and 2-lane roads with class I and class II designations. For a freeway and multilane highway segment, LOS is determined by the density of the traffic stream. For a two-lane road segment with class I designation, LOS for class I 2-lane roads is calculated according to percent time spent following an average travel speed. For the 2-lane road segment with class II designation, the LOS is computed in accordance with percent time spent following. For intersection-related mobility assessment, intersections are split into signalized and unsignalized intersections. The control delay per vehicle is estimated to correspond to different LOS levels.

When LOS is employed for establishing threshold performance levels for traffic mobility, a LOS D is typically considered to be the minimum acceptable mobility performance level. If any highway segment or intersection in the highway network experiences traffic mobility worse than LOS D in the AM and/or PM peak periods of a typical weekday, it is a candidate for mobility improvement treatment. The analysis can be expended to identify mobility improvement needs for the entire highway network.

(b) *Travel time index and travel time buffer index*: The LOS analysis of traffic mobility assessment is effective in identifying highway segments and intersections with mobility concerns. However, it lacks strength in correlating the segment- or intersection-specific

Table 6.7 Vehicular traffic mobility assessment using LOS analysis

Network component		*Performance measure*		*LOS*	*Description*
Highway segment	Freeway	Density, *D* (pcpmpl)	$D \leq 11$	A	Free flow
			$D > 11–18$	B	Reasonably free flow, effects of minor incidents still easily absorbed
			$D > 18–26$	C	At or near free-flow speed, queues may form behind any significant blockage
			$D > 26–35$	D	Approaching unstable flow
			$D > 35–45$	E	Unstable flow, operating at capacity
			$D > 45$	F	Forced or breakdown flow, queues form behind breakdown points
	Multilane	Density, *D* (pcpmpl)	$D \leq 11$	A	Free flow
			$D > 11–18$	B	Reasonably free flow, effects of minor incidents still easily absorbed
			$D > 18–26$	C	At or near free-flow speed, queues may form behind any significant blockage
			$D > 26–35$	D	Approaching unstable flow
			$D > 35–40–45$	E	Unstable flow, operating at capacity
	Two-lane, Class I	Percent-time-spent-following, *P* (%) Average travel speed, *V* (mph)	$P \leq 35, V \geq 55$	A	Low following, high speed
			$P > 35–50, V > 50–55$	B	Moderate following, relatively high speed
			$P > 50–65, V > 45–50$	C	Relatively high following, moderate speed
			$P > 65–80, V > 40–45$	D	High following, moderate to low speed
			$P > 80, V \leq 40$	E	Extremely high following, low speed
	Two-lane, Class II	Percent-time-spent-following, *P* (%)	$P \leq 40$	A	Low following
			$P > 40–55$	B	Moderate following
			$P > 55–70$	C	Relatively high following
			$P > 70–85$	D	High following
			$P > 85$	E	Extremely high following
Intersection	Signalized	Control delay, *d* (sec/veh)	$d \leq 10$	A	Free flow
			$d > 10–20$	B	Stable flow
			$d > 20–35$	C	Stable flow with acceptable delay
			$d > 35–55$	D	Approaching unstable flow with tolerable delay
			$d > 55–80$	E	Unstable flow, with intolerable delay
			$d > 80$	F	Forced or breakdown flow
	Unsignalized	Control delay, *d* (sec/veh)	$d \leq 10$	A	Free flow
			$d > 10–15$	B	Stable flow
			$d > 15–25$	C	Stable flow with acceptable delay
			$d > 25–35$	D	Approaching unstable flow with tolerable delay
			$d > 35–50$	E	Unstable flow, with intolerable delay
			$d > 50$	F	Forced or breakdown flow

mobility issues with the overall O–D path travel quality for individual travelers. In addition, the static or time-dependent traffic details used for the LOS analysis cannot examine traffic mobility impacts caused by schedule-varying traffic dynamics, such as providing tradeoff opportunities for travelers to leave earlier or later than the regular departure time that will lead to earlier or delayed arrival compared with the target arrival time.

Conversely, the travel time index could provide useful information on the travel quality associated with an O–D travel path comprising different classes of highway segments or an entire highway network. Further, travel time buffer index can be computed to incorporate reliability considerations for a route, corridor or an O–D travel path. The two indices are generally calculated as follows:

$$TTI_r = \frac{\sum_{l=1}^{L}\left[\left(V_{l,r}/V_{l,r}^{ff}\right)\cdot VMT_{l,r}^{peak}\right]}{\sum_{l=1}^{L}\left(VMT_{l,r}^{peak}\right)} \tag{6.1}$$

$$TTBI_r = \left[\frac{\left(TT_r^{95} - \overline{TT}_r\right)}{\overline{TT}_r}\right] \times 100\% \tag{6.2}$$

where

$$TT_r^{95} = \sum_{l=1}^{L}\left(TT_{l,r}^{95}\right)$$

where

TTI_r = Travel time index of route, corridor or O–D path r
$TTBI_r$ = Travel time buffer index of route, corridor or path r
$V_{l,r}^{ff}$ = Free-flow volume of link l of route, corridor or path r
$V_{l,r}$ = Traffic volume of link l of route, corridor path r
$VMT_{l,r}^{peak}$ = Peak period VMT of link l of route, corridor or path r
$TT_{l,r}^{95}$ = Travel time of link l of route, corridor or path r corresponding to the 95th percentile travel time of the path
TT_r^{95} = 95th percentile travel time of route, corridor path r
$\overline{TT}_r$ = Average travel time of route, corridor or travel path r experienced by N travelers
l = 1, 2, … , L as the total number of links on route, corridor or travel path r
n = 1, 2, … , N as the total number of travelers utilized route, corridor or travel path r

With travel time index and travel time buffer index in place, an upper bound value can be predefined for each index. The two indices can be used independently or jointly to identify routes, corridors or O–D paths with peak period traffic mobility issues in need of improvement as long as one or both index values exceeds the predefined upper bound values. If only considering one highway segment at a time, the above indices can also be employed to identify individual highway segments with traffic mobility concerns.

Transit quality of service needs assessment: Transit quality of service is viewed from availability and quality perspectives. The availability is concerned with spatial and temporal availability of transit service. With transit service being available, the quality of service can be assessed in terms of comfort and convenience experienced by transit riders. For both the availability and quality aspects, the assessment can be carried out separately by transit element, including transit stops/stations, route segments, and transit system with multiple routes providing service to a specified area.

(a) *LOS analysis of transit service availability*: Table 6.8 presents LOS-based performance analysis of transit service availability according to service frequency for stops/stations, service hours for route segments, and service coverage for the transit system.

(b) *LOS analysis of transit service quality*: Table 6.9 presents LOS-based performance analysis of available transit service quality from passenger loads for stops/stations, reliability for route segments, and travel time difference between transit and auto travel for the transit system, respectively.

The transit quality of service needs assessment first needs to focus on the availability of transit service. If transit service is available, the service quality needs to be examined. The assessment may focus on transit stops/stations, routes or the system. Transit improvements are required if the service level of any performance measure listed in Tables 6.8 and 6.9 falls below LOS C.

Shared use bicycle path mobility needs assessment: The LOS analysis can also be utilized for shared use bicycle path mobility assessment (Patten et al., 2006). The LOS ratings from A to F are determined based on bicycle-related LOS scores computed in the following

$$LOS_{score}^{Bike} = 5.446 - 0.00809E - 15.86/W - 0.287CL - 0.5DP \tag{6.3}$$

where

LOS_{score}^{Bike} = Bicycle traffic LOS score
E = Weighed events per minute, E = Meetings/min + 10 (active passing/min)
W = Bikeway width, ft
CL = 1, if the bikeway has centerline 0, otherwise
DP = min (delayed passing/min, 1.5)

Table 6.10 lists LOS scores corresponding to different LOS ratings. Practically, LOS C is used as the threshold level for shared use bicycle path mobility improvements.

Pedestrian walking mobility needs assessment: The LOS approach can also be utilized for mobility assessment of pedestrian walking on a link, crossing an intersection, and traveling on one or more arterial segments that cover multiple links and intersections (TRB, 2010). The LOS ratings from A to F are determined based on pedestrian-related LOS scores computed as below

$$LOS_{link}^{ped} = 6.0468 + F_{w,link} + F_{v,link} + F_{s,link} \tag{6.4}$$

$$LOS_{int}^{ped} = 0.5997 + F_{w,int} + F_{v,int} + F_{s,int} + F_{delay} \tag{6.5}$$

$$LOS_{seg\,l}^{ped} = F_{cd} \cdot \left(0.318LOS_{link}^{ped} + 0.220LOS_{int}^{ped} + 1.606\right) \tag{6.6}$$

$$LOS_{A}^{ped} = \left[\sum_{l=1}^{L}\left(LOS_{seg\,l}^{ped} \cdot L_{seg\,l}\right)\right] \Bigg/ \left(\sum_{l=1}^{L} L_{seg\,l}\right) \tag{6.7}$$

Table 6.8 Performance analysis of transit service availability

	Performance measure				
Element	*Frequency*	*Headway (min)*	*Veh/h*	*LOS*	*Description*
Stop	Urban scheduled transit	<10	>10	A	Highly frequent service, riders do not need schedules
		10–14	5–6	B	Frequent service, passengers consult schedules
		15–20	3–4	C	Max desirable time to wait if a bus/train is missed
		21–30	2	D	Service unattractive to selective riders
		31–60	1	E	Service available during hour
		>10	<1	F	Service unattractive to all riders
	Frequency	Access time (h)			
	Paratransit	0.0–0.5			Fairly prompt response
		0.6–1.0		B	Acceptable response
		1.1–2.0		C	Tolerable response
		2.1–4.0		D	Poor response, may need pre-trip planning
		4.1–24.0		E	Need pre-trip planning
		>24.0		F	Irregular service
	Frequency	Trips/day			
	Intercity scheduled transit	>15		A	Frequent daily service
		12–15		B	Mid-day and frequent peak hour service
		8–11		C	Mid-day or frequent peak hour service
		4–7		D	Minimum service to provide choice of travel times
		2–3		E	Possible daily round-trip
		0–1		F	Impossible daily round-trip
Route segment	Service hours	Hours/day			
		19–24		A	Night service provided
		17–18		B	Late evening service provided
		14–16		C	Early evening service provided
		12–13		D	Daily service provided
		4–11		E	Peak hour/limited mid-day service
		0–3		F	Very limited or no service
System	Service coverage	Transit supportive area of coverage (%)			
		90–100		A	
		80–89.9		B	
		70–79.9		C	
		60–69.9		D	
		50–59.9		E	
		<50		F	

Table 6.9 Performance analysis of transit service quality

Element	*Performance measure*			*LOS*	*Description*
Stop	Passenger loads	m²/p	p/seat		
	Bus	>1.20	0.00–0.50	A	No passengers need to sit next to each other
		0.80–1.19	0.51–0.75	B	Passengers can choose where to sit
		0.60–0.79	0.76–1.00	C	All passengers can sit
		0.50–0.59	1.01–1.25	D	Comfortable standee load for design
		0.40–0.49	1.26–1.50	E	Maximum schedule load
		<0.40	>1.50	F	Crush loads
	Rail	>1.85	0.00–0.50	A	No passengers need to sit next to each other
		1.30–1.85		B	Passengers can choose where to sit
		0.95–1.29	0.76–1.00	C	All passengers can sit
		0.50–0.94	1.01–2.00	D	Comfortable standee load for design
		0.30–0.49	2.01–3.00	E	Maximum schedule load
		<0.30	>3.00	F	Crush loads
Route segment	Reliability	On-time operation (%), varying by 5–10 min for fixed route, 20 min for paratransit			
		97.5–100			One late transit vehicle per month
		95.0–97.4		B	Two late transit vehicles per month
		90.0–94.9		C	One late transit vehicle per week
		85.0–89.9		D	
		80.0–84.9		E	One late transit vehicle per direction per week
		<80		F	
System	Transit/auto time	Travel time difference (min)			
		≤0		A	Faster by transit than by auto
		1–15		B	About as fast by transit as by auto
		16–30		C	Tolerable for selective transit riders
		31–45		D	Round-trip at least an hour longer by transit
		46–60		E	Tedious for all transit riders
		>60		F	Unacceptable to most riders

Table 6.10 Shared use bicycle path mobility assessment using los analysis

Network component	*Performance measure*		*LOS*	*Description*
Bikeway	LOS^{Bike}_{score}	>4	A	Optimondition, bikeway can accommodate more riders
		>3.5–4	B	Facility condition is good, bikeway can absorb more riders
		=3–3.5	C	Fair condition with marginal ability to absorb more riders
		=2.5–3	D	Reduction in bicycle travel speed
		=2–2.5	E	Significant reduction in travel speed with crowded demand
		≤2	F	Significant conflicts

with

$$F_{w,link} = -1.2276\ln(W_v + 0.5W_1 + 50p_{pk} + W_{buf} \cdot f_b + W_{aA} \cdot f_{sw})$$

$$F_{w,int} = \left[0.681\left(N_d^{0.514}\right)\right]$$

$$F_{v,link} = 0.0091[v_m/(4N_{th})]$$

$$F_{v,int} = 0.00569[(v_{rtor} + v_{lt,perm})/4] - N_{rtci,d} \cdot \left[0.0027\left((0.25/N_d) \cdot \left(\sum_{i \in m_d} v_i\right)\right) - 0.1946\right]$$

$$F_{s,link} = [4((S_r/100)^2)]$$

$$F_{s,int} = 0.00013\left[\left((0.25/N_d) \cdot \left(\sum_{i \in m_d} v_i\right)\right) \cdot S_{85,mj}\right]$$

$$F_{delay} = 0.0401[ln(((C - g_{walk,mi})^2)/2C)]$$

where

LOS_{link}^{ped} = LOS for pedestrians walking on a link
LOS_{int}^{ped} = LOS for pedestrians crossing an intersection
$LOS_{seg\,l}^{ped}$ = LOS for pedestrians walking on segment l with a link and an intersection
LOS_{A}^{ped} = LOS for pedestrians walking on an arterial containing multiple segments
$F_{w,link}$ = Cross-section factor
$F_{w,int}$ = Width of the crossing
$F_{v,link}$ = Motorized vehicle volume factor
$F_{v,int}$ = Volume of motor vehicles crossing the crosswalk
$F_{s,link}$ = Motorized vehicle speed factor
$F_{v,int}$ = Speed of motor vehicles on the street being crossed
F_{delay} = Pedestrian delay, sec
F_{cd} = Roadway crossing difficulty factor
W_v = Effective width of the outside through lane, bike lane, and shoulder, ft
W_1 = Effective width of the combined bicycle lane and shoulder, ft
p_{pk} = Proportion of on-street parking occupied
W_{buf} = Buffer width between roadway and sidewalk, t
f_b = Presence of a continuous barrier at least 3 ft high
W_{aA} = Adjusted available sidewalk width, ft
f_{sw} = Sidewalk coefficient
N_d = Number of traffic lanes crossed
v_m = Midblock demand flow rate, veh/hr
N_{th} = Number of through lanes on the street in the direction of travel being considered
v_{rtor} = Volume of right turns on red across subject crosswalk, veh/hr

Table 6.11 Pedestrian walking mobility needs assessment arterials using LOS analysis

Network component	*Performance measure*		*LOS*
Pedestrian walking on an arterial link	LOS_{link}^{Ped} score, Average pedestrian space, S^{ped} (ft²/p)	$LOS_{link}^{Ped} \le 2.0$, $S^{ped} > 60$	A
		$LOS_{link}^{Ped} = 2.0-2.75$, $S^{ped} = 40-60$	B
		$LOS_{link}^{Ped} = 2.75-3.5$, $S^{ped} = 24-40$	C
		$LOS_{link}^{Ped} = 3.5-4.25$, $S^{ped} = 15-24$	D
		$LOS_{link}^{Ped} = 4.25-5.0$, $S^{ped} = 8-15$	E
		$LOS_{link}^{Ped} > 5.0$, $S^{ped} \le 8$	F
Pedestrian walking on an arterial segment with a link and an intersection	LOS_{seg}^{Ped} score, S^{ped} (ft²/p)	$LOS_{seg}^{Ped} \le 2.0$, $S^{ped} > 60$	A
		$LOS_{seg}^{Ped} = 2.0-2.75$, $S^{ped} = 40-60$	B
		$LOS_{seg}^{Ped} = 2.75-3.5$, $S^{ped} = 24-40$	C
		$LOS_{seg}^{Ped} = 3.5-4.25$, $S^{ped} = 15-24$	D
		$LOS_{seg}^{Ped} = 4.25-5.0$, $S^{ped} = 8-15$	E
		$LOS_{seg}^{Ped} > 5.0$, $S^{ped} \le 8$	F

$v_{lt,perm}$ = Volume of left turns across the subject crosswalk that are concurrent with its pedestrian phase, veh/hr
$N_{rtci,d}$ = Number of right-turn channelization islands
v_i = Volumes that cross the crosswalk from movement, $i = 1, 2, \ldots, m_d$
S_r = Vehicle running speed, mph
$S_{85,mj}$ = 85th percentile speed on the major street, mph
C = Cycle length, sec
$g_{walk,mi}$ = Effective walk time when walking on the minor street, sec
d_{pd} = Pedestrian diversion delay, sec
d_{pw} = Pedestrian crossing delay, sec
L_{segl} = Length of segment l, $l = 1, 2, \ldots, L$, ft

Table 6.11 lists different pedestrian LOS ratings. Practically, LOS C can be used as the threshold level for pedestrian walking improvements.

6.3.3 Threshold performance levels for highway traffic safety

In general, highway safety performance analysis is separately conducted for segments and intersections. Based on highway functional classifications, highway segments can be split into rural/urban freeway, multilane divided and undivided highway, and two-lane road segments. Meanwhile, intersections can be classified into signalized and unsignalized intersections. For each classification, the constituent highway segments in a highway network can be grouped according to traffic volume ranges, number of lanes, lane widths, outside paved shoulder widths, interchange density as applicable, and so on. Similarly, intersections within the same classification can be further grouped by major and minor approach traffic volume ranges, number of through, left-turn, and right-turn lanes, intersection skew angles, number of approaches, and so forth.

For each group of highway segments or intersections, the average and SD of crash frequency in terms of number of crashes per highway segment or intersection per year can be computed using multi-year crash records or crash estimates. If the crash frequency of a specific highway segment or intersection is higher than the average crash frequency plus one SD, the highway segment or intersection can be considered as a candidate site for safety improvements.

6.3.4 Threshold performance levels for vehicle air emissions

In the U.S., standards for controlling vehicle air emissions have been established by EPA (2017). The standards cover primary and secondary stipulations. Primary standards are concerned with population health protection and secondary standards deal with public welfare, protection against animals, vegetation, and buildings. Table 6.12 lists threshold emission pollutant levels of EPA standards. Based on the traffic operating conditions associated with individual highway segments, the vehicle emission rates of individual air pollutants can be calculated using the vehicle emission models. If the computed air emission rates of vehicles running on a highway segment in any time period of the day exceed the EPA standards, the highway segment should be selected as a candidate site for air quality improvements.

6.3.5 Threshold performance levels for vehicle noise pollution

Vehicle noise pollutant thresholds are concerned with daytime and night time noise levels for both indoor and outdoor environments. Table 6.13 illustrates noise threshold levels established by WHO and EPA (Bhatia, 2014).

Table 6.12 Vehicle air pollutant threshold levels

Pollutant		*Standard*	*Duration*	*Level*	*Form*
CO		Primary	8 hours	9 parts-per-million (ppm)	Not to be exceeded more than once per year
			1 hour	35 ppm	
O_3		Primary and secondary	8 hours	0.075 ppm	Annual fourth-highest daily maximum 8-hour concentration, average over 3 years
SO_2		Primary	1 hour	75 ppm	99th percentile of one hour daily maximum concentrations, average over 3 years
		Secondary	3 hours	0.5 ppm	Not to be exceeded more than once per year
NO_X		Primary	1 hour	100 parts-per-billion (ppb)	98th percentile, average over 3 years
		Primary and secondary	Annual	35 ppb	Annual average
Particle matters (PM)	PM 2.5	Primary	Annual	12 μg/m^3	Annual mean, average over 3 years
		Secondary	Annual	15 μg/m^3	Annual mean, average over 3 years
		Primary and secondary	24 hours	35 μg/m^3	98th percentile, average over 3 years
	PM10	Primary and secondary	24 hours	150 μg/m^3	Not to be exceeded more than once per year

Table 6.13 Noise threshold levels

Standard	*Measure*	*Health protective threshold dB(A)*
WHO	L_{night}, Outdoor	40
EPA	L_{day}, Indoors	45

6.4 TRANSPORTATION PERFORMANCE IMPROVEMENT STRATEGIES AND MEASURES

When dealing with a multimodal transportation system, measures for performance improvements can be generally classified into three broad categories: travel demand management, capacity expansion, and efficient utilization of available capacity. Figures 6.1 through 6.3 depict typical performance improvement strategies within each category. Candidate projects associated with those demand management, capacity expansion and multimodal integration, and efficiency capacity utilization strategies can be proposed, evaluated, and prioritized for actual implementation to collectively achieve maximized overall performance improvements for a transportation system.

Table 6.14 lists some typical measures for physical transportation facility and system usage improvements.

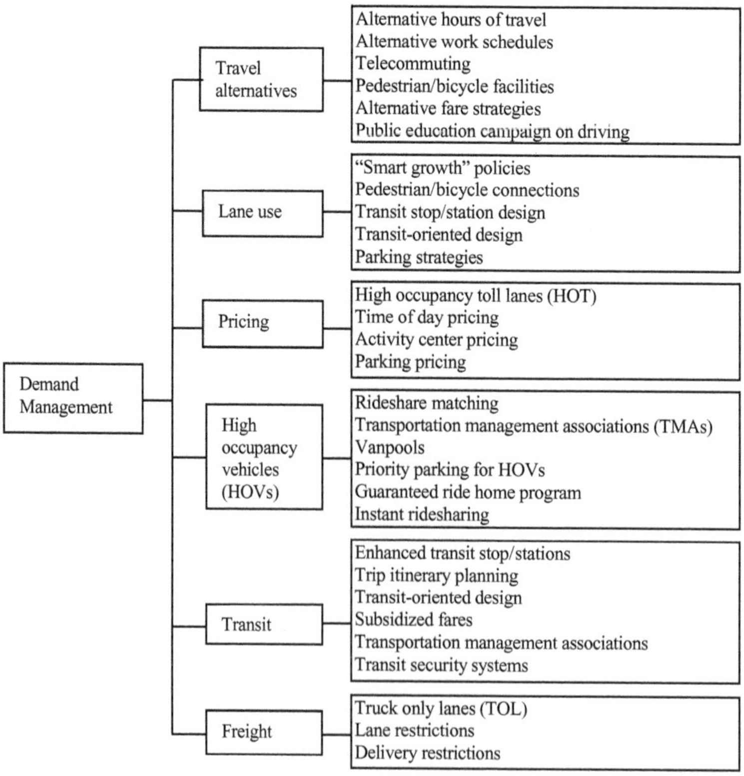

Figure 6.1 Travel demand management strategies.

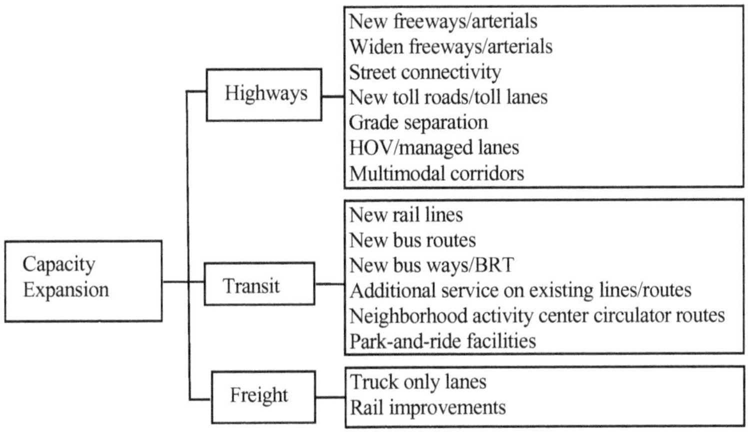

Figure 6.2 Transportation capacity expansion strategies.

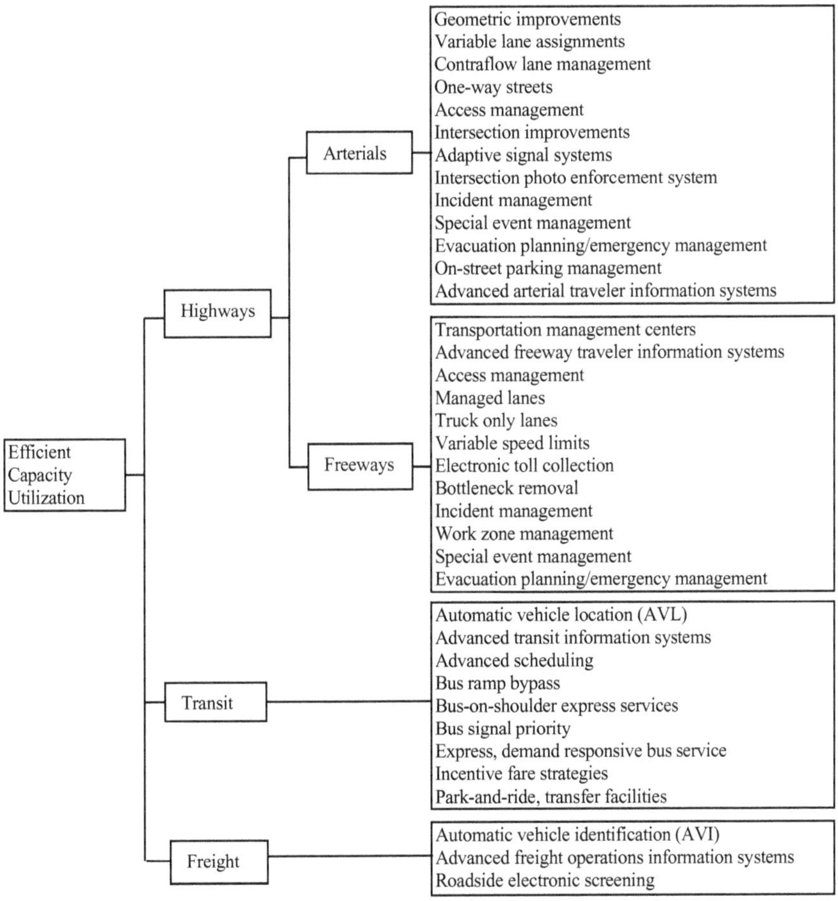

Figure 6.3 Efficient multimodal transportation capacity utilization strategies.

Table 6.14 Typical transportation performance/improvement measures

System component	*System needs*	*Improvement measure*
Pavements	Maintenance	Crack seal Crack filling Fog seal Sand seal Scrub seal Slurry seal Rejuvenator Micro-surfacing Ship seal Thin hot-mix overlays Full and partial-depth repair Slab stabilization Joint and crack resealing Dowel bar retrofit
	Rehabilitation	Cold milling Cold recycle in place
Bridges	Maintenance	Deck maintenance Superstructure maintenance Deck and superstructure maintenance Substructure maintenance
	Rehabilitation	Deck rehabilitation Deck rehabilitation and bridge widening Superstructure strengthening Superstructure strengthening and bridge widening Deck and superstructure rehabilitation Deck and superstructure rehabilitation and bridge widening Substructure rehabilitation
	Replacement	Deck replacement Deck replacement and bridge widening Superstructure replacement and bridge widening Deck replacement and superstructure rehabilitation Deck replacement, superstructure rehabilitation, and bridge widening Substructure replacement as part of bridge replacement
Traffic control and safety hardware	Replacement	This option is adopted based on the device and hardware useful life
	Retroreflectivity	Replace based on inspection
	Night time visual	Replace based on visual inspection
Transit	Facilities	Bus/rail transit right-of-way, travel lanes, and tracks Loading areas Stops/stations, transfer facilities, and park-and-ride facilities Terminals and garages
	Vehicles	Type, capacity, and amenities Maintenance and repair
	Operations	Advanced scheduling Vehicle control
Rail	Maintenance	Surface treatment Switches Level crossing Structure Spots

(*Continued*)

Table 6.14 (Continued) Typical transportation performance/improvement measures

System component	*System needs*	*Improvement measure*
		Tamping Ballast regulating Ballast stabilizing Rail grinding Joint straightening Ballast cleaning Spots
	Renewal	Parts Track continuous or panels Switches complete or parts Formation Structure
Mobility		Freeway ramp metering Freeway and arterial access management Congestion and incident management Special event management Emergency management Arterial surveillance and management Multimodal signal optimization Connected vehicles Advanced traveler information systems and assistance
Safety		Highway geometry, sight distance, pavement surface conditions Grade separated crossings Reduce speed limits Adaptive signal control Intersection red-light running photo enforcement Warning signs in problem areas Sobriety check points
Emissions		Control vehicle speed and speed changes Monitor vehicle status
Noise pollution		Noise barrier Monitor and regulate private refuse service vehicles

PROBLEMS

6.1 Transportation system needs are important in performance-based transportation asset management process. Describe the process of transportation needs assessment in terms of time and space.

6.2 A local agency wants to prepare a database for condition assessment of the local road pavement network within its authority. Assist the agency in identifying primary pavement performance measures and threshold levels for the assessment.

6.3 Three inbound lanes of an interstate highway have a capacity of 1800 passenger-cars per hour per lane. It has been determined that the following link travel time equation describes the congestion effects along a 1-mile segment: $t_1 = t_0[1 + 0.15(V/C)^4]$ minutes.

During the morning peak hour, it takes traffic an average of $1.75t_0$ minutes to travel that 1-mile segment. What is peak hour volume on this three-lane segment?

6.4 In the same context as question 6.3, suppose the current 3-lane volume on this segment is 4800 passenger-cars per hour and t_0 is 1 minute. If one of the three lanes were to be converted to a high-occupancy vehicle (HOV) lane and 25 percent of the current vehicles would be eligible to use it, what would be the travel time in the HOV lane over the 1-mile segment? What would be the travel time in the two remaining lanes?

6.5 A 15-mile long freeway segment has a free-flow speed of 55 mph, capacity $c = 2400$ vehicles per hour per lane, current flow rate $q = 1100$ vehicles per hour per lane, the level of service parameter $\tau = 0.1$, $\alpha = 0.474$, $\beta = 4$, $\gamma = 0.833$, and $T_0 = 15$ minutes. Apply (a) the BPR function; by (b) Davidson formula; and (c) Greenshields traffic stream model to find the travel time associated with the freeway segment.

a. BPR function: $T = T_0\ [1 + \alpha\ (v/(\gamma\ c))^{\beta}]$

b. Davidson's formula: $T = T_0\ [1 - (1 - \tau)\ (q/(\gamma\ c))]/[1 - (q/(\gamma\ c))]$

c. Greenshields model: $(v/o_f) + (D/D_o) = 1$, $vD = q$

6.6 You are tasked by the Mayor of a small city to identify needs and possible scenarios that could potentially maximize transit usage. Mention all the major steps involved with this assessment process.

Hint: start your scenario from evaluating the current transportation system performance and end by demand management, multimodal integration, and efficient transportation system usage.

6.7 Compare between the demand and the facility supply countermeasures. Use fixed guideway transit as an example.

Chapter 7

Fundamentals of project evaluation

7.1 PROJECT-LEVEL VERSUS NETWORK-LEVEL EVALUATION

Transportation project evaluation based on benefit–cost analysis frequently provides a quantitative basis for comparing and prioritizing investment alternatives proposed to improve the performance of physical transportation facilities and system usage from the needs assessment. When choosing a benefit–cost analysis method for project evaluation, tradeoffs must be considered between accuracy and simplicity of the method. In general, the methods fall into one of the following two categories: (i) project-level evaluation that uses standard assumptions to compute direct project benefits in the immediate project area and indirect benefits to project-affected areas; and (ii) network-level evaluation that estimates project benefits based upon the output of a regional planning model so as to capture project benefits associated with the overall impacts on the transportation network.

The ease or difficulty of real-world implementation is crucial in adopting project- versus network-level evaluation. As compared to project-level benefit–cost analysis, network-level analysis generally requires more time, data, and assumptions, necessitates the use of travel demand forecasting models such as the traditional four-step or demand forecasting model or even the agent-based demand forecasting model, and requires significantly more effort. For projects proposed to improve transportation system performance in the context of travel demand management, transportation system capacity expansion, and efficient capacity utilization, some are well suited for project-level evaluation and for others it is desirable to conduct the network-level evaluation. Table 7.1 illustrates some exemplary types of projects for the project- and network-level evaluations.

7.2 TRANSPORTATION COSTS

7.2.1 Total costs, variable costs, and variable social costs

The total transportation costs can be broken down into fixed costs and total variable costs. As seen in Table 7.2, total variable costs are classified as agency costs, user costs, externalities, and user charges. Variable social costs are the total variable costs that exclude user charges. These costs are defined as variable costs because they vary by system usage level normally measured by VMT. The average variable costs (AVCs), defined as the combined average unit costs per vehicle-mile social costs, might rise, decline or remain constant with vehicle volume. Most of the components of variable social costs vary slightly with volume due to congestion, but the one that varies by far the most is travel time. The marginal cost (MC) is the additional cost associated with the supply of an additional unit of travel.

Table 7.1 Exemplary types of projects suitable for project- and network-level evaluation

Project evaluation	*Decision criterion*	
	Project type	*Land area or facility type*
Project level	Resurfacing, restoration, and rehabilitation Safety improvements, including roadway geometry, lane, access, and roadside improvements Minor capacity improvement, such as addition of passing, auxiliary, and truck climbing lanes	Physical facilities with no alternative routes, such as bridges and tunnels Low-volume systems well under capacity Rural areas with relatively sparse roadway networks
Network level	New construction and significant capacity expansion such as added travel lanes and interchanges Addition of HOV or HOT lanes Corridor/area-wide adaptive signal systems Corridor/area-wide traffic control such as access management, intersection red-light photo enforcement New or improved park-and-ride facilities ITS projects, such as ramp metering, access management, traffic surveillance, and regional traveler information system	High-volume systems at or over capacity Urban areas with relatively dense roadway networks with alternative path choices

Table 7.2 Transportation variable cost categories and items

Variable cost category	*Variable cost item*	*Total variable costs*	*Variable social costs*	*Price*
Agency costs	Construction costs	√	√	
	• Repair costs	√	√	
	• Maintenance costs	√	√	
	• System operation costs	√	√	
User costs	Vehicle operating costs	√	√	√
	Travel time	√	√	√
	Vehicle crashes	√	√	√
Externalities	Vehicle air emissions	√	√	
	Vehicle noise pollution	√	√	
User charges	Fuel tax	√		√
	Tolls	√		√
	Parking charges	√		√

Price is the cost to road users and it includes user costs and user charges that vary with usage. The AVC, MC, and price as a function of the vehicle volume provide the information necessary for calculating user benefits (Lee, 2002).

7.2.2 Agency and user costs

Agency costs are those borne by transportation agencies, which generally include direct agency costs of facility construction and subsequent costs of maintenance and repair incurred during facility service life cycle. User costs are those incurred from transportation system usage. The primary user cost components include costs of vehicle operations, travel time, crashes, and air emissions. Each user cost component may be classified into normal operations condition with transportation system usage free of construction, maintenance, or rehabilitation activities that restrict the capacity, and work zone condition with added usage cost.

7.2.3 Average and MCs

Assuming a base case and one project alternative case, the physical characteristics of each case are given by the MC and AVC curves (excluding fixed costs and fixed charges such as the annual vehicle registration fee), while the price curve constitutes the policies affecting how the highway system is operated. The MC, AVC, and price are assumed to be converted into dollar values, referred to as generalized cost or generalized price, meaning that it combines money and in-kind components on the same scale. As shown in Figure 7.1, MC, AVC, and price to the user at any given vehicle volume are all different. MC and AVC are mathematically related, and will diverge if any component of cost varies with volume or volume-to-capacity ratio. That is, MC is unequal to AVC if the AVC goes up or down with volume. Because travel time for the unit distance of travel rises with congestion, for most volume levels MC lies above AVC.

Because users are faced with the average rather than the MC, it is frequently assumed that price and AVC are the same, but this usually is not true because of user charges, agency costs, and externalities. The MC and AVC functions are mathematically related, such that either one could be derived from the other, but only AVC can be observed empirically. Total variable costs can be measured either as the area under the MC curve (up to q_0) or as the AVC (ac) times the volume (q_0), the latter being a rectangle.

The price function in Figure 7.1a is shown as lying above AVC. This might be the case if variable user charges exceed variable agency costs and externalities. If the reverse is true, then the price curve lies below the AVC curve, as shown in Figure 7.1b. For congested conditions, it is unlikely that price will be above MC without a congestion-related toll, but the price could be above AVC. Whether the price is above or below AVC depends upon the magnitude and valuation of agency costs and externalities relative to user charges.

Vehicle volume could be determined by any of the three functions: by MC at p_{mc} for efficient pricing and first-best evaluation; by AVC at p_{ac}, which ignores actual user charges, agency costs, and externalities; or by the price function at p, which is the most general case. The inefficiency from not pricing at MC is given by the triangle bounded by p_{mc}, mc, and p.

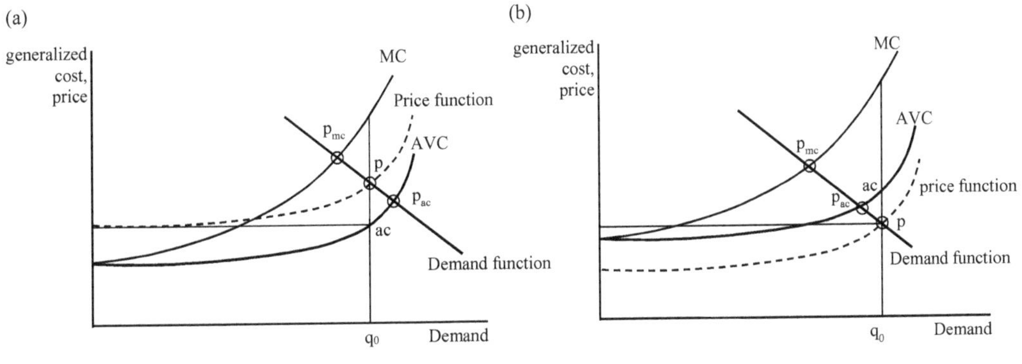

Figure 7.1 Illustration of marginal cost, average variable cost, and price curves. (a) Price function above average variable cost function. (b) Price function below average variable cost function. (Adapted from Li, Z. and P. Kaini. 2007. Optimal investment decision-making for highway transportation asset management under risk and uncertainty. Report MRUTC 07-10. Department of Civil, Architectural, and Environmental Engineering, Illinois Institute of Technology, Chicago, IL.)

7.3 TRANSPORTATION FACILITY LIFE-CYCLE AGENCY AND USER COSTS

7.3.1 Engineering economic basics

7.3.1.1 *Cash flow diagrams*

The cash flow diagram contains five variables: (i) present amount received or incurred in base year 0, such as the construction cost; (ii) future amount, that is, equivalent to a given present amount, which may be incurred (or paid) at end of analysis period or anytime within the analysis period; (iii) annual amount received or incurred every year; (iv) analysis period, that is, the total time over which economic evaluation is carried out; and (v) interest rate (or discount rate), depending on when the amount is available, which represents the extent to which money changes in value over time.

7.3.1.2 *Effective versus nominal interest rates*

The interest rate is either fixed or undergoes finite or infinite continuous compounding within the year. Simple fixed interest is generally better than compound interest from the borrower's perspective. Therefore, it is bad enough when the borrowed amount is increased at a compound rate even when the interest rate is fixed. There are three possible scenarios between the effective annual interest rate i_e and nominal interest rate r: (i) $i_e = r$, when interest rate is fixed throughout the year; (ii) $i_e = (1 + r/m)^m - 1$, when the nominal interest rate r is compounded m times a year throughout the year; and (iii) $i_e = e^r - 1$, when the nominal is compounded infinite number of times a year throughout the year.

7.3.1.3 *Current, constant, and discounted cost amounts*

The facility life-cycle activity profile forms the basis of deriving the facility life-cycle agency and user costs. In general, the agency costs of construction, repair, and maintenance work are directly estimated in monetary values. For individual user cost items, the quantities are first calculated and then converted to monetary values using the estimated qualities multiplied by corresponding monetized unit rates. In the calculation process, the monetary values of individual cost items incurred in different years of the facility service life cycle are adjusted to constant monetary values to remove the impact of inflation. With constant monetary values in place, they are discounted by accounting for the timing effect to a common reference year to arrive at the present worth of monetary values that are utilized for further analysis. Table 7.3 lists discounting formulas for discrete compounding.

EXAMPLE 7.1

Solve the following, using interest at 7% compounded annually:

a. What is the amount that will be accumulated in a sinking fund at the end of 20 years if \$500 is deposited in the fund at the beginning of each of the 20 years?
b. Uniform deposits are to be made on January 1 of 2001, 2002, 2003, and 2004 into a fund that is intended to provide \$2000 on January 1 of 2015, 2016, and 2017. What must be the size of these deposits?

Solution

a. $F = \left(500 \times \dfrac{(1+0.07)^{19} - 1}{0.07} + 500\right) \times (1+0.07)^{20} = \$21932.6$

Table 7.3 Cash flow relationships with end-of-period compounding

Type	*Find/given*	*Factor notation and formula*	*Relation*	*Sample cash flow diagram*
Single amount	*F/P* Compound amount	$(F/P, i, n) = (1+i)^n$	$F = P(F/P, i, n)$	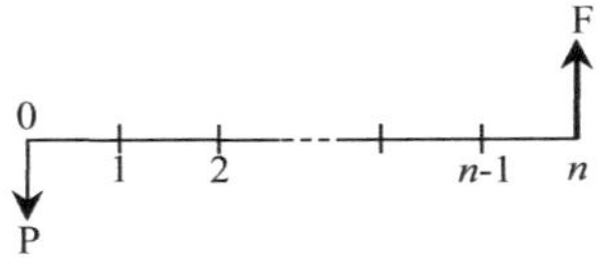
	P/F Present worth	$(P/F, i, n) = \frac{1}{(1+i)^n}$	$P = F(P/F, i, n)$	
Uniform series	*P/A* Present worth	$(P/A, i, n) = \frac{[(1+i)^n - 1]}{[i \cdot (1+i)^n]}$	$P = A(P/A, i, n)$	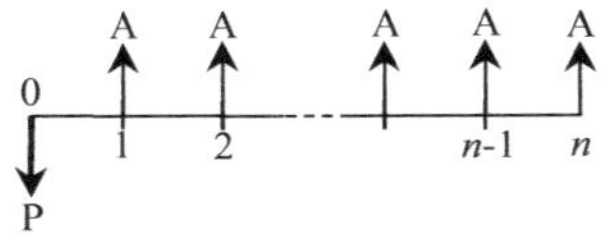
	A/P Capital recovery	$(A/P, i, n) = \frac{[i \cdot (1+i)^n]}{[(1+i)^n - 1]}$	$A = P(A/P, i, n)$	
	F/A Compound amount	$(F/A, i, n) = \frac{[(1+i)^n - 1]}{i}$	$F = A(F/A, i, n)$	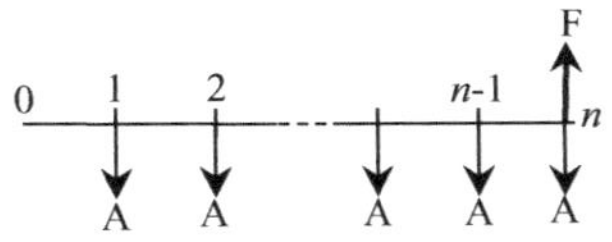
	A/F Sinking fund	$(A/F, i, n) = \frac{i}{[(1+i)^n - 1]}$	$A = F(A/F, i, n)$	
Arithmetic gradient	P_G/G Present worth	$(P/G, i, n) = \frac{[(1+i)^n - i \cdot n - 1]}{[i^2 \cdot (1+i)^n]}$	$P_G = G(P/G, i, n)$	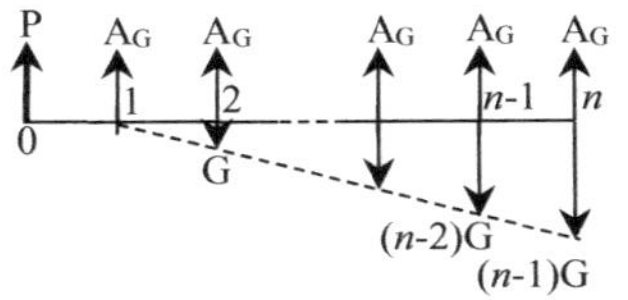
	A_G/G Uniform series	$(A/G, i, n) = \frac{1}{i} - \frac{n}{[(1+i)^n - 1]}$	$A_G = G(A/G, i, n)$	
Geometric gradient	P_g/A_1 and g Present worth	$P_g = \begin{cases} \frac{A_1 \cdot \left[1 - ((1+g)/(1+i))^n\right]}{(i-g)} \\ A_1 \cdot \left(\frac{n}{1+i}\right) \end{cases}$	$g \neq i$ $g = i$	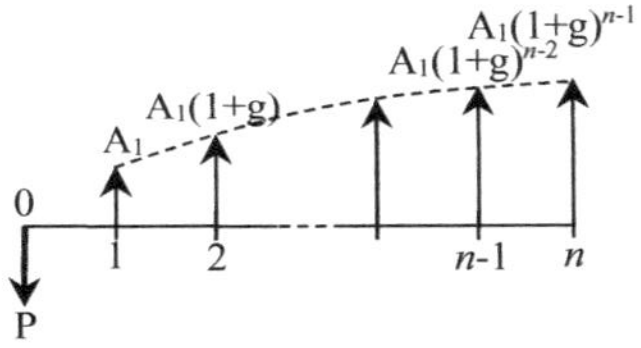

b. Set the beginning of 2000 as the base year

$$P_{2015} = \frac{2000}{(1+0.07)^{15}} = \$724.89$$

$$P_{2016} = \frac{2000}{(1+0.07)^{16}} = \$677.47$$

$$P_{2017} = \frac{2000}{(1+0.07)^{17}} = \$633.15$$

$$P_{2015} + P_{2016} + P_{2017} = \$2035.51$$

Therefore,

$$A = 2035.51 \times \frac{0.07(1+0.07)^4}{(1+0.07)^4 - 1} = \$600.94/\text{year}$$

EXAMPLE 7.2

a. What annual expenditure for 15 years is equivalent to spending \$1000 at the end of the first year, \$2000 at the end of the 5th year, and \$3000 at the end of the 9th year, if the interest is at 8% per annum?
b. What single amount paid at the beginning of the first year is equivalent to the series of unequal payments in the above, with interest at 8%?

Solution

$$P = 1000 \times \left[\frac{1}{(1+0.08)}\right] + 2000 \times \left[\frac{1}{(1+0.08)^5}\right] + 3000 \times \left[\frac{1}{(1+0.08)^9}\right] = \$2861.91$$

$$A = 2861.91 \times \left[\frac{0.08(1+0.08)^{15}}{(1+0.08)^{15} - 1}\right] = \$334.36/\text{year}$$

$$P = \$2861.91$$

EXAMPLE 7.3

What uniform annual payment for 40 years is equivalent to spending \$10000 immediately, \$10000 at the end of 10 years, \$10000 at the end of 20 years, and \$2000/year for the first 30 years? Assume an interest rate of 5%.

Solution

$$P = 10000 \times \left(1 + \frac{1}{(1+0.05)^{10}} + \frac{1}{(1+0.05)^{20}}\right) + 2000 \times \left[\frac{(1+0.05)^{30} - 1}{0.08(1+0.05)^{30}}\right] = \$39123.6$$

$$A = 39123.6 \times \left[\frac{0.05(1+0.05)^{40}}{(1+0.05)^{40} - 1}\right] = \$2280.05/\text{year}$$

EXAMPLE 7.4

A company leased a storage yard from a city and prepaid the rent for 5 years; the terms of the lease permit the company to continue to rent the site for 5 years more by payment of \$5000 at the beginning of each year of the second 5-year period. Two years of the prepaid period have expired and the city is in need of funds; it proposed to the company that it now prepays the rental that was to have been paid year by year in the second 5-year period. If interest is figured at 4%, what is a fair payment to be made now in lieu of these five annual payments?

Solution

Set the beginning of the second 5-year period as the base year.

$$P_5 = 5000 \times \left[\frac{(1+0.04)^5 - 1}{0.04(1+0.04)^5} \right] = \$22259.1$$

Convert the rental of the second 5-year period to the present worth in the second year.

$$P_2 = 22259.1 \times \left[\frac{1}{(1+0.04)^3} \right] = \$19788.3$$

EXAMPLE 7.5

A pre-stressed concrete bridge is proposed to be constructed. The estimated design service life is 35 years. The construction cost is \$3000000, followed by a deck rehabilitation of \$500000 in the 20th year, and it would have a salvage value of \$200000 at the end of its useful life. The yearly maintenance cost would be \$3000. What would be the total yearly cost? The interest rate is 5%.

Solution

$$P_{Construction} = \$3000000$$

$$P_{Rehabilitation} = \$500000 \times \left[\frac{1}{(1+0.05)^{20}} \right] = \$188445$$

$$P_{Salvage} = -\$200000 \times \left[\frac{1}{(1+0.05)^{35}} \right] = -\$36258.1$$

$$A = (3000000 + 188445 - 36258.1) \times \left[\frac{0.05(1+0.05)^{35}}{(1+0.05)^{35} - 1} \right] + 3000 = \$195509/\text{year}$$

EXAMPLE 7.6

A government decided to make a 25-year loan on \$2500000 at 10% interest to a Toll Highway Authority. Four years of the loan period was to be spent to build the road during which time the government agreed to "capitalize" the interest until the road is completed and toll collection has started in the 5th year. \$500000 was expected to be spent each of the first two years and the rest equally in the 3rd and 4th years. What annual payment by the Toll Authority would repay the loan (principal plus interest) over the remaining 21 years?

Solution

$$F_4 = 500000 \times \left[(1+0.1)^4 + (1+0.1)^3 \right] + 750000 \times \left[(1+0.1)^2 + (1+0.1) \right] = \$3130050$$

$$A = 3130050 \times \left[\frac{0.1(1+0.1)^{21}}{(1+0.1)^{21} - 1} \right] = \$361910/\text{year}$$

7.3.1.4 Perpetual costs

Perpetual means recurring forever. The transportation facility maintains a finite number of years of useful service life. Once the service life is ended, the facility is reconstructed or replaced completely and the second useful service life would begin. In the limit, the service life cycles will repeat into perpetuity horizon. The facility life-cycle agency costs in perpetuity can be derived from the facility life-cycle agency costs in one cycle as given below.

Denote

PW_T = Present worth of facility service life-cycle agency costs in a T-year useful service life
PW_∞ = Present worth of facility service life-cycle agency costs in perpetuity
EUA_∞ = Equivalent uniform annualized agency costs in perpetuity
T = Facility design useful service life in number of years
i = Discount rate

$$\begin{aligned} PW_\infty &= \mathrm{PW}_T + \mathrm{PW}_T\Big/\left[(1+i)^T\right] + \mathrm{PW}_T\Big/\left[(1+i)^{2T}\right] + \cdots + \mathrm{PW}_T\Big/\left[(1+i)^{(n-1)\cdot T}\right] \\ &= \mathrm{PW}_T \cdot \left[1 + 1\Big/\left[(1+i)^T\right] + 1\Big/\left[(1+i)^{2T}\right] + \cdots + 1\Big/\left[(1+i)^{(n-1)\cdot T}\right]\right] \\ &= \mathrm{PW}_T \cdot \left[\left((1+i)^T\right)\Big/\left((1+i)^T - 1\right)\right] \end{aligned} \tag{7.1}$$

$$EUA_\infty = i \cdot PW_\infty^M = (i \cdot PW_T) \cdot [((1+i)^T)/((1+i)^T - 1)] \tag{7.2}$$

7.3.1.5 Evaluation of alternative projects

A project is referred to as an undertaking that is associated with a cost and is expected to yield some benefits. Alternative projects imply that the projects are mutually exclusive. In other words, if two projects are described as alternatives, the occurrence of one project completely precludes the occurrence of another. Evaluation of a project means that the positive impacts (benefits) of the project are compared with the negative impacts (costs) to confirm whether the project is economically feasible as a candidate for possible implementation.

7.3.1.6 Common project evaluation methods

Benefit-to-cost ratio method: The benefit-to-cost ratio method compares the discounted benefits and costs for each project and then evaluates each alternative compared to the other.

Incremental benefit–cost analysis method: The benefits and costs considered for each project are not the totals, but rather the additional benefits achieved and costs incurred over the next effective project. This analysis considers, in effect, whether an investment necessary to achieve the next incremental step in the system can be justified in terms of the incremental benefits that would be achieved.

Net present worth method: The net present worth method uses the chosen discount rate to convert the project benefits and costs to equivalent present values and then compares these values. The present value of the benefits and costs is equal to the summation of the values of these effects multiplied by the present worth factor appropriate to the period over which the benefits and costs occur. The net present worth then equals the difference between the present-value benefits and costs.

Equivalent uniform annual cost method: The equivalent uniform annual cost method converts non-uniform series of project benefits and costs into equivalent uniform annualized amounts of benefits and costs, respectively. The annualized benefits and costs are then used to compare project alternatives on an equal basis.

Cost-effectiveness method: The effectiveness of a project alternative is usually represented as a scaled quantity relating to a specific goal. For instance, the number of car pools formed and reduction in vehicle air emission quantities. Cost-effectiveness ratios can thus be calculated to show the degree of goal attainment per dollar of net expenditure. This method is particularly useful when it is difficult to reach a consensus in unit values of user cost elements, such as values of travel time, vehicle crashes, and air emissions.

EXAMPLE 7.7

A proposal is being considered to improve an existing road connecting two small cities in the Midwest region to reduce transportation costs. The cost of the project is \$2000000. Present annual transportation cost for all traffic amounts to \$2540000 per year and would continue if no improvement is made. After the improvement, annual transportation costs are estimated to be \$2320000. Assume the life of the project is 15 years, and minimum attractive rate of return (MARR) is 12%. Should the project be undertaken? Evaluate this proposal by using the net present worth, benefit-to-cost, and internal rate of return (IRR) methods. Assume costs appear at the beginning of each year while benefits are at the end of a year.

Solution

$$\text{Cost} = \$2000000$$

$$\text{Benefit} = \text{Cost Reduction} = \$2540000 - \$2320000 = \$220000/\text{year}$$

$$P_{benefit} = 220000 \times \left[\frac{(1+0.12)^{15} - 1}{0.12(1+0.12)^{15}}\right] = \$1498390.$$

a. Net present worth method

$$\text{Cost} - \text{Benefit} = \$1498390 - \$2000000 = -\$501610$$

b. Benefit-to-cost ratio method

$$\$1498390/\$2000000 = 0.75$$

c. Internal rate of return method

$$2000000 = 220000 \times \left[\frac{(1+i)^{15} - 1}{i(1+i)^{15}}\right]$$

Solve for i, $i = 0.070 = 7.0\%$

Since Net present worth (−\$501610) < 0, Benefit/cost (0.75) < 1, and IRR (7.0%) < MARR (12%), the proposal should not be undertaken.

EXAMPLE 7.8

A County Highway Division is considering either replacing or rehabilitating a bridge by contract which crosses over a river. The cash flows associated with bridge replacement and rehabilitation options are shown below. For the bridge replacement option, it is assumed to recur infinite number of times into perpetuity horizon. For the bridge rehabilitation option, the same case flow as the replacement option will be followed after the service life of the rehabilitated bridge terminates. Determine the best choice using the equivalent uniform annual cost (EUAC) method. Consider an annually compounded interest rate of 6%.

Alternative I. Bridge replacement with 40-year service life

a. Replacement cost
   Initial = 3500000
   15th year = 80000
   20th year = 15000
   30th year = 60000
b. Annual maintenance cost = \$1000/year
c. Salvage of existing bridge beams = \$10000

Alternative II. Bridge rehabilitation with 20-year service life

a. Rehabilitation Costs
   Initial = 1200000
   10th and 20th year = 10000
b. Annual maintenance cost = \$2000/year
c. No salvage of existing bridge beams

Solution

Alternative I:

$$EUAC_I = \left(3500000 + \frac{80000}{(1+0.6)^{15}} + \frac{15000}{(1+0.6)^{20}} + \frac{60000}{(1+0.6)^{30}} - \frac{10000}{(1+0.6)^{40}}\right) \times \frac{0.06(1+0.06)^{40}}{(1+0.06)^{40}-1} + 1000 = \$233620/\text{year}$$

Alternative II:

$$EUAC_{II} = \left(1200000 + \frac{10000}{(1+0.6)^{10}} + \frac{10000}{(1+0.6)^{20}}\right) \times \frac{0.06(1+0.06)^{20}}{(1+0.06)^{20}-1} + 2000 = \$106629/\text{year}$$

Alternative II is the more cost-effective choice.

7.3.2 Basic analytical steps

The overall benefits of a transportation project in the facility service life cycle may be extracted from both the agency and user perspectives. With investments in the transportation system, it may reduce agency costs and result in savings of user costs in terms of vehicle operations, travel time, crashes, and air emissions in the facility service life cycle. Thus reductions in agency and user costs in the facility service life cycle are treated as project benefits. To estimate the change in life-cycle agency costs, the activity profiles in the facility service life cycle containing details of frequency, timing, and magnitude of facility construction, repair, and maintenance work for primary transportation facilities such as pavements and bridges need to be established. For pavement or bridge facilities, the performance deterioration trend and maintenance and repair treatments in the facility service life cycle differ

Table 7.4 Analytical steps for estimating agency and user costs in facility service life cycle

Analytical step	*Data and information needs*
Define base case and project alternatives	The network elements affected Engineering characteristics Project build-out schedule Project agency cost schedule Project user cost schedule
Determine level of details required	Types of benefits and costs Link versus corridor perspective Vehicle classes to be studied Hourly, daily, and seasonal details Time periods within a day to be explicitly modeled
Develop basic agency cost factors	Physical asset performance models Activity frequency, timing, and magnitude
Develop basic user cost factors	Vehicle operating unit costs Vehicle occupancy rates Values of travel time Vehicle crash rates and unit costs Vehicle air emission rates and unit costs
Select economic factors	Discount and inflation rates Analysis period Physical asset service life cycle assumptions Physical asset salvage values at the end of service life cycle
Obtain traffic data for base case and project alternatives for explicitly modeled periods	Travel demand and traffic assignment models Hourly, daily, and seasonal traffic volumes, speeds, and occupancy before and after improvement Traffic growth rate factors Volume-delay function factors Peak-spreading assumptions
Measure agency costs for base case and project alternatives	Project direct agency costs of construction Discounted life-cycle costs of maintenance and rehabilitation
Measure user costs for base case and project alternatives for affected links or networks	Operating, delay, crash, and emission costs during construction Life-cycle vehicle operating costs Life-cycle travel time costs (including delay costs) Life-cycle accident costs Life-cycle air emission costs

significantly by material type. Therefore, life-cycle agency cost profiles need to be developed for different types of pavements and bridges. For instance, different life-cycle activity profiles are needed for flexible, right, and composite pavements; and for concrete and steel bridges. Corresponding to facility life-cycle agency cost profiles, the life-cycle user cost profiles can be created for costs of vehicle operations, travel time, crashes, and air emissions, respectively. Table 7.4 lists the generic steps for estimating transportation agency and user costs in facility service life cycle (AASHTO, 1978, 2003).

7.3.3 Transportation facility service life cycle

The transportation facility service life cycle may be classified into design useful service life cycle and repair service life cycle. The design useful service life cycle is defined as the time interval between two consecutive construction interventions, whereas the repair service life cycle is defined as the time interval of construction to rehabilitation, rehabilitation to rehabilitation or rehabilitation to reconstruction/new construction during the design useful service life cycle. Therefore, one design useful service life cycle may cover multiple repair service life cycles. Within each repair service life cycle, maintenance treatments are typically implemented. The maintenance treatments may be preventive or corrective. Preventive maintenance treatments are applied before the onset of significant structural deterioration (O'Brien, 1989). Conversely, corrective maintenance treatments are typically carried out to address the existing performance deterioration issues. In project evaluation, tradeoffs exist between design and repair service life cycles affected by repair treatments, between repair service life cycle and maintenance treatments, and between corrective and preventive maintenance treatments (Labi and Sinha, 2002, 2005).

7.3.4 Transportation facility life-cycle activity profiles

The transportation facility life cycle activity profile refers to the frequency, timing, magnitude of construction, repair, and maintenance treatments in the facility design useful service life cycle. A typical life-cycle activity profile represents the most cost effective way of implementing strategically coordinated repair and maintenance treatments to achieve the intended design useful service life. In practice, transportation facility design useful service life-cycle activity profiles are determined using preset time intervals for treatments and condition triggers for treatments, respectively. In current practices, preset time intervals are largely used because of lack of consensus in facility condition deterioration measurements and condition trigger values (AASHTO, 1978, 2003; FHWA, 1987, 1991; Gion et al., 1993; INDOT, 2002). Figure 7.2 provides a schematic illustration of pavement facility life-cycle activity profile.

7.3.5 Facility life-cycle agency costs

7.3.5.1 Facility life-cycle agency cost components

A multimodal transportation system may deal with passenger travel and freight shipments. Typically, highway transportation is the primary mode to accommodate passenger travel. Of various types of highway facilities, pavements and bridges are the predominant ones. Both pavements and bridges are designed to provide travel services for a relatively

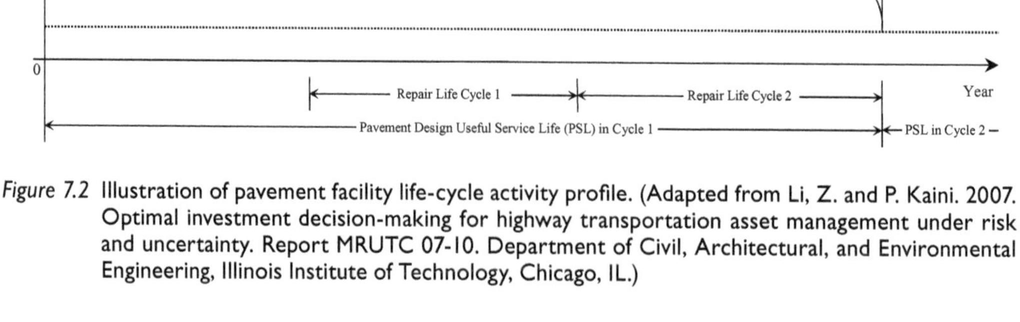

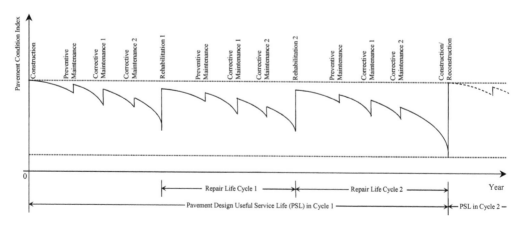

Figure 7.2 Illustration of pavement facility life-cycle activity profile. (Adapted from Li, Z. and P. Kaini. 2007. Optimal investment decision-making for highway transportation asset management under risk and uncertainty. Report MRUTC 07-10. Department of Civil, Architectural, and Environmental Engineering, Illinois Institute of Technology, Chicago, IL.)

long period. To achieve the design useful service life cycle, effective maintenance and repair treatments need to be implemented in a timely manner. In this respect, the facility life-cycle agency costs generally include costs of construction for facility delivery, and additional costs of maintenance and repair treatments to ensure that the facility condition is always kept above the threshold levels and no early termination of the actual service life occurs.

The repair treatments are implemented in multiyear intervals. The length of each interval is determined according to predefined time durations or by incorporating the facility condition deterioration models. The maintenance treatments may be classified into annual routine maintenance and periodic maintenance treatments applied in multiyear intervals. The annual routine maintenance may be reactive, that is, implemented in the spring season to correct the damages to the facility during the winter season or may be preventive, usually applied before the winter season to avoid excessive damages to the facility during the coming winter season that could be found in the next spring season.

7.3.5.2 Growths in annual traffic volumes, truck volumes, and axle loading

In the facility service life cycle, traffic volumes, truck traffic volumes, and truck axle loading in terms of ESALs are expected to grow. This will, in turn, lead to increases in cumulative travel and axle loading, resulting in accelerated facility condition deterioration. Therefore, facility repair and maintenance costs will increase accordingly to recover excessive damage to the facilities.

Denote

$AADT_0$ = Annual average daily traffic in base year 0
L = Total miles of travel in base year 0
$CVMT(t)$ = Cumulative vehicle miles of travel for t years from the base year 0
$CVMT(T)$ = Cumulative vehicle miles of travel in T-year facility service life cycle
$CVMT_\infty$ = Cumulative vehicle miles of travel in perpetuity
VMT = Average annual VMT in T-year facility service life cycle
$ESAL_0$ = Annual $ESAL$ value in base year 0
$ESAL'$ = Small constant increment of annual $ESAL$ value
$CESAL(t)$ = Cumulative $ESAL$ value in service for t years from the base year 0
$CESAL(T)$ = Cumulative $ESAL$ value in T-year facility service life cycle
$CESAL_\infty$ = Cumulative $ESAL$ value in perpetuity
$ESAL$ = Average annual $ESAL$ value in T-year facility service life cycle
g_V = Annual traffic volume growth rate
g_L = Annual $ESAL$ growth rate
T = Facility design service life
i = Discount rate

Cumulative and annual travel with growth in T-year life cycle and perpetuity: With the base year $AADT_0$ and total miles of travel L determined, we could consider a baseline annual growth rate of g_V that leads to a growth path of $AADT_0 \cdot L \cdot e^{(g_V \cdot t)}$. Hence, $CVMT$ in t years from the base year 0 can be calculated as

$$CVMT(t) = \int_0^t \left[AADT_0 \cdot L \cdot e^{(g_V \cdot \tau)}\right] d\tau = AADT_0 \cdot L \cdot \left[\frac{e^{(g_V \cdot t)} - 1}{g_V}\right] \tag{7.3}$$

Hence, the cumulative VMT in T-year facility service life cycle and in perpetuity, as well as average annual VMT become

$$CVMT(T) = \int_0^T [AADT_0 \cdot L \cdot e^{(g_V \cdot \tau)}] d\tau = AADT_0 \cdot L \cdot \left[\frac{e^{(g_V \cdot T)} - 1}{g_V}\right] \tag{7.4}$$

$$CVMT_\infty = \int_0^{n \cdot T} \left[AADT_0 \cdot L \cdot e^{(g_V \cdot \tau)}\right] d\tau = AADT_0 \cdot L \cdot \left[\frac{e^{(g_V \cdot n \cdot T)} - 1}{g_V}\right] (n \to \infty). \tag{7.5}$$

$$VMT = AADT_0 \cdot L \cdot \left[\frac{e^{(g_V \cdot n \cdot T)} - 1}{g_V \cdot n \cdot T}\right] \tag{7.6}$$

Cumulative and annual traffic loading with growth in T-year life cycle and perpetuity: For the calculation of annual $ESAL$, growth in annual traffic loading in the T-year facility service life cycle generally occurs due to increases in truck traffic volumes and axle loads. With the base year traffic loading $ESAL_0$ determined, we could consider a baseline annual growth rate of g_L that leads to a growth path of $ESAL_0 \cdot e^{(g_L \cdot t)}$ and a small constant increment $ESAL'$ per year from that baseline path. Hence, the cumulative traffic loading in t years can be calculated as

$$CESAL(t) = \int_0^t \left[ESAL_0 \cdot e^{(g_L \cdot \tau)} + ESAL'\right] d\tau = ESAL_0 \cdot \left[\frac{e^{(g_L \cdot t)} - 1}{g_L}\right] + ESAL' \cdot t \tag{7.7}$$

Hence, the cumulative traffic loading in T-year facility service life cycle and in perpetuity, as well as average annual traffic loading, become

$$CESAL(T) = \int_0^T [ESAL_0 \cdot e^{(g_L \cdot \tau)} + ESAL'] d\tau = ESAL_0 \cdot \left[\frac{e^{(g_L \cdot T)} - 1}{g_L}\right] + ESAL' \cdot T \tag{7.8}$$

$$CESAL_\infty = \int_0^{n \cdot T} [ESAL_0 \cdot e^{(g_L \cdot \tau)} + ESAL'] d\tau = ESAL_0 \cdot \left[\frac{e^{(g_L \cdot n \cdot T)} - 1}{g_L}\right] + ESAL' \cdot n \cdot T (n \to \infty) \tag{7.9}$$

$$ESAL = ESAL_0 \cdot \left[\frac{e^{(g_L \cdot n \cdot T)} - 1}{g_L \cdot n \cdot T}\right] + ESAL' \tag{7.10}$$

7.3.5.3 Growth in repair and routine maintenance costs in facility service life cycle

With a deteriorating trend of the facility condition that is generally followed, the amounts of subsequent repair treatments are expected to increase with the passage in time. For the same reason, the amount of annual routine maintenance treatment is envisioned to increase. For either the repair or annual routine maintenance cost, the amount is generally categorized into load-related and non-load-related portions. The load-related portion of the cost is estimated based on the MC to recover damages caused by traffic loading in terms of the ESAL value. The non-load-related portion of the cost is calculated in accordance with vehicle miles of travel. The increases in the load-related portion of the cost are attributable to growth in both truck traffic volumes and in axle loads for the same type of trucks over time. The increases in the non-load-related portion of the cost are caused by growth in total traffic volumes, especially truck traffic volumes over time. Thus the load-related and non-load-related portions of repair and annual routine maintenance costs can be estimated by considering geometric growth rates for the annual traffic volume, truck traffic volume, and axle loading for trucks about the base year traffic volume, truck traffic volume, and axle loading, respectively.

Present worth and equivalent uniform annualized repair and maintenance costs with annual growth: Assuming that the facility repair treatment is repetitively implemented in T-year intervals, repair cost increases from the first repair treatment in the Tth year with a geometric growth rate, and annual routine maintenance cost in the T-year repair interval also grows from the first year with a geometric growth rate.

Denote

- C_0^M = Annual facility maintenance cost discounted to base year 0, beginning in year 1
- C_t^M = Annual facility maintenance cost in year t
- $PWF^M(T)$ = Present worth factor of annual routine maintenance cost in T-year repair interval
- PWF_∞^M = Present worth factor of annual routine maintenance cost in perpetuity
- $PW^M(T)$ = Present worth of annual facility maintenance cost in T-year repair interval
- PW_∞^M = Present worth of annual facility maintenance cost in perpetuity
- EUA^M = Equivalent uniform annualized facility maintenance cost
- g^M = Growth rate of annual facility maintenance cost
- C_0^R = Facility repair cost discounted to base year 0, beginning in year T and recurring in every T years
- PW_∞^R = Present worth of facility repair cost in perpetuity
- EUA^R = Equivalent uniform annualized facility repair cost
- g^R = Growth rate of facility repair cost
- T = T-year repair interval
- i = Discount rate

The present worth and equivalent uniform annualized repair and maintenance costs are given as

$$C_t^M = C_0^M \cdot e^{(g^M \cdot t)}$$

$$PW^M(T) = \int_0^T \left[\left(C_0^M \cdot e^{\left(g^M \cdot \tau\right)} \right) \cdot e^{(-i \cdot \tau)} \right] d\tau = C_0^M \cdot \left[\frac{\left[e^{\left((g^M - i) \cdot T\right)} - 1 \right]}{(g^M - i)} \right] \tag{7.11}$$

$$PW_\infty^M = \int_0^{n \cdot T} \left[\left(C_0^M \cdot e^{\left(g^M \cdot \tau\right)} \right) \cdot e^{(-i \cdot \tau)} \right] d\tau = C_0^M \cdot \left[\frac{\left[e^{\left(\left(g^M - i\right) \cdot n \cdot T\right)} - 1 \right]}{(g^M - i)} \right] \tag{7.12}$$

$$EUA^M = i \cdot PW_\infty^M = \left(i \cdot C_0^M \right) \cdot \left[\frac{\left[e^{\left(\left(g^M - i\right) \cdot n \cdot T\right)} - 1 \right]}{(g^M - i)} \right] \tag{7.13}$$

Alternatively, given the annual routine maintenance costs C_0^M in base year 0 and $C_T^M = C_0^M \cdot e^{(g^M \cdot T)}$ in year T, with the effective geometric growth rate g^M, the following can be calculated:

$$PWF^M(T) = \frac{\left[e^{\left((g^M - i)\cdot T\right)} - 1\right]}{(g^M - i)} \tag{7.14}$$

$$PWF_\infty^M = \frac{\left[e^{\left((g^M - i)\cdot n\cdot T\right)} - 1\right]}{(g^M - i)} \tag{7.15}$$

$$PW^M(T) = C_0^M \cdot PWF^M(T) = C_0^M \cdot \frac{\left[e^{\left((g^M - i)\cdot T\right)} - 1\right]}{(g^M - i)}$$

$$PW_\infty^M = C_0^M \cdot PWF_\infty^M = C_0^M \cdot \frac{\left[e^{\left((g^M - i)\cdot n\cdot T\right)} - 1\right]}{(g^M - i)}$$

$$EUA^M = i \cdot PW^M = \left(i \cdot C_0^M\right) \cdot \left[\frac{\left[e^{\left((g^M - i)\cdot n\cdot T\right)} - 1\right]}{(g^M - i)}\right]$$

It shows that the two methods yield identical results.
Similarly,

$$\begin{aligned} PW_\infty^R &= \left[\left(C_0^R \cdot e^{(g^R)}\right) \cdot e^{(-T\cdot i)}\right] + \left[\left(C_0^R \cdot e^{(2g^R)}\right) \cdot e^{(-2T\cdot i)}\right] + \cdots + \left[\left(C_0^R \cdot e^{(n\cdot g^R)}\right) \cdot e^{(-n\cdot T\cdot i)}\right] \\ &= C_0^R \cdot \left[e^{(g^R - T\cdot i)} + e^{(2(g^R - T\cdot i))} + \cdots + e^{(n(g^R - T\cdot i))}\right] \\ &= C_0^R \cdot e^{(g^R - T\cdot i)} \cdot \left[\left(e^{(n(g^R - T\cdot i))} - 1\right) \Big/ \left(e^{(g^R - T\cdot i)} - 1\right)\right] \end{aligned} \tag{7.16}$$

$$EUA^R = i \cdot PW_\infty^R = \left(i \cdot C_0^R \cdot e^{(g^R - T\cdot i)}\right) \cdot \left[\left(e^{(n(g^R - T\cdot i))} - 1\right) \Big/ \left(e^{(g^R - T\cdot i)} - 1\right)\right] \tag{7.17}$$

Note that when $n = 1$, $EUA^R = i \cdot C_0^R \cdot e^{(g^R - T\cdot i)}$, which is just the annualized amount of present worth of the first repair cost allocated to the T-year repair interval.

7.3.6 Facility life-cycle user costs

7.3.6.1 User cost components and categories

User costs are incurred by highway users in the transportation facility service life cycle. User cost components mainly include costs of vehicle operations, travel time, crashes, and air emissions (AASHTO, 1978, 2003; FHWA, 2000). Each user cost component consists of two cost categories: user cost in normal operational conditions and excessive user costs due to work zones (FHWA, 1998b).

7.3.6.2 *Growth in annual traffic volumes in facility service life cycle*

The cumulative and annual travel with growth in *t*-year life cycle and perpetuity can be computed in the same way as the previous section.

7.3.6.3 *Unit costs of vehicle operations*

Vehicle operating costs refer to costs of fuel, oil, tire wear, maintenance and repair, and mileage-dependent vehicle depreciation and are measured in terms of dollars per VMT.

Transportation projects can affect vehicle operating costs directly by improving operating conditions such as fewer changes in speed, reduced grades, smoother pavements, and wider curves or indirectly by influencing traveler behavior including more frequent usage and more direct routing. The highway vehicle operating costs are affected by vehicle type, vehicle speed, speed changes, gradient, curvature, and road surface condition, as briefly described in Table 7.5. In addition, Table 7.6 provides a range of estimates used in several benefit–cost models for project evaluation.

7.3.6.4 *Unit costs of travel time*

Transportation improvement projects often lead to higher speeds and lower travel times for drivers, passengers, and freight. Since travel time can make up a major portion of total user costs, it is important to use an appropriate value of time when converting travel time into

Table 7.5 Factors affecting vehicle operating costs

Factor	*Description*
Vehicle type	Generally, cars have lower operating costs than trucks, due to lower fuel and oil consumption, and lower price of vehicle and parts, as well as maintenance and repairs. Since vehicle technology, fuel efficiency, and price/costs change over time, vehicle operating costs for various classes of vehicles will also change and must be periodically updated
Vehicle speed	Empirical research indicates that vehicle speed is the dominant factor in determining vehicle operating costs. They decrease as vehicle speed increases, reaching an optimum efficiency point at mid-range speeds of approximately 55 mph, after which point costs will increase as vehicle speed increases further. Obviously, vehicle operating costs will get higher under congested traffic conditions and when vehicles are idling at stop-controlled/signalized intersections, ramp meters, and railroad crossings
Speed changes	Empirical research indicates that vehicle operating costs increase with speed change cycles and the added cost of speed cycling is higher at higher speeds
Gradient	Driving a vehicle up a steep, positive grade requires more fuel than driving it along on a level road at the same speed, and the additional load on the engine imposes added costs of maintenance. Roadway sections with negative gradient would have an opposite effect. However, as the steepness of the downgrade increases, it may be necessary to apply the brakes and this also imposes an added operating cost burden
Curvature	Curves impose costs through the centrifugal force that tends to keep the vehicle following a tangent rather than a radial path. The force is countered by superelevation of the roadway and the side friction between the tire tread and the roadway surface. As a result, there is a greater usage of energy and more fuel is required to negotiate curved sections. In addition, the side friction increases tire wear and raises this component of operating costs
Road surface condition	The motion of a vehicle on a deteriorated pavement surface will experience greater rolling resistance, which requires more fuel consumption compared to traveling at a similar speed on a smooth surface. The roughness of road surface contributes to reduction of speed, additional tire wear, and influences the vehicle maintenance and repair expenses incurred in the operation of a vehicle

Table 7.6 Summary of vehicle operating cost estimation methods (2000 dollars)

Model	*Attribute*				
	VOC Items	*Factors considered*	*VOC Range $/veh-mile (year)*	*Vehicle Types*	*Source*
2003 AASHTO Red Book	Fuel, oil, tire, maintenance	Speed, speed cycling, grade, curvature, pavement condition	Auto: • 0.039–0.117 gal fuel/veh-mile Truck: • 0.158–0.503 gal fuel/veh-mile Car, SUV, and van: • $0.095–0.124/veh-mile	Car, sports utility vehicles (SUV), van, truck	American Automobile Association (AAA) (1999) Cohn et al. (1992)
CAL-B/C	Fuel, non-fuel	Speed (for fuel only)	Auto: • 0.033–0.182 gal fuel/veh-mile • $0.165/veh-mile for non-fuel cost Truck: • 0.008–0.511 gal fuel/veh-mile • $0.285/veh-mile for non-fuel cost	Auto, truck	U.S. Department of Transportation (USDOT) (1992)
HERS-ST	Fuel, oil, tire, maintenance and repair, depreciation	Speed, speed cycling, grade, curvature, and pavement condition	$0.203	2 car types, 5 truck types	Zaniewski et al. (1982)
STEAM	Fuel, tire, maintenance and repair	Speed (for fuel only)	$0.058–0.105	Car, truck	USDOT (1992)
StratBENCOST	Fuel, oil, tire, maintenance and repair, depreciation	Speed, speed cycling, grade, curvature, and pavement condition	$0.187–0.351	Car, truck, bus	Zaniewski et al. (1982)

Table 7.7 Factors affecting the value of travel time

Travel time component		*Factor*
Resource cost	Wage rate	It is generally thought that higher income groups value travel time at a higher price than lower income groups. Different wage rates are recommended to be used as the basis for calculating time values for truck drivers, air travelers, and travelers on surface passenger modes
	Trip purpose	There is consensus that on-the-clock work travel should be valued at the wage rate including fringe benefits, while other trip purposes should be valued at some fraction of the wage rate
	Amount of timing saving	There has been substantial disagreement in the literature on the value of small units of time. Some studies suggest that small increments of time have lower unit values than do larger increments of time. Others valued time savings at the same rate, regardless of the amount of time savings
Disutility cost	Congestion	Travel under congested conditions puts extra stress on the driver. As a result, reductions in travel time during peak periods, which are most likely to be congested, are likely to be valued more highly than reductions in travel time during off-peak periods
	Passenger and driver time	It is logical that stresses of driving may make travel time savings more important to drivers than to passengers with a higher value of time for drivers
	LOS, walking, and waiting time	There is disagreement about whether distinctions should be made between travel modes due to differences in comfort and other service attributes. It is generally accepted that time spent walking and waiting for a vehicle exposure to adverse weather has a higher value to the rider than time spent riding

monetary values. The time cost of travel generally includes two components: the resource cost reflecting the value to the traveler of an alternative use of time such as work and the disutility cost as the level of discomfort, boredom, or other negative aspects associated with time lost due to travel. Table 7.7 lists factors affecting the value of travel time.

The methods derived for measuring the value of travel time typically fall into five types of analyses: mode choice, route choice, speed choice, dwelling choice, and wage rate-based analyses, as briefly summarized in Table 7.8.

FHWA provides detailed factors for values of travel time computation in a different situation, as presented in Tables 7.9 through 7.11. Table 7.12 lists values of travel time used in some existing models for project evaluation.

7.3.6.5 Unit costs of vehicle crashes

Vehicle crashes can vary in severity and the number of individuals involved. By severity, vehicle crashes can be divided into fatal, injury, and property damage only (PDO) categories. Fatalities result in lost years of life, while injuries result in lost years of productive life and may also cause pain and suffering. In addition, all vehicle crashes result in property damages of varying severity. Tables 7.13 and 7.14 present methods evaluating vehicle crash losses and unit costs of vehicle crashes in existing models.

7.3.6.6 Unit costs of vehicle air emissions

Transportation investments affect the environment because of the construction process, impacts of the physical asset itself, and resulting changes in travel behavior. Vehicle emissions

Table 7.8 Methods for estimating the value of travel time

Method	*Description*
Mode choice	Mode choice analysis attempts to compare a fast, but expensive mode with an inexpensive, but a slow one. The difference in cost is presumably equal to the value of the difference in time. Most of these analyses compare automobiles with some sort of transit
Route choice	In route choice analysis, a slow and inexpensive route option is compared with a faster and more expensive route option for a single travel mode. The difference in cost is presumably equal to the value of the difference in time
Speed choice	Speed choice analysis is one attempt to supplement the results of route choice analysis. The analyses are based on the economic assumption that rational, utility maximizing individuals adopt driving speeds that minimize their total trip costs. While travel time is one component of the trip cost, there are other trip costs, such as vehicle operating costs and accident costs. Assuming that all costs are perceived by drivers and that the least cost speed is selected, the perceived time costs can then be determined
Dwelling choice	In this form of analysis, the value of time is calculated by comparing housing value against the time it takes to reach the work location. The analysis results can be used to corroborate other methods
Wage rate	For "off-the-clock" travel, the hourly wage rate is treated as a standard against which the value of time is measured. The concept underlying this approach is that travelers' hourly wages give the opportunity cost of their time. The percentage of wage rate appears to be a convenient metric to measure value of time associated with "off-the-clock" travel
	For the value of "on-the-clock" travel time, there is a general consensus that a driver's wage rate is the right measure of the value of time when highway travel is part of the person's work. Thus, the average labor cost for truck drivers is an appropriate value of time for truck traffic

Table 7.9 Travel purpose by modes

Trip purpose	*Nonwork travel (%)*	*Work travel (%)*
Local travel by surface mode	95.4	4.6
Intercity travel by surface mode	78.6	21.4
Intercity travel by air	59.6	40.4

Table 7.10 Recommended value of travel time (% wage rate)

Category	*Surface mode*	*Air and high-speed rail (%)*	*%*
Local	Nonwork	50	
	Work	100	
Intercity	Nonwork	70	70
	Work	100	100

Note: Operators are 80%–120% on all modes.

generally fall into two categories: vehicle emit pollutants such as CO, NO_X, VOC, PM, SO_2; and greenhouse gas emissions, mainly caused by CO_2. Air pollutants can cause damage to human health, building materials, and agriculture and vegetation, as well as limit visibility. Increasing concentrations of greenhouse gases in the atmosphere may be causing changes in the Earth's climate that could potentially impose substantial costs on society in terms of flooding, crop loss, and increased incidence of disease (Matthews et al., 2001). Table 7.15 presents factors that affect vehicle air emission quantities.

Table 7.11 Recommended value of travel time (2016 dollars)

Category	*Surface mode*	*Air and high-speed rail*	
Local	Nonwork	13.36	
	Work	25.49	
	All purpose	13.91	
Intercity	Nonwork	18.59	35.50
	Work	25.49	63.66
	All purpose	23.03	46.86
Truck drivers	27.49		
Bus drivers	27.27		
Transit rail operators	44.96		
Locomotive engineers	38.18		
Airline pilots and engineers	84.70		

The air emission unit costs are typically estimated based either on damage costs or control costs. Damage cost valuation involves estimating the actual value of the harm caused by air emissions, whereas control cost valuation simply examines the cost of the measures necessary to reduce air pollutant emissions. Damage cost valuation is preferable because studies that use control costs to value air pollution rely on the assumption that the controls placed on pollution are efficient. Table 7.16 lists steps involved with a damage cost valuation. Table 7.17 summarizes unit costs of vehicle air emissions established in existing models.

7.3.6.7 Annual and life-cycle user costs

With data traffic details of daily traffic volumes, vehicle mixes, speeds, and speed changes associated with a transportation project made available, the quantities of daily user costs for different user cost components including costs of vehicle operations, travel time, crashes, and air emissions can be computed. The user cost quantities can be multiplied by corresponding monetized unit costs to establish the daily monetary user costs that can be extrapolated to arrive at the annual monetary user costs.

Present worth and equivalent uniform annual user costs with annual growth: With growth in annual traffic volumes during the facility service life cycle, the total travel will grow accordingly. This may lead to a growing trend of annual user costs of vehicle operations, travel time, crashes, and air emissions. Without loss of generality, a geometric growth rate can be assumed for annual user costs over time in reference to the annual user costs in base year 0.

Denote

C_0^{UC} = Annual user costs in the base year 0
C_t^{UC} = Annual user costs in year t
$PW^{UC}(T)$ = Present worth of annual user costs in T-year analysis period
PW_{∞}^{UC} = Present worth of annual user costs in perpetuity
EUA^{UC} = Equivalent uniform annualized user costs
g^{UC} = Growth rate of annual user costs
T = T-year analysis period
i = Discount rate

Table 7.12 Summary of values of travel time in existing models (2000 dollars)

Model	*Auto*	*Bus*	*Truck*	*Source*
2003 AASHTO Red Book STEAM CAL-B/C	• 50% of the wage rate for driving alone commute • 60% of the wage rate for carpool driver commute • 40% of the wage rate for carpool passenger commute • 50% of the wage rate for personal local trip • 70% of the wage rate for personal intercity trip • 100% of total compensation for business	• 50% of the wage rate for in-vehicle commute • 50% of the wage rate for in-vehicle personal • 100% of the wage rate for nonbusiness waiting, walking, or transfer time • 100% of total compensation for business	• 100% of total compensation for in-vehicle business • 100% of total compensation for business waiting time	USDOT (1992)
HERS-ST	Work-related travel: $13.96/veh-hour	–	Work-related travel: • $15.82/veh-hour for 4-tire truck • $29.72/veh-hour for 6-tire truck • $33.97/veh-hour for 3–4 axle truck • $37.76/veh-hour for 4-axle comb truck • $37.98/veh-hour for 5-axle comb truck	USDOT (1992)
	Nonwork travel: 60% of the wage rate	–	Nonwork travel: 60% of the wage rate	Jack Faucett Associates (1991)
StratBENCOST	• Low: $12.04/veh-hour • Med: $12.93/veh-hour • High: $25.64/veh-hour	• Low: $84.78/veh-hour • Med: $91.03/veh-hour • High: $180.50/veh-hour	• Low: $33.00/veh-hour • Med: $35.43/veh-hour • High: $70.25/veh-hour	TTI (1990) Jack Faucett Associates (1991)

Note: The unit dollar values can be updated to current year dollars using relevant consumer price indices and producer price indices for non-trucks and trucks, respectively.

Table 7.13 Methods for valuating vehicle crash losses

Method	*Description*
Direct cost	This method measures only the easily measurable out-of-pocket costs of accidents, which include crash clean-up, injury treatment, property repair and replacement, accounting for workplace disruption, and insurance claims processing and related costs. The personal costs, emotional and physical, are ignored in the direct costs method
Human capital	This method calculates values as a function of salary. As a result, lower values are computed for women and children than for men. This method ignores pain, suffering, and lost quality of life. Human capital costs are useful to determine the dollars lost to injury and death, and form the basis for legal compensation awards
Years of loss plus direct cost	This method estimates two sets of costs: years of life lost to fatalities and productive life lost to nonfatal injuries, and the dollar value of the medical costs. Since medical costs for a serious injury are much higher than for a sudden death, the combined value could be misleading
Willingness to pay	This method involves evaluating the reduction of accident risk by estimating the amount people pay for small decreases in safety and health risks, often obtained through the analysis of safety equipment purchases made by individuals. The method places a value on people's behavior of exchanging money, time, comfort, and convenience for safety. Frequently, these values are added to results of the direct cost approach to obtain an overall crash value

Table 7.14 Summary of vehicle crash unit costs in existing models

Model	*Fatality*	*Injury*	*PDO*	*Source*
2003 AASHTO Red Book	Cost per fatal crash: $3366388 Delay component: $9148	Cost per injury crash: Critical: $2402997 Severe: $731580 Serious: $314204 Moderate: $157958 Minor: $15017 Delay component: Critical: $9148 Severe: $999 Serious: $940 Moderate: $846 Minor: $777	Cost per PDO crash: $3900	U.S. National Highway Traffic Safety Administration (NHTSA) (2000)
CAL-B/C	$3104738	$81572	$6850	U.S. National Safety Council (NSC) (1995)
HERS-ST	$2911243	Urban: $14556–26201 Rural: $24746–29112	Urban: $7278–8734 Rural: $5822–7278	Jack Faucett Assoc. (1991)
STEAM	$2925093	$64071	$3565	FHWA (1994)
StratBENCOST (1996 dollars)	Low: $887948 Med: $3864742 High: $8887021	Low: $16403 Med: $92024 High: $237829	Low: $1583 Med: $6372 High: $12863	FHWA (1994)

The present worth and equivalent uniform annualized user costs can be estimated as below:

$$C_t^{UC} = C_0^{UC} \cdot e^{(g^{UC} \cdot t)}$$

$$PW^{UC}(T) = \int_0^T \left[\left(C_0^{UC} \cdot e^{(g^{UC} \cdot \tau)}\right) \cdot e^{(-i \cdot \tau)}\right] d\tau = C_0^{UC} \cdot \left[\frac{\left[e^{((g^{UC}-i) \cdot T)} - 1\right]}{(g^{UC} - i)}\right] \tag{7.18}$$

$$PW_{\infty}^{UC} = \int_{0}^{n \cdot T} \left[\left(C_0^{UC} \cdot e^{(g^{UC} \cdot \tau)} \right) \cdot e^{(-i \cdot \tau)} \right] d\tau = C_0^{UC} \cdot \left[\frac{\left[e^{((g^{UC} - i) \cdot n \cdot T)} - 1 \right]}{(g^{UC} - i)} \right] \quad (7.19)$$

$$EUA^{UC} = i \cdot PW_{\infty}^{UC} = \left(i \cdot C_0^{UC} \right) \cdot \left[\frac{\left[e^{((g^{UC} - i) \cdot n \cdot T)} - 1 \right]}{(g^{UC} - i)} \right] \quad (7.20)$$

Table 7.15 Factors affecting vehicle air emission quantities

Factor	*Description*
Vehicle age	The engine fuel efficiency decreases with the increase of vehicle age. This accordingly will increase air emission rates
Vehicle speed	Speeds are essential to determine vehicle emission rates. In general, VOC emission rates tend to drop as speed increases, whereas NO_X and CO emission rates increase at higher speeds (above 55 miles/h)
Vehicle mix	Mix of vehicle types in the traffic stream and mix changes affect emission rates
Traffic condition	Emission rates are also higher during stop-and-go, congested traffic conditions than during free flow conditions at the same average speed
Ambient air temperature and cold-start trips	A cold-start vehicle generates additional emissions because a vehicle's emissions control equipment has not reached its optimal operating temperature

Table 7.16 Damage cost method for estimating the unit cost of vehicle air emissions

Step	*Description*
Impact of pollutant emissions on air quality	Ambient air pollution concentrations are the result of air pollutant dispersion, reaction, and residence, complicated by meteorology and topography. These processes result in nonlinear relationships between pollutant emissions and air concentrations that can be determined through computer modeling
Increase of health problems caused by air quality deterioration	The dose–response functions can be used to estimate the increased risk of developing a certain adverse health effect, such as headaches, chronic respiratory problems, or mortality, in response to increased air pollutant concentrations
Dollar costs per health effect	Health impacts in monetary terms can be quantified using a revealed preferences method that estimates costs based on people's behavior; and expressed preferences that ask people about the cost of an impact
Estimation of unit costs	Unit costs per ton of pollutants emitted can be estimated based on the above information

Table 7.17 Summary of vehicle air emission costs per ton in existing models (2000 dollars)

Model	*CO*	*NO_X*	*PM*	*SO_2*	*VOC*	*Source*
CAL-B/C HERS-ST STEAM StratBENCOST	Rural • Low: $11 • High: $55 Urban • Low: $22 • High: $110	Rural • Low: $1674 • High: $3978 Urban • Low: $2511 • High: $5968	Rural • Low: $1329 • High: $2646 Urban • Low: $2658 • High: $5292	Rural • Low: $1757 • High: $9220 Urban • Low: $2636 • High: $13831	Rural • Low: $1157 • High: $3023 Urban • Low: $1735 • High: $4534	McCubbin and Delucchi (1996)

7.4 FACILITY LIFE-CYCLE BENEFIT ANALYSIS

7.4.1 Life-cycle agency benefit analysis

The typical life-cycle activity profile of a physical transportation facility such as a pavement or bridge represents the most cost-effective investment strategy to manage the facility. If any needed treatment fails to be implemented on time as planned in the typical life-cycle activity profile, it may lead to significant increases in the needed repair cost, and an early termination of the facility service life could potentially happen. Thus, the typical life-cycle activity profile can be used as the base case activity profile and the case with early service-life termination can be considered as an alternative case activity profile. For each facility, the reduction in life-cycle agency costs of the base case activity profile compared with the alternative case activity profile can be computed as project-level life-cycle agency benefits of timing implementing the needed project in facility service life cycle. Similarly, the decrease in facility life-cycle user costs according to the base case activity profile against the alternative case activity profile can be estimated as the project-level life-cycle user benefits.

Figure 7.3 illustrates an example of base case and alternative case activity profiles for the steel-box beam bridge and the method for estimating project-level life-cycle agency benefits and user benefits by keeping the typical life-cycle activity profile for the bridge. For the base case life-cycle activity profile, agency costs in the T-year bridge service life consist of initial bridge construction cost C_{CON} in year 0, first deck rehabilitation cost $C_{DECK\ REH1}$ in year t_1, deck replacement cost $C_{DECK\ REP}$ in year t_2, second deck rehabilitation cost $C_{DECK\ REH2}$ in year t_3, and annual routine maintenance costs. The annual routine maintenance costs between two major treatments in the bridge life cycle will gradually increase over time due to the combined effect of higher traffic volume, truck traffic volume, and axle loading; aging materials; climate conditions; and other non-load-related factors. Different geometric gradient growth rates are used for intervals between year 0 and t_1, t_1 and t_2, t_2 and t_3, and t_3 and T, respectively.

For the alternative life-cycle activity profile, it is assumed that the deck replacement project (with the cost of $C_{PROJECT}$) is actually implemented y_1 years after year t_2 as the base case profile, namely, $C_{DECK\ REP}$ in year t_2 is replaced by $C_{PROJECT}$ in year $t_2 + y_1$. This will defer the second deck rehabilitation by y_1 years. Due to postponing deck replacement and the second deck rehabilitation, the bridge service life may experience an early termination of y_2 years. As for the annual routine maintenance costs, different geometric gradient growth

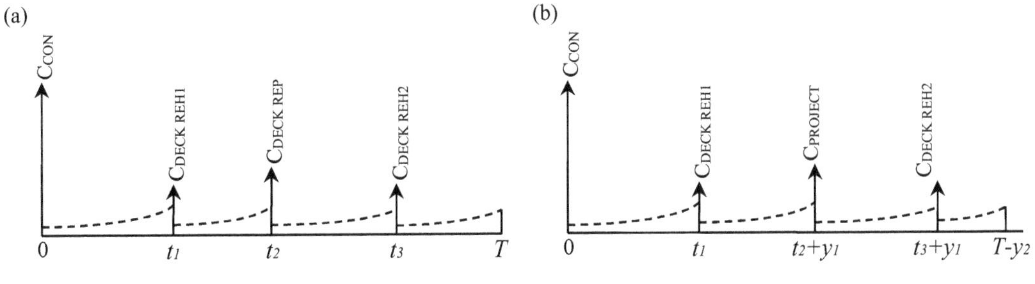

Figure 7.3 Illustration of base- and alternative-case life-cycles of a steel-box beam bridge. (a) Base case life-cycle. (b) Alternative case life-cycle. (Adapted from Li, Z. and P. Kaini. 2007. Optimal Investment Decision-Making for Highway Transportation Asset Management under Risk and Uncertainty. Report MRUTC 07-10. Department of Civil, Architectural, and Environmental Engineering, Illinois Institute of Technology, Chicago, IL. Li, Z. and S. Madanu. 2009. *ASCE Journal of Transportation Engineering* 135(8), 516–526.)

rates are used for intervals between year 0 and t_1, t_1 and $t_2 + y_1$, $t_2 + y_1$ and $t_3 + y_1$, and $t_3 + y_1$ and $T - y_2$, correspondingly. In particular, the annual routine maintenance cost profiles for the base case and alternative case profiles are identical from year 0 to year t_2. The project-level life-cycle agency benefits are estimated as the reduction in bridge life-cycle agency costs quantified according to the base case activity profile compared with the alternative case activity profile.

The primary user cost items include costs of vehicle operations, travel time, crashes, and air emissions. For each user cost item, the base case and alternative case annual user cost profiles in bridge life cycle follow a pattern that resembles the profile of annual routine maintenance costs in bridge life cycle. In either the base case profile or alternative case profile, the "first-year" user cost amounts immediately after the major treatments including bridge construction, first deck rehabilitation, deck replacement, and second deck rehabilitation are directly computed based on the unit user cost in constant monetized values per VMT and the annual VMT. The geometric growth rate is then applied to the "first-year" user cost amount for each interval between two major repair treatments to establish the annual user cost amounts for subsequent years within the interval. Additional work-zone-related costs can be estimated using the procedures in FHWA (1998b, 2000) and AASHTO (1978, 2003), and added to the annual user cost amounts for the years in which major treatments are implemented. This ultimately establishes the base case and alternative case annual user cost profiles for costs of vehicle operations, travel time, crashes, and air emissions, respectively.

For each user cost item, the annual user cost profiles for the base case and alternative case are identical from year 0 to t_2 and are different for the remaining years in the bridge life cycle. The travel demand in terms of annual VMT for a specific year after year t_2 could be different between the base case and alternative case due to the fact that the traffic volume, that is, AADT and/or travel distance associated with the bridge, might change for the two cases. The consumer surplus concept is employed to separately compute the user benefits by comparing the base case and alternative case annual user cost profiles for intervals from year t_2 to $t_2 + y_1$, $t_2 + y_1$ to t_3, t_3 to $t_3 + y_2$, $t_3 + y_2$, $T - y_2$, and $T - y_2$ to T. The total project-level life-cycle user benefits are the combination of individual user benefit items associated with reductions in costs of vehicle operations, travel time, crashes, and air emissions in the bridge life cycle. With equal weights assigned for agency benefits and user benefits, the total project-level life-cycle benefits by keeping the typical life-cycle activity profile for the bridge are established by directly adding the two sets of benefits.

7.4.2 Life-cycle user benefit analysis

7.4.2.1 *Concept of consumer surplus*

Benefits of a transportation project are normally an amalgamation of cost savings and additional travel. Figure 7.4 demonstrates the user benefits of the project. It assumes price locates above AVC. The project user benefits can be calculated in two ways: (i) using the combination of MC and demand curve and (ii) using the combination of AVC and demand curve.

Project user benefits estimated using MC and demand curves: As seen in Figure 7.4a, the area under MC curve up to the current volume q_0 represents total variable costs for the base case without project implementation. The corresponding area for the alternative case with project implementation is lower but extends out to volume q_1. The cost difference is an area of cost savings between the two curves up to q_0, and an area of additional costs under MC curve MC_1 from q_0 to q_1. The latter is offset by the (not necessarily equal) incremental benefits from the additional travel, illustrated by the area under the demand

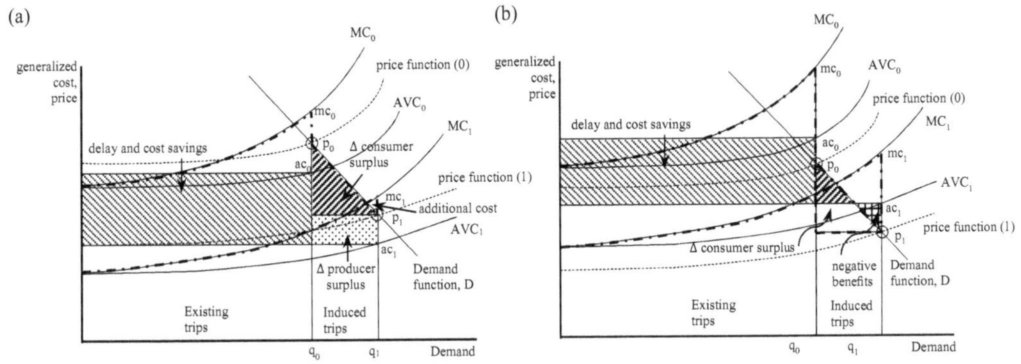

Figure 7.4 Illustration of user benefits of a transportation project. (a) Price function above average variable cost. (b) Price function below average variable cost. (Adapted from Li, Z. and P. Kaini. 2007. Optimal investment decision-making for highway transportation asset management under risk and uncertainty. Report MRUTC 07-10. Department of Civil, Architectural, and Environmental Engineering, Illinois Institute of Technology, Chicago, IL.)

curve from q_0 to q_1. The aggregated project user benefits are the area represented by the dot–dash line. It can be depicted as the area between the two MC curves and beneath the demand curve.

Where MC crosses above the demand curve, the area marked "additional costs" is negative. These negative benefits are a result of underpricing the project alternative, relative to MC pricing. If the new project is efficiently priced, the project user benefits could be increased by that much. Correspondingly, the project user benefits would be reduced if the inefficiency from underpricing the base case is not included.

Project user benefits estimated using AVC and demand curves: Areas under the MC curve can also be represented by rectangles constructed from the AVC curve using the following relation:

$$\int_0^q MC = q \cdot AVC_q \tag{7.21}$$

For the case where the price curve is above the AVC curve as illustrated in Figure 7.4a, the area under MC_0 up to q_0 is equal to the rectangle whose length and height are q_0 and ac_0, respectively. Similarly, the area under MC_1 up to q_1 is equal to the rectangle whose length and height are q_1 and ac_1, respectively. The difference between these two rectangles is the shaded area that is exactly equal to the outlined area generated from the marginal curves. This shaded are is labelled "delay and cost savings," minus the additional costs from q_0 to q_1, plus the area under the demand curve from q_0 to q_1.

In reality, by reducing in price from p_0 to p_1, a distinction between the base case old trips q_0, and new trips from q_0 up to q_1 is derived. A reason for making this distinction is the nature of the benefits to the two groups: existing old users have demonstrated their willingness to pay for their travel, and so the cost reductions over their previous generalized cost to them become the benefits. In contrast, new trip makers on this facility have not shown any willingness to pay. Their benefits must be calculated from the demand curve as additional consumer surplus and producer surplus over what they actually pay when using the improved facility.

Consumer surplus is the amount users would be willing to pay above what they actually pay, measured as an area under the demand curve between the "with" and "without" induced vehicle volumes and above the price. The incremental consumer surplus applies to induced "new" trips and the relevant volumes are q_0 (without improvement) and q_1 (with improvement). The incremental consumer surplus is a triangular area whose hypotenuse is the demand curve between p_0 and p_1, and whose legs are $(p_0 - p_1)$ and $(q_1 - q_0)$.

Producer surplus is an area under the demand curve—that is, below what users pay but above short-run variable cost. Normally, user fees are ignored in estimating benefits as they are considered as transfers, but here it is simply a part of the means for valuing induced travel. Similar to consumer surplus, it implies a willingness to pay for new trips. A congestion toll generates producer surplus, but only the portion on new trips is counted as a benefit; the portion applying to old trips is already counted in the cost and time savings on old trips. The net of toll revenues above incremental agency costs and externalities is producer surplus.

Figure 7.4b demonstrates a scenario where price curve is below the average variable curve for both the base case and project alternative case. The outline of project user benefits based on MC is typically the same, but the area defined by AVC curves to some extent has a different shape. Savings on old trips start above the current price, because the elimination of externalities in the base case is a benefit. Similarly, the benefits stop farther up, because some of the vehicle operating cost and travel time savings are offset by agency costs or externalities in the project alternative case. Cost savings would come down to p_1 were it not for the new externalities. The incremental consumer surplus is the same in both figures, but a share of it is offset by the negative producer surplus where the toll revenues are below incremental agency costs and externalities. In summary, the primary components of project user benefits are changes in consumer surplus on old trips, incremental consumer surplus on new trips, and producer surplus on new trips.

7.4.2.2 User benefits on a directly affected highway segment with shift in demand

Provided with a demand curve, the consumer surplus is the amount road users in the aggregate would have been willing to put on top of what they are actually asked to pay. The change in consumer surplus between a project alternative and the base case is considered as the user benefits associated with the project alternative. For a generalized case where the demand curve shifts upward because of a project improvement, the user benefits can be estimated as shown in Figure 7.5.

7.4.2.3 User benefits on an indirectly affected highway segment with shift in demand

In some cases, a backward shift in demand may happen at some indirectly affected segments where improvements cause traffic to shift to the improved segment. That is, the travel demand on the indirectly affected segments is reduced at every user cost. As illustrated in Figure 7.6, the change in consumer surplus is only related to the one measured on the directly affected segment. The approach can be applied to every affected link to account for all changes in consumer surplus.

7.4.2.4 Annual user benefits in facility service life cycle

The user benefits as changes in consumer surplus for projects directly and indirectly affected highway segments can be separately computed for costs of vehicle operations,

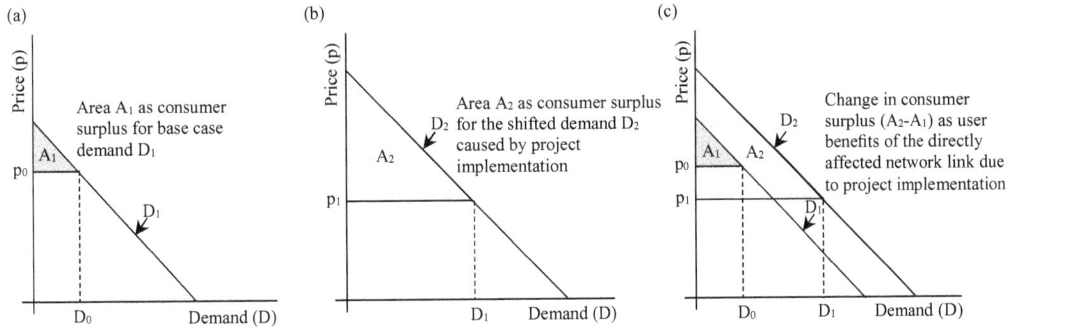

Figure 7.5 Illustration of user benefits on a directly affected network link with demand shift. (a) Demand curve for base case. (b) Upward shifted demand curve after project implementation. (c) Combined effect for a directly affected highway segment. (Adapted from Li, Z. and P. Kaini. 2007. Optimal investment decision-making for highway transportation asset management under risk and uncertainty. Report MRUTC 07-10. Department of Civil, Architectural, and Environmental Engineering, Illinois Institute of Technology, Chicago, IL.)

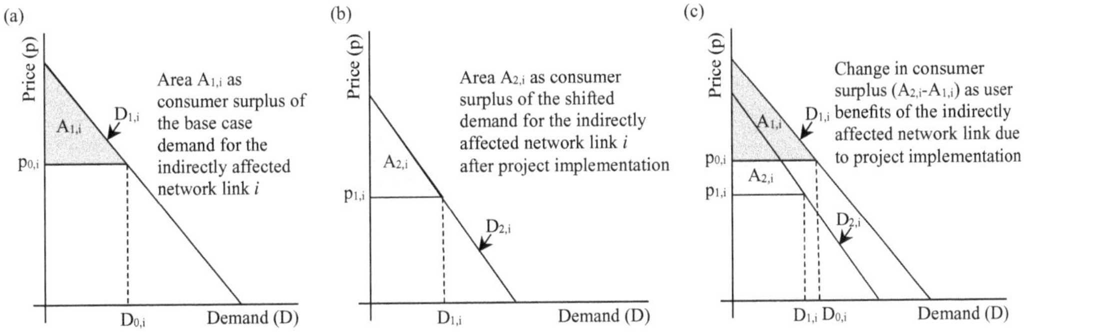

Figure 7.6 Illustration of user benefits on an indirectly affected network link with demand shift. (a) Demand curve for the indirectly affected segment in base case. (b) Backward shifted demand for the indirectly affected segment due to project implementation. (c) Combined effect for an indirectly affected highway segment. (Adapted from Li, Z. and P. Kaini. 2007. Optimal investment decision-making for highway transportation asset management under risk and uncertainty. Report MRUTC 07-10. Department of Civil, Architectural, and Environmental Engineering, Illinois Institute of Technology, Chicago, IL.)

vehicle travel time, crashes, and air emissions. The individual user benefits could be aggregated to establish the overall user benefits. The computation can be made based on daily travel to calculate the daily user benefits, which can be expanded to annual user benefits.

7.4.2.5 Life-cycle user benefits

After an estimation of the overall user benefits in the first year of the physical asset useful service life is obtained, the overall user benefits for the future years can be extrapolated by considering a geometric annual traffic growth rate. This will help determine the life-cycle user benefits for the project.

EXAMPLE 7.9

A rural 2-lane road connecting two towns currently has an alignment staying close to a river. The alignment consists of two tangent sections on both ends, four identical sharp curves (1430 ft radius, approximately 19° curvature) in the middles, and three tangent sections connecting the curved sections. Each tangent section is 1000 ft and the curve section is 250 ft long. The road will climb a +3% grade from one end to the other. The design speeds are 50 and 35 mph for the tangent and curved sections, respectively. During the remaining 15-year pavement design service life period, a pavement rehabilitation treatment that costs \$50000 is required in the 7th year and annual maintenance treatments are needed, which are expected to be \$1000 per year. However, the existing alignment suffers from an embankment erosion problem which requires a repair treatment of \$100000 after the road has been in service for 5 years.

Alternatively, the local Highway Department is intending to change the entire alignment that stays away from the river bank to completely avoid the embankment erosion problem. For the proposed alignment, it only consists of one 5000 ft long horizontal curve (6000 ft radius, approximately 1° curvature). The road will climb a +3% grade for the first 3000 ft and a +2% grade for the remaining 2000 ft. The design speed is at 55 mph for the entire alignment. For the 20-year design service life period, the initial construction cost is \$400000, and one pavement rehabilitation treatment is required in the 10th year that costs \$50000. The annual maintenance cost is expected to be \$500 annually.

The base year daily traffic is 2500 vehicles per day, with autos only in the traffic stream and 50/50 directional split. The annual traffic growth rate is at 2%. Use a discount rate of 5%. The daily traffic is assumed to be static after the realignment.

Assume that the new construction cost occurs in the base year and that all other costs including embankment erosion control, rehabilitation, maintenance, and highway user cost are counted at the end of each year. Compute the agency and user costs associated with vehicle operations and travel time at \$23.1/hour to determine which option should be chosen.

Average running speed (mph)	*Grade (%)*	*Tangent running cost on grades (dollar/1000 VMT)*	*Degree of curvature*	*Added running cost on curves (dollar/1000 VMT)*
35	+3	150.3	19	43.5
35	−3	115.9		
50	+3	155.7	–	–
50	−3	112.3		
55	+3	157.6	1	5.4
55	−3	117.7		
55	+2	153.9		
55	−2	121.3		

Solution

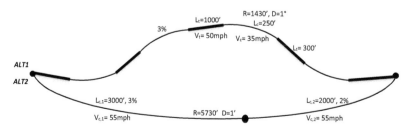

1. Agency costs
   a. Existing alignment

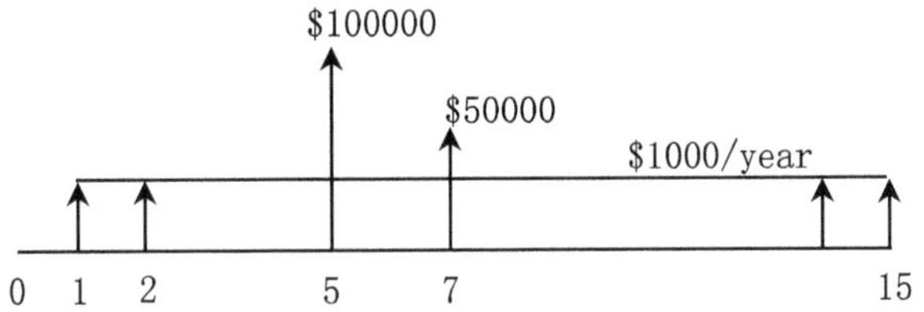

$$AC_0 = \left[100000 \times \frac{1}{(1+0.05)^5} + 50000 \times \frac{1}{(1+0.05)^7}\right] \times \left[\frac{0.05(1+0.05)^{15}}{(1+0.05)^{15}-1}\right] + 1000 = \$11972.1/\text{year}$$

   b. Proposed alignment

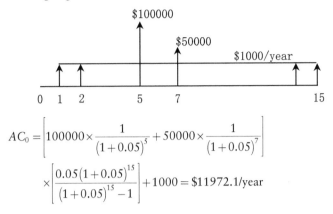

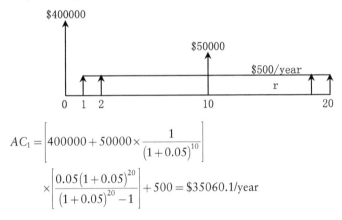

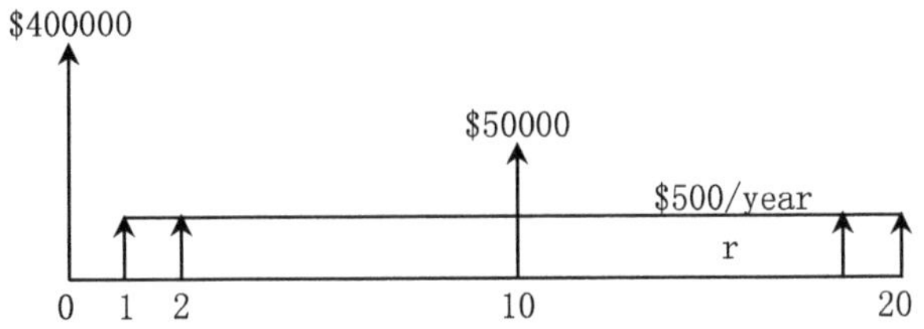

$$AC_1 = \left[400000 + 50000 \times \frac{1}{(1+0.05)^{10}}\right] \times \left[\frac{0.05(1+0.05)^{20}}{(1+0.05)^{20}-1}\right] + 500 = \$35060.1/\text{year}$$

2. *User costs and benefits*: Assume the average running speed is identical to the design speed for simplification purposes.
   In the base year (Year 0):
   Directional Daily Volume = 2500 × 50% = 1250 veh/day
   Directional Annual Volume = 1250 × 365 = 456250 veh/year
   a. Existing alignment
   Travel Time (TT) per kVMT on Tangents = 1000/50 = 20 h/kVMT
   Travel Time per kVMT on Curves = 1000/35 = 28.6 h/kVMT

		Tangents (50 mph)	*Curves (35 mph)*
VC: +3%	a) TT	20 h/kVMT × \$23.1/h = \$462/kVMT	28.6 h/kVMT × \$23.1/h = \$660/kVMT
	b) $VOC_{tangent}$	\$155.7/kVMT	\$150.3/kVMT
	c) VOC_{curve}	0	\$43.5/kVMT
	Sum	\$617.7/kVMT	\$793.8/kVMT
VC: −3%	a) TT	20 h/kVMT × \$23.1/h = \$462/kVMT	28.6 h/kVMT × \$23.1/h = \$660/kVMT
	b) $VOC_{tangent}$	\$112.3/kVMT	\$115.9/kVMT
	c) VOC_{curve}	0	\$43.5/kVMT
	Sum	\$574.3/kVMT	\$819.4/kVMT

Total unit cost ($/kVMT)	617.7 + 574.3 = $1192/kVMT	793.8 + 819.4 = $1613.2/kVMT
Total unit cost ($/kV)	1192 × 1000 × 5/5280 = $1128.79/kV	1613.2 × 250 × 4/5280 = $305.53/kV

The equivalent Unit Cost = 1128.79 + 305.53 = 1434.32$/kV
Total Usage = 456250 veh/year = 456.25 kV/year

b. Proposed alignment:
Travel Time per kVMT = 1000/55 = 18.2 h/kVMT

		Curve 1 (3000 ft)	*Curve 2 (2000 ft)*
VC: +3%	a) TT	18.2 h/kVMT × $23.1/h = $420.42/kVMT	
	b) $VOC_{tangent}$	$157.6/kVMT	
	c) VOC_{curve}	$5.4/kVMT	
	Sum	$583.42/kVMT	
VC: -3%	a) TT	$420.42/kVMT	
	b) $VOC_{tangent}$	$117.7/kVMT	
	c) VOC_{curve}	$5.4/kVMT	
	Sum	$543.52/kVMT	
VC: +2%	a) TT		$420.42/kVMT
	b) $VOC_{tangent}$		$153.9/kVMT
	c) VOC_{curve}		$5.4/kVMT
	Sum		$579.72/kVMT
VC: −2%	a) TT		$420.42/kVMT
	b) $VOC_{tangent}$		$121.3/kVMT
	c) VOC_{curve}		$5.4/kVMT
	Sum		$547.12/kVMT
Total unit cost ($/kVMT)		583.42 + 543.52 = $1126.94/kVMT	579.72 + 547.12 = $1126.84/kVMT
Total unit cost ($/kV)		1126.94 × 3000/5280 = $640.31/kV	1126.84 × 2000/5280 = $426.83/kV

The equivalent Unit Cost = 640.31 + 426.83 = 1067.14 $/kVMT
Total Usage = 456.25 kV/year

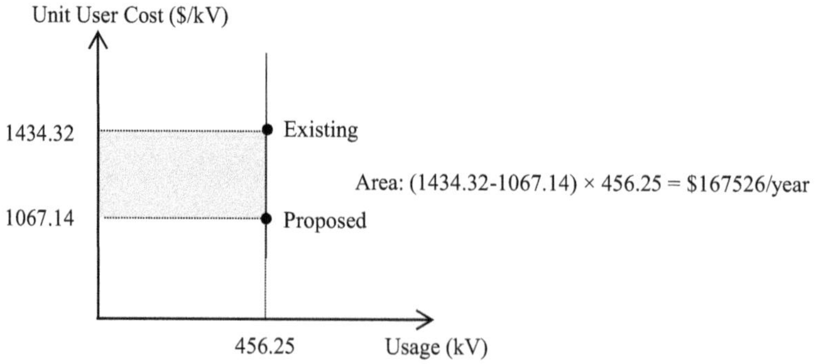

The user benefits (UB) measured by changes in consumer surplus are
UB(year 0) = Changes in consumer surplus = $167526
UB(year 1) = $167526 × (1 + 0.02) = $170877

The cash flow of the user benefit can be viewed as a geometric series because of the constant yearly growth factor. Therefore, the present worth of user benefits from year 1 to year 20 is approximately computed as follows:

$$PWF_{0,N} = \frac{1 - e^{-((i-r)N/(1+r))}}{i-r} = \frac{1 - e^{-(((0.05-0.02)20)/(1+0.02))}}{0.05-0.02} = 14.82$$

$$PW_{UB} = UB(year1) \times PWF_{0,20} = 170877 \times 14.82 = \$2532397$$

$$\text{UB} = 2532397 \times \left[\frac{0.05(1+0.05)^{20}}{(1+0.05)^{20} - 1}\right] = \$203206/\text{year}$$

3. *Conclusion*: Now we are considering two alternatives: 1) Keep using the existing infrastructure; 2) Use the proposed alignment. The net present worth (NPW) estimates are

$$NPW_o = \text{UB}_0 - AC_0 = 0 - 11972.1 = -\$11972.1/\text{year} < 0$$

$$NPW_1 = \text{UB}_1 - AC_1 = 203206 - 35060.1 = \$168146/\text{year} > 0$$

Since the proposed new alignment alternative has a positive and higher net present value, building a new alignment is recommended.

7.4.3 Risk and uncertainty considerations

In the estimation of project agency and user benefits, various factors related to facility service lives, current and future traffic volumes and mixes, agency costs, user costs, and discount factors to bring costs occurring at different time points to a common reference point in time, are involved in the analysis. For a specific factor in the above, it is not always the case that a single value is attached to it. That is, the factor may not always be deterministic or under certainty.

Alternatively, the factor may exhibit multiple possible values that can be separately treated as risk and uncertainty cases. For the factor under risk, a probability distribution can be assigned to the possible values. The probabilistic risk assessment can be performed to compute its mathematical expectation or expected value. The expected value of the input factor can then be used for analysis. For the factor under uncertainty, a reliable probability distribution may not be determined for the possible values or even the full range of possible values is unknown. Accordingly, the mathematical expectation cannot be established. Uncertainty-based assessment in place of probabilistic risk assessment needs to be conducted (Ang and Tang, 1984a,b; Grinstead and Snell, 2012; Haimes, 2015).

7.4.3.1 Factors under risk and uncertainty considerations

The agency costs of construction, repair, and maintenance may not remain unchanged. Traffic predictions in the facility service life cycle may not follow the projected values. The discount rate may fluctuate over time in the facility life cycle. Such variations will in turn result in changes in the overall project-level life-cycle agency and user benefits. Correspondingly, risk and uncertainty of these factors are incorporated into project life-cycle agency and user cost and benefit analysis.

7.4.3.2 Risk factor assessment

Selection of probability distribution: The minimum and maximum values of agency cost, traffic, and discount rate factors under risk considerations are constrained by nonnegative values. In addition, the distribution of possible values of each risk factor could be either symmetric or skewed. Such distribution characteristics can be readily modeled by the beta distribution that is continuous over a finite range and allows for practically any degree of skewness and kurtosis. The beta distribution has four parameters—lower bound (L), upper bound (H), and two shape parameters α and β, with density function given by

$$f(x|\alpha,\beta,L,H) = \frac{\Gamma(\alpha+\beta)\cdot\left[(x-L)^{\alpha-1}\right]\cdot\left[(H-x)^{\beta-1}\right]}{\Gamma(\alpha)\cdot\Gamma(\beta)\cdot\left[(H-L)^{\alpha+\beta-1}\right]}(L \le x \le H) \tag{7.22}$$

where the Γ-functions are used to normalize the distribution so that the area under the density function from L to H is exactly 1.

The mean and variance of the beta distribution are calculated as

$$\mu = \alpha/(\alpha+\beta) \quad \text{and} \quad \sigma^2 = (\alpha\cdot\beta)/[((\alpha+\beta)^2)\cdot(\alpha+\beta+1)] \tag{7.23}$$

Generating possible values for the risk factor using simulation analysis: Simulation is typically a rigorous extension of sensitivity analysis that uses randomly sampled values from an input probability distribution to achieve possible values of a random variable or factor. Two types of sampling techniques are commonly used to perform simulations. Monte Carlo sampling is a technique that uses random numbers to select values from the probability distribution. The second type is the Latin hypercube sampling technique where the probability scale of the cumulative distribution curve is divided into an equal number of probability ranges. The number of ranges used is equal to the number of iterations performed in the simulation. Therefore, the Latin hypercube sampling technique is likely to achieve convergence in fewer iterations as compared to those of the Monte Carlo sampling technique (FHWA, 1998b).

In the simulation analysis, each simulation run will contain multiple iterations and replicated simulation runs could be executed. This way, the sample mean and standard deviation values can be computed using data on the multiple values from each simulation run. The grand mean and standard values then can be obtained using the mean values obtained from replicated simulation runs. The grand mean value could be considered as the expected value for the factor to be used as the input value for project-level life-cycle benefit and cost analysis. To ensure the statistical significance of using the grand mean value to represent the expected value, the minimum number of iterations in each simulation run and the number of replicated simulation runs for substantial variance reductions need to be achieved.

$$F_{(E)} = \left(\sum_{m=1}^{M}\sum_{n=1}^{N} X_i\right)/(M\cdot N) \tag{7.24}$$

where

$F_{(E)}$ = Grand average of possible values obtained for a factor from replicated simulation runs
X_i = A simulation output representing a possible value for the factor
N = Number of iterations in each simulation run
M = Number of replicated simulation runs

7.4.3.3 *Uncertainty factor assessment*

When a factor is inherited with uncertainty, a reliable probability distribution may not be established to characterize possible values that it takes or even the full range of possible values is unknown. In this case, a meaningful mathematical expectation cannot be directly computed through probabilistic risk assessment. Among various methods for uncertainty-based analysis, Shackle's model (an in-depth description of Shackle's model is given in Chapter 17) is well suited to handle the case of a factor under uncertainty (Shackle, 1949).

In general, Shackle's model overcomes the limitation of inability to establish the mathematical expectation for a factor involving a range of possible values. The model relies on three pillars. First, a degree of surprise is used to assess uncertainty associated with the possible values in place of the probability distribution. Then, a priority index is introduced by jointly evaluating each known value and the associated degree of surprise. Next, it identifies two values of the factor maintaining the maximum priorities, one on the gain side and another on the loss side from the expected value $F_{(E)}$. The expected value could be the mean, median, or mode of all known possible values, but it is not the mathematical expectation as the corresponding probabilities are unknown. The two values need to be standardized to remove the associated degrees of surprise. The absolute deviations of two outcomes relative to the expected value are termed as standardized focus gain x_{SFG} and standardized focus loss x_{SFL} from the expected value $F_{(E)}$. This model yields a triple $\langle x_{SFL}, F_{(E)}, x_{SFG}\rangle$ for each factor under uncertainty. More details of Shackle's model are given in Chapter 17; Ford and Ghose (1998), Young (2001), Li and Sinha (2004, 2009a), and Li and Madanu (2009).

To simplify applications of Shackle's model for uncertainty-based analysis, the grand average of simulation outputs from replicated simulation runs where each run is executed in multiple iterations can be used as the expected value $F_{(E)}$ for a factor under uncertainty. If higher values are preferred for the factor, the absolute deviation of the average of simulation outputs that are lower than the expected value can used as standardized focus loss value x_{SFL} and the absolute deviation of the average value of simulation outputs that are equal to or higher than the expected value can used as standardized focus gain value x_{SFG}, which are computed by

$$x_{SFL} = \left| \left(\sum_{m=1}^{M} \sum_{n=1}^{N_r} X_i \right) / (M \cdot N_r) - F_{(E)} \right| \tag{7.25}$$

$$x_{SFG} = \left| \sum_{m=1}^{M} \left(\sum_{n=1}^{N} X_i - \sum_{n=1}^{N_r} X_i \right) / [M \cdot (N - N_r)] - X_{(E)} \right| \tag{7.26}$$

x_{SFL} = Standardized focus loss value from the expected value $F_{(E)}$ (that cannot be treated as mathematical expectation) for the factor

x_{SFG} = Standardized focus gain value from the expected value $F_{(E)}$ for the factor

N_r = Number of simulation outputs in the rth simulation run such that $X_i < F_{(E)}$ if a higher value is preferred for the factor

For some input factors such as the discount rate, lower outcome values are preferred. In these cases, the N_r for computing the standardized focus loss value x_{SFL} and the standardized focus gain value x_{SFG} refers to number of the simulation outputs in the rth simulation run such that $X_i > F_{(E)}$.

As an extension to Shackle's model, a decision rule is introduced to help compute a single value X for the input factor based on the triple $\langle x_{SFL}, F_{(E)}, x_{SFG}\rangle$ that can be used for estimating project benefits. Assuming that the decision maker only tolerates loss from the expected value for the factor under uncertainty by ΔX and if higher values are preferred, the decision rule is set as

$$X = \begin{cases} F_{(E)}, & \text{if } x_{SFL} \leq \Delta X \\ \left(F_{(E)} - x_{SFL}\right) / \left[1 - \left(\Delta X / F_{(E)}\right)\right], & \text{otherwise} \end{cases} \tag{7.27}$$

When lower values are preferred for the factor, the decision rule is revised to

$$X = \begin{cases} F_{(E)}, & \text{if } X_{SFL} \leq \Delta X \\ (F_{(E)} + X_{SFL}) / [1 + (\Delta X / F_{(E)})], & \text{otherwise} \end{cases} \tag{7.28}$$

If the standardized focus loss x_{SFL} from the expected value $F_{(E)}$ does not exceed the tolerance ΔX, the expected value will be utilized for the factor for the computation. This will produce an identical input factor value for both uncertainty- and risk-based analyses. If the standardized focus loss x_{SFL} from the expected value $F_{(E)}$ is greater than ΔX, a penalty is applied to derive a unique value for the input factor. Different tolerance levels ΔX's may be used for different factors inherited with uncertainty in project life-cycle agency and user benefit–cost analysis.

7.4.4 A generalized framework for facility life-cycle benefit analysis

Figure 7.7 shows a generalized framework for project-level facility life-cycle benefit analysis when input factors under mixed conditions of *certainty* (the factor is purely deterministic with a single value), *risk* (the factor has many possible values with a known probability distribution), and *uncertainty* (the factor has many possible outcomes with unknown probabilities). If a factor is under certainty, the single value of the factor can be used for analysis. If a factor is under risk, a mathematical expectation of the factor can be applied for analysis. If

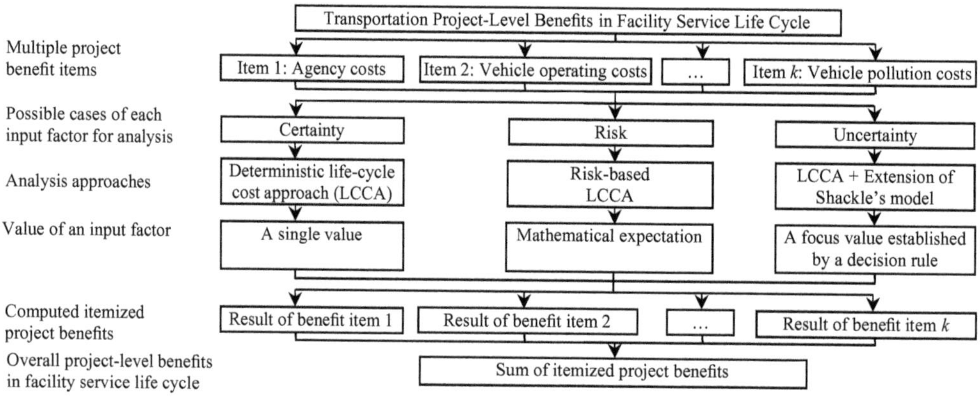

Figure 7.7 Framework of transportation project-level life-cycle benefit analysis under certainty, risk, and uncertainty. (Adapted from Li, Z. and P. Kaini. 2007. Optimal investment decision-making for highway transportation asset management under risk and uncertainty. Report MRUTC 07-10. Department of Civil, Architectural, and Environmental Engineering, Illinois Institute of Technology, Chicago, IL. Li, Z. and S. Madanu. 2009. *ASCE Journal of Transportation Engineering* 135(8), 516–526.)

a factor is under uncertainty, the single value of the factor determined according to the decision rule extended from Shackle's model can be adopted for analysis (Li and Madanu, 2009).

By using values of factors in the analysis determined under certainty, risk or uncertainty, the proposed framework helps estimate the project-level life-cycle agency and user benefits as reductions in agency costs, savings of vehicle operating costs, shortening of travel time, decreases in vehicle crashes, and cutbacks of vehicle air emissions. The combination of certainty, risk, and uncertainty cases for factors in the analysis may vary by benefit item for the same project and for different projects.

7.4.5 Facility life-cycle agency and user benefits in perpetuity

The project-level facility life-cycle benefits in perpetuity can be quantified based on the base case and alternative life-cycle activity profiles. Since the base case life-cycle activity profile is considered as the most cost-effective investment strategy, investment decisions are always made with the intention to follow the base case life-cycle activity profile. That is, the base case facility life-cycle activity profile in perpetuity will be represented by repeating the base case facility life-cycle activity profile an infinite number of times. For the alternative case life-cycle activity profile in perpetuity, early termination of service life may occur in the first life cycle, in the first and second life cycles or in the first several life cycles. After experiencing early service life terminations, the base case typical facility life cycle is expected to be resumed for the subsequent life cycles in perpetuity horizon. This is because the base case life cycle profile represents the most cost-effective investment strategy that the decision maker always aims to achieve.

Having created base case and alternative facility life-cycle activity profiles in perpetuity, the project-level facility life-cycle agency and annual user costs in perpetuity that follow the two activity profiles can be estimated. The reduction in project-level facility life-cycle agency costs according to the base case and alternative case activity profiles in perpetuity can be calculated and is treated as project-level facility life-cycle agency benefits in perpetuity. Similarly, reductions in project-level life-cycle user costs of vehicle operations, travel time, crashes, and air emissions based on the base case and alternative case activity profiles in perpetuity can be individually computed and then combined to establish project-level facility life-cycle user benefits in perpetuity. With equal weights assigned to agency and user benefits, they can be directly added together to establish overall project-level facility life-cycle benefits in perpetuity.

7.4.6 Annualized project-level facility life-cycle benefits and costs in perpetuity

To facilitate project evaluation, the project-level facility life-cycle agency costs of construction, repair, and maintenance according to the base case and alternative facility life-cycle activity profiles are first converted to present worth amounts and then to equivalent uniform annualized agency costs. The difference between the two amounts is treated as equivalent uniform annual agency benefits. Likewise, the project-level facility life-cycle annual user costs of vehicle operations, travel time, crashes, and air emissions according to the base case and alternative facility life-cycle activity profiles are first converted to present worth amounts and then to equivalent uniform annualized use costs. The difference between the two amounts is treated as equivalent uniform annual user benefits. The equivalent uniform annual agency and user benefits are summed up to establish the equivalent uniform annual benefits generated from project implementation. Meanwhile, a number of project costs are spread across the facility service life cycle to obtain the equivalent uniform annual costs. The equivalent

Table 7.18 Comparison of some benefit–cost analysis models

Name	*Source*	*Project type*	*Level of analysis*	*Special feature*	*Limitation*
2003 AASHTO Red Book	AASHTO	Highway operational improvements and safety projects	Project level	Travel time, VOC, and crash benefits of additional lanes, new highways, traffic control, signal systems, ITS improvements, pricing and regulatory policies; geometry, lane, access, and roadside safety improvements	Limited accounting for network effects; no accounting for modal interaction
Cal-B/C	California Department of Transportation (CALTRANS)	Highway, transit	Network level	Travel time, VOC, crash, and emission benefits of highway improvements, ITS, and transit improvements	No accounting for interaction between modes
HDM4	World Bank	Highway improvements	Network level	Includes 16 motorized and 8 nonmotorized vehicle types; includes roadway deterioration model for asphalt, concrete, gravel, and dirt roads; estimates emissions and energy consumption	No accounting for interaction between modes
IDAS	Cambridge Systematics	ITS improvements	Project level	Estimates benefits and costs for signals, ramp metering, incident management, electronic payment, traveler information, weigh-in-motion, and traffic surveillance	Evaluates ITS options only
MicroBENCOST	Texas Transportation Institute (TTI)	Highway improvements and safety projects	Project level	Includes intersection and interchange delay, bridges, railroad crossings, HOVs, and safety improvements; analyze emissions, construction delays; estimates discomfort costs based on road condition	Limited accounting for network effects; no accounting for interaction between modes
Roadside	AASHTO	Roadside improvements	Project level	Integrated with design tool	Only accounts for safety-related benefits
STEAM	FHWA	Highway, transit, TDM, tolls, multimodal	Network level	Accepts input from four-step models; separate analysis of peak and off-peak periods by trip purpose and mode; emissions; fuel consumption; revenue transfers	Some costs must be estimated outside model; requires trip tables and network from external travel demand model
StratBENCOST	HLB Decision Economics Inc.	Highway improvements	Network level	Risk analysis, environmental effects, separate modules for network-wide or single-roadway analysis; includes construction delays	No accounting for interaction between modes

uniform annual benefits and costs in one facility service life cycle can be extrapolated to perpetuity horizon that is used to assess the economic feasibility of project implementation.

In the agency cost estimation process, the facility repair and annual routine maintenance costs are expected to increase over time owing to facility condition deterioration, coupled with the impacts of traffic and axle loading growths. In addition, the annual user costs of vehicle operations, travel time, crashes, and air emissions are expected to increase due to traffic or travel growth. Particular to annual routine maintenance cost and annual user costs, geometric annual growth rates are generally employed to estimate the annual cost increases. The multiyear costs are discounted to establish the present worth values and then converted to equivalent uniform annual amounts.

7.4.7 Comparison of available benefit–cost analysis software tools

Table 7.18 lists the most popular software packages that are frequently used by analysts to estimate the benefits of transportation projects. The features of individual models in terms of the level of analysis, special features, and software limitations are summarized.

PROBLEMS

7.1 Why is life-cycle analysis important to transportation project evaluation? List all possible agency cost and user cost items for a typical highway project.

7.2 What is the concept of consumer surplus? Draw an illustrative figure to demonstrate the change in consumer surplus. How should this concept be employed for a transportation project evaluation?

7.3 Explain the differences between transportation network-level and project-level evaluation, and how to use quantitative and qualitative methods to conduct the evaluation.

7.4 An individual wishes to accumulate $1000000 in 30 years to invest in a transportation project. If 30 end-of-year deposits are made into an account that pays interest at a rate of 10% compounded annually, what size deposit is required to meet the stated objective?

7.5 It is expected that a truck will incur operating costs of $4000 the first year and that these costs will increase by $500 each year thereafter for the 10-year life of the truck. If money is worth 10% per year (i.e., the annual interest rate at 10%) to the truck company, what is the equivalent annual worth of the operating costs?

7.6 The present cost of a transportation project is $2.5 per consumer. In the future, it will be replaced by a new project which costs $1.75 per consumer. Since the reduction of cost, 20% more consumers will use the system. The annualized agency cost for investment is 40% of the present consumer cost. Maintenance and other costs remain unchanged. Compute the benefit-cost ratio.

7.7 Given the following two investment opportunities, determine which one is worthwhile to choose. Use net present worth method.

	Alternative A	*Alternative B*
First costs	$10000	$15000
Life	5 years	10 years
Salvage value	$2000	$0
Annual receipts	$5000	$7000
Annual disbursements	$2200	$4000
Interest rate	8%	8%
Study period	10 years	10 years

Note: The lowest common multiple of the lives is 10 years. Assume the service will be needed for at least that long and that what is estimated to happen in the first 5 years for project A will be repeated in the second 5 years.

7.8 Compute Internal Rate of Return for the following single investment opportunity.

	Cash flow
First costs	$10000
Project life	5 years
Salvage value	$2000
Annual receipts	$5000
Annual disbursements	$2200

7.9 Solve the following using an interest rate at 6% compounded annually:

a. What is the amount that will be accumulated in a sinking fund at the end of the 15th year if $1200 is deposited in the fund at the beginning of each of the 15 years?

b. What uniform annual payment for 20 years is equivalent to spending $15000 immediately, $11000 at the end of 5 years, $12000 at the end of 10 years, and $2000 a year for 20 years?

7.10 Given the following cash flow data:

	Cash flow	
Year end	*Alternative A*	*Alternative B*
0	−1000	600
1	600	500
2	600	−2000
3	600	400
4	600	200
5	−2100	−1000
6	700	1386

a. Solve for the net present worth of the alternatives A and B using interest rates from 0% upward.

b. Plot the curves on the resulting net present worth against interest rates.

c. What are the rates of return of the two alternatives?

d. What is the rate of return on the difference between the alternatives?

e. If your MARR is 12 percent, which alternatives would you select? Comment briefly on your results. What are some problems associated with the rate of return (ROR) method?

f. Assume the negative cash flows are costs and the positive ones are benefits. What are the Benefit Cost Ratios over the analysis period?

This problem requires many iterations. Use interest tables and/or write a computer program to solve it.

7.11 A proposed 10-mile, 4-lane road widening project is worth 3.5 million. The existing daily traffic is 20000 in the 3-hour peak period. After the widening, daily traffic will be increased by 20%. Assume 250 days per year. Average speed will increase from 40 to 50 mph. Energy use will go from an average of 28 mpg to 23 mpg. The annual maintenance cost is $125000. Assume that the time value is $20 per hour and gasoline costs $3 per gallon, the service life is 30 years and the discount rate is

5%. Compute the net present worth, benefit-to-cost ratio, and equivalent uniform annual benefits of the project.

7.12 A secondary road in a developing country, 30 miles long, is to be improved by surface treating the gravel surface without any change in length. The cost of the improvement is estimated at \$150000 per mile. A present annual transport cost for all traffic on the existing road (vehicle operating cost, maintenance cost, etc.) is estimated at \$200000 per mile. After improvement, this is expected to reduce to \$170000 per mile per annum. Reconstruction takes place in two years with equal expenditures in each year. Assume that in the second year of construction, transport cost on the improved road is equivalent to present costs in half the length and new transport cost on the other half. Would you undertake the project? (MARR = 8% and the project life is 20 years after reconstruction.) Resealing would be required 10 years after reconstruction at a cost of \$55000 per mile. Use the net present worth method.

Chapter 8

Economic analysis of highway pavement preservation

8.1 OVERVIEW

The proposed methodology considers agency and user costs in the service life cycle of a highway pavement. Agency costs comprise construction, repair, and maintenance costs. User costs contain costs of vehicle operation, travel time, crashes, and air emissions in normal operation and work zone conditions, as seen in Figure 8.1.

8.1.1 Pavement project direct costs

The project direct costs generally include direct agency costs and added user costs related to work zones. Direct agency cost items include costs of project land acquisition, design and engineering support, and construction. Extra user costs related to work activities include additional costs of vehicle operation and delays upstream of and within work zones.

8.1.2 Pavement life-cycle costs for agency and user

In life-cycle cost analysis, the overall agency costs include costs of construction, maintenance, and repair treatments incurred during pavement service life cycle. Major user cost items include costs of vehicle operation, travel time, crashes, and air emissions. Life-cycle user costs are calculated in accordance with those user items for each year of pavement service life cycle.

8.1.3 Project-level benefits in pavement service life cycle

The overall project-level benefits in pavement service life cycle may come from both the agency and user aspects. With the investment in a highway pavement, it may reduce the agency's pavement life-cycle costs and instigate savings of life-cycle costs for users. The decreases in agency and user costs are considered as the overall project-level benefits.

8.2 DETERMINISTIC PAVEMENT AGENCY LIFE-CYCLE COST ANALYSIS

8.2.1 Pavement types

Highway pavements can be generally categorized into flexible, rigid, and composite. Flexible pavements have a surface layer that consists entirely of an asphalt/aggregate mix laid over a granular treated or untreated base layer, and sometimes an untreated natural gravel subbase

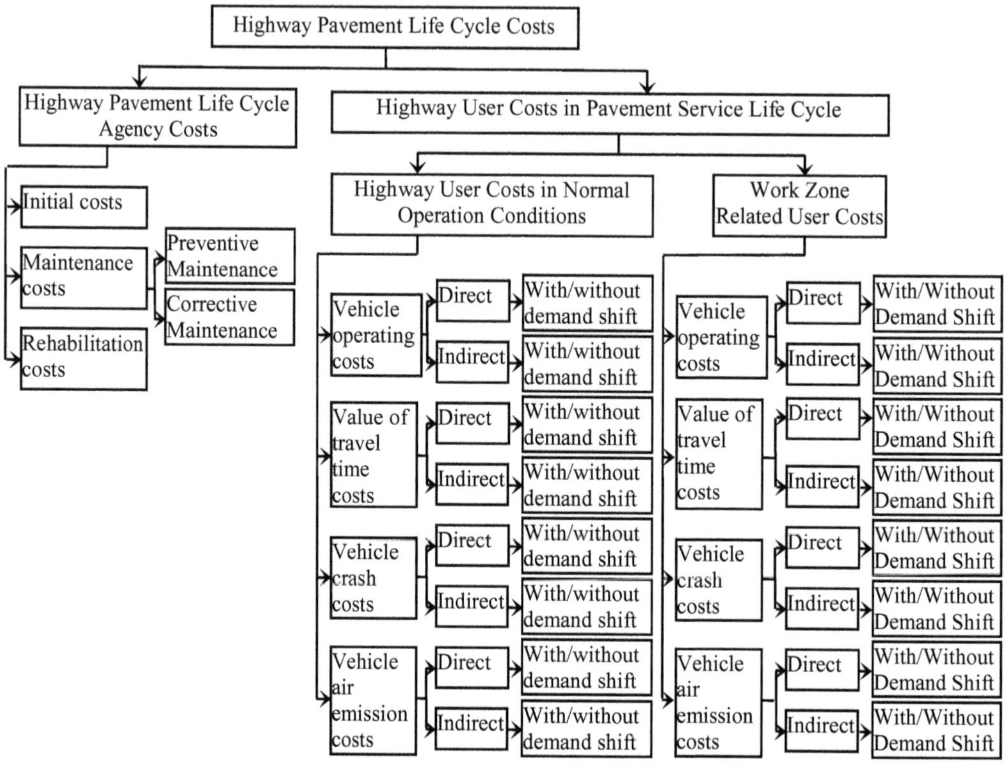

Figure 8.1 Agency and user cost categories in highway pavement service life cycle. (Adapted from Li, Z. and P. Kaini. 2007. Optimal investment decision-making for highway transportation asset management under risk and uncertainty. Report MRUTC 07-10. Department of Civil, Architectural, and Environmental Engineering, Illinois Institute of Technology, Chicago, IL.)

layer (AASHTO, 1993; INDOT, 2002). Rigid pavements are commonly classified as jointed Portland cement concrete, jointed reinforced concrete, and continuously reinforced concrete pavements. All rigid pavement types are typically constructed on a layer of untreated or treated granular subbase layer. In some cases, an additional but lower-quality natural gravel or crushed rock layer is used to separate the granular layer from the subgrade. Composite pavements are mainly constructed from existing flexible or rigid pavements with asphalt or concrete overlays.

8.2.2 Pavement design service life cycle

Within the pavement design lifespan, maintenance and rehabilitation treatments are implemented to ensure the expected design life could withstand the cumulative impacts of traffic loading and non-load factors. Table 8.1 lists service lives of some standard treatments recommended by the FHWA and state transportation agencies (FHWA, 1987; NYDOT, 1992, 1993, 1999; INDOT, 2002).

8.2.3 Pavement treatment categories

Pavement treatments are generally classified into maintenance and rehabilitation categories. In the maintenance category, it can be further divided into preventive and corrective

Table 8.1 Recommended pavement design service life cycle

Pavement type	*Treatment*	*Design life (year)*
Flexible	New full-depth *HMA* pavements	20
	Thin milling and resurfacing	8
	Micro-surfacing overlay	6
	Chip sealing	4
	Crack sealing	3
Rigid	New *PCC* pavements	30
	PCC pavement joint sealing	8
	Concrete pavement rehabilitation (*CPR*) techniques	7
Composite	Concrete pavements over existing pavements	25
	HMA overlay over rubblized *PCC* pavements	20
	HMA overlay over cracked and seated *PCC* pavements	15
	HMA overlay over *CRC* pavements	15
	HMA overlay over jointed concrete with sawed and sealed joints	15
	HMA overlay over jointed concrete	12
	HMA overlay over asphalt pavements	15
	Micro-surfacing overlay	6

Table 8.2 Typical maintenance and rehabilitation treatments

Pavement type	*Preventive maintenance*	*Corrective maintenance*	*Rehabilitation*
Flexible	Thin resurfacing Thin asphalt/concrete overlay Micro-surfacing Seal coating Localized crack sealing, bump grinding	Shallow patching Deep patching	Cold milling and resurfacing Hot or cold recycling
Rigid	Thin resurfacing Thin asphalt/concrete overlay Localized crack sealing Fault grinding Under-sealing Retrofitting	Shallow patching Deep patching	Resurfacing Rubblization followed by resurfacing Crack seating followed by resurfacing (Un)bonded concrete overlay Concrete pavement restoration
Composite	Thin resurfacing Thin asphalt overlay/inlay Micro-surfacing Ultrathin concrete overlay Seal coating Localized crack sealing, bump grinding, sawing, and sealing	Shallow patching Deep patching	Resurfacing Milling followed by resurfacing Milling followed by rubblization and resurfacing Milling followed by crack-and-seating and resurfacing

maintenance, dependent upon the treatment purpose, as listed in Table 8.2 (FHWA, 1991; Geoffroy, 1996; Labi et al., 2008).

8.2.4 Pavement life-cycle maintenance and rehabilitation strategies

8.2.4.1 Pavement service life cycle and rehabilitation life cycle

A *pavement service life cycle* is defined as the time interval between two consecutive construction interventions. Similarly, a *pavement rehabilitation life cycle* is defined as the time interval from construction to rehabilitation, rehabilitation to rehabilitation, rehabilitation to reconstruction, or new construction.

8.2.4.2 Maintenance and rehabilitation strategies

A strategy is regarded as a combination of treatments and their respective frequencies and timings. Within the pavement service life cycle, a *rehabilitation strategy* that involves a combination of rehabilitation activities such as pavement resurfacing and overlays and concrete restoration applied at various times can be created. Within pavement rehabilitation life cycle, a *maintenance strategy* comprising a combination of maintenance activities applied at various times can be established (Peshkin et al., 2005).

Pavement maintenance strategies typically include preventive treatments, such as crack sealing, chip sealing, and thin overlays. Such preventive treatments are applied before the onset of significant structural deterioration (O'Brien, 1989). In the past, corrective maintenance treatments have generally been excluded from strategy formulations because they are implemented to address distress that has already existed. They are considered in the current analysis as long as the effects of those treatments in a pavement maintenance strategy for correcting pavement distresses can be reliably predicted. Table 8.3 presents criteria for implementing preventive and corrective maintenance treatments (Labi et al., 2008; Anwaar et al., 2013).

8.2.5 Typical pavement life-cycle activity profiles

In a pavement service life cycle, maintenance and rehabilitation treatments are implemented in a coordinated manner to form a pavement repair strategy. Different strategies may be adopted, which could be determined by (i) preset time intervals according to pavement age

Table 8.3 Application criteria of preventive and corrective maintenance treatments

Pavement type	*Treatment*	*Average age at first application (year)*	*Average frequency of application (yearly interval)*	*Average received treatment life (year)*
Flexible	Crumb rubber sealing	2	N/I	N/I
	Crack sealing	3	4	3
	Chip sealing	7	5	6
	Sand sealing	12	4	5
	Micro-surfacing	15	N/I	3
	Thin hot mixed asphalt (HMA) overlay	17	11	11
Rigid	Underdrain maintenance	1	2	2
	Crack sealing	6	4	6
	Joint sealing	8	6	10
Composite	Underdrain maintenance	1	1	2
	Crumb rubber sealing	2	N/I	N/I
	Crack sealing	2	3	4
	Chip sealing	10	5	5
	Sand sealing	12	4	5
	Micro-surfacing	15	N/I	3
	Thin HMA overlay	20	11	9

Note: N/I—not indicated.

or cumulative traffic passage; and (ii) condition triggers for treatments using disaggregated condition measures such cracking, rutting, and faulting indices or aggregated measures such as PSI. Conversely, the condition trigger-based strategy requests for a specific maintenance or rehabilitation activity at any time a selected pavement condition measure or performance deteriorates to a certain threshold value. This approach is theoretically sound, but difficult to implement in practice. At present, preset time intervals are largely more popular than condition triggers to establish pavement repair strategies.

For using preset time intervals, pavement age is generally used to determine pavement repair strategies. This assumes that pavement condition follows a predictable pattern according to pavement age. This assumption is justified by two accounts: (i) age is considered a surrogate for combined impacts of cumulative traffic loading and non-load factors; and (ii) reliable data details on traffic loading and all other non-load factors are relatively difficult to collect. With the passage in time, the condition trigger approach could still be applicable.

In the current practice, 20- and 30-year design lives are generally assumed for new full-depth HMA and PCC pavements, respectively. The current practices also propose a 15-year design life for HMA overlays over in-service flexible pavements, a 12–15-year design life for HMA over in-service rigid pavements, and a 25-year design life for concrete over in-service flexible or rigid pavements. Given a new or in-service pavement, the expected design service life may be shortened without appropriate implementation of maintenance treatments. The portion of service life of a new pavement after applying the first rehabilitation treatment can be regarded as the service life of a composite pavement.

In the current analysis, a 40-year design useful service life is proposed for the life-cycle cost analysis of new flexible and rigid pavements. Within the 40-year period, one rehabilitation, one or more major preventive maintenance treatments, and regular or irregular routine or corrective maintenance treatments are planned. The timing of applying rehabilitation and preventive maintenance treatments is relative to pavement age, subject to slight modification according to cumulative traffic loading. If a higher level of cumulative traffic loading is expected, the treatments would be conducted earlier. Table 8.4 summarizes criteria used to create the typical pavement life-cycle activity profiles for flexible, rigid, and composite pavements, respectively. Figures 8.2 through 8.5 depict proposed life-cycle activity profiles for flexible and rigid pavements. Two sets of activity profiles are proposed for each pavement type to remain flexible. The life-cycle activity profiles for composite pavements are embedded into the profiles for flexible and rigid pavements.

Table 8.4 Criteria considered for establishing typical pavement life cycle activity profiles

	Pavement type		
Criterion	*Flexible*	*Rigid*	*Composite*
Extended service life	A fixed 40-year service life cycle for flexible pavements extended from 20 years of design lives	A fixed 40-year service life cycle for rigid pavements extended from 30 years of design lives	With passage of time, new flexible or rigid pavements will eventually become composite pavements depending on types of materials used for rehabilitation, that is, HMA over HMA, HMA over PCC or PCC over PCC
Repair frequency	Once for rehabilitation Once or twice for preventive maintenance	Once for rehabilitation Four times for preventive maintenance	
Repair timing	First treatment and treatment intervals first determined by pavement age The treatment timing then adjusted by traffic loading	First treatment and treatment intervals first determined by pavement age The treatment timing then adjusted by traffic loading	

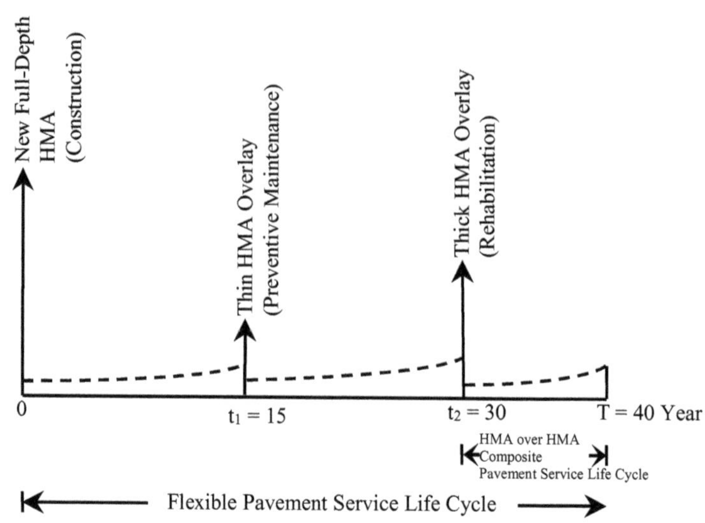

Figure 8.2 Typical life-cycle activity profile for flexible pavements—strategy I. (Adapted from Li, Z. and P. Kaini. 2007. Optimal investment decision-making for highway transportation asset management under risk and uncertainty. Report MRUTC 07-10. Department of Civil, Architectural, and Environmental Engineering, Illinois Institute of Technology, Chicago, IL.)

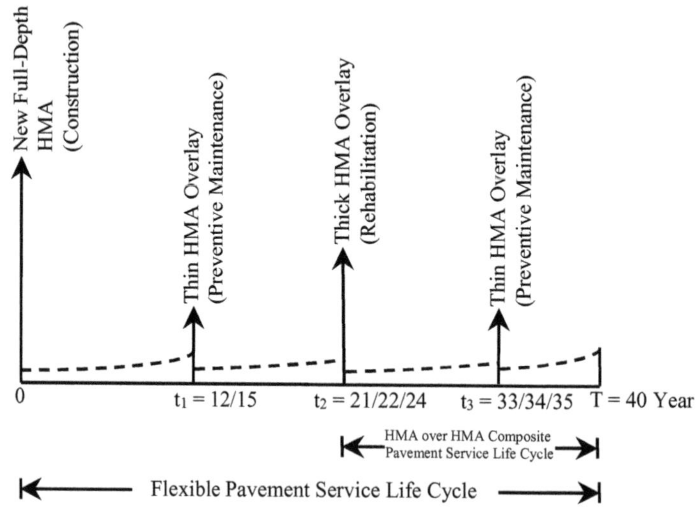

Figure 8.3 Typical life-cycle activity profile for flexible pavements—strategy II. (Adapted from Li, Z. and P. Kaini. 2007. Optimal investment decision-making for highway transportation asset management under risk and uncertainty. Report MRUTC 07-10. Department of Civil, Architectural, and Environmental Engineering, Illinois Institute of Technology, Chicago, IL.)

8.2.6 Pavement life-cycle agency cost analysis

8.2.6.1 Pavement life-cycle agency cost categories

In the pavement life-cycle costs analysis, all construction, rehabilitation, and maintenance costs incurred are considered. In the cost calculations, inflation is removed so that all cost

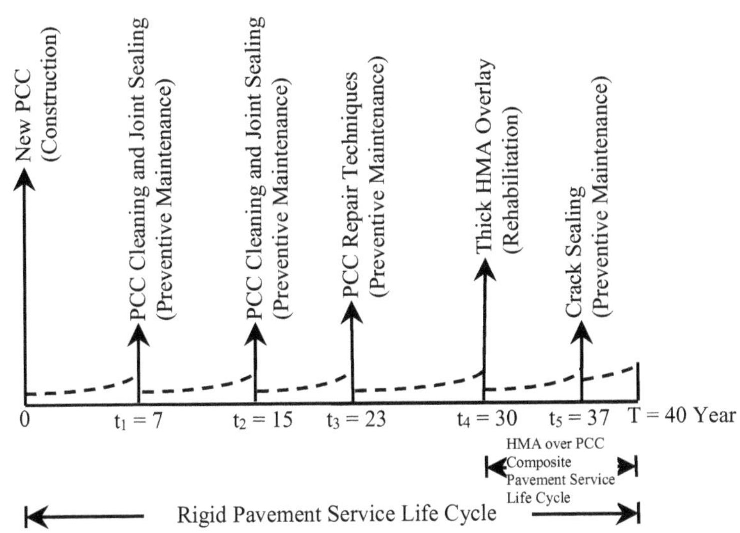

Figure 8.4 Typical life-cycle activity profile for rigid pavements—strategy I. (Adapted from Li, Z. and P. Kaini. 2007. Optimal investment decision-making for highway transportation asset management under risk and uncertainty. Report MRUTC 07-10. Department of Civil, Architectural, and Environmental Engineering, Illinois Institute of Technology, Chicago, IL.)

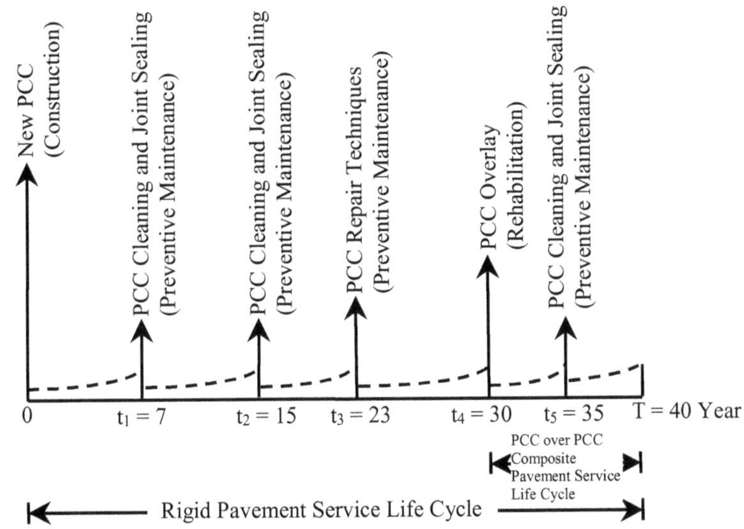

Figure 8.5 Typical life-cycle activity profile for rigid pavements—strategy II. (Adapted from Li, Z. and P. Kaini. 2007. Optimal investment decision-making for highway transportation asset management under risk and uncertainty. Report MRUTC 07-10. Department of Civil, Architectural, and Environmental Engineering, Illinois Institute of Technology, Chicago, IL.)

amounts are expressed in constant dollars. The constant dollars are then discounted to establish the present worth of life-cycle costs in a reference year (FHWA, 1998b, 2000; Labi and Sinha, 2002).

Pavement construction costs are the capital costs incurred in all phases of the design and construction, such as feasibility studies, surveying, geometric and pavement design services, right-of-way(ROW) acquisition, and construction of pavements. Construction costs may be reported by pavement section from the initial construction and the next reconstruction.

Pavement maintenance costs are incurred to maintenance treatments for preserving the conditions of highway pavements so that they can provide a satisfactory riding quality to highway users. Maintenance treatments could be for either preventive or corrective purposes on a routine or periodic basis. Maintenance costs may be expressed in average unit accomplishment costs per treatment or in average costs of all maintenance treatments per lane-mile by pavement type and age group.

Pavement rehabilitation costs are costs of subsequent rehabilitation activities. Like maintenance costs, rehabilitation costs may be reported by unit accomplishment costs per treatment activity or by pavement section considering all types of treatments the pavement receives for the period between initial construction and the next reconstruction.

8.2.6.2 Quantification of pavement life-cycle agency costs

The pavement life-cycle agency costs for one service life cycle can be quantified using the proposed pavement life-cycle activity profiles. Routine maintenance costs are expected to increase slightly in response to pavement condition deterioration. A geometric annual growth rate of maintenance costs can be used to facilitate estimation of pavement life-cycle agency costs. Different gradients can be adopted for different periods between major repair treatments. The service life cycle is expected to repeat infinite times. This will help determine the life-cycle agency costs in perpetuity. Tables 8.5 and 8.6 present pavement life-cycle agency cost analysis results in present worth and equivalent uniform annual amounts.

Denote

PW_{LCAC} = Present worth of pavement life-cycle agency costs
$PW_{LCAC\infty}$ = Present worth of pavement life-cycle agency costs in perpetuity
$EUAAC$ = Equivalent uniform annual pavement agency costs
$EUAAC_{\infty}$ = Equivalent uniform annual pavement agency costs in perpetuity
C_{CON} = Pavement construction cost
C_{REH} = Pavement rehabilitation cost
C_{PM1} = Pavement first preventive maintenance cost
C_{PM2} = Pavement second preventive maintenance cost
C_{PM3} = Pavement third preventive maintenance cost
C_{PM4} = Pavement fourth preventive maintenance cost
C_{MAIN1} = Annual routine maintenance cost incurred between construction and first major repair
C_{MAIN2} = Annual routine maintenance cost incurred between the first and second major repairs
C_{MAIN3} = Annual routine maintenance cost incurred between the second and third major repairs
C_{MAIN4} = Annual routine maintenance cost incurred between the third and fourth major repairs
C_{MAIN5} = Annual routine maintenance cost incurred between the fourth and fifth major repairs
C_{MAIN6} = Annual routine maintenance cost incurred between the fifth and sixth major repairs
g_{M1} = Growth rate of annual routine maintenance cost between construction and first major repair

Table 8.5 Computation of flexible pavement life-cycle agency costs

Strategy		*Computation*
Strategy I	Agency cost profile	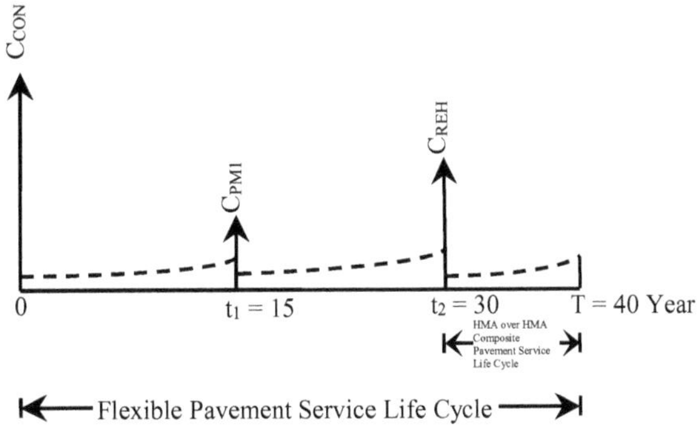
	PW_{LCAC}	$= C_{CON} + C_{PM1}/(1+i)^{t_1} + C_{REH}/(1+i)^{t_2}$ $+ (C_{MAIN1}(1-(1+g_{M1})^{t_1}(1+i)^{-t_1}))/(i-g_{M1})$ $+ ((C_{MAIN2}(1-(1+g_{M2})^{(t_2-t_1)}(1+i)^{-(t_2-t_1)}))/(i-g_{M2}))/(1+i)^{t_1}$ $+ ((C_{MAIN3}(1-(1+g_{M3})^{(T-t_2)}(1+i)^{-(T-t_2)}))/(i-g_{M3}))/(1+i)^{t_2}$
	$PW_{LCAC\infty}$	$= PW_{LCAC}/(1-(1/(1+i)^T))$
	$EUAAC$	$= PW_{LCAC} \cdot ((i(1+i)^T)/((1+i)^T-1))$
	$EUAAC_{\infty}$	$= PW_{LCAC\infty} \cdot i$
Strategy II	Agency cost profile	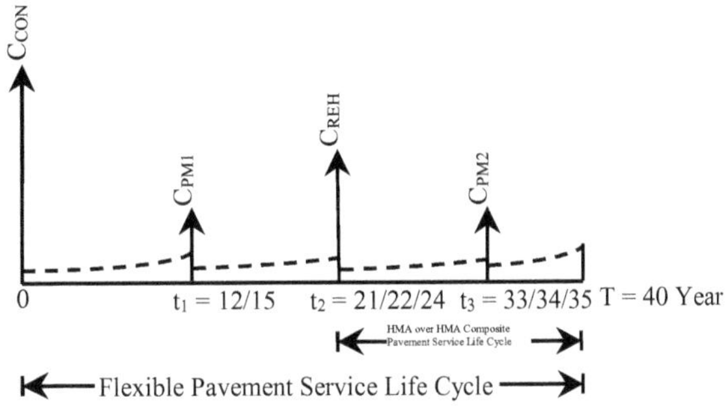
	PW_{LCAC}	$= C_{CON} + C_{PM1}/(1+i)^{t_1} + C_{REH}/(1+i)^{t_2} + C_{PM2}/(1+i)^{t_3}$ $+ (C_{MAIN1}(1-(1+g_{M1})^{t_1}(1+i)^{-t_1}))/(i-g_{M1})$ $+ ((C_{MAIN2}(1-(1+g_{M2})^{(t_2-t_1)}(1+i)^{-(t_2-t_1)}))/(i-g_{M2}))/(1+i)^{t_1}$ $+ ((C_{MAIN3}(1-(1+g_{M3})^{(t_3-t_2)}(1+i)^{-(t_3-t_2)}))/(i-g_{M3}))/(1+i)^{t_2}$ $+ ((C_{MAIN4}(1-(1+g_{M4})^{(T-t_3)}(1+i)^{-(T-t_3)}))/(i-g_{M4}))/(1+i)^{t_3}$
	$PW_{LCAC\infty}$	$= PW_{LCAC}/(1-(1/(1+i)^T))$
	$EUAAC$	$= PW_{LCAC} \cdot ((i(1+i)^T)/((1+i)^T-1))$
	$EUAAC_{\infty}$	$= PW_{LCAC\infty} \cdot i$

g_{M2} = Growth rate of annual routine maintenance cost between the first and second major repairs

g_{M3} = Growth rate of annual routine maintenance cost between the second and third major repairs

g_{M4} = Growth rate of annual routine maintenance cost between the third and fourth major repairs

Table 8.6 Computation of rigid pavement life-cycle agency costs

Strategy		*Computation*
Strategy I	Agency cost profile	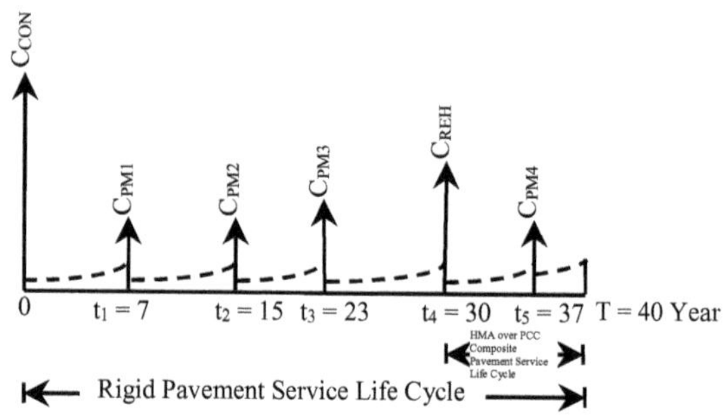
	PW_{LCAC}	$= C_{CON} + C_{REH}/(1+i)^{t4}$ $+ C_{PM1}/(1+i)^{t1} + C_{PM2}/(1+i)^{t2} + C_{PM3}/(1+i)^{t3} + C_{PM5}/(1+i)^{t5}$ $+ (C_{MAIN1}(1-(1+g_{M1})^{t1}(1+i)^{-t1}))/(i-g_{M1})$ $+ ((C_{MAIN2}(1-(1+g_{M2})^{(t_2-t_1)}(1+i)^{-(t_2-t_1)}))/(i-g_{M2}))/(1+i)^{t1}$ $+ ((C_{MAIN3}(1-(1+g_{M3})^{(t_3-t_2)}(1+i)^{(t_3-t_2)}))/(i-g_{M3}))/(1+i)^{t2}$ $+ ((C_{MAIN4}(1-(1+g_{M4})^{(t_4-t_3)}(1+i)^{-(t_4-t_3)}))/(i-g_{M4}))/(1+i)^{t3}$ $+ ((C_{MAIN5}(1-(1+g_{M5})^{(t_5-t_4)}(1+i)^{-(t_5-t_4)}))/(i-g_{M5}))/(1+i)^{t4}$ $+ ((C_{MAIN6}(1-(1+g_{M6})^{(T-t_5)}(1+i)^{-(T-t_5)}))/(i-g_{M6}))/(1+i)^{t5}$
	$PW_{LCAC\infty}$	$= PW_{LCAC}/(1-(1/(1+i)^T))$
	$EUAAC$	$= PW_{LCAC} \cdot ((i(1+i)^T)/((1+i)^T-1))$
	$EUAAC_{\infty}$	$= PW_{LCAC\infty} \cdot i$
Strategy II	Agency cost profile	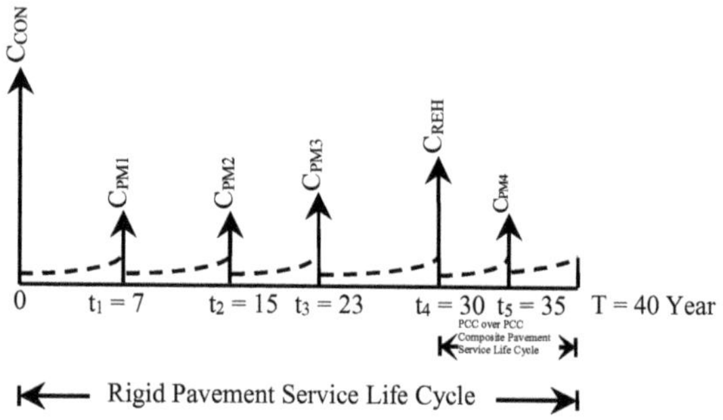
	PW_{LCAC}	$= C_{CON} + C_{REH}/(1+i)^{t4}$ $+ C_{PM1}/(1+i)^{t1} + C_{PM2}/(1+i)^{t2} + C_{PM3}/(1+i)^{t3} + C_{PM5}/(1+i)^{t5}$ $+ (C_{MAIN1}(1-(1+g_{M1})^{t1}(1+i)^{-t1}))/(i-g_{M1})$ $+ ((C_{MAIN2}(1-(1+g_{M2})^{(t_2-t_1)}(1+i)^{-(t_2-t_1)}))/(i-g_{M2}))/(1+i)^{t1}$ $+ ((C_{MAIN3}(1-(1+g_{M3})^{(t_3-t_2)}(1+i)^{-(t_3-t_2)}))/(i-g_{M3}))/(1+i)^{t2}$ $+ ((C_{MAIN4}(1-(1+g_{M4})^{(t_4-t_3)}(1+i)^{-(t_4-t_3)}))/(i-g_{M4}))/(1+i)^{t3}$ $+ ((C_{MAIN5}(1-(1+g_{M5})^{(t_5-t_4)}(1+i)^{-(t_5-t_4)}))/(i-g_{M5}))/(1+i)^{t4}$ $+ ((C_{MAIN6}(1-(1+g_{M6})^{(T-t_5)}(1+i)^{-(T-t_5)}))/(i-g_{M6}))/(1+i)^{t5}$
	$PW_{LCAC\infty}$	$= PW_{LCAC}/(1-(1/(1+i)^T))$
	$EUAAC$	$= PW_{LCAC} \cdot ((i(1+i)^T)/((1+i)^T-1))$
	$EUAAC_{\infty}$	$= PW_{LCAC\infty} \cdot i$

g_{M5} = Growth rate of annual routine maintenance cost between the fourth and fifth major repairs
g_{M6} = Growth rate of annual routine maintenance cost between the fifth and sixth major repairs
I = Discount rate
T = Time of year a major treatment is implemented
T = Number of years of service life

8.3 DETERMINISTIC HIGHWAY USER COST ANALYSIS IN PAVEMENT SERVICE LIFE CYCLE

8.3.1 Highway user cost components and categories

User costs are those incurred by highway users over the pavement service life cycle, depending on the highway improvements and associated repair strategies over the service life cycle. They comprise a substantial part of the total transportation costs for highway investments and can often be the decisive factor in life-cycle cost analysis. There are two dimensions of highway user costs: (i) user cost *components* including costs of vehicle operation, travel time, crashes, and air emissions (Zaniewski et al., 1982; FHWA, 2000; AASHTO, 2003); and (ii) user cost *categories* that are related to normal operation and work zone traffic conditions (FHWA, 1998b) (Figure 8.6).

8.3.1.1 Highway user cost components

Vehicle operating costs are mileage-dependent costs of running automobiles, trucks, and other motor vehicles on highways, including the expenses of fuel, tires, engine oil, maintenance, and the portion of vehicle mileage related depreciation. Factors affecting vehicle operating costs include vehicle type, engine type, age, vehicle speed, speed changes, gradient, curvature, and pavement surface conditions.

Travel time costs refer to the value of time spent in travel and include costs to businesses of time spent by their employees, vehicles, and goods, and costs to consumers of personal unpaid time spent on travel, including time spent parking and walking to and from a vehicle.

Vehicle crash costs are costs related to motor vehicle crashes classified into fatality, injury, and PDO categories.

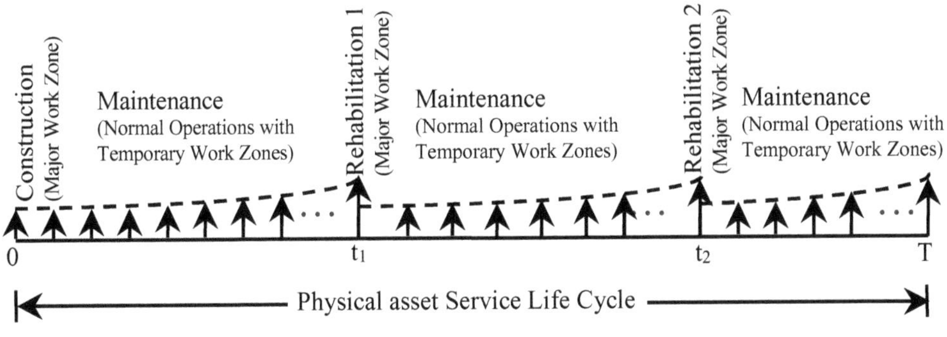

Figure 8.6 User cost trends in highway pavement service life cycle. (Adapted from Li, Z. and P. Kaini. 2007. Optimal investment decision-making for highway transportation asset management under risk and uncertainty. Report MRUTC 07-10. Department of Civil, Architectural, and Environmental Engineering, Illinois Institute of Technology, Chicago, IL.)

Vehicle air emission costs are external costs associated with major pollutants emitted by vehicles, including NMHC, CO, CO_2, NO_X, SO_2, and total suspended particles (TSP).

8.3.1.2 Highway user cost categories

Highway user costs in normal operation conditions reflect costs associated with highway usage during periods free of construction, maintenance or rehabilitation activities. Among individual user cost components in this category, vehicle operating costs vary considerably according to vehicle type, engine type, age, speed, speed changes, design features, and pavement conditions. During normal operations, little difference exists between delay and crash costs resulting from pavement design alternatives.

Work zone-related user costs are mainly related to increased costs of vehicle operation and delays to highway users in the presence of work zones. User costs in this category are a function of the work zone configuration, duration, timing, and scope, and depend on the volume and operating characteristics of the traffic traversing through work zones.

8.3.2 Highway normal operation user costs in pavement service life cycle

8.3.2.1 Annual highway normal operation user costs for individual user cost components

For highway segments with traffic operations affected by a specific project, the annual costs of vehicle operation, travel time, crashes, and air emissions for the initial year are separately calculated based on the corresponding VMT and respective per VMT unit rates of respective user cost components. The quantified user cost components are then converted to dollar values and summed up to arrive at the annual total user costs (Zaniewski et al., 1982; FHWA, 2000; AASHTO, 2003). Table 8.7 presents computational steps for establishing the annual user costs under normal operating conditions.

Denote

UC = Annual total highway user costs
VOC = Annual vehicle operating costs, dollars/year
TTC = Annual travel time costs, dollars/year
DC_{INT} = Annual vehicle intersection delay costs, dollars/year
DC_{RR} = Annual vehicle railroad crossing delay costs, dollars/year
DC_{IC} = Annual vehicle incident delay costs, dollars/year
VCC = Annual vehicle crash costs, dollars/year
VEC = Annual vehicle emission costs, dollars/year
VMT_i = Annual vehicle miles of travel for vehicle class i
$UVOC_{ik}$ = Unit cost of VOC component k for vehicle class i, dollars/VMT
$UTTC_i$ = Unit travel time value for vehicle class i, dollars/hr
$UVCC_p$ = Unit cost of vehicle crashes for crash severity p, dollars/crash
$UVEC_{iq}$ = Unit rate of vehicle emitted pollutant type q for vehicle class i, dollars/VMT
S_i = Average travel speed for vehicle class i
N_l = Number of intersections of type l
$\bar{D}_{INT,l}$ = Average delay at intersection of type l, hr/veh
K_m = Number of trains passing railroad crossing m per year
TSD_m = Total stopped delay time per train at railroad crossing m, hr/train
$\bar{D}_{IC,i}$ = Delay time per incident by vehicle class i, hr/veh

Table 8.7 Computation of initial-year normal operation user costs

Computational step		*Description*
Step 0	Determine inputs	Identify project-affected highway segments (directly affected segments, indirectly affected segments) Determine initial-year traffic demand of affected segments (AADT, directional hourly demand, vehicle composition) Determine normal operations characteristics (highway capacity, speed)
Step 1	Determine initial-year VMT for project directly and indirectly affected segments	Vehicle class *i* initial-year $VMT_{i,D}$ for project directly affected highway segment = 365 × (initial year *AADT*) × (% vehicle class *i*) × (length of the directly affected segment) Vehicle class *i* initial-year $VMT_{ij,ID}$ for project indirectly affected highway segment j = 365 × (initial year *AADT*) × (% vehicle class *i*) × (length of indirectly affected segment *j*)
Step 2	Compute vehicle operating costs	Determine initial-year VOC for project-affected segments $VOC_D = \sum_{i=1}^{13}\left[\left(\sum_{k=1}^{5} UVOC_{ik}\right)\cdot VMT_{i,D}\right]$ $VOC_{ID} = \sum_{i=1}^{13}\left[\left(\sum_{k=1}^{5} UVOC_{ik}\right)\cdot\left(\sum_{j=1}^{N} VMT_{ij,ID}\right)\right]$
Step 3	Calculate travel time cost	Determine initial-year travel time costs for project-affected segments $TTC_D = \sum_{i=1}^{13}\left[UTTC_i\cdot(VMT_{i,D}/S_i)\right]$ $TTC_{ID} = \sum_{i=1}^{13}\left[UTTC_i\cdot\left(\sum_{j=1}^{N}(VMT_{ij,ID}/S_{ij,ID})\right)\right]$
Step 4	Quantify delay costs	Determine initial-year delay costs for project-affected segments $DC_{INT,D/ID} = \sum_{i=1}^{13}\left[UTTC_i\cdot\left(\sum_{l=1}^{L}(N_l\cdot\bar{D}_{INT,l})\right)\right]$ $DC_{RR,D/ID} = \sum_{i=1}^{13}\left[UTTC_i\cdot\left(K_m\cdot\sum_{m=1}^{M} TSD_m\right)\right]$ $DC_{IC,D} = \sum_{i=1}^{13}\left[UTTC_i\cdot\left(\bar{D}_{IC,i}\cdot R_i^{IC}\cdot(VMT_{i,D}/1000000)\right)\right]$ $DC_{IC,ID} = \sum_{i=1}^{13}\left[UTTC_i\cdot\left(\bar{D}_{IC,i}\cdot R_i^{IC}\cdot\left(\sum_{j=1}^{N} VMT_{ij,ID}/1000000\right)\right)\right]$
Step 5	Establish vehicle crash costs	Determine initial-year vehicle crash costs for project-affected segments $VCC_D = \sum_{s=1}^{3}\left[UVCC_i\cdot R_s^{VC}\cdot\sum_{i=1}^{13}(VMT_{i,D}/1000000)\right]$

(Continued)

Table 8.7 (Continued) Computation of initial-year normal operation user costs

Computational step		*Description*
Step 6	Calculate vehicle air emission costs	Determine initial-year emission costs for project-affected segments $VCC_D = \sum_{s=1}^{3}\left[UVCC_i \cdot R_s^{VC} \cdot \sum_{i=1}^{13}(VMT_{i,D}/1\,000000)\right]$ $VCC_{ID} = \sum_{s=1}^{3}\left[UVCC_i \cdot R_s^{VC} \cdot \sum_{i=1}^{13}\left(\left(\sum_{j=1}^{N}VMT_{ij,ID}\right)/1\,000000\right)\right]$
Step 7	Determine initial-year total user costs	Determine initial-year total user costs for project-affected segments $UC_D = VOC_D + TTC_D + DC_{INT,D} + DC_{RR,D} + DC_{IC,D} + VCC_D + VEC_D$ $UC_{ID} = VOC_{ID} + TTC_{ID} + DC_{INT,ID} + DC_{RR,ID} + DC_{IC,ID} + VCC_{ID} + VEC_{ID}$

R_i^{IC} = Number of incidents per million VMT by vehicle class i
R_s^{VC} = Vehicle crashes of severity level s per million VMT, crashes/million VMT
R_p^{VE} = Quantities of pollutant type p emitted by vehicle class i, tons/VMT
i = Vehicle class 1–13
j = Number of project indirectly affected highway segments
k = VOC component 1–5 for fuel consumption, oil consumption, tire wear, vehicle depreciation, and maintenance and repair
l = Intersection type 1 to L
m = Railroad crossing 1 to M
s = Crash severity level 1–3
p = Pollutant type 1–6 for NMHC, CO, CO_2, NO_X, SO_2, and TSP

Note: Subscripts "*D*" and "*ID*" refer to project directly and indirectly affected highway segments.

8.3.2.2 Highway normal operation user costs in pavement service life cycle

The constant-dollar annual total user costs are expected to increase slightly in response to a traffic demand increase. A geometric annual growth rate can be used to calculate life-cycle user costs. Different gradients are proposed for different periods between major repair treatments. Table 8.8 lists the computation details.

Denote

PW_{LCUC} = Present worth of highway pavement life-cycle user costs
$PW_{LCUC\infty}$ = Present worth of highway pavement life-cycle user costs in perpetuity
$EUAUC$ = Equivalent uniform annual highway pavement user costs
$EUAUC_\infty$ = Equivalent uniform annual highway pavement user costs in perpetuity
C_{AUC2} = Annual pavement user cost incurred between the first and second major repairs
C_{AUC3} = Annual pavement user cost incurred between the second and third major repairs
C_{AUC4} = Annual pavement user cost incurred between the third and fourth major repairs
C_{AUC5} = Annual pavement user cost incurred between the fourth and fifth major repairs
C_{AUC6} = Annual pavement user cost incurred between the fifth and sixth major repairs
g_{AUC1} = Growth rate of annual pavement user cost between initial construction and first major repairs

Table 8.8 Computation of highway normal operation user costs in pavement service life cycle

Pavement type		*Computation*
Flexible strategy I	User cost profile	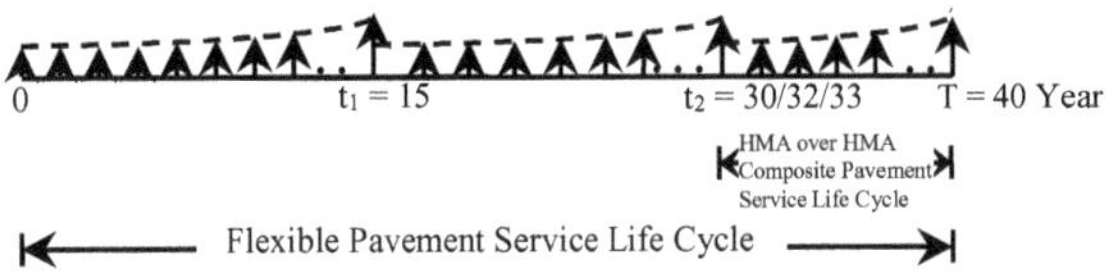
	PW_{LCUC}	$= (C_{AUC1}(1-(1+g_{AUC1})^{t_1}(1+i)^{-t_1}))/(i-g_{AUC1})$ $+((C_{AUC2}(1-(1+g_{AUC1})^{(t_2-t_1)}(1+i)^{-(t_2-t_1)}))/(i-g_{AUC1}))/(1+i)^{t_1}$ $+((C_{AUC3}(1-(1+g_{AUC1})^{(T-t_2)}(1+i)^{-(T-t_2)}))/(i-g_{AUC1}))/(1+i)^{t_2}$
	$PW_{LCUC\infty}$	$= PW_{LCUC}/(1-(1/(1+i)^T))$
	$EUAUC$	$= PW_{LCUC}\cdot((i(1+i)^T)/((1+i)^T-1))$
	$EUAUC_{\infty}$	$= PW_{LCUC\infty}\cdot i$
Flexible strategy II	User cost profile	
	PW_{LCUC}	$= (C_{AUC1}(1-(1+g_{AUC1})^{t_1}(1+i)^{-t_1}))/(i-g_{AUC1})$ $+((C_{AUC2}(1-(1+g_{AUC2})^{(t_2-t_1)}(1+i)^{-(t_2-t_1)}))/(i-g_{AUC2}))/(1+i)^{t_1}$ $+((C_{AUC3}(1-(1+g_{AUC3})^{(t_3-t_2)}(1+i)^{-(t_3-t_2)}))/(i-g_{AUC3}))/(1+i)^{t_2}$ $+((C_{AUC4}(1-(1+g_{AUC4})^{(T-t_3)}(1+i)^{-(T-t_3)}))/(i-g_{AUC4}))/(1+i)^{t_3}$
	$PW_{LCUC\infty}$	$= PW_{LCUC}/(1-(1/(1+i)^T))$
	$EUAUC$	$= PW_{LCUC}\cdot((i(1+i)^T)/((1+i)^T-1))$
	$EUAUC_{\infty}$	$= PW_{LCUC\infty}\cdot i$
Rigid strategy I	User cost profile	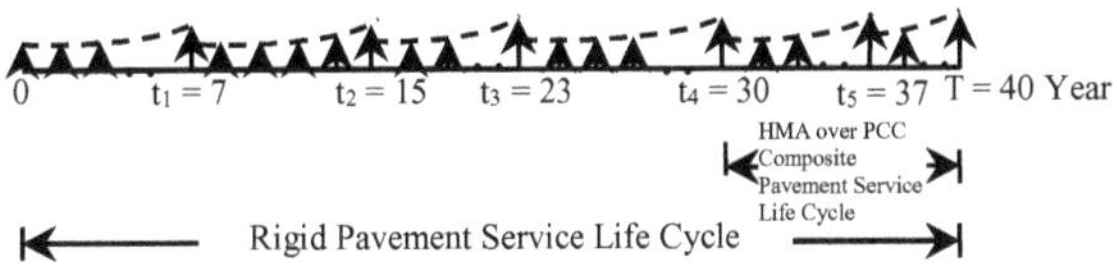
	PW_{LCUC}	$= (C_{AUC1}(1-(1+g_{AUC1})^{t_1}(1+i)^{-t_1}))/(i-g_{AUC1})$ $+((C_{AUC2}(1-(1+g_{AUC2})^{(t_2-t_1)}(1+i)^{-(t_2-t_1)}))/(i-g_{AUC2}))/(1+i)^{t_1}$ $+((C_{AUC3}(1-(1+g_{AUC3})^{(t_3-t_2)}(1+i)^{-(t_3-t_2)}))/(i-g_{AUC3}))/(1+i)^{t_2}$ $+((C_{AUC4}(1-(1+g_{AUC4})^{(t_4-t_3)}(1+i)^{-(t_4-t_3)}))/(i-g_{AUC4}))/(1+i)^{t_3}$ $+((C_{AUC5}(1-(1+g_{AUC5})^{(t_5-t_4)}(1+i)^{-(t_5-t_4)}))/(i-g_{AUC5}))/(1+i)^{t_4}$ $+((C_{AUC6}(1-(1+g_{AUC6})^{(T-t_5)}(1+i)^{-(T-t_5)}))/(i-g_{AUC6}))/(1+i)^{t_5}$
	$PW_{LCUC\infty}$	$= PW_{LCUC}/(1-(1/(1+i)^T))$
	$EUAUC$	$= PW_{LCUC}\cdot(i(1+i)^T)/((1+i)^T-1))$
	$EUAUC_{\infty}$	$= PW_{LCUC\infty}\cdot i$
Rigid strategy II	User cost profile	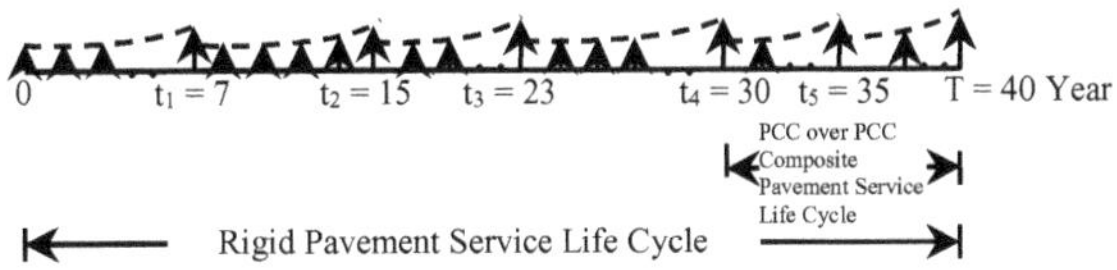

(Continued)

Table 8.8 (Continued) Computation of highway normal operation user costs in pavement service life cycle

Pavement type		*Computation*
	PW_{LCUC}	$=(C_{AUC1}(1-(1+g_{AUC1})^{t1}(1+i)^{-t1}))/(i-g_{AUC1})$
		$+((C_{AUC2}(1-(1+g_{AUC2})^{(t2-t1)}(1+i)^{-(t2-t1)}))/(i-g_{AUC2}))/(1+i)^{t1}$
		$+((C_{AUC3}(1-(1+g_{AUC3})^{(t3-t2)}(1+i)^{-(t3-t2)}))/(i-g_{AUC3}))/(1+i)^{t2}$
		$+((C_{AUC4}(1-(1+g_{AUC4})^{(t4-t3)}(1+i)^{-(t4-t3)}))/(i-g_{AUC4}))/(1+i)^{t3}$
		$+((C_{AUC5}(1-(1+g_{AUC5})^{(t5-t4)}(1+i)^{-(t5-t4)}))/(i-g_{AUC5}))/(1+i)^{t4}$
		$+((C_{AUC6}(1-(1+g_{AUC6})^{(T-t5)}(1+i)^{-(T-t5)}))/(i-g_{AUC6}))/(1+i)^{t5}$
	$PW_{LCUC\infty}$	$=PW_{LCUC}/(1-(1/(1+i)^{T}))$
	$EUAUC$	$=PW_{LCUC}\cdot((i(1+i)^{T})/((1+i)^{T}-1))$
	$EUAUC_{\infty}$	$=PW_{LCUC\infty}\cdot i$

g_{AUC2} = Growth rate of annual pavement user cost between the first and second major repairs
g_{AUC3} = Growth rate of annual pavement user cost between the second and third major repairs
g_{AUC4} = Growth rate of annual pavement user cost between the third and fourth major repairs
g_{AUC5} = Growth rate of annual pavement user cost between the fourth and fifth major repairs
g_{AUC6} = Growth rate of annual pavement user cost between the fifth and sixth major repairs
i = Discount rate
t = Time of year a major pavement treatment is implemented
T = Number of years of service life for a highway pavement

8.3.3 Additional work zone-related highway user costs

The HCM defines a work zone as an area of a highway in the presence of work activities that impinge on the number of lanes available to traffic or affect travelers traversing through the area (TRB, 2010). Highway work activities can significantly reduce the highway capacity and vehicle operating speed, which may lead to queue development and consequently in vehicle delays and operating cost increases. User costs for work zone operations are influenced by such factors as pavement type, work type, traffic volume, vehicle composition, and work zone characteristics. Table 8.9 lists various components of excessive vehicle operating costs and delays caused by work zones.

Table 8.10 provides a procedure for estimating the additional vehicle operating costs and delay costs owing to work zones.

8.4 PROJECT-LEVEL AGENCY AND USER BENEFITS IN PERPETUITY

8.4.1 Project-level agency benefits due to pavement service life change

The typical activity profiles for different types of pavements proposed in the previous sections represent the pavements in service in the ideal situation. Specifically, these reflect that

Table 8.9 Excessive vehicle operating costs and delays caused by work zones

	Existence of		*VOC component*	*Delay component*	
Flow characteristic	*Work zone*	*(WZ) queue*	*Work zone upstream*	*Work zone upstream*	*Within work zone*
Uncongested	Yes	No	Speed change	Speed change	WZ reduced speed
Congested	Yes	Yes	Speed change Stopping Queue idling	Speed change Stopping Queue reduced speed	WZ reduced speed
	No	Yes	Stopping Queue idling	Stopping Queue reduced speed	None

Table 8.10 Computation of additional work zone-related highway user costs

Step		*Description*
Step 0	Determine inputs	Determine project future-year traffic demand (AADT, directional hourly demand, vehicle composition) Determine normal operations characteristics (highway capacity, speed) Determine work zone characteristics (construction duration, work zone operation hours, and length)
Step 1	Determine future-year traffic demand	Vehicle class *i* future-year $AADT_i$ = (Base year *AADT*) × (% vehicle class *i*) × ((1 + class *i* annual growth)$^{\text{(Future year-base year)}}$)
Step 2	Calculate work zone directional hourly volume (DHV)	Vehicle class *i* directional hourly volume DHV_i = (Future year $AADT_i$) × (Directional distribution *i*) × (Hourly traffic distribution factor *i*)
Step 3	Determine roadway capacity	Determine pavement normal operations capacity using *HCM* Determine work zone capacity using *HCM*
Step 4	Identify user cost components	Identify various *VOC* and delay components for each hour as listed in Table 8.11
Step 5	Quantify traffic affected for each VOC and delay component	Hourly queue rate = *DHV* − normal/work zone capacity Hourly vehicles queued = cumulative hourly queue rates Vehicles traverse work zone = *DHV* with work zone Vehicles traverse queue = normal/WZ capacity with queue Vehicles stopped for the queue = *DHV* with queue Vehicles slowed down = *DHV* with work zone, no queue
Step 6	Compute queue reduced speed delay	Hourly volumes through queue and upstream of queue Hourly speeds through queue and upstream of queue Hourly densities through queue and upstream of queue *WZ* delay = *WZ* length/*WZ* speed − *WZ* length/upstream speed Queue delay = Queue length/queue speed − queue length/upstream speed Average hourly vehicles in queue = Arithmetic average of vehicles queued at the beginning and end of each hour Average hourly queue length = Average hourly vehicles in queue/(density in queue − density upstream of queue) Average queue delay per vehicle = Average hourly queue length/queue speed
Step 7	Select added VOC rates	Select added VOC rates due to speed change, stopping, and queue idling by vehicle class
Step 8	Select added delay time and hourly time values	Select added delay time due to speed change, stopping, queue reduced speed, and work zone reduced speed by vehicle class Select hourly time values by vehicle class Compute added delay costs
Step 9	Assign traffic to vehicle classes	Distribute respective number of vehicles affected by speed change, stopping, queue, and traversing work zone to each vehicle class
Step 10	Compute work zone user costs	Compute total added VOC costs for the construction duration Compute total added delay costs for the construction duration

the design service lives are achieved when the pavement repair strategies are effectively implemented. Early termination of a service life may occur if inappropriate types of treatments are implemented and/or optimal timings are missed. As such, the typical service life cycle can be used as the base case scenario and the cases with early terminations of a service life can be used as alternative scenarios. To maintain the typical service life cycle, projects as treatments of a strategy need to be implemented. The resulting difference in pavement life-cycle agency costs between the base case and an alternative scenario can be treated as the overall project-level life-cycle agency benefits of maintaining the base case activity profile.

8.4.2 Project-level user benefits due to pavement service life change

Similar to the calculation of project-level agency benefits in one service life cycle, the reduction in life-cycle user costs is considered as project-level life-cycle user benefits of maintaining the base case activity profile.

8.4.3 Project-level user benefits due to demand shift

Other than the life-cycle user benefits due to pavement service life cycle shift as compared with the base case service life, life-cycle user benefits may be achieved because of traffic demand shift after implementing the highway project. The upward shift of the demand curve related to highway segments directly affected by the project may reduce traffic demand on alternative routes (AASHTO, 2003). The calculation of the change in annual total highway user costs due to demand shift as user benefits is explained below.

8.4.3.1 Annual project-level user benefits for a directly affected road segment due to demand shift

The annual user benefits from implementing a highway project are captured by the concept of consumer surplus. Given a demand curve, the consumer surplus is the difference between what highway users collectively would have been willing to pay and what they are actually paying as far as the user costs are concerned. The change in consumer surplus before and after project execution is considered to define the user benefits generated by the project. For a generalized case where the demand curve shifts upward as a result of a project improvement, the user benefits can be calculated as depicted in Figure 8.7a. The annual user benefits associated with the highway segment directly affected by the project could be related to vehicle operating costs, travel time, crashes, and air emissions.

8.4.3.2 Annual project-level user benefits for an indirectly affected road segment due to demand shift

If the implementation of a highway project causes traffic to shift to the improved segment, other indirectly affected segments may see a demand reduction. That is, the travel demand on the indirectly affected segments is reduced for every user cost. As shown in Figure 8.7b, the change in consumer surplus is consistent with the change in consumer surplus that is measured on the directly affected segment. The approach can be applied to each indirectly affected highway segment to capture all changes in consumer surplus. The annual user benefits for the highway segments indirectly affected by the project could be relevant to vehicle operating costs, travel time, vehicle crashes, and emissions.

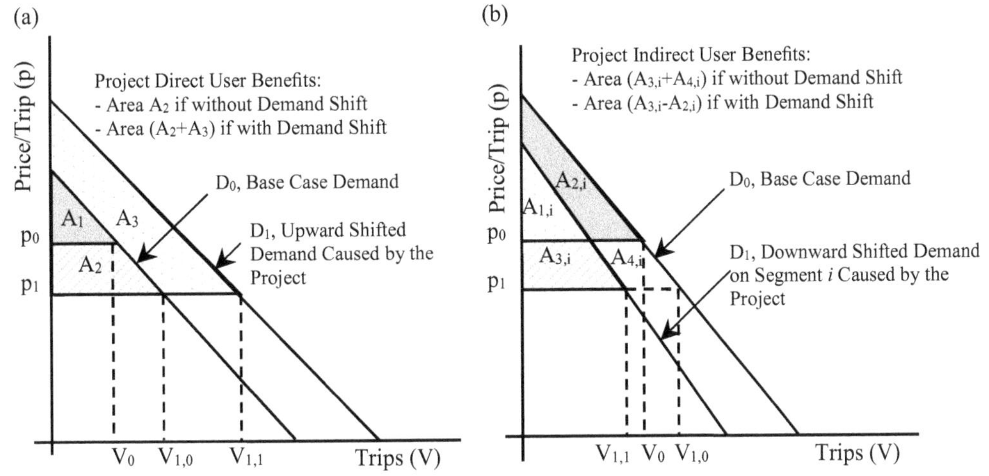

Figure 8.7 Project-level user benefits due to demand shift. (a) Directly affected highway segment. (b) Indirectly affected highway segment *i*. (Adapted from Li, Z. and P. Kaini. 2007. Optimal investment decision-making for highway transportation asset management under risk and uncertainty. Report MRUTC 07-10. Department of Civil, Architectural, and Environmental Engineering, Illinois Institute of Technology, Chicago, IL.)

8.4.4 Project-level user benefits due to demand shift in pavement service life cycle

The annual total life-cycle user benefits of a highway project due to demand shift are the aggregation of changes in all consumer surpluses for all user cost components related to all directly and indirectly affected highway segments. After obtaining annual total user benefits for the base year with demand shift, the annual total user benefits involving demand shifts for the future years in a typical service life cycle can be extrapolated using a linear or geometric gradient annual growth rate. With all annual total user benefits for the typical service life cycle being calculated, the overall project life-cycle user benefits due to demand shift can be determined and discounted to a common base year.

8.4.5 Project-level user benefits due to both demand shift and pavement service life change

The computation includes three steps: (i) establishing the annual total user benefits for the base year with demand shift using the concept of consumer surplus; (ii) extrapolating the annual total user benefits for future years in a reduced service life cycle because of service life cycle shift; and (iii) determining the project overall life-cycle user benefits with demand shift and service life cycle shift by discounting future-year benefits to base-year values.

8.4.6 Project-level agency and user benefits in perpetuity

Compared with the base case that maintains the typical pavement service life cycle for an infinite number of times, an alternative scenario may have an early service life termination for the first life cycle and then follow the typical service life cycle from the second cycle to perpetuity. It is also possible that the alternative case will have early service life terminations for the first and second life cycles, and then follow the typical service life cycle from the third cycle onward to perpetuity, and so forth. The difference in life-cycle agency and user costs in perpetuity between the base case and an alternative case is considered as the overall project-level agency and user benefits in perpetuity horizon. Table 8.11 illustrates the computation details.

Denote

$PW_{LCAC\infty}$ = Present worth of pavement life-cycle agency costs in perpetuity
$PW_{LCUC\infty}$ = Present worth of pavement life-cycle user costs in perpetuity
$EUAAC_{\infty}$ = Equivalent uniform annual agency costs in perpetuity
$EUAUC_{\infty}$ = Equivalent uniform annual user costs in perpetuity
PW_{AB} = Present worth of project agency benefits in perpetuity
PW_{UB} = Present worth of project user benefits in perpetuity
PW_{B} = Present worth of overall project benefits in perpetuity
EUA_{AB} = Equivalent uniform annual project agency benefits in perpetuity
EUA_{UB} = Equivalent uniform annual project user benefits in perpetuity
EAU_{B} = Equivalent uniform annual overall project benefits in perpetuity
i = Discount rate
T = Number of years of service life for a highway pavement

8.5 DISTRIBUTION OF PROJECT COSTS IN PERPETUITY

The total costs of a pavement project consist of project direct costs and work zone-related user costs during project execution. Since the typical pavement service life cycle is treated as the base case, the total project costs are distributed in accordance with the base case service life cycle repeated into perpetuity horizon, as illustrated in Table 8.12.

Denote

PW_{PC} = Present worth of project direct construction, maintenance, or rehabilitation costs
PW_{WZUC} = Present worth of excessive work zone user costs caused by the project
$PW_{C,\infty}$ = Present worth of the total of project direct costs and work zone user costs in perpetuity
$EUA_{C,\infty}$ = Total of annualized project direct costs and work zone user costs in perpetuity
i = Discount rate
T = Number of years of service life for a highway pavement

8.6 PROJECT-LEVEL BENEFIT AND COST ANALYSIS

With the project-level benefits and costs established, the values can be compared to determine the economic feasibility of the proposed pavement project. The computational procedure is illustrated in Table 8.13.

8.7 RISK AND UNCERTAINTY CONSIDERATIONS IN PAVEMENT PROJECT-LEVEL BENEFIT–COST ANALYSIS

The total benefits of a highway pavement project are established by assessing its impacts on pavement life-cycle agency and user costs in pavement service life cycle, either for one

Table 8.11 Computation of project-level agency and user benefits in perpetuity

Case		*Computation*
Base case: No early termination	Agency cost profile	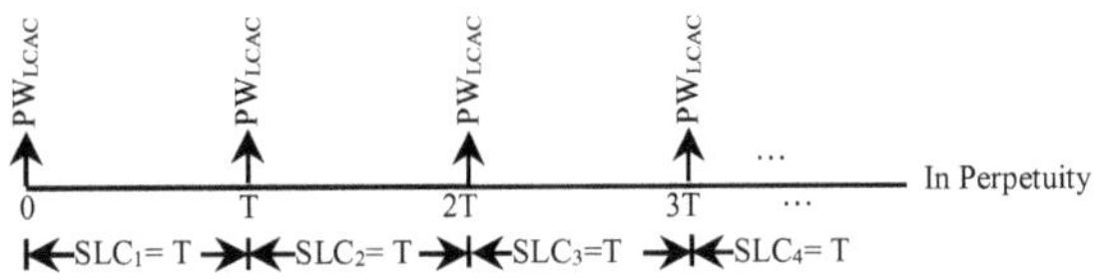
	User cost profile	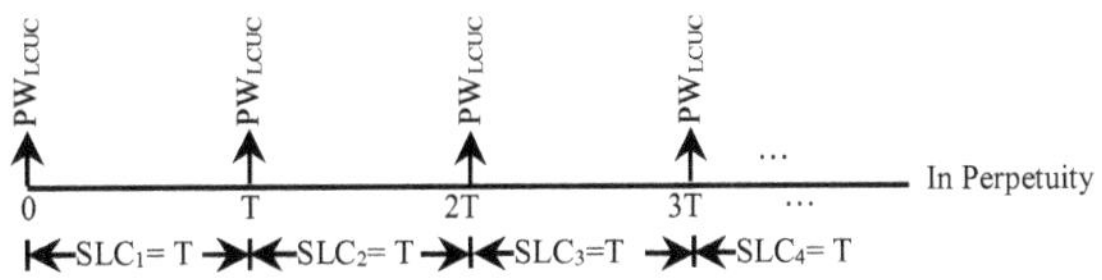
	Present worth	$PW_{LCAC\infty,0} = PW_{LCAC}/(1-(1/(1+i)^{T}))$ $PW_{LCUC\infty,0} = PW_{LCUC}/(1-(1/(1+i)^{T}))$
	Annual worth	$EUAAC_{\infty,0} = PW_{LCAC\infty,0} \cdot i, EUAUC_{\infty,0} = PW_{LCUC\infty,0} \cdot i$
Case 1: Early termination in cycle 1	Agency cost profile	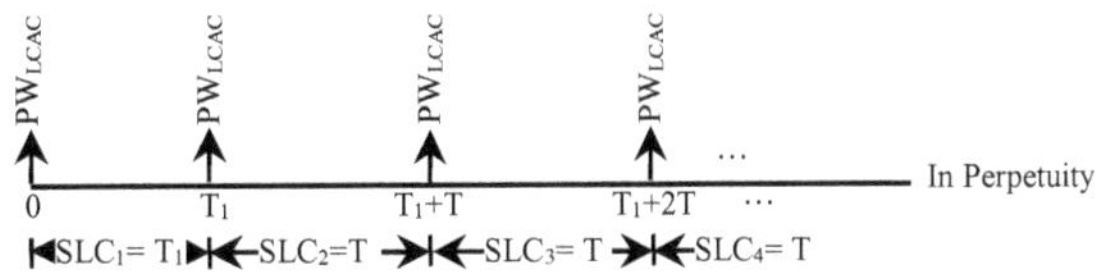
	User cost profile	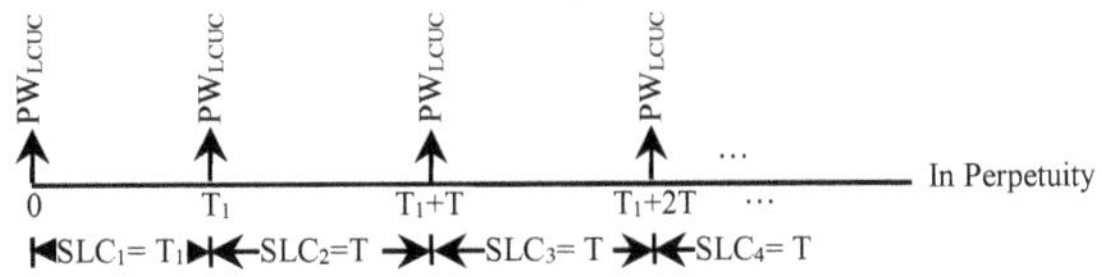
	Present worth	$PW_{LCAC\infty,1} = PW_{LCAC1} + (PW_{LCAC}/(1-(1/(1+i)^{T})))/(1+i)^{T_1}$ $PW_{LCUC\infty,1} = PW_{LCUC1} + (PW_{LCUC}/(1-(1/(1+i)^{T})))/(1+i)^{T_1}$
	Annual worth	$EUAAC_{\infty,1} = PW_{LCAC\infty,1} \cdot i, EUAUC_{\infty,1} = PW_{LCUC\infty,1} \cdot i$
	Base case benefits	Agency Benefits : $PW_{AB} = PW_{LCAC\infty,1} - PW_{LCAC\infty,0}$ $EUA_{AB} = EUAAC_{LCAC\infty,1} - EUAAC_{LCAC\infty,0}$ User Benefits : $PW_{UB} = PW_{LCUC\infty,1} - PW_{LCUC\infty,0}$ $EUA_{UB} = EUAUC_{LCAC\infty,1} - EUAUC_{LCUC\infty,0}$ Overall Benefits : $PW_{B} = PW_{AB} + PW_{UB}, EUA_{B} = EUA_{AB} + EUA_{UB}$
Case 2: Early termination in cycles 1 and 2	Agency cost profile	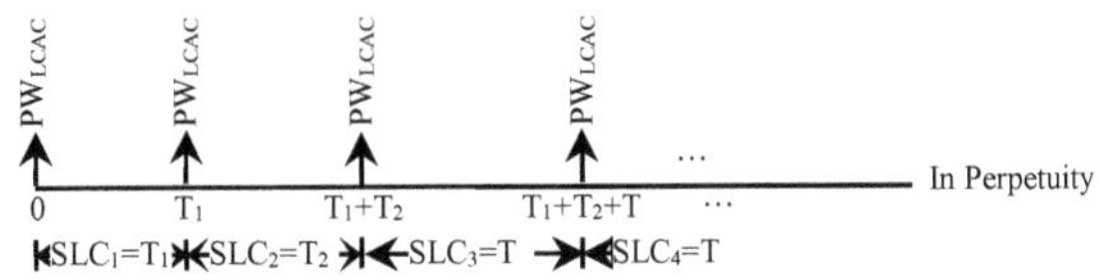
	User cost profile	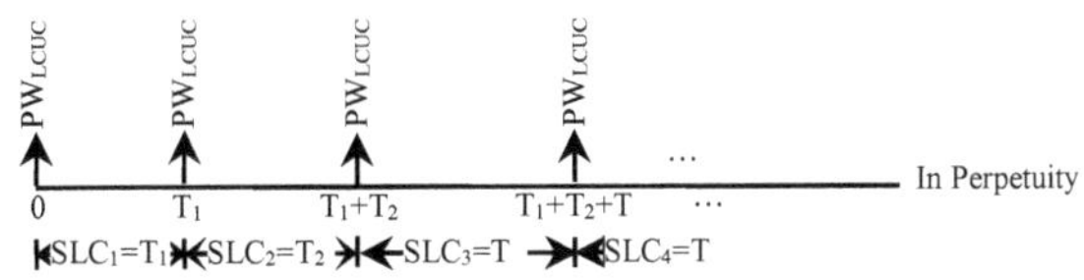

(*Continued*)

Table 8.11 (Continued) Computation of project-level agency and user benefits in perpetuity

Case		*Computation*
	Present worth	$PW_{LCAC\infty,2} = PW_{LCAC1} + PW_{LCAC2}/(1+i)^{T_1} + (PW_{LCAC}/(1-(1/(1+i)^{T})))/(1+i)^{(T_1+T_2)}$ $PW_{LCUC\infty,2} = PW_{LCUC1} + PW_{LCUC2}/(1+i)^{T_1} + (PW_{LCUC}/(1-(1/(1+i)^{T})))/(1+i)^{(T_1+T_2)}$
	Annual worth	$EUAAC_{\infty,2} = PW_{LCAC\infty,2} \cdot i$, $EUAUC_{\infty,2} = PW_{LCUC\infty,2} \cdot i$
	Base case benefits	Agency Benefits : $PW_{AB} = PW_{LCAC\infty,2} - PW_{LCAC\infty,0}$ $EUA_{AB} = EUAAC_{LCAC\infty,2} - EUAAC_{LCAC\infty,0}$ User Benefits : $PW_{UB} = PW_{LCUC\infty,2} - PW_{LCUC\infty,0}$ $EUA_{UB} = EUAUC_{LCAC\infty,2} - EUAUC_{LCUC\infty,0}$ Overall Benefits : $PW_{B} = PW_{AB} + PW_{UB}, EUA_{B} = EUA_{AB} + EUA_{UB}$

Table 8.12 Distribution of project costs in perpetuity horizon

Case		*Computation*
Base case: No early termination	Project direct cost profile	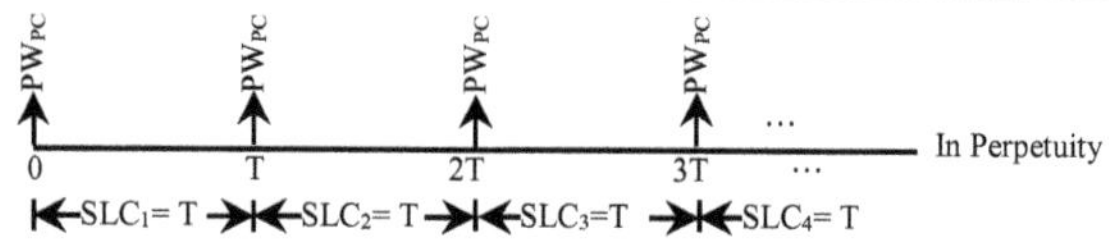
	Project work zone-related user cost profile	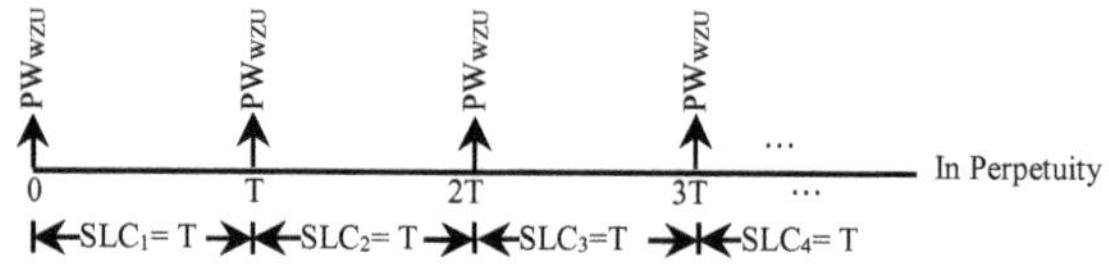
	Present worth	$PW_{C,\infty,0} = (PW_{PC} + PW_{WZUC})/(1-(1/(1+i)^{T}))$
	Annual worth	$EUA_{C,\infty,0} = PW_{C,\infty,0} \cdot i$

service life cycle or in perpetuity (Sinha and Fwa, 1988). Quite a few input factors are used to estimate the benefit spanning the pavement service life cycle. For some input factors, historical data are used to derive the values for computation. For other input factors, predicted values are utilized for calculation. The historical data are inherited with accuracy and precision issues, and predicted values may not be fully accurate either. In association with the historical data and prediction values with the pavement project for benefit and cost analysis, the issues of accuracy and precision of the adopted values get carried over.

Therefore, using a single value for each input factor to calculate the benefits or costs for the pavement project may be questionable. Often multiple possible values may be considered for some input factors for the computation. The analysis can be classified into two cases. For the first case, the range of multiple possible values is known and a reliable probability distribution can be assigned. Thus, a mathematical expectation (or expected value) via probabilistic risk analysis can be calculated and used for pavement project-level benefit and cost calculation. For the second case, the full range of multiple possible values may be unknown, and even if it is known a reliable probability distribution cannot be established. Hence a mathematical expectation (or expected value) cannot be calculated. For any pavement project-level life-cycle benefit and cost analysis, the combination of input factors under

Table 8.13 Pavement project benefit–cost analysis in perpetuity

Case		*Present worth*		*Annual worth*	
		Agency costs	*User costs*	*Agency costs*	*User costs*
Base case: No early termination	Life-cycle costs	$PW_{LCAC\infty,0}$	$PW_{LCUC\infty,0}$	$EUAAC_{\infty,0}$	$EUAUC_{\infty,0}$
Case 1: Early termination In cycle 1	Life-cycle costs	$PW_{LCAC\infty,1}$	$PW_{LCUC\infty,1}$	$EUAAC_{\infty,1}$	$EUAUC_{\infty,1}$
	Base case benefits	$PW_{AB,\infty} = PW_{LCAC\infty,1} - PW_{LCAC\infty,0}$ $PW_{UB,\infty} = PW_{LCUC\infty,1} - PW_{LCUC\infty,0}$		$EUA_{AB,\infty} = EUAAC_{LCAC\infty,1} - EUAAC_{LCAC\infty,0}$ $EUA_{UB,\infty} = EUAUC_{LCAC\infty,1} - EUAUC_{LCUC\infty,0}$	
	Base case project costs	$PW_{C,\infty,0}$		$EUA_{C,\infty,0}$	
	Benefit/cost analysis	$NPW = (PW_{AB,\infty} + PW_{UB,\infty}) - PW_{C,\infty,0}$ $B/C = (PW_{AB,\infty} + PW_{UB,\infty})/PW_{C,\infty,0}$		$NPW = (EUA_{AB,\infty} + EUA_{UB,\infty}) - EUA_{C,\infty,0}$ $B/C = (EUA_{AB,\infty} + EUA_{UB,\infty})/EUA_{C,\infty,0}$	
Case 2: Early termination In cycles 1 and 2	Life-cycle costs	$PW_{LCAC\infty,2}$	$PW_{LCUC\infty,2}$	$EUAAC_{\infty,2}$	$EUAUC_{\infty,2}$
	Base case benefits	$PW_{AB,\infty} = PW_{LCAC\infty,2} - PW_{LCAC\infty,0}$ $PW_{UB,\infty} = PW_{LCUC\infty,2} - PW_{LCUC\infty,0}$		$EUA_{AB,\infty} = EUAAC_{LCAC\infty,2} - EUAAC_{LCAC\infty,0}$ $EUA_{UB,\infty} = EUAUC_{LCAC\infty,2} - EUAUC_{LCUC\infty,0}$	
	Base case project costs	$PW_{C,\infty,0}$		$EUA_{C,\infty,0}$	
	Benefit/cost analysis	$NPW = (PW_{AB,\infty} + PW_{UB,\infty}) - PW_{C,\infty,0}$ $B/C = (PW_{AB,\infty} + PW_{UB,\infty})/PW_{C,\infty,0}$		$NPW = (EUA_{AB,\infty} + EUA_{UB,\infty}) - EUA_{C,\infty,0}$ $B/C = (EUA_{AB,\infty} + EUA_{UB,\infty})/EUA_{C,\infty,0}$	

Note: NPW, Net present worth; B/C, benefit-to-cost ratio

certainty, risk, and uncertainty varies. In some instances, all input factors for analysis may be under certainty and deterministic analysis is involved. In other instances, the input factors may be mixed with certainty and risk or certainty, risk, and uncertainty. Correspondingly, risk-based or uncertainty-based analysis is involved. They should be handled on a case-by-case basis.

8.7.1 Primary input factors involving risk and uncertainty considerations

Of the primary input factors for project-level life-cycle cost and benefit analysis, the costs of construction, repair, and maintenance, traffic demand, and discount rate are identified for risk and uncertainty considerations in the life-cycle cost analysis.

Construction and repair costs may not remain unchanged from the point in time when the estimates are made to the time when the pavement project is constructed. However, the construction and repair costs may vary due to market and political influences.

Traffic demand forecasted in the pavement service life span may not follow the projected path. The increased use of pavements by heavy vehicles will potentially reduce pavement useful service life cycles. Therefore, the project-level benefits and costs in pavement service life cycle will change accordingly.

The *discount rate* may fluctuate over time during the pavement service life cycle.

8.7.2 Probabilistic risk analysis

8.7.2.1 Selection of input probability distributions

Strictly speaking, construction and repair costs, traffic demand, and discount rate are discrete variables. Their minimum and maximum values of possible outcomes are bounded by nonnegative values. In addition, the distributions of the possible outcomes could be either symmetrical or skewed. Such characteristics can be modeled by the beta distribution that is continuous over a finite range and allows for virtually any degree of skewness and kurtosis. The general beta distribution has four parameters: lower range (L), upper range (H), and two shape parameters referred to as α and β. The beta density function is given by

$$f(x|\alpha,\beta,L,H) = \frac{\Gamma(\alpha+\beta)\cdot(x-L)^{\alpha-1}\cdot(H-x)^{\beta-1}}{\Gamma(\alpha)\cdot\Gamma(\beta)\cdot(H-L)^{\alpha+\beta-1}}(L \leq x \leq H)$$

where the Γ-function factors serve to normalize the distribution so that the area under the density function from L to H is exactly one. The mean and variance for the beta distribution are given as $\mu = \alpha/(\alpha+\beta)$ and $(\alpha\cdot\beta)/[((\alpha+\beta)^2)\cdot(\alpha+\beta+1)]$.

It is seen that the distribution mean is a weighted average of L and H such that when $0 < \alpha < \beta$, the mean is closer to L and the distribution is skewed to the right; whereas for $\alpha > \beta > 0$, the mean value is closer to H and the distribution is skewed to the left. When $\alpha = \beta$, the distribution is symmetric. For a given α/β ratio, the mean value is constant and the variance varies inversely with the absolute value of $\alpha + \beta$. If α and β are increased proportionately, the variance may decrease while keeping the mean value unchanged. Practically, the skewness and variance (kurtosis) can be considered as high, medium or low depending on the values of α and β. Table 8.14 presents the combinations of skewness and variance (kurtosis) for beta distributions as a guide to conduct risk analysis.

8.7.2.2 Determination of distribution-controlling parameters

For state-maintained highways, historical data on highway construction and repair costs, traffic volumes, and discount rates are generally available. The data can be processed for risk-based analysis.

8.7.2.3 Execution of simulation runs

As a rigorous extension of sensitivity analysis, simulation employs randomly sampled values from the input probability distribution to calculate discrete results. Two types of sampling techniques are commonly used. The first one is Monte Carlo sampling that utilizes random

Table 8.14 Approximate values of shape parameters for beta distributions

Combination type	*Skewness*	*Variance (kurtosis)*	α	β
1	Skewed to the left	High	1.50	0.50
2	Symmetric	High	1.35	1.35
3	Skewed to the right	High	0.50	1.50
4	Skewed to the left	Medium	3.00	1.00
5	Symmetric	Medium	2.75	2.75
6	Skewed to the right	Medium	1.00	3.00
7	Skewed to the left	Low	4.50	1.50
8	Symmetric	Low	4.00	4.00
9	Skewed to the right	Low	1.50	4.50

numbers to select values from specific probability distributions. The second one is Latin hypercube sampling that divides the probability scale of the cumulative distribution curve into an equal number of probability ranges consistent with the number of iteration runs. This helps achieve convergence at a faster pace comparatively to the Monte Carlo simulation (FHWA, 1998b; Reigle, 2000).

8.7.3 Uncertainty-based analysis

The probabilistic risk analysis becomes ineffective in dealing with the case when an input factor is under uncertainty. Shackle's model is one of the alternative approaches that overcomes limitations of risk-based analysis approaches. The model is based on three pillars. First, it uses degree of surprise as a measure of uncertainty of the input factor for computing the project-level agency or user benefits instead of probability distribution. Second, it introduces a priority weight by jointly assessing each known value of the input factor and its degree of surprise pair. It eventually identifies and standardizes the focus gain and focus loss values relative to an expected value (sample mean, median or mode established for the possible values) from maximum priority weights (Shackle, 1949; Ford and Ghose, 1998; Young, 2001). The triple of standardized gain, loss, and expected value is established for the input factor from the Shackle model analysis.

Next, a decision rule is used to establish a single value based on the triple that will be eventually used for project-level benefit and cost analysis. Among the triple of standardized gain, loss, and expected values, the decision maker will likely assign higher weight to standardized focus loss and expected values since standardized focus gain value has exceeded the expected value. If the difference between standardized focus loss and expected values is within the decision-maker's tolerance level, the expected value can be adopted for the calculation. Otherwise, a compromised value between the standardized focus loss and expected values that is acceptable from the decision maker can be calculated and used for project-level benefit and cost analysis (Li and Sinha, 2004, 2009a; Li and Madanu, 2009).

In general, if input is under certainty, the single value could be used for the computation. If an input factor is under risk, the mathematical expectation can be used for calculation. If an input factor is under uncertainty, the expected value or a compromised value between the expected value and standardized focus loss can be used for analysis.

Case Study: Flexible pavement rehabilitation project evaluation

A 15-million-dollar pavement rehabilitation project was proposed in the year 2010 for a 10-mile, 2-lane rural principal arterial road. The road maintains a flexible pavement constructed in 1978 as the base year with a 40-year design useful service life, indicating that it has been in service for 32 years. The AADT in 2010 is 7380 with a 2% annual growth rate from the base year. The average vehicle operating speed is 59.37 mph.

For the typical flexible pavement life-cycle activity profile, rehabilitation treatments need to be implemented in the 15th and 30th years. For the alternative activity profile that reflects the actual situation of investments, it shows that the first rehabilitation treatment was appropriately implemented in the 15th year. However, the second rehabilitation is delayed by 2 years from the 30th year to the 32nd year, leading to a reduction in the useful service life by 2 years from 40 years to 38 years. Assuming that the annual incremental gradients are 3% for annual routine maintenance cost and 2% for individual user cost components of vehicle operation, travel time, crashes, and air emissions, with delayed implementation of the second pavement rehabilitation treatment, the incremental gradients for maintenance costs and user costs are expected to be 10% higher than the original gradients. A 4% discount rate is used for analysis.

Based on historical data, factors related to pavement agency costs of construction, rehabilitation, and maintenance; traffic growth rate; annual maintenance and user cost gradients; and discount rate are identified for risk and uncertainty analysis. The details are given below:

Cost item	μ	σ	*L*	*H*
Construction ($/lane-mi)	1353536.53	694614.00	588385.00	3165840.00
Rehabilitation ($/lane-mi)	155287.00	509879.00	29147.00	1119863.00
Preventive maintenance ($/lane-mi)	4120.00	6544.00	186.00	21999.00
Annual maintenance ($/lane-mi)	137.97	499.00	4.00	2186.00
Annual traffic growth rate r	2%	1%	1%	3%
Annual maintenance gradient g_{M1}	3%	1%	1%	5%
Annual maintenance gradient g_{M2}	3%	1%	1%	5%
Annual maintenance gradient g_{M3}	3%	1%	1%	5%
Annual user cost gradient r_1	2%	1%	1%	3%
Annual user cost gradient r_2	2%	1%	1%	3%
Annual user cost gradient r_3	2%	1%	1%	3%
Discount rate i	4%	1%	3%	5%

The unit user costs of vehicle operation, travel time, crashes, and air emissions in the year 2000 constant dollars are computed by the following formulas (Li et al., 2010):

$$\text{Vehicle operating costs } (\$/VMT) = 0.3523 - 0.0022 \times \text{Speed}$$
$$\text{Travel Time cost } (\$/VMT) = 0.40327 - 0.004245 \times \text{Speed}$$
$$\text{Crash cost } (\$/VMT) = -0.1483 + 0.004877 \times \text{Speed}$$
$$\text{Air emission costs } (\$/VMT) = 0.2059 + 0.00006256 \times \text{Speed}$$

Conduct project evaluation using the life-cycle cost analysis approach without and with considerations of risk and uncertainty of the above input factors.

SOLUTION

1. Deterministic project evaluation
   1A. Base case life-cycle agency cost analysis using the typical activity profile
      a. Base case agency cost items

Agency cost item	*Unit cost ($/lane-mi)*	*Project-related agency cost*
Construction	1353536.53	$= 1353536.53 \times 10 \times 2 = 27070730.60$
Rehabilitation	155287.00	$= 155287.00 \times 10 \times 2 = 3105740.00$
Preventive maintenance	4120.00	$= 4120.00 \times 10 \times 2 = 82400.00$
Annual maintenance	138.00	$= 138.00 \times 10 \times 2 = 2760.00$

b. Base case life-cycle agency costs

Agency cost profile

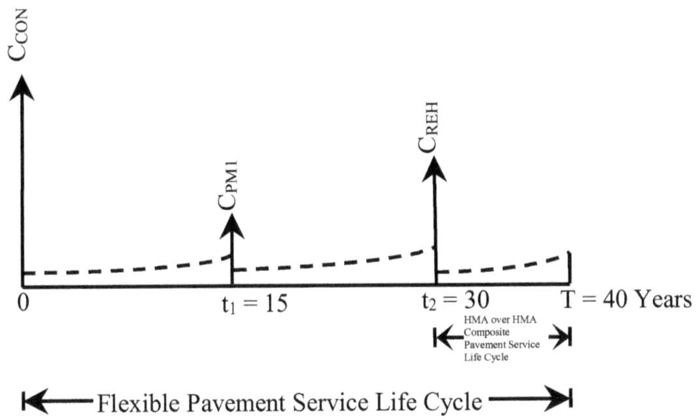

$$
\begin{aligned}
PW_{LCAC} &= C_{CON} + C_{PMI}/(1+i)^{t_1} + C_{REH}/(1+i)^{t_2} + (C_{MAIN1}(1-(1+g_{M1})^{t_1}(1+i)^{-t_1}))/(i-g_{M1}) \\
&\quad +((C_{MAIN2}(1-(1+g_{M2})^{(t_2-t_1)}(1+i)^{-(t_2-t_1)}))/(i-g_{M2}))/(1+i)^{t_1} \\
&\quad +((C_{MAIN3}(1-(1+g_{M3})^{(T-t_2)}(1+i)^{-(T-t_2)}))/(i-g_{M3}))/(1+i)^{t_2} \\
&= 27070730.60 + 82400.00/(1+4\%)^{15} + 3105740.056/(1+4\%)^{30} \\
&\quad +(2760.00(1-(1+3\%)^{15}(1+4\%)^{-15}))/(4\%-3\%) \\
&\quad +((2760.00(1-(1+3\%)^{(30-15)}(1+4\%)^{-(30-15)}))/(4\%-3\%))/(1+4\%)^{15} \\
&\quad +((2760.00(1-(1+3\%)^{(40-30)}(1+4\%)^{-(40-30)}))/(4\%-3\%))/(1+4\%)^{30} \\
&= 28139792.32 \\
PW_{LCAC\infty} &= PW_{LCAC}/(1-(1/(1+i)^T)) = 28139792.32/(1-(1/(1+4\%)^{40})) = 35543012.42 \\
EUAAC &= PW_{LCAC}((i(1+i)^T)/((1+i)^T-1)) \\
&= 28139792.32((4\%(1+4\%)^{40})/((1+4\%)^{40}-1)) = 1421720.49 \\
EUAAC_{\infty} &= PW_{LCAC\infty}\cdot i = 35543012.42\times 4\% = 1421720.49
\end{aligned}
$$

1B. Base case life-cycle user cost analysis using the typical activity profile

a. Base year user costs

i. Base year $AADT$: $7380/[(1+2\%)^{32}] = 3916$ (An annual growth rate of 2%)

ii. Base year annual user costs:

$$\text{Vehicle opt. costs (\$/VMT)} = 0.3523 - 0.0022\times\text{Speed} = 0.3523 - 0.0022\times 59.37 = 0.2217$$

$$\text{Travel time cost (\$/VMT)} = 0.40327 - 0.004245\times\text{Speed} = 0.4033 - 0.004245\times 59.37 = 0.1534$$

$$\text{Crash cost (\$/VMT)} = -0.1483 + 0.004877\times\text{Speed} = -0.1483 + 0.004877\times 59.37 = 0.1413$$

$$\text{Air emission cost (\$/VMT)} = 0.2059 + 0.00006256\times\text{Speed} = 0.2059 + 0.00006256\times 59.37 = 0.2096$$

iii. Base year annual user costs:

Annual vehicle operating costs	$= 0.2217 \times 391610 \times 365 = 3168846.78$
Annual travel time cost	$= 0.1534 \times 391610 \times 365 = 2192607.56$
Annual vehicle crash cost	$= 0.1413 \times 391610 \times 365 = 2019657.42$
Annual vehicle air emission cost	$= 0.2096 \times 391610 \times 365 = 2995896.64$
Total	$= 10377008.40$

b. Base case life-cycle user costs

User cost profile

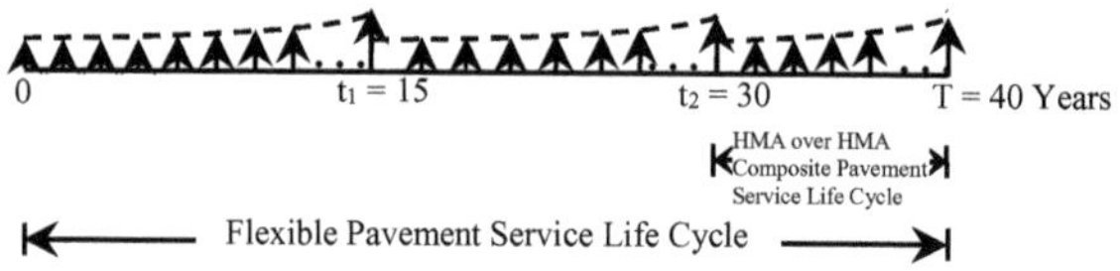

PW_{LCUC}

$$= (C_{AUC1}(1 - (1 + r_1)^{t_1}(1 + i)^{-t_1}))/(i - r_1)$$
$$+ ((C_{AUC2}(1 - (1 + r_2)^{(t_2-t_1)}(1 + i)^{-(t_2-t_1)}))/(i - r_2))/(1 + i)^{t_1}$$
$$+ ((C_{AUC3}(1 - (1 + r_3)^{(T-t_2)}(1 + i)^{-(T-t_2)}))/(i - r_3))/(1 + i)^{t_2}$$
$$= (10377008.40(1 - (1 + 2\%)^{15}(1 + 4\%)^{-15}))/(4\% - 2\%)$$
$$+ ((10377008.40(1 - (1 + 2\%)^{(30-15)}(1 + 4\%)^{-(30-15)}))/(4\% - 2\%))/(1 + 4\%)^{15}$$
$$+ ((10377008.40(1 - (1 + 2\%)^{(40-30)}(1 + 4\%)^{-(40-30)}))/(4\% - 2\%))/(1 + 4\%)^{30}$$
$$= 232139250.65$$

$PW_{LCUC\infty}$

$$= PW_{LCUC}/(1 - (1/(1 + i)^T))$$
$$= 232139250.65/(1 - (1/(1 + 4\%)^{40})) = 293212123.80$$

$EUAUC$

$$= PW_{LCUC}((i(1 + i)^T)/((1 + i)^T - 1))$$
$$= 232139250.65((4\%(1 + 4\%)^{40})/((1 + 4\%)^{40} - 1)) = 11728484.95$$

$EUAUC_{\infty}$

$$= PW_{LCUC\infty} = i = 293212123.80 \times 4\% = 11728484.95$$

2A. Alternative case life-cycle agency cost analysis with early termination
Reduction in pavement useful service life
Two years of reduction in pavement useful service life
Conversion of project cost to 2000 constant dollars

$$Dollar_{2000} = Dollar_{2010}/(1 + i)^{2010\text{-}2000} = 15000000/(1 + 4\%)^{10} = \$10133462.53$$

Pavement maintenance gradient adjustment.
Due to late implementation of the rehabilitation treatment compared with the timing in the base case profile, the annual gradient of annual routine maintenance cost is expected to be higher by 10%. The adjusted annual gradient becomes

$$g'_{M3} = g_{M3} + (10\% \times ((1 + g_{M2})^{(32-15)} - 1)) = 3\% + (10\% \times ((1 + 3\%)^{(32-15)} - 1)) = 9.528\%$$

a. Alternative case agency cost items

Agency cost item	*Unit cost ($/lane-mi)*	*Project-related agency cost*
Project cost		10133462.53
Construction	1353536.53	27070730.60
Rehabilitation	155287.00	3105740.00
Preventive maintenance	4120.00	82400.00
Annual maintenance	138.00	2760.00

b. Alternative case life-cycle agency costs

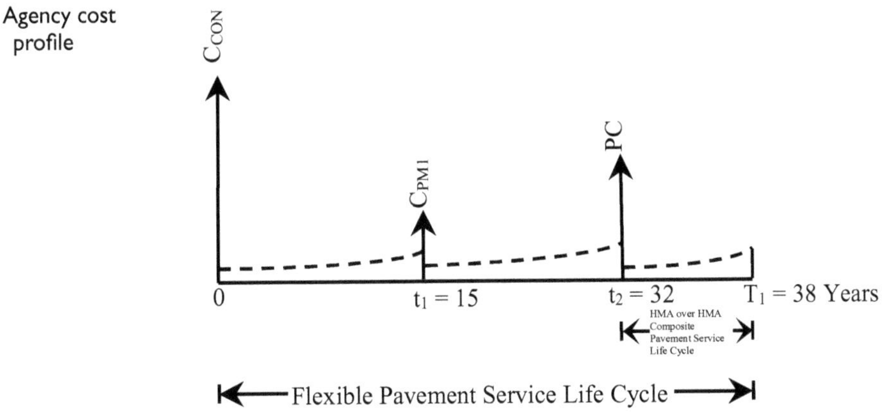

PW_{LCAC} $= C_{CON} + C_{PMI}/(1+i)^{t_1} + PC/(1+i)^{t_2}$
$+ (C_{MAIN1}(1-(1+g_{M1})^{t_1}(1+i)^{-t_1}))/(i-g_{M1})$
$+ ((C_{MAIN2}(1-(1+g_{M2})^{(t_2-t_1)}(1+i)^{-(t_2-t_1)}))/(i-g_{M2}))/(1+i)^{t_1}$
$+ ((C_{MAIN3}(1-(1+g'_{M3})^{(T_1-t_2)}(1+i)^{-(T_1-t_2)}))/(i-g'_{M3}))/(1+i)^{t_2}$
$= 27070730.60 + 82400.00/(1+4\%)^{15} + 10133462.53/(1+4\%)^{32}$
$+ (2760.00(1-(1+3\%)^{15}(1+4\%)^{-15}))/(4\%-3\%)$
$+ ((2760.00(1-(1+3\%)^{(32-15)}(1+4\%)^{-(32-15)}))/(4\%-3\%))/(1+4\%)^{15}$
$+ ((2760.00(1-(1+9.528\%)^{(38-32)}(1+4\%)^{-(38-32)}))/(4\%-9.528\%))/(1+4\%)^{32}$
$= 30070745.45$

$EUAAC$ $= PW_{LCAC}((i(1+i)^{T'_1})/((1+i)^{T'_1}-1))$
$= 30070745.45((4\%(1+4\%)^{38})/((1+4\%)^{38}-1)) = 1552610.30$

2B. Alternative case life-cycle user cost analysis with early termination

a. Base year user costs

Base year annual user costs:

Annual vehicle operating costs	3168846.78
Annual travel time cost	2192607.56
Annual vehicle crash cost	2019657.42
Annual vehicle air emission cost	2995896.64
Total	10377008.40

b. Alternative case life-cycle user costs

Annual user cost gradient adjustment

Due to late implementation of the rehabilitation treatment compared with the timing in the base case profile, the annual gradient of user costs is expected to be higher by 10%. The adjusted annual gradient becomes

$$r_3' = r_3 + (10\% \times ((1+r_2)^{(32-15)} - 1)) = 2\% + (10\% \times ((1+2\%)^{(32-15)} - 1)) = 6\%$$

User cost profile

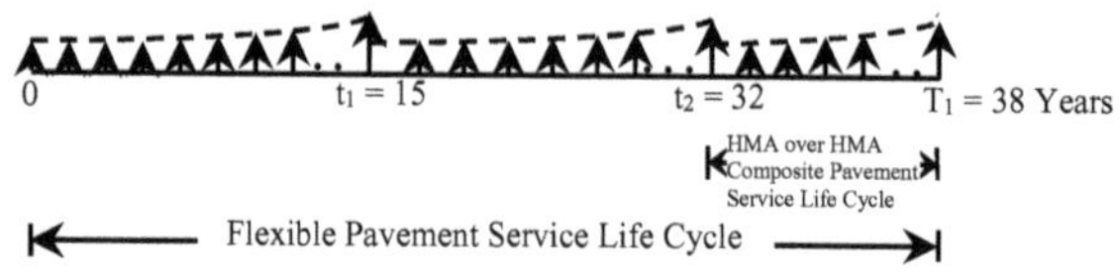

PW_{LCUC}

$$= (C_{AUC1}(1 - (1 + r_1)^{t_1}(1 + i)^{-t_1}))/(i - r_1)$$
$$+ ((C_{AUC2}(1 - (1 + r_2)^{(t_2-t_1)}(1 + i)^{-(t_2-t_1)}))/(i - r_2))/(1 + i)^{t_1}$$
$$+ ((C_{AUC3}(1 - (1 + r_3')^{(T_1-t_2)}(1 + i)^{-(T_1-t_2)}))/(i - r_3'))/(1 + i)^{t_2}$$
$$= (10377008.40(1 - (1 + 2\%)^{15}(1 + 4\%)^{-15}))/(4\% - 2\%)$$
$$+ ((10377008.40(1 - (1 + 2\%)^{(32-15)}(1 + 4\%)^{-(32-15)}))/(4\% - 2\%))/(1 + 4\%)^{15}$$
$$+ ((10377008.40(1 - (1 + 6\%)^{(38-32)}(1 + 4\%)^{-(38-32)}))/(4\% - 6\%))/(1 + 4\%)^{32}$$
$$= 230015425.70$$

$EUAUC$

$$= PW_{LCUC}((i(1 + i)^{T_1})/((1 + i)^{T_1} - 1))$$
$$= 230015425.70\ ((4\%(1 + 4\%)^{38})/((1 + 4\%)^{38} - 1)) = 11876137.86$$

3. Project life-cycle overall benefits in perpetuity
   Since the typical life-cycle activity profile warrants the lowest life-cycle agency and user costs, the pavement manager would always like to revert to the typical activity profile after experiencing early termination. Here, early termination is only assumed for the first life cycle and the typical service life cycle as in the base case is followed for the rest of the cycles into perpetuity.

Base case: No early Termination

Agency cost profile

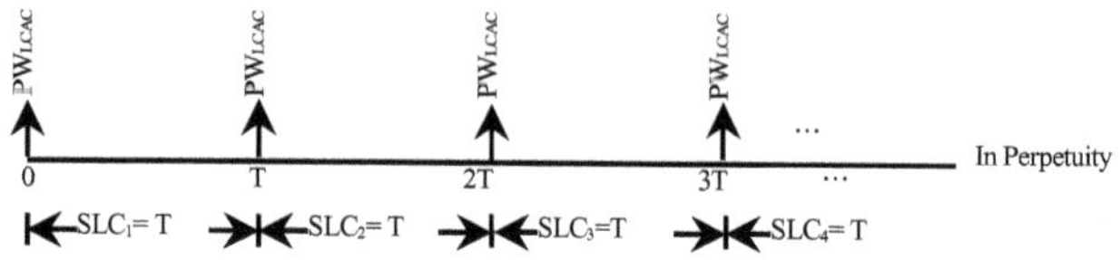

User cost Profile

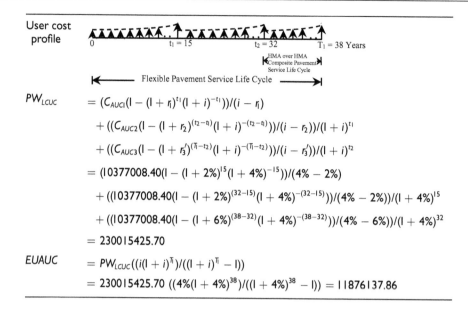

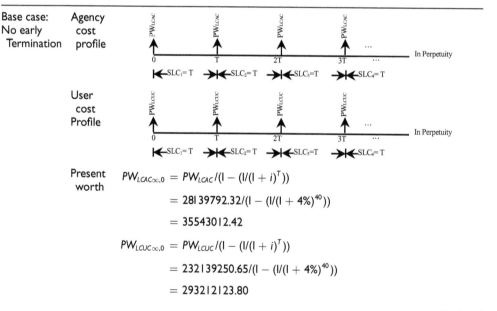

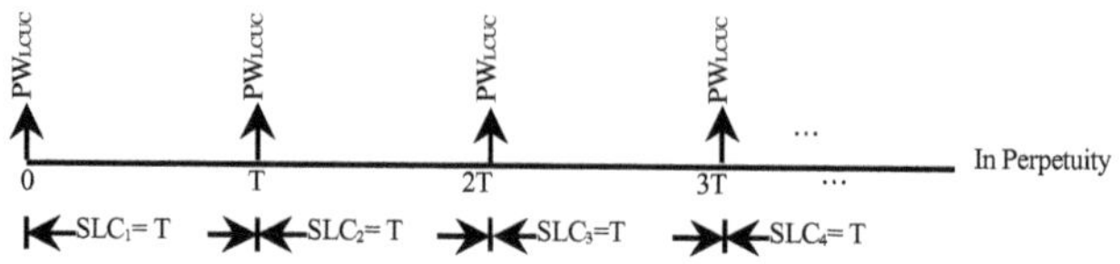

Present worth

$$PW_{LCAC\infty,0} = PW_{LCAC}/(1 - (1/(1 + i)^T))$$
$$= 28139792.32/(1 - (1/(1 + 4\%)^{40}))$$
$$= 35543012.42$$

$$PW_{LCUC\infty,0} = PW_{LCUC}/(1 - (1/(1 + i)^T))$$
$$= 232139250.65/(1 - (1/(1 + 4\%)^{40}))$$
$$= 293212123.80$$

Continued

Alternative Case: Early termination in cycle 1

Agency cost profile

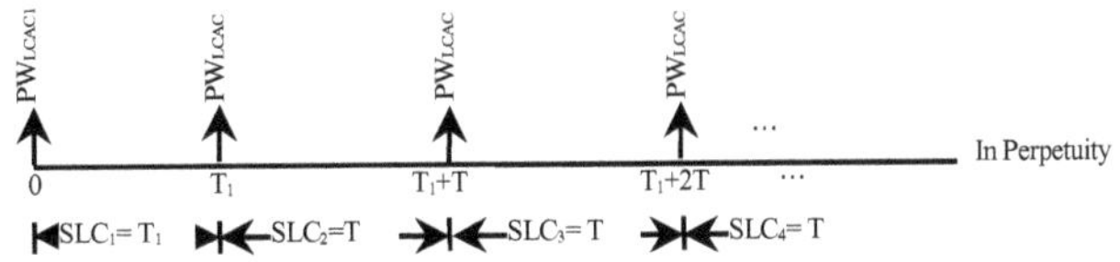

User cost Profile

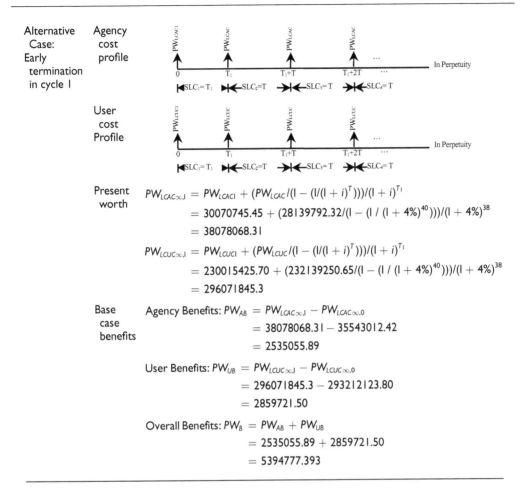

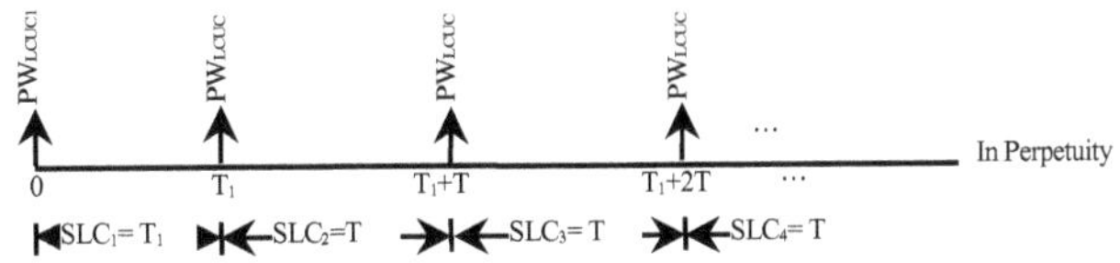

Present worth

$$PW_{LCAC\infty,1} = PW_{LCAC1} + (PW_{LCAC}/(1 - (1/(1+i)^T)))/(1+i)^{T_1}$$
$$= 30070745.45 + (28139792.32/(1 - (1\ /\ (1+4\%)^{40})))/(1+4\%)^{38}$$
$$= 38078068.31$$

$$PW_{LCUC\infty,1} = PW_{LCUC1} + (PW_{LCUC}/(1 - (1/(1+i)^T)))/(1+i)^{T_1}$$
$$= 230015425.70 + (232139250.65/(1 - (1\ /\ (1+4\%)^{40})))/(1+4\%)^{38}$$
$$= 296071845.3$$

Base case benefits

Agency Benefits: $PW_{AB} = PW_{LCAC\infty,1} - PW_{LCAC\infty,0}$
$$= 38078068.31 - 35543012.42$$
$$= 2535055.89$$

User Benefits: $PW_{UB} = PW_{LCUC\infty,1} - PW_{LCUC\infty,0}$
$$= 296071845.3 - 293212123.80$$
$$= 2859721.50$$

Overall Benefits: $PW_B = PW_{AB} + PW_{UB}$
$$= 2535055.89 + 2859721.50$$
$$= 5394777.393$$

II. Project evaluation under certainty and risk.

Input factors considered for risk-based analysis include pavement agency costs of construction, maintenance, and rehabilitation; traffic growth rates; and discount rates. Monte Carlo simulations were simultaneously performed for those factors using beta distributions. Ten simulation runs and with 1000 iterations each were used. For each input factor involving risk consideration, the grand average was established based on the average values of 10 simulation runs, each of which was determined in accordance with the 1000 iteration outcomes. In the end, the grand average values for all factors considered for probabilistic risk analysis were used for computing the expected project life-cycle benefits incorporating risk. The analytical procedure used is identical to that for deterministic life-cycle benefit analysis.

It should be noted that the analysis incorporating risk is essentially an analysis under a case of mixed certainty and risk. This is because apart from the agency costs, traffic growth rates, and discount rate, the remaining input factors such as useful service life and project size are still treated under certainty.

The following values are used for project evaluation with risk considerations:

Case	*Cost component*	*Cost item*	*Input value*
Factors under risk	Agency costs	Construction	1492234.51
		Rehabilitation	417362.84
		Preventive maintenance	7954.30
		Annual maintenance	732.84
		Annual maintenance gradient g_{M1}	2.998%
		Annual maintenance gradient g_{M2}	2.998%
		Annual maintenance gradient g_{M3}	2.998%
	Vehicle operating costs	Annual traffic growth rate r	2.011%
	Travel time	Annual user cost gradient r_1	1.9959%
	Vehicle crashes	Annual user cost gradient r_2	1.9959%
	Vehicle air emissions	Annual user cost gradient r_3	1.9959%
		Discount rate i	4.01%

Project life-cycle overall benefits under risk in perpetuity.

Early termination is only assumed for the first life cycle, and the typical service life cycle used in the base case is followed for the following cycles into perpetuity.

Base case: No early termination

Agency cost profile

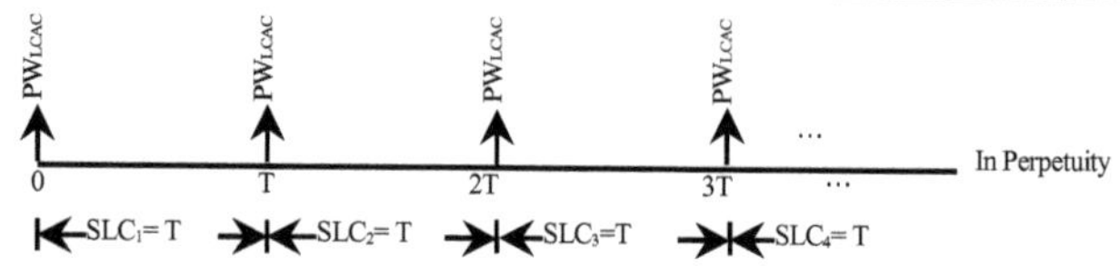

User cost profile

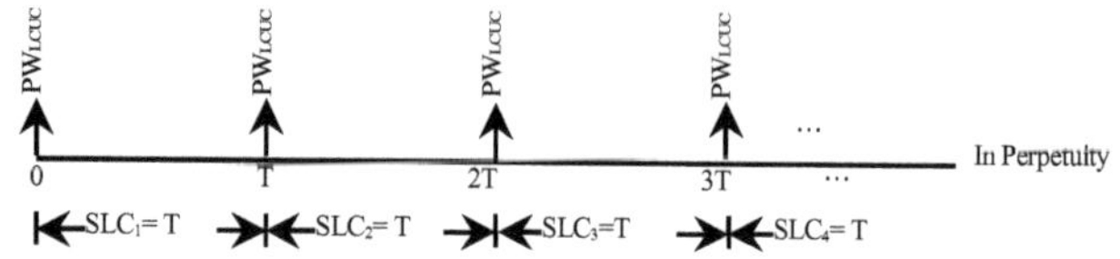

Present worth

$$PW_{LCAC\infty,0} = PW_{LCAC}/(1 - (1/(1+i)^T)) = 32847688.62/(1 - (1/(1+4.01\%)^{40})) = 41447638.91$$

$$PW_{LCUC\infty,0} = PW_{LCUC}/(1 - (1/(1+i)^T)) = 230960466.15/(1 - (1/(1+4.01\%)^{40})) = 291428907.31$$

Alternative Case: Early termination in cycle 1

Agency cost profile

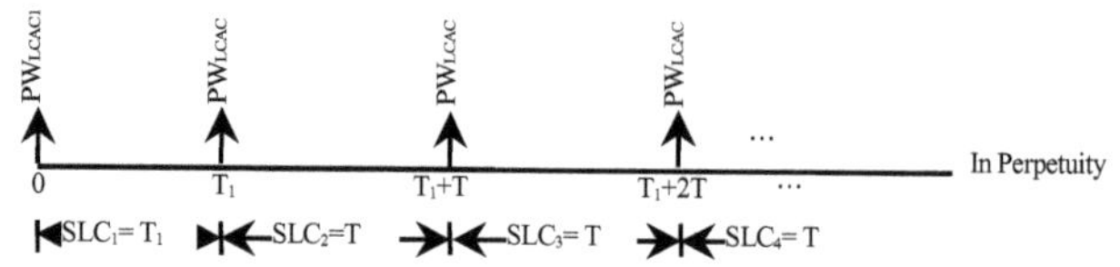

User cost Profile

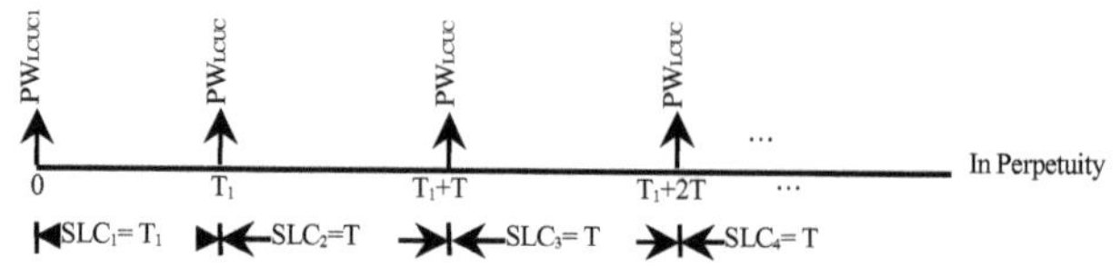

Continued

Present worth	$PW_{LCAC\infty,1} = PW_{LCAC1} + (PW_{LCAC}/(1-(1/(1+i)^T)))/(1+i)^{T_1}$
	$= 33157868.79$
	$+ (32847688.62/(1-(1/(1+4.01\%)^{40})))/(1+4.01\%)^{38}$
	$= 42504847.8$
	$PW_{LCUC\infty,1} = PW_{LCUC1} + (PW_{LCUC}/(1-(1/(1+i)^T)))/(1+i)^{T_1}$
	$= 228847472.20$
	$+ 230960466.15/(1-(1/(1+4.01\%)^{40})))/(1+4.01\%)^{38}$
	$= 294568462.6$
Base case benefits	Agency Benefits: $PW_{AB} = PW_{LCAC\infty,1} - PW_{LCAC\infty,0}$
	$= 42504847.80 - 41447638.91$
	$= 1057208.89$
	User Benefits: $PW_{UB} = PW_{LCUC\infty,1} - PW_{LCUC\infty,0}$
	$= 294568462.6 - 291428907.31$
	$= 3139555.30$
	Overall Benefits: $PW_B = PW_{AB} + PW_{UB}$
	$= 1057208.89 + 3139555.30$
	$= 4196764.19$

III. Project evaluation under certainty, risk, and uncertainty.

It the process of conducting risk-based project evaluation, benefits associated with reductions in agency costs, vehicle operating costs, and air emission costs are found to be relatively stable with limited variations, whereas decreases of travel time and crash costs vary considerably. As such, travel time and crash costs are further selected for uncertainty-based analysis. The risk-based project benefits related to reductions in agency costs, vehicle operating costs, and air emission costs are combined with uncertainty-based benefits in terms of decreases of travel time and vehicle crashes to arrive at the overall project life-cycle benefits under certainty, risk, and uncertainty.

The following values are used for project evaluation with risk considerations:

Case	*Cost component*	*Cost item*	*Input value*
Factors under risk	Agency costs	Construction	1492234.51
		Rehabilitation	417362.84
		Preventive maintenance	7954.30
		Annual maintenance	732.84
		Annual maintenance gradient g_{M1}	2.998%
		Annual maintenance gradient g_{M2}	2.998%
		Annual maintenance gradient g_{M3}	2.998%
	Vehicle operating costs	Annual traffic growth rate r	2.011%
	Vehicle air emissions	Annual user cost gradient r_1	1.9959%
		Annual user cost gradient r_2	1.9959%
		Annual user cost gradient r_3	1.9959%
		Discount rate i	4.01%

For uncertainty-based analysis of travel time saving and vehicle crash decrease benefits, the values of input factors are adjusted according to the following decision rules as an extension of Shackle's model:

$$\text{If higher values are preferred, } X = \begin{cases} F_{(E)}, & \text{if } x_{SFL} \leq \Delta X \\ \dfrac{(F_{(E)} - x_{SFL})}{[1-(\Delta X/F_{(E)})]}, & \text{otherwise} \end{cases}$$

$$\text{If lower values are preferred, } X = \begin{cases} F_{(E)}, & \text{if } x_{SFL} \leq \Delta X \\ \dfrac{(F_{(E)} + x_{SFL})}{[1+(\Delta X/F_{(E)})]}, & \text{otherwise} \end{cases}$$

For annual traffic growth rate *r*, considering that higher values are preferred

$$\text{Since } |x_{SFL} - F_{(E)}| = |1\% - 2.011\%| = 1.011\% > \Delta X = 39\% \times 2\% = 0.78\%$$
$$r = x_{SFL}/[1 - \Delta X / F_{(E)}] = 1\%/[1-(0.78\%/2.011\%)] = 1.63\%$$

For annual user cost gradients r_1, r_2, and r_3, supposing that higher values are preferred

$$\text{Since } |x_{SFL} - F_{(E)}| = |1\% - 1.995\%| = 0.995\% > \Delta X = 39\% \times 2\% = 0.78\%$$
$$r_1 = r_2 = r_3 = x_{SFL}/[1 - \Delta X/X_{(E)}] = 1\%/[1-(0.78\%/1.995\%)] = 1.64\%$$

For discount rate *i*, assuming that lower values are preferred

$$\text{Since } |x_{SFL} - F_{(E)}| = |5\% - 4.01\%| = 0.99\% > \Delta X = 20\% \times 4\% = 0.8\%$$
$$i = x_{SFL}/[1 + \Delta X/F_{(E)}] = 5\%/[1+0.8\%/4.01\%] = 4.17\%$$

The following values are used for project evaluation with uncertainty considerations:

Cost component	*Cost item*	*Average, $F_{(E)}$ (%)*	*x_{SFL} (%)*	*x_{SFG} (%)*	*Tolerance (ΔX) (%)*	*Input value (%)*
Travel time Vehicle crashes	Annual traffic growth rate *r*	2.011	1	3	39% of $\mu = 2$	1.63
	Annual user cost gradient r_1	1.9959	1	3	39% of $\mu = 2$	1.64
	Annual user cost gradient r_2	1.9959	1	3	39% of $\mu = 2$	1.64
	Annual user cost gradient r_3	1.9959	1	3	39% of $\mu = 2$	1.64
	Discount rate *i*	4.01	5	3	20% of $\mu = 4$	4.17

Project life-cycle overall benefits under certainty, risk, and uncertainty in perpetuity.

Early termination is only assumed for the first life cycle, and the typical service life cycle used in the base case is followed for the following cycles into perpetuity.

Base case: No early termination

Agency cost profile

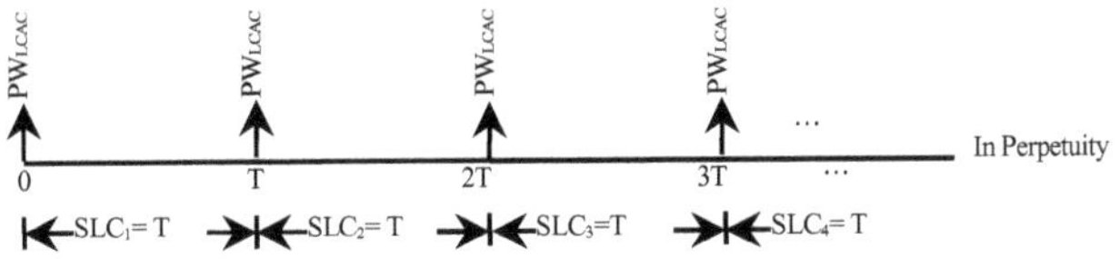

User cost profile

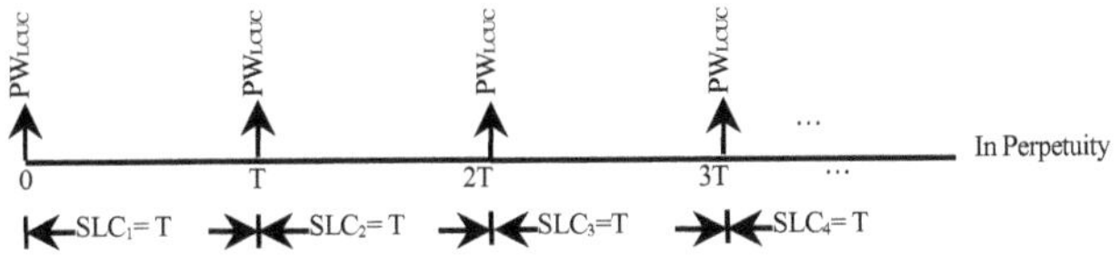

Present worth

$$PW_{LCAC\infty,0} = PW_{LCAC}/(1 - (1/(1+i)^{T}))$$
$$= 32847688.62/(1 - (1/(1 + 4.01\%)^{40}))$$
$$= 41447638.91$$

$$PW_{LCUC\infty,0} = PW_{LCUC(VOC+VAE)}/(1 - (1/(1+i)^{T}))$$
$$+ PW_{LCUC(VTT+VCC)}/(1 - (1/(1+i)^{T}))$$
$$= 137208332.14/(1 - (1/(1 + 4.01\%)^{40}))$$
$$+ 100911545.69/(1 - (1/(1 + 4.17\%)^{40}))$$
$$= 298505312.6$$

Alternative case: Early termination in cycle 1

Agency cost profile

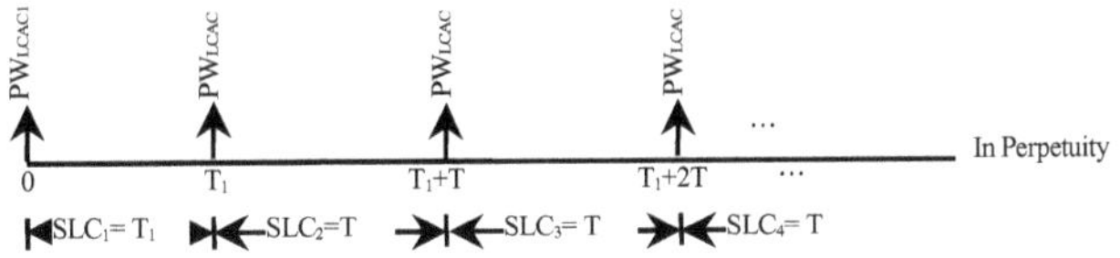

User cost profile

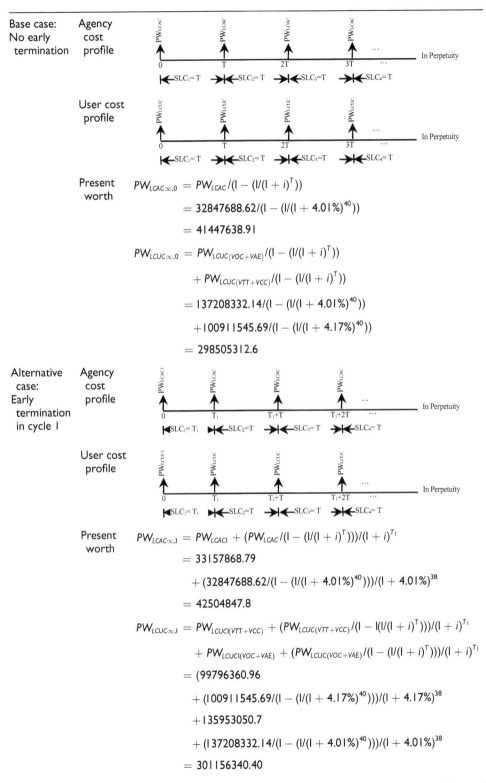

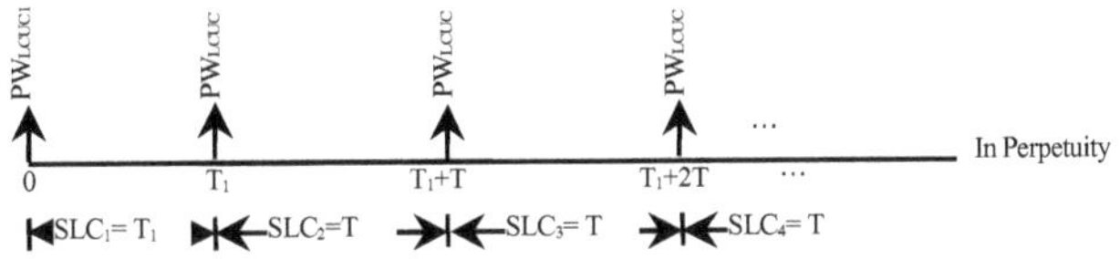

Present worth

$$PW_{LCAC\infty,1} = PW_{LCAC1} + (PW_{LCAC}/(1 - (1/(1+i)^{T})))/(1+i)^{T_1}$$
$$= 33157868.79$$
$$+ (32847688.62/(1 - (1/(1 + 4.01\%)^{40})))/(1 + 4.01\%)^{38}$$
$$= 42504847.8$$

$$PW_{LCUC\infty,1} = PW_{LCUC1(VTT+VCC)} + (PW_{LCUC(VTT+VCC)}/(1 - 1(1/(1+i)^{T})))/(1+i)^{T_1}$$
$$+ PW_{LCUC1(VOC+VAE)} + (PW_{LCUC(VOC+VAE)}/(1 - (1/(1+i)^{T})))/(1+i)^{T_1}$$
$$= (99796360.96$$
$$+ (100911545.69/(1 - (1/(1 + 4.17\%)^{40})))/(1 + 4.17\%)^{38}$$
$$+ 135953050.7$$
$$+ (137208332.14/(1 - (1/(1 + 4.01\%)^{40})))/(1 + 4.01\%)^{38}$$
$$= 301156340.40$$

Continued

Base case benefits	Agency Benefits: PW_{AB}	$= PW_{LCAC\infty,I} - PW_{LCAC\infty,0}$ $= 42504847.80 - 41447638.91 = 1057208.89$
	User Benefits: PW_{UB}	$= PW_{LCUC\infty,I} - PW_{LCUC\infty,0}$ $= 301156340.40 - 298505312.6 = 2651027.79$
	Overall Benefits: PW_B	$= PW_{AB} + PW_{UB}$ $= 1057208.89 + 2651027.79 = 3708236.68$

IV. Study summary

1. Input data

a. Project evaluation under certainty

Data item	*AC*	*VOC*	*TTC*	*VCC*	*VEC*
Project cost	10133462.53				
Construction cost	27070730.60				
Rehabilitation cost	3105740.00				
Preventive maintenance	82400.00				
Annual routine maintenance	2760.00				
Maint. gradient $g_{ml}, g_{m2}, g_{m3}/g'_{m3}$	3%/9.528%				
Discount rate i	4%	4%	4%	4%	4%
Base year AADT	3916	3916	3916	3916	3916
Annual traffic growth r	2%	2%	2%	2%	2%
Annual UC gradient $r_1, r_2, r_3/r'_3$		2%/6%	2%/6%	2%/6%	2%/6%
Service life for base case	40	40	40	40	40
Service life for alternative case	38	38	38	38	38

b. Project evaluation under certainty and risk

Data item	*AC*	*VOC*	*TTC*	*VCC*	*VEC*
Project cost	10123723.969				
Construction cost	29844690.20				
Rehabilitation cost	8347256.80				
Preventive maintenance	159086.00				
Annual routine maintenance	14656.80				
Maint. gradient $g_{ml}, g_{m2}, g_{m3}/g'_{m3}$	2.998%/9.52%				
Discount rate i	4.01%	4.01%	4.01%	4.01%	4.01%
Base year *AADT*	3903	3903	3903	3903	3903
Annual traffic growth r	2.011%	2.011%	2.011%	2.011%	2.011%
Annual *UC* gradient $r_1, r_2, r_3/r'_3$		1.995%/ 5.9887%	1.995%/ 5.9887%	1.995%/ 5.9887%	1.995%/ 5.9887%
Service life for base case	40	40	40	40	40
Service life for alternative case	38	38	38	38	38

c. Project evaluation under certainty, risk, and uncertainty

Data item	*AC*	*VOC*	*TTC*	*VCC*	*VEC*
Project cost	10123723.969				
Construction cost	29844690.20				
Rehabilitation cost	8347256.80				
Preventive maintenance	159086.00				
Annual routine maintenance	14656.80				
Maintenance gradient $g_{m1}, g_{m2}, g_{m3}/g'_{m3}$	2.998%/9.52%				
Discount rate i	4.01%	4.01%	4.17%	4.17%	4.01%
Base year *AADT*	3903	3903	4399	4399	3903
Annual traffic growth r	2.011%	2.011%	1.63%	1.63%	2.011%
Annual *UC* gradient $r_1, r_2, r_3/r'_3$		1.995%/ 5.9887%	1.64%/ 4.825%	1.64%/ 4.825%	1.995%/ 5.9887%
Service life for base case	40	40	40	40	40
Service life for alternative case	38	38	38	38	38

2. Estimated project benefits

Benefit category	*Certainty*	*Certainty and risk*	*Certainty, risk, and uncertainty*
PWAgency	2535055.89	1057208.89	1057208.89
PWUser	2859721.50	3139555.30	2651027.79
PWTotal	5394777.39	4196764.19	3708236.68

3. Project costs in perpetuity

Case	*Computation*
Project cost profile	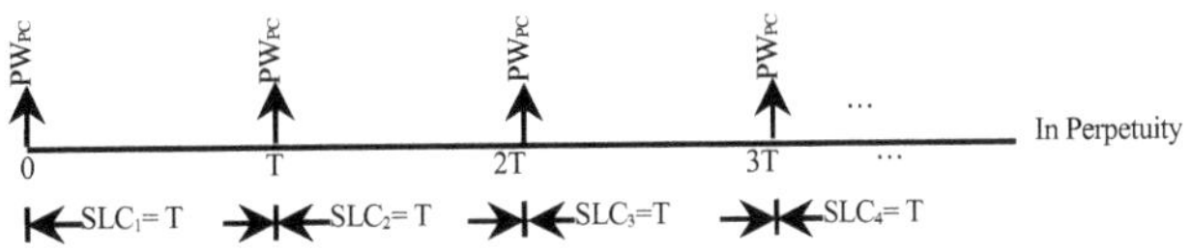
Certainty ($i = 4\%$)	$PW_{LCPC} = \text{Project Cost}/((1 + i)^{(2010-2000)+\text{Project Timing}})$ $= 15000000/((1 + 4\%)^{(2010-2000)+32}) = 2888623.955$ $PW_{LCPC\infty} = PW_{LCPC}/(1 - (1/(1 + i)^T))$ $= 2888623.955/(1 - (1/(1 + 4\%)^{40})) = 3648584.039$
Certainty and risk ($i = 4.01\%$)	$PW_{LCPC} = \text{Project Cost}/((1 + i)^{(2010-2000)+\text{Project Timing}})$ $= 15000000/((1 + 4.01\%)^{(2010-2000)+32}) = 2876982441$ $PW_{LCPC\infty} = PW_{LCPC}/(1 - (1/(1 + i)^T))$ $= 2876982441/(1 - (1/(1 + 4.01\%)^{40})) = 3630213.703$
Certainty, risk, and uncertainty ($i = 4.01\%$)	$PW_{LCPC} = 2876982441$ $PW_{LCPC\infty} = 3630213.703$

Note: LCPC, Life-cycle project costs

4. Project evaluation results

Item	*Certainty*	*Certainty and risk*	*Certainty, risk, and uncertainty*
NPW	= 5394777.39 – 3648584.039 = 1746193.351	= 4196764.19–3630213.703 = 566550.487	= 3708236.68 – 3630213.703 = 78022.98
B/C	= 5394777.39/3648584.039 = 1.48	= 4196764.19/3630213.703 = 1.16	= 3708236.68/3630213.703 = 1.02

PROBLEMS

8.1 A flexible pavement construction project is proposed. The estimated service life is 20 years. The initial cost is estimated at \$200000, and it would have a salvage value of \$35000 at the end of its useful life. The yearly maintenance is estimated at \$2500. What would be the present worth of life-cycle costs? What will be the user cost saving amounts annually to justify the pavement construction? Use interest rate is 5%.

8.2 A proposal is being considered to improve an existing roadway condition by resurfacing the road to reduce transportation costs. The cost of the project is \$1500000. Present annual transportation cost for all traffic amounts to \$1950000 per year and would continue if no improvement is made. After the improvement, annual transportation costs are estimated to be \$1700000. Assume the life of the pavement design is 20 years and MARR is 10%. Evaluate this proposal by using the net present worth, benefit-to-cost ratio, and internal rate of return methods. Assume costs appear at the beginning of each year while benefits appear at the end of each year.

8.3 It is proposed to reconstruct a 10-mile, 4-lane concrete pavement highway. The initial cost per lane-mile is \$80000 and it would have zero salvage value at the end of service life of 25 years. The annual maintenance cost is \$2500 per lane-mile, with 5% increase annually. Assume a discount rate of 4%. What would be the net present agency cost?

8.4 A proposal for rehabilitating the existing highway using an asphalt overlay is made for a 10-mile, 6-lane highway. The initial rehabilitation cost per lane-mile is \$50000 with a zero salvage value at the end of its service life of 15 years. The annual maintenance cost is expected to be \$5500 per lane-mile, with 4% increase annually. Assume a discount rate of 6%. Calculate annual agency cost over the life cycle of new pavement construction.

Chapter 9

Economic analysis of highway bridge preservation

9.1 OVERVIEW

The proposed methodology considers all agency and user costs in the service life cycle of a highway bridge. Agency costs mainly comprise capital costs associated with project construction and discounted future costs of bridge component repair, replacement, and maintenance treatments. User costs consist of cost components of vehicle operations, travel time, crashes, and air emissions in normal operation and work zone conditions, as seen in Figure 9.1.

9.1.1 Bridge project direct costs

The project direct costs generally include direct agency costs and additional user costs caused by work zones. Direct agency cost elements mainly comprise capital costs of project land acquisition, design and engineering support, and construction. User costs related to construction work include additional costs of vehicle operations and delays immediately upstream of and within work zones.

9.1.2 Bridge life-cycle agency and user costs

In life-cycle cost analysis, the overall agency costs contain direct agency costs of construction and subsequent costs of rehabilitation, replacement, and maintenance of different bridge components in the bridge service life cycle. User costs include cost components of vehicle operations, travel time, crashes, and air emissions. Life-cycle user costs are estimated based on these user components for each year in the bridge service life cycle.

9.1.3 Project-level benefits in bridge service life cycle

The overall bridge project-level benefits in bridge service life cycle may be obtained from the agency and the user perspectives. The investment in a highway bridge may lead to decreases in bridge life-cycle agency costs and instigate savings of life-cycle user costs. The potential reductions in life-cycle agency and user costs from project implementation are considered as the overall project-level benefits. To estimate the change of life-cycle agency costs, the activity profiles represented by the frequency, timing, and magnitude of construction, maintenance, and repair treatments for bridge components need to be established. For bridges, the life-cycle benefits in perpetuity can be quantified assuming that the predetermined life-cycle activity profiles are repeated ad infinitum.

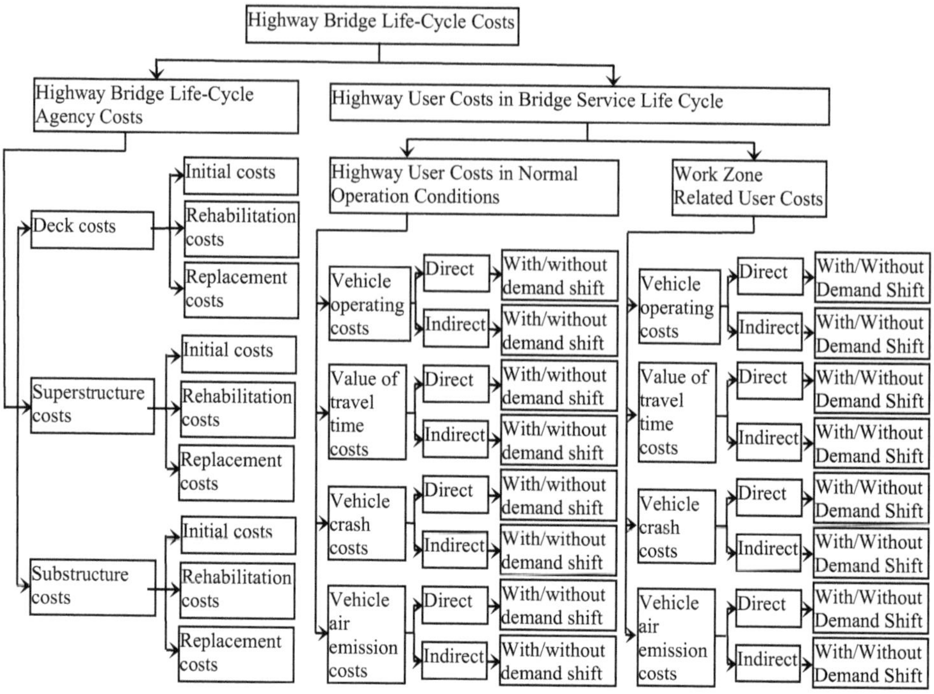

Figure 9.1 Agency and user cost categories in highway bridge service life cycle. (Adapted from Li, Z. and P. Kaini. 2007. Optimal Investment Decision-Making for Highway Transportation Asset Management under Risk and Uncertainty. Report MRUTC 07-10. Department of Civil, Architectural, and Environmental Engineering, Illinois Institute of Technology, Chicago, IL.)

9.2 DETERMINISTIC BRIDGE LIFE-CYCLE AGENCY COST ANALYSIS

9.2.1 Categorization of bridge types

A bridge is composed of a deck, superstructure, and substructure components. Reinforced concrete and composite materials are commonly used for bridge decks. Five types of superstructure (reinforced concrete slab or box-beam, concrete I-beam, steel beam, and steel girder) and two types of substructure (solid stem and pile type) are commonly encountered in practice.

9.2.2 Bridge design service life cycle

A *bridge service life cycle* is defined as the time interval between two consecutive bridge replacements or new bridge construction interventions. A *bridge repair life cycle* is defined as the time interval from construction to subsequent bridge component rehabilitation/replacement, consecutive bridge component rehabilitation/replacement or bridge component rehabilitation/replacement to new construction (Gion et al., 1993; Hawk, 2003). Table 9.1 lists typical bridge design lives by bridge type.

9.2.3 Typical bridge treatment types

Maintenance, rehabilitation, and replacement treatments can be applied to a specific bridge component or be jointly applied to multiple bridge components (Gion et al., 1993; Hawk, 2003; FHWA, 2011). Table 9.2 presents typical bridge treatment types.

Table 9.1 Recommended bridge design service life cycle

Superstructure material	*Superstructure type*	*Design life (year)*
Concrete	Channel beam	35
	T-beam	70
	Slab	60
	Girder	70
Prestressed concrete	Box-beam	65
	Segmental box girder	50
Steel	Box-beam	70
	Girder	70
	Truss	80

Table 9.2 Typical bridge maintenance, rehabilitation, and replacement treatments

Bridge component	*Maintenance*	*Rehabilitation*	*Replacement*
Deck	Deck maintenance	Deck rehabilitation Deck rehabilitation and bridge widening	Deck replacement Deck replacement and bridge widening
Superstructure	Superstructure maintenance	Superstructure strengthening Superstructure strengthening and bridge widening	Superstructure replacement and bridge widening
Deck and superstructure	Deck and superstructure maintenance	Deck and superstructure rehabilitation Deck and superstructure rehabilitation and bridge widening	Deck replacement and superstructure rehabilitation Deck replacement, superstructure rehabilitation, and bridge widening
Substructure	Substructure maintenance	Substructure rehabilitation	Substructure replacement as part of bridge replacement

9.2.4 Bridge life-cycle repair strategies and activity profiles

Of various types of treatments to bridge components, deck rehabilitation is the most commonly used. Typically, the first deck rehabilitation is implemented 20, 25, or 30 years after the initial construction. Depending on bridge superstructure types, deck replacement or superstructure replacement is scheduled 15 years after the first deck rehabilitation. In the bridge service life cycle, the second deck rehabilitation is carried out 20 or 25 years after deck or superstructure replacement. The recommended life-cycle activity profiles for various types of bridges are shown in Figures 9.2 through 9.10.

9.2.5 Bridge agency life-cycle cost analysis

9.2.5.1 Bridge agency cost categories

Bridge agency costs are those related to design and construction, routine maintenance, deck and superstructure rehabilitation and replacement, and bridge replacement. Design costs include all the costs related to the engineering design, field tests and related equipment, and human resource costs. Construction costs include the costs of materials, equipment, and labor. Administrative costs associated with the bridge project are also included. Routine maintenance costs are a function of type of material, weather conditions, location, and traffic levels. Routine maintenance costs are normally estimated for individual bridge components. Rehabilitation costs of bridge components include major repair activities and require

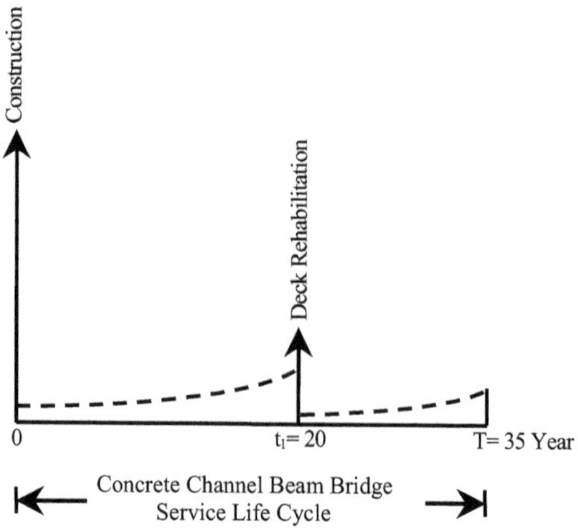

Figure 9.2 Typical life-cycle activity profile of concrete channel beam bridges. (Adapted from Li, Z. and P. Kaini. 2007. Optimal Investment Decision-Making for Highway Transportation Asset Management under Risk and Uncertainty. Report MRUTC 07-10. Department of Civil, Architectural, and Environmental Engineering, Illinois Institute of Technology, Chicago, IL.)

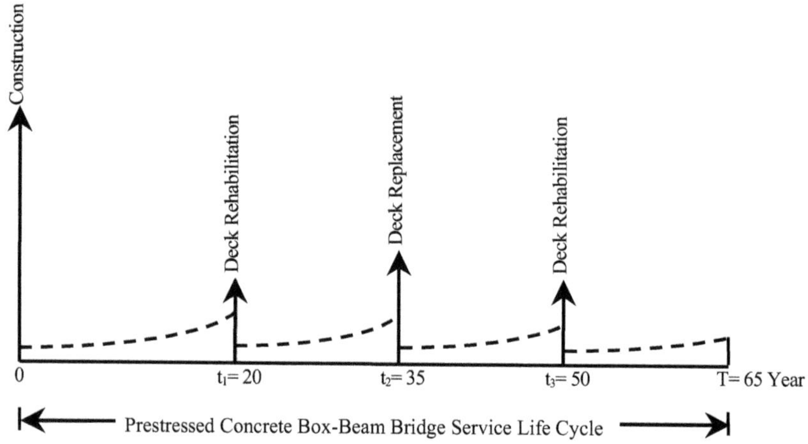

Figure 9.3 Typical life-cycle activity profile of prestressed concrete box-beam bridges. (Adapted from Li, Z. and P. Kaini. 2007. Optimal Investment Decision-Making for Highway Transportation Asset Management under Risk and Uncertainty. Report MRUTC 07-10. Department of Civil, Architectural, and Environmental Engineering, Illinois Institute of Technology, Chicago, IL.)

engineering analysis. The costs associated with the replacement of any bridge component when its service life ends are taken as the component replacement costs. The analysis follows the same procedure adopted for component rehabilitation costs.

9.2.5.2 Estimation of bridge element costs

Accurate cost estimations provide firmer support for a comparative analysis of various alternatives. Cobb–Douglas production functions can be utilized to estimate bridge component replacement costs as a function of bridge length, deck width, and vertical clearance.

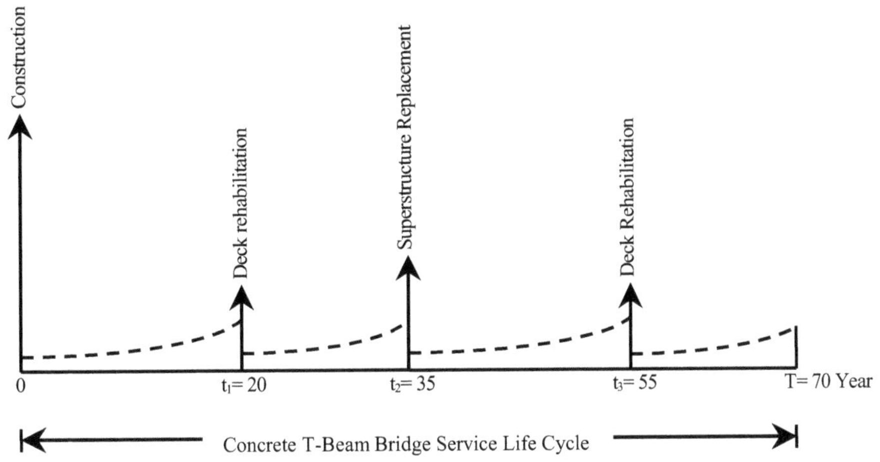

Figure 9.4 Typical life-cycle activity profile of concrete T-beam bridges. (Adapted from Li, Z. and P. Kaini. 2007. Optimal Investment Decision-Making for Highway Transportation Asset Management under Risk and Uncertainty. Report MRUTC 07-10. Department of Civil, Architectural, and Environmental Engineering, Illinois Institute of Technology, Chicago, IL.)

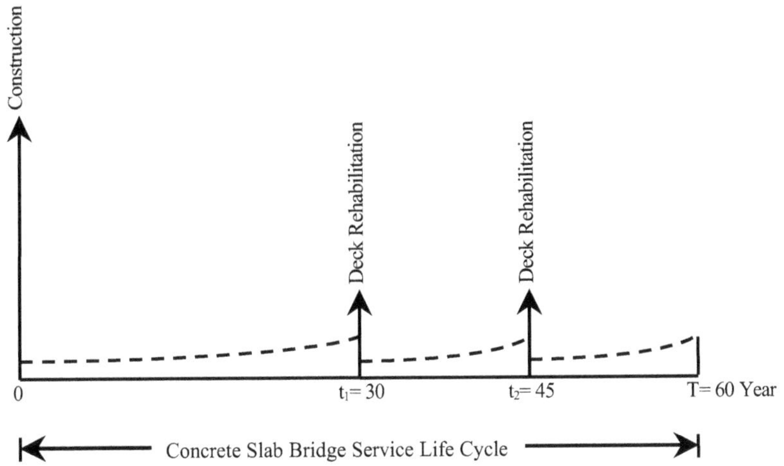

Figure 9.5 Typical life-cycle activity profile of concrete slab bridges. (Adapted from Li, Z. and P. Kaini. 2007. Optimal Investment Decision-Making for Highway Transportation Asset Management under Risk and Uncertainty. Report MRUTC 07-10. Department of Civil, Architectural, and Environmental Engineering, Illinois Institute of Technology, Chicago, IL.)

9.2.5.3 Typical bridge life-cycle agency cost analysis

The bridge life-cycle agency costs can be quantified based on the typical bridge life-cycle activity profile. The costs of bridge component routine maintenance are expected to increase slightly in response to bridge condition deterioration. A geometric gradient annual growth rate can be utilized to calculate bridge life-cycle agency costs. Different gradients can be proposed for different bridge repair intervals. Given that such service life cycle will repeat ad infinitum, the bridge life-cycle agency costs in perpetuity can be calculated. Tables 9.3 and 9.4

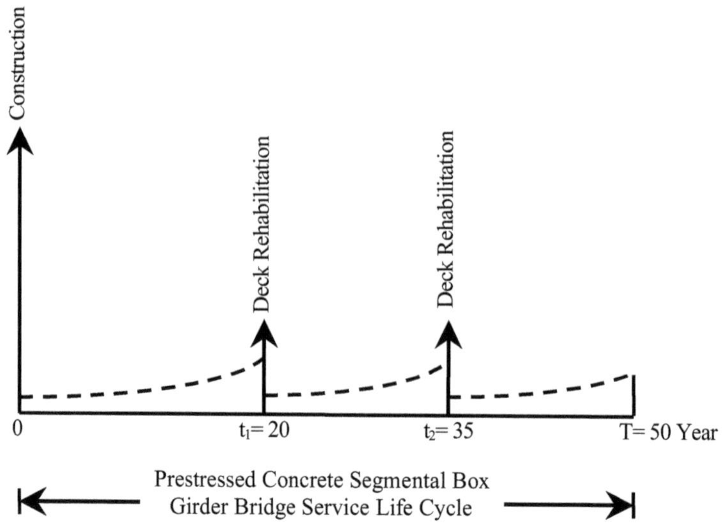

Figure 9.6 Typical life-cycle activity profile of prestressed concrete segmental box girder bridges. (Adapted from Li, Z. and P. Kaini. 2007. Optimal Investment Decision-Making for Highway Transportation Asset Management under Risk and Uncertainty. Report MRUTC 07-10. Department of Civil, Architectural, and Environmental Engineering, Illinois Institute of Technology, Chicago, IL.)

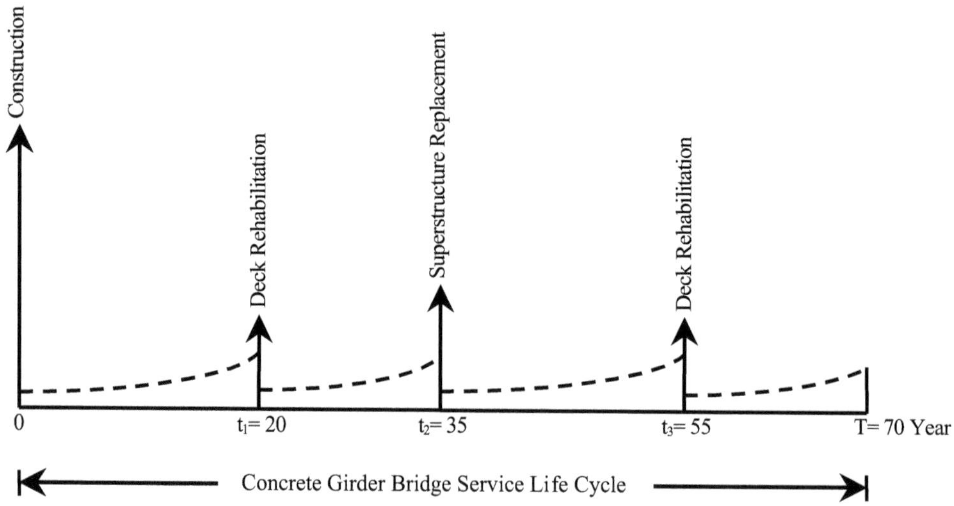

Figure 9.7 Typical life-cycle activity profile of concrete girder bridges. (Adapted from Li, Z. and P. Kaini. 2007. Optimal Investment Decision-Making for Highway Transportation Asset Management under Risk and Uncertainty. Report MRUTC 07-10. Department of Civil, Architectural, and Environmental Engineering, Illinois Institute of Technology, Chicago, IL.)

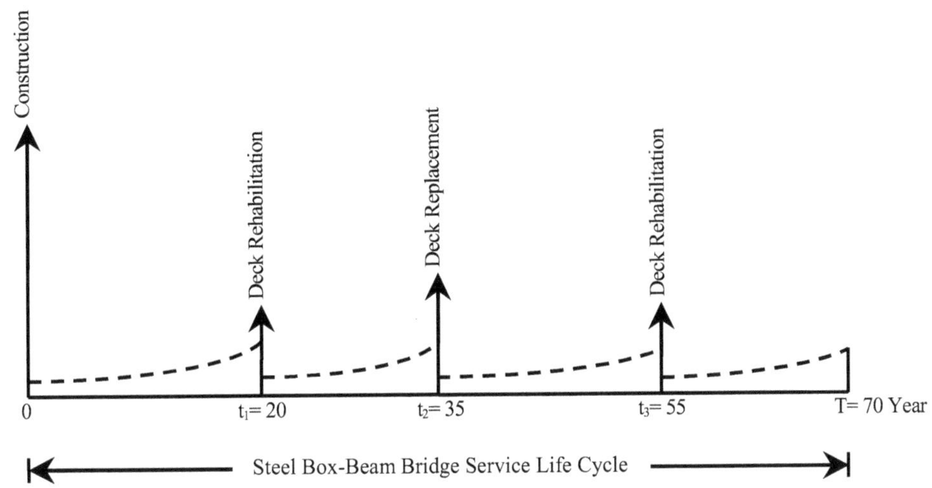

Figure 9.8 Typical life-cycle activity profile of steel box-beam bridges. (Adapted from Li, Z. and P. Kaini. 2007. Optimal Investment Decision-Making for Highway Transportation Asset Management under Risk and Uncertainty. Report MRUTC 07-10. Department of Civil, Architectural, and Environmental Engineering, Illinois Institute of Technology, Chicago, IL.)

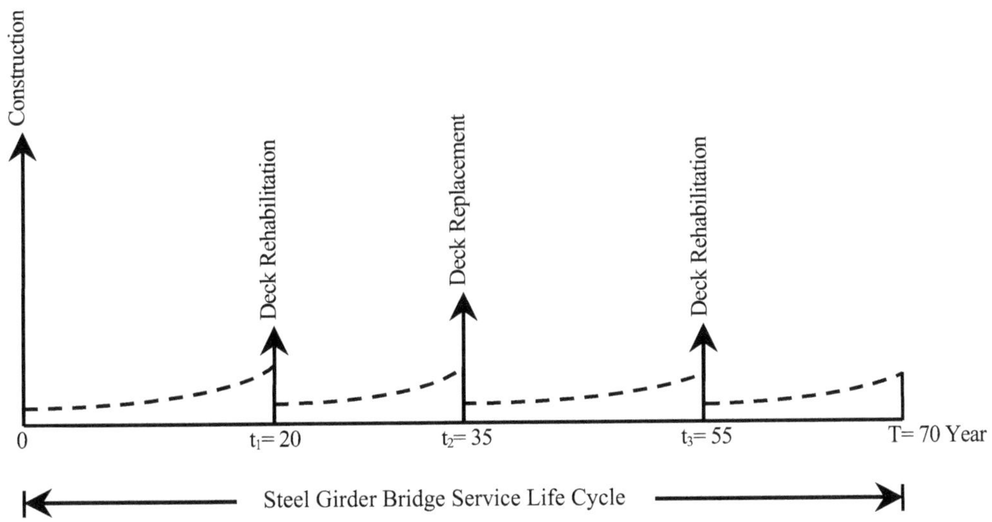

Figure 9.9 Typical life-cycle activity profile of steel girder bridges. (Adapted from Li, Z. and P. Kaini. 2007. Optimal Investment Decision-Making for Highway Transportation Asset Management under Risk and Uncertainty. Report MRUTC 07-10. Department of Civil, Architectural, and Environmental Engineering, Illinois Institute of Technology, Chicago, IL.)

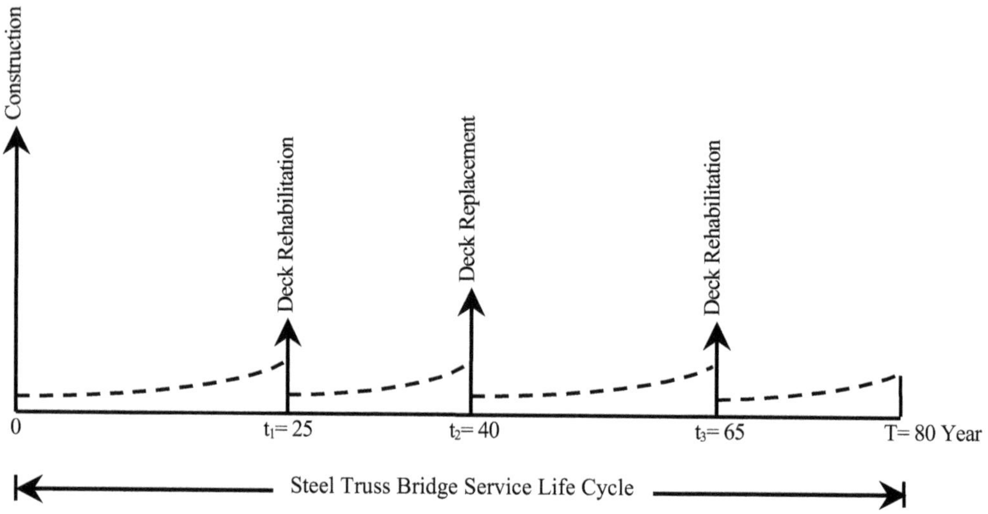

Figure 9.10 Typical life-cycle activity profile of steel truss bridges. (Adapted from Li, Z. and P. Kaini. 2007. Optimal Investment Decision-Making for Highway Transportation Asset Management under Risk and Uncertainty. Report MRUTC 07-10. Department of Civil, Architectural, and Environmental Engineering, Illinois Institute of Technology, Chicago, IL.)

Table 9.3 Computation of concrete bridge life-cycle agency costs

Bridge type	*Computation*	
Concrete channel beam	Agency cost profile	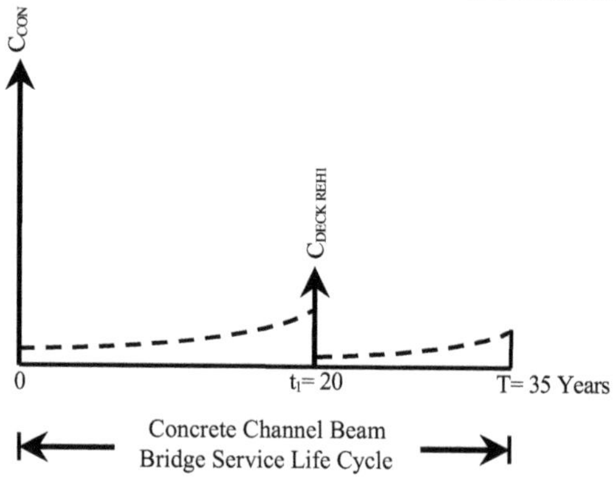
	PW_{LCAC}	$= C_{CON} + C_{DECK\ REHI}/(1+i)^{t_1} + (C_{MAIN1}(1-((1+g_{M1})^{t_1}(1+i)^{-t_1}))/(i-g_{M1}) + ((C_{MAIN2}(1-(1+g_{M2})^{(T-t_1)}(1+i)^{-(T-t_1)}))/(i-g_{M2}))/(1+i)^{t_1}$
	$PW_{LCAC\infty}$	$= PW_{LCAC}/(1-(1/(1+i)^{T}))$
	$EUAAC$	$= PW_{LCAC}\cdot((i(1+i)^{T})/((1+i)^{T}-1))$
	$EUAAC_{\infty}$	$= PW_{LCAC\infty}\cdot i$

(Continued)

Table 9.3 (Continued) Computation of concrete bridge life-cycle agency costs

Bridge type	*Computation*	
Prestressed concrete box-beam	Agency cost profile	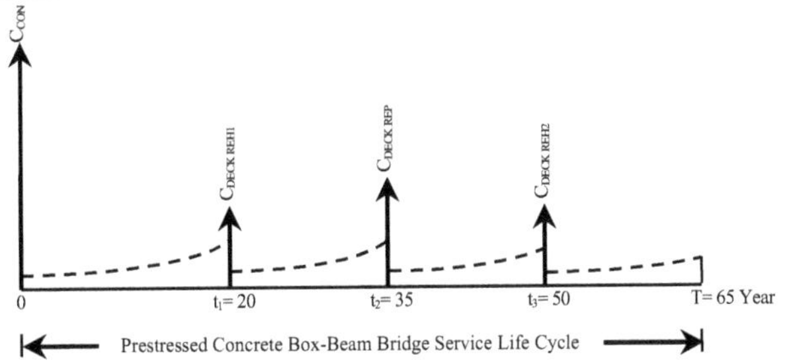
	PW_{LCAC}	$= C_{CON} + C_{DECK\ REHAB1}/(1+i)^{t_1} + C_{DECK\ REP}/(1+i)^{t_2} + C_{DECK\ REH2}/(1+i)^{t_3}$ $+ (C_{MAIN1}(1-(1+g_{M1})^{t_1}(1+i)^{-t_1}))/(i-g_{M1})$ $+ ((C_{MAIN2}(1-(1+g_{M2})^{(t_2-t_1)}(1+i)^{-(t_2-t_1)}))/(i-g_{M2}))/(1+i)^{t_1}$ $+ ((C_{MAIN3}(1-(1+g_{M3})^{(t_3-t_2)}(1+i)^{-(t_3-t_2)}))/(i-g_{M3}))/(1+i)^{t_2}$ $+ ((C_{MAIN4}(1-(1+g_{M4})^{(T-t_3)}(1+i)^{-(T-t_3)}))/(i-g_{M4}))/(1+i)^{t_3}$
	$PW_{LCAC\infty}$	$= PW_{LCAC}/(1-(1/(1+i)^T))$
	$EUAAC$	$= PW_{LCAC}\cdot((i(1+i)^T)/((1+i)^T-1))$
	$EUAAC_{\infty}$	$= PW_{LCAC\infty}\cdot i$
Concrete T-beam	Agency cost profile	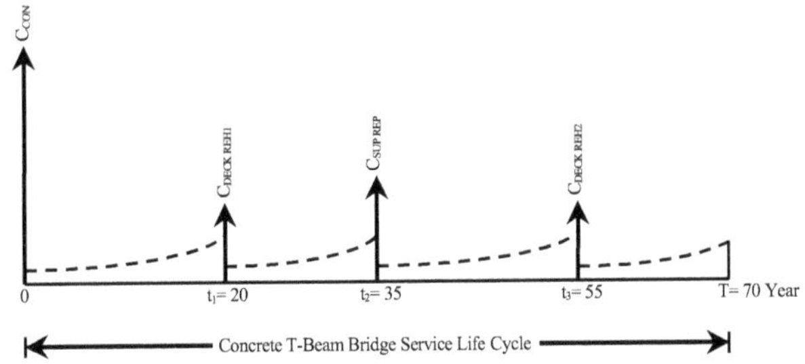
	PW_{LCAC}	$= C_{CON} + C_{DECK\ REHAB1}/(1+i)^{t_1} + C_{SUP\ REP}/(1+i)^{t_2} + C_{DECK\ REH2}/(1+i)^{t_3}$ $+ (C_{MAIN1}(1-(1+g_{M1})^{t_1}(1+i)^{-t_1}))/(i-g_{M1})$ $+ ((C_{MAIN2}(1-(1+g_{M2})^{(t_2-t_1)}(1+i)^{-(t_2-t_1)}))/(i-g_{M2}))/(1+i)^{t_1}$ $+ ((C_{MAIN3}(1-(1+g_{M3})^{(t_3-t_2)}(1+i)^{-(t_3-t_2)}))/(i-g_{M3}))/(1+i)^{t_2}$ $+ ((C_{MAIN4}(1-(1+g_{M4})^{(T-t_3)}(1+i)^{-(T-t_3)}))/(i-g_{M4}))/(1+i)^{t_3}$
	$PW_{LCAC\infty}$	$= PW_{LCAC}/(1-(1/(1+i)^T))$
	$EUAAC$	$= PW_{LCAC}\cdot((i(1+i)^T)/((1+i)^T-1))$
	$EUAAC_{\infty}$	$= PW_{LCAC\infty}\cdot i$
Concrete slab	Agency cost profile	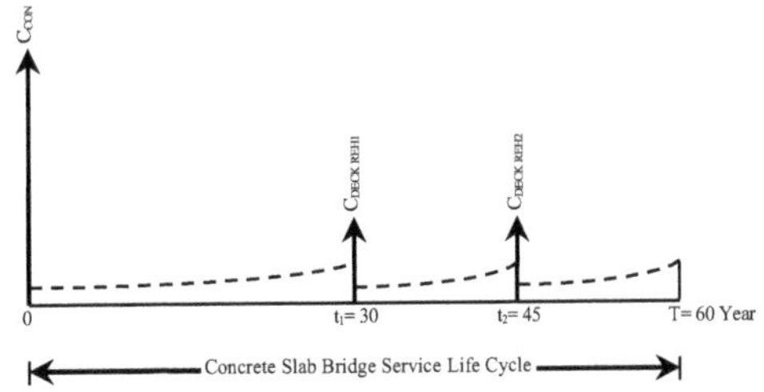

(*Continued*)

Table 9.3 (Continued) Computation of concrete bridge life-cycle agency costs

Bridge type		*Computation*
	PW_{LCAC}	$= C_{CON} + C_{DECK\ REH1}/(1+i)^{t_1} + C_{DECK\ REH2}/(1+i)^{t_2}$
		$+ (C_{MAIN1}(1-(1+g_{M1})^{t_1}(1+i)^{-t_1}))/(i-g_{M1})$
		$+ ((C_{MAIN2}(1-(1+g_{M2})^{(t_2-t_1)}(1+i)^{-(t_2-t_1)}))/(i-g_{M2}))/(1+i)^{t_1}$
		$+ ((C_{MAIN3}(1-(1+g_{M3})^{(T-t_2)}(1+i)^{-(T-t_2)}))/(i-g_{M3}))/(1+i)^{t_2}$
	$PW_{LCAC\infty}$	$= PW_{LCAC}/(1-(1/(1+i)^T))$
	$EUAAC$	$= PW_{LCAC} \cdot ((i(1+i)^T)/((1+i)^T-1))$
	$EUAAC_{\infty}$	$= PW_{LCAC\infty} \cdot i$
Prestressed concrete segmental box girder	Agency cost profile	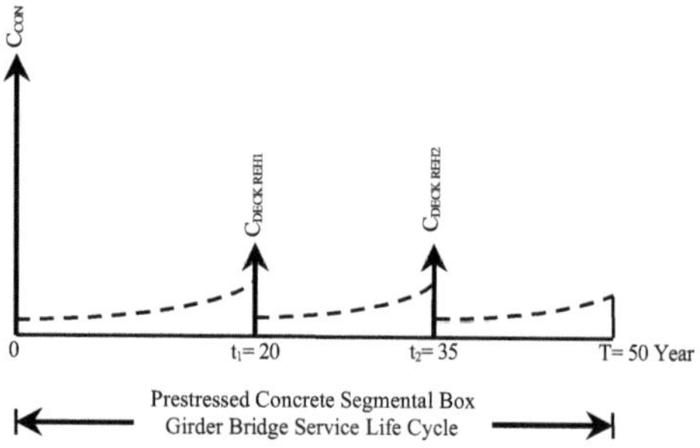

Bridge type		*Computation*
	PW_{LCAC}	$= C_{CON} + C_{DECK\ REH1}/(1+i)^{t_1} + C_{DECK\ REH2}/(1+i)^{t_2}$
		$+ (C_{MAIN1}(1-(1+g_{M1})^{t_1}(1+i)^{-t_1}))/(i-g_{M1})$
		$+ ((C_{MAIN2}(1-(1+g_{M2})^{(t_2-t_1)}(1+i)^{-(t_2-t_1)}))/(i-g_{M2}))/(1+i)^{t_1}$
		$+ ((C_{MAIN3}(1-(1+g_{M3})^{(T-t_2)}(1+i)^{-(T-t_2)}))/(i-g_{M3}))/(1+i)^{t_2}$
	$PW_{LCAC\infty}$	$= PW_{LCAC}/(1-(1/(1+i)^T))$
	$EUAAC$	$= PW_{LCAC} \cdot ((i(1+i)^T)/((1+i)^T-1))$
	$EUAAC_{\infty}$	$= PW_{LCAC\infty} \cdot i$
Concrete girder	Agency cost profile	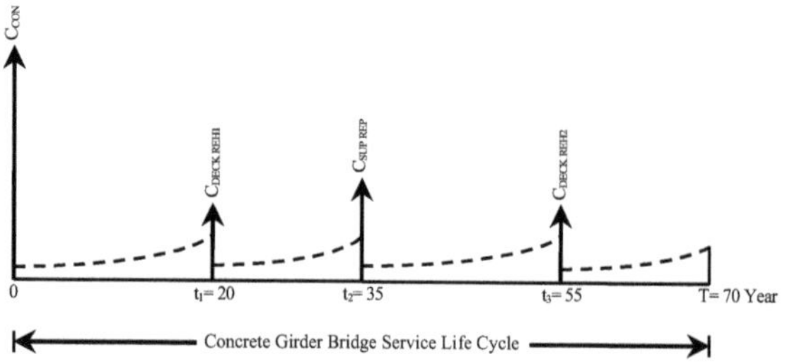

Bridge type		*Computation*
	PW_{LCAC}	$= C_{CON} + C_{DECK\ REH1}/(1+i)^{t_1} + C_{SUP\ REP}/(1+i)^{t_2} + C_{DECK\ REH2}/(1+i)^{t_3}$
		$+ (C_{MAIN1}(1-(1+g_{M1})^{t_1}(1+i)^{-t_1}))/(i-g_{M1})$
		$+ ((C_{MAIN2}(1-(1+g_{M2})^{(t_2-t_1)}(1+i)^{-(t_2-t_1)}))/(i-g_{M2}))/(1+i)^{t_1}$
		$+ ((C_{MAIN3}(1-(1+g_{M3})^{(t_3-t_2)}(1+i)^{-(t_3-t_2)}))/(i-g_{M3}))/(1+i)^{t_2}$
		$+ ((C_{MAIN4}(1-(1+g_{M4})^{(T-t_3)}(1+i)^{-(T-t_3)}))/(i-g_{M4}))/(1+i)^{t_3}$
	$PW_{LCAC\infty}$	$= PW_{LCAC}/(1-(1/(1+i)^T))$
	$EUAAC$	$= PW_{LCAC} \cdot ((i(1+i)^T)/((1+i)^T-1))$
	$EUAAC_{\infty}$	$= PW_{LCAC\infty} \cdot i$

Table 9.4 Computation of steel bridge life-cycle agency costs

Bridge type		*Computation*
Steel box-beam	Agency cost profile	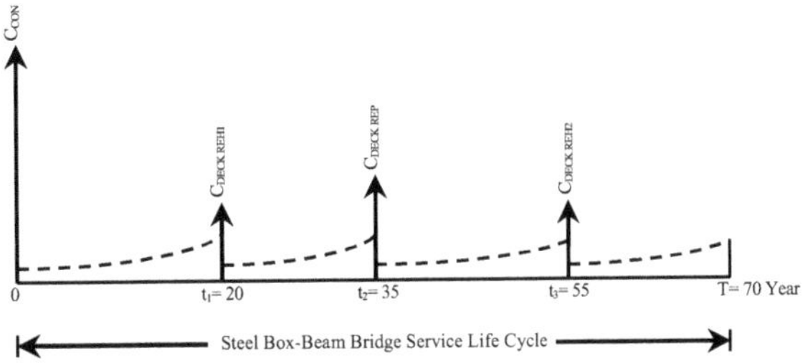
	PW_{LCAC}	$= C_{CON} + C_{DECK\ REH1}/(1+i)^{t_1} + C_{SUP\ REP}/(1+i)^{t_2} + C_{DECK\ REH2}/(1+i)^{t_3}$ $+ (C_{MAIN1}(1-(1+g_{M1})^{t_1}(1+i)^{-t_1}))/(i-g_{M1})$ $+ ((C_{MAIN2}(1-(1+g_{M2})^{(t_2-t_1)}(1+i)^{-(t_2-t_1)}))/(i-g_{M2}))/(1+i)^{t_1}$ $+ ((C_{MAIN3}(1-(1+g_{M3})^{(t_3-t_2)}(1+i)^{-(t_3-t_2)}))/(i-g_{M3}))/(1+i)^{t_2}$ $+ ((C_{MAIN4}(1-(1+g_{M4})^{(T-t_3)}(1+i)^{-(T-t_3)}))/(i-g_{M4}))/(1+i)^{t_3}$
	$PW_{LCAC\infty}$	$= PW_{LCAC}/(1-(1/(1+i)^T))$
	$EUAAC$	$= PW_{LCAC} \cdot ((i(1+i)^T)/((1+i)^T-1))$
	$EUAAC_{\infty}$	$= PW_{LCAC\infty} \cdot i$
Steel girder	Agency cost profile	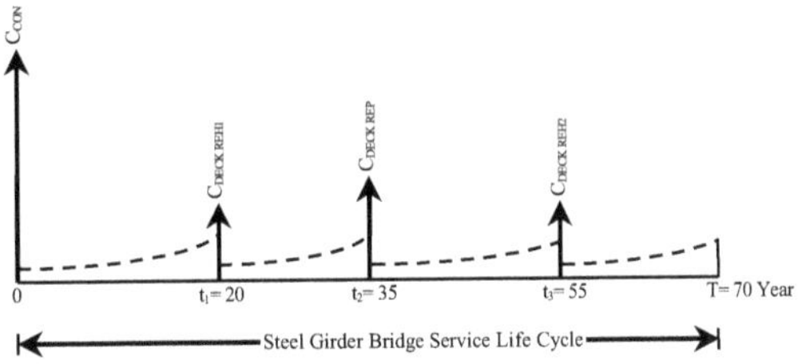
	PW_{LCAC}	$= C_{CON} + C_{DECK\ REH1}/(1+i)^{t_1} + C_{DECK\ REP}/(1+i)^{t_2} + C_{DECK\ REH2}/(1+i)^{t_3}$ $+ (C_{MAIN1}(1-(1+g_{M1})^{t_1}(1+i)^{-t_1}))/(i-g_{M1})$ $+ ((C_{MAIN2}(1-(1+g_{M2})^{(t_2-t_1)}(1+i)^{-(t_2-t_1)}))/(i-g_{M2}))/(1+i)^{t_1}$ $+ ((C_{MAIN3}(1-(1+g_{M3})^{(t_3-t_2)}(1+i)^{-(t_3-t_2)}))/(i-g_{M3}))/(1+i)^{t_2}$ $+ ((C_{MAIN4}(1-(1+g_{M4})^{(T-t_3)}(1+i)^{-(T-t_3)}))/(i-g_{M4}))/(1+i)^{t_3}$
	$PW_{LCAC\infty}$	$= PW_{LCAC}/(1-(1/(1+i)^T))$
	$EUAAC$	$= PW_{LCAC} \cdot ((i(1+i)^T)/((1+i)^T-1))$
	$EUAAC_{\infty}$	$= PW_{LCAC\infty} \cdot i$
Steel truss	Agency cost profile	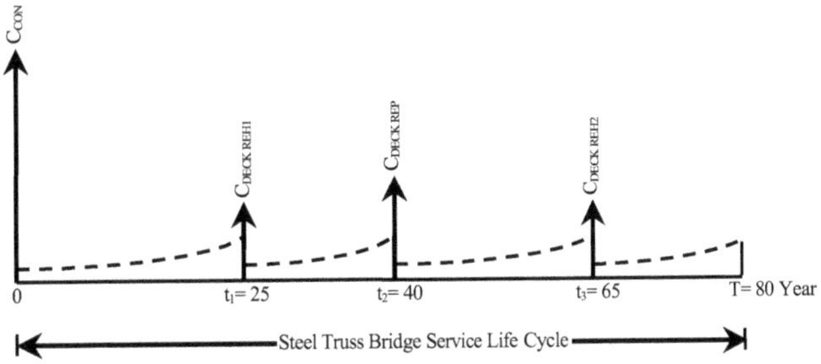

(Continued)

Table 9.4 (Continued) Computation of steel bridge life-cycle agency costs

Bridge type		*Computation*
	PW_{LCAC}	$= C_{CON} + C_{DECK\ REH1}/(1+i)^{t_1} + C_{DECK\ REP}/(1+i)^{t_2} + C_{DECK\ REH2}/(1+i)^{t_3}$ $+ (C_{MAIN1}(1-(1+g_{M1})^{t_1}(1+i)^{-t_1}))/(i-g_{M1})$ $+ ((C_{MAIN2}(1-(1+g_{M2})^{(t_2-t_1)}(1+i)^{-(t_2-t_1)}))/(i-g_{M2}))/(1+i)^{t_1}$ $+ ((C_{MAIN3}(1-(1+g_{M3})^{(t_3-t_2)}(1+i)^{-(t_3-t_2)}))/(i-g_{M3}))/(1+i)^{t_2}$ $+ ((C_{MAIN4}(1-(1+g_{M4})^{(T-t_3)}(1+i)^{-(T-t_3)}))/(i-g_{M4}))/(1+i)^{t_3}$
	$PW_{LCAC\infty}$	$= PW_{LCAC}/(1 - (1/(1+i)^T))$
	$EUAAC$	$= PW_{LCAC} \cdot ((i(1+i)^T)/((1+i)^T - 1))$
	$EUAAC_{\infty}$	$= PW_{LCAC\infty} \cdot i$

present bridge life-cycle agency cost analysis results in present worth and equivalent uniform annual amounts.

Denote:

PW_{LCAC} = Present worth of bridge life-cycle agency costs
$PW_{LCAC\infty}$ = Present worth of bridge life-cycle agency costs in perpetuity
$EUAAC$ = Equivalent uniform annual bridge agency costs
$EUAAC_{\infty}$ = Equivalent uniform annual bridge agency costs in perpetuity
C_{CON} = Bridge construction cost
$C_{DECK\ REH1}$ = First bridge deck rehabilitation cost
$C_{DECK\ REH2}$ = Second bridge deck rehabilitation cost
$C_{DECK\ REP}$ = Bridge deck replacement cost
$C_{SUP\ REP}$ = Bridge superstructure replacement cost
C_{MAIN1} = Annual bridge maintenance cost incurred between construction and first major repair
C_{MAIN2} = Annual bridge maintenance cost incurred between the first and second major repairs
C_{MAIN3} = Annual bridge maintenance cost incurred between the second and third major repairs
C_{MAIN4} = Annual bridge maintenance cost incurred between the third and fourth major repairs
g_{M1} = Growth rate of annual bridge maintenance cost between construction and first major repair
g_{M2} = Growth rate of annual bridge maintenance cost between the first and second major repairs
g_{M3} = Growth rate of annual bridge maintenance cost between the second and third major repairs
g_{M4} = Growth rate of annual bridge maintenance cost between the third and fourth major repairs
i = Discount rate
t = Time of year a major treatment is implemented
T = Number of years of service life

9.3 DETERMINISTIC HIGHWAY USER COST ANALYSIS IN BRIDGE SERVICE LIFE CYCLE

9.3.1 General

The highway user cost analysis for a bridge preservation project in the bridge service life cycle is conducted similarly to pavement-related life-cycle user cost analysis, as detailed in Section 8.3 of Chapter 8.

Two dimensions are considered for the user cost analysis: (i) user cost *components*, including costs of vehicle operations, travel time, crashes, and air emissions; and (ii) user cost *categories*, dealing with normal operation and work zone-affected traffic conditions. Vehicle operating costs contain mileage-dependent costs related to expenditures for fuel, tires, engine oil, maintenance, and mileage-dependent vehicle depreciation. Travel time costs include costs to businesses of time by their employees, vehicles, and goods, and costs to consumers of personal unpaid time spent on travel. Vehicle crash costs are costs related to motor vehicle crashes classified into fatal, injury, and PDO severity levels. Vehicle air emission costs are external costs associated NMHC, CO, CO_2, NO_X, SO_2, and TSP emitted from vehicle use.

The individual user cost components can be assessed in normal operation and work zone conditions. Highway user costs in normal operating conditions reflect highway user costs associated with travelers using a bridge free of work activities that restrict the bridge capacity. Among the individual user cost components in this category, vehicle operating costs vary considerably according to vehicle type, speed, engine type, age, speed changes, design features, and bridge conditions. During normal operations, little difference exists between delay and crash costs resulting from bridge design alternatives. Work zone-related user costs mainly contain the increased vehicle operating costs and delays to highway users resulting from work activities. User costs in this category are a function of the work zone configuration, duration, timing, and scope, and depend on the volume and operating characteristics of the traffic traversing the work zones.

9.3.2 Highway user costs in bridge service life cycle

For highway segments with traffic operations affected by a specific bridge project, the annual costs of vehicle operations, travel time including delays, crashes, and air emissions for the initial year are separately calculated based on the corresponding VMT and respective per VMT unit rates of individual user cost components.

With the initial-year costs of individual user cost components estimated, the bridge life-cycle activity profile can be followed to derive the bridge life-cycle user costs. Additional work zone related user costs need to be calculated for the annual user costs at which the repair and maintenance treatments are implemented for bridge components. The annual total user costs are assumed to increase slightly in consecutive years in response to traffic volume increases. A geometric gradient annual growth rate can be used for calculating the life-cycle user costs. Different gradients can be proposed for different periods between major repair treatments. Tables 9.5 and 9.6 list computation details.

Denote:

PW_{LCUC} = Present worth of highway bridge life-cycle user costs

$PW_{\text{LCUC}\infty}$ = Present worth of highway bridge life-cycle user costs in perpetuity

$EUAUC$ = Equivalent uniform annual highway bridge user costs

$EUAUC_{\infty}$ = Equivalent uniform annual highway bridge user costs in perpetuity

C_{AUC1} = Annual bridge user cost incurred between initial construction and first major repair

Table 9.5 Computation of highway normal operation user costs in concrete bridge service life cycle

Bridge type		*Computation*
Concrete channel beam	User cost profile	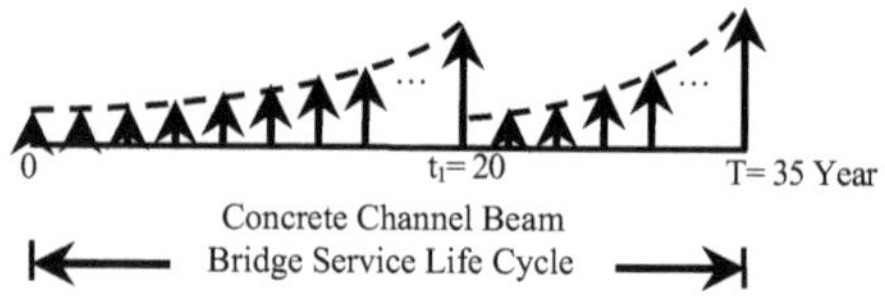
	PW_{LCUC}	$= (C_{AUC1}(1-((1+g_{AUC1})^{t_1}(1+i)^{-t_1}))/(i-g_{AUC1})$
		$+ ((C_{AUC2}(1-((1+g_{AUC2})^{(T-t_1)}(1+i)^{-(T-t_1)}))/(i-g_{AUC2}))/(1+i)^{t_1}$
	$PW_{LCUC\infty}$	$= PW_{LCUC}/(1-(1/(1+i)^T))$
	$EUAUC$	$= PW_{LCUC}\cdot((i(1+i)^T)/((1+i)^T-1))$
	$EUAUC_{\infty}$	$= PW_{LCUC\infty}\cdot i$
Prestressed concrete box-beam	User cost profile	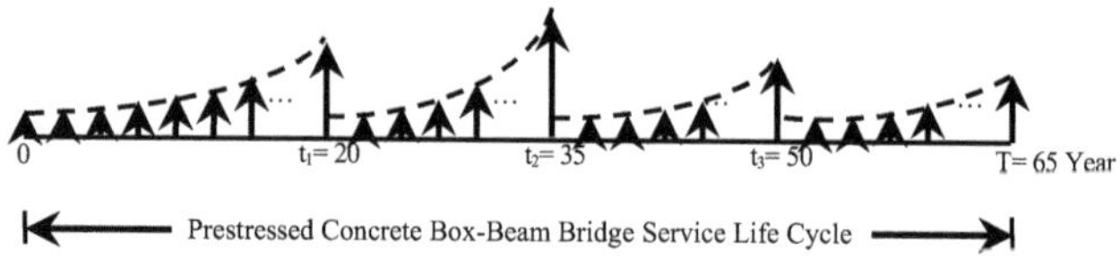
	PW_{LCUC}	$= (C_{AUC1}(1-(1+g_{AUC1})^{t_1}(1+i)^{-t_1}))/(i-g_{AUC1})$
		$+ ((C_{AUC2}(1-(1+g_{AUC2})^{(t_2-t_1)}(1+i)^{-(t_2-t_1)}))/(i-g_{AUC2}))/(1+i)^{t_1}$
		$+ ((C_{AUC3}(1-(1+g_{AUC3})^{(t_3-t_2)}(1+i)^{-(t_3-t_2)}))/(i-g_{AUC3}))/(1+i)^{t_2}$
		$+ ((C_{AUC4}(1-(1+g_{AUC4})^{(T-t_3)}(1+i)^{-(T-t_3)}))/(i-g_{AUC4}))/(1+i)^{t_3}$
	$PW_{LCUC\infty}$	$= PW_{LCUC}/(1-(1/(1+i)^T))$
	$EUAUC$	$= PW_{LCUC}\cdot((i(1+i)^T)/((1+i)^T-1))$
	$EUAUC_{\infty}$	$= PW_{LCUC\infty}\cdot i$
Concrete T-beam	User cost profile	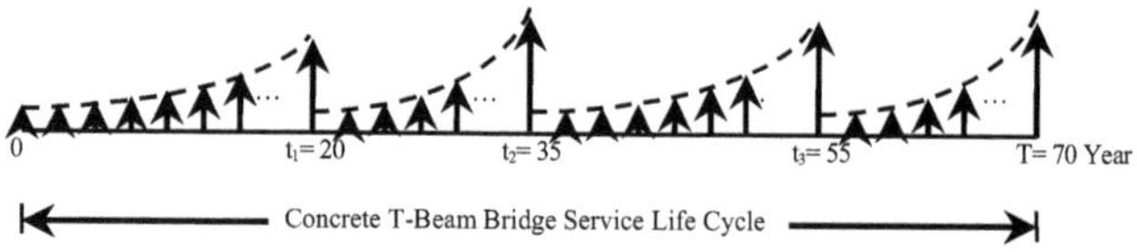
	PW_{LCUC}	$= (C_{AUC1}(1-(1+g_{AUC1})^{t_1}(1+i)^{-t_1}))/(i-g_{AUC1}))$
		$+ ((C_{AUC2}(1-(1+g_{AUC2})^{(t_2-t_1)}(1+i)^{-(t_2-t_1)}))/(i-g_{AUC2}))/(1+i)^{t_1}$
		$+ ((C_{AUC3}(1-(1+g_{AUC3})^{(t_3-t_2)}(1+i)^{-(t_3-t_2)}))/(i-g_{AUC3}))/(1+i)^{t_2}$
		$+ ((C_{AUC4}(1-(1+g_{AUC4})^{(T-t_3)}(1+i)^{-(T-t_3)}))/(i-g_{AUC4}))/(1+i)^{t_3}$
	$PW_{LCUC\infty}$	$= PW_{LCUC}/(1-(1/(1+i)^T))$
	$EUAUC$	$= PW_{LCUC}\cdot((i(1+i)^T)/((1+i)^T-1))$
	$EUAUC_{\infty}$	$= PW_{LCUC\infty}\cdot i$
Concrete slab	User cost profile	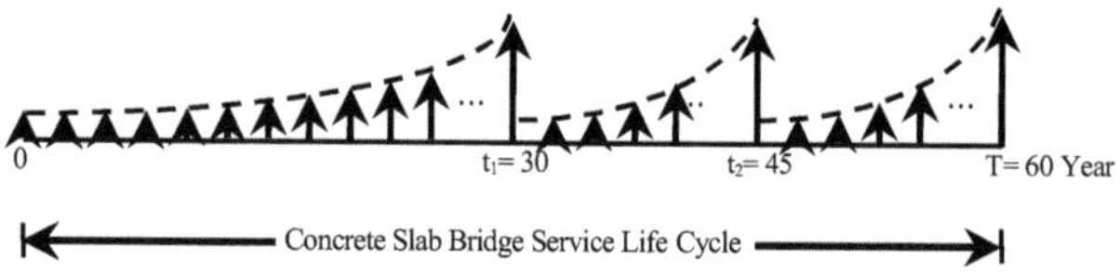

(Continued)

Table 9.5 (Continued) Computation of highway normal operation user costs in concrete bridge service life cycle

Bridge type	*Computation*	
	PW_{LCUC}	$= (C_{AUC1}(1-(1+g_{AUC1})^{t_1}(1+i)^{-t_1}))/(i-g_{AUC1}) + ((C_{AUC2}(1-(1+g_{AUC2})^{(t_2-t_1)}(1+i)^{-(t_2-t_1)}))/(i-g_{AUC2}))/(1+i)^{t_1} + ((C_{AUC3}(1-(1+g_{AUC3})^{(T-t_2)}(1+i)^{-(T-t_2)}))/(i-g_{AUC3}))/(1+i)^{t_2}$
	$PW_{LCUC\infty}$	$= PW_{LCUC}/(1-(1/(1+i)^T))$
	$EUAUC$	$= PW_{LCUC}\cdot((i(1+i)^T)/((1+i)^T-1))$
	$EUAUC_{\infty}$	$= PW_{LCUC\infty}\cdot i$
Prestressed concrete segmental box girder	User cost profile	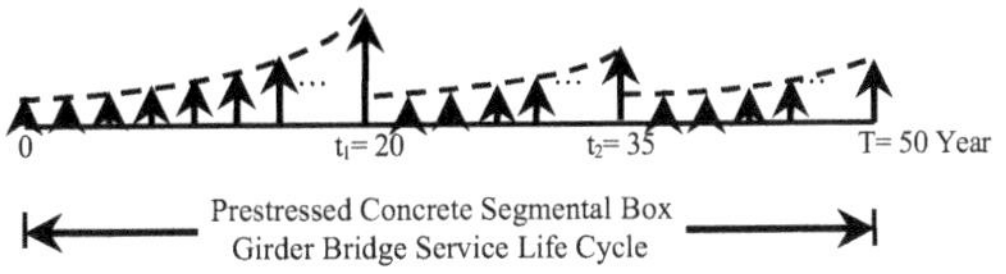
	PW_{LCUC}	$= (C_{AUC1}(1-(1+g_{AUC1})^{t_1}(1+i)^{-t_1}))/(i-g_{AUC1}) + ((C_{AUC2}(1-(1+g_{AUC2})^{(t_2-t_1)}(1+i)^{-(t_2-t_1)}))/(i-g_{AUC2}))/(1+i)^{t_1} + ((C_{AUC3}(1-(1+g_{AUC3})^{(T-t_2)}(1+i)^{-(T-t_2)}))/(i-g_{AUC3}))/(1+i)^{t_2}$
	$PW_{LCUC\infty}$	$= PW_{LCUC}/(1-(1/(1+i)^T))$
	$EUAUC$	$= PW_{LCUC}\cdot((i(1+i)^T)/((1+i)^T-1))$
	$EUAUC_{\infty}$	$= PW_{LCUC\infty}\cdot i$
Concrete girder	User cost profile	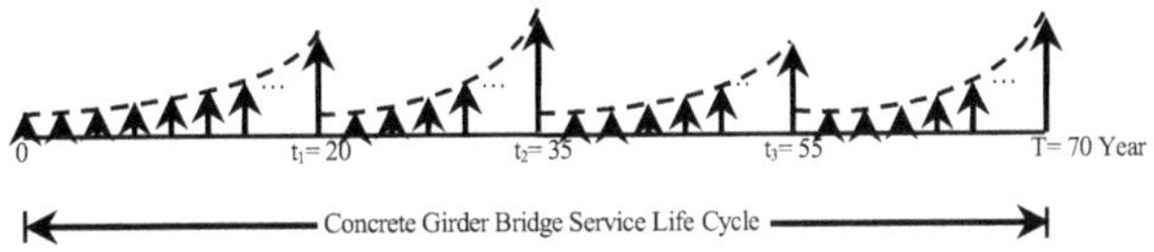
	PW_{LCUC}	$= (C_{AUC1}(1-(1+g_{AUC1})^{t_1}(1+i)^{-t_1}))/(i-g_{AUC1}) + ((C_{AUC2}(1-(1+g_{AUC2})^{(t_2-t_1)}(1+i)^{-(t_2-t_1)}))/(i-g_{AUC2}))/(1+i)^{t_1} + ((C_{AUC3}(1-(1+g_{AUC3})^{(t_3-t_2)}(1+i)^{-(t_3-t_2)}))/(i-g_{AUC3}))/(1+i)^{t_2} + ((C_{AUC4}(1-(1+g_{AUC4})^{(T-t_3)}(1+i)^{-(T-t_3)}))/(i-g_{AUC4}))/(1+i)^{t_3}$
	$PW_{LCUC\infty}$	$= PW_{LCUC}/(1-(1/(1+i)^T))$
	$EUAUC$	$= PW_{LCUC}\cdot((i(1+i)^T)/((1+i)^T-1))$
	$EUAUC_{\infty}$	$= PW_{LCUC\infty}\cdot i$

C_{AUC2} = Annual bridge user cost incurred between the first and second major repairs
C_{AUC3} = Annual bridge user cost incurred between the second and third major repairs
C_{AUC4} = Annual bridge user cost incurred between the third and fourth major repairs
C_{AUC5} = Annual bridge user cost incurred between the fourth and fifth major repairs
C_{AUC6} = Annual bridge user cost incurred between the fifth and sixth major repairs
g_{AUC1} = Growth rate of annual bridge user cost between initial construction and first major repair
g_{AUC2} = Growth rate of annual bridge user cost between the first and second major repairs
g_{AUC3} = Growth rate of annual bridge user cost between the second and third major repairs

Table 9.6 Computation of highway normal operation user costs in steel bridge service life cycle

Bridge type	*Computation*	
Steel box-beam	User cost profile	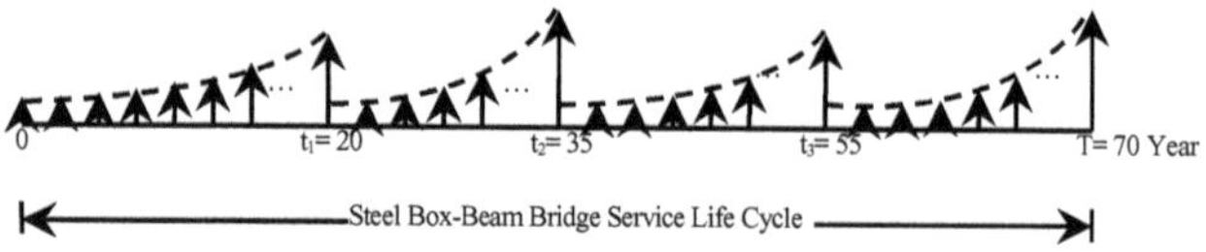
	PW_{LCUC}	$= (C_{AUC1}(1-(1+g_{AUC1})^{t_1}(1+i)^{-t_1}))/(i-g_{AUC1})$ $+ ((C_{AUC2}(1-(1+g_{AUC2})^{(t_2-t_1)}(1+i)^{-(t_2-t_1)}))/(i-g_{AUC2}))/(1+i)^{t_1}$ $+ ((C_{AUC3}(1-(1+g_{AUC3})^{(t_3-t_2)}(1+i)^{-(t_3-t_2)}))/(i-g_{AUC3}))/(1+i)^{t_2}$ $+ ((C_{AUC4}(1-(1+g_{AUC4})^{(T-t_3)}(1+i)^{-(T-t_3)}))/(i-g_{AUC4}))/(1+i)^{t_3}$
	$PW_{LCUC\infty}$	$= PW_{LCUC}/(1-(1/(1+i)^T))$
	$EUAUC$	$= PW_{LCUC}\cdot((i(1+i)^T)/((1+i)^T-1))$
	$EUAUC_\infty$	$= PW_{LCUC\infty}\cdot i$
Steel girder	User cost profile	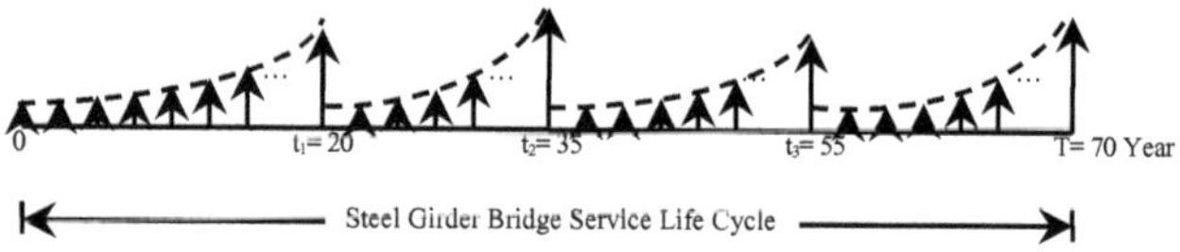
	PW_{LCUC}	$= (C_{AUC1}(1-(1+g_{AUC1})^{t_1}(1+i)^{-t_1}))/(i-g_{AUC1})$ $+ ((C_{AUC2}(1-(1+g_{AUC2})^{(t_2-t_1)}(1+i)^{-(t_2-t_1)}))/(i-g_{AUC2}))/(1+i)^{t_1}$ $+ ((C_{AUC3}(1-(1+g_{AUC3})^{(t_3-t_2)}(1+i)^{-(t_3-t_2)}))/(i-g_{AUC3}))/(1+i)^{t_2}$ $+ ((C_{AUC4}(1-(1+g_{AUC4})^{(T-t_3)}(1+i)^{-(T-t_3)}))/(i-g_{AUC4}))/(1+i)^{t_3}$
	$PW_{LCUC\infty}$	$= PW_{LCUC}/(1-(1/(1+i)^T))$
	$EUAUC$	$= PW_{LCUC}\cdot((i(1+i)^T)/((1+i)^T-1))$
	$EUAUC_\infty$	$= PW_{LCUC\infty}\cdot i$
Steel truss	User cost profile	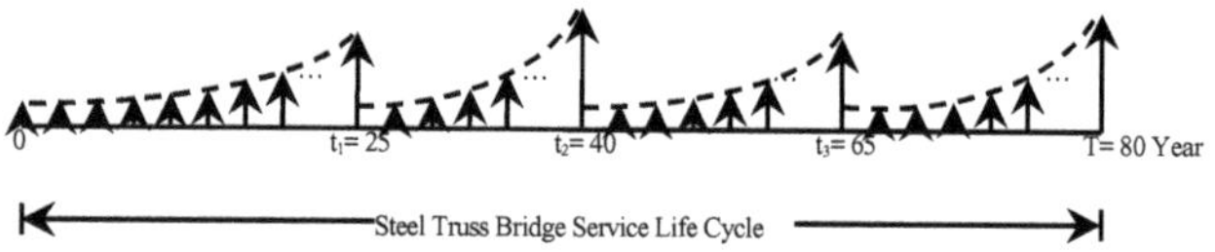
	PW_{LCUC}	$= (C_{AUC1}(1-(1+g_{AUC1})^{t_1}(1+i)^{-t_1}))/(i-g_{AUC1})$ $+ ((C_{AUC2}(1-(1+g_{AUC2})^{(t_2-t_1)}(1+i)^{-(t_2-t_1)}))/(i-g_{AUC2}))/(1+i)^{t_1}$ $+ ((C_{AUC3}(1-(1+g_{AUC3})^{(t_3-t_2)}(1+i)^{-(t_3-t_2)}))/(i-g_{AUC3}))/(1+i)^{t_2}$ $+ ((C_{AUC4}(1-(1+g_{AUC4})^{(T-t_3)}(1+i)^{-(T-t_3)}))/(i-g_{AUC4}))/(1+i)^{t_3}$
	$PW_{LCUC\infty}$	$= PW_{LCUC}/(1-(1/(1+i)^T))$
	$EUAUC$	$= PW_{LCUC}\cdot((i(1+i)^T)/((1+i)^T-1))$
	$EUAUC_\infty$	$= PW_{LCUC\infty}\cdot i$

g_{AUC4} = Growth rate of annual bridge user cost between the third and fourth major repairs

g_{AUC5} = Growth rate of annual bridge user cost between the fourth and fifth major repairs

g_{AUC6} = Growth rate of annual bridge user cost between the fifth and sixth major repairs

i = Discount rate

t = Time of year a major bridge treatment is implemented

T = Number of years of service life for a highway bridge

9.4 PROJECT-LEVEL AGENCY AND USER BENEFITS IN PERPETUITY

9.4.1 Project-level agency benefits due to bridge service life change

The project-level agency benefits are mainly achieved from a service life change perspective. The typical activity profiles for various types of bridges proposed in the previous sections represent the bridge service lives under the ideal situation. In particular, the recommended service lives are achievable provided that repair and maintenance treatments are implemented in a timely fashion to individual bridge components. Early termination of a service life may occur if inappropriate treatments are implemented and/or the activity timing fails the effective time window for treatments—either too early or too late. As such, the typical service life cycle can be used as the base case scenario and the cases with early terminations can be used as alternative scenarios. To maintain the typical service life cycle, a current project needs to be implemented at the right timing. The resulting difference in bridge life-cycle agency costs between the base case and an alternative scenario can be treated as the overall bridge project-level life-cycle agency benefits due to service life cycle shift.

9.4.2 Project-level user benefits due to bridge service life change and demand shift

The project-level user benefits in the bridge service life cycle can be quantified from bridge service life change and traffic demand shift aspects. The bridge life-cycle user costs are quantified on the basis of the typical and alternative bridge life-cycle activity profiles. The reductions in life-cycle user costs associated with the base case typical and alternative profiles are considered as life-cycle user benefits due to service life cycle change.

In addition, life-cycle user benefits may be achieved due to demand shift after bridge project implementation. The impacts of traffic demand shift may occur for highway segments within the physical range of the bridge preservation project and outside of the project physical range. If the total travel demand remains unchanged, increases in the traffic volumes on the highway segments directly affected by the project and within the project physical range will lead to decreases in the traffic volumes on the highway segments indirectly affected by the project and outside the project physical range. For both parts, user benefits are expected if the unit user costs in terms of user costs per VMT get reduced after project implementation. For each highway segment that is directly or indirectly affected by the bridge preservation project, the user benefits can be computed as changes in consumer surplus (AASHTO, 2003). The user benefits of all highway segments directly and indirectly affected by the project can be aggregated to establish the total user benefits due to demand shift. In the analysis process, the overall project-level user benefits resulting from bridge facility service life change and traffic demand shift can be addressed simultaneously in annual user cost calculation and be extrapolated to one bridge service life cycle. The project-level life-cycle user benefits can be expressed in present worth amounts about the base year.

9.4.3 Project-level agency and user benefits in perpetuity

As compared to the base case, which maintains the typical bridge service life cycle ad infinitum, an alternative scenario may have an early service life termination for the first life cycle and then follow the typical service life cycle from the second cycle to perpetuity. The alternative case could possibly have early service life terminations for the first two life cycles and then follow the typical service life cycle from the third cycle onward to perpetuity, and so on. The difference in life-cycle agency and user costs in perpetuity between the base case and an alternative case defines the overall project-level agency and user benefits in perpetuity. Table 9.7 illustrates the computation details.

Denote:

$PW_{LCAC\infty}$	= Present worth of bridge life-cycle agency costs in perpetuity
$PW_{LCUC\infty}$	= Present worth of bridge life-cycle user costs in perpetuity
$EUAAC_{\infty}$	= Equivalent uniform annual agency costs in perpetuity
$EUAUC_{\infty}$	= Equivalent uniform annual user costs in perpetuity
PW_{AB}	= Present worth of project agency benefits in perpetuity
PW_{UB}	= Present worth of project user benefits in perpetuity
PW_{B}	= Present worth of overall project benefits in perpetuity
EUA_{AB}	= Equivalent uniform annual project agency benefits in perpetuity
EUA_{UB}	= Equivalent uniform annual project user benefits in perpetuity
EAU_{B}	= Equivalent uniform annual overall project benefits in perpetuity
i	= Discount rate
T	= Number of years of service life for a highway bridge

Table 9.7 Computation of project-level agency and user benefits in perpetuity

Case		*Computation*
Base case: no early termination	Agency cost profile	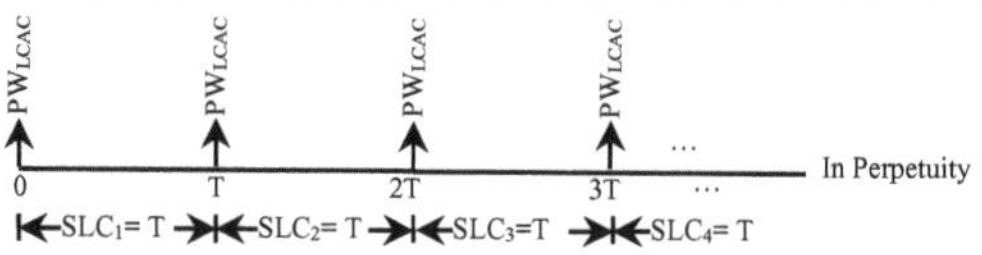
	User cost profile	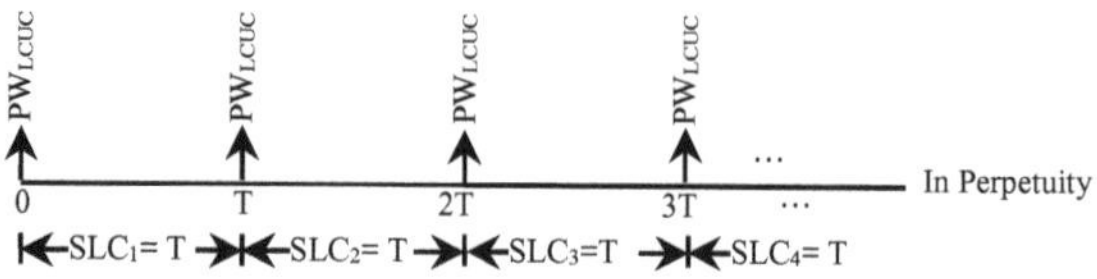
	Present worth	$PW_{LCAC\infty,0} = PW_{LCAC}/(1-(1/(1+i)^T))$ $PW_{LCUC\infty,0} = PW_{LCUC}/(1-(1/(1+i)^T))$
	Annual worth	$EUAAC_{\infty,0} = PW_{LCAC\infty,0} \cdot i$, $EUAUC_{\infty,0} = PW_{LCUC\infty,0} \cdot i$
Case 1: early termination in cycle 1	Agency cost profile	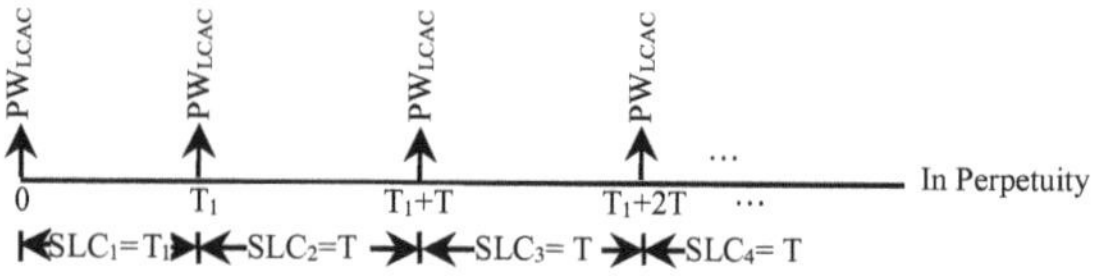
	User cost profile	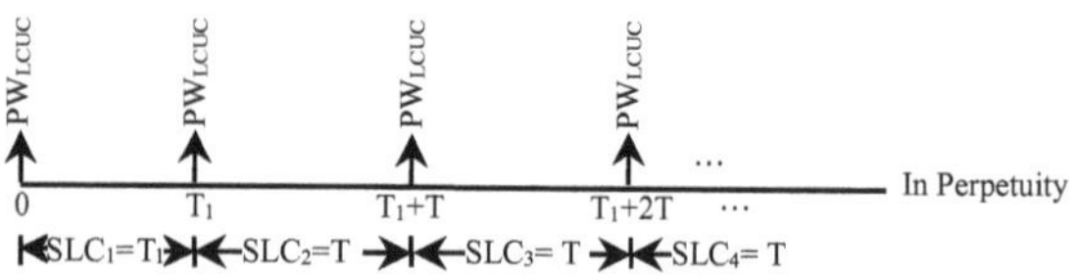
	Present worth	$PW_{LCAC\infty,1} = PW_{LCAC1} + (PW_{LCAC}/(1-(1/(1+i)^T)))/(1+i)^{T_1}$ $PW_{LCUC\infty,1} = PW_{LCUC1} + (PW_{LCUC}/(1-(1/(1+i)^T)))/(1+i)^{T_1}$
	Annual worth	$EUAAC_{\infty,1} = PW_{LCAC\infty,1} \cdot i$, $EUAUC_{\infty,1} = PW_{LCUC\infty,1} \cdot i$

(Continued)

Table 9.7 (Continued) Computation of project-level agency and user benefits in perpetuity

Case		*Computation*
	Base case benefits	Agency benefits: $PW_{AB} = PW_{LCAC\infty,1} - PW_{LCAC\infty,0}$ $EUA_{AB} = EUAAC_{LCAC\infty,1} - EUAAC_{LCAC\infty,0}$ User benefits: $PW_{UB} = PW_{LCUC\infty,1} - PW_{LCUC\infty,0}$ $EUA_{UB} = EUAUC_{LCAC\infty,1} - EUAUC_{LCUC\infty,0}$ Overall benefits: $PW_B = PW_{AB} + PW_{UB}, EUA_B = EUA_{AB} + EUA_{UB}$
Case 2: early termination in cycles 1 and 2	Agency cost profile	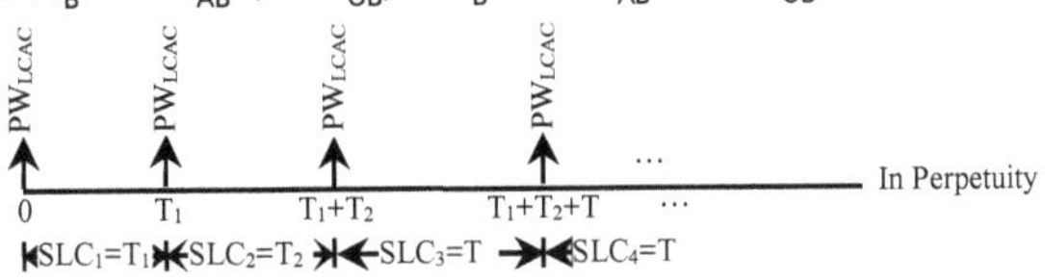
	User cost profile	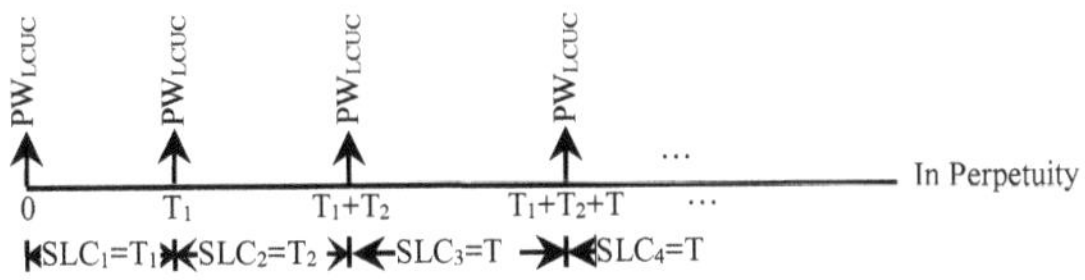
	Present worth	$PW_{LCAC\infty,2} = PW_{LCAC1} + PW_{LCAC2}/(1+i)^{T_1} + (PW_{LCAC}/(1-(1/(1+i)^T)))/(1+i)^{(T_1+T_2)}$ $PW_{LCUC\infty,2} = PW_{LCUC1} + PW_{LCUC2}/(1+i)^{T_1} + (PW_{LCUC}/(1-(1/(1+i)^T)))/(1+i)^{(T_1+T_2)}$
	Annual worth	$EUAAC_{\infty,2} = PW_{LCAC\infty,2}\cdot i, EUAUC_{\infty,2} = PW_{LCUC\infty,2}\cdot i$
	Base case benefits	Agency benefits: $PW_{AB} = PW_{LCAC\infty,2} - PW_{LCAC\infty,0}$ $EUA_{AB} = EUAAC_{LCAC\infty,2} - EUAAC_{LCAC\infty,0}$ User benefits: $PW_{UB} = PW_{LCUC\infty,2} - PW_{LCUC\infty,0}$ $EUA_{UB} = EUAUC_{LCAC\infty,2} - EUAUC_{LCUC\infty,0}$ Overall benefits: $PW_B = PW_{AB} + PW_{UB}, EUA_B = EUA_{AB} + EUA_{UB}$

9.5 DISTRIBUTION OF PROJECT COSTS IN PERPETUITY

The total costs of a highway bridge project consist of project direct costs and added work zone-related user costs during project execution. Since the typical bridge service life cycle is treated as the base case, the total project costs are distributed according to the typical bridge service life cycle in perpetuity, as illustrated in Table 9.8.

Denote:

PW_{PC} = Present worth of project direct construction, maintenance or rehabilitation costs

PW_{WZUC} = Present worth of excessive work zone user costs caused by the project

Table 9.8 Distribution of project costs in perpetuity horizon

Case		*Computation*
Base case: no early termination	Project direct cost profile	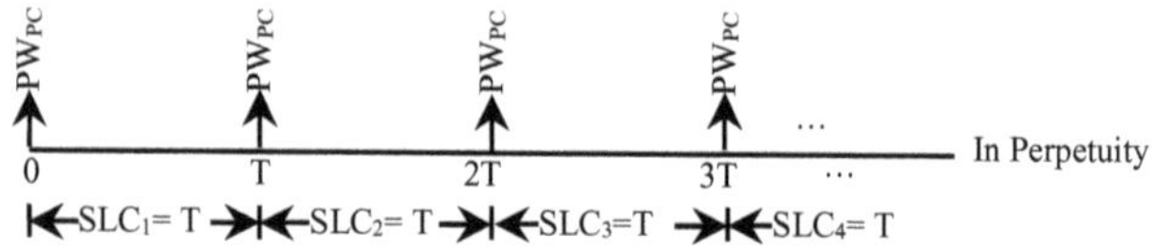
	Project work zone-related user cost profile	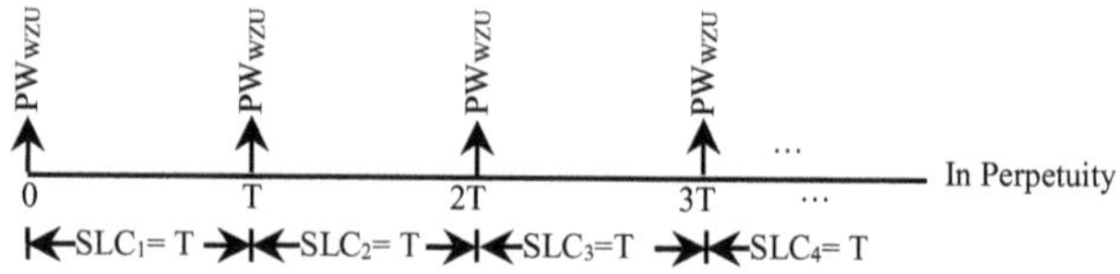
	Present worth	$PW_{C,\infty,0} = (PW_{PC} + PW_{WZUC})/(1 - (1/(1+i)^T))$
	Annual worth	$EUA_{C,\infty,0} = PW_{C,\infty,0} \cdot i$

$PW_{C,\infty}$ = Present worth of the total of project direct costs and work zone user costs in perpetuity

$EUA_{C,\infty}$ = Total of annualized project direct costs and work zone user costs in perpetuity

i = Discount rate

T = Number of years of service life for a highway bridge

9.6 PROJECT-LEVEL BENEFIT AND COST ANALYSIS

With the project-level life-cycle benefits and costs established, these amounts can be compared to evaluate economic feasibility of the proposed bridge project, as illustrated in Table 9.9.

9.7 RISK AND UNCERTAINTY CONSIDERATIONS IN BRIDGE PROJECT-LEVEL BENEFIT–COST ANALYSIS

Similar to the estimation of pavement project-level benefits and costs in pavement service life cycle, the bridge project-level benefit and cost analysis involves the use of a large number of input factors. In bridge service life cycle, agency costs of bridge component construction, and repair and maintenance treatments are estimated using historical cost data. Predicted traffic volumes are used for user cost calculations, and the discount rate may fluctuate over time. Therefore, using a single value for each of the above input factors to calculate the benefits or costs for the bridge project may not produce reliable results. Conversely, risk and uncertainty of the above input factors for the analysis need to be taken into consideration.

For the input factor under risk considerations, it is assumed that the range of multiple possible values of the factor is known and a reliable probability distribution can be assigned. Thus a mathematical expectation can be calculated via probabilistic risk assessment (FHWA, 1998b; Reigle, 2000). The mathematical expectation can be used for bridge project-level benefit and cost calculation. Practically, the beta distribution can be employed for analysis. With historical data made available, the lower (L) and upper (H) bounds, sample mean, and sample variance can be computed. These values can help derive the two shape parameters

Table 9.9 Bridge project benefit–cost analysis in perpetuity

Case		*Present worth*		*Annual worth*	
		Agency costs	*User costs*	*Agency costs*	*User costs*
Base case: no early termination	Life-cycle costs	$PW_{LCAC\infty,0}$	$PW_{LCUC\infty,0}$	$EUAAC_{\infty,0}$	$EUAUC_{\infty,0}$
Case 1: early termination in cycle 1	Life-cycle costs	$PW_{LCAC\infty,1}$	$PW_{LCUC\infty,1}$	$EUAAC_{\infty,1}$	$EUAUC_{\infty,1}$
	Base case benefit	$PW_{AB,\infty} = PW_{LCAC\infty,1} - PW_{LCAC\infty,0}$ $PW_{UB,\infty} = PW_{LCUC\infty,1} - PW_{LCUC\infty,0}$		$EUA_{AB,\infty} = EUAAC_{LCAC\infty,1} - EUAAC_{LCAC\infty,0}$ $EUA_{UB,\infty} = EUAUC_{LCAC\infty,1} - EUAUC_{LCUC\infty,0}$	
	Base case project costs	$PW_{C,\infty,0}$		$EUA_{C,\infty,0}$	
	Benefit/cost analysis	$NPW = (PW_{AB,\infty} + PW_{UB,\infty}) - PW_{C,\infty,0}$ $B/C = (PW_{AB,\infty} + PW_{UB,\infty})/PW_{C,\infty,0}$		$NPW = (EUA_{AB,\infty} + EUA_{UB,\infty}) - EUA_{C,\infty,0}$ $B/C = (EUA_{AB,\infty} + EUA_{UB,\infty})/EUA_{C,\infty,0}$	
Case 2: early termination in cycles 1 and 2	Life-cycle costs	$PW_{LCAC\infty,2}$	$PW_{LCUC\infty,2}$	$EUAAC_{\infty,2}$	$EUAUC_{\infty,2}$
	Base case benefit	$PW_{AB,\infty} = PW_{LCAC\infty,2} - PW_{LCAC\infty,0}$ $PW_{UB,\infty} = PW_{LCUC\infty,2} - PW_{LCUC\infty,0}$		$EUA_{AB,\infty} = EUAAC_{LCAC\infty,2} - EUAAC_{LCAC\infty,0}$ $EUA_{UB,\infty} = EUAUC_{LCAC\infty,2} - EUAUC_{LCUC\infty,0}$	
	Base case project costs	$PW_{C,\infty,0}$		$EUA_{C,\infty,0}$	
	Benefit/cost analysis	$B/C = (PW_{AB,\infty} + PW_{UB,\infty})/PW_{C,\infty,0}$		$NPW = (EUA_{AB,\infty} + EUA_{UB,\infty}) - EUA_{C,\infty,0}$ $B/C = (EUA_{AB,\infty} + EUA_{UB,\infty})/EUA_{C,\infty,0}$	

referred to as α and β. The four parameter values for each input factor can be used as input values for Monte Carlo simulation runs using Latin square sampling method and each run with multiple iterations. The grand average of replicated simulation run outputs can be computed and treated as the mathematical expectation value. In the end, the grand average values for all input factors considered for probabilistic risk analysis can be used for estimating the expected bridge project-level benefits in the bridge facility service life cycle incorporating risk.

For the input factor under uncertainty considerations, the full range of multiple possible values of the factor is unknown or if it is known a reliable probability distribution cannot be established. Hence, a mathematical expectation cannot be computed. In this case, Shackle's model can be employed for calculation (Shackle, 1949; Ford and Ghose, 1998; Young, 2001).

First, the grand average of replicated simulation run results can be used as the expected value for the factor under uncertainty. If higher values are preferred for the factor, the absolute deviation of the average of simulation outputs that are lower than the expected value can be used as standardized focus loss value, and the absolute deviation is equal to or higher than the expected value can be used as standardized focus gain value.

Next, a decision rule can be followed to establish a single value in accordance with the triple of expected value, standardized focus loss, and standardized focus gain that will eventually be used for bridge project-level benefit and cost analysis. Among the triple values, the decision maker will likely assign more weight to standardized focus loss and expected values since standardized focus gain value has exceeded the expected value. If the difference between standardized focus loss and expected values is within the tolerance level, the expected value can be adopted for computation. Otherwise, a compromise value between the standardized focus loss and expected values that is acceptable from the decision maker can be calculated and used for project-level benefit and cost analysis (Li and Sinha, 2004, 2009a; Li and Madanu, 2009).

Finally, the single value for an input factor under certainty, the mathematical expectation for an input factor under risk, and the compromise value for an input factor under uncertainty can be used to estimate the bridge project-level life-cycle agency and user costs and benefits in perpetuity horizon.

Case Study: Steel truss bridge deck replacement project evaluation

A 3.39-million-dollar project is proposed in year 2011 for deck replacement of a steel truss bridge on an interstate highway. The bridge is 246.6 feet long with a total project length of 0.934 mile. The bridge was constructed in 1978 as the base year with an 80-year design useful service life, indicating that it has been in service for 33 years. The AADT in 2002 is 108210 with a 2% annual growth rate from the base year. The average vehicle operating speed is 56.78 mph.

For the typical steel truss bridge life-cycle activity profile, deck rehabilitation treatments need to be implemented in the 25th and 65th years, and deck replacement needs to be implemented in the 40th year. The alternative activity profile, which reflects the actual situation of investments, shows that the first deck rehabilitation treatment was implemented in a timely fashion in the 15th year. However, the deck replacement is earlier by 7 years from the 40th year to 33rd year, leading to a reduction in the useful service life. Assume the annual incremental gradients are 3% for annual routine maintenance costs and 2% for individual user cost components of vehicle operations, travel time, crashes, and air emissions. With earlier implementation of the bridge deck replacement treatment, the incremental gradients for maintenance costs and user costs are expected to be 5% higher than the original gradients. A 4% discount rate is used for analysis.

Based on historical data, input factors related to bridge agency costs of deck, superstructure, substructure, and surface treatments; traffic growth rate; annual maintenance and user cost gradients; and discount rate are identified for risk and uncertainty analysis. The details are given below:

Cost item	μ	σ	L	H
Deck (\$/ft^2)	62.019	42.00	0.10	387.00
Superstructure (\$/ft^2)	109.617	82.00	0.20	372.00
Substructure (\$/ft^2)	114.597	92.00	0.10	372
Surface (\$/ft^2)	62.019	42.00	0.10	387.00
Annual traffic growth rate r	2%	1%	1%	3%
Annual maintenance gradient g_{M1}	3%	1%	1%	5%
Annual maintenance gradient g_{M2}	3%	1%	1%	5%
Annual maintenance gradient g_{M3}	3%	1%	1%	5%
Annual user cost gradient r_1	2%	1%	1%	3%
Annual user cost gradient r_2	2%	1%	1%	3%
Annual user cost gradient r_3	2%	1%	1%	3%
Discount rate i	4%	1%	3%	5%

The bridge deck area is estimated as 246.6 × ((6 × 12-ft lane) + (2 × 6-ft shoulder)) = 20714.40 ft^2; the construction, deck rehabilitation and replacement, and maintenance treatments costs are estimated as

Bridge construction cost = (Deck + Superstructure + Substructure + Surface) × Deck area
Deck rehabilitation cost = Deck × Deck area; Deck replacement cost = (Deck + Surface) × Deck area
Maintenance cost = 2.5% (Construction/Design life)

The unit user costs of vehicle operations, travel time, crashes, and air emissions in year 2000 constant dollars are computed by the following formulas (Li et al., 2010):

Vehicle operating costs (\$/VMT) $= 0.8156 - 0.0077 \times$ Speed
Travel Time cost (\$/VMT) $= 0.4595 - 0.00531 \times$ Speed
Crash cost (\$/VMT) $= -0.22 + 0.005612 \times$ Speed
Air emission cost (\$/VMT) $= 0.2037 + 0.000013 \times$ Speed

Conduct project evaluation using the life-cycle cost analysis approach without and with considerations of risk and uncertainty of the above input factors.

SOLUTION

1. Deterministic project evaluation

1A. Base case life-cycle agency cost analysis using the typical activity profile

a. Base case agency cost items

Agency cost item	*Unit cost (\$/ft²)*	*Project-related agency cost*
Construction	348.25	$= 348.25 \times 20714.4 = 7213789.80$
Deck rehabilitation	62.019	$= 62.019 \times 20714.4 = 1284686.37$
Deck replacement	124.04	$= 124.04 \times 20714.4 = 2569414.18$
Annual maintenance	–	$= (2.5\% \times 7213789.80)/80 = 2254.31$

b. Base case life-cycle agency costs

Agency cost profile

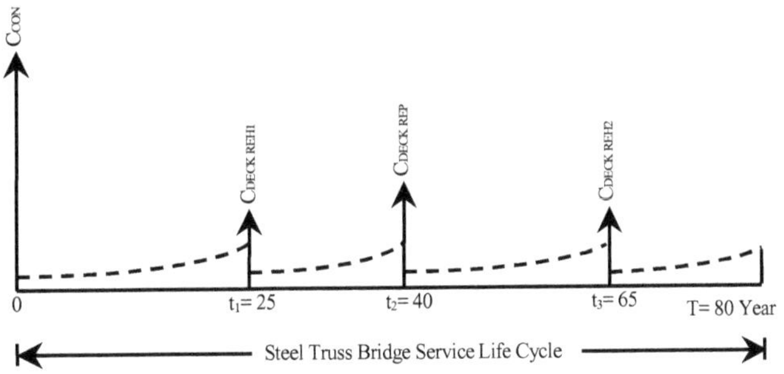

PW_{LCAC}

$$
\begin{aligned}
&= C_{CON} + C_{DREH1}/(1+i)^{t_1} + C_{DREP}/(1+i)^{t_2} + C_{DREH2}/(1+i)^{t_1} \\
&\quad + (C_{MAIN1}(1-(1+g_{M1})^{t_1}(1+i)^{-t_1}))/(i-g_1) \\
&\quad + ((C_{MAIN2}(1-(1+g_{M2})^{(t_2-t_1)}(1+i)^{-(t_2-t_1)}))/(i-g_2))/(1+i)^{t_1} \\
&\quad + ((C_{MAIN3}(1-(1+g_{M3})^{(t_3-t_2)}(1+i)^{-(t_3-t_2)}))/(i-g_3))/(1+i)^{t_2} \\
&\quad + ((C_{MAIN4}(1-(1+g_{M4})^{(T-t_3)}(1+i)^{-(T-t_3)}))/(i-g_4))/(1+i)^{t_3} \\
&= 7213789.80 + 1284686.37/(1+4\%)^{25} \\
&\quad + 2569414.18/(1+4\%)^{40} \\
&\quad + 1284686.37/(1+4\%)^{65} \\
&\quad + (2254.31(1-(1+3\%)^{25}(1+4\%)^{-25}))/(4\%-3\%) \\
&\quad + ((2254.31(1-(1+3\%)^{(40-25)}(1+4\%)^{-(40-25)}))/(4\%-3\%))/(1+4\%)^{25} \\
&\quad + ((2254.31(1-(1+3\%)^{(65-40)}(1+4\%)^{-(65-40)}))/(4\%-3\%))/(1+4\%)^{40} \\
&\quad + ((2254.31(1-(1+3\%)^{(80-65)}(1+4\%)^{-(80-65)}))/(4\%-3\%))/(1+4\%)^{65} \\
&= 8403486.02
\end{aligned}
$$

Continued

$PW_{LCAC\infty}$	$= PW_{LCAC}/(1-(1/(1+i)^{T}))$
	$= 8403486.02/(1-(1/(1+4\%)^{80}))$
	$= 8784599.97$
$EUAAC$	$= PW_{LCAC}((i(1+i)^{T})/((1+i)^{T}-1))$
	$= 8403486.02((4\%(1+4\%)^{80})/((1+4\%)^{80}-1))$
	$= 351384.00$
$EUAAC_{\infty}$	$= PW_{LCAC\infty}\cdot i = 8784599.97\times 4\% = 351384.00$

1B. Base case life-cycle user cost analysis using the typical activity profile

a. Base year user costs

i. Base year AADT: 108210/[(1 + 2%) 24] = 67276 (An annual growth rate of 2%)

ii. Base year annual user costs:

Vehicle opt. costs (\$/VMT)	$= 0.8156 - 0.0077\times \text{Speed} = 0.8156 - 0.0077\times 56.78 = 0.3784$
Travel time cost (\$/VMT)	$= 0.4595 - 0.00531\times \text{Speed} = 0.4595 - 0.00531\times 56.78 = 0.1582$
Crash cost (\$/VMT)	$= -0.22 + 0.005612\times \text{Speed} = -0.22 + 0.005612\times 56.78 = 0.0986$
Air emission cost (\$/VMT)	$= 0.2037 + 0.000013\times \text{Speed} = 0.2037 + 0.000013\times 56.78 = 0.2038$

iii. Base year annual user costs:

Annual vehicle operating costs	$= 0.3784\times 67276\times 0.934$ miles $\times 365 = 8678627.14$
Annual travel time cost	$= 0.1582\times 67276\times 0.934$ miles $\times 365 = 3628326.67$
Annual vehicle crash cost	$= 0.0986\times 67276\times 0.934$ miles $\times 365 = 2261397.03$
Annual vehicle air emission cost	$= 0.2038\times 67276\times 0.934$ miles $\times 365 = 4674165.46$
Total	19242516.30

b. Base case life-cycle user costs

User cost profile

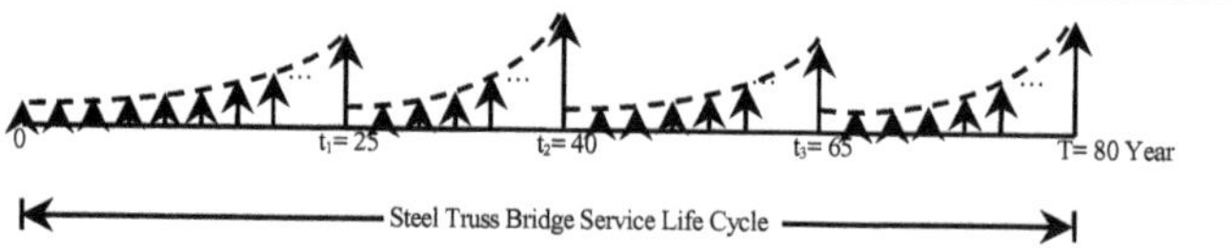

PW_{LCUC}

$$
\begin{aligned}
&= (C_{AUC1}(1-(1+r_1)^{t_1}(1+i)^{-t_1}))/(i-r_1)\\
&\quad + ((C_{AUC2}(1-(1+r_2)^{(t_2-t_1)}(1+i)^{-(t_2-t_1)}))/(i-r_2))/(1+i)^{t_1}\\
&\quad + ((C_{AUC3}(1-(1+r_3)^{(t_3-t_2)}(1+i)^{-(t_3-t_2)}))/(i-r_3))/(1+i)^{t_2}\\
&\quad + ((C_{AUC4}(1-(1+r_4)^{(T-t_3)}(1+i)^{-(T-t_3)}))/(i-r_4))/(1+i)^{t_3}\\
&= (19242516.30(1-(1+2\%)^{25}(1+4\%)^{-25}))/(4\%-3\%)\\
&\quad + ((19242516.30(1-(1+2\%)^{(40-25)}(1+4\%)^{-(40-25)}))/(4\%-2\%))/(1+4\%)^{25}\\
&\quad + ((19242516.30(1-(1+2\%)^{(65-40)}(1+4\%)^{-(65-40)}))/(4\%-2\%))/(1+4\%)^{40}\\
&\quad + ((19242516.30(1-(1+2\%)^{(80-65)}(1+4\%)^{-(80-65)}))/(4\%-2\%))/(1+4\%)^{65}\\
&= 557278148.56
\end{aligned}
$$

Continued

$PW_{LCUC\infty}$	$= PW_{LCUC}/(1-(1/(1+i)^T))$
	$= 557278148.56/(1-(1/(1+4\%)^{80})) = 582551764.29$
$EUAUC$	$= PW_{LCUC}((i(1+i)^T)/((1+i)^T-1))$
	$= 557278148.56((4\%(1+4\%)^{80})/((1+4\%)^{80}-1)) = 23302070.57$
$EUAUC_\infty$	$= PW_{LCUC\infty} \cdot i$
	$= 582551764.29 \times 4\% = 23302070.57$

2A. Alternative case life-cycle agency cost analysis with early termination

Reduction in pavement useful service life

Seven years of reduction in pavement useful service life

Conversion of project cost to 2000 constant dollars

$$Dollar_{2000} = Dollar_{2011}/(1+i)^{2011-2000} = 3390000/(1+4\%)^{11} = \$2202079.35$$

Pavement maintenance gradient adjustment

The annual routine maintenance cost is assumed to increase at a faster pace for the period after project implementation as

$$g'_{M3} = (g_{M3} + 5\% \times (1+g_2)^{(33-25)} - 1) = 4.334\%$$
$$g'_{M4} = (g_{M4} + 5\% \times (1+g_3)^{(58-33)} - 1) = 8.469\%$$

a. Alternative case agency cost items

Agency cost item	*Unit cost (\$/ft²)*	*Project-related agency cost*
Project cost	–	2202079.35
Construction	348.25	= 348.25 × 20714.4 = 7213789.80
Deck rehabilitation	62.019	= 62.019 × 20714.4 = 1284686.37
Deck replacement	124.04	= 124.04 × 20714.4 = 2569414.18
Maintenance cost	–	= (2.5% × 7213789.80)/80 = 2254.31

b. Alternative case life-cycle agency costs

With deck replacement implemented earlier by 7 years from the 40th year to 33rd year, the useful service life is assumed to reduce by 7 years from 80 years to 73 years.

Agency cost profile

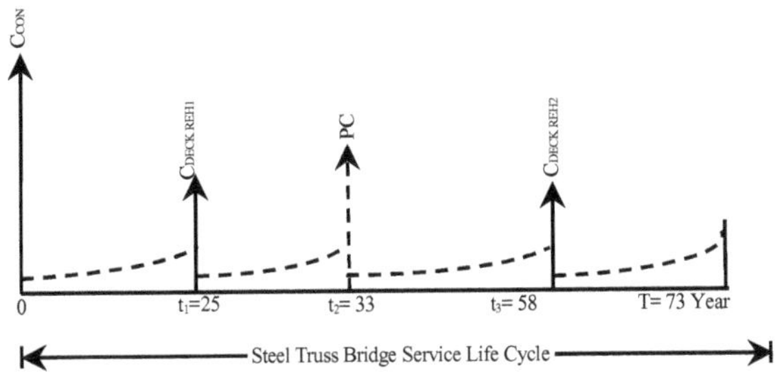

Continued

PW_{LCAC}

$$= C_{CON} + C_{DECK\ REH1}/(1+i)^{t_1} + PC/(1+i)^{t_2} + C_{DECK\ REH2}/(1+i)^{t_3}$$
$$+ (C_{MAIN1}(1-(1+g_{M1})^{t_1}(1+i)^{-t_1}))/(i-g_{M1})$$
$$+ ((C_{MAIN2}(1-(1+g_{M2})^{(t_2-t_1)}(1+i)^{-(t_2-t_1)}))/(i-g_{M2}))/(1+i)^{t_1}$$
$$+ ((C_{MAIN3}(1-(1+)^{(t_3-t_2)}(1+i)^{-(t_3-t_2)}))/(i-))/(1+i)^{t_2}$$
$$+ ((C_{MAIN4}(1-(1+)^{(T-t_3)}(1+i)^{-(T-t_3)}))/(i-))/(1+i)^{t_3}$$
$$= 7213789.80 + 1284686.37/(1+4\%)^{25} + 2202079.36/(1+4\%)^{33}$$
$$+ 1284686.37/(1+4\%)^{58}$$
$$+ (2254.31(1-(1+3\%)^{25}(1+4\%)^{-25}))/(4\%-3\%)$$
$$+ ((2254.31(1-(1+3\%)^{(33-25)}(1+4\%)^{-(33-25)}))/(4\%-3\%))/(1+4\%)^{25}$$
$$+ ((2254.31(1-(1+4.334\%)^{(58-33)}(1+4\%)^{-(58-33)}))/(4\%4.334\%\%))/(1+4\%)^{33}$$
$$+ ((2254.31(1-(1+8.469\%)^{(73-58)}(1+4\%)^{-(73-58)}))/(4\%-8.469\%))/(1+4\%)^{58}$$
$$= 8506025.26$$

$EUAAC$

$$= PW_{LCAC}((i(1+i)^{T_i})/((1+i)^{T_i}-1))$$
$$= 8506025.26((4\%(1+4\%)^{73})/((1+4\%)^{73}-1))$$
$$= 360841.76$$

2B. Alternative case life-cycle user cost analysis with early termination

a. Base year user costs

Base year annual user costs:

Annual vehicle operating costs	8678627.14
Annual travel time cost	3628326.67
Annual vehicle crash cost	2261397.03
Annual vehicle air emission cost	4674165.46
Total	19242516.30

b. Alternative case life-cycle user costs

Annual user cost gradient adjustment

The costs of annual vehicle operating costs, travel time, vehicle crashes, and vehicle emissions are assumed to increase at a faster pace by 5% in excess of the 2% annual gradient for the period after project implementation as

$$r_3' = (r_3 + 5\% \times (1+r_2)^{(33-25)} - 1) = 2.858\%$$
$$r_4' = (r_4 + 5\% \times (1+r_3)^{(33-25)} - 1) = 5.203\%$$

User cost profile

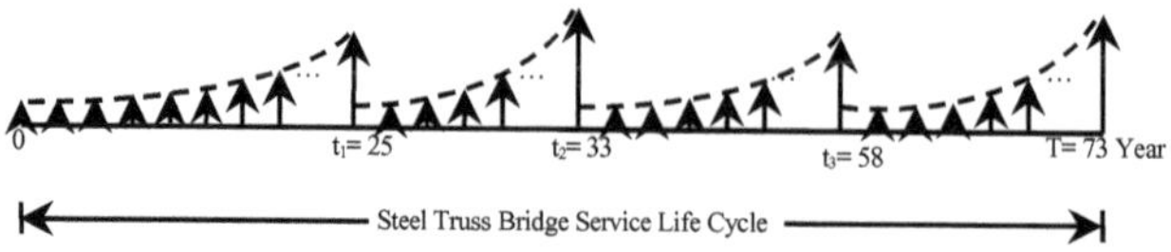

Continued

$$
\begin{aligned}
PW_{LCUC} &= (C_{AUC1}(1-(1+r_1)^{t_1}(1+i)^{-t_1}))/(i-r_1) \\
&\quad + ((C_{AUC2}(1-(1+r_2)^{(t_2-t_1)}(1+i)^{-(t_2-t_1)}))/(i-r_2))/(1+i)^{t_1} \\
&\quad + ((C_{AUC3}(1-(1+r_3')^{(t_3-t_2)}(1+i)^{-(t_3-t_2)}))/(i-r_3'))/(1+i)^{t_2} \\
&\quad + ((C_{AUC4}(1-(1+r_4')^{(T-t_3)}(1+i)^{-(T-t_3)}))\ /\ (i-r_4'))/(1+i)^{t_3} \\
&= (19242516.30(1-(1+2\%)^{25}(1+4\%)^{-25}))/(4\%-2\%) \\
&\quad + ((19242516.30(1-(1+2\%)^{(33-25)}(1+4\%)^{-(33-25)}))/(4\%-2\%))/(1+4\%)^{25} \\
&\quad + ((19242516.30(1+2.858\%\%)^{(58-33)}(1+4\%)^{-(58-33)}))/(4\%-2.858\%))/(1+4\%)^{33} \\
&\quad + ((19242516.30\ (1-(1+5.203\%)^{(73-58)}(1+4\%)^{-(73-58)}))/(4\%-5.203\%))/(1+4\%)^{58} \\
&= 564316892.21
\end{aligned}
$$

$$
\begin{aligned}
EUAUC &= PW_{LCUC}((i(1+i)^{T_1})/((1+i)^{T_1}-1)) \\
&= 564316892.21((4\%(1+4\%)^{73})/((1+4\%)^{73}-1)) \\
&= 23939395.24
\end{aligned}
$$

3. Project life-cycle overall benefits in perpetuity
   Since the typical life-cycle activity profile confers the lowest life-cycle agency and user costs, the pavement manager would always like to revert back to the typical activity profile after experiencing early termination. Here, early termination is only assumed for the first life cycle, and the typical service life cycle used in the base case is followed for following cycles into perpetuity.

Base case: no early termination

Agency cost profile

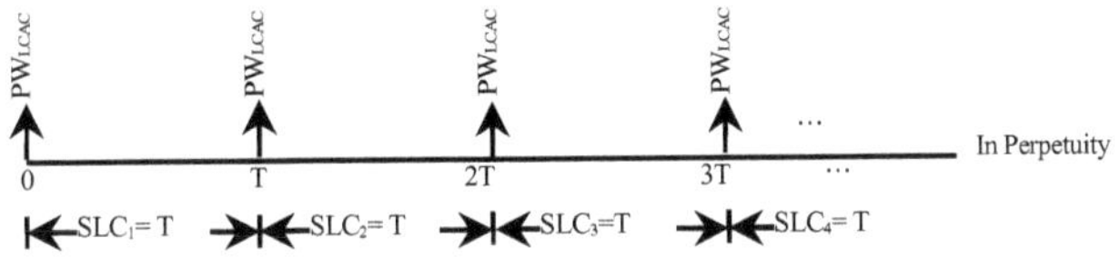

User cost profile

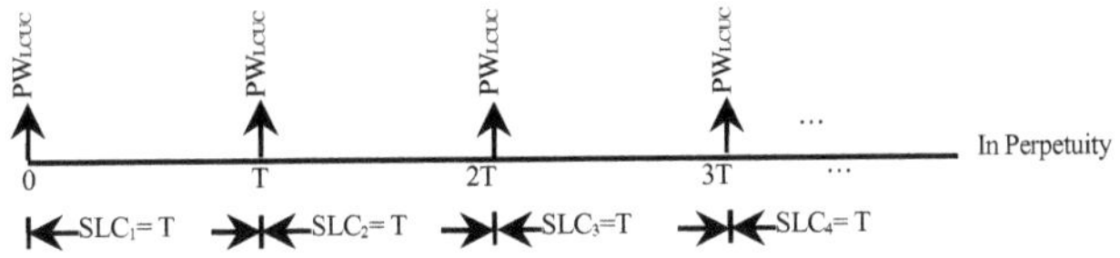

Present worth

$$
\begin{aligned}
PW_{LCAC\infty,0} &= PW_{LCAC}/(1-(1/(1+i)^{T})) \\
&= 8403486.02/(1-(1/(1+4\%)^{80})) = 8784599.97 \\
PW_{LCUC\infty,0} &= PW_{LCUC}/(1-(1/(1+i)^{T})) \\
&= 557278148.56/(1-(1/(1+4\%)^{80})) = 582551764.29
\end{aligned}
$$

Alternative case: early termination in cycle 1

Agency cost profile

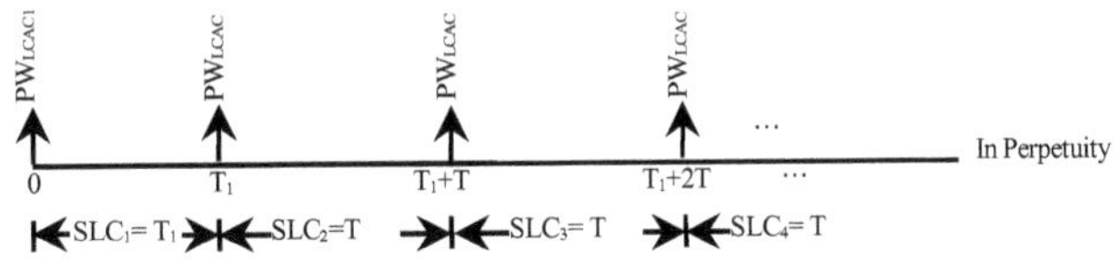

User cost profile

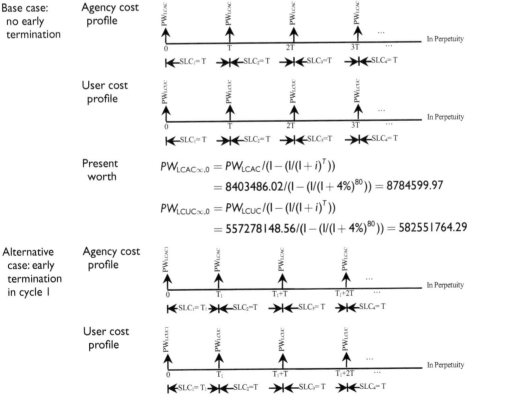

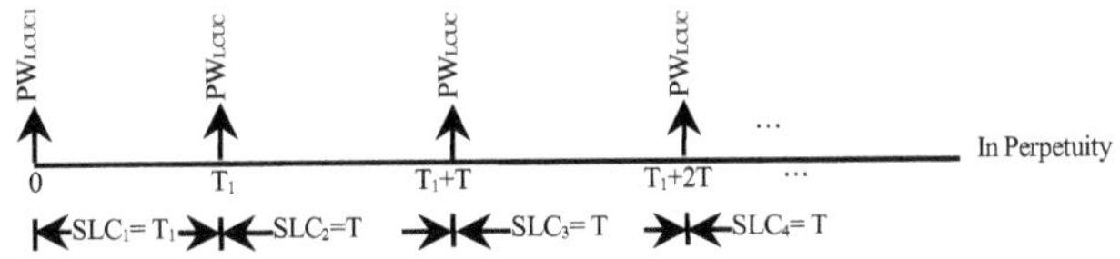

Continued

Present worth

$$PW_{LCAC\infty,I} = PW_{LCACI} + (PW_{LCAC}/(1-(1/(1+i)^{T})))/(1+i)^{T_I}$$
$$= 8506025.26 + (8403486.02/(1(1/(1+4\%)^{80})))/(1+4\%)^{73}$$
$$= 9007545.22$$

$$PW_{LCUC\infty,I} = PW_{LCUCI} + (PW_{LCUC}/(1-(1/(1+i)^{T})))/(1+i)^{T_I}$$
$$= 564316892.21 + 557278148.56(/((1(1/(1+4\%)^{80})))/(1+4\%)^{73}$$
$$= 597575246.32$$

Base case benefits

Agency benefits:

$$PW_{AB} = PW_{LCAC\infty,I} - PW_{LCAC\infty,0}$$
$$= 9007545.22 - 8784597.78 = 222945.24$$

User benefits:

$$PW_{UB} = PW_{LCUC\infty,I} - PW_{LCUC\infty,0}$$
$$= 597575246.32 - 582551764.29$$
$$= 15023482.03$$

Overall benefits:

$$PW_{B} = PW_{AB} + PW_{UB}$$
$$= 222945.24 + 15023482.03$$
$$= 15246427.27$$

II. Project evaluation under certainty and risk

Input factors considered for risk-based analysis include bridge agency costs of construction, deck rehabilitation, deck replacement, and routine maintenance; traffic growth rates; and discount rates. Monte Carlo simulations were simultaneously performed for those factors using beta distributions. Ten simulation runs were used, each with 1000 iterations. For each input factor involving risk consideration, the grand average was established based on the average values of 10 simulation runs, each of which was determined in accordance with the 1000 iteration outcomes. The grand average values for all factors considered for probabilistic risk analysis were used for computing the expected project life-cycle benefits incorporating risk. The analytical procedure used is identical to that for deterministic life-cycle benefit analysis. With remaining input factors treated as deterministic, the analysis incorporating risk is essentially an analysis under a mixed case of certainty and risk.

The following values are used for project evaluation with risk considerations:

Case	*Cost component*	*Cost item*	*Input value*
Factors under risk	Agency costs	Construction	10893081.53
		Deck rehabilitation	270193.48
		Deck replacement	5508166.10
		Routine maintenance	3404.08
		Annual maintenance gradient g_{M1}	2.998%
		Annual maintenance gradient g_{M2}	2.998%
		Annual maintenance gradient g_{M3}	2.998%
		Annual maintenance gradient g_{M4}	2.998%
	Vehicle operating costs Travel time Vehicle crashes Vehicle air emissions	Annual traffic growth rate r	2.011%
		Annual user cost gradient r_1	1.9959%
		Annual user cost gradient r_2	1.9959%
		Annual user cost gradient r_3	1.9959%
		Discount rate i	4.01%

Project life-cycle overall benefits under risk in perpetuity

With deck replacement implemented earlier by 7 years from the 40th year to 33rd year, the useful service life in risk-based analysis is assumed to decrease by 5 years from 80 years to 75 years. Early termination is only assumed for the first life cycle, and the typical service life cycle used in the base case is used for following cycles into perpetuity.

Base case: no early termination

Agency cost profile

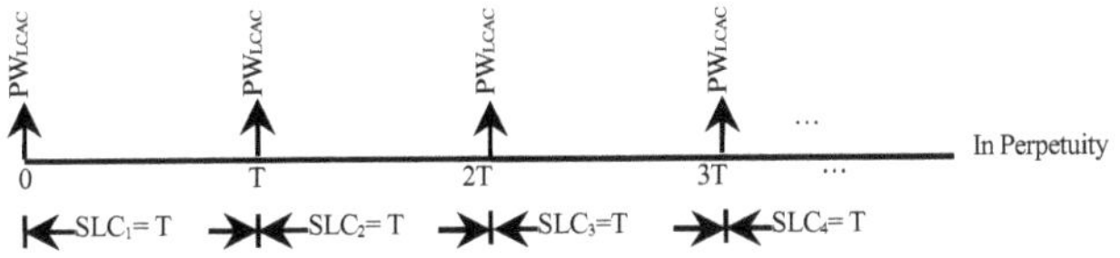

User cost profile

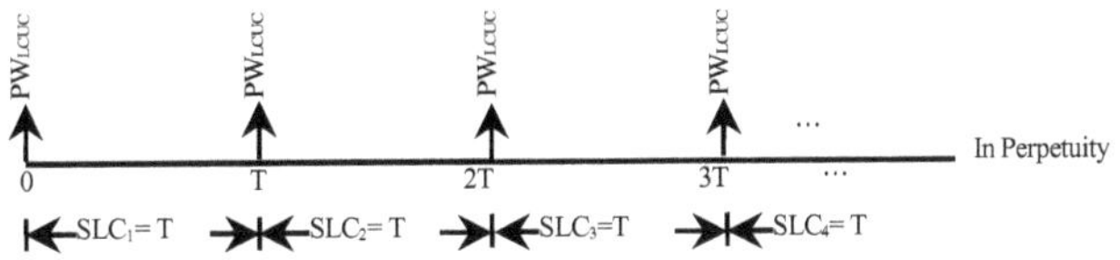

Present worth

$$PW_{LCAC\infty,0} = PW_{LCAC}/(1-(1/(1+i)^{T}))$$
$$= 13365800.29/(1-(1/(1+4.01\%)^{80}))$$
$$= 13967110.90$$

$$PW_{LCUC\infty,0} = PW_{LCUC}/(1-(1/(1+i)^{T}))$$
$$= 718104044.70/(1-(1/(1+4.01\%)^{80}))$$
$$= 750410646.36$$

Alternative case: early termination in cycle 1

Agency cost profile

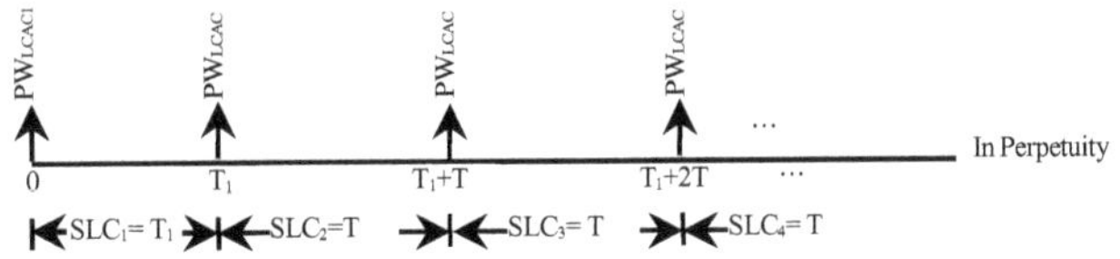

User cost profile

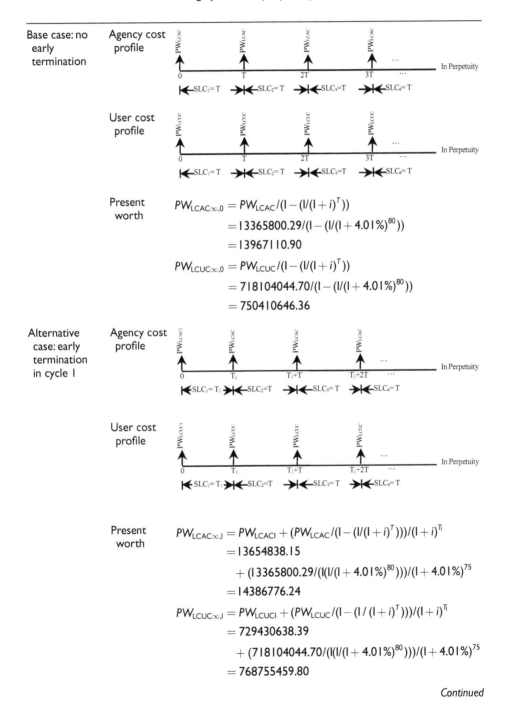

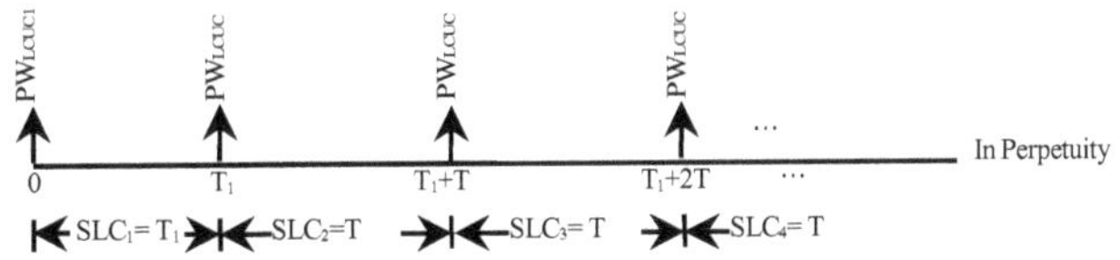

Present worth

$$PW_{LCAC\infty,1} = PW_{LCAC1} + (PW_{LCAC}/(1-(1/(1+i)^{T})))/(1+i)^{T_1}$$
$$= 13654838.15$$
$$+ (13365800.29/(1(1/(1+4.01\%)^{80})))/(1+4.01\%)^{75}$$
$$= 14386776.24$$

$$PW_{LCUC\infty,1} = PW_{LCUC1} + (PW_{LCUC}/(1-(1/(1+i)^{T})))/(1+i)^{T_1}$$
$$= 729430638.39$$
$$+ (718104044.70/(1(1/(1+4.01\%)^{80})))/(1+4.01\%)^{75}$$
$$= 768755459.80$$

Continued

Base case benefits	Agency benefits: $$PW_{AB} = PW_{LCAC\infty,1} - PW_{LCAC\infty,0} = 14386776.24 - 13967110.90 = 419665.34$$ User benefits: $$PW_{UB} = PW_{LCUC\infty,1} - PW_{LCUC\infty,0} = 768755459.80 - 750410646.36 = 18344813.42$$ Overall benefits: $$PW_{B} = PW_{AB} + PW_{UB} = 419665.34 + 18344813.42 = 18764478.76$$

III. Project evaluation under certainty, risk, and uncertainty

In the process of conducting risk-based project evaluation, benefits associated with reductions in agency costs, vehicle operating costs, and air emission costs are found to be relatively stable, with limited variations except for decreases in travel time and crash costs. Thus these costs are further selected for uncertainty-based analysis. The project level life-cycle benefits are estimated under mixed cases of certainty, risk, and uncertainty.

The following values are used for project evaluation with risk considerations:

Case	*Cost component*	*Cost item*	*Input value*
Factors under risk	Agency costs	Construction	10893081.53
		Deck rehabilitation	2702193.48
		Deck replacement	5508166.10
		Routine maintenance	3404.08
		Annual maintenance gradient g_{M1}	2.998%
		Annual maintenance gradient g_{M2}	2.998%
		Annual maintenance gradient g_{M3}	2.998%
		Annual maintenance gradient g_{M4}	2.998%
	Vehicle operating costs	Annual traffic growth rate r	2.011%
	Vehicle air emissions	Annual user cost gradient r_1	1.9959%
		Annual user cost gradient r_2	1.9959%
		Annual user cost gradient r_3	1.9959%
		Annual user cost gradient r_4	1.9959%
		Discount rate i	4.01%

For uncertainty-based analysis of travel time saving and vehicle crash decrease benefits, the values of input factors are adjusted according to the following decision rules as an extension of Shackle's model:

$$\text{If higher values are preferred, } X = \begin{cases} F_{(E)}, \textit{if } x_{SFL} \leq \Delta X \\ \left(F_{(E)} - x_{SFL}\right) / \left[1 + \left(\Delta X / F_{(E)}\right)\right], \textit{ otherwise} \end{cases}$$

$$\text{If lower values are preferred, } X = \begin{cases} F_{(E)}, \textit{if } x_{SFL} \le \Delta X \\ \left(F_{(E)} - x_{SFL}\right) / \left[1 + \left(\Delta X / F_{(E)}\right)\right], \textit{otherwise} \end{cases}$$

For annual traffic growth rate *r*, considering that higher values are preferred

$$\text{Since } |x_{SFL} - F_{(E)}| = |1\% - 2.011\%| = 1.011\% > \Delta X = 39\% \times 2\% = 0.78\%$$

$$r = x_{SFL} / [1 - \Delta X / F_{(E)}] = 1\% / [1 - (0.78\%/2.011\%)] = 1.63\%$$

For annual user cost gradients r_1, r_2, r_3, and r_3, supposing that higher values are preferred

$$\text{Since } |x_{SFL} - F_{(E)}| = |1\% - 1.995\%| = 0.995\% > \Delta X = 39\% \times 2\% = 0.78\%$$

$$r_1 = r_2 = r_3 = r_4 = x_{SFL}/[1 - \Delta X / X_{(E)}] = 1\%/[1 - (0.78\% / 1.995\%)] = 1.64\%$$

For discount rate *i*, assuming that lower values are preferred

$$\text{Since } |x_{SFL} - F_{(E)}| = |5\% - 4.01\%| = 0.99\% > \Delta X = 20\% \times 4\% = 0.8\%$$

$$i = x_{SFL} / [1 + \Delta X / F_{(E)}] = 5\%/[1 + 0.8\%/4.01\%] = 4.17\%$$

The following values are used for project evaluation with uncertainty considerations:

Cost component	*Cost item*	*Average, $F_{(E)}$*	x_{SFL}	x_{SFG}	*Tolerance ($X\Delta$)*	*Input value*
Travel time vehicle crashes	Annual traffic growth rate r	2.011%	1%	3%	39% of $\mu = 2\%$	1.63%
	Annual user cost gradient r_1	1.9959%	1%	3%	39% of $\mu = 2\%$	1.64%
	Annual user cost gradient r_2	1.9959%	1%	3%	39% of $\mu = 2\%$	1.64%
	Annual user cost gradient r_3	1.9959%	1%	3%	39% of $\mu = 2\%$	1.64%
	Annual user cost gradient r_3	1.9959%	1%	3%	39% of $\mu = 2\%$	1.64%
	Discount rate *i*	4.01%	5%	3%	20% of $\mu = 4\%$	4.17%

Project life-cycle overall benefits under certainty, risk, and uncertainty in perpetuity.
With deck replacement implemented earlier by 7 years from the 40th year to 33rd year, the useful service life in risk-based analysis is assumed to reduce by 5 years from 80 years to 75 years. Early termination is only assumed for the first life cycle, and the typical service life cycle used in the base case is used for the following cycles into perpetuity.

Base case: no early termination

Agency cost profile

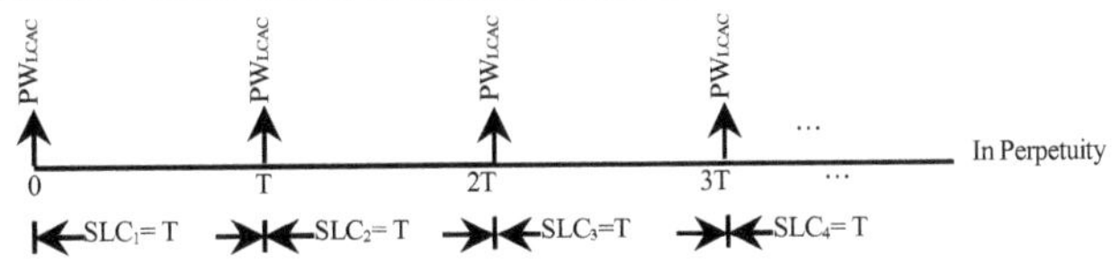

User cost profile

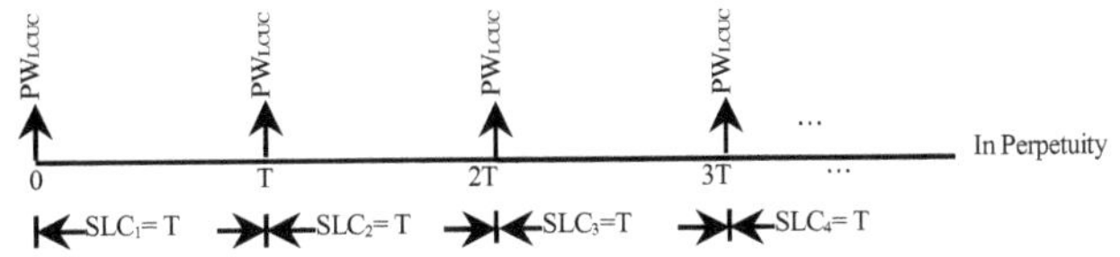

Continued

Present worth

$$PW_{LCAC\infty,0} = PW_{LCAC}/(1-(1/(1+i)^{T}))$$
$$= 13365800.29/(1-(1/(1+4.01\%)^{80})) = 13967110.90$$

$$PW_{LCUC\infty,0} = PW_{LCUC(VOC+VAE)}/(1-(1/(1+i)^{T})) + PW_{LCUC(VTT+VCC)}/(1-(1/(1+i)^{T}))$$
$$= 498307717.12/(1(1/(1+4.01\%)^{80})) + 213609916.30/(1(1/(1+4.17\%)^{80})) = 742789926.55$$

Alternative case: early termination in cycle 1

Agency cost profile

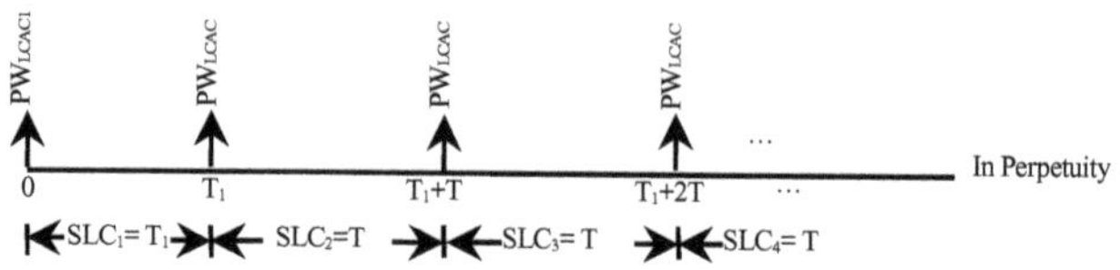

User cost profile

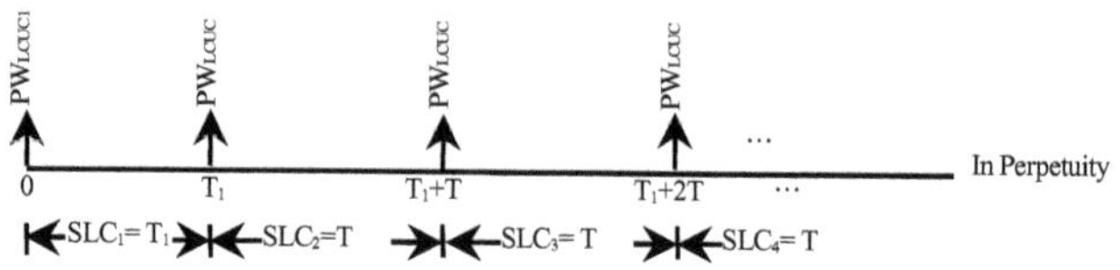

Present worth

$$PW_{LCAC\infty,1} = PW_{LCACI} + (PW_{LCAC}/(1-(1/(1+i)^{T})))/(1+i)^{T_1}$$
$$= 13654838.15 + (13365800.29/(1-(1/(1+4.01\%)^{80})))/(1+4.01\%)^{75}$$
$$= 14386776.24$$

$$PW_{LCUC\infty,1} = PW_{LCUCI(VOC+VAE)} + (PW_{LCUC(VOC+VAE)}/(1-(1/(1+i)^{T})))/(1+i)^{T_1} + PW_{LCUCI(VTT+VCC)} + (PW_{LCUC(VTT+VCC)}/(1-(1/(1+i)^{T})))/(1+i)^{T_1}$$
$$= 506167482.13 + (498307717.12/(1-(1/(1+4.01\%)^{80})))/(1+4.01\%)^{75} + 215479142.15 + (213609916.30/(1-(1/(1+4.01\%)^{80})))/(1+4.01\%)^{75}$$
$$= 759304945.60$$

Base case benefits

Agency benefits:

$$PW_{AB} = PW_{LCAC\infty,1} - PW_{LCAC\infty,0}$$
$$= 14386776.24 - 13967110.90 = 419665.34$$

User benefits:

$$PW_{UB} = PW_{LCUC\infty,1} - PW_{LCUC\infty,0}$$
$$= 759304945.60 - 742789926.55 = 16515019.15$$

Overall benefits:

$$PW_{B} = PW_{AB} + PW_{UB}$$
$$= 419665.34 + 16515019.15 = 16934684.49$$

IV. Study summary

1. Input data

a. Project evaluation under certainty

Data item	*AC*	*VOC*	*TTC*	*VCC*	*VEC*
Project cost	2202079.35				
Construction cost	7213789.80				
Deck rehabilitation cost	1284686.37				
Deck replacement cost	2569414.18				
Annual routine maintenance	2254.31				
Maintenance gradient g_1, g_2, g_3, g_4/ Maintenance gradient g_3', g_4'	3%/4.334% 8.469%				
Discount rate i	4%	4%	4%	4%	4%
Base year AADT	67276	67276	67276	67276	67276
Annual traffic growth r	2%	2%	2%	2%	2%
Annual UC gradient r_1, r_2, r_3, r_4/ Annual UC gradient r_3', r_4'		2%/2.858% 5.203%	2%/2.858% 5.203%	2%/2.858% 5.203%	2%/2.858% 5.203%
Service life for base case	80	80	80	80	80
Service life for alt case	73	73	73	73	73

b. Project evaluation under certainty and risk

Data item	*AC*	*VOC*	*TTC*	*VCC*	*VEC*
Project cost	2199751.58				
Construction cost	10893081.53				
Deck rehabilitation cost	2702193.48				
Deck replacement cost	5508166.104				
Annual routine maintenance	3404.08				
Maintenance gradient g_1, g_2, g_3, g_4/ Maintenance gradient g_3', g_4'	2.998%/5.68% 4.855% 6.637%				
Discount rate i	4.01%	4.01%	4.01%	4.01%	4.01%
Base year AADT	86926	86926	86926	86926	86926
Annual traffic growth r	2.011%	2.011%	2.011%	2.011%	2.011%
Annual UC gradient r_1, r_2, r_3, r_4/ Annual UC gradient r_3', r_4'		1.995%/3.608% 3.143% 4.122%	1.995%/3.608% 3.143% 4.122%	1.995%/3.608% 3.143% 4.122%	1.995%/3.608% 3.143% 4.122%
Service life for base case	80	80	80	80	80
Service life for alt case	75	75	75	75	75

c. Project evaluation under certainty, risk, and uncertainty

Data item	*AC*	*VOC*	*TTC*	*VCC*	*VEC*
Project cost	2199751.58				
Construction cost	10893081.53				
Deck rehabilitation cost	2702193.48				
Deck replacement cost	5508166.104				
Annual routine maintenance	3404.08				
Maintenance gradient g_1, g_2, g_3, g_4/ Maintenance gradient g_3', g_4'	2.998%/5.68% 4.855% 6.637%				
Discount rate i	4.01%	4.01%	4.17%	4.17%	4.01%
Base year AADT	86926	86926	90579	90579	86926
Annual traffic growth r	2.011%	2.011%	1.63%	1.63%	2.011%
Annual UC gradient r_1, r_2, r_3, r_4/ Annual UC gradient r_3', r_4'		1.995%/3.608% 3.143% 4.122%	1.64%/2.920% 2.560% 3.311%	1.64%/2.920% 2.560% 3.311%	1.995%/3.608% 3.143% 4.122%
Service life for base case	80	80	80	80	80
Service life for alt case	75	75	75	75	75

2. Estimated project benefits

Benefit category	*Certainty*	*Certainty and risk*	*Certainty, risk, and uncertainty*
PW_{Agency}	222945.24	419665.34	419665.34
PW_{User}	15023482.03	18344813.42	16515019.15
PW_{Total}	15246427.27	18764478.76	16934684.49

3. Project costs in perpetuity

Case	*Computation*
Project cost profile	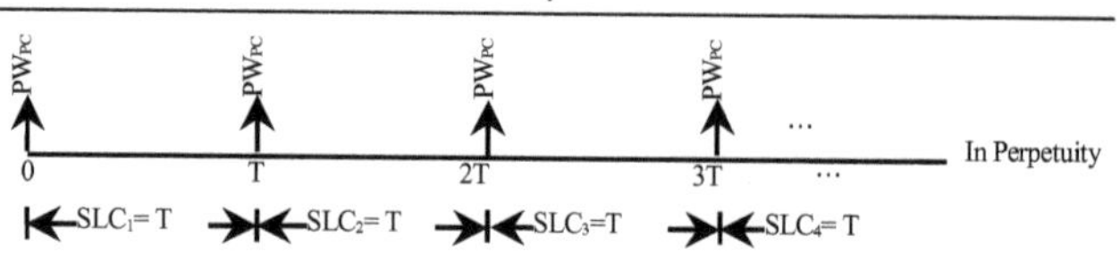
Certainty ($i = 4\%$)	$PW_{LCPC} = \text{Project Cost}/((1+i)^{(2011-2000)+\text{Project Timing}})$ $= \$3390000/((1+4\%)^{(2011-2000)+33}) = 603577.1208$ $PW_{LCPC\infty} = PW_{LCPC}/(1-(1/(1+i)^T))$ $= 603577.1208/(1-(1/(1+4\%)^{80})) = 630950.48$

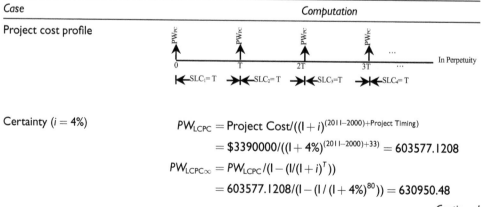

Continued

Certainty and risk ($i = 4.01\%$)	$PW_{LCPC} = \text{Project Cost}/((1+i)^{(2011-2000)+\text{Project Timing}})$
	$= \$3390000/((1+4.01\%)^{(2011-2000)+33}) = 601029.0418$
	$PW_{LCPC\infty} = PW_{LCPC}/(1-(1/(1+i)^T))$
	$= 601029.0418/(1-(1/(1+4.01\%)^{80})) = 628068.59$
Certainty, risk, and uncertainty ($i = 4.01\%$)	$PW_{LCPC} = 601029.0418$
	$PW_{LCPC\infty} = 628068.59$

4. Project evaluation results

Item	*Certainty*	*Certainty and risk*	*Certainty, risk, and uncertainty*
NPW	= 15246427.27	= 18764478.76	= 16934684.49
	−630950.48	−628068.59	−628068.59
	= 14615476.79	= 18136410.17	= 16306615.90
B/C	= 15246427.27/	= 18764478.76/	= 16934684.49/
	630950.48	628068.59	628068.59
	= 24.16	= 29.87	= 26.96

PROBLEMS

9.1 Two designs have been proposed for a short span bridge in a rural area, as shown in Table P9.1. The first design option is to construct the bridge in two phases (Phase I now and Phase II in 30 years). The second proposal is to complete the construction of the bridge in one phase. Assume the annual interest rate is 8%, determine which alternative is preferred using present worth method.

Table P9.1 Alternative details

Alternative	*Construction costs (dollars)*	*Annual maintenance costs (dollars)*	*Service period (years)*
I (Phase I)	$17300000	$120000	1–30
I (Phase II)	$14200000	$40000	31–60
II	$28000000	$160000	1–60

9.2 Two alternatives are under consideration for maintenance of a bridge. Select the most cost-effective alternative using equivalent uniform annual cost method. Assume an interest rate of 8% per year and a design life of 70 years for each alternative.
Alternative I consists of annual maintenance costs of $10000 per year for design life except for:

Year 30, in which bridge deck repairs will cost $30000
Year 40, in which a deck overlay and structural repairs will cost $150000

Alternative II comprises annual maintenance costs of $6000 per year for design life except for:

Year 30, in which bridge deck repairs will cost $45000
Year 40, in which a deck overlay and structural repairs will cost $125000

9.3 The initial investment for constructing a median and guardrails for a prestressed concrete bridge is \$1000000. Maintaining these facilities is expected to cost \$2000 annually for the first five years of service and \$10000 for the next five years of service. It is expected that these facilities will be rehabilitated at the end of the tenth year at a cost of \$60000, after which the maintenance costs are expected to decrease to \$6000 per year. What is equivalent uniform annual cost over a 15-year period of service if the interest rate is 6% per year?

9.4 Your company has been presented with an opportunity to invest in a bridge project. The facts on the project are presented below:

Investment required	*\$60000000*
Salvage value after 10 years	None
Gross income expected from the project	\$20000000/year
Operating costs	
Labor	\$2500000/year
Materials, licenses, insurance, etc.	\$1000000/year
Fuel and other costs	\$1500000/year
Maintenance costs	\$500000/year

The project is expected to operate as shown for 10 years. If your management expects to make 25% on its investments before taxes, would you recommend this project?

Chapter 10

Economic analysis of highway traffic control and safety hardware preservation

10.1 OVERVIEW

Traffic control and safety hardware facilities, including traffic signs, lighting, signals, pavement markings, crash cushions, and guardrails, are an essential part of physical highway facilities. Their overall conditions contribute significantly to highway safety performance. Owing largely to differences in geometric design features and traffic movement characteristics associated with highway segments and intersections, economic analysis of proposed projects is conducted separately to preserve traffic control and safety hardware associated with highway segments and intersections. In the analysis process, highway segments are classified by land area and highway functional classification. Likewise, intersections are categorized by land area, traffic control type, and geometric design feature. Economic analysis is then conducted for each class of highway. Figure 10.1 proposes a framework for economic analysis of traffic control and safety hardware projects. For a typical highway segment (or an intersection), the types of crash occurrences can be achieved from historical crash data. For each type of crash recorded, the crash frequency is mainly affected by detailed safety issue items of the highway segment (or the intersection) in relation to geometric design, consistency of design standards, pavement conditions, traffic control and safety hardware conditions, and roadside features. The crash severity is mostly caused by roadside features. The safety index (SI) of the highway segment (or intersection), measured by fatal, injury, and PDO crash frequencies per segment (or per intersection) per year, can be estimated by considering total traffic exposure of the segment (or intersection), crash occurrence probabilities for different types of crashes, crash frequency for each type of crash, and crash severity for different types of crashes. This SI can be called current-case SI.

If a traffic control and safety hardware project is implemented to improve the condition of one or more traffic control and safety hardware facilities installed in the highway segment (or within the intersection), the crash frequency and crash severity for different types of crashes that are potentially affected by the traffic control and safety hardware facilities are expected to be reduced. Hence a lower SI for the highway segment by crash severity level is expected. It is reasonable to assume that the traffic control and safety hardware issues could be improved to allow crash frequency and crash severity to decrease to the system-wide average levels. The lower-valued SI obtained from the implementation of the traffic control and safety hardware project can be called the base-case SI.

The difference in SI values of the current-case SI before implementing the traffic control and safety hardware project and the base-case SI after implementing the project is considered as the potential for safety improvement (PSI). In reality, the useful service lives of different types of traffic control and safety hardware facilities vary considerably, and the traffic control and safety hardware service lives are often shorter than the design useful service life of a highway segment (or an intersection). To compute the cumulative PSIs for fatal, injury,

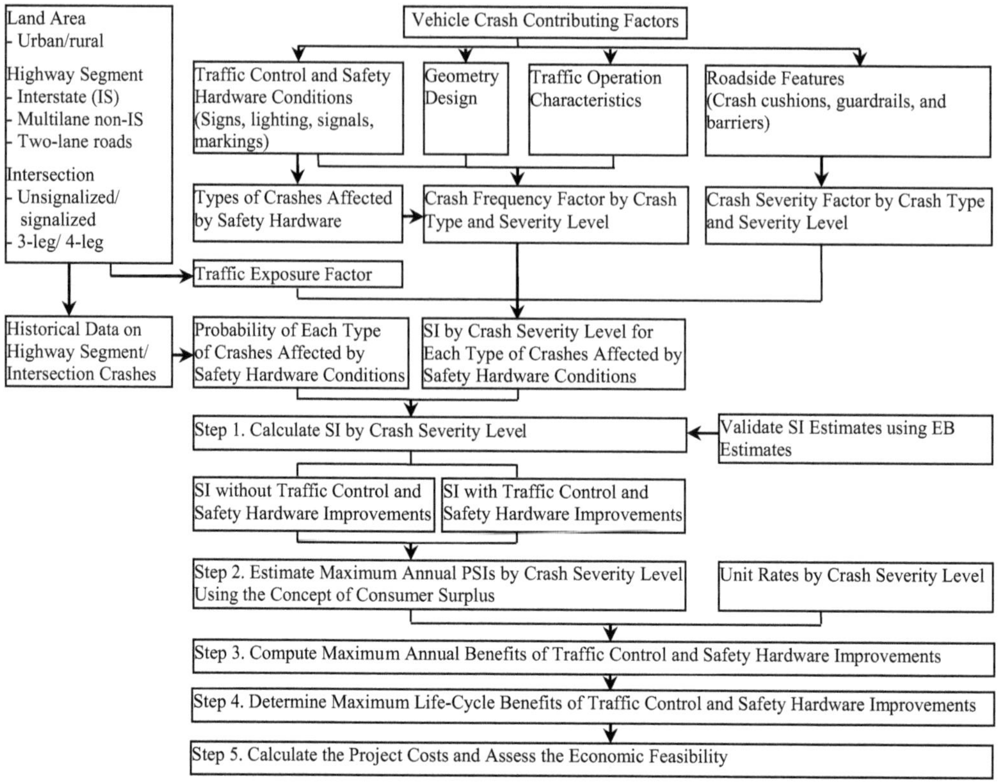

Figure 10.1 Framework of economic analysis of traffic control and safety hardware preservation. (Adapted from Li, Z. and S. Madanu. 2008. Report MRUTC 08-06. Department of Civil, Architectural, and Environmental Engineering, Illinois Institute of Technology, Chicago, IL; Madanu, S. et al. 2010. *ASCE Journal of Transportation Engineering* 136(2), 129–140; Li, Z. et al. 2010. *TRB Journal of Transportation Research Record* 2160, 1–11.)

and PDO crash severity levels in the highway facility design useful service life cycle based on the estimated first-year PSI and the annual traffic growth rate, it is essential to assume that traffic control and safety hardware projects must be implemented in phases so that the multiple service lives of traffic control and safety hardware would match the design service life cycle of the highway segment (or intersection). By that assumption, the PSI level will be consistent throughout the useful service life cycle of the highway segment (or intersection). Having computed the life-cycle cumulative PSIs for fatal, injury, and PDO crash severity levels, and by applying the respective unit rates for fatal, injury, and PDO crashes, the life-cycle cumulative PSIs can be converted to life-cycle overall vehicle crash cost reductions. The crash cost reductions are essentially the traffic control and safety hardware project benefits. The benefits can be compared with the project costs to assess the economic feasibility of the proposed projects.

10.2 SI FOR A HIGHWAY SEGMENT OR AN INTERSECTION

The SI is used to measure the relative safety performance of a highway segment or an intersection (Ogden, 1996; De Leur and Sayed, 2002; Transfund New Zealand, 2003;

TAC, 2004; Montella, 2005; Cafiso et al., 2006; Lamm et al., 2006; Li et al., 2010, 2017; Madanu et al., 2010). It is formulated as a function of the traffic exposure factor (TEF) (the exposure of highway users to safety hazards), crash frequency factor (the probability of a vehicle being involved in a crash), and crash severity factor (the consequence of a crash) as

$$SI_s = TEF_s \cdot \sum_{r=1}^{R} [P(Crash\ Type\ r) \cdot CFF_{rs} \cdot CSF_{rs}] \tag{10.1}$$

where

- SI_s = Safety index of a highway segment or an intersection for crash severity level s
- TEF_s = Overall TEF for crash severity level s
- $P(Crash\ Type\ r)$ = Probability of type r crash occurrences
- CEF_{rs} = Overall crash frequency factor for a highway segment or an intersection measuring the risk of increasing the crash frequency for crash type r in crash severity level s
- CSF_{rs} = Overall crash severity factor for a highway segment or an intersection measuring the risk of increasing the crash severity for crash type r in crash severity level s

10.2.1 Safety issue items contributing to crash frequency and crash severity risks

Tables 10.1 and 10.2 present detailed safety issue items identified as contributing to highway segment or intersection crashes. These detailed safety issue items are associated with geometric design, consistency of design standards, pavement conditions, safety hardware conditions, and roadside features.

10.2.2 TEF by crash severity level

TEF gauges the exposure of highway users to safety hazards and it is calculated by crash severity level as

$$TEF_{seg,s} = \alpha_{seg,0} \cdot AADT^{\alpha_{seg,1}} \cdot L \tag{10.2}$$

$$TEF_{int,s} = \alpha_{int,0} \cdot AADT_{major}^{\alpha_{int,1}} \cdot AADT_{minor}^{\alpha_{int,2}} \tag{10.3}$$

where

- $AADT$ = Annual average daily traffic on a highway segment, veh/day
- L = Length of the highway segment under consideration, mi
- $AADT_{major}$ = Intersection major road AADT, veh/day
- $AADT_{minor}$ = Intersection minor road AADT, veh/day
- $\alpha_{seg,0}$, $\alpha_{seg,1}$ = Model coefficients of applicable safety performance function (SPF) to a highway segment with $\alpha_{seg,1} < 1$ to consider nonlinearity of crashes with traffic volume
- $\alpha_{int,0}$, $\alpha_{int,1}$, $\alpha_{int,2}$ = Model coefficients of applicable SPF to an intersection with $\alpha_{int,1}$, $\alpha_{int,2} < 1$ to consider nonlinearity of crashes with traffic volume
- s = Crash severity level, including fatal, injury, and PDO crashes

Table 10.1 Detailed safety issue items contributing to highway segment crashes

Category	*Type*	*Detailed safety issue item*
Geometric design	Alignment	Very severe curve realignment needed
		Inadequate sight distance on horizontal curves
		Inadequate sight distance on vertical curves
	Cross section/ Lane width	• very narrow • narrow
	Shoulder width	• no shoulder • very narrow • narrow • medium
	Passing lane	Missing passing lane
	Climbing lane	Missing climbing lane
	Rumble strips	Missing audible edge-lines
		Missing audible center-lines
	Ped crosswalks	Missing or ineffective crosswalks
	Access control	Dangerousness of accesses
		Excessive density of uncontrolled accesses
Consistency of design standards	Curve sections	Curve section design consistency level
	Tangent sections	Tangent section design consistency level
Pavement conditions	Skid resistance	Inadequate skid resistance
	Unevenness	Potholes, rutting, shoving
Traffic control and safety hardware conditions	Signs	Missing or invisible curve warning signs
		Missing or invisible no-passing signs
		Unmet night time retroreflectivity
	Lighting	No roadway lighting
		Roadway lighting off
	Pavement markings	Missing or inadequate edge-line markings
		Missing or inadequate centerline markings
		Missing or inadequate no-passing markings
		Unmet night time retroreflectivity
	Delineation	Chevron missing or ineffective on curves
Roadside features	Embankment	Unshielded embankment
	Guardrails and barriers	Embankment shielded by ineffective barrier
		Ineffective barrier for overpass bridges
		No breakaway barrier terminals
		Missing transition
		Missing or damaged reflectors
	Bridges	Inadequate bridge rails
	Ditches	Ditches in presence nearby
	Trees	Trees in presence nearby
	Utility poles	Utility poles in presence nearby
	Obstacles	Obstacles in presence nearby

10.2.3 Crash frequency factor by crash type and severity level

The crash frequency factor indicates the risk of increasing the crash frequency. The crash frequency factor by crash type r and severity level s for a highway segment, CFF_{rs}, is computed as the production of crash frequency factors by crash type r and severity level s for

Table 10.2 Detailed safety issue items contributing to highway intersection crashes

Category	*Type*	*Detailed safety issue item*
Geometric design	Alignment	Intersection angle
		Horizontal curvature index
		Vertical curve grade index
	Cross section	Major road outside shoulder width
		Major road median width
		Major road number of through lanes
		Minor road number of through lanes
		Presence of right-turn lane on the major road
	Channelization	Major road right-turn channelization
		Major road left-turn channelization
	Speed limit	Major road design speed
	Terrain	Level/rolling/mountainous
	Access control	Number of driveways on the major-road legs within the intersection range
Traffic operation characteristics	Turning traffic	Percentage of minor-road left-turn traffic during the AM and PM peak hours combined
	Truck traffic	Percent trucks during the peak hours
Traffic control and safety hardware conditions	Signs	Sign control type
		Unmet night time retroreflectivity
	Signals	Signal phasing indicator variable
		Presence of protected left-turn signal phase
	Lighting	Intersection lighting indicator variable
	Pavement markings	Missing or inadequate edge-line markings
		Missing or inadequate centerline markings
		Unmet night time retroreflectivity
Roadside features	Embankment	Unshielded embankment
	Guardrails and barriers	Embankment shielded by ineffective barrier
	Ditches	Ditches in presence nearby
	Trees	Trees in presence nearby
	Utility poles	Utility poles in presence nearby
	Obstacles	Obstacles in presence nearby
	Roadside hazard rating	Roadside hazard rating within 250 ft of the intersection on the major road

multiple safety issue types i concerning geometric design, design consistency, pavement conditions, and safety hardware conditions as below (Proctor et al., 2001; Transfund New Zealand, 2003):

$$CFF_{rs} = \prod_{i=1}^{I} CFF_{rsi} \tag{10.4}$$

with

$$CFF_{rsi} = 1 + \sum_{j=1}^{J} (W_{ij} \cdot \Delta CF_{rsij} \cdot P_{rsij})$$

where

CEF_{rs} = Overall crash frequency factor for a highway segment or an intersection with respect to crash type r and severity level s

CEF_{rsi} = Crash frequency factor concerning safety issue type i for a highway segment or an intersection with respect to crash type r and severity level s

W_{ij} = Safety issue weighting factor as the percentage of a highway segment or intersection approaches affected by detailed safety issue item j under safety issue type i

ΔCF_{rsij} = Relative increase in the crash frequency due to detailed safety issue item j under safety issue type i with respect to crash type r and severity level s

P_{rsij} = Crash frequency proportion factor as the proportion of crashes on the highway segment or at an intersection affected by detailed safety issue item j under safety issue type i with respect to crash type r and severity level s

Table 10.3 demonstrates an example of relative increase in the crash frequency caused by highway facility, and traffic control and safety hardware issues for rural two-lane roads (Hassan et al., 1996; RDMTD, 1999; Harwood et al., 2000; Hauer, 2002; Shen et al., 2004; NCHRP, 2005; Rose and Carlson, 2005; Lamm et al., 2006).

10.2.3.1 Crash frequency weighting factor

For a highway segment, the safety issue weighting factor W_{ij} for each detailed safety issue item j under safety issue type i represents the proportion of the highway segment affected by the safety issue item j. The weighting factors could be established from the following:

a. *Weighting factors for safety issue items related to geometric design, pavement conditions, and safety hardware conditions*: The weighting factor W_{rsij} for each detailed safety issue item j under safety issue type i represents the proportion of the highway segment affected by the safety issue item. First, data on individual safety items that reflect the geometric design, pavement conditions, and safety hardware conditions along the highway segment alignment need to be collected. The highway segment can be divided into multiple sections of approximately one-tenth of a mile in length. Then 0/1 values can be assigned to each section representing absence/presence of a specific safety issue item in the section. This helps determine the actual fraction of the length of the highway segment exposed to each safety issue. The ratio between the length affected by the safety issue item and the total length of the highway segment is exactly the weighting factor for the safety issue item.
b. *Weighting factors for safety issue items associated with the consistency of design standards*: For a typical highway segment, different design elements are involved with the curved sections and tangent sections in the segment. The weighting factors for curved sections and tangent sections may not be identical. Thus they are estimated separately for curved sections and tangent sections.
   1. *Weighting factor for the curved section-induced safety issue*: A consistent design of a highway will provide uniform driving conditions and thereby produce good driving performance. Also, drivers tend to change speeds according to the posted speed limits and alignment changes. If a highway segment under safety evaluation has more than one curved section, the weighting factor for the curved section-induced safety issue is computed as

$$W_{rsij\_CURVE} = \left[\sum_{m=1}^{M}(W_{rsij\_m,CURVE} \cdot L_{m,CURVE})\right] / \left(\sum_{m=1}^{M} L_{m,CURVE}\right) \tag{10.5}$$

Table 10.3 Relative increase in the crash frequency caused by highway facility, and traffic control and safety hardware issues for rural two-lane roads

Category	*Safety issue type (i)*	*Detailed safety issue item (j)*	*Affected crash type (r)*	*Related effect*	ΔCF_{rsij} (%)
Geometric design	Alignment	Very severe curve realignment needed	All	650 ft	100
		Inadequate sight distance on horizontal curves	All	650 ft	5
		Inadequate sight distance on vertical curves	All	650 ft	50
	Cross section	Lane width • Very narrow < 9 ft • Narrow < 11 ft	Run-off-road Head-on, sideswipe	Segment Segment	5–50 2–30
	Shoulder attributes	Shoulder width • No shoulder • Very narrow < 1 ft • Narrow 1–4 ft • Medium 5–8 ft	Run-off-road Run-off-road Head-on, sideswipe Head-on, sideswipe	Segment Segment Segment Segment	9–40 9–40 6–20 6–20
	Passing lane	Missing passing lane	All	Segment	33
	Climbing lane	Missing climbing lane	All	Segment	33
	Rumble strips	Missing audible edge-lines	Run-off-road	Segment	40
		Missing audible centerlines	Head-on	Segment	11
	Pedestrian crosswalks	Missing or ineffective crosswalks	Hit pedestrian	Segment	60
	Access control	Dangerousness of accesses	All	Segment	135
		Excessive density of uncontrolled accesses	All	Segment	75
Consistency of design standards	Curve sections	Curve section design consistency level	All	Segment	700
	Tangent sections	Tangent section design consistency level	All	Segment	700
Pavement conditions	Skid resistance	Inadequate skid resistance	Wet	Segment	30
	Unevenness	Potholes, rutting, shoving	All	Segment	10
Safety hardware conditions	Signs	Curve warning missing or not visible	All	200 m	10
		No-passing sign missing or not visible	Head-on	Segment	50
		Night time retroreflectivity not met	All	Segment	50
	Lighting	No roadway lighting	All	Segment	10
		Roadway lighting off	All	Segment	10
	Pavement markings	Edge-lines marking missing or inadequate	All	Segment	8
		Centerline marking missing or inadequate	All	Segment	13
		No-passing line marking missing or inadequate	Head-on	Segment	50
		Night time retroreflectivity not met	All	Segment	50
	Delineation	Chevron missing or ineffective on severe curve	All	650 ft	20

with

$$W_{rsij\_m,CURVE} = \sum_{n=1}^{N_{SC}} W_{rsij\_m,n,CURVE} / N_{SC}$$

where

W_{rsij_CURVE} = Weighting factor for safety issue item j under safety issue type i induced by all curve sections in the highway segment with respect to crash type r and severity level s

$W_{rsij\_m,CURVE}$ = Weighting factor for safety issue item j under safety issue type i induced by curve section m in the highway segment with respect to crash type r and severity level s

$W_{rsij\_m,n,CURVE}$ = Weighting factor for safety issue item j under safety issue type i induced by safety criterion n of curve section m in the highway segment with respect to crash type r and severity level s

$L_{m,CURVE}$ = Length of the mth curve section in the highway segment, ft

m = Number of curve sections in the highway segment, 1, 2, ..., M

n = Number of safety criteria considered for the evaluation, 1, 2, ..., N_{SC}

i. *Criteria for evaluating the crash risk affected by a curve section*: The following four safety criteria are utilized for evaluating the weighting factor for the crash risk affected by a curve section:
   - Curve length consistency: comparing the curve length L with the acceptable minimum curve length L_{min} and maximum curve length L_{max} for each curve section.
   - Design speed consistency: comparing the difference between the 85th percentile speed V_{85} and the design speed V_D on each curve section.
   - Operating speed consistency: comparing the difference in V_{85}s between successive curve sections in the highway segment.
   - Driving dynamic consistency: comparing the difference between the design speed V_D-based side friction $f_{s,D}$ and 85th percentile speed V_{85}-based side friction $f_{s,85}$ on each curve section. Table 10.4 depicts the three criteria.

ii. Weighting scales for $W_{rsij\_m,n,CURVE}$. The following weighting scales can be used:
   - A weight of "+1" is given to the curve section indicating the good design class
   - A weight of "0" is given to the fair design class
   - A weight of "−1" is given to the poor design class

The weight obtained from all the safety criteria $W_{rsij\_m,n,CURVE}$ are summed up and then averaged to get the weighting factor $W_{rsij\_m,CURVE}$ for each curve section. $W_{rsij\_m,CURVE} \geq 0.75$ indicates a good design class for curve sections of a highway segment; $-0.75 < W_{rsij\_m,CURVE} < 0.75$ represents a fair design class; and $W_{rsij\_m,CURVE} \leq -0.75$ shows a poor design class.

2. Weighting factor for tangent section-induced safety issue items
   i. Criteria for evaluating the crash risk caused by a tangent section. Tangent sections in a highway segment also need to be evaluated for consistency of design standards. Two criteria may be used for the evaluation:
      - Maximum length of tangent section, L_{max}
      - Minimum length of tangent section, L_{min}

For the maximum length, the tangent section must have a length less than 115 times the design speed: $TL_{max} < 115V_D$, where TL in ft and V_D in mph.

Table 10.4 Criteria for evaluating the crash risk affected by a curve section

Consistency criterion	*Design class*		
	Good	*Fair*	*Poor*
Curve length	$(3L_{min} + L_{max})/4 < L < (L_{min} + 3\ L_{max})/4$	$L_{min} < L < (3\ L_{min} + L_{max})/4$ or $(L_{min} + 3\ L_{max})/4 < L < L_{max}$	$L < L_{min}$ or $L > L_{max}$
Design speed	$\lvert V_{85,i} - V_{D,i}\rvert \leq 6.5$ mph	6.5 mph $< \lvert V_{85,i} - V_{D,i}\rvert \leq 13$ mph	$\lvert V_{85,i} - V_{D,i}\rvert > 13$ mph
Operating speed	$\lvert V_{85,i} - V_{85,i+1}\rvert \leq 6.5$ mph	6.5 mph $< \lvert V_{85,i} - V_{85,i+1}\rvert \leq 13$ mph	$\lvert V_{85,i} - V_{85,i+1}\rvert > 13$ mph
Vehicle dynamics	$f_{s,D} - f_{s,85} \geq +0.01$	$-0.04 \leq f_{s,D} - f_{s,85} < +0.01$	$f_{s,D} - f_{s,85} < -0.04$

Where
L = Length of the curve, ft
L_{min} = Minimum curve length, $L_{min} = 100 + 10(V_D - 30)$, ft
L_{max} = Maximum curve length, $L_{max} = 1$ mile for multilane, 0.5 mile for high speed two-lane, and 0.25 mile for low speed two-lane curve section
V_D = Design speed of curve section, mph
V_{85} = 85th percentile speed on curve, $V_{85} = 63.67 + 0.0007L - (18,341 + 3.56L)/R$, mph
R = Radius of the curve, ft
$f_{s,D}$ = Design speed V_D-based side friction, $f_{s,D} = 0.32745 - 0.001674V_D + 0.00000324V_D^2$
$f_{s,85}$ = 85th percentile speed V_{85}-based side friction, $f_{s,85} = \left[\left(V_{85}^2\right)/(15R)\right] - e$
e = Superelevation, $e = tan\beta$ with β being the superelevation angle of the curve

The minimum length of the tangent section needs to be checked for reverse curves that are closely spaced with deflections in opposite directions. There are two possible cases for this situation: in the first case, there is a normal crown section between the reverse curves; in the second case, there is no normal section and the pavement is continuously rotated in a plane about its axis. The minimum length of the tangent section for each case is calculated using the following (INDOT, 2002):

$$TL_{min} = \begin{cases} 0.67L_A + 0.67L_B + TR_A + 2.94V_D + TR_B, \text{normal crown} \\ 0.67L_A + 0.67L_B, \text{continuously rotated in a plane} \end{cases} \quad (10.6)$$

where
TL_{min} = Minimum length of a tangent section measured between point of tangency (PT) of the first curve and point of curvature (PC) of the section curve, ft
L_A = Superelevation runoff length for the first curve, ft
L_B = Superelevation runoff length for the second curve, ft
TR_A = Tangent runout length for the first curve, ft
TR_B = Tangent runout length for the second curve, ft

Table 10.5 classifies the superelevation runoff length and tangent runout length for different design speeds recommended by AASHTO (2001).

ii. The weighting factor for the tangent section length that potentially increases the crash risk. The weighting factor W_{rsij} for the potential safety issue item of tangent section length under the safety issue type of consistency of design standards represents the proportion of the tangent sections in the highway segment possessing inadequate lengths such that the crash risk is increased. First, the length data of each tangent section and the requirements of minimum tangent

Table 10.5 Superelevation runoff length and tangent runout length for different design speeds

Design speed (mph)	Superelevation runoff length (ft)	Tangent runout length (ft)				
		Superelevation rate (%)				
		2	4	6	8	10
15	44	44	–	–	–	–
20	59	59	30	–	–	–
25	74	74	37	25	–	–
30	88	88	44	29	–	–
35	103	103	52	34	26	–
40	117	117	59	39	29	–
45	132	132	66	44	33	–
50	147	147	74	49	37	–
55	161	161	81	54	40	–
60	176	176	88	59	44	–
65	191	191	96	64	48	38
70	205	205	103	68	51	41
75	220	220	110	73	55	44
80	235	235	118	78	59	47

length and maximum tangent length for each tangent section need to be collected. By comparing the actual length to the minimum length and maximum length for each tangent section, the list of tangent sections with inadequate length can be identified. The ratio between total length of all inadequate tangent sections and total length of all tangent sections in the highway segment is the weighting factor for the tangent length safety issue item.

For an intersection, the weighting factor W_{ij} is the ratio of the crash risk factor affected length over the total influence length of the highway intersection which can be taken as 250 ft on each intersection direction.

10.2.3.2 Relative increase in crash frequency

For safety issue items related to geometric design, pavement conditions, and safety hardware conditions, the relative increase in crash frequency ΔCF_{rsij} is related to the crash modification factor CMF_{rsij} and it can be computed as $\Delta CF_{rsij} = CMF_{rsij} - 1$, where the CMF_{rsij} can be achieved from SPFs applicable to the highway segment or intersection. For safety issue items related to the consistency of design standards, the increase in crash risk on poorly curved sections and inadequate tangent sections as compared to adequate tangents is reported as 700% (Lamm et al., 2006).

10.2.3.3 Crash frequency proportion factor

The proportion of crashes P_{rsij} affected by a detailed safety issue item j under safety issue type i by crash type r and severity level s can be established in a similar manner as the weighting factor for each detailed safety issue item.

For a highway segment, historical crash data and crash locations associated with the highway segment need to be collected. The highway segment can be divided into multiple

Table 10.6 Proportion of crashes affected by the safety issue type of consistency of design standards

Safety issue item	*Crash type affected*	*Criterion*		P_{rsij} (%)
Curve section consistency	Run-off-road, partially head-on, sideswipe	Curve length Design speed Operating speed Vehicle dynamics	Good	20
		Curve length Design speed Operating speed Vehicle dynamics	Fair	50
		Curve length Design speed Operating speed Vehicle dynamics	Poor	100
Tangent section consistency	Run-off-road	Minimum length		10
		Maximum length		10

sections. 0/1 values can be assigned to each section representing the absence/presence of a specific safety issue item in the section. If a crash by type and by severity level occurred in a highway section that is also affected by a specific safety issue item, this section would be considered affected by the specific safety issue item. The ratio between the cumulative number of sections affected by the safety issue item and the total number of sections is the proportion of crashes affected by the safety issue item.

For an intersection, historical crash data associated with the highway intersection need to be collected. 0/1 values can be assigned respective to the absence/presence of a detailed crash risk factor in each intersection approach. The ratio between the cumulative number of crashes affected by a crash risk factor and the total number of intersection crashes is the proportion of crashes affected by the detailed crash risk factor.

Table 10.6 lists ratios of crashes by type and severity affected by curve sections and tangent design standards under consistency of design standards (Cafiso et al., 2006).

10.2.4 Crash severity factor by crash type and severity level

The crash severity factor portrays the consequences of being involved in a vehicle crash. The vehicle operating speed and roadside features of a highway segment or an intersection play a major role in calculating this factor. The crash severity factor by crash type r and by crash severity level s, CSF_{rs}, is computed as the production of crash severity factors by crash type r and by crash severity level s for multiple safety issue types i concerning roadside features. The crash severity factor needs to be adjusted to represent the operating conditions by taking the ratio between the 85th percentile speed and the speed limit. For each identified roadside safety issue type i, multiple detailed safety issue items j might be involved. The crash severity factor related to roadside safety issues type i is calculated by combining the safety issue weighting factor (W), relative change in the crash severity (ΔCS), and crash proportion factor (P) for all constituent safety issues items j.

$$CSF_{rs} = (V_{85}/V_{SL}) \cdot \prod_{i=1}^{I} CSF_{rsi\_RSH} \tag{10.7}$$

with

$$CSF_{rsi\_RSH} = 1 + \sum_{j=1}^{J} (W_{ij\_RSH} \cdot \Delta CS_{rsij\_RSH} \cdot P_{rsij\_RSH})$$

where

CSF_{rs} = Overall crash severity factor for a highway segment or an intersection with respect to crash type r and severity level s

V_{85} = 85th percentile of speed distribution weighted along the highway segment or intersection approaches, mph

V_{SL} = Speed limit distribution weighted along the segment or intersection approaches, mph

CSF_{rsi_RSH} = Roadside safety issue type i induced crash severity factor for the highway segment or intersection with respect to crash type r and severity level s

W_{ij_RSH} = Roadside safety hazard-related safety issue weighting factor as the percentage of highway segment or intersection approaches affected by detailed safety issue item j under roadside safety issue type i

ΔCS_{rsij_RSH} = Relative increase in the crash severity risk caused by detailed safety issue item j under roadside safety issue type i with respect to crash type r and severity level s

P_{rsij_RSH} = Crash severity proportion factor as the proportion of crashes on the highway segment or intersection approaches affected by safety issue item j under roadside safety issue type i with respect to crash type r and severity level s

In the analysis process, the injury crash severity factor for each crash type is first computed. The fatal crash severity factor and PDO crash severity factor for each roadside safety issue item for each type of crash are derived based on the injury crash severity factor for the same detailed safety issue item j under roadside safety issue type i as below:

$$CSF_{rIi\_RSH} = 1 + \sum_{j=1}^{J} (W_{ij\_RSH} \cdot \Delta CS_{rIij\_RSH} \cdot P_{rIij\_RSH}) \tag{10.8}$$

$$CSF_{rF/Pi\_RSH} = 1 + \sum_{j=1}^{J} [W_{ij\_RSH} \cdot (\Delta CS_{rIij\_RSH} \cdot (1 + \Delta S_{rF/Pij\_RSH})) \cdot P_{rIij\_RSH}] \tag{10.9}$$

where

CSF_{rIi_RSH} = Roadside safety issue type i induced injury crash severity factor for a highway segment or an intersection with respect to crash type r

CSF_{rF/Pi_RSH} = Roadside safety issue type i induced fatal/PDO crash severity factor for a highway segment or an intersection with respect to crash type r

W_{ij_RSH} = Roadside safety hazard-related safety issue weighting factor as the percentage of highway segment or intersection approaches affected by detailed safety issue item j under safety issue type i

ΔCS_{rIij_RSH} = Relative increase in injury crash risks for crash type r caused by detailed safety issue item j under roadside safety issue type i

Table 10.7 Relative increases in crash severity for rural two-lane roads caused by roadside issues

Roadside safety issue type (i)	Detailed safety issue item (j)	Affected crash type (r)	Related effect	Increase in injury crash severity (%) ΔCS_{rIij_RSH}	Additional increase in fatal/PDO crash severity (%) ΔCS_{rFij_RSH}	ΔCS_{rPij_RSH}
Embankment	Unshielded embankment					
	• 9 < height < 18 ft, grade > 0.5%	Run-off-road	Segment	80	800	800
	• height > 18 ft, grade > 0.5%	Run-off-road	Segment	100	1400	1400
Guardrails and barriers	Embankment shielded					
	• 9 < height < 18 ft, grade > 0.5%	Run-off-road	Segment	10	70	100
	• height > 18 ft, grade > 0.5%	Run-off-road	Segment	11	100	100
	Ineffective barrier for overpass bridges	Run-off-road	30 m	60	70	100
	No breakaway barrier terminals and transitions	Run-off-road	30 m	60	300	300
	Missing transition	Run-off-road	30 m	60	300	300
	Reflectors missing or damaged	Run-off-road	30 m	8	0	100
Bridges	Inadequate bridge rails	Run-off-road	30 m	6	2000	2000
Ditches	Ditches located less than 9 ft	Run-off-road	Segment	50	150	150
Trees	Trees located < 9 ft	Run-off-road	60 m	90	1000	1000
Utility poles	Utility poles located < 9 ft	Run-off-road	60 m	90	1000	1000
Obstacles	Obstacles located < 9 ft	Run-off-road	60 m	90	1000	1000

P_{rIij_RSH} = Crash frequency proportion factor as the proportion of injury crashes on the highway segment or intersection affected by detailed safety issue item j under roadside safety issue type i with respect to crash type r

$\Delta S_{rF/Pij\_RSH}$ = Net increase in fatal/PDO crash risk over injury crash risk for crash type r caused by detailed safety issue item j under roadside safety issue type i

Table 10.7 details an example of relative increase in the crash severity risk as a result of roadside safety issues for rural two-lane roads (Harwood et al., 2000; Hauer, 2002; Shen et al., 2004; Lamm et al., 2006).

10.2.4.1 Roadside hazard-related crash severity weighting factor

The roadside hazard-related safety issue weighting factor W_{ij_RSH} represents the proportion of the highway segment affected by each detailed safety issue item j under roadside safety issue type i. It is determined similarly as the W_{ij} for the crash frequency factor.

10.2.4.2 Relative increase in injury and fatal/PDO crash risks ΔCS_{rIij_RSH} and $\Delta S_{rF/Pij\_RSH}$

The relative increase in injury crashes ΔCF_{rIij} caused by roadside safety issue items is related to the crash modification factor for injury crashes CMF_{rsij} and it can be computed as $\Delta CF_{rIij} = CMF_{rIij} - 1$, where the CMF_{rIij} can be established from SPFs applicable to the highway segment or intersection.

Because the vehicle crash rate for a highway segment in terms of crashes per VMT (or vehicle crash rate for an intersection in terms of crashes per entering vehicle) reflects the crash risk encountered by each traveler making a unit distance of travel (or entering an intersection), it can therefore be used to calculate the relative increase in fatal and PDO crash risks over the injury crash risk. The relative increase in fatal crash risk ΔS_{rFij_RSH} over the injury crash risk due to a roadside safety issue item can be measured as the percentage change in the fatal and injury crash rate over injury crash rate using only historical crash data. Similarly, the relative increase in PDO crash risk ΔS_{rPij_RSH} over the injury crash risk can be computed as the percentage change in the injury and PDO crash rate over injury crash rate using only historical crash data.

10.2.4.3 Injury crash proportion factor

The injury crash proportion factor P_{rIij_RSH} represents the proportion of injury crashes in the highway segment affected by a detailed safety issue item j under roadside safety issue type i. It is computed similarly to the P_{rsij} for the crash frequency factor.

10.2.5 SI by crash type and by severity

The SI is computed as the production of TEF, crash frequency factor, and crash severity factor. The TEF for a highway segment or an intersection measures the extent of highway users exposed to road hazards when traversing the highway segment or intersection.

The crash frequency factor by crash type and by severity level considers detailed safety issue items under various safety issue types that can be grouped into safety issue categories regarding geometric design, consistency of design standards, pavement conditions, and safety hardware conditions. The crash frequency factors for multiple safety issue items under a specific safety issue type can be added to obtain the safety issue type-specific crash frequency factor. The overall crash frequency factor for the highway segment or intersection by crash type and by crash severity is established as the production of crash frequency factors corresponding to multiple safety issue types.

The crash severity factor by crash type and by severity level considers detailed safety issue items in the roadside safety issue category. The crash severity factors for multiple safety issue items under a specific roadside safety issue type can be added to obtain the roadside safety issue type-specific crash severity factor. The overall crash severity factor for the highway segment or intersection by crash type and by crash severity is established as the production of crash severity factors corresponding to multiple roadside safety issues types can be computed to establish the overall crash severity factor for the highway segment or intersection by crash type and by crash severity.

10.2.6 Crash occurrence probability by crash type

The production of estimated TEF, crash frequency factor, and crash severity factor by crash type and crash severity level is the SI for the highway segment or intersection by crash type and crash severity level. To establish the SI for the highway segment or intersection only by severity level, the probability of vehicle crash occurrences by crash type needs to be determined.

The distribution of vehicle crashes on a highway segment or within an intersection by crash type can be readily calculated using historical crash data. Such a distribution reflects the crash occurrences in the past and only represents the crash possibility distribution. Thus it cannot be directly used as the crash probability distribution to predict specific types of crashes that can be expected in the future. To estimate the SI value for a highway segment or an intersection by fatal, injury, and PDO crash severity level, the crash possibility distribution needs to be transformed into crash probability distribution. The transformation can be achieved based on the uncertainty invariance principle in information theory (Klir and Parviz, 1992; Klir, 2006). Specifically, the generalized Hartley measure of the possibility profile is set equal to the Shannon entropy of the derived probability distribution. The commonly used possibility–probability transformation approach is log-interval transformation discussed as follows:

Given the distribution of crash occurrences by crash type $CT = \{CT_1, CT_2, \ldots, CT_n\}$, the Mobius function $m = \langle m_1, m_2, \ldots, m_n \rangle$ is defined as

$$m_i = \sum_{k=1}^{i} CT_k / \sum_{l=1}^{n} \sum_{k=1}^{l} CT_k \ (i, k, l = 1, 2, \ldots, n) \tag{10.10}$$

The possibility profile defined as $r = \langle r_1, r_2, \ldots, r_n \rangle$ $(1 = r_1 \geq r_2 \geq \cdots \geq r_n \geq 0)$ can be computed by

$$r_i = \sum_{k=i}^{n} m_k \tag{10.11}$$

The possibility profile $r = \langle r_1, r_2, \ldots, r_n \rangle$ can be used to derive a probability distribution $p = \langle p_1, p_2, \ldots, p_n \rangle$ using possibility–probability transformation approaches. Without loss of generality, the log-interval scaling transformation can be employed to accomplish this effort (Klir, 2006). The transformation covers the following computational steps:

- Compute generalized Hartley measure $GH(r) = \sum_{i=2}^{n} \left[r_i \cdot log_2^{(i/(i-1))} \right]$
- Determine α from $GH(r) = S(p)$, where Shannon entropy $S(p) = -\sum_{i=1}^{n} (p_i \cdot log_2^{p_i})$ and $p_i = r_i^{(1/\alpha)} / \sum_{l=1}^{n} (r_l^{(1/\alpha)})$
- Obtain probability distribution $p_i = r_i^{(1/\alpha)} / \sum_{l=1}^{n} (r_l^{(1/\alpha)})$ using the derived α value

EXAMPLE 10.1

The possibility distribution of four types of crashes is given as 30% head-on, 20% sideswipe, 40% fixed object, 10% run-off-road, reorder it to $CT = \{CT_1 = 40\%, CT_2 = 30\%, CT_3 = 20\%, CT_4 = 10\%\}$. Perform possibility–probability transformation using the log-interval scaling transformation.

Solution

The Mobius function $m = \langle m_1, m_2, m_3, m_4 \rangle$ is obtained as

$$m_1 = \sum_{k=1}^{1} CT_k / \sum_{l=1}^{4} \sum_{k=1}^{l} CT_k$$

$$= \frac{40\%}{[40\% + (40\% + 30\%) + (40\% + 30\% + 20\%) + (40\% + 30\% + 20\% + 10\%)]} = 0.1333$$

$$m_2 = \sum_{k=1}^{2} CT_k / \sum_{l=1}^{4}\sum_{k=1}^{l} CT_k$$

$$= \frac{(40\% + 30\%)}{[40\% + (40\% + 30\%) + (40\% + 30\% + 20\%) + (40\% + 30\% + 20\% + 10\%)]} = 0.2333$$

$$m_3 = \sum_{k=1}^{3} CT_k / \sum_{l=1}^{4}\sum_{k=1}^{l} CT_k$$

$$= \frac{(40\% + 30\% + 20\%)}{[40\% + (40\% + 30\%) + (40\% + 30\% + 20\%) + (40\% + 30\% + 20\% + 10\%)]} = 0.3000$$

$$m_4 = \sum_{k=1}^{4} CT_k / \sum_{l=1}^{4}\sum_{k=1}^{l} CT_k$$

$$= \frac{(40\% + 30\% + 20\% + 10\%)}{[40\% + (40\% + 30\%) + (40\% + 30\% + 20\%) + (40\% + 30\% + 20\% + 10\%)]} = 0.3333$$

The possibility profile $r = \langle r_1, r_2, ..., r_n \rangle$ can be computed as

$$r_1 = m_1 + m_2 + m_3 + m_4 = 0.1333 + 0.2333 + 0.3000 + 0.3333 = 1.0$$

$$r_2 = m_2 + m_3 + m_4 = 0.2333 + 0.3000 + 0.3333 = 0.8667$$

$$r_3 = m_3 + m_4 = 0.3000 + 0.3333 = 0.6333$$

$$r_4 = m_4 = 0.3333$$

with the possibility profile $r = \langle 1, 0.8667, 0.6333, 0.3333 \rangle$ in place, the log-interval scaling transformation can be accomplished as follows:

$$GH(r) = \sum_{i=2}^{4}\left[r_i \cdot log_2^{(i/(i-1))}\right] = 0.8667 log_2^{(2/1)} + 0.6333 log_2^{(3/2)} + 0.3333 log_2^{(4/3)} = 1.3755$$

with

$$S(p) = -[(1/s) \cdot log_2^{(1/s)} + (0.8667^{(1/\alpha)}/s) \cdot log_2^{(0.8667^{(1/\alpha)}/s)} + (0.6333^{(1/\alpha)}/S) \cdot log_2^{(0.6333^{(1/\alpha)}/s)} + (0.3333^{(1/\alpha)}/s) \cdot log_2^{(0.3333^{(1/\alpha)}/s)}] = 1.3755$$

where $s = 1 + 0.8667^{(1/\alpha)} + 0.6333^{(1/\alpha)} + 0.3333^{(1/\alpha)}$, the value of α is then obtained as $\alpha = 0.181$.

Hence, $s = 1 + 0.8667^{(1/0.181)} + 0.6333^{(1/0.181)} + 0.3333^{(1/0.181)} = 1.5362$. This helps derive the corresponding probabilities as

$$p_1 = 1/1.5362 = 0.6509$$

$$p_2 = 0.8667^{(1/0.181)}/1.5362 = 0.2953$$

$$p_3 = 0.6333^{(1/0.181)}/1.5362 = 0.0522$$

$$p_4 = 0.3333^{(1/0.181)}/1.5362 = 0.0015$$

Therefore, the probabilities for head-on, sideswipe, fixed object, and run-off-road crashes are 0.2953, 0.0522, 0.6509, and 0.0015.

10.2.7 SI by crash severity level

The traffic exposure factor TEF_s, probability distribution of crash occurrences by crash type P (Crash Type r), crash frequency factor by crash type and severity level CFF_{rs}, and crash severity factor by crash type and severity level CSF_{rs} can be synthesized using Equation 10.1 to compute the SI by crash severity level SI_s.

10.2.8 Validation of the estimated SI value

Compared with the empirical Bayesian (EB) method for crash predictions, the SI method is a disaggregated method that simultaneously considers crash contributing factors associated with highway geometric design, consistency of geometric design standards, pavement conditions, and traffic control and safety hardware conditions. It could capture the subtle impacts on safety performance caused by changes in conditions of different types of traffic control and safety hardware. Therefore, it is well suited for estimating the PSIs resulting from traffic control and safety hardware project implementation.

However, the SI estimate needs to be validated before being used for traffic control and safety hardware project evaluation. The benchmark crash frequency estimate is the EB crash estimate calculated as a weighted sum of multiyear crash records and crash frequencies predicted by a highway segment SPF in conjunction with crash modification factors to address site-specific conditions. The EB estimate corrects for regression-to-mean bias for crash predictions, making it a reliable benchmark value for validating the SI estimate (Hauer, 1997).

$$EB_{seg,i} = w_{seg,i} \times CF_{seg,i,P} + (1 - w_{seg,i}) \times CF_{seg,i,O} \tag{10.12}$$

$$EB_{int,i} = w_{int,i} \times CF_{int,i,P} + (1 - w_{int,i}) \times CF_{int,i,O} \tag{10.13}$$

with

$$w_{seg,i} = \frac{1}{1 + \left[k_{seg} \times (T \times L_{seg,i}) \times \sum_{t=1}^{T} CF_{seg,i,P,t} \right]}$$

$$w_{int,i} = \frac{1}{1 + \left[k_{int} \times T \times \sum_{t=1}^{T} CF_{int,i,P,t} \right]}$$

where

$EB_{seg,i}$, $EB_{int,i}$ = EB-adjusted multiyear crash frequency for segment or intersection i

$w_{seg,i}$, $w_{int,i}$ = Weighting factor between SPF predicted and field observed multiyear crash frequencies for segment or intersection site i

k_{seg}, k_{int} = Overdispersion parameter of the crash frequency for highway segment per mile per year or for intersection per year

$CF_{seg,i,P,t}$, $CF_{int,i,P,t}$ = SPF-predicted crash frequency for segment or intersection site i in year t

$CF_{seg,i,P}$, $CF_{int,i,P}$ = SPF-predicted multiyear crash frequency with further adjustments according to the crash modification factors for segment or intersection site i

$CF_{seg,i,O}$, $CF_{int,i,O}$ = Field observed multiyear crash frequency for segment or intersection site i

t = 1, 2, …, T

Table 10.8 Statistical measure and tests for validating the SI estimates

Statistical measure/test	*Description*
Root mean square error	$$RMSE_{SI} = \sqrt{\sum_{i=1}^{N}[(SI_i - CF_{i,O})^2]/N}$$ $$RMSE_{EB} = \sqrt{\sum_{i=1}^{N}[(EB_i - CF_{i,O})^2]/N}$$ where, $RMSE_{SI}$ = Root mean square error of *SI* estimates compared with field observed crashes for multiple highway segments or intersections; $RMSE_{EB}$ = Root mean square error of *EB* estimates compared with field observed crashes for multiple highway segments or intersections; SI_i, EB_i, $CF_{i,O}$ = *SI* value, *EB* estimate, and field observed crash frequency of highway segment or intersection *i*; and N = Number of highway segments or intersections Conclusion: A smaller RMSE value confirms a higher level of consistency between the estimated and field observed crashes
Spearman's rank correlation test	Null hypothesis H_0: The ranks of SI and EB crash estimates sorted in descending order are not correlated. Reject H_0 if standard normal z-statistic $\geq z\alpha_{,(N-1)}$ $$z = \left[1 - \frac{6\sum_{i=1}^{N} D_i^2}{N\cdot(N^2-1)}\right]\cdot\sqrt{N-1}$$ where, D_i = The difference between ranks of SI and EB crash estimates for highway segment or intersection *i* sorted in descending order; and N = Number of highway segments or intersections
Mann–Whitney *U* test	Null hypothesis H_0: The two samples of EB and SI crash frequency estimates came from the same population. Reject H_0 if standard normal z-statistic $\lvert z\rvert \geq z\alpha_{,(N-1)}$ $$z = \frac{(T-\mu)\pm 0.5}{\sigma}\ (\text{“}-0.5\text{” if } T \geq \mu,\ \text{“}+0.5\text{” if } T < \mu)$$ with $$T = min\left(T_{SI} = \sum_{i=1}^{N} Rank(SI_i), T_{EB} = \sum_{i=1}^{N} Rank(EB_i)\right)$$ $$\mu = [N\cdot S(2N+1)]/2,\ \sigma = N\cdot\sqrt{(2N+1)/12}$$ where, T_{SI} = Summation of ranks of all *SI* crash frequency estimates; T_{EB} = Summation of ranks of all *EB* crash frequency estimates, SI_i, EB_i = *SI* and *EB* crash frequency estimates for highway segment or intersection *i*; and N = Number of highway segments or intersections
Chi-square test	Null hypothesis H_0: The difference in EB and SI estimates is statistically insignificant. Reject H_0 if χ^2-statistic $\geq \chi^2_{\alpha,(N-1)}$ $$\chi^2 = \sum_{i=1}^{N}\left[\frac{(SI_i - EB_i)^2}{EB_i}\right]$$ where SI_i, EB_i = *SI* and *EB* crash frequency estimates for highway segment or intersection *i*; and N = Number of highway segments or intersections

With SI and EB estimates calculated for multiple highway segments or intersections, statistical methods can be adopted to assess the level of agreements between the two sets of crash frequency estimates. In general, the root mean square error (RMSE) measure and three statistical tests can be utilized to examine the consistency of SI and EB estimates for validation: Spearman's rank correlation test, Mann–Whitney U test, and Chi-square test. The RMSE is a frequently used measure that serves to aggregate differences between values predicted by a model or an estimator and the values actually observed from the thing being modeled or estimated into a single measurement of predictive power (Kutner et al., 2004). Spearman's rank correlation coefficient is a measure of association between the ranking of SI and EB crash estimates calculated on each highway segment (Siegel and Castellan, 1988). The Mann–Whitney U test (also known as the Mann–Whitney–Wilcoxon test, Wilcoxon rank-sum test, or Wilcoxon–Mann–Whitney test) is a nonparametric test for assessing whether two samples of observations come from the same distribution (McCullagh and Nelder, 1989). The Chi-square test is a statistical test commonly used to compare observed data with data one would expect to obtain under the hypothesis of no significant difference between the expected and observed values (Kutner et al., 2004). Table 10.8 shows the RMSE calculation and test details.

10.3 TRAFFIC CONTROL AND SAFETY HARDWARE PROJECT EVALUATION

10.3.1 Annual PSIs

Once the SI values are validated, the proposed method for computing the current-case SI values can then be readily used for traffic control and safety hardware project evaluation. After project implementation to a highway segment or an intersection, the traffic control and safety hardware condition is improved to the system-wide average level. The crash frequency factor and the crash severity factor concerning the detailed traffic control and safety hardware issue items are expected to decrease from the existing values of greater than one to a value of one. This will eventually reduce the overall crash frequency factor and the crash severity factor for the entire highway segment or intersection, leading to reductions in SI values for fatal, injury, and PDO crashes.

The reduction in the SI values after implementing the traffic control and safety hardware project (base-case SI) from the SI values associated with the existing traffic control and safety hardware (current-case SI) by crash severity level is treated as the $PSIs$. The current-case and base vehicle crash rates for a highway segment or an intersection can be calculated as

$$CR_{seg,i,s}^{current} = \frac{SI_{seg,i,s}^{current}}{\left(365 \times 10^{-6} \times AADT_{seg,i}^{current}\right)} \tag{10.14}$$

$$CR_{seg,i,s}^{base} = \frac{SI_{seg,i,s}^{base}}{\left(365 \times 10^{-6} \times AADT_{seg,i}^{base}\right)} \tag{10.15}$$

$$CR_{int,i,s}^{current} = \frac{SI_{int,i,s}^{current}}{\left[365 \times 10^{-6} \times \left(AADT_{int,major,i}^{current} + AADT_{int,minor,i}^{current}\right)\right]} \tag{10.16}$$

$$CR_{int,i,s}^{base} = \frac{SI_{int,i,s}^{base}}{\left[365 \times 10^{-6} \times \left(AADT_{int,major,i}^{base} + AADT_{int,minor,i}^{base}\right)\right]} \tag{10.17}$$

where

Annual $PSI_{seg,i,s}$ = Annual potential for safety improvements for highway segment i by crash severity level s, crashes/seg/year

Annual $PSI_{int,i,s}$ = Annual potential for safety improvements for intersection i by crash severity level s, crashes/int/year

$CR_{seg,i,s}^{current}$ = Current-case crash rate for highway segment i by crash severity level s, crashes/million VMT

$CR_{seg,i,s}^{base}$ = Base-case crash rate for highway segment i by crash severity level s, crashes/million VMT

$CR_{int,i,s}^{current}$ = Current-case crash rate for intersection i by crash severity level s, crashes/million entering vehicles (EV)

$CR_{int,i,s}^{base}$ = Base-case crash rate for intersection i by crash severity level s, crashes/million EV

$SI_{seg,i,s}^{current}$ = Current-case SI value for highway segment i by crash severity level s, crashes/seg/year

$SI_{seg,i,s}^{base}$ = Base-case SI value for highway segment i by crash severity level s, crashes/seg/year

$SI_{int,i,s}^{current}$ = Current-case SI value for intersection i by crash severity level s, crashes/int/year

$SI_{int,i,s}^{base}$ = Base-case SI value for intersection i by crash severity level s, crashes/int/year

$AADT_{seg,i}^{current}$ = Current-case annual average daily traffic ($AADT$) for highway segment i, veh/day

$AADT_{seg,i}^{base}$ = Base-case $AADT$ for highway segment i, veh/day

$AADT_{int,major,i}^{current}$ = Current-case $AADT$ on major approaches for intersection i, veh/day

$AADT_{int,minor,i}^{current}$ = Current-case $AADT$ on minor approaches for intersection i, veh/day

$AADT_{int,major,i}^{base}$ = Base-case $AADT$ on major approaches for intersection i, veh/day

$AADT_{int,minor,i}^{base}$ = Base-case $AADT$ on minor approaches for intersection i, veh/day

L = Length of highway segment i, mi

$MVMT_{seg,i}^{current}$ = Current-case annual million VMT for highway segment i

$MVMT_{seg,i}^{base}$ = Base-case annual million VMT for highway segment i

$MEV_{int,i}^{current}$ = Current-case annual million EVs for intersection i

$MEV_{int,i}^{base}$ = Base-case annual million EVs for intersection i

s = Crash severity category, including fatal, injury, and PDO crashes

As seen in Figure 10.2, the annual PSIs for a highway segment or an intersection achieved by addressing all needed traffic control and safety hardware improvements can be computed using the concept of changes in consumer surplus (Miller, 1992; AASHTO, 2003).

$$\text{Annual } PSI_{seg,i,s} = \frac{1}{2} \times \left(CR_{seg,i,s}^{current} - CR_{seg,i,s}^{base}\right) \times \left(MVMT_{seg,i}^{current} + MVMT_{seg,i}^{base}\right) \tag{10.18}$$

$$\text{Annual } PSI_{int,i,s} = \frac{1}{2} \times \left(CR_{int,i,s}^{current} - CR_{int,i,s}^{base}\right) \times \left(MEV_{int,i}^{current} + MEV_{int,i}^{base}\right) \tag{10.19}$$

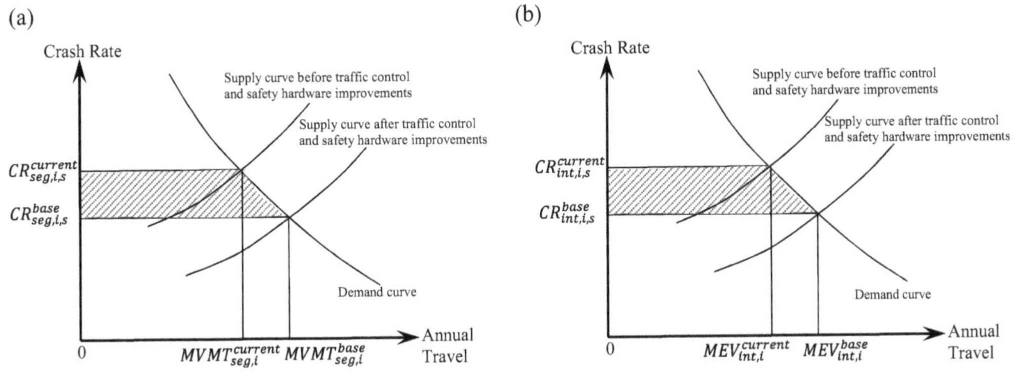

Figure 10.2 Illustration of annual PSI calculation. (a) Annual PSIs for a highway segment. (b) Annual PSIs for an intersection. (Adapted from Li, Z. and S. Madanu. 2008. Report MRUTC 08-06. Department of Civil, Architectural, and Environmental Engineering, Illinois Institute of Technology, Chicago, IL; Madanu, S. et al. 2010. *ASCE Journal of Transportation Engineering* 136(2), 129–140; Li, Z. et al. *TRB Journal of Transportation Research Record* 2160, 1–11.)

with

$$MVMT^{current}_{seg,i,s} = 365 \times 10^{-6} \times AADT^{current}_{seg,i} \times L$$

$$MVMT^{base}_{seg,i,s} = 365 \times 10^{-6} \times AADT^{base}_{seg,i} \times L$$

$$MEV^{current}_{int,i,s} = 365 \times 10^{-6} \times \left(AADT^{current}_{int,major,i} + AADT^{current}_{int,minor,i}\right)$$

$$MEV^{base}_{int,i,s} = 365 \times 10^{-6} \times \left(AADT^{base}_{int,major,i} + AADT^{base}_{int,minor,i}\right)$$

10.3.2 Project-level annual safety improvement benefits

The annual PSIs for fatal, injury, and PDO crashes reflect the highest extent of crash reductions that could be achieved per year by implementing all required traffic control and safety hardware improvements for a specific highway segment or an intersection. The annual PSIs for fatal injury and PDO crashes could then be converted to dollar values using the respective unit crash costs and then combined to arrive at the annual benefits of traffic control and safety hardware improvements for the highway segment or intersection.

$$\text{Annual } B^{TCSH}_{seg,i} = \sum_{s=1}^{3} (\text{Annual } PSI_{seg,i,s} \cdot UCC_s) \tag{10.20}$$

$$\text{Annual } B^{TCSH}_{int,i} = \sum_{s=1}^{3} (\text{Annual } PSI_{int,i,s} \cdot UCC_s) \tag{10.21}$$

where

Annual $B^{TCSH}_{seg,i}$ = Annual safety benefits from traffic control and safety hardware improvements for highway segment i, dollars/seg/year

Annual $B^{TCSH}_{int,i}$ = Annual safety benefits from traffic control and safety hardware improvements for intersection i, dollars/int/year

UCC_s	= Unit crash cost for crash severity level s, dollars/crash
s	= Vehicle crash severity level, including fatal, injury, and PDO crashes

10.3.3 Project-level safety improvement benefits in traffic control and safety hardware life cycle

In the service life cycle of the longest-lasting traffic control and safety hardware installed on a highway segment or within an intersection, a geometric annual growth rate of safety benefits for project implementation may be assumed to keep abreast of the annual traffic growth. The project-level safety benefits can be computed as

$$PW_{LCB_{seg,i}^{TCSH}} = \text{Annual } B_{seg,i}^{TCSH} \cdot \left[\frac{(e^{(g-i)\cdot N} - 1)}{(g-i)}\right] \quad (10.22)$$

$$PW_{LCB_{int,i}^{TCSH}} = \text{Annual } B_{int,i}^{TCSH} \cdot \left[\frac{(e^{(g-i)\cdot N} - 1)}{(g-i)}\right] \quad (10.23)$$

$$EUA_{LCB_{seg,i}^{TCSH}} = PW_{LCB_{seg,i}^{TCSH}} \cdot \left[\frac{i\cdot(1+i)^N}{(1+i)^N - 1}\right] \quad (10.24)$$

$$EUA_{LCB_{int,i}^{TCSH}} = PW_{LCB_{int,i}^{TCSH}} \cdot \left[\frac{i\cdot(1+i)^N}{(1+i)^N - 1}\right] \quad (10.25)$$

where

$PW_{LCB_{seg,i}^{TCSH}}$	= Present worth of project-level life-cycle safety benefits for highway segment i, dollars
$PW_{LCB_{int,i}^{TCSH}}$	= Present worth of project-level life-cycle safety benefits for intersection i, dollars
$EUA_{LCB_{seg,i}^{TCSH}}$	= Equivalent uniform annualized project-level life-cycle safety benefits for highway segment i, dollars/year
$EUA_{LCB_{int,i}^{TCSH}}$	= Equivalent uniform annualized project-level life-cycle safety benefits for intersection i, dollars/year
g	= Geometric gradient of annual safety benefit growth
i	= Discount rate
N	= Service life-cycle of the longest lasting traffic control and safety hardware

10.3.4 Traffic control and safety hardware project costs

The useful service lives of different types of traffic control and safety hardware vary considerably. The useful service lives of some types of traffic control and safety hardware could be much shorter than the useful service life of the longest-lasting traffic control and safety hardware. For instance, the service life of pavement markings is much shorter than that of guardrails. For the same type of hardware, the service life is affected significantly by material type. For example, the useful service life of pavement markings varies greatly with the use of different types of materials. In addition, different parts of the hardware could maintain highly different life spans. For instance, the sheeting, post, and concrete foundation of a traffic sign vary widely in their longevity. Moreover, some types of hardware, such as traffic signs, crash cushions, and guardrails, may be damaged whenever vehicle crashes occur, leading to early termination of service lives.

When estimating the safety hardware improvement costs in the service life of the longest-lasting traffic control and safety hardware, multiple rounds of safety hardware improvement costs might need to be considered. The multiple project costs occurring at different points in time of the longest-lasting hardware service life can be treated on a case-by-case basis. The multiple costs can be discounted to establish the present worth cost. The present worth cost can be spread across the longest-lasting hardware service life to establish the equivalent uniform annualized costs.

10.3.5 Traffic control and safety hardware project benefit-cost analysis

The project-level equivalent uniform annualized safety improvement benefits and project costs can be utilized to compute the benefit-to-cost ratio and net present worth to justify the economic feasibility of the traffic control and safety hardware projects from project efficiency and effectiveness perspectives.

Case Study 1: Evaluation of highway segment traffic control and safety improvements

A county in the U.S. Midwest region is approximately 70% urban. It has a good mix of highways in different functional classes. The state-maintained highway network in the county is divided into 193 highway segments according to the homogeneity of geometric design and traffic characteristics. For the data set provided for the period 2002–2006, data details are available for crashes, segment length, traffic volume, geometric design, posted speed limit, cross section, pavement and bridge conditions, conditions of traffic control and safety hardware including signs, lighting, pavement markings, and guardrails, and roadside features. Table 10.9 provides some basic information on the data.

a. *SI estimates for individual highway segments*: For each of the 193 highway segments, the SI value is separately computed for each year. Applicable SPFs are identified and the model coefficients are adopted for estimating the TEFs. Detailed safety issue items for the highway segments are identified based on available data. The safety issue item weighting factors, increases in the crash frequency and severity risks, and crash proportion factors are used for estimating the crash frequency factors and crash severity factors. Table 10.10 illustrates a detailed SI computation for a rural interstate highway segment.
b. *Consistency comparisons of SI and EB estimates*: To validate the SI estimates established by the proposed method, the EB estimates for fatal, injury, and PDO crashes are also computed for each highway segment as the weighted sum of observed and predicted crashes. The EB and SI estimates for the 193 highway segments are then compared for the level of agreements. As shown in Table 10.11, the Spearman's rank correlation test results for each five-year analysis period maintain the lowest z-statistic values of 11.26 for fatal, 12.58 for injury, and 12.76 for PDO crashes. This indicates that the ranking of SI and EB estimates for all highway segments agrees at 99% confidence level. The Mann–Whitney test generally produces similar results, with the one exception that the ranking of fatal crashes in 2003 had a lower confidence level. The level of agreements in the magnitude of EB and SI estimates was further evaluated by the Chi-square test and regression analysis. Further listed in Table 10.5, the highest Chi-square statistical values for each five-year analysis period are 1.15 for fatal crashes, 13.71 for injury crashes, and 45.62 for PDO crashes. All values are lower than the Chi-square critical value of 240.50 for 99% confidence level. This demonstrates that the differences in EB and SI estimates for fatal, injury, and PDO crashes are insignificant.

Source: Adapted from Li, Z. et al. 2010. *TRB Journal of Transportation Research Record* **2160**, 1–11.

Table 10.9 Case study 1 data summary statistics

Data item		*Urban area* Interstate	*Urban area* Multilane divided	*Urban area* Multilane undivided	*Urban area* Two-lane	*Rural area* Interstate	*Rural area* Multilane divided	*Rural area* Two-lane
Segments	No.	31	58	3	45	22	27	7
Segment lengths (std. dev.)	Mile	0.9581 (0.6625)	0.3355 (0.3522)	0.5700 (0.4034)	0.4402 (0.3577)	1.1564 (0.3028)	0.8930 (0.3965)	0.9657 (0.4179)
AADT, veh/day (std. dev.)	2002	46928 (11164)	16435 (5770)	13348 (6628)	10126 (3065)	26907 (1141)	9735 (910)	7469 (421)
	2003	47536 (11280)	16604 (5843)	13480 (6696)	10213 (3112)	27186 (1239)	9836 (925)	7553 (440)
	2004	48153 (11399)	16774 (5918)	13613 (6765)	10302 (3159)	27468 (1341)	9939 (943)	7638 (460)
	2005	48778 (11519)	16947 (5994)	13748 (6835)	10390 (3209)	27754 (1446)	10043 (962)	7724 (480)
	2006	49411 (11641)	17121 (6072)	13884 (6906)	10480 (3259)	28043 (1556)	10148 (983)	7811 (501)
Fatal crashes	2002	0	1	0	1	0	0	0
	2003	0	1	0	1	0	0	0
	2004	0	1	0	1	0	0	0
	2005	0	1	0	1	0	0	0
	2006	0	1	0	1	0	0	0
	Total	0	6	0	6	0	0	0
Injury crashes	2002	18	32	7	30	7	1	3
	2003	18	32	4	31	4	4	2
	2004	26	30	6	33	10	7	5
	2005	24	39	8	35	13	5	3
	2006	22	31	6	29	9	3	4
	Total	108	164	31	158	43	20	17
PDO crashes	2002	58	101	23	97	22	5	8
	2003	57	102	14	99	12	14	6
	2004	82	94	19	105	32	23	16
	2005	77	126	24	112	42	15	10
	2006	72	100	20	91	28	10	11
	Total	346	523	100	504	136	67	51

Case Study 2: Evaluation of intersection traffic control and safety improvements

The same 70% urban American Midwest county has a combination of intersections with different geometric configurations and traffic control. There are 226 intersections on the state-maintained highways in the county in total. A dataset provided for the period 2002–2006 with the details include vehicle crashes by type and severity level, classified turning counts on intersection major and minor approaches, geometric design standards, field speeds and posted speed limits, cross section, terrain, access control, turning lanes, channelization, traffic control and safety hardware conditions, and road-side features. Table 10.12 presents data summary statistics.

a. *SI values for individual intersections*: For each of the 226 intersections, the SI value is separately computed for each year. Applicable SPFs are identified and the model coefficients are adopted for estimating the TEFs. Detailed safety issue items for the intersections are identified based on available data. The safety issue item weighting factors, increases in the crash frequency and severity risks,

Table 10.10 An example of computing the SI for a rural interstate segment

I. Traffic exposure factor (TEF_s)

$TEF_F = e^{-2.1827} \times \text{Length} \times [(AADT/1000)^{-0.1009}]$, $TEF_I = e^{-0.2489} \times \text{Length} \times [(AADT/1000)^{0.1878}]$, $TEF_{TOTAL} = e^{0.9112} \times \text{Length} \times [(AADT/1000)^{0.2143}]$

2006 AADT	26585 veh/day
Segment length	1.48 miles

$TEF_F = e^{-2.1827} \times 1.48 \times 26.585^{-0.1009}] = 0.1198$, $TEF_I = e^{-0.2489} \times 1.48 \times 26.585^{0.1878} = 2.1365$, $TEF_P = e^{0.9112} \times 1.48 \times 26.585^{0.2143}–0.1198–2.1365 = 5.1789$

II. Crash frequency factor (CFF_{rs}): $CFF_{rs} = \prod CFF_{rsi}$, $CFF_{rsi} = 1 + \Sigma W_{ij} \times \Delta CF_{rsij} \times P_{rsij}$

Safety issue category	*Safety issue type (i)*		*Detailed safety issue item (j)*		W_{ij}	ΔCF_{rsij} (%)	P_{rsij} (%)	CFF_{rsi}
Geometric design	Alignment CFF_{rs1}	($i = 1$)	Realignment needed for very severe curve	($j = 1$)	0.1	87	10.54	1.0092
	Cross section CFF_{rs2}	($i = 2$)	Lane width	($j = 1$)	1.0	100	10.00	1.1000
	Shoulder attributes CFF_{rs3}	($i = 3$)	Medium shoulder width	($j = 1$)	0.5	20	4.94	1.0049 for head-on and sideswipe, 1.00 for other crash types
	Passing lane CFF_{rs4}	($i = 4$)	Missing passing lane	($j = 1$)	1.0	20	10.00	1.0200
	Terrain CFF_{rs5}	($i = 5$)	Rolling terrain	($j = 1$)	0.5	50	10.00	1.0250
	Access control CFF_{rs6}	($i = 6$)	Excessive density of uncontrolled accesses	($j = 1$)	0.1	75	100.00	1.0750
Consistency of design standards	Tangent sections $CFF_{r,s,7}$	($i = 7$)	Tangent section design consistency level	($j = 1$)	0.25	87	10.54	1.0229
Pavement conditions	Skid resistance $CFF_{r,s,8}$	($i = 8$)	Inadequate skid resistance	($j = 1$)	1.0	45	20.00	1.0900
Safety hardware conditions	Signs, $CFF_{r,s,9}$	($i = 9$)	Unmet night time retroreflectivity	($j = 1$)	0.5	10	100.00	1.0500
	Pavement markings $CFF_{r,s,10}$	($i = 10$)	Missing edge-line markings	($j = 1$)	0.25	8	100.00	1.0525
			Missing centerline markings	($j = 2$)	0.25	13	100.00	

$CFF_{head-on,\,I/F/P} = CFF_{sideswipe,\,I/F/P} = CFF_{rs1} \times CFF_{rs2} \times \cdots \times CFF_{r,s,10} = 1.5449$
$CFF_{fixed\ object,\,I/F/P} = CFF_{run\text{-}off\text{-}road,\,I/F/P} = CFF_{rs1} \times CFF_{rs2} \times \cdots \times CFF_{r,s,10} = 1.5374$

(*Continued*)

Table 10.10 (Continued) An example of computing the SI for a rural interstate segment

III. Crash severity factor (CFF_{rs}): $CFF_{rs} = (V_{85}/V_{SL}) \times \prod CSF_{rsi}$, $CSF_{rli} = 1 + \Sigma W_{ij\_RSH} \times \Delta CS_{rlij} \times P_{rlij}$, $CSF_{r,F/P,i} = 1 + \Sigma W_{ij\_RSH} \times [\Delta CS_{rlij} \times (1 + \Delta S_{r,F/P,i,j})] \times P_{rlij}$

Safety issue category	Safety issue type (i)		Detailed safety issue item (j)		W_{ij_RSH}	ΔCS_{rlij} (%)	P_{rlij} (%)	ΔS_{rFij} (%)	ΔS_{rPij} (%)	CSF_{rli}	CSF_{rFi}	CSF_{rPi}
Roadside features	Embankment CSF_{rs1}	$(i = 1)$	Unshielded embankment	$(j = 1)$	0.05	80	5	800	800	1.002	1.018	1.018
	Guardrails	$(i = 2)$	Damaged guardrails	$(j = 1)$	0.1	35	13	70	100	1.0046	1.0077	1.0091
	Ditches, CSF_{rs4}	$(i = 3)$	Ditches located < 9 ft	$(j = 1)$	0.5	40	4.78	150	150	1.0096	1.0239	1.0239
	Trees, CSF_{rs5}	$(i = 4)$	Trees located < 9 ft	$(j = 1)$	0.5	90	3.46	1000	1000	1.0156	1.1713	1.1713
	Utility poles, CSF_{rs6}	$(i = 5)$	Utility pol. located < 9 ft	$(j = 1)$	0.5	90	1.32	1000	1000	1.0059	1.0653	1.0653
	Obstacles, CSF_{rs7}	$(i = 6)$	Obstacles located < 9 ft	$(j = 1)$	0.1	90	1.32	1000	1000	1.0012	1.0131	1.0131

$CSF_{head\text{-}on,F/I/P} = CSF_{sideswipe,I/F/P} = CSF_{fixed\ object,I/F/P} = (V_{85}/V_{SL}) \times 1.00 = (73/65) \times 1.00 = 1.1231$
$CSF_{run\text{-}off\text{-}road,F} = (V_{85}/V_{SL}) \times (CSF_{rF1} \times CSF_{rF2} \times \cdots \times CSF_{rF6}) = (73/65) \times 1.018 \times 1.0077 \times \cdots \times 1.0131 = 1.4912$
$CSF_{run\text{-}off\text{-}road,I} = (V_{85}/V_{SL}) \times (CSF_{rI1} \times CSF_{rI2} \times \cdots \times CSF_{rI6}) = (73/65) \times 1.002 \times 1.0046 \times \cdots \times 1.0012 = 1.1674$
$CSF_{run\text{-}off\text{-}road,P} = (V_{85}/V_{SL}) \times (CSF_{rP1} \times CSF_{rP2} \times \cdots \times CSF_{rP6}) = (73/65) \times 1.018 \times 1.0091 \times \cdots \times 1.0131 = 1.4933$

IV. Probability distribution of crash types

$p_F = \{p_F(\text{Head-on}), p_F(\text{Sideswipe}), p_F(\text{Fixed object}), p_F(\text{Run-off-road})\} = \{0.2886, 0.0761, 0.6286, 0.0068\}$
$p_I = \{p_I(\text{Head-on}), p_I(\text{Sideswipe}), p_I(\text{Fixed object}), p_I(\text{Run-off-road})\} = \{0.2825, 0.0691, 0.0054, 0.6430\}$
$P_P = \{P_P(\text{Head-on}), P_P(\text{Sideswipe}), P_P(\text{Fixed object}), P_P(\text{Run-off-road})\} = \{0.0052, 0.2773, 0.0641, 0.6534\}$

V. SI for the highway segment: $SI_s = TEF_s \times \Sigma[p(\text{Crash Type } r) \times CFF_{rs} \times CSF_{rs}]$

Head-on	r = Head-on, $s = F, I, P$	$SI_{head-on,F} = 0.2079$, $SI_{head-on,I} = 3.7070$, $SI_{head-on,P} = 8.9858$
Sideswipe	r = Sideswipe, $s = F, I, P$	$SI_{sideswipe,F} = 0.2079$, $SI_{sideswipe,I} = 3.7070$, $SI_{sideswipe,P} = 8.9858$
Fixed object	r = Fixed object, $s = F, I, P$	$SI_{fixed\ object,F} = 0.2069$, $SI_{fixed\ object,I} = 3.6890$, $SI_{fixed\ object,P} = 8.9422$
Run-off-road	r = Run-off-road, $s = F, I,$	$SI_{run-off-road,F} = 0.2746$, $SI_{run-off-road,I} = 3.8345$, $SI_{run-off-road,P} = 11.8897$
Total SI	Fatal	$SI_F = \sum[p(\text{Crash Type } r) \times SI_{r,F}] = 0.2886 \times 0.2079 + 0.0761 \times 0.2079 + 0.6286 \times 0.2069 + 0.0068 \times 0.2746 = 0.2077$
	Injury	$SI_I = \sum[p(\text{Crash Type } r) \times SI_{r,I}] = 0.2825 \times 3.7070 + 0.0691 \times 3.7070 + 0.0054 \times 3.6890 + 0.6430 \times 3.8345 = 3.7889$
	PDO	$SI_P = \sum[p(\text{Crash Type } r) \times SI_{r,P}] = 0.0052 \times 8.9858 + 0.2773 \times 8.9858 + 0.0641 \times 8.9422 + 0.6534 \times 11.8897 = 10.8804$

Table 10.11 Comparisons of SI and EB estimates by crash severity level

		Spearman's rank test		*Mann–Whitney U test*			*Chi-square test*
Crashes	*Year*	ρ_s	*z-statistic*	U_{SI}	U_{EB}	*z-statistic*	χ^2*-statistic*
Fatal	2002	0.89	12.31	16547	20713	−1.90	0.67
	2003	0.81	11.26	20805	15682	−2.68	0.92
	2004	0.89	12.39	16393	20866	−2.04	0.69
	2005	0.93	12.83	16545	20712	−1.90	1.15
	2006	0.93	12.82	17055	20203	−1.44	0.90
Injury	2002	0.91	12.58	16077	21180	−2.32	13.71
	2003	0.98	13.63	16522	20739	−1.92	8.85
	2004	0.92	12.75	16631	20630	−1.82	7.09
	2005	0.94	13.03	16516	20741	−1.92	10.60
	2006	0.95	13.22	16573	20508	−1.71	7.08
PDO	2002	0.93	12.83	20262	16580	−1.87	27.15
	2003	0.92	12.76	17131	20127	−1.37	21.59
	2004	0.99	13.78	17353	19907	−1.17	21.06
	2005	0.97	13.47	16996	20261	−1.49	33.84
	2006	0.99	13.70	17381	19881	−1.15	45.62
Critical value		$z_{0.005} = 2.576$		$z_{0.005} = 2.576$			$\chi^2_{0.01\ 192} = 240.50$

and crash proportion factors are used for estimating the crash frequency factors and crash severity factors. Table 10.13 illustrates an example of SI computation for an urban 4-leg stop-controlled intersection.

b. *Consistency tests of SI and EB estimates*: To validate the SI estimates established by the proposed risk-based approach, the EB estimates for fatal, injury, and PDO crashes are computed for each intersection as the weighted sum of observed and predicted crashes. The SI and EB estimates for fatal, injury, and PDO crashes for the 226 intersections are compared separately for the level of agreements. As indicated in Table 10.14, both SI and EB estimates maintain relatively low RMSE values. In most cases, the RMSE values for SI estimates are slightly lower. The Spearman's rank correlation test results for each five-year analysis period maintain the lowest z-statistic values of 12.16 for fatal crashes, 13.87 for injury crashes, and 14.50 for PDO crashes. This indicates that the ranking of SI and EB estimates for all highway intersections agrees at 99% confidence level. The Mann–Whitney test generally produces similar results, with the exception of the ranking of fatal crashes in 2002. The level of agreements in the magnitude of EB and SI estimates was further evaluated by the Chi-square test and regression analysis. As shown in Table 10.4, the highest Chi-square statistical values for each five-year analysis period for fatal crashes, injury crashes, and PDO crashes are 4.15, 5.53, and 11.17, respectively. All values are lower than the Chi-square critical value of 277.27 for 99% confidence level. This demonstrates that the differences in EB and SI estimates for fatal, injury, and PDO crashes are insignificant.
c. *Project-level benefits*: As the example in Table 10.13 shows, the CFF_{rs7}, CFF_{rs8}, and CFF_{rs9} will all be reduced from 1.1080, 1.0435, and 1.1300 to 1.00 if we improve the conditions of intersection traffic control and safety hardware including signs, pavement markings, and lighting to reduce frequencies of all types of crashes. This will help reduce the overall crash severity factors by crash pattern and severity level. As a result, the SI estimates for fatal, injury, and PDO crashes will be reduced from 0.0942, 1.4958, and 2.0336 crashes per year to 0.0721, 1.1449, and 1.5565, respectively. The annual PSIs as percentage reductions in fatal, injury, and PDO crashes per year by improving all types of traffic control and safety hardware are 23.5%, 23.4%, and 23.5%, correspondingly.

Table 10.12 Case study 2 data summary statistics

Data item		*Urban area*				*Rural area*				
		4-leg signalized	*3-leg signalized*	*4-leg STOP*	*3-leg STOP*	*4-leg signalized*	*4-leg STOP*	*3-leg STOP*	*Off-ramp*	*On-ramp*
Intersections	Year	36	1	47	106	5	7	11	4	9
AADT, veh/day (std. dev.)	2002	9813	9728	9643	8591	9282	8881	9438	8042	8958
		(3059)		(3744)	(2552)	(215)	(756)	(637)	(1245)	(805)
	2003	10149	9826	10158	9136	9735	9400	9739	8508	9477
		(3079)		(3594)	(2475)	(213)	(914)	(616)	(1284)	(852)
	2004	10506	9945	10735	9747	10218	9955	10052	9001	10027
		(3129)		(3486)	(2485)	(511)	(1128)	(645)	(1311)	(901)
	2005	10886	10245	11387	10434	10731	10547	10378	9523	10608
		(3223)		(3465)	(2637)	(878)	(1393)	(734)	(1384)	(953)
	2006	11291	10814	12123	11209	11278	11179	10719	10075	11223
		(3377)		(3595)	(2982)	(1293)	(1703)	(880)	(1411)	(1008)
Fatal crashes	2002	0	0	0	0	0	1	0	0	0
	2003	0	0	0	1	0	0	0	0	0
	2004	0	0	0	0	0	0	1	0	0
	2005	0	0	1	1	0	0	0	0	1
	2006	0	0	0	0	0	0	0	0	0
	Total	0	0	1	2	0	1	1	0	1
Injury crashes	2002	12	2	27	10	3	2	2	0	2
	2003	2	0	3	24	0	0	4	0	1
	2004	10	0	11	16	0	0	1	2	2
	2005	11	0	8	11	0	2	2	0	4
	2006	3	0	4	10	0	0	0	0	2
	Total	38	2	53	71	3	4	9	2	11
PDO crashes	2002	73	4	71	70	9	11	5	2	4
	2003	22	0	30	76	0	1	4	1	6
	2004	48	0	36	58	2	0	10	0	0
	2005	28	0	20	46	0	2	6	2	7
	2006	19	0	13	31	0	2	2	0	7
	Total	190	4	170	281	11	16	27	5	24

For the same example, to calculate the annual life-cycle safety improvement benefits expressed in present worth and equivalent uniform annual amount, the following assumptions are made: 5% increase in the daily total traffic on intersection major and minor approaches with safety hardware improvements, which brings the daily entering traffic volume from 11660 to 12243; $4008900, $82588, and $7400, respectively, for each fatal, injury, and PDO crash saved (AASHTO, 2010a,b); 3% annual traffic growth rate; 20-year overall intersection safety hardware service life cycle; and 4% discount rate. Based on the SI values of 0.0942 (0.0721), 1.4958 (1.1449), and 2.0336 (1.5565) for fatal, injury, and PDO crashes before and after intersection traffic control and safety hardware improvements, the annual safety benefits could be calculated as $143420 for the first year of the 20-year overall intersection safety hardware service life cycle period. The present worth and equivalent uniform annual amounts of the maximum safety benefits for the 20-year overall intersection safety hardware service life cycle could be estimated as $2588715 and $190482, respectively.

Source: Adapted from Madanu, S. et al. 2010. *ASCE Journal of Transportation Engineering* **136**(2), 129–140.

Table 10.13 An example of computing SI for an urban 4-leg stop-controlled intersection

I. Traffic exposure factor (TEF)

$TEF_F = (2.459\times10^{-5})\times(AADT_{major})^{0.499}\times(AADT_{minor})^{0.430}$, $TEF_I = (4.015\times10^{-4})\times(AADT_{major})^{0.499}\times(AADT_{minor})^{0.430}$,

$TEF_P = (5.421\times10^{-4})\times(AADT_{major})^{0.499}\times(AADT_{minor})^{0.430}$

2006 AADT — *Major approach: 10959 veh/day; minor approach: 701 veh/day*

$TEF_F = 2.459\times10^{-5}\times10959^{0.499}\times701^{0.430} = 0.0427$; $TEF_I = 4.015\times10^{-4}\times10959^{0.499}\times701^{0.430} = 0.6969$; $TEF_P = 5.421\times10^{-4}\times10959^{0.499}\times701^{0.430} = 0.9409$

II. Crash frequency factor (CFF_{rs}): $CFF_{rs} = \prod CFF_{rsi}$, $CFF_{rsi} = 1 + \sum W_{ij}\times\Delta CF_{rsij}\times P_{rsij}$

Category	*Crash risk factor (i)*		*Detailed crash risk factor (j)*		*Affected crash type (r)*	W_{ij}	ΔCF_{rsij} (%)	P_{rsij} (%)	CFF_{rsi}
Geometric design	Alignment CFF_{rs1}	$(i = 1)$	Presence of horizontal and vertical curves	$(j = 1)$	All	0.5	100.00	10.54	1.0527
	Cross section CFF_{rs2}	$(i = 2)$	Major road outside shoulder width	$(j = 1)$	Sideswipe, Fixed object	0.5	8.67	10.54	1.071 for HO, 1.088 for SS, 1.057 for RE, AGL, and 1.048 for FO
			Major road median width	$(j = 2)$	Head-on, Sideswipe	1.0	27.00	10.00	
			Major road number of lanes	$(j = 3)$	All	1.0	20.00	11.00	
			Major road through lanes	$(j = 4)$	All	1.0	20.00	11.00	
			Presence of right-turn lane on the major road	$(j = 5)$	Rear-end, sideswipe, angle	1.0	26.00	5.00	
	Channelization CFF_{rs3}	$(i = 3)$	Major road right-turn channelization	$(j = 1)$	Rear-end, sideswipe, angle	1.0	26.00	2.00	1.0337 for RE, AGL, SS, 1.00 for others
			Major road left-turn channelization	$(j = 2)$	Rear-end, sideswipe, angle	1.0	19.00	15.00	
	Terrain CFF_{rs4}	$(i = 4)$	Rolling terrain	$(j = 1)$	All	1.0	100.00	10.00	1.1000
	Access control, CFF_{rs5}	$(i = 5)$	Excessive density of uncontrolled accesses	$(j = 1)$	All	1.0	10.60	10.00	1.0106

(Continued)

Table 10.13 (Continued) An example of computing SI for an urban 4-leg stop-controlled intersection

Traffic operations	Truck traffic CFF_{rs}	$(i = 6)$	Presence of trucks during peak hour	$(j = 1)$	All	0.5	30.25	8.84	1.0134
Safety hardware (SH)	Signs, CFF_{rs7}	$(i = 7)$	Unmet night,time retroreflectivity	$(j = 1)$	All	1.0	30.00	36.00	1.1080
	Pavement markings, CFF_{rs9}	$(i = 8)$	Missing edge-line markings	$(j = 1)$	All	1.0	29.00	10.00	1.0435
			Missing centerline markings	$(j = 2)$	All	0.5	29.00	10.00	
	Lighting, CFF_{rs10}	$(i = 9)$	Intersection lighting indicator variable	$(j = 1)$	All	1.0	52.00	25.00	1.1300

$CFF_{head\text{-}on,F/I/P} = CFF_{rs1} \times CFF_{rs2} \times \cdots \times CFF_{rs,9} = 1.7153$; $CFF_{sideswipe,}$ $CFF_{sideswipe,F/I/P} = CFF_{rs1} \times CFF_{rs2} \times \cdots \times CFF_{r,s,9} = 1.8022$
$CFF_{rear\text{-}end,F/I/P} = CFF_{angle,F/I/P} = CFF_{rs1} \times CFF_{rs2} \times \cdots \times CFF_{rs,9} = 1.7499$; $CFF_{fixed\ object,F/I/P} = CFF_{rs1} \times CFF_{rs2} \times \cdots \times CFF_{rs,9} = 1.6794$

III. Crash severity factor (CSF_{rs}): $CSF_{rs} = (V_{85}/V_{SL}) \times \Pi CSF_{rsi}$, $CSF_{rIi} = 1 + \Sigma W_{ij\_RSH} \times \Delta CS_{rIij} \times P_{rIij}$, $CSF_{r,F/P,i} = 1 + \Sigma W_{ij\_RSH} \times [\Delta CS_{rIij} \times (1 + \Delta S_{r,F/P,i,j})] \times P_{rIij}$

Category	*Crash risk factor*	*(i)*	*Detailed crash risk factor*	*(j)*	*Crash type*	W_{ij_RSH}	ΔCS_{rsij} (%)	P_{rsij} (%)	ΔS_{rFij} (%)	ΔS_{rPij} (%)	CSF_{rIi}	CSF_{rFi}	CSF_{rPi}
Roadside features	Roadside hazard rating	$(i = 1)$	Roadside hazard rating within 250 ft of the intersection on the major road	$(j = 1)$	Fixed object	0.5	10	18	1000	1000	1.009	1.099	1.099

$CSF_{head\text{-}on,F/I/P} = CSF_{sideswipe,F/I/P} = CSF_{rear\text{-}end,F/I/P} = CSF_{angle,F/I/P} = (V_{85}/V_{SL}) \times 1.00 = (43/35) \times 1.00 = 1.2286$
$CSF_{fixed\ object,F} = (V_{85}/V_{SL}) \times (CSF_{rF1}) = (43/35) \times 1.099 = 1.3502$, $CSF_{fixed\ object,I} = (V_{85}/V_{SL}) \times (CSF_{rI1}) = (43/35) \times 1.009 = 1.2396$,
$CSF_{fixed\ object,P} = (V_{85}/V_{SL}) \times (CSF_{rP1}) = (43/35) \times 1.099 = 1.3502$

(*Continued*)

Table 10.13 (Continued) An example of computing SI for an urban 4-leg stop-controlled intersection

IV. SI for the highway intersection: $SI_s = TEF_s \times \Sigma[p(\text{Crash Type } r) \times CFF_{rs} \times CSF_{rs}]$		
$p_F = \{p_F(\text{Head-on}), p_F(\text{Sideswipe}), p_F(\text{Rear-end}), p_F(\text{Angle}), p_F(\text{Fixed object})\} = \{0.1312, 0.0331, 0.0027, 0.3257, 0.5072\}$ $p_I = \{p_I(\text{Head-on}), p_I(\text{Sideswipe}), p_I(\text{Rear-end}), p_I(\text{Angle}), p_I(\text{Fixed object})\} = \{0.1125, 0.0248, 0.3202, 0.5406, 0.0018\}$ $p_P = \{p_P(\text{Head-on}), p_P(\text{Sideswipe}), p_P(\text{Rear-end}), p_P(\text{Angle}), p_P(\text{Fixed object})\} = \{0.0021, 0.1261, 0.3345, 0.5090, 0.0283\}$		
Head-on	r = Head-on, $S = F, I, P$	$SI_{head\text{-}on,F} = 0.0900$, $SI_{head\text{-}on,I} = 1.4687$, $SI_{head\text{-}on,P} = 1.9829$
Sideswipe	r = Sideswipe, $S = F, I, P$	$SI_{sideswipe,F} = 0.0945$, $SI_{sideswipe,I} = 1.5431$, $SI_{sideswipe,P} = 2.0833$
Rear-end	r = Rear-end, $S = F, I, P$	$SI_{rear\text{-}end,F} = 0.0918$, $SI_{rear\text{-}end,I} = 1.4983$, $SI_{rear\text{-}end,P} = 2.0229$
Angle	r = Angle, $S = F, I, P$	$SI_{angle,F} = 0.0918$, $SI_{angle,I} = 1.4983$, $SI_{angle,P} = 2.0229$
Fixed object	r = Fixed object, $S = F, I, P$	$SI_{fixed\ object,F} = 0.0968$, $SI_{fixed\ object,I} = 1.4508$, $SI_{fixed\ object,P} = 2.1335$
Total SI	Fatal	$SI_F = \sum[p(\text{Crash Type } r) \times SI_{r,F}]$ $= 0.1312 \times 0.0900 + 0.0331 \times 0.0945 + 0.0027 \times 0.0918 + 0.3257 \times 0.0918 + 0.5072 \times 0.0968 = 0.0942$
	Injury	$SI_I = \sum[p(\text{Crash Type } r) \times SI_{r,I}]$ $= 0.1125 \times 1.4687 + 0.0248 \times 1.5431 + 0.3202 \times 1.4983 + 0.5406 \times 1.4983 + 0.0018 \times 1.4508 = 1.4958$
	PDO	$SI_P = \sum[p(\text{Crash Type } r) \times SI_{r,P}]$ $= 0.0021 \times 1.9829 + 0.1261 \times 2.0833 + 0.3345 \times 2.0229 + 0.5090 \times 2.0229 + 0.0283 \times 2.1335 = 2.0336$

Table 10.14 Comparisons of SI and EB estimates by crash severity level

		Root mean squared error		Spearman's rank correlation test		Mann–Whitney U test			Chi-square test
Crashes	*Year*	*SI estimate*	*EB estimate*	ρ_s	*z-statistic*	U_{SI}	U_{EB}	*z-statistic*	χ^2*-statistic*
Fatal	2002	0.11	0.14	0.98	14.78	21425	29651	−2.96	0.70
	2003	0.16	0.28	0.99	14.87	22303	28773	−2.33	2.98
	2004	0.04	0.05	0.99	14.93	22790	28286	−1.98	1.53
	2005	0.19	0.33	0.99	14.91	22896	28180	−1.90	4.15
	2006	0.06	0.08	0.81	12.16	21915	29161	−2.61	0.40
Injury	2002	0.14	0.22	0.98	14.79	26336	24740	−0.57	2.76
	2003	0.11	0.09	0.92	13.87	23858	27218	−1.21	3.34
	2004	0.14	0.13	0.99	14.92	23020	28056	−1.81	5.35
	2005	0.13	0.18	0.99	14.90	26313	24763	−0.56	1.42
	2006	0.06	0.08	0.99	15.00	27596	23480	−1.48	5.53
PDO	2002	0.34	0.19	0.96	14.53	23797	27279	−1.25	11.17
	2003	0.15	1.06	0.97	14.56	25548	25528	−0.57	4.80
	2004	0.44	0.39	0.96	14.50	24736	26340	−0.58	4.45
	2005	0.20	0.09	0.99	14.90	23999	27077	−1.11	2.64
	2006	0.07	0.10	0.98	14.76	23221	27855	−1.67	7.32
Critical value					$z_{0.005} = 2.58$			$z_{0.005} = 2.58$	$\chi^2_{0.01,225} = 277.27$

PROBLEMS

10.1 For the exemplary calculation of SI value for the rural interstate highway segment in Case Study 1, assume that the annual traffic growth rate is 5% and it is proposed to re-apply pavement markings every 2 years, replace traffic signs every 10 years, and re-install guardrails every 20 years. With the guardrail useful service life being consider as the longest, compute the SI value for the highway segment.

10.2 For the exemplary calculation of SI value for the urban 4-leg stop controlled in Case Study 2, assume that the annual traffic growth rate is 6% and it is proposed to re-apply pavement markings every 2 years, replace traffic signs every 10 years, and fully eliminate all roadside hazards. With the longest useful service life cycle for the traffic control and safety hardware estimated as 10 years, compute the SI value for the intersection.

Chapter 11

Economic analysis of transit facility preservation

The transit system is a dynamic transportation sector. The usage of this system over time shows trends that depend on car ownership, travel cost, fuel cost, and service provided to the customer. Apart from densely populated urban areas, where transit may compete with driving for convenience and levels of service, in most developed countries transit primarily serves captive riders who cannot afford the expense of a private automobile. The system is affected by the transit agencies' investments, operational innovations, method of financing, and the level of public commitment to transit as an environmental or congestion mitigation strategy. This chapter covers methods for transit project evaluation.

11.1 TRANSIT IMPACTS ANALYSIS

Providing a safe, reliable, and attractive transit service requires efficient physical systems, competent personnel, and good organizations. Considering the size and characteristics of the transit service area, there are considerable differences in organizational, functional, and operational practices among transit agencies and companies. Passengers and agency ownership should be fully considered during the process of establishing funding requirements, operating improvements, and capital planning. Essential to the performance of an urban transit investment is the connectivity of public transit with the other modes. The economic and population growth from improved transit development can provide travelers more flexibility and choice, cost savings, ubiquity of service, and a more attractive transportation network with higher frequencies and better connectivity across an urban area.

11.1.1 Transit performance measures

Transit agencies have been assisted by appropriate definitions for performance measures. Performance measures are needed to evaluate transportation provisions in a quantitative or qualitative manner, which could be directly or indirectly affected by the degree to which results meet the needs and expectations of the users. Two categories of performance measures are usually used: efficiency and effectiveness. Performance efficiency refers to the relationship between inputs and outputs or what is generally known as "productive" or "technical" efficiency in economics. Efficiency includes resource utilization conditions, such as labor utilization and vehicle utilization, and financial performance, such as cost per unit output, revenue per unit output, and the ratio of revenue and input. On the other hand, effectiveness refers to the use of outputs to achieve objectives or service consumption. It includes safety, accessibility, and percentage of on-time arrival and service equity. Equity is the percentage of elderly and differently abled people serviced by the transit facilities, and needs to be considered as an integral part of the service performance measures. Hatry (1980) argued that

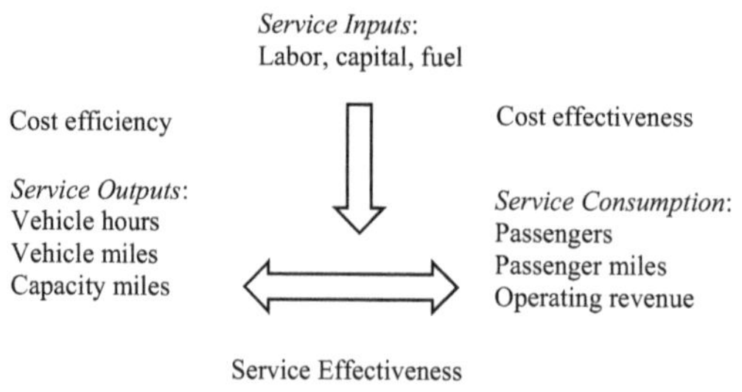

Figure 11.1 Conceptual framework for transit performance analysis.

in public agencies, efficiency should be considered separately from effectiveness. Efficiency is the relationship between inputs and outputs or what is referred to as "productive" or "technical" efficiency in the economics literature. Effectiveness, on the other hand, refers to the use of outputs to achieve objectives or service consumption. Fielding (1987) explained this further by using a triangle with service input at the top and service output and service consumption as the base, as shown in Figure 11.1.

Effectiveness is also considered as the ratio of outputs to inputs, but it is related to the results achieved by a given input. Therefore, it includes both the quality and quantity of the service provided. For instance, the percentage of the population served is a measure of effectiveness. Serving a higher percentage of the population could help attract population growth in the surrounding area. In addition, quality of service is an important measure in transit system evaluation, especially for commuters during peak-hour travel. As with effectiveness ratio, it is usually desirable to maximize the coefficient of utilization. However, in many cases there are trade-offs to high utilization ratios, such as higher maintenance costs and lower passenger comfort in heavily loaded vehicles, and fewer employee social benefits when the labor utilization ratio is high.

Since labor and vehicle operating costs represent most of transit operating costs, efficient performance should take resource utilization into account. Higher values for vehicle per driver, vehicle hours per bus, and percent peak vehicle use with same or better levels of service means better resource utilization or improved efficiency of the transit agency. Furthermore, since urban mass transit may exist for different reasons in some areas, it is only reasonable to consider various policy issues in light of the goals and objectives the system is expected to achieve. For example, the percentage of elderly and differently abled serviced may be used to reflect the extent to which a transit system serves special groups. Failure to recognize such special population groups may result in inaccurate assessment of transit performance. Generally, it would seem reasonable to assume that those transit systems that provide a high level of service to the elderly and the differently abled would incur higher unit operating costs than do those systems in cities with a small elderly and differently abled population.

11.1.2 Transit economic analysis framework

There are many forms and types of ownership of transit organizations. It could be one of a city's departments or a semi-independent agency that belongs to a county, state or other governmental unit. The sources of financing also vary greatly, including various forms of user charges, special funds, federal funds or allocation from general tax revenues.

Figure 11.2 depicts a framework for transit economic analysis. In the analysis, the life-cycle cost defined as the present worth of all future expenditures over the transit facility

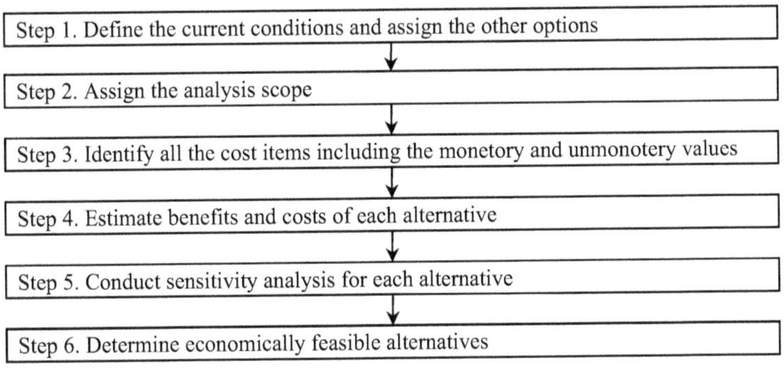
Step 1. Define the current conditions and assign the other options
Step 2. Assign the analysis scope
Step 3. Identify all the cost items including the monetory and unmonotery values
Step 4. Estimate benefits and costs of each alternative
Step 5. Conduct sensitivity analysis for each alternative
Step 6. Determine economically feasible alternatives

Figure 11.2 Framework of transit economic analysis.

service lifespan for each alternative needs to be determined. Benefits include both user and nonuser benefits. Costs cover initial capital investment, life-cycle routine maintenance, and major period maintenance costs. Present worth amounts of benefits and costs are calculated using the standard discount rate along with sensitivity analysis. Then, the benefit and cost estimates are compared for each alternative to justify its economic viability. Several economic analysis methods are used for this purpose, the predominant ones include net present worth (NPW), benefit-to-cost ratio (BCR), and internal rate of return (IRR).

Financial performance of transit agencies, which is the most significant component of efficiency performance, is composed of two major categories: cost per unit output and revenue per unit output. Figure 11.3 shows these elements with their categorizations.

The American Public Transit Association (APTA) indicates that transit investments should consider both capital and operating investments (APTA, 1991). Capital investments mainly comprise the "hardware" of the transit systems, like vehicles, maintenance facilities, track in the case of rail transit, and other system components, as well as expense for land acquisition and hardware modernization. Table 11.1 provides a good classification of transit asset management.

Typically, operating expenditures include labor, maintenance, and supplies. Operating expenditures provide direct benefits to the local economy since salaries and wages typically comprise two-thirds of total operating expenditures. At the initiation of an economic analysis study, the general managers and other representatives of transit systems would gather feedback from transit operators, the existing condition of the systems would be evaluated, specific objectives and performance indicators would be checked, and the need for increased transit performance and accountability emphasized.

If all transit systems operated in identical environments and under similar policy constraints, their performance in different locations could then be explained only in terms of variations in levels of service. However, in some cases, the major difficulty of the operating performance of one system cannot be directly compared with other systems. Besides, certain

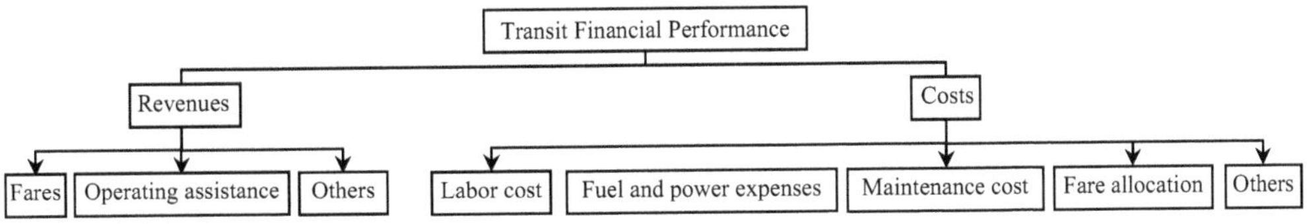
Transit Financial Performance
Revenues
Costs
Fares
Operating assistance
Others
Labor cost
Fuel and power expenses
Maintenance cost
Fare allocation
Others

Figure 11.3 Transit financial performance factors.

Table 11.1 Classical transit asset management system inventory structure

Transit asset component	*Transit element*
Guideway element	Guideway
	Track work
	Speed structure
	Bus/BRT guideway
Station	Rail transit station
	Bus/BRT stop
	Ferry dock
Facility	Buildings
	Storage yard
	Equipment
	Major shops
	Control center
Vehicle	Revenue vehicles
	Non-revenue vehicles
	Equipment/parts
System	Train control
	Roadway traffic signals
	Electrification
	Communications
	Security
	Revenue collection
	Utilities
	ITS

inherent biases exist in transit performance measurements because of the effect of several environmental variables beyond the control of transit operators. Population density, congestion, and network configuration are most often cited as intruding environmental and policy effects. These and other elements outside the transit operator's control can have a significant effect on certain performance indicators that are used to describe system productivity. For example, bus travel speed and ridership are affected by each of the above factors, and the conditions for bus travel in a small city are quite different from those in a large city like New York. As congestion may decrease the vehicle's average speed, the failure to consider such an effect will result in misleading productivity measures. Consequently, several environmental and policy factors that appear to constrain the performance of transit systems are given below and discussed in the following sections. Therefore, a stratification analysis is employed: the integrated transit system would be subdivided into a number of groups so that the systems within a group would have substantially similar operating environments. After the working session with representatives from the transit operator, the final evaluation criteria and definition of desirable options will be chosen (Table 11.2).

11.1.3 Transit impacts

11.1.3.1 Efficiency impacts

Efficiency impacts refer to changes in efficiency from people shifting from automobile travel to transit travel. Efficiency benefits include vehicle cost savings, avoided chauffeuring,

Table 11.2 Stratification factors, variables, and data sources

Stratification factor	*Variable*	*Data source*
Congestion	Average vehicle operating speed	System wide measure derived from APTA operating report will also be available from Financial and Reporting Elements (FARE) system
Wage rate	Average wage per driver	System wide measure derived from APTA data and calculated by dividing compensation to operators by operator person hours; wage will be available from FARE
Population	Number of people in urban area in which system operates	Obtained from APTA data in FARE system, reported by metropolitan planning organization (MPO)
Population density	Population per square mile of land area in urbanized area, central city, and service area	System route maps and available census population data provided by MPO
Organization type	Qualitative distinction (e.g., municipal transit authority, contract management)	Obtained directly from transit management, highly susceptible to error due to varying definitions of management types
Network configuration	Qualitative distinction (e.g., radial, grid, circumferential)	Derived from route maps of transit systems, also susceptible to error due to definitional inconsistencies
Local transit policy	Percentage of trips by elderly, percentage of work trips, percentage of elderly population	Elderly population available from census reports; trip distribution by rider characteristics (age, sex, income, handicap) and trip purpose (work, shop) will be provided by MPOs from transit user surveys

congestion reductions, avoided costs of roadway expansions, parking cost savings, increased safety and health, and energy conservation. As a city grows, these benefits play an increasingly critical role in alleviating traffic congestion and parking problems, especially in major centers of commercial, employment, and other activity. Whereas benefits of automobile transportation systems tend to decline with higher use due to congestion, the higher capacity and inherent space efficiency of transit allows the user and public benefits to grow with higher use. The efficiency benefits of transit can create a positive feedback loop in which service improvement attracts more riders, which leads to more frequent service and possibly more priority in traffic, which in turn attracts more riders.

Ideally, transit systems want to achieve both equity and efficiency. In practice, most transit systems in large cities are designed mainly to provide basic mobility to the public rather than efficiency. Buses also operate at times and locations where demand is low, particularly in areas where origin/destination is dispersed. Achieving equity goals means that average occupancy is lower than it could be, and average operating cost, energy consumption, or pollution emissions per passenger-mile are higher than they could be. On the other hand, travel demand is more likely to be concentrated on the major corridors, where congestion and parking problems are also concentrated. Transit service along these major corridors tends to achieve great efficiency benefits.

Vehicle cost savings: Shifting from automobile to transit provides vehicle cost savings to travelers. Shifting from driving to transit saves fuel and oil, the cost of which depends on fuel price and travel miles. In addition, depreciation, insurance, and parking costs are partly variable, since increased driving increases the probability of vehicle repair and maintenance, reduces vehicle salvage value, and increases the risks of crashes, traffic, and parking citations. Savings may be greater in cases where transit users avoid parking fees or road tolls.

The greatest vehicle savings is realized when shifting from driving to transit allows households to own or lease fewer automobiles.

Specifically, the category of potential vehicle cost savings includes:

- *Vehicle operating costs*: Fuel, oil, and tire wear costs. They are estimated by per-mile costs times mileage reduced.
- *Long-term mileage-related costs*: Vehicular mileage-related depreciation mileage lease fee, potential crash costs, and a part of insurance. Similarly, this is indicated by the product of per-mile costs and mileage reduced.
- *Special costs*: Parking fee, tickets, and toll fees. This type of cost is always determined by market survey and the result varies from place to place.
- *Vehicle ownership costs*: Vehicle purchase cost, insurance premiums, license, and taxes. This is obtained by simply taking the product of reduced vehicle ownership and unit vehicle ownership cost.
- *Residential parking*: Residential parking costs due to reduced vehicle ownership. Where parking is included in residential cost or rent, often mandated by zoning, the cost of providing parking is bundled with the cost of housing and can constitute a substantial portion of household spending. This value is sensitive to the location, varying from \$100 to over \$1000 per vehicle per year. For residents who do not need the parking, decoupling the cost of parking from the cost of housing and eliminating parking minimums for residential development can result in substantial saving. It is measured by the product of reduced vehicle ownership and cost per reduced residential parking space.

Energy conservation: The transportation industry is one of the major energy users in the U.S. Sharing rides through transit can directly save fuel. It can also reduce the need for constructing more transportation infrastructure facilities and manufacturing new vehicles, which also consume a great deal of energy. Congestion relief from transit also saves fuel that is otherwise wasted when vehicles are stopped in congested situations.

Avoided chauffeuring: Chauffeuring is the marginal automobile travel specifically to carry a passenger. It can also include taxi and rideshare trips. Ridesharing is not included. Common examples include chauffeuring children to school and other activities, friends to the airport, and elderly relatives on errands. Such trips can be extremely inefficient if the effective trip is only a small portion of the total trip. For example, driving a family member to the airport requires the driver to make an empty return trip (Litman, 2015).

Drivers sometimes enjoy chauffeuring as they view this kind of activity as a great opportunity to interact with friends and family members. However, chauffeuring is always an undesirable burden for transportation networks and this appears to be true for drivers on some occasions. For example, the driver is willing to spend time on other important activities. Transit provides an alternative option to solve this undesirable chauffeuring problem.

Congestion reduction: Traffic congestion impact consists of the additional vehicular delay, road rage, total travel time, vehicle operating costs, and emission that each additional vehicle imposes on other road users. Congestion occurs when traffic demand approaches the capacity of roads or intersections. Once the traffic demand approaches capacity, the severity of congestion is very sensitive to the volume.

Congestion reduction benefits can be difficult to assess because urban traffic is dynamic, and tends to maintain equilibrium. Specifically, traffic volume (demand) grows until travel time cost is unacceptable to the additional traveler. The corresponding phenomenon found in practice is that expanded road capacity is always rapidly filled with induced travel demand and the demand shifted from other roadways. In other words, there is a near-perfect elasticity of demand for road space, and it dramatically limits the ability of capacity-expanding construction investments to deliver lasting congestion relief. However, transit service can reduce the equilibrium point by diverting some trips from automobile trips, consequently easing congestion severity. Transit services tend to perform effectively in alleviating congestion under the following conditions:

- Transit is able to offer high-quality service in terms of travel time, comfortable riding experience, timely operation, and nice station
- Transit routes are able to cover a major share of urban trips. Transit routes along major corridors generally attract high ridership
- Transit routes have the exclusive right of way (e.g., a dedicated rail transit or bus lane). These transit operations will be much less impacted by congestion, particularly if they are granted transit signal priority at intersections. This is a huge advantage when compared with driving under congested conditions
- The transit fare is acceptable

There are several approaches to measuring congestion reduction benefits resulting from reduced automobile trips. One approach is to simulate the transportation behavior with and without transit service, and calculate the variation in travel time and vehicle operating cost. These variations are then added together and converted into a dollar value. This approach requires a high-fidelity, large-scale simulation software. Another approach is to investigate the costs of expanding roadway capacity such that a given congestion reduction is achieved, and then divide that cost by the number of peak-period vehicle-miles to achieve a unit rate. This method also requires a system that could model travelers' choices under each circumstance, which is the bottleneck of implementing these two approaches.

Parking cost savings: Automobile-to-transit shifts could reduce not only auto travel demand, but also parking demand, and so reduce parking costs. Specifically, reduced vehicle ownership reduces residential parking demand, including parking lot requirements and on-street parking demand in residential areas. And non-residential parking demand reduction results due to reduced automobile trips. This benefit is considered as user cost savings in terms of parking fee savings, reduced parking congestion, and increased convenience. This benefit could also be viewed as agency cost savings in terms of savings of the investment in building, managing, and subsidizing parking facilities. Note that savings in the amount of land needed for parking facilities is also significant especially in the area where land is scarce such as downtown, allowing more compact development. The cost of providing parking is bundled with the cost of goods and services, and consequently parking cost savings accrue to consumers. For example, if a grocery store is served by a new transit line and can sell its parking lot to a developer, the avoided cost of providing the parking and the revenue from the property development mean lower prices for customers, more profit for the storeowner or both.

Parking cost savings can be divided into two parts: residential parking cost savings and non-residential parking cost savings. Residential parking cost savings is calculated by simply taking the product of reduced vehicle ownerships and average cost per residential parking space. Non-residential parking cost savings is calculated by multiplying the reduced automobile round trips to the average cost per commercial parking space. Parking tends to be expensive in times and locations where supply is less than demand, such as peak hours in urban areas, where automobile-to-transit shifts are most common. In these situations, transit could provide plenty of parking cost savings.

Traffic safety: Transit is a relatively low crash-risk mode of transportation as heavy transit vehicles are much safer in collisions than private automobiles, and the drivers are professionals. Historical data across multiple regions indicate that the fatality rate per passenger-mile of transit passengers is much less than that of car occupants. Traffic safety benefits directly result from the prevention of potential crashes and could include the further negative impact caused by crashes such as property damage, time wasted in dealing with the crash, and unexpected congestion due to one lane being occupied by the crash.

Personal security: Personal security refers to freedom from assault, theft, sexual attacks, robbery, larceny, harassment, and abuse. The crime risks faced by transit users vary from place to place. In general, per capita crime rates on transit tend to decline as ridership increases. The potential reasons include improved surveillance, better emergency response, and police dispatch strategy.

Health impacts: Extensive medical studies imply that regular physical activities are associated with important health benefits, including reduced risk for premature death, cardiovascular disease, ischemic stroke, type two diabetes, colon and breast cancers, and depression. Less active and less fit people have a greater risk of developing high blood pressure and other diseases. Furthermore, lack of physical activity can contribute to feelings of anxiety and depression. It is widely accepted that increased walking and cycling activities could help increase public fitness and health. Public transportation users walk more frequently, and for longer distances on average, and tend to meet minimum physical activity goals more readily than drivers. Meanwhile, the health benefit also comes from reduced air pollution, which will be addressed in following sections.

Community impacts: Community impacts refer to the effects transit development has on adjacent neighborhoods and communities, beyond those intentionally affecting travelers and direct users of the transportation facilities. This is also known as social impacts. Transit affects neighborhood livability (the quality of the local environment as experienced by people who live, work, or visit there) as a result of changes in visual quality, walking environment, and community cohesion. Specifically, community cohesion refers to residents who live in the same community working toward a society in which there is a common vision and sense of belonging, and strong and positive relationships are being developed between people from different backgrounds in the workplace, in schools, and within neighborhoods. Many people consider cohesion an important community attribute, as it tends to improve quality of life. More importantly, cohesion tends to encourage community members to help and protect one another, increasing neighborhood safety and security. Transit and transit-oriented development can enhance community cohesion. Transit service provides opportunities for community members to interact at stations and on board vehicles, which helps improve community cohesion.

11.1.3.2 Mobility impacts

Mobility benefits refer to the improved mobility provided to travelers of all modes who benefit from the increased capacity and flexibility conferred by an efficient transit system. Public transit trips generally have longer travel time than automobile trips, since transit travel requires extra access, transfer and waiting time, and dwell time at stops. Some these time components can be minimized by more organized and timely operations, assigning exclusive right of way and transit signal priority at intersections, well-planned transit network, improved access to stations, advanced electronic payment and fare management system, and more frequent services. Moreover, transit development can attract employment and commercial centers near transit stations, which means more trip destinations are close to the stations. This could also help reduce the walking distances and increase accessibility.

These mobility benefits are particularly important to transportation disadvantaged travelers, who cannot drive due to physical, economic, or social constraints. There are several categories of mobility benefits as described below. Note that some of these categories may overlap.

User benefits: These are the direct benefits to transit users from improved service quality. For example, less walking time and waiting time, shorter trip time, a less crowded environment, and more comfortable seats. Moreover, user benefits also include the indirect benefit from increased access to services and activities. For example, economic benefits and enjoyment from the extra accessibility provided by transit to social and recreational activities, and commercial benefits from bringing consumers to shop at stores far from their origins. By enlarging people's activity range, more education and job opportunities are offered, which could further increase economic opportunities. Generally, this is measured by rider surveys, investigating how much users rely on the transit service, the trip purposes, and how much they value these trips.

Public service support: Transit can facilitate government activities in offering mobility. For example, transit services can help people reach medical and education services. Transit service, especially equity-oriented transit service, can enable the elderly and differently abled to live independently, which could help reduce care facility costs. Therefore, a part of transit subsidies could be recovered by reduction of these government budgets. This type of benefit is generally obtained through the consultation with public agency officials, and surveys of clients, to determine the role transit plays in supporting public service goals.

Equity benefits: Many people who could afford to drive and park wherever they go choose to take transit because it is faster or more convenient for some trips or for economic or ethical reasons. Nevertheless, transit has a mission to help achieve social equity objectives. By expanding people's activity range, transit service offers extra educational, economic, and social opportunities for disadvantaged groups, especially for those who are economically, physically, and socially disadvantaged. Transit helps reduce the relative disability of someone who is not able to drive. The equity benefit is indicated by the percentage of transit users who belong to disadvantaged groups, and the value that society places on the resulting increased equity.

Option value: Option value refers to the value people place on having an additional available travel mode option even if they do not currently use it (ECONorthwest and PBQD, 2002). Transit is always a backup option for travelers even if some motorists hardly use it. But this kind of travel option seems to be critical when these travelers encounter some emergencies where auto trips are restricted, such as with vehicular mechanical failure. This is similar to lifeboats on a ship, whose value does not depend on how frequently they are used.

Increased productivity: Increased economic productivity is created by improved mobility for non-drivers, including improved access to education and employment. Business is generally more productive when a workplace is readily accessible to workers who prefer not to or are unable to drive to work. Improved access is measured by demographic mapping or transit user surveys aimed at investigating the portion of transit riders that rely on transit for school and work purposes. Other indicators include transit service quality and the value society places on work and school.

Reduced high-risk drivers: Transit offers high-risk drivers another travel option—for example, an alternative to drinking and driving. The lack of transit service will compel high-risk drivers to drive, which could potentially result in crashes. This is measured by consulting experts and the public, investigating the relationship between inadequate travel options and high-risk driving.

11.1.3.3 Land use impacts

Transit planning decisions affect land use directly by determining which land is devoted to transit facilities such as bus/BRT stops, rail transit stations and tracks, parking lots for park-and-ride, and other facilities. They affect land use indirectly by affecting land use and transit-oriented development (TOD) due to the improved mobility and, in some cases, reduced parking requirements. TOD is the connection between transit stops or stations and economic activity on the adjacent land, such as shopping, entertainment, residential development, and employment. Proximity to high-quality transit service can add value to property and sparks new development. Some of the value added from TOD can be captured through a variety of legislative and administrative mechanisms and returned to the public as a revenue stream for paying debt service on the capital and operating costs of transit system investments.

Since transit can help reduce the amount of land devoted to transportation infrastructure and parking as a result of the reduction of automobile travel, it can help achieve more compact urban development or other land use planning objectives. Transit plays an important role in optimizing the current land use pattern with limited resources.

It can be difficult to evaluate the exact land-use impacts of a particular transit project implementation, especially the indirect, long-term impacts. No model exists that could predict the change of land use after implementing a certain transit project. Impacts are affected by so many factors that even a tiny change from one factor could affect the land use significantly in the long term. For example, local commercial policy could encourage the development of a shopping center near a particular transit station. These factors can also affect each other. For example, a transportation policy could affect the local travel demand, which could further affect transit planning decisions.

11.1.3.4 Economic impacts

Economic impacts refer to increased productivity, reduced business activity costs, employment, and property values. Economic impacts analysis is not the same as benefit–cost analysis. Economic impacts analysis focuses specifically on measurable changes in cash flow from the perspective of household and businesses.

Direct impacts: Transit itself is a labor-intensive industry. Transit provides a lot of jobs and local business activity. Specifically, capital investment in transit development supports purchases of equipment and facilities, including tracks, rights-of-way, communication equipment, land, and so on. Daily operations also play an important role in stimulating economic development. Operation-related investment supports all associated jobs and purchases of supplies needed for continuing operations, including fuel, electric power, maintenance parts, and materials. Therefore, investment in transit (both development and services) can directly help economic development by providing job positions, as well as public-sector purchases that further indirectly affect market and industry activity, and thus support jobs in supplier industries.

Business productivity gains: Transit helps improve business productivity in many ways. A shift from automobile to transit will facilitate increased economic productivity and competitiveness for cities. Transit helps alleviate traffic congestion and shorten travel time, so that time and money can be saved and further be invested in business. For example, a delivery company may be more productive if transit reduces traffic congestion. Transit expands the activity range of individuals. From the perspective of business, transit gives companies access to a wider labor market by making job centers available to workers who do not drive, which may lead to a reduction in the wages needed to attract workers. From the perspective of individuals, improved mobility gives them more options in the job market. Overall, transit helps markets optimize resource allocation. For example, a non-driving resident of a small village can only find a job in his village without transit service if he is not willing to move, but he can commute to the large city nearby every workday by using transit services.

Property values: Transit can increase the value of real estate, especially near high-capacity rail transit stations. The amount of value increase depends highly on the local regulatory and commercial environment, transit connections, and land use. Property value increases near rail transit stations for two reasons. First, improved mobility and access makes housing and commercial property more attractive. Second, and more importantly, proximity to a rail transit station can trigger a reduction in the parking requirement and increase in allowable density (through a TOD zoning overlay, for example), which allows much more profitable development (more apartments, less parking), making the property worth more. Essentially, the increase in property values near a transit station is another form of mobility improvement benefit, including access cost savings and travel-time savings due to transit service. Therefore, considering this value in the local or regional economic impacts study will lead to "double counting," since the value of mobility improvement is already included in travel time impact analysis. However, analyzing the change of property values due to transit is helpful because it provides a fresh perspective to demonstrate and calibrate the value transit provides in the local market.

11.1.3.5 Environmental impacts

Air emissions: Transit can eliminate many automobile trips, especially in dense urban areas, replacing many separate emissions-producing vehicles with fewer transit vehicles that generally reduce pollution per passenger-mile. Measuring emission impacts of an automobile-to-transit shift is not easy because there exist so many possible combinations of vehicle models, engine types, and driving conditions. Current simulation software is unable to accurately model these complicated situations, but can still give a rough result.

Noise pollution: Traffic noise is always a problem in urban areas. Conventional diesel buses are quite noisy because of the large engines and low power-to-weight ratio. The noise produced by a typical diesel bus is equivalent to that produced by 5–15 average automobiles, depending on conditions (Delucchi and Hsu, 1998). Hybrid and electric buses can run much more quietly than conventional diesel buses. If a bus can reduce noisy vehicle trips such as those of a motorcycle, it can help reduce noise in the community.

Water pollution: Vehicles generally contribute little to water pollution in normal operations. But fuel leaks from engines could pollute water. Transit is more likely to contribute less to water pollution since fewer vehicles are needed, and these vehicles are properly maintained. On the other hand, air pollution fallout is also another source of water pollution. In this process, water serves as a solvent for several air pollutants, notably acid deposits. The most notable and destructive fallouts are sulfuric and nitric acids, which may alter the pH of water if they are present in sufficient concentrations. Fallouts are accelerated and concentrated in an area by rainy conditions.

11.1.4 Evaluating transit benefits

There are two major categories of transit benefits:

Efficiency oriented: which result when transit substitutes for automobile travel
Equity oriented: that are resulted from the availability and use of transit by disadvantaged people

Although the goal of transit development is to provide both efficiency and equity, some transit services are justified more by one than the other. For example, transit services arranged in an area with low transit trip demand are primarily equity-justified, since they tend to provide basic mobility for that area even though demand is low and does little to reduce traffic congestion, facility costs, or pollution emissions. Vanpooling, express bus, and commuter rail services are mainly efficiency-justified, since they have higher fares with the potential of serving only middle- and higher-income riders, and tend to ease traffic congestion and other negative traffic impacts.

In general, transit service in rural areas and small cities is primarily equity-justified, while bus and rail transit services in large cities provide both benefits. Efficiency-justified transit service tends to have a higher cost recovery and consequently lower subsidy per passenger-mile due to the relatively high ridership. The difference between equity- and efficiency-justified subsidies is often critical for transit evaluation (Walker, 2008).

Transit can bring about a variety of benefits, many of which are partly or completely overlooked in conventional transport economic analysis, as summarized below:

User benefits: Increased convenience, speed, and comfort to users from transit service improvements. This is usually taken into consideration.

Transportation diversity: Improved transportation options, particularly for non-drivers. This type of benefit is sometimes considered qualitatively.

Consumer savings: Reduced consumer transportation costs, including reduced vehicle operating and ownership costs. The vehicle operating cost is always considered in practice, but the ownership cost is not.

Congestion reduction: Reduced traffic congestion. This benefit is sometimes considered indirectly.

Facility cost savings: Reduced road and parking facility costs. This type of benefit is generally not taken into account.

Road safety: Reduced per capita traffic crash rates. This is considered directly.

Environmental benefits: Reduced air and noise pollution emissions and habitat degradation. This type of benefit is increasingly considered today.

Land use efficiency: More compact development and reduced sprawl. This impact is sometimes considered.

Economic development: Increased productivity and agglomeration efficiencies. This is generally not considered or considered indirectly.

Community cohesion: Positive interactions among people in a community. This type of benefit is generally not considered.

Public health: Increased physical activity (particularly walking). This type of benefit is generally not considered.

Tourism: Cities that are easy to get around in without driving and parking are much more attractive to tourists.

Competition with suburbs and other cities: Increasingly, high-quality transit is one of the main features cities use to compete with suburbs and other cities for dynamic companies and talented workers.

11.1.5 Transit project evaluation and selection

The benefit-cost analysis method was developed decades ago for transportation project evaluation, and the method is still a powerful tool in practice. Benefit–cost analysis is briefly introduced in this section. A more detailed discussion can be found in Chapter 7.

In project evaluation, life cycle of the planned facility is an important input factor. It is always estimated as the shortest of the functional life, economic life, and physical life. The benefit–cost analysis is performed within the life cycle.

If the benefits obtained from the planned project in year t are B_t, then the total present worth of the benefit is computed as follows:

$$PW_B = B_0 + \frac{B_1}{1+i} + \frac{B_2}{(1+i)^2} + \cdots + \frac{B_t}{(1+i)^t} + \cdots + \frac{B_L}{(1+i)^L} + \frac{SB}{(1+i)^L} = \sum_{t=0}^{L} \frac{B_t}{(1+i)^t} + \frac{SB}{(1+i)^L} \tag{11.1}$$

where

PW_B = Total present worth of life-cycle benefits
SB = Salvage benefits
i = Discount rate
L = Facility service life-cycle

Similarly, the total present worth of the benefit can be obtained by the same computation.

$$PW_C = C_0 + \frac{C_1}{1+i} + \frac{C_2}{(1+i)^2} + \cdots + \frac{C_t}{(1+i)^t} + \cdots + \frac{C_L}{(1+i)^L} + \frac{SC}{(1+i)^L} = \sum_{t=0}^{L} \frac{C_t}{(1+i)^t} + \frac{SC}{(1+i)^L} \tag{11.2}$$

where

PW_C = The present worth of life-cycle benefits
SC = Salvage cost
C_t = Cost spent on the project in year t

The discount rate measures the time value of money. Compared with the previous year, even with the same amount of money, the worth in this year is $1/(1 + i)$ times of that in the previous year. This factor has a lot of names depending on the case of application. In the context of a bank loan, this becomes the loan interest. If we could find a value of this discount rate such that the present worth of benefit is identical to the present worth of cost, this value is called the IRR. Another name of the discount rate is MARR, which is the minimum attractive rate of return on a project at which the decision-maker will accept the project, with its potential risk and the opportunity cost given. If the discount rate is less than MARR, the project is not attractive.

The NPW is a single economic performance indicator of a project. This value measures the present worth of the net benefit result from the project. It is computed or estimated into a quantitative value. The project is economically feasible only if a positive NPW is achieved. In the case with multiple alternatives, the one with the highest NPW is most attractive because of its highest economic effectiveness.

$$NPW = PW_B - PW_C \tag{11.3}$$

Although the NPW could give a rough picture of the gain or loss of implementing a certain project, it fails to reflect the efficiency, which is the amount of benefit that can be obtained per dollar investment. A complement to NPW is the BCR. The BCR gives a percentage gain relative to the costs for the project. The project is not economically feasible as long as the

BCR is less than one. In the case with multiple alternatives, the one with the highest BCR tends to be the most attractive because of its highest economic efficiency.

$$BCR = \frac{PW_B}{PW_C} \tag{11.4}$$

Another commonly used procedure in project evaluation is the IRR method. As mentioned, IRR is the value of discount rate such that the NPW is zero. Therefore, IRR is solved by setting NPW of a project equal to zero. If the current discount rate is less than the IRR, the project is viable.

Note that the economic project evaluation methods introduced above can be used only for projects with similar characteristics and life span. When comparing two or more alternatives with different life cycles, equivalent uniform annual cost and benefit should be used to replace PW_c and PW_B in Equations 11.3 and 11.4, respectively. These two values are computed as follows:

$$EUAC = PW_C \times CRF(L,i) \tag{11.5}$$

$$EUAB = PW_B \times CRF(L,i) \tag{11.6}$$

where

$EUAB$ = Equivalent uniform annual benefits
$EUAC$ = Equivalent uniform annual costs
CRF = Cost recovery factor, $CRF(n, i) = (i(1 + i)^n)/((1 + i)^n - 1)$

EXAMPLE 11.1

A 20-mile long bus-rapid transit (BRT) system is planned. The construction cost of this project is estimated at \$150000 per mile. Specifically, \$80000 per mile at year 0 and \$70000 per mile at year 1. During the 2-year construction, all traffic needs to detour. The detour may cause an additional network-wide cost (user cost) which is estimated to be \$2500000 per annum. After the BRT system is built, users can save \$1200000 per annum. The project has a 20-year life span after reconstruction and zero salvage value. The operating cost during the life cycle is \$150000 per annum. The benefits such as fare and advertisement revenues generated from this BRT system amount to \$50000 per annum. Maintenance is required 10 years after the construction at a cost of \$50000 per mile. Determine whether the agency should undertake this project if the discount rate is 7%.

Solution

Compute construction cost at first two years

$$\text{Year } 0 = \$80000 \times 20 = \$1600000$$

$$\text{Year } 1 = \$70000 \times 20 = \$1400000$$

Compute maintenance cost at year 12

$$\text{Year } 12 = \$50000 \times 20 = \$1000000$$

Compute annual cost during the life cycle

$$\text{Annual cost} = \$150000 - \$50000 = \$100000$$

The following table summarizes the user benefits and agency costs along the construction years and the 20-year life cycle.

Year	*0*	*1*	*2*	*3–11*	*12*	*13–22*
User benefits	0	−$2500000	−$2500000	$1200000	$1200000	$1200000
Agency costs	$1600000	$1400000	0	$100000	$1100 k	$100 k

$$PW_B = 0 + \frac{-2500}{1+7\%} + \frac{-2500}{(1+7\%)^2} + \frac{1200}{(1+7\%)^3} + \cdots + \frac{1200}{(1+7\%)^{22}} = \$6583.821\ k$$

$$PW_C = 1600 + \frac{1400}{1+7\%} + \frac{0}{(1+7\%)^2} + \frac{100}{(1+7\%)^3} + \cdots + \frac{100}{(1+7\%)^{22}} + \frac{1000}{(1+7\%)^{12}}$$
$$= \$4277.745\ k$$

$$NPW = PW_B - PW_C = \$2306076 > 0$$

$$BCR = \frac{PW_B}{PW_C} = 1.539 > 1$$

Therefore, this project is economically feasible.

11.2 BUS AND BRT PROJECT EVALUATION

The primary source of societal benefits associated with transit is its impacts on travel behavior. Benefits are realized when transportation users realize that their travel cost has been reduced. If a transit improvement does not change the cost of travel perceived by users (of all modes), it cannot affect users' travel behavior. BRT has emerged as a viable option to enhance transportation capacity and provide increased levels of mobility and accessibility. BRT systems vary from one application to another but all provide a higher level of service than traditional bus transportation. Service on BRT systems is generally faster than regular bus service because the buses make fewer stops and may run as often as comparable rail transit systems during peak travel times.

Bus transit has a lot of advantages. Empirical studies indicate that high-quality bus service such as BRT can attract high ridership and stimulate TOD. Generally, a bus service is much cheaper to build and more flexible than rail transit that generally serve the highest-ridership routes. Moreover, due to the flexibility of bus transit, it is much easier to design bus routes and schedules to balance both equity and efficiency. Ultimately, it is desirable to match various modes to demand by route, not choose one mode for all routes of an entire system, to ensure efficiency.

11.2.1 Basic performance evaluation elements

The following procedure provides an objective assessment of bus system performance by calculating a set of variables affecting the choice of transportation mode. The basic elements include round-trip operating speed, in-vehicle travel time (IVTT), waiting time and walking time, ridership, passenger miles, vehicle miles, and vehicle hours, number of buses, number of drivers, revenue, operation costs, and user costs.

11.2.1.1 Introduction

On any given bus route, the operational policy variables may be grouped into the following categories:

Resource variables: Number of buses, number of drivers, operator costs, and user costs
Service variables: Average operating speed, vehicle travel time, walking time, and waiting time
Output variables: Ridership, passenger miles, vehicle miles, vehicle hours, and revenue
Performance measures: Cost efficiency (total cost per hour, total cost per vehicle mile, total cost per passenger, total cost per passenger mile), revenue efficiency (revenue per operating cost, revenue per passenger), vehicle utilization efficiency (annual vehicle miles per vehicle, passenger per vehicle), ridership efficiency (passengers per vehicle mile, passengers per vehicle hour, passengers per dollar of operating cost, and passenger miles per seat mile), and other measures (e.g., deficit per passenger).

Figures 11.4 and 11.5 present the relationship of these variables. These changes occur as a result of the inherent elastic nature of demand in response to changes in the level of service characteristics. Figure 11.4 indicates how the number of stops could ultimately affect ridership. On the one hand, increasing the number of stops without altering other supply elements could reduce the distance between two successive bus stops and consequently reduce the walking time, which could further attract more ridership. On the other hand, more stops means more time spent dwelling, which ultimately reduce the average speed and increases IVTT, potentially shedding ridership.

As illustrated in Figure 11.5, an increase in fare will generally reduce ridership as a direct result of the negative elasticity of demand with respect to fare. However, this reduction in ridership might improve the average operating speed and cause a reduction in the vehicle travel time, consequently inducing an increase in ridership. Yet, the net change in ridership may still be negative on account of the relative magnitudes of these opposing changes. On the other hand, an increase in number of available stops for transfer would decrease the waiting time to the next transfer stop and decrease the average passenger travel time, which will result in an increase in ridership.

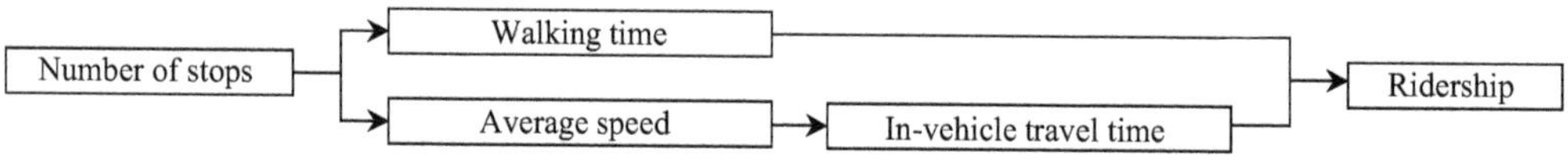

Figure 11.4 Transit ridership estimation.

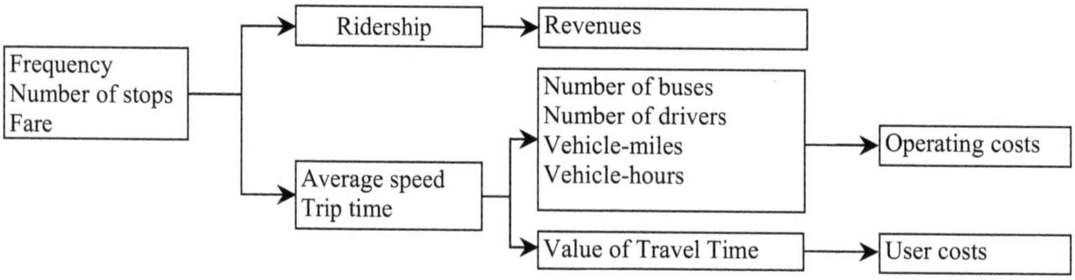

Figure 11.5 Transit system costs and revenues.

11.2.1.2 Round-trip operating speed

Before establishing computational models, the underlying assumptions are made to simplify the computation. These assumptions are listed as follows:

Origins and destinations (each stops) are uniformly distributed along the route.
The probability distribution of number of passengers that board a bus at a given stops follows a Poisson distribution.
Passengers are equally likely to get off at any stop. Multiple people have multiple brains to make multiple decisions.

Service frequency: Equation 11.7 provides the optimal service frequency with respect to passenger waiting time and agency operating time.

$$X = \sqrt{0.5 \times \omega V \times \alpha / OC \times T} \tag{11.7}$$

with

$$X = 60/H \quad \text{if } X \leq 60/H$$
$$X = Q_{peak}/k \quad \text{if } X > 60/H$$

where

X = Frequency of service, buses/hr
ωV = Dollar value of excess time including walking time and waiting time (ω ranges generally from 2 to 3)
α = Coefficient parameters for certain route
OC = Operating costs, dollars/bus/hr
T = Round-trip time, hr
H = Policy headway (60 min)
Q_{peak} = Peak passenger volume, passengers/hr
K = Design capacity for a certain route

EXAMPLE 11.2

A certain route has a peak passenger volume of 750 passengers per hour. The total travel time including layover is 105 min. The operating cost for this route is $250 per bus per hour, and the design capacity for the route is 1200 passengers per hour. Assume that it has 5 as a coefficient parameter, and = 2.5. Estimate the service frequency.

Solution

$$X = \sqrt{0.5 \times 2.5 \times 5/250 \times 1.75}$$

$X = 12.5$ min, using 15 min (frequency always expressed by multiply of 5).

Round-trip operating time: The total round-trip time expression developed by H. Mohring is expressed by Equation 11.8.

$$T = \frac{L}{S} = \frac{L}{S''} + \frac{2QE}{X} + \xi YL(1 - e^{-(2Q/XYL)}) \tag{11.8}$$

where

T	= Round-trip time, hr
X	= Frequency of service, buses/hr
L	= Round trip route length, mi
S	= Average operating speed over the entire route, mph
Q	= Average hourly demand, passengers/route/hr
S''	= Bus running speed of bus, mph
T''	= Running time (i.e., the time spent when the bus is in motion), hr
Y	= Stop density, stops/mi
E	= Time spent per passenger in boarding or alighting from a bus, hr
ξ	= Time spent in a stopping and starting maneuver, hr
$2QE/X$	= Time spent in loading and unloading passengers, hr
ξYL	= Time spent in starting and stopping maneuvers, hr
$1 - e^{-(2Q/XYL)}$	= Possibility that a given stop has passengers alighting or boarding demand

In the above expression, the total round-trip time, is composed of three components: running time, loading/unloading time, and starting/stopping time. The total time passengers spent between the bus's first acceleration from each stop to the last braking to the next one is called running time. It is usually dependent on the average bus running speed of the route. Loading and dwell time is the time taken by the passengers to get on and off the buses, which depends mainly on the passenger demand for each of the bus service. The total time that the bus takes in applying the acceleration and brake is accounted as the bus starting and stopping time. Owing to the probability distribution of number of passengers boarding a bus at a given stop following a Poisson distribution, there would be a number of factors influencing the total starting and stopping time. Factors influencing the total starting and stopping time are: frequency of service in buses per hour, number of stops per mile, round trip route length, average hourly demand, and time spent in a stopping and starting maneuver for each stop.

Round-trip operating speed: The average round-trip operating speed can be calculated by dividing L by the round-trip time as

$$S = \frac{L}{T} = \frac{L}{(L/S'') + (2Q\varepsilon/X) + \xi YL(1 - e^{-(2Q/XY)})} \tag{11.9}$$

EXAMPLE 11.3

A certain route is 22.8 miles (round trip length), with running speed of 28 mph. This route has on average 3 stops per mile, accommodated by 6 buses per hour. In one day, each bus spends 1.1 h for passenger boarding and alighting along the route, with 0.4 h for bus decelerating, acceleration, and mixing with the traffic. The estimated hourly demand for this route is 550 passengers per hour. Compute the round-trip time and average round-trip operating speed.

Solution

$$T = \frac{L}{S''} + \frac{2Q\varepsilon}{X} + \xi YL(1 - e^{-(2Q/XYL)}) = \frac{22.8}{28} + 1.1 + 0.4\left(1 - e^{-\frac{2\times 550}{3\times 6\times 22.8}}\right) = 2.28\,\text{h}$$

$$S = \frac{L}{T} = \frac{22.8}{2.28} = 10\,\text{mph}$$

11.2.1.3 IVTT per passenger

The largest perceived cost of using transit is travel time, which is composed of time spent walking to transit, waiting, riding, and transferring between routes. The IVTT per passenger can be calculated by dividing the average trip length by the average speed presented in following:

$$\text{IVTT} = 60\ \text{M/S} \tag{11.10}$$

where
$IVTT$ = In vehicle travel time, min
M = Average trip length per passenger, mi
S = Operating speed, mph

11.2.1.4 Waiting and walking time

The waiting time (WTT) is accounted for by considering the basic assumption that no user would wait for over 1 h to use the service. The average waiting time is assumed to be equal to one-half of the headway for a headway time less than 30 min and varies linearly between 15 and 22 min for a headway time over 30 min.

$$\begin{aligned} WTT &= \frac{30}{X} \quad \text{if } X \geq 2 \\ WTT &= 8 + \frac{14}{X} \quad \text{if } X < 2 \end{aligned} \tag{11.11}$$

where
WTT = Walking time, min
X = Frequency of service, buses/hr

The specific knowledge about the population distribution and the corresponding walking distance distribution is difficult to acquire. Therefore, the maximum walking distance is assumed to be one-half the distance between the stops. For the sake of the present study, the average walking distance is assumed to be $1/2Y$ miles. And since walking occurs at both ends of the trip, the total walking distance per trip is assumed as $1/Y$. The walking distance can then be converted to walking time (WKT) as below.

$$WKT = \frac{(2 \times (1/2Y))}{(1.466V_w)} = \frac{0.6818}{(Y \times V_w)} \tag{11.12}$$

where
Y = Stop density, stops/mi
V_w = Transit rider average walking speed, ft/sec

11.2.1.5 User costs of travel time

The user costs are traditionally defined as the monetary value that transit users place on their travel. The function built to estimate this value is shown as follows:

$$UC_i = V(IVTT)_i + \omega V(WKT + WTT)_i \tag{11.13}$$

where

UC_i = The user costs per passenger in the period i
V = Dollar value of vehicle travel time
ωV = Dollar value of excess time (ω ranges generally from two to three)

11.2.1.6 Ridership

The effect of the change in various influencing parameters on the ridership is quantified by calculating the change in the ridership numbers before and after any change to the system. The demand function is assumed to be the product form with constant elasticity so that the new ridership level (Q_1) after a small change (Δ) in the service variables is obtained from the equation below

$$Q_1 = Q_0\left\{1+\alpha\left[\frac{\Delta(IVTT)}{IVTT_0}\right]+\beta\left[\frac{\Delta(WKT+WTT)}{WKT_0+WTT_0}\right]+\gamma\left[\frac{\Delta(FARE)}{FARE_0}\right]\right\} \tag{11.14}$$

where

α, β, γ = Demand elasticity with respect to in vehicle travel time, excess travel time, and fare

The fare, change in vehicle travel time, and excess travel time are divided into N smaller components to have better accuracy according to a different period of time, and the equations are solved for N time. It is assumed that there are n distinct periods, and i denotes such period, where $i = 1, 2, 3, \ldots, n$. The hourly impact variable in for i^{th} period is then computed as follows:

$$Q_{i1} = Q_{i0}\left\{1+\alpha\left[\frac{\Delta(IVTT_i)}{IVTT_{i0}}\right]+\beta\left[\frac{\Delta(WKT_i+WTT_i)}{WKT_{i0}+WTT_{i0}}\right]+\gamma\left[\frac{\Delta(FARE_i)}{FARE_{i0}}\right]\right\} \tag{11.15}$$

11.2.1.7 Passenger miles

The total distances traveled by transit passengers within a unit time interval is a direct measure of transit service output quantity. This is always recorded using passenger miles per hour ($PMPH$). The cumulative sum of distances traveled by transit passengers in the ith hour on a certain route is termed as passenger miles and it can be quantified as below

$$PMPH = Q_i \times M_i \tag{11.16}$$

where

Q_i = Average hourly demand in the ith hour
M_i = Average trip length per passenger in miles in the ith hour

11.2.1.8 Vehicle miles

The total number of miles that all vehicles travel during the reporting period, from the time they pull out to go into revenue service to the time they pull in from revenue service, is termed vehicle miles. This term could also be documented on an hourly basis, which is vehicle miles per hour ($VMPH$)

$$VMPH_i = X_i \times L \tag{11.17}$$

where

L = Round trip route length, mi

X_i = Frequency of service in buses per hour in the ith hour

11.2.1.9 Vehicle hours

Just as vehicle miles is the cumulative measure of transit service in the space domain, vehicle hours is the cumulative measure in the time domain. The key concept in this context is platform hours—that is, the number of hours buses are on the road for a given route, including:

Revenue time: The number of hours buses are operating scheduled trips for a given route. This time does not include layover or deadhead time.

Deadhead time: The scheduled time spent on driving to and from the base or between trips on different routes. Passengers may not be conveyed on deadheading trips.

Layover/recovery time: The scheduled time spent at a route's terminal between consecutive trips on a single bus. For example, the schedule arranges for a bus to arrive at its terminal at 4:10 PM and to leave this terminal at 4:30 PM. The "layover" or "recovery" time for this bus would be 20 min. Layover or recovery time is designed for the purpose of allowing bus drivers a break and setting a time cushion in case the preceding trip is delayed.

Vehicle hours, referring to platform hours, can be considered the total time that vehicles are scheduled to or actually travel, from the time they pull out of the garage to go into service to the time they pull in from service. This term can also be documented into an hourly rate known as vehicle hours per hour (*VHPH*)

$$VHPH_i = X_i \times (L/S_i)(1 + LOF) \tag{11.18}$$

where

X_i = Frequency of service in buses in the ith hour

S_i = Operating speed of buses in the ith hour

L = Round trip route length, mi

LOF = The layover time factor as a fraction of round trip time

11.2.1.10 Number of buses

One important resource variable is number of buses. The number of buses needed during any period i is computed as follows. Note that the computed value represents the lower bound of the actual number of buses required.

$$N_{bus,i} = X_i \times (L/S_i) \tag{11.19}$$

where

$N_{bus,i}$ = Number of buses in the ith hour

X_i = Frequency of service in buses in the ith hour

S_i = Operating speed of buses in the ith hour, mph

L = Round trip route length, mi

11.2.1.11 Number of drivers

The number of drivers required for daily operation is a function of run cutting, labor rules, and the peak to off-peak service ratios. Direct estimating is very difficult and unnecessary. Therefore, several simplifying assumptions are made to estimate the value with a reasonable accuracy level. For example, denote a certain average ratio of the number of pay hours to platform hours as *R*, assuming that a driver is paid for *N* hours per day on average. Then, the number of drivers required on any day can be estimated by using Equation 11.20.

$$N_{drivers} = (\mathrm{VHPD} \times \mathrm{R})/\mathrm{N} \tag{11.20}$$

where

$N_{drivers}$ = Number of drivers per day
VHPD = Vehicle hours per day
R = Average ratio of the number of paid hours to platform hours
N = Average number paid hours of a driver

11.2.2 Agency costs

11.2.2.1 Revenue

The fare collected from transit riders is considered the operating revenue. Usually, fares consist of direct payments by passengers (fare-box revenues), as well as revenues from the sale of weekly and monthly passes.

$$REV_i = Q_i \times FARE_i \tag{11.21}$$

where

REV_i = Revenue in the ith hour
Q_i = Ridership in the ith hour
$FARE_i$ = Fare in the ith hour

11.2.2.2 Operating costs

Operating cost is a major agency cost in daily transit operations. This term is estimated on an hourly basis as shown in Equation 11.22

$$OC_i = \theta \times VH_i + \mu \times VMPH_i \tag{11.22}$$

where

OC_i = The operating cost in the period *i*
θ = Unit costs of bus operation per vehicle hour for a bus of a particular size, fuel price at certain time, and operation scenario
μ = Unit costs of bus operation per vehicle mile for a bus of a particular size, fuel price at certain time, and operation scenario
VH_i = Vehicle hours in the period *i*
$VMPH_i$ = Vehicle miles per hour in the period *i*

11.2.2.3 Total costs

The total cost refers to total agency cost, including capital cost, operating cost, and maintenance cost. The capital cost includes bus purchase price, startup costs, fueling station price, and so forth. The operating cost covers driver wages, fuel and energy costs, administrative costs, and marketing costs. The maintenance cost refers to preventative maintenance cost, corrective maintenance cost, and mid-cycle overhaul cost. Therefore, the total cost could be obtained by taking the sum of these three types of cost and considering the discount rate.

Another expression for computing the total cost can be based on the unit rate for each cost item. The final estimated total cost is the computed by adding together all individual costs.

EXAMPLE 11.4

An agency has developed formulas to allocate bus operating costs, and this formula is expressed as

$$C=11.13H+0.28M+20059.22V+0.06R$$

where

C = Annual costs of system operation
H = Annual vehicle-hours of service
M = Annual vehicle-miles of service
V = Peak vehicle needs
R = Annual system revenue

Assume that this agency has operated 750000 annual vehicle hours, with 0.5 million vehicle miles of travel. The peak service needed is 5 h daily and this agency operates this peak service only on weekdays (with using 250 weekdays per year), and the annual system revenue is 1.5 million dollars. Find the total agency costs.

Solution

$$\begin{aligned}\text{Total cost} &= 11.13H + 0.28M + 20059.22V + 0.06R \\ &= 11.13\times 750000 + 0.28\times 0.5 + 20059.22\times 5\times 250 + 0.06\times 1.5 \\ &= 33651525\end{aligned}$$

11.3 RAIL TRANSIT PROJECT EVALUATION

Rail transit systems play a key role in TOD. In this type of system, rail transit stations are designed to serve as a hub for local economic development. Presumably, communities surrounding rail transit stations tend to receive economic benefits as result of increased commuter access to the area. This may include increased commercial activity, including an increase in the number of permanent jobs as employers relocate to areas with better accessibility. In addition, the residents in the neighborhood of the rail transit stations could benefit from the system because of the resulting increased property values for real estate nearby.

11.3.1 On-line travel time

Station to station travel time: The time interval between two adjacent stations includes running time and station standing time. As illustrated in Figure 11.6, the running time is the time the train spent moving (represented by t_1 and $t_3 + t_4 + t_5$), and dwell time is the duration of train standing at a station for passenger boarding and alighting purpose (represented by t_2). Therefore, the travel time is computed as follows:

$$T_{si} = t_{ri} + t_{si} \tag{11.23}$$

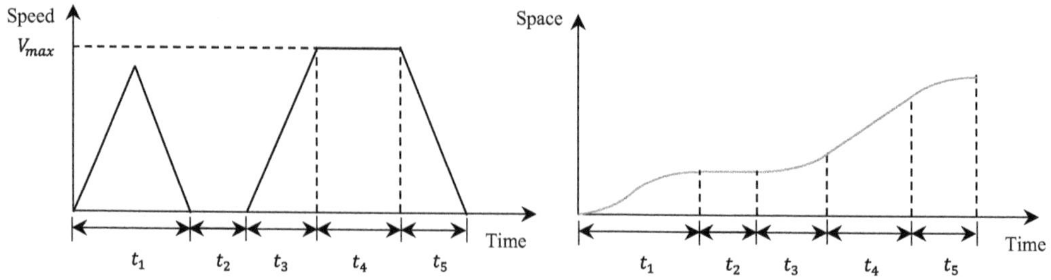

Figure 11.6 Rail transit travel time.

where

T_{si} = Station to station travel time in minutes on any spacing i
t_{ri} = Running time on any spacing i
t_{si} = Station standing time on any spacing i

Station to station running time: The running time should include three time components: acceleration time, running time at maximum speed, and deceleration time. Sometimes the train cannot reach the maximum speed because of the short station spacing. In this case, the running only consists of acceleration and deceleration. The running time is computed as follows:

$$t_{ri} = t_a + t_b = \sqrt{\frac{2(a+b)S'}{ab}} \tag{11.24}$$

where

a, b = Average acceleration and deceleration rates, m/s^2
S' = Distance between two consecutive stations, in m

If the station spacing is long enough to enable the train to reach the maximum speed, the running time is computed as follows:

$$t_{ri} = \frac{3.6S}{V_{max}} + \frac{V_{max}}{7.2}\left(\frac{1}{a} + \frac{1}{b}\right) \tag{11.25}$$

where

S = Distance between two consecutive stations, in m
V_{max} = The maximum speed in km/hr

Operating time: The operating time is the scheduled time interval between departure of a train from one terminal and its arrival at the other terminal on the line. The operating time is therefore the sum of station-to-station travel time for all stations i between terminals, as shown by Equation 11.26

$$T_0 = \sum_i T_{si} = \sum_i (t_{ri} + t_{si}) \tag{11.26}$$

where

T_0 = Operating time of a rail transit line

Terminal time: Terminal time is the time a train (or bus) spends at a line terminal (strictly, after a t_s is subtracted). Terminal time is extremely important for a successful transit operation. It gives time to turn vehicles, rest/change the driver, adjust the schedule, and recover delays incurred in travel. Therefore, the terminal time should be long enough to meet these purposes. The way to measure terminal time is slightly different from other time. We use a percentage to designate the terminal time as represented by Equation 11.27

$$\gamma = \frac{t_{t1} + t_{t2}}{2T_0} \times 100\% \tag{11.27}$$

where

t_{t1} = Terminal time at one terminal
t_{t2} = Terminal time at another terminal

The value of γ usually varies between 10% and 30%, the typical average being around 15%. On lines with long uniform headways, cycle times (introduced later), being integer multiples of headways, must sometimes include long terminal times, resulting in $\gamma > 30\%$. It is intuitive that the transit routes with such characteristics are more likely to have longer delay and put more stress to drivers. Thus, more terminal time is needed. For example, bus transit routes always have peak-hour terminal times longer than off-peak terminal times.

Cycle time: Cycle time refers to the total round-trip time on a line or the interval between two consecutive times a transit unit in regular service leaves the same terminal. It consists of operating times for both directions and both terminal times

$$T = T_{01} + T_{02} + t_{t1} + t_{t2} \approx 2(T_0 + t_t) \tag{11.28}$$

where

T_{01} = Operating time of one direction
T_{02} = Operating time of another direction

Deadhead time: Deadhead time is the portion of travel time during which the transit unit is not in passenger service. It includes travel from the depot to the line and back or between lines when a transit unit is reassigned. Deadhead time is not directly productive, and so needs to be minimized.

Platform time: Platform time is the total time a transit unit is in operation. When a transit unit makes k round trips on a line and has both deadhead times equal, its platform time is

$$T_p = kT + 2t_d \tag{11.29}$$

where

T_p = Platform time
T = Cycle time
t_d = Deadhead time

Platform time sometimes is also affected by a portion of terminal times for schedule adjustments or if the transit unit is assigned to other lines with different cycle time. Platform time represents the net on-duty time for computing labor requirement.

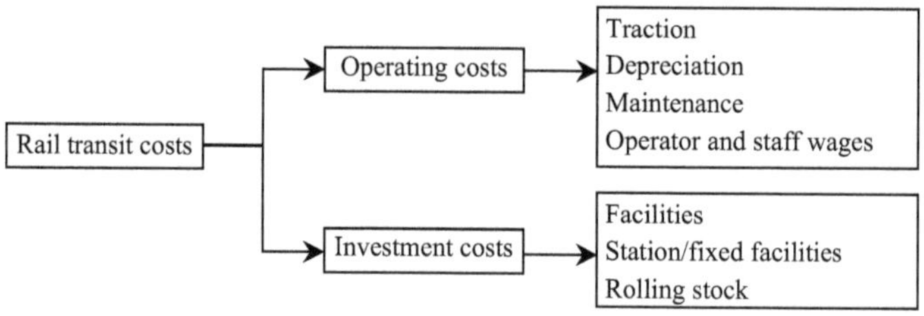

Figure 11.7 Rail transit cost components.

11.3.2 Agency costs

Rail transit cost items are categorized mainly into two broad categories: investment cost and operating cost. The investment cost (also known as capital cost) is the expenditure for realization or the purchase of components of rail transit construction (e.g., infrastructure, rolling stock, installations, etc.). And the operating cost (also known as management cost) contains the costs for the operation and maintenance of transit service. Figure 11.7 depicts these cost components.

11.3.2.1 Rail facility costs

The physical transit facility costs are the sum of all cost items. These cost items are studies, land and right cost, infrastructure building, track built, electric traction equipment, and controlling system

$$C_{PTF} = C_{STUD} + C_{LAND} + C_{BUILD} + C_{TRACK} + C_{ELEC} + C_{SIGN} \tag{11.30}$$

where

C_{PTF} = The cost spent on feasibility study, preliminary study, and the project
C_{STUD} = The cost spent on land and right
C_{BUILD} = The cost spent on main infrastructure building works
C_{TRACK} = The cost spent on trackage
C_{SIGN} = The cost spent on traction equipment
C_{ELEC} = The cost spent on signaling system

Specifically, the study cost includes costs for feasibility study, preliminary study, and project study. This cost is related to the transit route length. The cost for land and rights is related to the characteristics of the location, such as population density. Infrastructure building cost includes preparation of the ground, embankments, drainage, structures, fences, noise-protection equipment, and so on. It is always expressed in per unit length cost. Rail transit track cost includes ballast, sleepers or crossters, rail fastening, rails, welds or fish-platings, laying, and initial additional maintenance. Traction equipment cost includes cost of substations and the cost of catenary. The signaling cost includes cables, automatic block system, spot repetition of signal, cab signal, communication channel between the control center and the train, and the level crossing with light and acoustic signals and automatic barriers.

Fixed equipment investment cost: The fixed equipment cost includes stations, locomotive service and repair facilities, maintenance shops for rolling stock, track, and so on.

Rolling stock investment cost: Rolling stock investment cost refers to the purchase of vehicles. This cost is related to the number of trains needed for the normal operation with breakdowns and maintenance considered.

11.3.2.2 Rail operating costs

Operating costs refer to the expenditures for operating a rail transit service. It includes five cost components: traction, capital depreciation or leasing costs of rolling stock, maintenance of rolling stock, crew salary, and tolls for the use of infrastructures. Thus the operating cost model can be represented as the sum of all cost items

$$C_{OP} = C_{TR} + C_{DEP} + C_{MAN} + C_{SAL} + C_{ACC} \tag{11.31}$$

where

C_{OP} = The operating cost
C_{TR} = The traction cost
C_{DEP} = The rolling stock depreciation cost
C_{MAN} = The rolling stock maintenance cost
C_{SAL} = The salary cost
C_{ACC} = The access fee charged by the rail owner

Traction cost is the cost related to energy consumption for traction. The actual value depends on the weight of the train, type of engine, route alignment, its speed profile, and length of the route. The rolling stock depreciation cost is the difference between rolling stock purchase price and the cumulative depreciation. If the rolling stock is leased from elsewhere, the rolling stock lease cost should be considered instead of rolling stock depreciation cost. Salary cost is the amount of cost paid as the salary of crews including driving crew and onboard crew. The access cost is the access fee for using the rail if it is not owned. Otherwise, the rail maintenance cost should be considered instead (Gattuso and Restuccia, 2014).

11.4 NONMOTORIZED TRANSPORTATION PROJECT EVALUATION

Nonmotorized transportation refers to human-powered transportation modes, including walking and biking. From the transportation agency's perspective, providing the public with more environment-friendly options to travel is an important strategy to reduce energy consumption and air pollution, which contributes to building sustainable transportation.

Data trends indicate that Americans are seeking ways to drive less. Therefore, U.S. cities are adopting more innovative facilities like many European cities. One example of recently emerging innovative facilities is the protected bike lane or cycle track, is a bike lane that is physically protected from encroachment by motorized vehicles by some form of barrier. Another is bike share systems. Transit facilities and nonmotorized transportation facilities affect each other. Transit often serves as the primary mode in a trip, while nonmotorized transportation serves the last mile. High-quality transit service will encourage more nonmotorized transportation activities. Specifically, if transit is used as a primary commute mode, the resulting reduced need for automobile ownership may result in increased walking and biking. In addition, better nonmotorized transportation facilities could also help increase the ridership of transit. Both modes can enable the public to drive less, save money, and may also save time.

11.4.1 Impact categories

When evaluating a certain project, one must analyze the issue from multiple aspects. For example, FHWA initiated the Nonmotorized Transportation Pilot Program (NTPP) in August 2005. Up until the end of 2013, this program had provided over $25 million in contract authority to four pilot communities (Columbia, Missouri; Marin County, California; Minneapolis area, Minnesota; and Sheboygan County, Wisconsin) for pedestrian and bike infrastructure and nonmotorized programs. The evaluation of this program analyzed its effectiveness from four aspects: mode shift, access and mobility, environment and energy, safety and public health (Lyons et al., 2014).

11.4.1.1 User benefits

Implementing an nonmotorized transportation facility improvement project can directly benefit both existing nonmotorized travelers and induced former drivers. Indirectly, it can bring additional cost savings to users, such as reduced congestion, automobile ownership cost savings, and so on, as mentioned before.

The user benefits can be evaluated using the following methods:

Avoided costs: This method assesses the user benefits by evaluating the resulting reduced consumer expenditures on transportation cost (fuel, automobile ownership cost, etc.), travel time cost, and so on.

Contingent valuation: This method directly surveys the community to determine their willingness to pay for certain nonmotorized transportation projects.

Hedonic pricing: Similar to the hedonic pricing method using for transit projects, this method investigates the effects of a project on nearby property values. The value can be obtained from the local housing market.

11.4.1.2 Equity benefits

Equity (also called justice and fairness) refers to the distribution of impacts (benefits and costs) and whether that distribution is considered fair and appropriate (Litman, 2002). There are three general types of equity:

Horizontal equity (also called fairness and egalitarianism): This type concerns the distribution of impacts between individuals and groups considered equal in ability and need. It requires that users bear the costs of their transportation activities.

Vertical equity (also called social justice): This type concerns the distribution of impacts between individuals and groups of differing race, ethnic, income level, or education background. It requires transportation policies to favor lower-income people to compensate for overall inequities.

Vertical equity with regard to transportation ability and needs: This is a special type of vertical equity, only concerning impact distribution among individuals and groups of differing mobility ability and need. It requires that transportation policies favor mobility-disadvantaged people (such as people with disabilities).

Improving biking and walking conditions basically provides a fair share of resource to all travelers, including both motorist and non-motorist travelers. More specifically, the nonmotorized transportation improvement project provides basic mobility for the "real" public, including economically and socially disadvantaged groups. Therefore, it can help achieve equity objectives.

11.4.1.3 Barrier effect

The barrier effect refers to the travel delay in nonmotorized transportation caused by vehicular traffic. The barrier effect can be considered the congestion of non-motorists while traditional congestion analysis usually ignores this part. The barrier effect reduces the mobility of nonmotorized modes, and further encourage nonmotorized-to-motorized mode shifts, which could result in additional costs such as vehicle ownership cost, congestion cost, and parking cost. Generally, all projects that improve vehicular traffic without considering pedestrians and cyclists may enhance the barrier effect (Litman, 2015).

Case Study 1: Evaluation of a rapid rail operation project

A city plans to build a 20-mile long rapid rail line from the international airport to city center. The rail line contains double tracks and eight stations with 30-second dwell time at each station. Of the eight stations of the rail line running from the airport to the city center, three are located at the airport, three are located in suburban area, and two are located in the urban area. The distances between stations are as follows (Table 11.3):

The data details for the two best types of rail cars are shown below:

Item	*Type I*	*Type II*
Maximum speed	70 mph	110 mph
Constant acceleration rate	4 ft/sec^2	5 ft/sec^2
Constant braking rate	−5 ft/sec^2	−6 ft/sec^2
Investment cost	\$600000/car	\$800000/car
Base operating cost (including driver)	\$180/train-hr	\$150/train-hr
Additional operating cost	\$0.30/car-mi	\$0.20/car-mi
Life time	300000 miles/car	250000 miles/car

According to the airport passenger ridership analysis, it is planned to dispatch 8-car trains during rush hours (6–10 AM, 4–8 PM) with headways of 6 min. For the reminder of the time from 4 AM to midnight, it is planned to run 4-car trains with headways of 12 min. Determine which type of car should be select.

1. Calculate the average train speed from the time the doors close at Station 1 to the time it arrives Station 8 and opens its doors for each type of train.

   Three cases of speed profiles may exist for a train running between two successive stations. For the first case, the distance between two stations is so short that the train cannot reach the maximum speed. For the second case, the distance between two stations is long enough that allows the train to reach the maximum speed but cannot keep the top speed any longer before the need for speed reductions. For the third case, the train could reach and sustain the maximum speed for a certain period. The second case could be considered as the threshold case for the remaining two cases. The three cases of training operation profiles are summarized as follows (Table 11.4):

Table 11.3 Section length data information

Section	*Stations 1–2*	*Stations 2–3*	*Stations 3–4*	*Stations 4–5*	*Stations 5–6*	*Stations 6–7*	*Stations 7–8*
Length (mile)	0.5	0.5	6.0	6.0	6.0	0.5	0.5

Table 11.4 Three cases of train operation profiles

Case	*1*	*2*	*3*
Speed profile	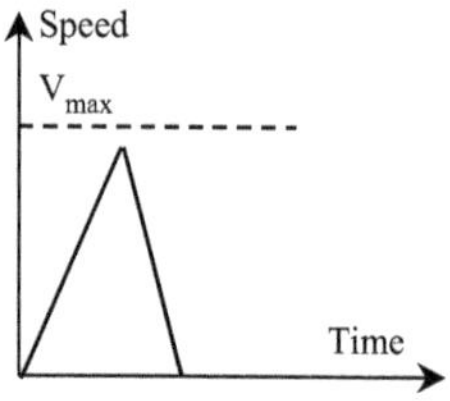	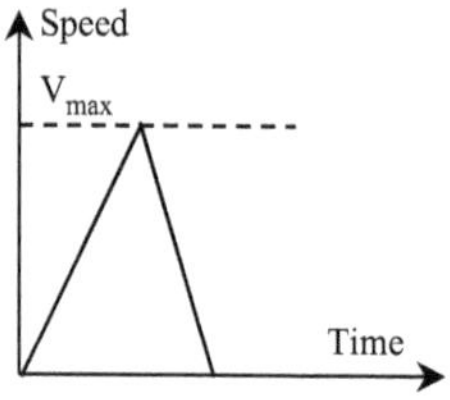	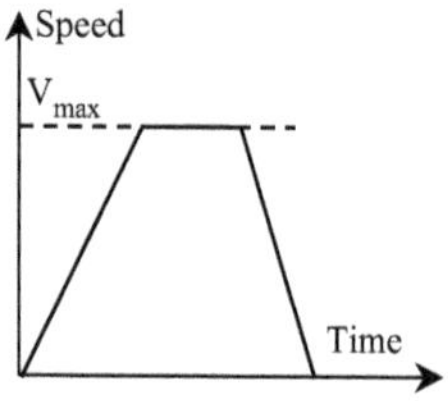
Threshold distance L_s		$\frac{1}{2}\cdot\left(\frac{1}{a}+\frac{1}{b}\right)\cdot V_{max}^2$	
Precondition	$L < L_s$	$L = L_s$	$L > L_s$
Maximum speed	$\sqrt{\frac{2abL}{a+b}}$	V_{max}	V_{max}
Acceleration time	$\sqrt{\frac{2bL}{a(a+b)}}$	V_{max}/a	V_{max}/a
Deceleration time	$\sqrt{\frac{2aL}{b(a+b)}}$	V_{max}/b	V_{max}/b
Running time at V_{max}	0	0	$(L - L_s)/V_{max}$
Running time	$\sqrt{\frac{2bL}{a(a+b)}}+\sqrt{\frac{2aL}{b(a+b)}}$	$\left(\frac{1}{a}+\frac{1}{b}\right)\cdot V_{max}$	$\left(\frac{1}{a}+\frac{1}{b}\right)\cdot V_{max}+\frac{L-L_s}{V_{max}}$

a. Type I train car

$$L_s = \frac{1}{2}\cdot\left(\frac{1}{a}+\frac{1}{b}\right)\cdot V_{max}^2 = \frac{1}{2}\times\left(\frac{1}{4}+\frac{1}{5}\right)\times\left(70\times\frac{5280}{3600}\right)^2 = 2371.6\,\text{ft} = 0.44\,\text{mile} < 0.5\,\text{mile}$$

For sections with 0.5 mile of distance, we could use the Case 3 formula. Then, the running time on these sections can be computed by

$$RT_{0.5} = \left(\frac{1}{a}+\frac{1}{b}\right)\cdot V_{max}+\frac{L-L_s}{V_{max}} = \left(\frac{1}{4}+\frac{1}{5}\right)\times 70\times\frac{5280}{3600}+\frac{0.5\times 5280-2371.6}{70\times(5280/3600)} = 48.8\,\text{sec}$$

For sections with 6 miles of distance, we also could use the Case 3 formula

$$RT_6 = \left(\frac{1}{a}+\frac{1}{b}\right)\cdot V_{max}+\frac{L-L_s}{V_{max}} = \left(\frac{1}{4}+\frac{1}{5}\right)\times 70\times\frac{5280}{3600}+\frac{6\times 5280-2371.6}{70\times(5280/3600)} = 331.7\,\text{sec}$$

The total travel time of one-way operation is the sum of total running time and total dwell time

$$TT_1 = (4\times 48.8+3\times 331.7)+6\times 30 = 1370.3\,\text{sec}$$

The average travel speed is computed as

$$V_I = \frac{20 \times 3600}{1370.3} = 52.54\,\text{mph}$$

b. Type II train car

$$L_s = \frac{1}{2} \cdot \left(\frac{1}{a} + \frac{1}{b}\right) \cdot V_{max}^2 = \frac{1}{2} \times \left(\frac{1}{5} + \frac{1}{6}\right) \times \left(110 \times \frac{5280}{3600}\right)^2 = 4771.9\,\text{ft} = 0.90\,\text{mile} > 0.5\,\text{mile}$$

For sections with 0.5 mile of distance, we could use the Case 1 formula. Then, the running time on these sections can be computed by

$$RT_{0.5} = \sqrt{\frac{2bL}{a(a+b)}} + \sqrt{\frac{2aL}{b(a+b)}} = \sqrt{\frac{2 \times 5 \times 0.5 \times 5280}{6 \times (5+6)}} + \sqrt{\frac{2 \times 6 \times 0.5 \times 5280}{5 \times (5+6)}} = 44.0\,\text{sec}$$

For sections with 6 miles of distance, we could use the Case 3 formula. Then, the running time on these sections can be computed as

$$RT_6 = \left(\frac{1}{a} + \frac{1}{b}\right) \cdot V_{max} + \frac{L - L_s}{V_{max}} = \left(\frac{1}{5} + \frac{1}{6}\right) \times 110 \times \frac{5280}{3600} + \frac{6 \times 5280 - 4771.9}{110 \times (5280/3600)} = 225.9\,\text{sec}$$

$$TT_{II} = (4 \times 44.0 + 3 \times 225.9) + 6 \times 30 = 1033.7\,\text{sec}$$

The average travel speed is computed as

$$V_{II} = \frac{20 \times 3600}{1033.7} = 70.0\,\text{mph}$$

2. Determine the number of cars needed by determining those needed for rush hour operation, considering 10 minutes of extra time for each train to turn around at the end of the line and 10% spare time to cover car downtime.
   a. Compute the cycle time of each train

   Type I: $\left(\frac{1370.3}{60} + 10\right) \times 2 = 65.7$ min

   Type II: $\left(\frac{1033.7}{60} + 10\right) \times 2 = 54.5$ min

   b. Compute the number of trains needed, considering the rush hour operation (the headway is 6 minutes)

   Type I: $\frac{65.7}{6} \times (1 + 10\%) = 12.05$ use 13

   Type II: $\frac{54.5}{6} \times (1 + 10\%) = 9.99$ use 10

   c. Compute the number of cars needed
   Type I: 13 × 8 = 104 cars
   Type II: 10 × 8 = 80 cars
3. Determine the train hours of operation per day (excluding the turn-around time) and per year, considering 325 days of full operations per year to account for reduced weekend and holiday operations.
   a. Compute the number of batches per day

   Rush hours: 8 hrs/day with 6 minutes of headway

Non-rush hours: 12 hours/day with 12 minutes of headway

Number of batches: $8 \times \frac{60}{6} + 12 \times \frac{60}{6} = 140$ batches/day

b. Compute the train hours of operation per day and per year

Operating time per batch

Type I: $\frac{1370.3}{60} \times 2 = 45.68$ min/batch $= 0.761$ hr/batch

Type II: $\frac{1033.7}{60} \times 2 = 34.46$ min/batch $= 0.574$ hr/batch

Train hours per day

Type I: $140 \times 0.761 = 106.54$ hrs/day

Type II: $140 \times 0.574 = 80.36$ hrs/day

Adjusted train hours per day considering the 325 days of full operations per year

Type I: $106.54 \times 325/365 = 94.86$ train-hrs/day

Type II: $80.36 \times 325/365 = 71.55$ train-hrs/day

4. Determine which type of cars should be selected for the rail line operations, using 5% discount rate.

a. Compute the operating length per day for each car

Rush hours: 8 hrs/day, 6 minutes of headway, 8 cars/train

$$8 \text{ cars} \times \frac{8 \text{ hrs} \times 60}{6 \text{ min}} = 640 \text{ car-batches}$$

Non-rush hours: 12 hrs/day, 12 minutes of headway, 4 car/train

$$4 \text{ cars} \times \frac{12 \text{ hours} \times 60}{12 \text{ min}} = 240 \text{ car-batches}$$

Total car-miles per day: $(640 + 240) \times 20 \times 2 = 35200$ car-miles/day

Type I: $\frac{35200}{104} = 338.46$ miles/car-day

Type II: $\frac{35200}{80} = 440.0$ miles/car-day

b. Compute the life cycle

Type I: $\frac{300000 \text{ miles/car}}{338.46 \text{ miles/car-day}} = 886$ days

Type II: $\frac{250000 \text{ miles/car}}{440 \text{ miles/car-day}} = 568$ days

c. Compute all cost items

Type I:

Investment cost: $104 \times \$600000 = \$62400000$

Base operating cost (per day): $94.86 \times \$180 = \$17074.8$

Additional operating cost (per day): $35200 \times \$0.30 = \$10560$

Total operating cost (per day): $\$17074.8 + \$10560 = \$27634.8$

Type II:

Investment cost: $80 \times \$800000 = \$64000{,}000$

Base operating cost (per day): $71.55 \times \$150 = \$10732.5$

Additional operating cost (per day): $35200 \times \$0.20 = \$7040$

Total operating cost (per day): $\$10732.5 + \$7040 = \$17772.5$

d. Compute the equivalent uniform daily costs (EUDC)

$$i_{day} = (1 + i_{year})^{1/365} - 1 = (1 + 5\%)^{1/365} - 1 = 1.337 \times 10^{-4}$$

Type I:

$$EUDC_I = 62400000 \times \left[\frac{i_{day}(1 + i_{day})^{886}}{(1 + i_{day})^{886} - 1} \right] + 27634.8 = \$102322/day$$

Type II:

$$EUDC_{II} = 64000000 \times \left[\frac{i_{day}(1 + i_{day})^{568}}{(1 + i_{day})^{568} - 1} \right] + 17772.5 = \$134788/day$$

11.5 CONCLUSION

Since $EUDC_I < EUDC_{II}$, the transit agency should select type I cars.

Case Study 2: Evaluation of an express bus-on-shoulder transit improvement project

A regional transit authority plans to open a new express bus line by authorizing buses running on shoulders of an expressway that connects a southwestern suburb of the metropolitan area to its central business district. The express bus-on-shoulder service is to operate in the AM and PM peak hours and freeway lanes are not affected. However, during the peak periods freeway traffic is expected to reduce because of modal shift. Former freeway users and potentially induced new travelers will be able to use the 50-mph express bus service.

The traffic and facility data assembled for the analysis are presented in Tables 11.5 through 11.8. The affected freeway section is designated "AB." "Alternative I" represents the existing freeway travel lane only option and "Alternative II" represents the freeway travel lane and the proposed express bus-on-shoulder option. Only one general class of bus trip is considered.

a. Assumptions
   - Without express bus line service, all bus users travel by car (2 miles on arterial at 25 mph average running speed and 10 miles on the freeway)
   - The 2-mile arterial maintains +2% grade, zero curvature, with \$169.2/kVMT tangent running cost on +2% grade

Table 11.5 Freeway facility data

Section	*Alternative*	*Section length*	*Design speed*	*Speed limit*	*Number of one-way lanes*	*Grade*	*Curvature*
AB	I	10	70	55	3	0%	0°
	II	10	70	55	3	0%	0°

Table 11.6 Express bus line facility data

System name	*Service speed*	*Seats per bus*	*Driver's wage*
Express	50 mph	50	\$20/hr

Table 11.7 Transit traffic data

Section	Alternative	Study year	Period	*Traffic volume (veh/hr)* One-way volume	One-way capacity	v/c ratio
AB	I	I	Peak hour	5400	7200	0.75
	II	I	Peak hour	4320	7200	0.60
	I	20	Peak hour	6120	7200	0.85
	II	20	Peak hour	4680	7200	0.65

Table 11.8 Bus transit traffic data

Alternative	*Study year*	*Period*	*Hourly patronage*	*Fare per trip*	*IVTT*	*OVTT*	*Annual bus miles*
I	I						
II	I	Peak	2200	\$5.00	30	6.0	1.5×10^6
I	20						
II	20	Peak	3200	\$10.00	30	6.0	1.5×10^6

- Discount rate $i = 5\%$ and analysis period is 25 years
- Extra annual expressway shoulder maintenance cost due to express bus-on-shoulder service = \$1000 per lane-mi;
- Use the following table to conduct freeway user cost computation

V/C	*Design speed (mph)*	*Speed limit (mph)*	*Tangent running cost (\$/kVMT)*	*Travel time (h/kVMT)*	*Added speed change cost (\$/kVMT)*
0.6	70	55	155.2	18.5	4.6
0.65	70	55	154.4	18.7	4.6
0.75	70	55	153.1	19.0	4.6
0.85	70	55	148.9	19.6	4.6

- Average occupants per vehicle = 1.2 and assume that walking and waiting time is valued at two times in-vehicle travel time per person hour
- Bus service is available on 5 weekdays for 52 weeks a year
- All data found in the tables are for peak period hourly values. Carry out the analysis only for AM and PM peak periods consisting of 6 hours
- Use the value of travel time \$20/hr.

b. Transit costs

- Transit user costs are given by the formula

$$TU = 1000 \times \left[V \times (IVTT/60 + (OVTT/60 \times w) + F\right]$$

where

V = Value of in-vehicle travel time, \$8/person
$IVTT$ = In-vehicle travel time, 30 min
$OVTT$ = Out of vehicle travel time, 6 min
w = 2
F = Transit fare, dollars

- Annual transit operating cost in million dollars are given as

$$TOC = 2.37 \times Q^{1.013} \times H^{0.785} \times S^{-0.862} \times e^{0.002P}$$

where
Q = Annual bus mileage, million miles
H = Bus driver's hourly wage, dollars/hr
S = Service speed, mph
P = Bus size

c. Expected benefits from transit improvement
- Change in user costs associated with freeway auto lanes (25 mile 3-lane freeway versus 25-mile 3-lane freeway, with auto traffic reductions)
- Change in user costs related to express bus-on-shoulder line (2-mile arterial +25 mile one-freeway shoulder lane versus 10 mile one-freeway shoulder lane)
- Transit revenue

d. Costs associated with transit improvement
- Arterial user costs
- Freeway user costs
- Extra maintenance costs of freeway shoulders
- Transit operating costs

SOLUTION

i. Compute user benefits for auto travelers on the expressway
  a. Compute the change in user costs associated with freeway auto lanes (10-mile 3-lane freeway with vehicle volume changes after introducing the express bus-on-shoulder service). The highway user cost items are summarized in terms of dollar per thousand vehicle miles of travel (kVMT) (Table 11.9).
  b. Compute the total kVMT per year (Table 11.10).
  c. Compute the highway auto user benefit by computing change in consumer surplus (Figure 11.8).
  User Benefit at Year 1: $UB_1 = (537.7 - 529.8) \times (67392 + 84240)/2 = \$598946$
  User Benefit at Year 20: $UB_{20} = (545.5 - 533.0) \times (73308 + 95472)/2 = \$1054875$
  d. Compute the present worth of total highway user benefits for the entire 20 years.

Table 11.9 Highway user cost change for peak hours

Alternative	*Cost item*	*Year 1*	*Year 20*
I	V/C	0.75	0.85
	TT	19.0 h/kVMT × \$20/h = \$380/kVMT	19.6 h/kVMT × \$20/h = \$392/kVMT
	$VOC_{tangent}$	\$153.1/kVMT	\$148.9/kVMT
	$VOC_{speed\ change}$	\$4.6/kVMT	\$4.6/kVMT
	Sum	\$537.7/kVMT	\$545.5/kVMT
II	V/C	0.6	0.65
	TT	18.5 h/kVMT × \$20/h = \$380/kVMT	18.7 h/kVMT × \$20/h = \$380/kVMT
	$VOC_{tangent}$	\$155.2/kVMT	\$154.4/kVMT
	$VOC_{speed\ change}$	\$4.6/kVMT	\$4.6/kVMT
	Sum	\$529.8/kVMT	\$533.0/kVMT

Table 11.10 VMT change

Alternative	*Year 1*	*Year 20*
I	5400 veh/hour × 6 hours × 10 miles × 52 weeks/year × 5 days/ week/1000 = 84240 kVMT/year	6120 veh/hour × 6 hours × 10 miles × 52 weeks/year × 5 days/ week/1000 = 95472 kVMT/year
II	4320 veh/hour × 6 hours × 10 miles ×52 weeks/year × 5 days/ week/1000 = 67392 kVMT/year	4680 veh/hour × 6 hours × 10 miles × 52 weeks/year × 5 days/ week/1000 = 73008 kVMT/year

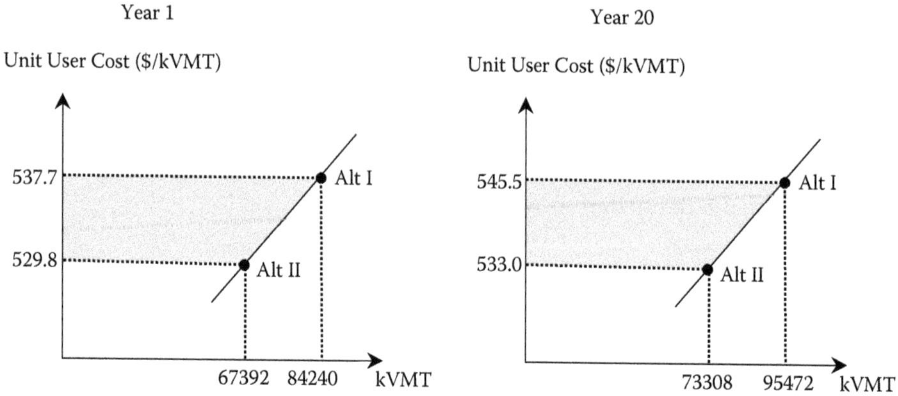

Figure 11.8 Changes in consumer surplus.

Denote

B_x = The payment at the end of the year x
B_y = The payment at the end of the year y
n = Analysis period in years
r = The constant annual growth/decline rate of benefit
i = Interest rate
PW = Present worth of the entire cash flow
PWF = Present worth factor

The following formulas are used to approximately will yield a present worth factor that can convert a geometric cash flow into a present worth for the analysis period, assuming continuous compounding of interest.

$$\alpha = \frac{B_x}{B_y}$$

$$r = \frac{\ln(\alpha)}{x - y}$$

$$PWF = \frac{1 - e^{-[(i-r)n]/(1+r)}}{i - r}$$

$$PW = B_1 \times PWF$$

In this case, we assume the benefit of each year is received at the beginning of the year.

$$\alpha = \frac{B_{20}}{B_1} = \frac{1054875}{598946} = 1.76$$

$$r = \frac{ln(\alpha)}{20-1} = \frac{\ln(1.76)}{19} = 0.0298$$

$$PWF = \frac{1-e^{-[(i-r)n]/(1+r)}}{i-r} = \frac{1-e^{-[(0.05-0.0298)25]/(1+0.0298)}}{0.05-0.0298} = 19.19$$

$$PW_{Auto} = UB_1 \times PWF = 598946 \times 19.19 = \$11493773$$

ii. Compute transit user costs

a. Alterative I:

The following cost items are related to traveling on the 2-mile arterial.

Travel time on arterial = 1000/25 = 40 h/kVMT

Travel time cost = 40 h/kVMT × \$20/h = \$800/kVMT

Tangent running cost on 2% grade = \$169.2/kVMT

Added speed change cost = \$5.1/kVMT

Total per kVMT cost = 800 + 169.2 + 5.1 = \$974.3/kVMT

Total per kV cost = \$974.3/kVMT × 2 miles = \$1948.6/kV

The transit user costs include the cost of 10-mile freeway traveling which has already been computed previously.

Assume that the arterial user cost will remain the same as for year 1. Therefore, one-way user cost would be

Alternative I in year 1: \$537.7/kVMT × 10 miles + \$1948.6/kV = \$7325.6/kV

Equivalently: 7325.6/1.2 = \$6104.67 per 10^3 person trips = \$6169.67/kp

Alternative I in year 20: \$545.5/kVMT × 10 miles + \$1948.6/kV = \$7403.6/kV

Equivalently: 7403.6/1.2 = \$6169.67/kp

b. Alternative II:

Transit user costs in this case are given by

$$TU_1 = 1000 \times \left(V \cdot \left(\frac{IVTT}{60} + \frac{OVTT}{60} \cdot w \right) + F \right) = 1000 \times \left(8 \times \left(\frac{30}{60} + \frac{6}{60} \times 2 \right) + 5 \right) = \$10600/kp$$

$$TU_{20} = 1000 \times \left(V \cdot \left(\frac{IVTT}{60} + \frac{OVTT}{60} \cdot w \right) + F \right) = 1000 \times \left(8 \times \left(\frac{30}{60} + \frac{6}{60} \times 2 \right) + 10 \right) = \$15600/kp$$

iii. Compute the transit user benefits using consumer surplus (Table 11.11)

User benefits in year 1: UB_1 = (7325.6 − 10600) × ((0 + 2.2)/2) × 6 × 5 × 52 = −\$5618870

User benefits in year 20: UB_{20} = (7403.6 − 15600) × ((0 + 3.2)/2) × 6 × 5 × 52 = −\$20458214

iv. Compute the present worth of total transit user benefit for the entire 25 years

v. $\alpha = \dfrac{-20458214}{-5618870} = 3.64$

Table 11.11 Transit user costs summary

Alternative	*Year*	*Mode*	*Period*	*One-way user cost*
I	1	Auto	Peak	$7325.6/kp
II	1	Bus	Peak	$10600/kp
I	20	Auto	Peak	$7403.6/kp
II	20	Bus	Peak	$15600/kp

vi. $r = \dfrac{\ln(3.64)}{20-1} = 0.0680$

vii. $PWF = \dfrac{1-e^{-[(i-r)n]/(1+r)}}{i-r} = \dfrac{1-e^{-[(0.05-0.068)25]/(1+0.068)}}{0.05-0.068} = 29.11$

viii. $PW_{Transit} = UB_1 + PWF = \$5618870 \times 29.11 = -\$163565305$

Compute the transit agency costs

a. Using the given transit operating cost formula, we have:

$$TOC_1 = 2.37 \times Q^{1.013} \times H^{0.785} \times S^{-0.862} \times e^{0.002P} = 2.37 \times 1.5^{1.013} \times 20^{0.785} \times 50^{-0.862} \times e^{0.002\times 50} = \$1423490$$

$$TOC_{20} = 2.37 \times Q^{1.013} \times H^{0.785} \times S^{-0.862} \times e^{0.002P} = 2.37 \times 2^{1.013} \times 20^{0.785} \times 50^{-0.862} \times e^{0.002\times 50} = \$1905099$$

$$\alpha = \frac{1905099}{1423490} = 1.34$$

$$r = \frac{\ln(1.34)}{20-1} = 0.0154$$

$$PWF = \frac{1-e^{-[(i-r)n]/(1+r)}}{i-r} = \frac{1-e^{-[(0.05-0.0154)25]/(1+0.0154)}}{0.05-0.0154} = 16.57$$

$$PW_{TOC} = TOC_1 \times PWF = \$1423490 \times 16.57 = \$23587229$$

b. The present worth of total shoulder maintenance cost for express bus lane at 25-year analysis period

$$PW_{MC} = 1000 \times 1 \times 10 \times \left[\frac{e^{0.05\times 24}-1}{e^{0.05\times 24}(e^{0.05}-1)} + 1\right] = \$146296$$

c. The total agency costs

$$PW_{AC} = PW_{TOC} + PW_{MC} = \$23587229 + \$146296 = \$23733525$$

ix. Compute the total user benefits

$$\text{Total benefits} = PW_{Auto} + PW_{Transit} = \$11493773 - \$163565305 = -\$152071532$$

x. Conclusion

The net present worth is $NPV = -\$152071532 - \$23733525 = -\$175805057 < 0$

Therefore, the proposal is not economically feasible.

PROBLEMS

11.1 List the major objectives and corresponding performance measures for a bus transit asset management project.

11.2 Help a transit agency select a project among the three alternatives. The data is given as follows: Comment on each project in terms of net present worth, benefit-to-cost ratio, and internal rate of return values.

Table P11.1 Alternative benefit and cost details

		Benefit		*Cost*	
Alternative	*Life span*	*Year 0*	*Year 1,2,..., 7*	*Year 0*	*Year 1,2,..., 7*
1	7 years	0	2545	18215	1005
2	7 years	0	3817	24707	1501
3	7 years	0	2428	14325	1907

11.3 The following table documents the detailed passenger travel information on a one-way operation of the bus rapid transit line. Its route consists of 6 stations connecting two rail transit stations. Based on the information given, answer the following question.

Table P11.2 Detailed BRT usage

Number of passengers	*Boarding station*	*Alighting station*	*Segment distance (mile)*	*Segment running time (min)*
2	1	2	4	7.2
7	1	3		
5	1	4		
14	1	5		
2	1	6		
4	2	3	3.5	6.5
7	2	4		
10	2	5		
8	2	6		
2	3	4	2.5	4.8
4	3	5		
2	3	6		
1	4	5	2	3.6
5	4	6		
1	5	6	4	8.2

a. What is the total passenger miles within this one-way BRT service?

b. What are the passenger loads on each section, and find the section with the maximum passenger loads?

c. Determine the number of boarding/alighting passengers at each station. Suppose this bus has two doors, one of which is for boarding only and another for alighting only. Use the following formula to determine the dwell time at each station.

Assume $t_a = 2s$ and $t_a = t_b = 1.5$ s/passenger, $t_d = p_a t_a + p_b t_b + t_{oc}$

where

t_d = The average dwell time, s
p_a/p_b = The alighting/boarding passengers per bus through the busiest door, respectively
t_a/t_b = The alighting/boarding time per passenger
t_{oc} = The time lost prior to opening and after the closing of doors

d. Compute the total and average in-vehicle travel time and cost using a unit rate of 22.0/hour
e. Compute the operating time of this bus, assuming terminal time at both terminals is 20 min and the total travel time of running the other direction is 29 min.

11.4 What are the main operating expenditures in daily rail transit operations?

11.5 How do transit and nonmotorized (active) transportation improvements contribute to fairness development?

Chapter 12

Economic analysis of rail facility preservation

Rail transportation consists of rail transit and freight rail, which is one of the primary modes of transportation supporting both people and goods movements. The economic analysis of rail facility preservation begins with brief descriptions of major rail facility cost items in the facility service life span. It then introduces cost estimation models, followed by descriptions of track maintenance and renewal preservation treatments. Some examples are provided for rail facility cost calculation, rail vehicle selection, and shortline rail preservation.

12.1 RAIL FACILITY LIFE-CYCLE COSTS

Rail tracks are the primary infrastructure of the rail mode. They also constitute the principal cost in rail construction. The term of life-cycle cost was first introduced by the FHWA in 1990s for analysis and design of highway pavements. It considers the total cost of constructing, maintaining, and operating the transportation facility in its service life cycle. The life-cycle cost based analysis could help evaluate the cost-efficiency of various types of transportation investment alternatives.

12.1.1 Rail track components

As seen in Figure 12.1, rail track components can be categorized into two groups: (i) superstructures consisting of sleepers, rails, ballast, and joints; and (ii) substructures that include sub-ballast and subgrade (AREMA, 2003).

Rails: Rails are the crucial element in railway structure. They direct train vehicles and transfer train vehicle loads to the sleepers. The functions of rails are distributing the train vehicle acceleration, maintaining adhesion for braking force, and providing a smooth surface transition.

Sleepers: Sleepers (or ties) are the transverse elements that connect two rails together. The main objectives of these elements are distributing transverse loads from rails to ballast and maintaining track gauge by holding the fasting system.

Fastener System: Fastener systems are the elements that are used to fix rails to railroad ties. Fasteners are used to keep the train on the track by resisting vertical, lateral, and longitudinal railway movements.

Ballast: Ballast is the crushed stone layer under sleepers. The main purpose of using ballast is to carry load from sleepers to subgrade so as to keep the track strong against vertical, longitudinal, and lateral movements. These movements are absorbed by the irregular shape of ballast materials. In addition, basalt helps with water drainage under rails and sleepers.

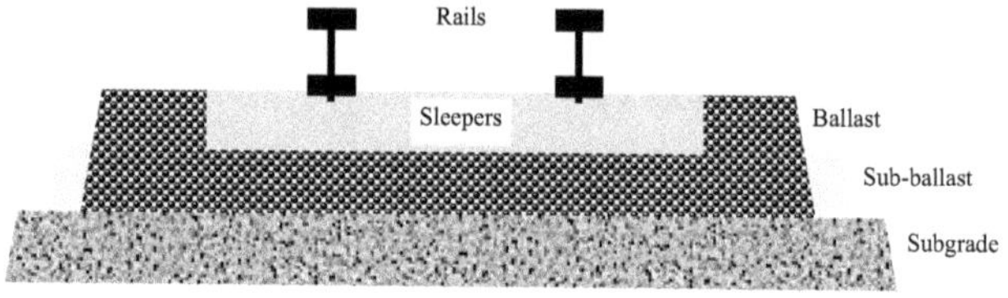

Figure 12.1 Illustration of rail track components.

Sub-ballast: Sub-ballast is a lower quality crushed stone layer between the ballast layer and the fine-graded subgrade. This layer is mainly used to assist ballast for water proofing along the side of the track.

Subgrade: Subgrade is the earth material layer that is used to offer support for track, distribute the train vehicle loads, and facilitate drainage under the rail track.

12.1.2 Rail facility life-cycle costs

Rail facility service life cycle refers to the time duration between two construction interventions. The costs incurred are classified into agency and user cost categories.

12.1.2.1 Agency costs

As depicted in Figure 12.2, agency costs consist of construction, repair, and maintenance costs, which are generally expressed in dollars per track mile or kilometer.

Construction cost: Rail facility construction cost mainly covers cost components of early stage studies, civil works, rail track facilities, intermodal facilities, and miscellaneous. Earth works forms a key portion of civil engineering works that are related to subgrade construction. The construction of rail track facilities includes elements of rails, sleepers, ballast, and sub-ballast. Depending upon the terrain and railway geometric design, rail facility construction may involve additional civil works associated with structures such as bridges, viaducts, and tunnels, as well as drainages. It also contains electrification and safety systems. Table 12.1 summarizes typical life spans of primary railway components (AREMA, 2003; Voelker and Clark, 2008; Popović et al., 2014).

Maintenance and renewal costs: Rail maintenance and renewal are defined as actions that aim to keep or restore railway components or elements to their acceptable performance functions. Maintenance treatments are classified into preventive, corrective, and renewal maintenance. These costs are estimated separately for different rail geometric designs, axle loads and vibration of rail vehicles, and cumulative utilization. In general, the rail element

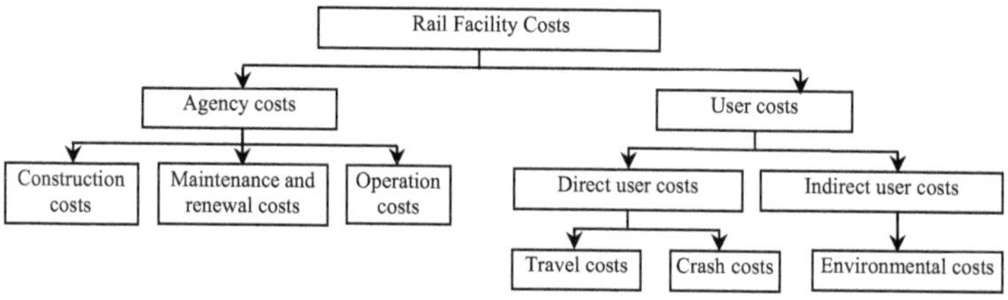

Figure 12.2 Rail facility cost components.

Table 12.1 Typical life spans of primary railway components

Component	*Element*	*Life span (year)*
Earth works	Subgrade—Soft ground	50
	Subgrade—Stable ground	100
Rail tracks	Rail—Jointed or bolted	40–60
	Rail—Continuously welded rail (CWR)	60–100
	Sleepers—Timber	20–40
	Sleepers—Concrete	50–55
	Ballast	25–60
	Switches and crossings—Inclined	25–50
	Switches and crossings—Vertical	30–65
	Electrification system	10–50
	Safety system	10–50
Structures	Bridges, viaducts, and tunnels	80–100
	Drainages	10–50
Intermodal	Habitability	2–10
	Aesthetics	1–5

with the shortest life span is the ballast, which requires partial renovation without changing rails or sleepers.

Rail operation costs: Rail operation costs are costs incurred from rail operations concerning vehicle fuel consumption, signal equipment, electrical machines, lighting, and crew compensation.

12.1.2.2 User costs

Rail user cost components mainly include travel costs, which are the ticket fees, crash costs, and environmental impacts costs.

Of the agency and user costs, the agency costs are generally considered as predominant and are therefore included in economic analysis.

12.2 RAIL FACILITY COST ESTIMATION

12.2.1 Rail facility construction cost

Rail facility construction cost constitutes a predominant portion of the total railway investments. The cost depends on several factors, including railway geometric layout, number of tracks, number of switches and crossings, number of electrical substations, safety installations, and labor costs. This cost is categorized into fixed and variable costs. Fixed cost refers to cost that is invariant to the amount of rail traffic. On the other hand, variable cost varies by traffic. As an example, Table 12.2 lists the distribution of various construction cost components (Profillidis, 2014).

Civil works cost: Cost of civil engineering work depends on the land topography; as the topography becomes difficult, the necessity of constructing tunnels and bridges becomes higher. The average cost for the construction cost activity (formation of the subgrade, tunnels, and bridges) depends on the number of tracks as well the topography of these elements (Profillidis, 2014; Table 12.3).

Table 12.2 Distribution of rail construction cost components

Component	*Share (%)*
Studies	7
Civil works (site preparation, subgrade, structures, and drainages)	57
Track (rails, sleepers, ballast, sub-ballast, switches, and crossings)	14
Electrification, signalization, and communication	17
Others	5

Table 12.3 Civil engineering works cost for new rail line construction

		Cost (million U.S. dollars/km)		
Type of track	*Maximum speed (km/h)*	*Easy topography*	*Average topography*	*Difficult topography*
Single	100	1.2–3.8	3.8–19.0	19.0–50.8
Double	100	1.2–5.1	3.8–25.4	25.4–63.4
Double	300	2.5–7.6	7.6–38.4	25.4–63.4

Table 12.4 Rail signaling system elements and cost rates (U.S. dollars)

Element	*Function*	*Unit cost*
Cables	For signaling and communicating	0.06–0.12 million
Automatic block system	Regulates the succession of trains	0.19–0.38/block section
Automatic or advanced train protection	Operate train network safely	0.025–0.038 million/unit
Cab signal	Automatic train control with transmission by track circuits or by cables	0.40 million/block section
Radio links	Data transmission along the track	0.025–0.05 million/km
Level crossing	Eliminate the conflict movement between adjacent crossing traffic with the train	Automatic half barrier 0.38 million/unit Four automatic barriers 0.88 million/unit

Track construction cost: Average construction cost per kilometer of track, including rails, sleepers, ballast, switches, and crossings, is estimated to be about $0.38–0.63 million per kilometer.

Electric traction construction cost: Electric traction is the means of applying the force needed to move the train. It contains substation and catenary cost components, which range from $0.25 to $0.38 million and $0.19 to $0.25 million per kilometer, respectively.

Signaling system construction cost: A rail signaling system consists of cables, automatic block system, automatic or advanced train protection, cab signal, radio link, and level crossing as applicable. Table 12.4 presents functions of individual elements and cost rates (Profillidis, 2014).

Light rail construction cost: Using data on as-built capital costs of 27 federally funded light rail transit projects completed over the last 20 years, Catalina (2016) calibrated a general model to estimate light rail construction as below:

$$C_{LR} = 10^{(8.2719299 - 0.00219 \times L^{0.5} + 0.2533808 \times ST^{0.5})} \tag{12.1}$$

where

C_{HR} = Light-rail construction cost, 2015 dollars
L = System length, mi
ST = Number of underground stations

EXAMPLE 12.1

A light rail system is planned to serve an area with two options proposed. Option one is to build a light rail line of 28 miles long with 12 stations on the north side of the city. Option two is to construct a light rail line of 16 miles in length with 10 stations on the south side. Using the lowest average cost per mile per station as the decision criterion, determine which option should be chosen?

For the north side option, the total and average costs are computed as

$$C_{LR}^{North} = 10^{(8.2719299-0.00219\times L^{0.5}+0.2533808\times ST^{0.5})}$$
$$= 10^{(8.2719299-0.00219\times 28^{0.5}+0.2533808\times 12^{0.5})}$$
$$= 1374289898 \text{ dollars}$$

Average $\text{cost}_{\text{North}} = 1374289898/(28 \times 12 \times 10^6) = 4.09$ million dollars per line-mile per station

For the south side option, the total and average costs are calculated as

$$C_{LR}^{South} = 10^{(8.2719299-0.00219\times L^{0.5}+0.2533808\times ST^{0.5})}$$
$$= 10^{(8.2719299-0.00219\times 16^{0.5}+0.2533808\times 10^{0.5})}$$
$$= 1159926160 \text{ dollars}$$

Average $\text{cost}_{\text{South}} = 1159926160/(16 \times 10 \times 10^6) = 7.25$ million dollars/line-mi/station

The north line option should be chosen because of a lower average cost.

Heavy rapid rail construction cost: As seen in Sinha and Labi (2011), a general model was proposed to estimate heavy rapid rail construction cost, which is of the following specification

$$C_{HR} = 3.9\times \mathrm{L}^{-0.702}\times \mathrm{U}^{1.08}\times \mathrm{ST}^{-0.36} \quad (12.2)$$

where

C_{HR} = Heavy rapid rail construction cost, million dollars/line-mi/station
L = Number of line-miles
U = Underground fraction of the system, in percent
ST = Number of stations

EXAMPLE 12.2

A city decided to extend one single-track line of the heavy rapid rail system to accommodate a 20-mile long new rail line with 3 stations. 20% of the new rail is underground. Estimate the total cost for the rail line extension project.

$$C_{HR} = 3.9\times L^{-0.702}\times U^{1.08}\times ST^{-0.36}$$
$$= 3.9\times 20^{-0.702}\times 20^{1.08}\times 3^{-0.36}$$
$$= 8.15 \text{ million dollars/line-mi/station}$$

Station and yard construction cost: The station construction cost varies considerably depending on whether the fixed guideway system is a light rail system or an underground subway or metro system. Table 12.5 lists station and yard construction costs for five U.S. cities.

Rolling stock cost: Rolling stock is vehicles used for rail operational services. Rolling stock cost depends on characteristics of the rail operators, vehicle sizes, and vehicle types. Tables 12.6 and 12.7 show rail vehicle purchase and operating costs (APTA, 1991).

Table 12.5 Station and yard costs of recently constructed rails (1000 U.S. dollars)

Facility	*Type*	*Los Angeles*	*Pittsburgh*	*Portland*	*Sacramento*	*San Jose*	*Average*
Station	At-grade center	1656		831		263	917
	At-grade side	1401	3248	910	636	312	1302
	Elevated	4493					4493
	Subway	42473	11491				26982
Yard and shop	Capital cost	67817	72323	22549	6900	31846	
	Capacity (vehicles)	54	97	100	50	50	
	Cost per unit of capacity	1256	746	226	138	637	600

Table 12.6 Rail vehicle purchase costs (million U.S. dollars)

Rail type	*Location*	*Year*	*Quantity*	*Total cost*	*Unit cost*	*Average unit cost*
Heavy rail	Chicago	1991	256	350.49	1.37	2.3
	Los Angeles	1989	54	106.70	1.98	
	New York	1990	19	66.35	3.49	
	San Francisco	1989	150	385.44	2.57	
	Washington, D.C.	1989	68	140.64	2.08	
Light rail	Boston	1991	86	222.86	2.58	2.6
	San Diego	1991	75	205.97	2.75	
	St. Louis	1990	31	76.65	2.46	
Commuter rail	Florida	1990	6	9.96	1.65	2.4
	Los Angeles	1990	40	86.10	2.16	
	New Jersey	1991	50	76.31	1.52	
	New York	1990	39	153.64	3.93	
	Indiana	1991	17	46.43	2.74	

Table 12.7 Rail vehicle operating costs (million U.S. dollars)

Rail type	*Location*	*Year*	*Vehicle type*	*Quantity*	*Total cost*	*Unit cost*	*Average unit cost*
Heavy rail	New York	1991	R33 Subway	494	339.35	0.69	0.84
	New York	1991	R44 Subway	280	250.54	0.89	
	New York	1990	R44 Subway	64	60.78	0.95	
Commuter rail	Maryland	1990		35	11.82	0.34	0.98
	New Jersey	1991		230	376.66	1.64	

12.2.2 Rail facility maintenance and renewal cost

Rail facility maintenance activities are mainly concerned with maintenance or renewal of tracks and platforms, electrification, lighting, and communication systems, and structures such as bridges and tunnels (Zarembski, 2000, 2015). Table 12.8 lists some of the rail facility maintenance and renewal cost models.

Table 12.8 Rail facility maintenance and renewal cost models

Treatment or activity	*Specification*
Rail grinding cost	$\sum_{i=1}^{k}\sum_{j=1}^{N-1}[(T_{g_i} \times C_L \times L_i \times n_{g_i} + C_{eg} \times T_{gi} \times L_i \times n_{gi}) \times (m/m_{gi})]/[(1+r)^j]$
Rail lubrication cost	$\sum_{i=1}^{k}\sum_{j=1}^{N-1}(T_{clu} \times C_L \times n_{li})/[(1+r)^j]$
Rail renewal cost	$\sum_{i=1}^{k}\sum_{j=1}^{N-1}[(T_r \times l_i + C_l \times T_{rri} \times l_i + C_{err} \times T_{rri} \times l_i) \times (m/m_{rri})]/[(1+r)^j]$
Track tamping cost	$\sum_{i=1}^{k}\sum_{j=1}^{N-1}[(T_{ta_i} \times C_L \times L_i + C_{eta} \times T_{tai} \times L_i) \times (m/m_{tai})]/[(1+r)^j]$
Ballast cleaning cost	$\sum_{i=1}^{k}\sum_{j=1}^{N-1}[(T_{b_i} \times C_L \times L_i + C_{eba} \times T_{bi} \times L_i) \times (m/m_{bi})]/[(1+r)^j]$
Track inspection cost	$\sum_{i=1}^{k}\sum_{j=1}^{N-1}[(T_t \times C_L \times L + C_{et} \times T_t \times l) \times (m/m_t)]/[(1+r)^j]$

where

C_{eb} = Equipment cost for ballast cleaning
C_{ebr} = Equipment cost for ballast renewal
C_{eg} = Equipment cost for grinding
C_{err} = Equipment cost for rail renewal
C_{et} = Equipment cost for track inspection
C_{eta} = Equipment cost for tamping
C_{clu} = Cost of lubrication material for each lubricator per year
C_L = Average labor cost
j = Analysis time
L_i = Length of ith curve
m = Gross tonnage per year, tons
m_{bi} = Interval for ballast cleaning for i^{th} curve, million gross tons (MGT)
m_{gi} = Interval for grinding for i^{th} curve, MGT
m_{rr} = Interval for rail renewal for i^{th} curve, MGT
m_{tai} = Interval for track taming for i^{th} curve, MGT
m_t = Interval for track inspection
N = Track service life-cycle, years
n_{gi} = Number of grinding passes on i^{th} curve
n_{li} = Number of wayside lubricators i^{th} curve
r = Discount rate
T_{tai} = Mean time to tamp for i^{th} curve
T_{bi} = Mean time to clean ballast for i^{th} curve
T_{gi} = Mean time to grind for i^{th} curve, hr/km
T_r = Mean time to renew track
T_{rri} = Mean time for rail renewal for i^{th} curve
T_t = Mean time to inspect track

EXAMPLE 12.3

A rail track inspection is carried out. There are five curves along the track, with each curve being 500 ft in length. The mean times for tamping vary along the curves, which are 0.7, 0.6, 0.9, 0.8, and 0.5 hour, respectively. The average labor cost is $16/hr, the equipment cost for tamping is $200/hr-ft, the gross tonnage per year is 15 MGT, and the load of tamping is 8 MGT after 5 years in service. Calculate the track tamping cost. Assume a discount rate of 3%.

The temping cost is calculated by

$$\sum_{i=1}^{k}\sum_{j=1}^{N-1}\frac{[(T_{ta_i}\times C_L\times L_i+C_{eta}\times T_{tai}\times L_i)\times(m/m_{tai})]}{[(1+r)^i]}$$

$$=\frac{\begin{array}{c}((0.7\times16\times500)+(200\times0.7\times500)+(0.6\times16\times500)+(200\times0.6\times500)+\\(0.9\times16\times500)+(200\times0.9\times500)+(0.8\times16\times500)+(200\times0.8\times500)+\\(0.5\times16\times500)+(200\times0.5\times500))\times(8/15)\end{array}}{(1+0.03)^5}$$

$$=\$171584$$

12.2.3 Rail system operation cost

Rail system operation cost mainly includes cost of managing the operator, driver, and labor, which constitutes more than 90% of the total operating cost, as well the cost of operations scheduling that accounts for 8%–10% of the total operating cost. This cost is estimated at approximately $1.4 per train-km per year (Zarembski, 2000, 2015).

12.3 RAIL FACILITY PRESERVATION TREATMENTS

Rail facility preservation mainly deals with the rail track component and vehicles that rely on tracks for operational service. Factors affecting the track condition can be grouped into load and non-load categories, which typically include the following:

- Track geometry
- Track stiffness
- Subgrade strength
- Moisture content
- Overall drainage

Track condition is bending, shear, contact, and thermal stresses. Vehicle performance is largely affected by the speed of the train and vehicle condition itself and is affected by vertical, lateral, and dynamic stresses. When conducting track condition assessment, focus can be on rail geometric measurements, such as rail alignment, deformation and twist, gauge, extent of wear and corrugation, and presence of faults or failures visible or detectable by non-destructive devices. Further, condition inspections can be made on sleepers, ballast, and sub-ballast. In the presence of structures and drainages, condition inspections are also needed. Table 12.9 lists some typical track maintenance and renewal treatments (Guler, 2013). The mechanized treatments are conducted for certain defects, as listed in Table 12.10.

Table 12.9 Typical railway maintenance and renewal treatments

Category		*Treatment*
Maintenance	Manual	Track surface treatments Switches Level crossings Structures
	Mechanical	Ballast cleaning Ballast regulating Ballast stabilizing Tamping Rail grinding Joint straightening Track straight
Renewal	Manual	Parts
	Mechanical	Track continuous or panels Switches complete or parts Formation Structures

Table 12.10 Railway defects and treatments

Mechanized treatment	*Defect*
Tamping machine	Correct track level, superelevation, and alignment
Rail grinding machine	Remove corrugations and grind welds
Strait	Straighten welds
Ballast regulator	Correct ballast profile
Ballast cleaner	Clean ballast bed

Case Study: Evaluation of a city beltway rail line preservation

A 30-mile long beltway rail on the west side of a large city provides local freight shipments for two suburban towns. The beltway runs in the north-south direction. The north side town is located 10 miles west from the intersection of the beltway mainline, which has a population of 5000 and is mainly involved with a large food processing business. According to business statistics, the beltway rail ships 80% of the food products. The south side of town is situated at the end of the beltway rail with 20000 residents working for one assembly factory. The economy of this town is dependent upon the factory that uses the beltway rail extensively for receiving auto parts and shipping cars to be sold nationwide.

Owing to a slower pace of economic growth, the use of beltway rail by food processing and auto businesses has been declining, making it hard to upkeep the beltway rail operations. Rail closure becomes increasingly thought of as an option to fix the deficit problem. However, the local governors are highly concerned that if the freight shipment demand in the two towns could not be met, the food processing and auto assembly businesses would relocate elsewhere. Conversely, all businesses indicated that they would stay if adequate transportation service were to be maintained. Meanwhile, a local entrepreneur has petitioned the state for loans with a lower interest rate to purchase the beltway right-of-way, immediately repair tracks, and buy new freight trains to operate it as a Class III rail. If the negotiation is successful, the current beltway rail could still be kept in use.

In response to the beltway rail closure, an alternative transportation option is proposed by the state transportation agency. That is, to reconstruct a 35-mile long, two-lane suburban arterial connecting the two towns as a truck route to provide transportation service to the businesses. The reconstruction is expected to be completed within one year.

Considering a 20-year analysis period, loan interest rate at 4.5%, and discount rate at 6%, determine which option is more preferred using the following data sources.

Alternative I: Upgrading the beltway rail to a class III rail

a. Beltway rail purchase cost
   Beltway rail right-of-way purchase cost: $2 million
   Immediate track repair cost: $15000/mi
   Locomotive purchase cost: $150000
b. Beltway rail maintenance cost
   Annual derailment cost: $4000/year
   Annual track maintenance cost: $6000/mi
c. Beltway rail operating cost
   Annual engine maintenance cost: $6000 + 0.05 × (yearly mileage)
   Annual engine operating cost: $4000 + 2 × (yearly mileage)
   Annual crew and miscellaneous cost: $40000 + 20 × (yearly man-hrs)
   Round trip travel time: 4 hours
   Frequency: 2 days/week
   Crew size: 2 operators
d. Freight shipment demand
   Food processing factory: 40 cars/month, 100 tons/car
   Auto assembly factory: 1000 cars/month, 50 tons/car
   Carloads are assumed to increase by 2% per year
e. Rail shipping charge
   10 cents per ton-mile

Alternative II: Converting the suburban arterial to a truck route

a. Arterial reconstruction and maintenance cost
   Reconstruction cost: $250000/mi
   Annual maintenance cost: $500/lane-mi
b. Freight shipment demand
   Food processing factory (5 miles from Interstate): 1000 trucks/year
   Auto assembly factory (25 miles from Interstate): 20000 trucks/year
   Truck volumes are assumed to increase by 3 percent per year
c. Truck shipping service charge
   5 cents per vehicle mile of travel

SOLUTION

1. Alternative I: Upgrading the beltway rail to a Class III rail

a. Cost items	
Cost of purchasing right-of-way:	$2000000
Immediate track repair:	$15000/mi × 30 miles = $450000
Locomotive purchase cost:	$200000
Total capital cost:	$2000000 + $450000 + $200000 = $2650000

Continued

Annual derailment cost: $4000/year

Annual track maintenance cost: $6000/mi × 30 miles = $180000/year

Annual engine maintenance cost: $6000 + 0.05 × (30 miles × 2 × 2 days/week × 52 weeks/year) = $6312/year

Annual engine operating cost: $4000 + 2 × (30 × 2 × 2 × 52) = $16480/year

Annual crew and miscellaneous cost: $40000 + 20 × (4 hours × 2 operators × 2 × 52) = $56640/year

Total annual cost: $4000 + $180000 + $6312 + $16480 + $56640 = $263432/year

b. Annual revenue

Food processing factory: 40 cars/month × 100 tons/car × 12 months/year × 10 miles × $0.1/ton-mi =$48000/year

Auto assembly factory: 1000 cars/month × 50 tons/car × 12 months/year × 30 miles × $0.1/ton-mi =$1800000/year

Total annual revenue: $48000 + $1800000 = $1848000/year

c. Net present worth

Note that the carloads are assumed to increase by 2% per year

$$PWF = \frac{1 - e^{-((i-r)N)/(1+r)}}{i-r} = \frac{1 - e^{-((0.045-0.02)20)/(1+0.02)}}{0.045-0.02} = 15.50$$

$$NPW = -2650000 - 263432 \times \left[\frac{(1+0.045)^{20} - 1}{0.045 \times (1+0.045)^{20}}\right] + 1848000 \times (1+0.02) \times 15.50$$

$$= \$23140173$$

2. Alternative II: Converting the suburban arterial to a truck route

a. Cost items

Reconstruction cost: $250000/mi × 35 miles = $8750000

Yearly maintenance cost: $500/lane-mi × 35 miles × 2 lanes = $35000/year

b. Annual revenue

Food processing factory: 1000 trucks/year × (5 × 2) mile/roundtrip × $0.05/VMT = $500/year

Auto assembly factory: 20,000 trucks/year × (25 × 2) mile/roundtrip × $0.05/VMT = $50000/year

Total annual revenue: $500 + $50000 = $50500/year

c. Net present worth

Similarly, the truck volumes are assumed to increase by 3% per year

$$PWF = \frac{1 - e^{-((i-r)N)/(1+r)}}{i-r} = \frac{1 - e^{-((0.065-0.03)20)/(1+0.03)}}{0.065-0.03} = 14.09$$

$$NPW = -8750000 - 35000 \times \left[\frac{(1+0.065)^{20} - 1}{0.065 \times (1+0.065)^{20}}\right] + 50500 \times (1+0.03) \times 14.09 = -\$8402756$$

3. Conclusion

Since Alternative I has a higher and positive NPW, upgrading the existing beltway rail is more preferred.

PROBLEMS

12.1 What is the definition of rail transportation? List the essential features of rail transportation.

12.2 What are the main performance measures for track condition assessment?

12.3 Why is life-cycle analysis important to evaluate a rail transportation project? What is the difference between life-cycle analysis of a highway and a rail transportation project?

12.4 What are the related railway project impacts on economy and environment?

12.5 Given information listed in Table P12.1, determine the equivalent uniform annual cost of a railway construction project that has the following components. Use a discount rate of 5%.

Table P12.1 Construction project component information

Component	*Cost (dollars)*	*Life (years)*	*Salvage value*
Land	6.2 million	100	6.2 million
Grading, drainage, subbase, and base	4.1 million	40	2 million
Rail and ties	9.2 million	25	20% of initial value
Signals and controls	1.7 million	10	0

12.6 What typical items are included in maintenance and renewal of rail facilities?

12.7 A 17-mile long new rail with 9 stations is going to be constructed, the line is one single-track line of the heavy rail system and 40% of the line is underground. Make an estimation for the cost of the project.

12.8 A 25-mile long new light-rail transit service along 16 stations is proposed. Estimate the construction cost of this project.

Chapter 13

Economic analysis of inland waterway system preservation and usage

13.1 INLAND WATERWAY TRANSPORTATION SYSTEM COMPONENTS

The inland waterway transportation system is one of the oldest transportation modes, and is still widely used for people and goods movements in many countries. Table 13.1 provides a summary of waterway transportation in the U.S. in 2015 (USACE, 2015).

Figure 13.1 illustrates a typical inland waterway transportation system that consists of facilities, vehicles, and user/nonusers. Facilities can be classified into navigable channels, dams, and locks. The component of navigable channels contains dredging, dikes, banks, and pumping components. The component of dams and locks generally covers chambers, two systems, signals, and communication systems. The vehicle component is concerned with different types of boats, ships, and vessels for passenger and freight movements. The system may deal with users or nonusers.

13.2 PRIMARY INLAND WATERWAY TRANSPORTATION FACILITIES

13.2.1 Waterway channel

A navigable waterway channel is an important component of an inland waterway transportation system. To maintain safe and efficient vessel movements, it is necessary to determine the adequate channel size with the least slope. It is assumed that water at the downstream end is freely discharged and the slope of the channel is uniform. The consideration for achieving a uniform slope for the waterway channel is mainly to avoid the risk of phenomenon change, which can yield a hydraulic jump. Water flow for the channel design is determined by Manning's formula as

$$Q = V \cdot A = (1.49/n) \cdot R^{2/3} \cdot S^{1/2} \cdot A \tag{13.1}$$

where

Q = Flow rate, ft^3/s
V = Velocity, ft/s
A = Flow area, ft^2
n = Roughness coefficient
R = Hydraulic radius, ft
S = Slope, ft/ft

Note that hydraulic radius (R) is the cross-sectional area (ft^2) of the channel divided by the wetted perimeter (ft), and roughness coefficient (n) is a constant determined by channel surface (Tables 13.2 and 13.3).

Table 13.1 Passenger and freight shipments in the U.S. in 2015 (million tons)

Type	*Arkansas*	*Columbia*	*Illinois*	*Mississippi*	*Ohio*	*Tennessee*
Recreational/passenger	0.0	0.0	0.2	0.1	4.5	0.3
Coal, lignite, and coal coke	4.0	0.0	7.6	29.6	400.0	15.5
Petroleum and petroleum products	4.3	5.1	22.8	16.8	99.7	3.0
Chemicals and related products	30.0	0.5	22.7	95.7	68.1	7.2
Crude materials, inedible, except fuels	14.1	3.5	26.3	59.8	199.6	23.6
All primary manufactured goods	11.7	0.0	19.4	38.4	65.7	8.0
Food and farm products	25.9	13.7	22.0	258.3	75.0	11.2
Manufactured equipment and machinery	0.3	0.2	0.5	1.4	2.2	0.3
Waste material	0.0	0.9	0.0	0.1	1.2	0.0

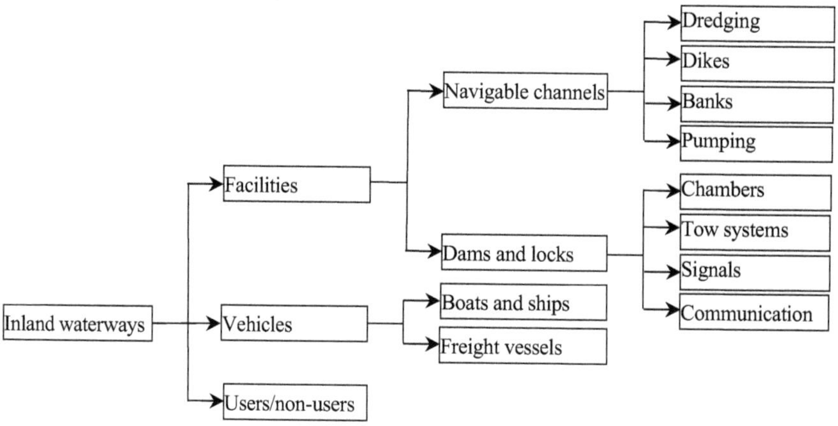

Figure 13.1 Inland waterway system components.

Table 13.2 Hydraulic radii

Type	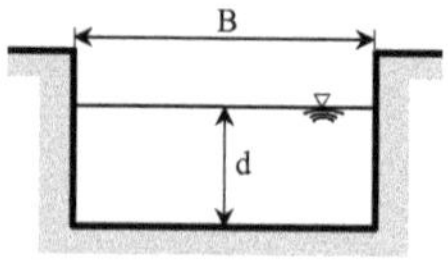	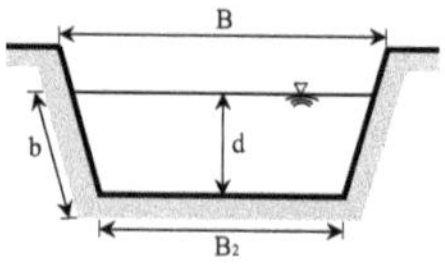
Area (A)	$B \cdot d$	$\frac{d(B_1 + B_2)}{2}$
Wetted perimeter (P)	$B + 2d$	$B_2 + 2b$
Hydraulic radius (A/P)	$\frac{Bd}{B + 2d}$	$\frac{d(B_1 + B_2)}{2(B_2 + 2b)}$

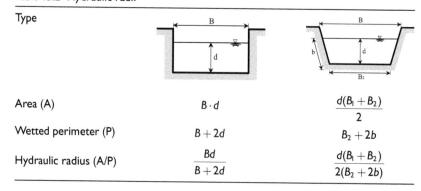

Table 13.3 Roughness coefficients

Channel surface material	*Roughness coefficient (N)*
Earth, free from weed, straight alignment, stones up to 250 ft	0.02 ~ 0.025
Ditto with poor alignment	0.03 ~ 0.05
Ditto with weeds and poor alignment	0.05 ~ 0.15
Gravel, 3–6 inches, free from weed, straight alignment	0.03 ~ 0.04
Ditto with poor alignment	0.04 ~ 0.08
Ditto with 6-inch stones	0.04 ~ 0.07
Concrete	0.012 ~ 0.017
Hand-placed pitching	0.025 ~ 0.035
Dressed stone, jointed	0.013 ~ 0.02
Cast iron	0.01 ~ 0.014

EXAMPLE 13.1

Consider a 100-ft wide rectangular concrete channel with a minimum slope of 0.005 that carries a water flow up to 500000 ft^3/s. Determine the height of the channel.

Solution

Referring to Table 13.2, roughness coefficient is 0.0145.

When the height of the channel is d, flow rate is obtained by

$$Q = V \cdot A = 500000 = \left(\frac{1.49}{0.0145}\right) \times \left(\frac{100d}{100+2d}\right)^{\frac{2}{3}} \times 0.005^{\frac{1}{2}} \times (100d)$$

Therefore, d is 72.05 ft.

13.2.2 Sediment control

Sediment control is an essential element for successful inland waterway channel management. Sediments are materials settled at the bottom of a channel, and are divided into two types: organic and inorganic materials. Organic sediments are depositions from deceased animals and plants, and inorganic sediments consist of coarse- and fine-grained compositions.

In a wide range of waterway management, sediment management includes deposition extraction (dredging) for navigation purposes, ecological conservation, flooding protection, agricultural uses, water supply, and hydropower generation. In waterway transportation, dredging technology is an especially critical factor to retain a sustainable corridor for shipping activities. Furthermore, sediment extraction has a direct impact on the economy and the environment, including water quality and habitats in nearby areas and contaminant distribution.

Sediment management plays an important role in determining channel alignments, depth, and width, which is correlated to flow conditions such as water capacity and velocity in channel. For example, excessive sediment leads to higher maintenance cost and frequency, decreases hydraulic capacity of the channel, and destroys wildlife habitat. On the other hand, insufficient sediment interrupts activity of vessels and yields the loss of wetlands. Also, high velocity flow with stream can wash a soft-bottom channel bank, which results in damage to channel structures and raises the risk of flooding. Therefore, it is important to determine proper quantity of sediment excavation by balancing shipping activities and environmental effects. Finally, sediment should be managed by all regions in a potential range of waterway usage since a local solution for sediment management on an upstream site can have an adverse influence on downstream sites. For instance, installing bulkheads along a bank in the upper river can prevent erosion of the riverbed on the site, but it can also reduce flow capacity, causing erosion downstream or even have negative effects upstream of the site. Therefore, channel configuration, sediment type, and flow condition should be considered holistically in all areas along the river.

13.2.3 Dredging

Dredging is the process of excavation from the bed or banks of a river. Since natural rivers generally do not provide a sufficiently passable navigation channel for safe movement of vessels, dredging is necessary not only to help keep the water clean but also to guarantee vessels can pass though safely and efficiently. Dredging is periodically implemented to maintain a clean and safe navigation system. Occasionally, it is required to additionally excavate sediment to accommodate big commercial vessels by enlarging and deepening the navigation channel. According to the USACE, dredging takes place around more than 400 ports and 25000 miles of navigation channels for effective waterway transport operation in the U.S. (USACE, 2015). Also, dredging work has seasonal restrictions because of environmental concerns about the dredging work. Dredging type and size depend on type of sediment, channel alignment, environmental restraints, and site restrictions. The two main types of dredging methods typically used are mechanical dredging and hydraulic dredging.

Mechanical dredging: A mechanical dredge digs or gathers sediment by using a dipper, clamshell, grab, dragline bucket, or backhoe, and materials removed from the channel bottom are placed on the waiting barge known as a scow, and then moved to disposal area. Generally, while mechanical dredging fixed by anchoring is implemented, excavated material is transported by scow to a disposal site. This dredging type is commonly utilized in many industries. Mechanical dredging is classified by bucket size. The most common types of mechanical dredge are bucket dredges and dipper dredges. Mechanical dredging is most efficient in a confined area or protected channel, in cases where digging is required near bridges, breakwater structures, wharfs, piers, or docks. Moreover, it is well suited to consolidated or hard-packed material and sediments consisting of rocks and debris. On the other hand, mechanical dredging is inefficient in areas where traffic volume is high and geographical terrain is rough.

Hydraulic dredging: Hydraulic dredging removes material by sucking the water through a temporary pipeline and then pumping it into another location. This dredging works like a giant vacuum and removes sediment very precisely. Hopper and pipeline dredging methods are typically used. A hopper dredge utilizes ships with a large containment area called a hopper, which is a storing area for dredged materials. Strong pumps suck deposits on the bottom of the channel, and they are piled up in the hoppers through long pipes called drag arms. Hopper dredging works efficiently for unconsolidated material or heavy sands, and can be operated in corridors with high traffic volume or rough area; also, it can move into another spot easily and quickly. However, it is difficult to use in confined or shallow river channels.

Another type of hydraulic dredging is pipeline dredging, which sucks sediments in water and discharges them into a disposal site directly through a pipeline. Dredge types are classified according to the intake method into cutter-head, suction, and dustpan dredges. Dredging work operates constantly because it is not required to convey dredged material into a disposal site. This dredge is effective for densely packed material and even rock removal, and it is possible to apply it in large areas with deep shoals. Table 13.4 lists dredge classifications and characteristics.

Table 13.4 Dredging classification and characteristics

Classification	*Dredge type*	*Advantage*	*Disadvantage*
Mechanical dredge	Dipper dredge Bucket dredge	Work well in consolidated or hard packed Work in rocks and debris Deliver a product having low water content	The production rate is low Fine-grained material leak when raised from deep underwater Generating turbidity
Hydraulic dredge	Hopper dredge Dustpan dredge Cutter-head dredge	Low cost High rate of production Long-distance pumping Movable easily Operate continuously	Dewatering required Not used in shallow channel
Other	Siphon dredges Pneumatic dredges	Effective in reservoir sediment dredging	Lack of reliability of system Limited hydraulic gradient

13.2.4 Dams and locks

A dam is a barrier that blocks a river stream and retains water for irrigation, generating electricity, flood prevention, or navigation. The purpose of a lock is to lift and lower ships or vessels between different water levels on a waterway channel. In other words, locks provide waterway transports with easier navigation and enable ships to be landed easily and safely. Among various inland waterway facilities, dams and locks require the most attention for maintenance work.

Dams and locks consist of signal, communication system, chamber, and tow system. In general, several barges are operated for towing. The number and size of chamber and towing devices are decided by the amount of shipping, vessel size, and river size, and so on. Furthermore, chamber size and towing size and distributions determine service time of the dams and locks.

13.2.5 Dikes

A dike, also known as a levee, is a structure to protect and manage the banks and channel of a river stably and is installed along the bank line. The purpose of dike is to prevent bank erosion or cutoff of the riverbed, which can cause the shape of a waterway to be modified. Dikes also help water current to concentrate on a main direction of progress. In most cases, stone or timber materials are used for dike construction. There are different types of dikes depending on the purpose of use: spur, longitudinal, kicker, L-head, ring, vane, setback, and closure dikes (Table 13.5).

Table 13.5 Dike design parameters

Parameter	*Description*
Dike length	Determined by the desired contracted channel width
Dike height	Based on the design flood level (usually decided based on past maximum flood level) Have a significant impact on the flow in stream and downstream A minimum top elevation of dike is about 2 or 3 feet above water surface
Dike angle	The angle between dike direction and river bank line Decide the location and amount of scrub of flow
Dike spacing	Determined by flow velocity and channel alignment An effective minimum spacing is two-thirds of the length of the upstream dike
Crest width	Determined by the method of construction Affected by the amount and pressure of flow, debris, and ice A general range of crest width is about 5–20 ft

13.2.6 Pipe flow and back-pumping

Proper flow level of channel should be maintained to establish stable waterway. In particular, overflowing from flooding can impair the designed channel, dike, and banking system. To manage the water level, a pumping system through pipes is utilized. Flow in a pipe is proportional to the square root of the pipe head. Friction losses and inlet and outlet losses at the pipe head are considered to calculate the amount of pumping. Velocity of flow through a fully submerged pipe is calculated by using Manning's formula as

$$V = (0.591/n) \cdot D^{2/3} \cdot S^{1/2} \cdot A \tag{13.2}$$

where

V = Velocity, ft/s
n = Roughness coefficient
D = Pipe diameter, ft
S = Hydraulic gradient, ft/ft
A = Flow area, ft^2

13.3 INLAND WATERWAY TRANSPORTATION SYSTEM USAGE

13.3.1 Water channel and lock delay and unavailability

Delay and unavailability of service at a waterway channel or a lock have an impact on reliability of the inland waterway system. A delay is the time that a vessel should wait to move, and unavailability time is defined as the duration caused by closure or outage because of system failure. Delay can occur when a vessel to be towed is too large because it results in increasing lock processing time or cargos load into vessels and land, as well as when the volume of vessels carrying cargo exceeds the capacity of lock usage. Both vessel traffic congestion and inefficient lock operations control cause delays, which degrade efficiency of the waterway system. On a river water channel, delays can also occur if water flow becomes relatively high or low due to flooding or drought. Unavailability time can result from a scheduled or unscheduled outage of waterway system because of system break, crash, flooding, or unexpected disaster.

For inland waterway transportation, lost time (T_{Lost}) yields cost to operators and users, which is defined as the sum of delay (t_d) and unavailability time (t_u).

$$T_{\text{Lost}} = t_{\text{delay}} + t_{\text{unavailability}} \tag{13.3}$$

Delay accounts for a predominant portion of total lost time. Most of unavailability time is related to scheduled unavailability (TRB, 2015). Unavailability time resulting from system failure can be calculated based on total hours of system outage. In an inland waterway system, delay largely fluctuates depending on weather and seasonal effects.

A major portion of delays occurs at locks. Thus delay is defined as an average waiting time for vessels to pass a lock. The average waiting time is obtained by using Marshall's formula, which is based on queue theory, which is expressed in terms of the arrival, departure, and service time variance as

$$t_w = \frac{\sigma_A^2 + 2\sigma_S^2 - \sigma_D^2}{2t_a(1-\rho)} \tag{13.4}$$

where

t_w = Average waiting time
σ_A^2 = Variance of inter-arrival times
σ_S^2 = Variance of lock service times
σ_D^2 = Variance of inter-departure time
ρ = V/C ratio

Chamber size and tow size primarily affect service time. Transit time at a lock is a vital consideration to determine service time, and it also determines lock performance. Transit time includes the following processes (TRB, 2015): (i) Time required for a tow to move from an arrival point to the lock chamber; (ii) time to close the gates and fill or empty the chamber; and (iii) time to open the gates and for the tow to exit from the chamber. Table 13.6 depicts average towing and lock delays of different waterways in the U.S. (USACE, 2015).

13.3.2 Safety

Compared to other travel modes such as auto and transit, inland waterway maintains relatively low frequency of crash occurrences. However, once a waterway crash occurs, the consequence could be significant, often leading to injuries, loss of life, cargo destruction, environmental damage, and obstruction of waterways. Of various types of waterway crashes, activities of recreational boating form the largest portion and pose high concerns to waterborne safety. Waterway crash contributing factors are mainly related to waterway channel and vessel characteristics. Inclement weather conditions can have an influence on the crash risk as well. In addition, waterway crashes are sensitive to types of goods being shipped. For instance, hazardous material shipping tends to have a much greater extent of adverse impacts on the waterway system directly and other associated systems (Table 13.7).

Table 13.6 Towing and lock processing delays of some major waterways in the U.S.

Inland waterway	*Average tow delay (hours)*		*Average processing delay (hours)*	
Year	2015	2014	2015	2014
Arkansas	0.20	0.21	0.88	0.85
Columbia	0.4	0.37	0.59	0.60
Illinois	2.38	2.71	0.92	0.92
Mississippi	1.81	2.72	0.60	0.61
Ohio	2.33	1.59	0.84	0.81
Tennessee	7.20	4.62	0.83	0.82
TENN–Tombigbee	0.31	0.19	0.61	0.60

Table 13.7 Inland waterway crash contributing factors

Waterway characteristic	*Vessel characteristic*	*Environmental condition*
Channel width and depth	Vessel type and size	Visibility information
Channel alignment	Vessel speed and direction	Wind speed
Navigable radius	Vessel clearance	Current velocity
Presence of locks	Types of cargo	Water level
Presence of bridges	The amount of cargo (overloading)	
	Presence of barge trains	

13.3.3 Environmental impacts

Compared with highway and rail transportation, inland waterway transportation is generally fuel efficient and environmentally friendly. However, modal share by inland waterway transportation to handle passenger travel and good movements is relatively low. Mainly for this reason, technologies and emission regulations adopted by waterway facility preservation and system operations generally fall behind those in use by surface transportation such as highways. Thus there is still room for improving waterway fuel efficiency and emission reductions.

The combustion of fuels for vessel activities produces air pollutants including NMHC, CO, CO_2, SO_2, NO_X, and particle matters (PM). The air pollutants emitted by inland waterway transportation are affected by vessel type, engine type, vessel loading capacity, shipping volume, and vessel cruising speed, as well as regional conditions where the vessel navigates, such as wind direction and speed, and water flow. At present, vessel air emissions are estimated per the IPCC Guidelines for National Greenhouse Gas Inventories (IPCC, 2006). Basically, emission (E) is calculated by the amount of fuel consumed (FC) multiplied by an appropriate emission factor (EF) as

$$E = FC \cdot EF \tag{13.5}$$

Here, emission factor is determined by technology type corresponding to fuel type and specific pollutant (IPCC, 2006). The following methods are generally used for quantifying vessel air emissions.

Vessel emission method 1 (Tier 1): The Tier 1 method is the simplest to estimate vessel air emissions using data on fuel consumption. It is a top-down method that total emissions from all fuel sources of combustion are calculated from the quantities of fuel consumed and average emission factors. In other words, the method uses quantities of fuel consumed for vessel activities from nationally collected data, and assumes emission factors based on fuel quality used for each pollutant. The emission factors depend on each type of vessel fuels such as bunker fuel oil, diesel oil, gas oil, and gasoline.

$$E_i = \sum_j (FC_j \cdot EF_{ij}) \tag{13.6}$$

where

E_i = Emission of pollutant i, kg
FC_j = Amount of fuel type j consumed or sold in navigation, tons
EF_{ij} = Fuel consumption emission factor of pollutant i and fuel type j, kg/ton
i = Vessel air pollutant i
j = Fuel type j

Vessel emission method (Tier 2): The Tier 2 method uses data not only on fuel consumption by fuel type, but also on specific vessel engine type for estimating emissions. Especially, non-CO_2 gases are far more affected by the engine characteristics than CO_2 emissions. Therefore, in the case of estimating non-CO_2 emissions, engine type can be a primary factor.

$$E_i = \sum_j \left(\sum_k FC_{jk} \cdot EF_{ijk} \right) \tag{13.7}$$

where

E_i = Emission of pollutant i, kg
FC_{jk} = Amount of fuel type j consumed by vessels with engine type k, tons
EF_{ijk} = Fuel consumption emission factor of pollutant i for vessels with fuel type j and engine type k, kg/ton
i = Vessel air pollutant i
j = Fuel type j
k = Engine type k

Vessel emission method 3 (Tier 3): The Tier 3 method relies on data on ship movements for individual vessels. While the Tier 3 method is more sophisticated than Tier 1 or Tier 2 method, it requires detailed data on ship movements and techniques. Emissions from vessel activities are estimated by adding together trips of all vessels. The emission for individual trips is estimated as

$$E_{\text{Trip},i} = E_{\text{Hotelling}} + E_{\text{Manouvering}} + E_{\text{Cruising}} = \sum_p (FC_{pjk} \cdot EF_{pijk}) \tag{13.8}$$

where

E_i = Emission of pollutant i over a trip, kg
FC_{pjk} = Amount of fuel type j consumed by vessels with engine type k in the trip phase p, tons
EF_{pijk} = Fuel consumption emission factor of pollutant i for vessels with fuel type j and engine type k in the trip phase p, kg/ton
i = Vessel air pollutant i
j = Fuel type j
k = Engine type k
p = Different phase of a trip p, including cruising, hoteling, and maneuvering

Method 4: For estimating emissions from small boats such as recreational vessels, factors related to boat type, fuel type, engine type, and boat technology layer and hours of use are considered. The emission for a specific fuel type from small boats is calculated as

$$E_i = \sum_v \sum_k \sum_t (N_{vkt} \cdot T_{vkt} \cdot P_{vkt} \cdot LF_{vkt} \cdot EF_{vkti}) \tag{13.9}$$

where

E_i = Emission of pollutant i over a trip, kg
N_{vkt} = Number of vessels
T_{vkt} = Average operation time with vessel type v, engine type k, and technology layer t, hr/vessel
P_{vkt} = Engine power with vessel type v, engine type k, and technology layer t, kW/hr
LF_{vkt} = Engine load factor with vessel type v, engine type k, and technology layer t, %
EF_{vkti} = Emission factor with vessel type v, engine type k, and technology layer t, kg/kW
i = Vessel air pollutant i
v = Vessel type v, including cabin boat, sailing, etc.
k = Engine type k
t = Technology layer t

13.4 ESTIMATION OF WATERWAY FACILITY AND SYSTEM USAGE COSTS

13.4.1 Navigable water channel cost

Dredging costs account for most of the navigable channel cost. The cost covers items ranging from pre-investigation for planning dredging and implementation of excavation to transporting and disposing excavated materials. Dredging cost is determined by dredging method and the amount of required excavation depends on the desired channel length, depth and width, type of sediment, environmental restriction, and the location of disposal area, as seen in Table 13.8. For example, in a waterway channel where mechanical dredging is required, the total cost of the dredging is calculated based on operating time about hydro-survey boat, usage of dredging equipment, and movement disposal barge. In other words, dredging cost is affected by quantity and type of dredging, geographic condition, and weather condition. Tables 13.9 and 13.10 list methods for dredging cost estimation and some examples of dredging quantities and costs in the U.S. (USACE, 2015).

To calculate dredging cost, disposal operation rate and excavation production rate need to be applied. The disposal operation rate, also called hauling production rate, is time for excavated material to be transported to the disposal location and disposal time. Dredged materials are transported using dump scow or through pumping station into disposal area. Therefore, this rate is determined by distance to disposal area, travel speed, and the size, type, speed, and number of scow. The excavation production rate is actual dredging volume per hour. Table 13.11 shows processes of computing excavating production rate.

Table 13.8 Factors contributing to dredging cost estimation

Factor	*Description*
Dredging quantity	It is a major component to determine dredging cost Typically works requiring dredging of large quantity have lower unit cost than small quantity jobs
Material type	The production rate of the dredging work is affected by material type (e.g., fine-grained material is lost from the bucket)
Geographic condition	Geographic characteristics near working area affect a decision of cost (e.g., plant availability, distance from dredging to disposal area, movement distance)
Seasonal or weather effect	Jobs in winter can increase cost because occurrence of ice interrupts dredging works Strong wind and wave affect dredging

Table 13.9 Methods for dredging cost estimation

Method	*Description*
Historical data	Rely on dredging records that similar work was performed Currently, the simplest and most reliable approach for estimating product It is appropriate for the geographical feature, which guides estimate production rate, dredging type, and effective operation time on the area
Computer simulation	Computer simulation modelling based on investigated geographic data is used to determine working method and quantity to be dredging
Combined method	A combination of the method is used when adjustment is required because the geographical or flow feature near dredging area is changed from prior conditions, or there is no historical data.

Table 13.10 Examples of dreading quantity and cost in 2015

Project location	*Dredging type*	*Quantity (yard³)*	*Costs*	*Working days*
Callout #1	Hooper	2438898	\$8777263	55.3
Chetco River (Y906)	Hooper	20088	\$387600	4.56
Columbia Lower William (E802)	Hooper	1127669	\$4642400	33.16
Mouth of Columbia River (E813)	Hooper	893129	\$3395000	24.25
Southwest Pass	Hooper	943023	\$6753542	62
Mississippi Valley Division St. Louis District (MVS) Dredging	Pipeline	239231	\$2494800	42
St. Louis River Way	Dustpan	233436	\$545289	5.7

Table 13.11 Process of computing the excavation production rate

Step	*Factor*	*Description*
1.	Determine bucket size	It is affected by material type to be excavated
2.	Compute the number of excavation per hour from bucket cycle time	It includes dredging work time from lowering of bucket into the river floor to landing dredged material into a waiting barge
3.	Determine bucket fill factor	It is affected by the type of materials (e.g., low-density material occupies more bucket area compare to high-density material filling the bucket completely)
4.	Compute bank factor	It is the ratio of the actual depth of bank cut depth divided by the optimum depth of bank cut (e.g., if actual bank cut depth available for a bucket is 7 ft and optimum bank cut depth is 10 ft, the bank factor would be 10/7 = 1.43)
5.	Compute effective working time rate	It is a rate of time actual dredging operation time to planned operation time
6.	Determine cleanup factor	

EXAMPLE 13.2

A sand type channel with the following cross-sectional dimension is planned to be dredged. The total length is about 10000 ft. Bucket size is 10 yd³ and cycle time is 50 seconds. Historical data reveal that bucket fill factor at the dredging area is 0.75. Also, optimum depth of bank cut is 10 ft and effective working time rate is 0.85. When 10% additional time is required to clean excavation area and dredging equipment plans to be operated 300 hours/month, determine total direct cost and unit cost for dredging. Equipment and labor costs are at \$200000/month and \$150000/month, respectively. Bulking factor = 1 and indirect cost is not considered.

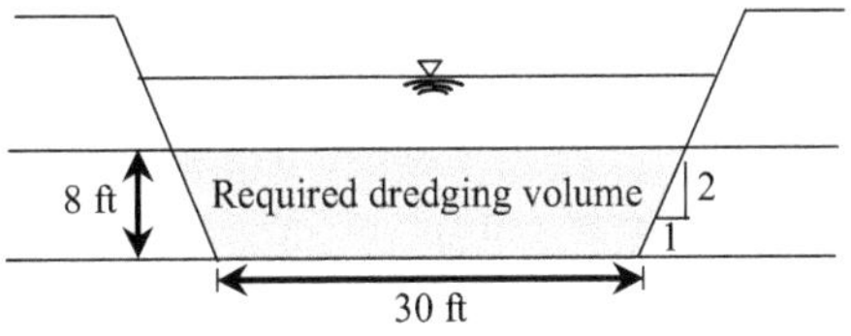

Solution

$$\text{Volume of excavation} = \frac{(30+38)\times 8}{2}\times 10000 = 2720000\,\text{ft}^3\ (=100741\,\text{yd}^3)$$

Production rate
When cycle time is 50 s and optimum depth of bank is 10 ft, the number of dredging per hour and bank factor are obtained by

$$\text{The number of excavation} = \frac{3600}{\text{cycle time}} = \frac{3600}{50} = 72 \text{ bucket/hr}$$

$$\text{Bank factor} = \frac{\text{average depth of bank cut}}{\text{optimum depth of bank cut}} = \frac{8}{10} = 0.8$$

Given bucket size = 10 yd^3, bucket fill factor = 0.75, effective working time = 0.85, and cleanup factor 0.9 (10% additional time), production rate = 10 × 72 × 0.75 × 0.8 × 0.8 5 × 0.9 = 330.48 yd^3/hr

$$\text{Time for dredging work} = \frac{100741\,\text{yd}^3}{330.48\frac{\text{yd}^3}{\text{hr}} \times 300\frac{\text{hrs}}{\text{month}}} \approx 1.0 \text{ month}$$

$$\text{Total direct dredging cost} = (200000 + 150000) \times 1 = \$350000$$

$$\text{Unit cost for direct dredging cost} = \frac{100741}{350000} = \$0.29/\text{yd}^3$$

13.4.2 Dam and lock costs

Dam and lock costs consist of labor, material, and equipment costs, which are dependent on the construction design and schedule. In general, the project design and the magnitude or impact of a particular item has an influence on the cost estimates. Work place, weather, and environmental conditions should be considered for assessing the dam and lock costs.

Labor costs include costs of equipment operators, ironworkers, carpenters, laborers, and so on. It can be calculated by wage ranges required by law for the work based on skill levels on each craft and area where dams and locks are located. Additionally, overtime costs anticipated for the project need to be taken into consideration since inland waterway projects have variable schedules to cope with weather and environmental conditions.

Material costs are estimated by items used for construction or maintenance for dams and locks. The time for manufacture and delivery, location factors or quantity are included in the costs. Historical data records or cost estimate models can be used to assess quantity and cost of materials.

Equipment costs are generated from the use of instruments required for construction of dams or locks, which include equipment ownership costs and operating costs. Equipment ownership costs are capital cost, cost of facility capital, and equipment overhead. The cost of facility capital is the cost invested in the equipment by considering the purchase price of the equipment, the economic life of the equipment, the salvage value, and the interest rate. Equipment operating costs include field repair, ground engagement, fuel expense, and electricity costs incurred while operating pieces of equipment. The costs are affected by project conditions, specification requirements, and knowledge of equipment capabilities.

Table 13.12 Waterway transportation unit crash cost in Netherland

Crash type		*Cost ($/case)*
Damage to waterway facility		47090
Damage to ship		120143
Damage to cargo		19091
Damage to human	Fatality	2269224
	Major injury	402173
	Minor injury	20363

13.4.3 Crash costs

The crash cost is assessed by combining crash risk, frequency, and unit crash costs determined by crash type of vessels. The type of vehicle, time of day, driver characteristics, and location involved with a vessel crash have an impact on estimating the crash cost.

The crash risk differs by type of waterway and type of vessel. The crash risk is separately assessed by crash severity level, including fatal, injury, and PDO and is distinguished between injurers and victims. The crash frequency represents the likelihood of a waterway crash occurrence, which can be determined based on historical crash records or through model predictions. The waterway crash occurrence is affected by type of waterway, type of vessels, location, and navigation condition. The unit crash cost includes the value for life and social costs such as administrative, material, production losses. In other words, the unit crash cost consists of all costs associated with a crash causing fatalities, injuries, and property damages. In general, the costs can be estimated by the willingness-to-pay/avoid method or net of insurance cost. Table 13.12 shows unit crash costs derived in the Netherlands (ECORYS and METTLE, 2005).

EXAMPLE 13.3

Data records indicate that 120 vessel crashes occurred in an inland waterway system. On the waterway facility and vehicle damages side, 20 crashes caused damages to the waterway channels, 30 resulted in vessel damages, and 30 led to cargo losses. As related to human damages, there were 12 fatal and 48 injury crashes with 20 major injuries and 28 minor injuries. Estimate total crash cost.

Solution

Cost of damage to infrastructure $= 20 \times 47090 = 941800$
Cost of damage to ship $= 30 \times 120143 = 3604290$
Cost of damage to cargo $= 30 \times 190091 = 572730$
Cost of damage to human $= 12 \times 2269224 + 20 \times 402173 + 28 \times 20363 = 35844312$

Crash type		*Number of crashes*	*Unit crash cost ($/case)*	*Crash cost ($)*
Damage to waterway facility		20	47090	941800
Damage to ship		30	120143	3604290
Damage to cargo		30	19091	572730
Damage to Human	Death	12	2269224	27230688
	Major injury	20	402173	8043460
	Minor injury	28	20363	570164
Total				$40963132

13.4.4 Air emission costs

The following procedure could be used to estimate the air emission cost. First, sources of airborne pollutants in waterway transportation, including NMHC, CO, CO_2, SO_2, NO_X, and PM, need to be identified and categorized. Then the quantity of vessel air emissions from sources of pollutants needs to be estimated by considering vessel types, passenger demand, cargo tonnage, engine operating time, and vessel speed. Next, unit costs of vessel air emissions can be estimated for different types of airborne pollutants. This can generally be done based on the willingness-to-pay/avoid method with due cognizance of differences in population density or land characteristics exposed to vessel air emissions. Finally, the vessel air emission cost per vessel can be estimated for different types of pollutants as

$$C_i^{VEC} = EF_i \cdot UC_i^{VEC} \tag{13.10}$$

where

C_i^{VEC} = Vessel emission cost of pollutant type i, dollars/vessel/mi
EF_i = Emission factor of pollutant type i, kg/vessel/mi
UC_i^{VEC} = Unit vessel emission cost of pollutant type i, dollars/kg
i = Vessel air pollutant i

13.4.5 Noise costs

Noise pollution is another type of impact on people and the environment caused by inland waterway transportation. Although inland waterways are typically distant to residential areas, noise pollution can still adversely influence human activities and wildlife in natural habitats near rivers. In most cases, noise impact is generated by vessel activities at locks and terminals. In order to calculate the noise cost of inland waterway transportation, the following procedure can be followed: (i) Determine cut-off value of noise level that cannot affect human and ecological systems; (ii) estimate number of households or people who are exposed to a certain noise level generated by vessel traffic above the cut-off level; (iii) establish values of unit cost for exposed noise intensity based on the number of vessels in the terminal and the impact of a certain type of vessels on noise intensity; and (iv) calculate noise cost based on noise intensity and the values of unit cost for waterway noise pollution.

EXAMPLE 14.4

Among 500 households suffering from waterway transportation noise, there are 200 households under 65–70 dBA, 100 households under 60–65 dBA, 70 households under 55–60 dBA, 50 households under 50–55 dBA, and 80 households under 45–50 dBA. With the cut-off value at 50 dBA, noise cost per year is at $27 per dB-household, determine the noise cost.

Solution

With cut-off value = 50 dBA, extra dBA values that affect households are +17.5 dBA for 200 households, +12.5 dBA for 100 households, +7.5 dBA for 70 households, +2.5 dBA for 50 households, and no effect for 80 households

The noise cost = (200 × 17.5 dBA + 100 × 12.5 dB + 70 × 7.5 dB + 50 × 2.5 dB) × $27 = $145800 per year

13.5 PERFORMANCE-BASED EVALUATION OF WATERWAY TRANSPORTATION PRESERVATION

13.5.1 Waterway transportation facility service life cycle

Comparable to pavement and bridge facilities in highway transportation, the inland waterway transportation facility service life cycle generally includes design useful service life cycle and repair service life cycle. The design useful service life cycle is defined as the time interval between two consecutive construction interventions. The repair service life cycle is the time interval of consecutive repair treatments. Because of differences in the function, design, materials used, and reliability, as well as usage, waterway channels, dams, and locks maintain different service lifespans.

13.5.2 Waterway transportation treatments and strategies

To achieve the waterway facility design useful service life cycle, effective maintenance and repair treatments need to be implemented in a timely manner. In this respect, the facility life-cycle agency costs generally include costs of facility construction and additional costs of maintenance and repair treatments to ensure that the facility condition is always kept above the threshold levels and no early termination of the actual service life occurs.

For waterway facilities such as waterway channels, dams, and locks, repair treatments are implemented in multi-year intervals. The length of each interval is determined according to predefined time durations or by incorporating the facility condition deterioration models. The maintenance treatments may be classified into annual routine maintenance and periodic maintenance treatments applied in repair intervals. The annual routine maintenance may be reactive to correct the damages or may be preventive usually applied to avoid excessive damage. In addition, treatments may be implemented to address or mitigate issues of system usage such as delays, vessel crashes, air, and noise pollution. A strategy pertains to multiple treatments implemented in waterway facility service life cycle in a coordinated way to help achieve the lowest annualized costs.

For assessing the condition of waterway facilities and determining the timing of facility treatments such as dams, two methods may be used. The first method is related to facility safety action classification. The second method relies on facility condition index level. A treatment to waterway dams, gates, and walls is required once the safety action classification reaches level three, denoted as high priority. Also, a treatment to waterway channels, dams, and locks is needed if the condition index level reaches level C. For the same facility such as a dam, a treatment needs to be implemented in time, whichever of the above two assessment criteria demands a treatment. Tables 13.13 and 13.14 present the descriptions of the two methods.

For assessing the waterway system usage performance, the concept of level of service (LOS) used in highway capacity analysis can be introduced (TRB, 2015). For an inland waterway system, LOS can be defined based on lost time or lock processing time for assessing performance because these factors affect users perceiving the service. These service levels could describe the degree associated with delays and levels of risk of system failure or unplanned outages that the people or goods movements could encounter. A treatment is needed once an LOS C is reached. Table 13.15 presents the LOS analysis of the waterway usage performance and determines the timing for a usage performance treatment.

Table 13.13 Safety action classification for dams, gates, and walls

Level	*Condition*	*Description*
1	Urgent and compelling	Indicates extremely high risk of failing Requires immediate action Will fail within a few years
2	Urgent	Indicates unsafe or potentially unsafe Failure will begin in normal operations or natural event Requires instant emergency action measures
3	High priority	Conditionally unsafe Moderate to high risk of failure Requires heightened monitoring and evaluation
4	Priority	Indicates a low risk of failure Requires monitoring and evaluation No plan for risk reduction measures
5	Normal	Indicates normal and safe Normal operation and maintenance

Table 13.14 Condition index levels for waterway facility assessment

Level	*Condition*	*Description*
A	Adequate	Limited probability of failure
B	Probably adequate	Low probability of failure
C	Probably inadequate	Moderate probability of failure
D	Inadequate	High probability of failure
E	Failed	The feature has failed

Table 13.15 Level of service for waterway transportation usage performance assessment

Level of service		*Waterway transportation usage impact*
A	Minimal delays (no unplanned outages)	Delays and outages will be in line with best service levels historically; there will be minor queuing
B	Moderate delays (no unplanned outages)	Queues, delays, and outages are expected, but the average is kept within a certain variance from historical best conditions
C	Significant delays (possible unplanned outages)	Delays and outages are unpredictable and windows of service may be constrained
D	Severe delays (high potential for unplanned outages)	Delays are expected to be lengthy and windows of service will be constrained, which could allow for an imminent risk of failure

13.5.3 Waterway transportation facility life-cycle agency costs

The life-cycle activity profile of a specific type of waterway facilities such as waterway channels, dams, and locks refers to the frequency, timing, and magnitude of construction, repair, and maintenance treatments in the facility design useful service life cycle. A typical life-cycle activity profile represents the most cost-effective way of implementing strategically coordinated repair and maintenance treatments to achieve the intended design useful service life. However, the typical facility life-cycle activity profile may not be followed in real-world conditions. This may trigger two consequences: the increased magnitude of repair costs to recover the further deteriorated facility condition and early termination of design useful service life. Comparing the equivalent uniform annualized life-cycle agency costs for the typical and actual facility life-cycle activity profiles, the annualized costs corresponding to the typical profile tend to be lower. The difference of the two annualized agency costs is the life-cycle agency benefits of implementing the needed repair and maintenance treatments and strategies in a timely manner.

13.5.4 Waterway transportation system usage costs

Waterway transportation system usage costs are those related to vessel operation, delays, crashes, air emissions, and noise pollution. According to the typical and actual facility activity profiles, the life-cycle use cost profiles can be established. The equivalent uniform annualized user costs corresponding to the two profiles can be computed. The reduction in equivalent uniform annualized user costs based on the typical user cost profile compared with equivalent uniform annualized user costs related to the actual user cost profile is treated as the facility life-cycle user benefits.

Because the vessel traffic in the facility useful service life cycle may follow a growing path from year to year, an annual vessel traffic growth rate needs to be taken into consideration in estimation of life-cycle user costs. With the base year average daily vessel traffic $ADVT_0$ determined, we could consider a baseline annual growth rate of g_V that leads to a traffic growth path of $ADVT_0 \cdot e^{(g_V \cdot t)}$ for any year t.

13.5.5 Risk and uncertainty considerations

In the process of computing waterway facility life-cycle agency and user costs, a number of input factors are used for the analysis. Some factors may not always follow a single value in the facility service life-cycle, which calls for risk and uncertainty considerations and analysis analogous to the risk and uncertainty-based analysis for highway pavement and bridge facilities.

When an input factor is under risk, it is assumed that the range of multiple possible values of the factor is known and a reliable probability distribution can be assigned. A mathematical expectation can be calculated via probabilistic risk assessment and used for the calculation. When an input factor is under uncertainty, either the full range of multiple possible values of the factor is unknown or if even though it is known a reliable probability distribution cannot be established. Hence, a mathematical expectation cannot be computed. An alternative approach such as Shackle's model, incorporating a decision rule as described in Chapter 7, needs to be employed to derive a single value for subsequent computation (Shackle, 1949; Ford and Ghose, 1998; Young, 2001; Li and Madanu, 2009). The following factors are recommended for risk and uncertainty considerations.

Dredging operation: Dredging operation can involve many special risks. Dredging has a harmful impact on the environment such as fish and mammals, which can yield substantial increases in environmental costs in some cases. Therefore, turbidity monitoring and control are required to reduce the risk of additional cost occurrence related to dredging. Also, dredging can have potential risks that can result in a high potential for storms, leading to strong winds and waves. This unexpected environmental change can cause vessel crashes, major equipment breakdowns, and encounter unexpected rock layers. Therefore, safety and environmental risks need to be thoroughly assessed when dredging is performed in a waterway channel.

High water flow (flooding): High water flow due to excessive water supply in channel from flooding would disrupt operating waterway transportation system. Therefore, an assessment of safety risk is required by using hydrograph analysis and flooding routing. Flooding flow should be accurately analyzed to prevent disconnecting waterway service and enable safe shipping and sufficient capacity to be provided in hydraulic structures.

Low water flow (drought): Low water flow due to restriction in water supply will disrupt shipping activities, which yields negative consequences. The frequency and extent of drought should be forecasted to prevent the risk of navigation restriction and to provide the proper amount of water available for vessel usage.

Occurrence of ice: Unlike other travel modes, waterway transportation is vulnerable to freezing weather. In winter season, the occurrence of ice on a river is one of the primary problems to cause interruption in navigation. On navigable rivers, there are sections with favorable and unfavorable navigation conditions at occurrence of ice. For example, if ice occurs on a channel that conveys a large number of ships, a significant socioeconomic loss is caused. Therefore, it is essential to manage the risk of occurrence of ice to provide successful operation of navigation system.

13.5.6 Methods for economic analysis of waterway transportation improvements

The implementation of a candidate project will potentially lead to reductions in waterway facility life-cycle agency and user costs, which are treated as benefits. A certain amount of project cost will occur because of project implementation. Ultimately, the waterway transportation improvement project-level benefits and costs are established. In general, the benefit-to-cost (B/C) ratio and incremental benefit-to-cost ratio methods could be employed to assess the cost-efficiency or economic feasibility of the proposed waterway project. The net present worth method can be adopted to evaluate the cost-effectiveness of the proposed project. These methods are described as follows:

Benefit–cost analysis: One of the most commonly used economic analysis approaches is the benefit–cost analysis. The benefit-to-cost (B/C) ratio is simply the equivalent benefits of a project divided by the equivalent costs of that project during the waterway facility service life. Benefit–cost comparisons are possible when the benefits of a waterway transportation improvement project can be assigned a monetary value. If the benefits of the project exceed its costs, the project is economically justifiable or cost-efficient. Furthermore, the ratio of each project provides a convenient basis for comparison, providing a means of the dollars of expected benefits of a project for each dollar spent on that project.

Incremental benefit–cost analysis: If candidate projects being analyzed build upon each other in terms of costs, quantities, complexities, and so forth of components that meet the waterway transportation management goals, it may be more appropriate to consider an incremental benefit–cost analysis. For this approach, the benefits and costs considered for each project are not the total, but rather the additional benefits achieved and costs incurred over the next effective project. This analysis considers whether an investment necessary to achieve the next incremental step in the system can be justified in terms of the incremental benefits that would be achieved.

Net present worth method: Computation of a candidate project's net present worth involves a conversion of all the project-level costs and benefits that are incurred at its initiation and throughout the waterway facility useful life cycle to an equivalent current value. The current value of equivalent costs is subtracted from the current value of equivalent benefits of the project. If the benefits exceed the costs, the project can be justified economically or is proven to be cost-effective. Further, comparisons among candidate projects can be carried out and the candidate project providing the greatest additional benefits over costs maintains the greatest net present worth.

PROBLEMS

13.1 You have been asked to evaluate the performance of an in-land waterway transportation system. What performance measures would you consider in the evaluation?

13.2 What are the main cost items of in-land waterway transportation? Because of severe flooding in an area, it is required to do dredging to keep constant water flow for vessels. Explain
   a. What type of dredging methods can be used
   b. How to estimate dredging cost.

13.3 A dam on Illinois river has a volume-to-capacity (V/C) ratio of 0.8, and variances of inter-arrival time, lock service time, and inter-departure time are 30, 20, and 10 mins, respectively. Estimate average waiting time ($t_a = 10$ mins).

13.4 Although in-land waterway transportation is fuel-efficient and environmentally friendly, technologies and regulation for environment issues fall behind those in use by surface transportation, either highway or rail transportation:
   a. What are the environmental impacts of waterway transportation?
   b. What factors can affect producing pollutants and how to estimate the pollutants emitted by in-waterway transportation?

13.5 Develop a maintenance and rehabilitation strategy for locks and dams that support an in-land waterway transportation.

13.6 Propose a life-cycle cost analysis method for evaluating waterway transportation projects.

13.7 What factors are considered under risk and uncertainty for managing the in-land waterway system?

Chapter 14

Economic analysis of mobility improvements

14.1 GENERAL

Mobility is having both the ability to travel from one location to another and the knowledge of how to do so using a multimodal transportation system (FHWA, 2004a). Access to accurate information on the expected travel times dramatically enhances travelers' perceptions of their trips because it allows them to make informed decisions, gives them more control over their journey, and makes the travel experience more pleasant. When travel information is unreliable or unavailable, the uncertainty generates anxiety about the duration, range, intensity, and evolution of the traffic congestion and delays they will face, causing them to perceive congestion and delays as worse than they really are. It may also lead to more erratic driving behavior and undermine public opinion of the transportation system's performance and management.

Congestion is defined as excessive travel time or delays compared to the normal travel time established by an agreed-upon norm (Lomax et al., 1997; FHWA, 2004b, 2005c; TRB, 2010). Thus, excessive travel time or delays will occur in the presence of congestion. The agreed-upon norm may vary by travel mode or sub-mode, geographic location, and time of day. Congestion is generally classified as: (i) recurrent; (ii) nonrecurrent; and (iii) emergency event-triggered congestion. Recurrent congestion forms when the travel demand exceeds the available capacity of the transportation system, and typically occurs during morning and evening peak periods. Recurrent congestion can be mitigated through measures such as reducing demand intensity, expanding capacity or both. Nonrecurrent congestion is caused by traffic incidents, including vehicle crashes and disablements, debris in the traveled way, work zones, and inclement weather conditions. To mitigate nonrecurrent congestion, it is essential to minimize the time required to detect an incident, respond, clear the traveled way, and restore the system. The impacts of an emergency event on congestion depend on the nature of the event—the range, duration, and intensity can vary considerably, and need to be addressed proactively and handled case by case. The key to mitigating these impacts is developing evacuation plans or emergency management strategies to respond effectively to emergency events well in advance so they can be implemented rapidly. To remain effective, these plans or strategies must be updated regularly before, during, and after the emergency events.

14.2 PEOPLE MOBILITY VERSUS TRAFFIC MOBILITY AND TYPICAL PERFORMANCE MEASURES

For a multimodal transportation system, a mobility analysis can be based on the mobility of people or the mobility of traffic—it can be based on person trips or vehicle trips. The analysis could assess the mobility of people between O–D pairs by individual travelers using multiple travel modes, including auto, transit, bike, and pedestrian walking. Alternatively, a traffic mobility analysis could assess the movements of vehicles on various levels of the transportation network, such as highway segments, intersections, corridors, subarea networks, or the entire network. The traffic mobility analysis could focus on a single type of vehicle, different types of vehicles of a single travel mode or all types of motorized and non-motorized vehicles of multiple travel modes. One example is the multimodal traffic mobility analysis for a complete urban street network whose streets are shared by auto, transit, bike, and pedestrian walking traffic. Further, the trip-based or traffic-based mobility can be conducted for morning and evening peak periods or for a 24-hour period. Tables 14.1 and 14.2 list typical performance measures utilized in a mobility analysis and their applicability for evaluating various mobility improvement projects (NCHRP, 2008).

Table 14.1 Typical performance measures for mobility analysis

Measure	*Description*
Travel time (person-hours)	Actual travel speed · volume · vehicle occupancy · length
Annual delays per traveler	(Actual travel time—free-flow travel time) · 250 weekdays/year
Travel time index	Actual travel time/free-flow travel time
Travel time buffer index	(95th percentile travel time—average travel time)/average travel time
Total delays	(Actual travel time—free-flow time) · volume · vehicle occupancy
Congested travel	$\sum$(Volume · congested segment length)
Percent of congested travel	$\dfrac{\sum(\text{Actual travel time—free-flow travel time})\cdot(\text{Volume}\cdot\text{vehicle occupancy}\cdot\text{length})}{\sum(\text{Actual travel rate}\cdot\text{volume}\cdot\text{vehicle occupancy}\cdot\text{length})}$
Congested mileage	$\sum$Congestion segment length
Accessibility	$\sum$Objective fulfillment opportunities where travel time $\leq$ target travel time

Table 14.2 Applicability of performance measures for evaluating different types of mobility improvement projects

Project type	*Travel time*	*Annual delays per traveler*	*Travel time index*	*Travel time buffer index*	*Total delays*	*Congested travel*	*Percent of congested travel*	*Congested roadway*	*Accessibility*
Individual location	Minor	Minor	Major	Major	Minor				
Short road segment	Major	Minor	Major	Major	Minor				
Long road, transit route		Minor	Major	Major	Minor				
Corridor		Minor	Major	Major	Major				Minor
Subarea network		Major	Major	Major	Major	Major	Major	Major	Major
Regional network		Major	Major	Major	Major	Major	Major	Major	Major
Multimodal analysis		Major	Major	Major	Major				Major

14.3 MOBILITY IMPROVEMENT MEASURES AND STRATEGIES

Congestion mitigation measures and strategies fall into three broad categories: demand management, capacity expansion, and efficient system utilization. A congestion mitigation measure aims to reduce the incidence of congestion and mitigate the adverse impacts when an incident does occur. A congestion mitigation strategy is a series of congestion mitigation measures identified and implemented in a coordinated manner to collectively maximize the ability to prevent congestion incidents and to minimize their impacts.

Demand management measures and strategies are concerned with reducing the total travel demand and reducing the intensity of demand, especially for the peak periods by effectively spreading the peak. Capacity expansion measures and strategies seek to increase the transportation supply with added system capacity in order to meet a travel demand by adding highway travel lanes and transit routes, and by enhancing multimodal integration to potentially trigger modal shifts from auto to transit. This can be achieved by integration of auto and transit modes by promoting ride share, integration of transit sub-modes, such as bus, BRT, fixed guideway transit, and taxi, in conjunction with non-motorized mode such as bike and walking. Efficient system utilization measures and strategies focus on improving the balance between demand and supply in fine-grained time and space domains.

14.3.1 Travel demand management

Providing auto travelers with more efficient travel options during peak travel periods can reduce congestion at very low cost. Reducing single occupancy vehicle trips by encouraging practices like park and ride, carpooling or vanpooling can reduce highway congestion. Alternative travel options can be implemented by public agencies that determine a variable pricing system for road travel and parking or by individual employers who choose to participate in telecommuting and compressed workweek scheduling.

Flexible work hours: Flexible work hours allow employees to work an 8-hour shift that typically begins between 6:00 and 9:00 AM and ends between 3:00 and 6:00 PM. Providing a range of start and end times reduces the number of workers traveling to and from work during the peak volume times while still allowing typical office hours. Staggered work hour programs are a variation of flextime that alternates the arrival of groups of employees, though the employees may not be able to choose their working hours. Prime targets include government agencies and manufacturing plants in heavily urbanized areas that employ a large number of people working typical shifts and that may want to alleviate crowding at entrances/exits, elevators, and parking areas. Flexible work hours can help employers improve productivity by allowing employees to work at times that fit their lifestyle needs. These programs distribute peak-hour traffic to less-congested hours, mitigating peak-hour congestion. Flexible work hour programs may be voluntary or mandatory, and may be used to satisfy trip reduction ordinances and air quality regulations. Whether a flexible work hour program is used and managed effectively, however, is up to a business's administration and its individual workers.

For some businesses, work schedules can easily be altered to achieve the goals of the flexible work hour technique. Adequate planning, enforcement, and coordination are required to make the technique successful at mitigating congestion. It is important that the plan include a voluntary adoption clause that also values employee input. The flexible work hour program should consider the distance workers must travel and external factors such as family and other circumstances that affect the hours they can work. Starting earlier or ending later may cause hardship for employees who travel long distances.

Carpooling: Carpooling programs are designed to promote ridesharing by identifying riders with similar origins and destinations. Using a database of interested riders, employers or regional agencies can promote carpooling for an entire region. Carpooling can increase the person-throughput of any roadway. But the biggest impacts can be expected when people with longer trips carpool: as the trip length increases, so do the positive impacts. The target market includes congested corridors during peak hours and activity centers with limited parking spaces. It can be paired with managed lanes that offer a price savings for carpools and with park-and-ride lots.

Carpooling itself is a simple concept, but several related issues can have significant impact on its effectiveness. Guaranteed ride home programs, priority carpool parking, vanpool vehicles, high-occupancy vehicle lanes, park-and-ride facilities, and real-time ridesharing technologies are all important techniques to consider.

Vanpooling: Vanpools usually consist of 5–15 riders that pay to commute for long distances into a city or to a transit facility. The service may allow patrons to ride transit for free or at a discount. Employers and local governments sponsor vanpools by providing incentives to employees for riding, including vouchers for transit, cost subsidies, and discounted parking. Vanpools work best in areas with little transit service and inadequate parking. This service is best paired with managed lanes that offer a price savings for vanpools and in areas with park-and-ride lots. The vanpool operator may be the transit agency or a third party may provide vans and administer the vanpool program. They are especially valuable in supplying commuting services to people who do not have a personal vehicle or where the public transit system does not reach a particular job center. Vanpools have the potential to offer a less expensive alternative to driving alone and to reduce costs to employers by reducing the amount of parking they need to provide.

Providing compelling information on how vanpools benefit employers and employees is essential to the success of the program. The initial push for these programs is typically done by the local government, but employers can also promote the benefits to their workers. Promoters should display the numerous benefits of not driving, including the economic savings to a household of owning one fewer vehicles and the community benefits of reducing the number of cars on the road. The overall management of the vanpool program depends on the available budget and preference of the local government. The vanpool program can be owned and administered by the transit authority or by a third-party entity. Vanpools require a support system to maintain adequate operations for the service area. Typically, this system includes collecting fares, purchasing or leasing new vans, finding room for new riders, scheduling and conducting van maintenance, paying for tolls and gas, and acquiring the necessary insurance. These programs are usually prevalent in trip patterns with inadequate transit options owing largely to low transit ridership.

Variable road pricing: Variable pricing manages travel demand to mitigate roadway congestion with carefully constructed pricing structures that encourage motorists to use the available roadway more efficiently. Transportation agencies use variable pricing on congested facilities, often utilizing existing toll components included in roads, bridges or tunnels. Variable road pricing can be classified into two types: (i) time-of-day pricing; and (ii) dynamic pricing. For time-of-day pricing, the administering agency varies the toll with the time of day according to a schedule that is set to generally, but not exactly, correlate with the level of congestion on the road: the times of day with higher congestion are tolled more than those with less congestion. Dynamic pricing resembles time-of-day pricing, but it provides an increased level of precision and adds technological complexity. Administering agencies adjust the toll in real time based on the current level of congestion on the road. If the average speed decreases, the vehicle count increases or some other measure of congestion increases, the toll increases in corresponding increments. Usually the agency establishes a minimum below which the toll cannot drop and a maximum above which it cannot rise. Variable pricing is most commonly used on roadways that are either currently congested or forecasted to be congested in the future. The strategy can be applied either to roads that are not currently tolled or to tolled facilities that currently have fixed pricing.

To implement variable pricing, the managing agency needs to first assess the needs of their facility and then select a suitable strategy to address those needs. If the toll component is new, tolling infrastructure will need to be constructed, software developed, and hardware selected for motorists' on-board units. Using all electronic tolling is vital for maximizing throughput. To implement variable pricing on a facility with existing tolling equipment, the agency needs to adjust the software algorithms to calculate and charge the new toll rates, especially with dynamic pricing. Signs that inform travelers of the new toll will need to be constructed. As with any major project, the managing agency should develop a concept of operations that analyzes the relevant constraints and considerations, such as level of demand, willingness to pay, infrastructure needed, level of public support, and availability of transportation alternatives for implementing pricing schemes.

Parking management: Managing the supply and price of parking can reduce congestion in surrounding corridors, especially in dense, high-activity locations where a significant portion of background traffic is for seeking parking spots. Encouraging drivers to use transit, carpools, or vanpools reduces the numbers of vehicles on major arterial and local city streets. The set of strategies that is ultimately deployed must have broad support from the affected businesses, workers, and communities. Typical parking management measures and strategies include: (i) increasing parking fees; (ii) reducing or eliminating minimum parking requirements; and (iii) parking cash-out. Increasing parking fees is typically applied in high-density areas to persuade commuters and visitors to reconsider their mode choice when commuting to major business or activity centers. Altering zoning ordinances to reduce or eliminate minimum parking requirements for certain areas can limit the parking availability and increase transit and carpool usage. The parking cash-out strategy provides stipends for employees to pay for parking, transit tickets, or to save for personal use. This method allows employers to give their employees money in lieu of providing parking, which can reduce the availability of parking and provide an incentive to use alternate modes of transportation. Parking management can be deployed for downtown or other high-density activity centers where transit service is adequate or parking is in short supply.

14.3.2 Capacity expansion

Expanding capacity is the most frequently adopted approach in congestion mitigation. Additional capacity is achieved by adding travel lanes, building new roadways, introducing managed lanes, or other design improvements. These types of improvements are implemented by public agencies, private entities, or public-private partnerships. For densely populated urban areas with land scarcity, adding more travel lanes may not be feasible due to prohibitively high cost and work zone constraints. In addition, long-term benefits of capacity expansion via adding more travel lanes or building new highways could be marginal. Alternatively, multimodal integration that consists of integration of auto and transit modes and transit sub-modes can be adopted. Center to multimodal integration is to reduce auto traffic via modal shifts while the riding quality of travelers diverted to transit is still maintained at a certain level.

Adding new lanes or roads: Adding new lanes to an existing road or building new roads can reduce congestion and provide alternate routes for travelers. Constructing additional lanes or a new highway, however, is a major undertaking that requires a significant funding commitment and typically takes a long time to implement. It is commonly applied only when needed to reduce serious congestion along a corridor or roadway. The planning and design aspects of building new lanes and roads are complex, and the strategy confers uncertainty and risk, the primary elements of which are the need to acquire the ROW and the negative impacts on the environment and the local community. The typical benefits conferred by reducing congestion and providing more reliable travel conditions include reduced commute times, improved freight and delivery schedules, reduced emissions and fuel consumption, and increased productivity and economic development. New lanes and roads are most beneficial in urban corridors that already own the ROW needed for the construction. The road or area considered for this project must be very congested to justify the effort and expense of new construction. Projects to add lanes and new roads should incorporate other congestion mitigation strategies into the design and operation, including strategies that manage demand and provide alternatives to driving to collectively improve mobility performance.

Roadway projects of this scale are often subject to various planning and environmental requirements as a prerequisite to securing funding. An environmental impact statement (EIS) must be completed if the project is expected to significantly impact the local environment. A plan may also be required to minimize the project's negative impact on air quality, depending on the jurisdiction and local laws where the project is located. When adding new lanes or roadways, the engineers, planners, and designers should consider both current and future demand. A project on the scale of adding new lanes or a new roadway will impact more than just the immediate corridor; it can change traffic patterns and operations throughout the surrounding area. Moreover, adding more lanes for longer lengths could potentially attract auto trips. The increased vehicular traffic may regress congestion to former levels and exacerbate bottleneck situations outside the project area. These issues should be explicitly addressed when considering adding travel lanes or building new roads.

Roadway widening: The number of lanes on a highway segment influences congestion and safety. In some bottleneck situations, adding more lanes for just a short distance can eliminate these constraints. These auxiliary lanes can be used for speed change, turning, weaving, truck climbing, merging and exiting traffic, and other purposes supplementary to through-traffic movement, which will balance the traffic load and maintain a more uniform level of service on the highway system.

Managed lanes: A managed lane is a broad term that refers to any lane or corridor whose use is controlled according to vehicle type, eligibility, pricing, or access management. They are deployed due to the inability to build enough lanes to address congestion during peak periods; the desire to encourage alternatives to driving, such as transit, in a congested corridor; the need to address funding issues and generate revenue; the desire to increase the effectiveness of existing HOV lanes; and the need to separate large vehicles from general traffic. Common types of managed lanes include: (i) high-occupancy vehicle (HOV); (ii) high-occupancy toll (HOT); (iii) express toll; and (iv) exclusive lanes. HOV lanes allow those cars with two, three, or more passengers to use lanes separated from the main traffic lanes. The lanes are managed by eligibility, in that only high-occupancy vehicles are allowed in the lane. HOV lanes typically provide travel-time savings and more reliable trip times, offering an incentive for ridesharing. HOT lanes, by contrast, allow lower-occupancy cars access to HOV lanes for a fee. The cost of purchasing access to HOT lanes is motivated by higher speed and more reliable trip time. The toll is adjusted to maintain free-flow conditions. This produces an alternative to congestion and generates revenue that can offset the cost of implementing the strategy. Express toll lanes similarly charge a toll that can be adjusted to maintain free-flow speeds, but differ from HOT lanes in that they charge all vehicles—high-occupancy vehicles are not exempted. Exclusive lanes restrict certain lanes to only buses, trucks, or other slower moving vehicles. A truck-specific lane typically separates the slower, less agile vehicles from the main lanes and allows higher speeds in the adjacent lanes for passenger cars. Bus lanes exclude passenger cars, reducing the effects of congestion on transit passengers and providing an incentive for drivers to take transit instead. As appropriate, HOT and bus lanes could be combined to let cars pay a toll to use the lane and keep the toll high enough so buses never get stuck in traffic. Managed lanes provide travel alternatives to the single-occupancy vehicle, giving users the flexibility to choose the best method of travel for each trip. This choice reduces congestion by making more efficient use of existing capacity. If flow can be maintained by pricing, eligibility or access, the throughput of vehicles and people in the corridor can be increased.

Selecting the type of lane and its design and operating rules should depend upon the primary goals for the lane—such as maintaining free-flow speed, maximizing person-moving capacity, maximizing vehicle throughput, revenue needs, freight-moving capacity, and so on. Managed lanes can boost the efficiency of both the current transportation network and any new or alternative network, such as dedicated transit or freight traffic facilities. Given the variety of managed lane strategies, and because no two facilities are alike, there is no single set of guidelines that can be applied across all circumstances. However, several planning, design, and operational factors should be considered: (i) the type of managed lane should be chosen to support the regional transportation vision and the goals for the specific corridor; (ii) the physical and operational characteristics of the corridor are vital in the development of managed lanes; and (iii) managed lanes are intended to promote a more efficient utilization of the existing system capacity.

Grade separations: Intersections that handle high volumes of vehicular and pedestrian traffic (and in some cases rail traffic) limit the capacity of the approaching roads. Grade-separating these points of conflict allows uninterrupted traffic flow while also eliminating the safety threat posed by crossing traffic. Three primary roadway improvement objectives are accomplished using grade-separated intersections: (i) increased capacity and uninterrupted flow; (ii) increased safety; and (iii) reduced auto-train conflicts and delays. Grade separation can be implemented for high-volume roadway intersections or highway-railroad crossings. Grade-separated intersections are expensive to build, thus they should be used only when other techniques such as signal timing and lane additions are unsuccessful in reducing congestion to an acceptable level or were impossible due to external constraints. Grade separation works well when one or more directions continue to experience heavy traffic and high congestion even after other methods have been exhausted.

Unconventional intersections: The search for innovative methods to improve traffic channelization at intersections begins with assessing existing unconventional intersection designs. The existing designs have primarily focused on improving left-turn traffic movements to flow and safety. They can be classified into two broad categories: unconventional at-grade intersection designs and unconventional overpass/underpass and interchange designs. Typical unconventional at-grade intersection design options include doublewide, continuous flow, median U-turn, and superstreet intersections.

Unconventional overpasses/underpasses: Overpasses/underpasses are used to mitigate the impact on traffic flow of vehicles entering and exiting the highway and the merging and weaving associated with interchanges. Improvements can be made to increase the capacity and safety of the weaving sections and the ramps that make up the interchange. The operation of the weaving sections is influenced by the weaving configuration, length, width, and entering and existing volumes. Parameters that influence the operations of ramp-highway junctions include the number of lanes, lane width, lateral clearances, terrain and grades, curvature, the length and type of acceleration or deceleration lanes, sight distance, speed, lane distribution, and free-flow speeds of upstream highway traffic. Unconventional overpass/underpass design options for interchanges include center-turn underpasses and tight diamond, single point, echelon, and median U-turn interchanges.

One-way streets: One-way streets are often used in high-volume situations, such as a downtown area with closely spaced intersections, or highways with heavy directional flows. One-way regulations are often incorporated into the original street design for new activity centers such as shopping centers, sports arenas or industrial parks. In some cases, reversible lanes can be implemented on highways or major streets to improve traffic movement in one direction for a time.

Multimodal integration: In some urban highway corridors, metropolitan planning organizations (MPOs) or other local planners may opt to add capacity to busy corridors by constructing multimodal transportation features such as dedicated bus-only and carpool lanes or commuter or light rail service parallel to the flow of automobile traffic. Transit that improves speed and reliability could potentially increase the flow of people in the transportation network in that it can carry many more passengers per vehicle. Alternate modal options might include express bus, bus-only lanes, HOT lanes, BRT, light rail transit (LRT), heavy rail transit (HRT), and commuter rail (CR). Modes that have a dedicated ROW that excludes car traffic, such as BRT, LRT, HRT, and CR, may also readily allow for additional capacity gains by adding more buses or trains or by adding additional rail cars to existing trains. Drivers who choose to use the alternative mode generally experience less congestion by avoiding roadway chokepoints. However, convincing a significant portion of drivers to switch to an alternative mode may prove difficult if transit station locations do not adequately serve urban destinations or if congestion relief or capacity expansion projects continue to be applied to the existing roadways. Shifting some drivers to another mode may provide some relief in the rate of growth of congestion on the problem highway facility. Multimodal integration may also be considered for high-traffic, congested urban transportation corridors and existing rail/highway corridors where population growth is expected.

14.3.3 Efficient system utilization

Reversible travel lanes: Reversible travel lanes mitigate congestion by borrowing available lane capacity from the less uncongested direction. Reversible lanes are also an effective way to reduce congestion when a special traffic event happens, such as crash blocking a lane during the morning or evening peak, or when there is construction or maintenance activity on the road. The lanes of multiple-lane roads can be reassigned as needed to make the road one-way in one direction, one-way in the other direction, or any combination in between. These adjustments, indicated by variable message signs and/or arrows, can be made at specified times of day or when volume exceeds certain limits. Roads with highly directional congestion, such as those leading to and from special event centers, and emergency evacuation routes are potential targets for reversible traffic lanes.

Ramp management and control: The geometric design of a highway ramp can have a positive or negative influence on the operation of the ramp itself, and on the operation of the highway at or upstream of the merge point. Highway design standards generally address those considerations. Ramp control, on the other hand, uses control devices such as traffic signals, signing, and gates at ramps to regulate the number of vehicles entering or leaving the highway to balance demand and capacity, prevent operational breakdowns, and enhance safety.

Ramp metering is the most common ramp control strategy, and can be installed at the entrance to or exit from a highway mainline. Metering on entrance ramps involves measuring highway flow rates, speeds or occupancies upstream and downstream of the ramp, and then determining the metering rate needed to effectively manage the flow of vehicles. The toll may be pre-timed, traffic responsive, or manually set. Exit ramp traffic flow may be blocked by closing the exit ramp or it may be enhanced on the ramp and on the nearby highway mainline by preferential signal timing at the intersection of the exit ramp and surface streets. Ramp metering strategies may use restrictive or non-restrictive metering rates, and a variety of such strategies may be applied in appropriate combinations locally or system-wide. Restrictive ramp metering sets the metering rate below the non-metered ramp volume, while non-restrictive ramp metering sets the metering rate equal to the average ramp arrival volume. The choice between the two primarily determines whether significant queues will build up on the ramps, and the extent to which traffic desiring ramp access is diverted onto the mainline, causing mainline congestion. Local ramp metering is employed when only the local conditions of the ramp are used to develop the metering rates. System-wide metering is employed when metering rates are established in a coordinated fashion based on criteria that consider a specific highway corridor or entire highway system as well as the surface streets that will be affected by metered traffic.

Traffic control and safety hardware: Traffic control and safety hardware mainly include traffic signs, lighting, signals, pavement markings, crash cushions, guardrails, barriers, and rumble strips (FHWA, 2004a,b, 2005c). Improving this hardware can be a less expensive and more practical alternative for enhancing the efficiency, reliability, and safety of the existing system.

One of the most common regulatory signs is the speed limit sign. When establishing speed zoning, factors such as spot speed distribution and variation, vehicle collision experience, traffic volume, interchange frequency and spacing, alignment, and roadside environment and potential distractions need to be considered. The use of warning signs should be kept to a minimum to avoid the tendency of breeding disrespect for all signs, and they should be placed to allow adequate time from perceiving to reaction. Guiding signs should be used frequently to promote safe and efficient operations by keeping highway users informed of their location.

Adding or enhancing roadway lighting can enhance visibility at night, thereby improving safety and traffic flow and reducing the incidence of crime and vandalism at night. For a given roadway or interchange condition to be lighted optimally, various combinations of lamps, luminaire type, mounting height, spacing and positioning, and energy consumption must be analyzed to determine a preferred design.

Pavement markings provide information on guidance, regulations, and warnings to highway users. Major marking types include pavement and curb markings, object markers, delineators, barricades, channelizing devices, and islands. In some cases, markings are used to supplement other traffic control devices such as signs and signals. As they wear, markings must be maintained and replaced on a regular basis because snow, debris, and water further reduce their visibility. The durability of markings depends on their material characteristics and traffic volumes.

Rumble strips may be considered another form of pavement markings—raised or grooved patterns installed on the pavement surface of a travel lane or a shoulder to cause a vibration and sound that alerts inattentive drivers that their vehicles are leaving the travel lane. Studies have demonstrated the benefits of shoulder rumble strips in reducing death and serious injury caused by inattentive drivers in run-off-road (ROR) collisions due to distraction and fatigue. However, they are not effective in eliminating ROR collisions by excessive speed, sudden turns to avoid on-road collisions or high-angle encroachments.

Signal operation and management: Improvements to signals are among the most common, readily available, and cost-effective strategies to alleviate vehicle or pedestrian delays. They involve a combination of technology and institutional cooperation. There are four primary categories of signal strategies that can increase travel speed and reduce delays: (i) upgrading signal equipment; (ii) improving signal timing; (iii) coordinating and interconnecting signals; and (iv) removing signals.

Older hardware and software can be updated to more efficient systems with improved vehicle sensing technologies and communications. The latest traffic signal controllers have more flexibility in signal timing and allow more adaptive, traffic-responsive control of signals. Newer control equipment also offers an improved interface with area-wide signal control systems, which agency staff can use to monitor and adjust signal timing in real time. Improving signal timing and coordination involves optimizing timing plans to optimize the flow of traffic by giving the main traffic flows green time when they need it most. Coordinating signals can create a green wave with minimal stops or slowdowns as traffic moves between signalized intersections along a street. Signals are interconnected and linked in time to ensure the integrity of the timing plan. State-of-the-art signal management systems facilitate the exchange of traffic flow information between signals, allowing for automated, real-time signal coordination. Technological advances now allow signals both to learn from historical patterns and to incorporate real-time data. As traffic patterns shift, signals that are no longer needed may be removed to reduce unnecessary delay and stops at an intersection. Potential candidates for signal improvements are arterial streets and major activity centers and central business districts.

Traffic incident and emergency management: A traffic incident is a nonrecurring event on or near the roadway that causes a reduction in roadway capacity or an abnormal increase in demand. Such events may be caused by vehicle collisions, vehicle disablements, roadway debris, work zones, and inclement weather conditions. Emergency events, either with notice or no-notice, such as natural disasters and terrorist attacks can also reduce capacity or increase demand. Because traffic incidents are unpredictable and dynamic, the responding agencies must effectively plan and coordinate their incident and emergency management. Effective traffic incident management can be achieved by shortening the time required to detect and verify an incident, timely motorist information and appropriate response, site and traffic management, rapid clearance, and managed recovery. Quickly clearing a traffic incident will also reduce the occurrence of secondary incidents in upstream traffic because of the unexpected queuing of cars. Effective emergency management relies on identifying potential emergency events, estimating the event characteristics in terms of possibility, duration, intensity, and affected population of occurrences, and developing emergency management plans to mitigate adverse impacts.

Work zone traffic control: Providing temporary signage, channelizing devices, barriers, pavement markings or working vehicles can maintain safe traffic flow within a work zone. Temporary traffic control plans play a vital role in providing continuity of safe and efficient traffic flow through a work zone, and their degree of detail depends on the complexity of the situation.

Advanced traveler information systems: An advanced traveler information systems (ATIS) can effectively disseminate many types of information requested by travelers and combines multimodal information in an effective and efficient manner. Information may be provided in static and real-time fashions. Static information comes from such sources as transit schedules, planned work zones, and known road closures. Real-time information comes from a variety of sources including roadway-based sensors, surveillance equipment, and drivers. Both the static and real-time information can help travelers select their mode of travel, route, and departure times both pre-trip and en-route. Pre-trip traveler information can provide the traveler with current roadway and transit information prior to deciding the time, mode, and route of their trip. This can help relieve congestion by giving the traveler the information needed to reroute, postpone start of the trip, change modes or avoid traveling altogether. En-route traveler information can provide the traveler with current roadway and transit information needed to avoid unexpected congestion along the way.

Information is typically provided via devices deployed along the side of the roadway or from devices mounted on the dashboard of the vehicle. Regardless of how it is provided, to be effective traveler information must be timely, complete, accurate, credible, available on demand,

and perceived by travelers as relevant to their needs and valuable when followed. The information could be disseminated via televisions, online, telephone-based call-in systems, smart phones, variable message signs, highway advisory radio, and in-vehicle navigation systems.

14.4 EVALUATION OF MOBILITY IMPROVEMENTS

14.4.1 Economic analysis methods for mobility improvement projects

The implementation of a candidate project can lead to reduced costs for public agencies and private users, which are treated as benefits. Constructing and deploying the project will have a certain cost. The economic analysis establishes the benefits and costs for the project. From an economic efficiency standpoint, the analysis should use the B/C ratio and incremental B/C ratio methods. From an economic effectiveness perspective, the net present worth method can be adopted. These methods are described below.

Benefit-to-cost ratio method: One of the most commonly used economic analysis approaches is the benefit–cost ratio. The B/C ratio is simply the equivalent benefits of a project divided by the equivalent costs of that project during the service life of the facility. Benefit–cost comparisons are possible when the benefits of a mobility improvement project can be assigned a monetary value. If the benefits of the project exceed its costs, the improvement is economically justifiable or cost-efficient. The method can be used to estimate the dollars of expected benefits for each dollar spent on that project. Furthermore, establishing a B/C ratio for several competing candidate projects provides a convenient basis for comparison.

Incremental benefit-cost analysis method: If different candidate projects being analyzed build one upon another in terms of costs, quantities, complexities, and so on of components that meet the transportation mobility goal, it may be more appropriate to consider an incremental benefit–cost analysis. For this approach, the benefits and costs considered for each project are not the totals, but rather the additional benefits achieved and costs incurred over the next effective project. This analysis considers, in effect, whether an investment necessary to achieve the next incremental step in the system can be justified in terms of the incremental benefits that would be achieved.

Net present worth method: To compute a candidate project's net present worth, all the project-level costs and benefits that are incurred at its initiation and throughout the useful life cycle of the facility are converted into an equivalent current value. The current value of equivalent costs is subtracted from the current value of equivalent benefits. If the benefits exceed the costs, the project can be justified economically or is shown to be cost-effective. Furthermore, comparisons among candidate projects are straightforward: the candidate project that provides the greatest additional benefits over costs is said to have the greatest net present worth.

14.4.2 Traffic mobility improvements from demand management

Demand management measures and strategies concentrate on reducing the demand intensity for travel by private autos on a congested facility. These strategies aim to reduce the total demand as much as possible and to cut demand intensity, especially during peak periods, via effective peak spreading—that is, by triggering changes in departure time, travel mode, destination location and arrival time, and travel path. The following example illustrates mobility improvements in the context of demand management involving land use/modal shift/parking management strategies.

EXAMPLE 14.1

A high-tech corporation headquartered in the edge of a large city's central business district (CBD) that maintains 4500 employees faces a major space problem. The 30 acres of parking space at the standard of 110 cars per acre have reached capacity. Those arriving a little late might need to circle for 15 minutes before finding a parking stall. Meanwhile, the corporation plans to hire 1500 additional employees soon and 5 acres of parking space will be taken away to build new office buildings for them. This poses more challenges to address the further deteriorating parking problem. If no feasible solution could be found, the corporation would choose to relocate elsewhere. The average home-to-work commuting distance is 15 miles and most employees come on one of the two freeways by single occupancy auto, 3-person carpool, and commuter rail. The following three alternatives are proposed:
Alternative I: Use some open parking space to build a 5-level square garage sufficient to park 4050 cars in total

- A total of 25 acres available for open parking and parking garage after reserving 5 acres for new office buildings
- Parking garage costs about \$22.5 per ft^2 or about \$9000 per parking stall
- Fence and remarking of the parking lot cost \$6.00 per linear foot + \$1.5 per ft^2
- Auto/carpool cost is \$0.12 per mile and commuter rail cost is \$2.0 per ride
- Charge parking cost of \$2.0 per auto and \$0.50 per carpool
- Consider 2 trips per day, 5 days per week, 50 weeks per year
- Assume 20 years of service life for the garage and 10 years of service life for fence and markings

Alternative II: No change in the configuration of the existing parking lots, add a new travel mode of 7-person vanpools, and offer free commuter rail riding service

- Remark the existing parking lot cost \$1.5 per ft^2
- Allow carpool and vanpool employees to use express lanes of the two freeways with no user charge
- Auto/carpool cost is \$0.12 per mile, vanpool cost is \$0.15 per ride
- Introduce 7-person vanpools with costs of vans undertaken by the corporation, where each van involves purchase cost of \$30000, annual maintenance cost of \$200 per year, annual operating cost of \$500, driver's compensation of \$50 per week, and \$15000 salvage value after 3 years in service
- Offer free commuter rail riding service with the \$2.0 per ride covered by the corporation
- Charge parking cost of \$2.0 per auto, \$0.50 per carpool, and free for vans
- Consider 2 trips per day, 5 days per week, 50 weeks per year
- Assume 10 years of service life for markings

Alternative III: Build a new satellite parking lot near the cooperation headquarters and provide free shuttle bus service between the parking lot and headquarters

- Purchase 20 acres of land made available from a relocated hospital 2 miles before reaching the headquarters at \$10000 per acre to facilitate parking for one-third of the employees (i.e., 2000 employees)
- The 20-acre land can be converted to a satellite parking lot via grading, paving, drainage, fence, markings, and security service. Grading, paving, and drainage cost is about \$0.2 per ft^2; fence and marking cost \$6.00 per linear foot and an additional \$1.5 per ft^2; and annual security service and maintenance cost of \$50000
- Purchase 20 shuttle buses with 50 seats per bus by the corporation, where each bus involves purchase cost of \$150000, annual maintenance cost of \$1000 per year,

annual operating cost of \$1000, driver's annual compensation of \$10000, and salvage value of 50% of the purchase cost after 5 years in service

- For the 2000 affected employees, the travel modes are auto and carpool from home to satellite parking lot, as well as shuttle bus from the lot to headquarters. Auto/carpool cost is \$0.12 per mile (for 13 miles to the satellite parking lot), shuttle bus riding (2 miles to the headquarter) is free, and parking for autos, carpools, and shuttle buses is free
- For the 4000 unaffected employees, the travel modes are autos, carpools, and commuter transit. Auto/carpool cost is \$0.12 per mile, commuter rail cost is free to the riders, but the \$2.0 per ride is covered by the corporation, and parking cost is \$2.0 per auto and \$0.50 per carpool
- Consider 2 trips per day, 5 days per week, 50 weeks per year
- Assume 10 years of service life for fence and markings at the satellite and existing parking lots

The mode split of home-to-work travelers is determined by multinomial Logit model as

Utility function: $U_m = B_m - 0.05IVTT_m - 0.15OVTT_m - 0.2OPC_m$

where

U_m = Utility of travel mode m
$IVTT_m$ = In-vehicle travel time of travel mode m, min
$OVTT_m$ = Out-of vehicle travel time of travel mode m, min
OPC_m = Out-of-pocket costs of travel mode m, dollars/trip
m = Travel mode m

According to multinomial Logit choice model, the probability of choosing mode i, $P(i) = e^{U_i} / \sum_{i=1}^{M} e^{U_i}$.

Alternative I	B_m	*IVTT access*	*IVTT in route*	*OVTT initial*	*OVTT access*	*OPC*
Auto (A)	2.1	6	35	0	6	0.12 × 15
Carpool (CP)	1.5	2	30	4	3	(0.12 × 15)/3
Commuter rail (R)	0	0	29	15	1	2.0

Alternative II	B_m	*IVTT access*	*IVTT in route*	*OVTT initial*	*OVTT access*	*OPC*
Auto (A)	2.1	7	35	0	6	0.12 × 15 + (2.0/2)
Carpool (CP)	1.5	3	30	4	3	[0.12 × 15 + (0.5/2)]/3
Vanpool (VP)	1.4	1	30	8	1	0.15
Commuter rail (R)	0.4	0	29	15	1	0

Note: As compared to carpools, the vans will be allowed to park right next to the building, reducing the "OVTT Access" to about 1 min.

Alternative III (affected)	B_m	*IVTT access*	*IVTT in route*	*OVTT initial*	*OVTT access*	*OPC*
a. 13 miles to the satellite parking						
Auto (A)	2.1	7	30	0	6	0.12 × 13
Carpool (CP)	1.5	3	25	4	3	(0.12 × 13)/3
b. 2 miles from the satellite parking to headquarter						
Shuttle bus (B)	–	1	5	–	–	0

Alternative III (unaffected)	B_m	*IVTT access*	*IVTT in route*	*OVTT initial*	*OVTT access*	*OPC*
Auto (A)	2.1	7	35	0	6	0.12 × 15 +(2.0/2)
Carpool (CP)	1.5	3	30	4	3	[0.12 × 15 + (0.5/2)]/3
Commuter rail (R)	0.4	0	29	15	1	0

Assuming that parking lots and the garage have a square shape (1 acre = 43560 per ft²), a discount rate of 5%, and changes in employees' vehicle operating cost and time value are negligible, determine the best alternative.

Solution

1. *Alterative I*: Based on the given utility function, $U_A = -1.21$, $U_{CP} = -1.27$, $U_R = -4.24$

$$P_A = \frac{e^{-1.21}}{e^{-1.21} + e^{-1.27} + e^{-4.25}} = 50.3\%$$

$$P_{CP} = \frac{e^{-1.27}}{e^{-1.21} + e^{-1.27} + e^{-4.25}} = 47.3\%$$

$$P_R = \frac{e^{-4.24}}{e^{-1.21} + e^{-1.27} + e^{-4.25}} = 2.4\%$$

No. of autos = 6000 × 0.503 = 3018 veh
No. of carpools = 6000 × 0.473/3 = 946 veh
No. of commuter rail riders = 6000 × 0.024 = 144
No. of parking spaces needed = 3018 + 946 = 3964 < 4050

Assume X acres will be used to build the parking garage.5 × 110$X \geq 4050$, and $X \geq 7.36$. Therefore, 7.5 acres will be reserved for the parking garage
7.5 acres × 43560 ft²/acre = 326700 ft², meaning it is a 572 ft × 572 ft square parking lot
Parking garage: \$22.5/ft² × 326700ft² = \$7350750 or 7.5 × 110 × 9000 = \$7425000, use \$7425000

Fence: \$6/ft × 4 × 572ft = \$13728
Remarking: \$1.5/*ft*² × (25 − 7.5) acre × 43560 *ft*²/acre = \$1143450
Total annualized cost:

$$7425000 \times \frac{0.05(1+0.05)^{20}}{(1+0.05)^{20}-1} + (13728+1143450) \times \frac{0.05(1+0.05)^{10}}{(1+0.05)^{10}-1}$$
$$= \$745661/\text{year}$$

Annual auto/carpool and rail ride cost to employees:

(\$0.12/mile × 15 miles × 3964 + \$2/ride × 144) × 2 × 5 × 50 = \$3711600/year

Annual Parking cost to employees = (3018 × \$2 + 946 × \$0.50) × 50 × 5 = \$1627250/year
Annual cost to employees = 3711600 + 1627250 = \$5338850/year
Annual cost to the employer: \$745661/year

2. *Alternative II*: Based on the given utility function, $U_A = -1.46$, $U_{CP} = -1.386$, $U_{VP} = -1.53$, $U_R = -3.45$

$$P_A = \frac{e^{-1.46}}{e^{-1.461} + e^{-1.386} + e^{-1.53} + e^{-3.45}} = 31.8\%$$

$$P_{CP} = \frac{e^{-1.386}}{e^{-1.461} + e^{-1.386} + e^{-1.53} + e^{-3.45}} = 34.2\%$$

$$P_{VP} = \frac{e^{-1.53}}{e^{-1.461} + e^{-1.386} + e^{-1.53} + e^{-3.45}} = 29.6\%$$

$$P_{R} = \frac{e^{-3.45}}{e^{-1.461} + e^{-1.386} + e^{-1.53} + e^{-3.45}} = 4.4\%$$

No. of autos = 6000 × 0.318 = 1908
No. of carpools = 6000 × 0.342/3 = 684
No. of vanpools = 6000 × 0.296/7 = 254
No. of commuter rail riders = 6000 × 0.044 = 264
No. of parking spaces needed = 1908 + 684 + 254 = 2846
Remarking: $1.5/ft^2 × 25 acre × 43560ft^2/acre = $1633500

Van cost to the employer (purchase + driver's wage − salvage):

$$\$30000 \times \frac{0.05(1+0.05)^3}{(1+0.05)^3 - 1} + (\$200 + \$50 \times 50) - 15000 \times \frac{0.05}{(1+0.05)^3 - 1} = \$8958/\text{year}$$

Commuter rail cost to the employer = 2 × 264 × 2 × 5 × 50 = $264000/year
Auto/carpool cost to employees = $0.12 × 15 × (1908 + 684) × 2 × 5 × 50 = $2332800/year
Vanpool cost to employees = $0.15 × 254 × 7 × 2 × 5 × 50 = $133350/year
Parking cost to employees = ($2 × 1908 + $0.5×684) × 5 × 50 = $1039500/year
Annual cost to the employer:

$$\$1633500 \times \left[\frac{0.05(1+0.05)^{10}}{(1+0.05)^{10} - 1}\right] + \$8958/\text{year} + \$264000/\text{year} = \$484504/\text{year}$$

Annual cost to employees: $2332800 + $133350 + $1039500 = $3505650/year

3. *Alternative III*
   a. Unaffected 4000 employees
   Based on the given utility function, $U_A = -1.46$, $U_{CP} = -1.39$, $U_R = -3.45$

$$P_A = \frac{e^{-1.46}}{e^{-1.46} + e^{-1.39} + e^{-3.45}} = 45.2\%$$

$$P_{CP} = \frac{e^{-1.39}}{e^{-1.46} + e^{-1.39} + e^{-3.45}} = 48.6\%$$

$$P_R = \frac{e^{-3.45}}{e^{-1.46} + e^{-1.39} + e^{-3.45}} = 6.2\%$$

No. of autos = 4000 × 0.452 = 1808
No. of carpools = 4000 × 0.486/3 = 648
No. of commuter rail riders = 4000 × 0.062 = 248
Auto/carpool cost to employees = $0.12 × 15 × (1808 + 648) × 2 × 5 × 50 = $2210400/year
Parking cost to employees = ($1.5 × 1808 + $0.50 × 648) × 5 × 50 = $83712/year
Commuter rail cost to the employer = $2 × 248 × 2 × 5 × 50 = $248000/year

b. Affected 2000 employees
Based on the given utility function, $U_A = -0.962$, $U_{CP} = -1.054$

$$P_A = \frac{e^{-0.962}}{e^{-0.962} + e^{-1.054}} = 52.3\%$$

$$P_{CP} = \frac{e^{-1.054}}{e^{-0.962} + e^{-1.054}} = 47.7\%$$

No. of autos = 2000 × 0.523 = 1046
No. of carpools = 2000 × 0.477/3 = 318
Auto/carpool cost to employees = \$0.12 × 13×(1046 + 318) × 2 × 5 × 50 = \$1063920/year

$$\text{Shuttle bus cost to the employer} = \left[\$150000 \times \left(\frac{0.05(1+0.05)^5}{(1+0.05)^5 - 1}\right) + \$1000 + \$10000 - \$150000 \times 0.5 \times \left(\frac{0.05}{(1+0.05)^5 - 1}\right)\right] \times 20$$
$$= \$641462/\text{year}$$

Land purchase cost to the employer = 20 × \$10000 = \$200000

$$\text{Satellite parking lot cost to the employer} = (\$0.2 + \$1.5) \times 20 \times 43560 + \$6 \times 4 \times \sqrt{20 \times 43560} = \$1705050$$

Security service and maintenance cost to the employer = \$50000/year

$$\text{Annual cost to the employer: } \$248000 + \$641462 + \$50000 + (\$200000 + \$1705050) \times \left[\frac{0.05(1+0.05)^{10}}{(1+0.05)^{10} - 1}\right]$$
$$= \$1186170/\text{year}$$

Annual cost to employees: \$2210400 + \$83712 + \$1063920 = \$3358032/year

4. Summary of annual costs

	Alternative I	*Alternative II*	*Alternative III*
Employer cost	745661	484504	1186170
Employee cost	5338850	3505650	3358032
Total cost	6084511	3990154	4544202

5. Conclusion
Alternative II has the lowest annual total cost and is the best choice.

14.4.3 Traffic mobility improvements from multimodal integration

For a multimodal transportation system, capacity expansion does not always mean adding new travel lanes, new roads or transit lines. Instead one can implement strategies that promote the integration of multiple travel modes, including auto and transit modes and transit sub-modes, which mainly include bus, BRT, streetcar/light rail, heavy rail, commuter rail, taxi, ride share, bike, and pedestrian walking. Each layer of integration involves facility and system operations. The following example illustrates the implementation and integration of alternative travel modes to improve traffic mobility.

EXAMPLE 14.2

In a coastal city, a significant number of travelers needs to cross a bay area. Travelers may choose to use the ferry system operated by the municipal transit authority or drive on a bypass to avoid the ferry. For the ferry users, the average daily traffic is 30000 vehicles, consisting of 80% cars and 20% trucks. The crossing takes 15 minutes and an average wait for the ferry is 15 minutes. The value of time is estimated to cost $20.00 per hour for cars and $50.00 per hour for trucks. The stop-and-start cost is estimated to be 2.0 cents for cars and 5.0 cents for trucks (3 stop-and-start operations are assumed), and the operating costs are 12 cents and 40 cents per mile, respectively. The transit authority charges $2.00 per vehicle for using the ferry system. The annual operating disbursements of the ferry system mainly related to vessel fuel use, maintenance and repair, and operator's wages are $400000. If the current ferry system gets sold to a private owner, it has a value of $1000000. Otherwise, it still has a remaining service life of 30 years with zero salvage value. For the bypass users, the daily traffic is 10000 vehicles with 100% cars in the traffic stream. The bypass is 5 miles long and the average running speed is 30 mph. The entire route includes 2 signalized intersections, on average each car gets 10 seconds of delay per intersection. The annual traffic is expected to grow by 2% and the average wait time is expected to increase by one percent per year compound, respectively.

Alternatively, a tunnel is proposed to eliminate the ferry system and potentially avoid the bypass route usage for mobility improvements. All travelers by cars and trucks will use the 1.5-mile tunnel at 50 mph without stopping. The tunnel involves construction cost $35000000, annual maintenance and repair cost of $60000, and annual operating disbursements of $90000. It is designed for a 30-year useful service life and zero salvage value. The toll charges for the tunnel are $0.50 per car and $1.0 per truck. Assume that the discount rate is 6%, determine whether the tunnel alternative is a better choice.

Solution

1. Annualized growth factors
   a. Annualized factor (traffic × waiting time)

$$g_{twt} = (1+0.01)(1+0.02) - 1 = 0.030$$

Equivalent uniform annual cost factor (EUACF) is

$$EUACF_{twt} = \frac{1-e^{-[(i-g_{twt})n]/(1+g_{twt})}}{i-g_{twt}} \times \left[\frac{0.06(1+0.06)^{30}}{(1+0.06)^{30}-1}\right] = 1.411$$

   b. Annualized factor (traffic)

$$g_t = 0.02$$

$$EUACF_t = \frac{1-e^{-[(i-g_t)n]/(1+g_t)}}{i-g_t} \times \left[\frac{0.06(1+0.06)^{30}}{(1+0.06)^{30}-1}\right] = 1.256$$

2. *Alternative 0*: Using the existing ferry and bypass
   a. Annual ferry agency cost

$$AAC_0 = 400 \text{ K/year}$$

b. Autos using ferry system

$$\text{Crossing time} = 20 \times \frac{15}{60} = \$5/\text{veh}$$

Stop and start = 0.02 × 3 = \$0.06/veh
Toll = \$2/veh
Annual user cost for cars using ferry system:
Waiting time cost: 5 × 30000 × 80% × 365 × (1 + 0.03) × 1.411 = \$63655.9 K/year
Other costs: (5 + 0.06 + 2) × 30000 × 80% × 365 × (1 + 0.02) × 1.256 = \$79231.6 K/year
Total: 63655.9 + 79231.6 = \$142887.5 K/year

c. Trucks using ferry system

$$\text{Waiting time} = 20 \times \frac{15}{60} = \$5/\text{veh}$$

$$\text{Crossing time} = 20 \times \frac{15}{60} = \$5/\text{veh}$$

Stop and start = 0.05 × 3 = \$0.15/veh
Toll = \$2/veh
Annual user cost for trucks using ferry system:
Waiting time cost: 12.5 × 30000 × 20% × 365 × (1 + 0.03) × 1.411 = \$39784.9 K/year
Other costs: (12.5 + 0.15 + 2) × 30000 × 20% × 365 × (1 + 0.02) × 1.256 = \$41102.8 K/year
Total: 39784.9 + 41102.8 = \$80887.7 K/year

d. Private vehicles (by driving 5 miles around the bay at an average speed of 30 mph)

$$\text{Travel time} = \$20/\text{hour} \times \left(\frac{5\,\text{mile}}{30\,\text{mph}} + \frac{2 \times 10}{3600}\right) = \$3.44/\text{veh}$$

Stop and start = 0.02 × 2 = \$0.04/veh
Operating = \$0.12/mile × 5 miles = \$0.6/veh
Total = 3.44 + 0.04 + 0.6 = \$4.08/veh
Annual user cost for cars using ferry system:
4.08 × 10000 × 365 × (1 + 0.02) × 1.256 = \$19078.4 K/year
Therefore, AUC_0 = 142887.5 + 80887.7 + 19078.4 = \$242853.6 K/year

3. Alternative I: Building a new tunnel

a. Annualized tunnel cost

$$\textit{Tunnel}: AAC_1 = (35000 - 1000) \times \left[\frac{0.06(1+0.06)^{30}}{(1+0.06)^{30} - 1}\right] + 60 + 90 = 2620.1\ \text{K/year}$$

b. Auto using tunnel system

$$\text{Time} = \$20/\text{hour} \times \frac{1.5\,\text{mile}}{50\,\text{mph}} = \$0.6/\text{veh}$$

Toll = \$0.5/veh
Operating = \$0.12/mile × 1.5 miles = \$0.18/veh/day

Annual user cost for trucks using tunnel system:
$(0.6+0.5+0.18)\times(30000\times 80\%+10000)\times 365\times(1+0.02)\times 1.244=$
\$20155.9 K/year

c. Trucks using tunnel system

$$\text{Time} = \$50/\text{hour}\times\frac{1.5\,\text{miles}}{50\,\text{mph}} = \$1.5/\text{veh}$$

Toll = \$1/veh
Operating = \$0.4/mile × 1.5miles = \$0.6/veh
Annual user cost for trucks using tunnel system:
$(1.5+1+0.6)\times 30000\times 20\%\times 365\times(1+0.02)\times 1.244=\$8614.4$ K/year
Therefore, $AUC_1 = 20155.9+8614.4=\$28770.3$ K/year

4. Conclusion

$$\frac{B}{C}=\frac{AUC_0-AUC_1}{AAC_1-AAC_0}=\frac{242853.6-28770.3}{2620.1-400}=96.4$$

The B/C ratio is greater than 1, so the tunnel alternative is a better choice.

14.4.4 Freeway mobility improvements

In freeway mobility improvement project evaluation, typical performance measures used for analysis are travel time index, travel time buffer index, and total delays (FHWA, 2003). The value of each performance measure is estimated before and after project implementation. The economic feasibility of the project is evaluated by comparing the above estimated values.

It is crucial that life-cycle costs of a freeway program be determined in terms of its complete implementation and operating schedule, recognizing that a freeway program will likely entail many separate steps, with elements that are deployed and become operational at different points in time. In developing life-cycle costs, the activity profiles of construction, repair, and maintenance costs need to be determined. Since multiple facilities are involved in a typical freeway system, only by applying life-cycle cost analysis can we achieve an equitable comparison of project costs and expected benefits.

14.4.4.1 Project economic feasibility screening

For a specific freeway corridor, a series of minor projects can be proposed as the do-minimum scenario. A number of major candidate projects can also be recommended based on alternative strategies. Without loss of generality, these strategies may be termed as low cost, truck issue, and new capacity strategies. The minor projects are those that must be implemented to maintain a minimum acceptable freeway operation and they will not go through a screening process for economic justification. However, a screening process must be undertaken for major projects to ensure that the final candidates considered for the corridor are all economically feasible.

The benefit–cost, incremental benefit–cost or net present worth analysis provides an objective means of comparing the quantifiable and monetary benefits of a candidate project to its costs. By applying any one of these analytical approaches to a specific project, its economic feasibility can be determined. This will further help to screen out those economically infeasible major candidate projects, ensuring that the remaining candidates are economically justified.

The first step in the screening process is to define the economic analysis period, which is accomplished by specifying a base year and future year. The next step involves quantifying agency and user benefits achieved from the candidate project. Savings of agency costs during project-related facility service life-cycle are regarded as agency benefits. Similarly, user benefits can be established by comparing user costs between the base year and future year with assumptions made for the intermediate years. Reductions in total user costs in future year are considered as user benefits. This process is illustrated by the following example.

Suppose that a major project P_i is proposed for a freeway corridor to improve truck mobility. The project is to be implemented in year t from the base year ($t=0$) and is going to last for 2 years for implementation with costs of C_{i1} and C_{i2}. Additionally, three minor projects are arranged for years 1, $t+2$, and $t+n$, and the costs of the minor projects are c_1, c_2, and c_3, respectively. Agency costs in terms of EUAC resulting from the combined effects of all four projects are expected to decrease by ΔEUA_{AC} in each year, and the reduction in user costs in future year ($t=T$) compared to cost in the base year is ΔUC_T, see Figure 14.1. The total project benefits and costs for a T-year analysis period could, therefore, be established as shown in Table 14.3.

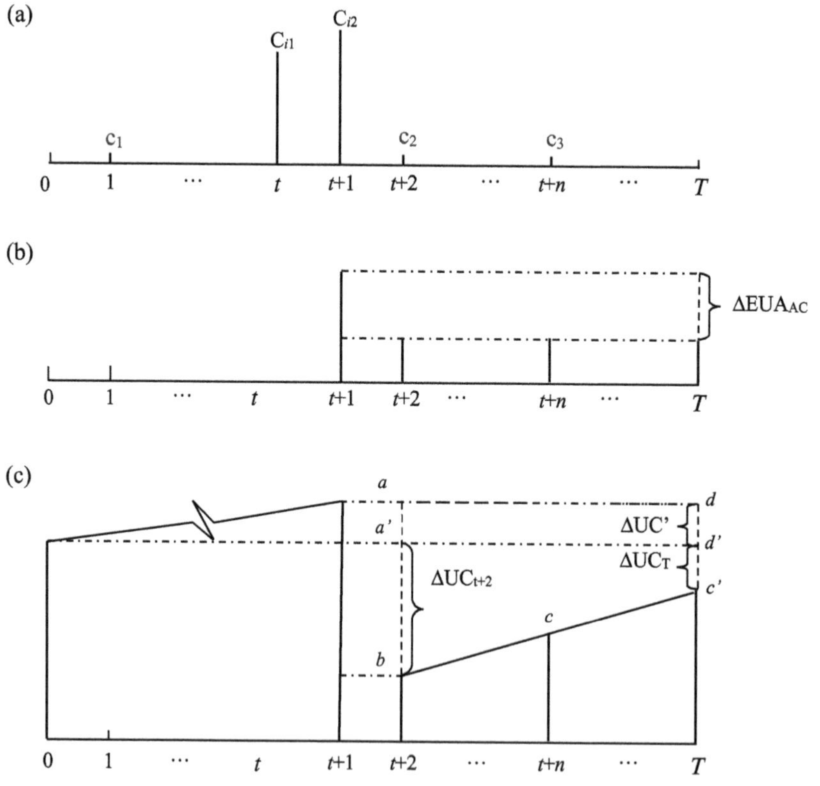

Figure 14.1 Illustration of freeway project-level agency and user benefits in a *T*-year analysis period. (a) Freeway agency cost profile. (b) Agency benefits of a freeway project. (c) User benefits of a freeway project. (Adapted from Li Z. 2004. Wisconsin Freeway Operations Assessment Guidelines. Department of Civil and Environmental Engineering, University of Madison, Madison, WI.)

Table 14.3 Tabulated method for a single major project screening

Designation		*Equation*	*Remark and assumption*
Agency benefits	Annual	$\Delta AC_{t+2} = \cdots = \Delta AC_T = \Delta EUA_{AC}$	1. Time period Base year: 0 Future year: T Benefits starting year: $t+2$ 2. Constant dollars used 3. User costs increase by k_1 each year before major project implementation 4. User benefits decrease linearly by k_2 each year after project implementation 5. Trapezoid rule applied for the computation of user benefits
	Total	$PW^{AC}_{t+1,T-t-1} = \Delta EUA_{AC} \cdot \left[\frac{i \cdot (1+i)^{T-t-1}}{\left((1+i)^{T-t-1} - 1\right)}\right]$	
User benefits	Annual	$\Delta UC_{t+2} = \Delta UC_T + (T - t - 2) \cdot k_2$ $\Delta UC' = (t+1) \cdot k_1$ $\Delta EUA_{UC} = (t+1) \cdot k_1 + \Delta UC_T + \left[\frac{(T-t-1) \cdot (1+i)^{T-t-1}}{\left((1+i)^{T-t-1} - 1\right)} - 1 - \frac{1}{i}\right] \cdot k_2$	
	Total	$PW^{UC}_{t+1,T-t-1} = \mathrm{AREA}(a - b - c' - d)$ $= \Delta EUA_{UC} \cdot \left[\frac{i \cdot (1+i)^{T-t-1}}{((1+i)^{T-t-1} - 1)}\right]$	
Total benefits		$PW_{\Delta B} = PW^{AC}_{t+1,T-t-1} + PW^{UC}_{t+1,T-t-1}$	
Total costs		$PW_{\Delta C} = c_1 \cdot (1+i)^t + \frac{c_2}{1+i} + \frac{c_3}{(1+i)^{n-1}} + C_{i1} \cdot (t+1) + C_{i2}$	
Screening criterion		Project is feasible if $PW_{\Delta B} > PW_{\Delta C}$	

Note: *AC*, Agency costs; *UC*, User costs; *B*, Benefits; *C*, Costs; and *PW*, Present worth.

14.4.4.2 Multi-project benefits and costs by freeway corridor

After the screening process, three sets of candidate projects based on the low cost, truck issue, and new capacity strategies can be established for each corridor. Individual projects within each set are economically feasible. However, some freeway project benefits are not readily quantified and not all quantifiable benefits can be converted to monetary values. Also, benefits resulting from multiple projects within the same corridor may not be directly additive due to interdependency effects. Because of this, alternative analyses are often needed to help assess which corridors meet the freeway system mobility goal most cost-effectively.

The method for a corridor-based multi-project evaluation by alternative mobility improvement strategy is similar to that used for single project screening. The given conditions include the base year and future year for economic analysis, costs, and implementation periods of individual projects proposed within the corridor, as well as the agency and user benefits based on the comparison of the outcome values of performance measures between the base year and future year. Since projects are recommended based on three strategies, there could be as many as three sets of candidate projects for each corridor. The time and space distribution of the corridor implementation plan is shown in Figure 14.2. It should be noted that different total lengths of the corridor are used for the base year (L_1) and future year (L_2) to reflect possible changes in the alignment.

As shown in Figure 14.2, the agency and user benefits will be generated after the completion of individual projects. The corridor-wide agency benefits can be quantified as the sum of agency benefits from individual projects in terms of respective products of EUAC changes and number of years of benefits counted in the economic analysis period. For instance, the agency benefits relevant to Project 1 are the product of net reduction in equivalent uniform annual agency costs times $(T-t_{11})$ years of benefits.

To quantify the corridor-wide user benefits, the information gathered includes user benefits associated with each project within the corridor during the screening process as well as corridor-wide user benefits that represent the combined effects of the multiple projects. In particular, we have an accurate assessment of the initial user benefits after the completion of the very first project and the overall user benefits of all projects within the corridor at the end of the analysis period. However, pertinent information on user benefits for the intermediate years is missing. This is explained in the following section.

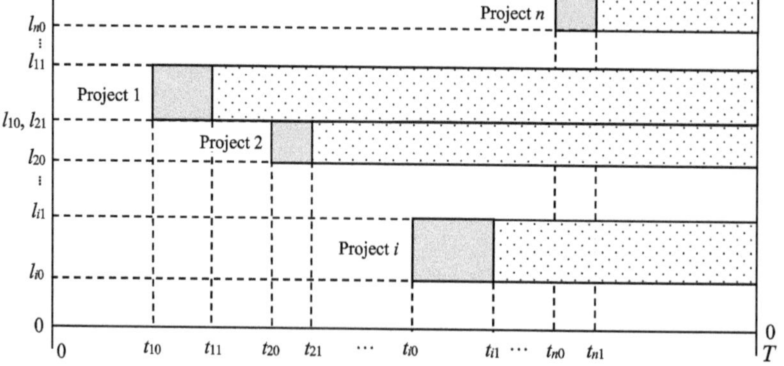

Figure 14.2 Time and space distribution of multiple projects implemented in a freeway corridor. (Adapted from Li Z. 2004. Wisconsin Freeway Operations Assessment Guidelines. Department of Civil and Environmental Engineering, University of Madison, Madison, WI.)

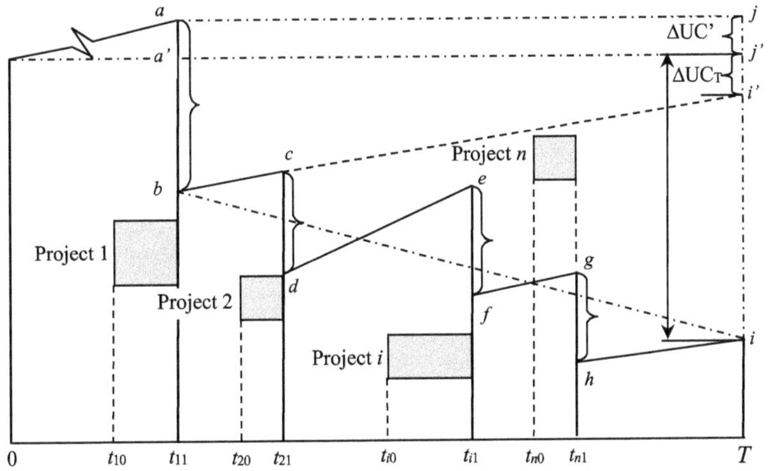

Figure 14.3 User benefits of multiple projects implemented in a freeway corridor for a T-year analysis period. (Adapted from Li Z. 2004. Wisconsin Freeway Operations Assessment Guidelines. Department of Civil and Environmental Engineering, University of Madison, Madison, WI.)

As illustrated in Figure 14.3, the irregular polygon bounded by a–b–c–d–e–f–g–h–i–j represents the user benefits generated by implementing *n* projects along a freeway corridor for a *T*-year analysis period. User benefits for Project 1 in year *T* are represented by line $i'j'$ and user benefits of the entire corridor resulting from all the *n* projects in year *T* are represented by line ij'. This is true because for both cases we are comparing changes in system usage performance between the base year (year 0) and a future year (year *T*). However, the total user benefits in year *T* for either case should be slightly higher as the system performance deteriorates further from the base year to the point where the very first project is implemented. Hence, the total user benefits from Project 1 and all projects within the corridor in year *T* should both be augmented by $\Delta UC'$, which are described by line $i'j$ and ij, respectively.

By comparing changes in system usage performance before and after project implementation, the total user benefits of Project 1 and all the *n* projects within the corridor in the *T*-year analysis period can be quantified. The analysis can be conducted in two steps. Step 1 is to establish user benefits generated by Project 1 in year t_{11} using linear extrapolation, assuming a constant diminishing rate for user benefits that only accrue from it. In this way, line ab can be extrapolated from line $i'j$. Since the exact user benefits in the interim years are unknown, they can be interpolated by assuming a linear diminishing rate between year t_{11} when the user benefits first begin as a result of implementing Project 1 and the future year *T* when the analysis period ends. Following this concept, the overall user benefits associated with n projects implemented along the freeway the corridor for the *T*-year analysis period, as represented by polygon a–b–c–d–e–f–g–h–i–j, can then be quantified using the approximated trapezoid a–b–i–j. With this simplified treatment, the corridor based multi-project user benefits can be quantified.

For each alternative mobility improvement strategy, the multi-project agency and user benefits could be summed up to arrive at the total benefits. The multi-project costs could also be aggregated to establish the total costs. The multi-project benefits and costs could be expressed as the present worth or equivalent uniform annual amounts. As summarized in Table 14.4, the corridor-based multi-project benefits and costs could be compared to assess the economic feasibility of alternative mobility improvement strategies. The benefits and costs of all economically feasible corridor-based mobility improvement strategies can

Table 14.4 Tabulated method for corridor-based multi-project evaluation by alternative freeway mobility improvement strategy

Strategy	*Benefits*	*Costs*	*Screening criterion*
Required strategy (k = 0 do minimum)	$B_{j,0} = \Delta B_{\left(\sum_{i=1}^{N_{j,0}} p_{i,j,0}\right)}$ $B_{j,0} \neq \Delta B_{(p_{1,j,0})} + \cdots + \Delta B_{\left(p_{N_{j,0},j,0}\right)}$	$C_{j,0} = \sum_{i=1}^{N_{j,0}} c_{i,j,0}$	All pass
Alternative strategies (k = 1 low cost k = 2 truck issue k = 3 new capacity)	$B_{\left(\sum_{i=1}^{N_{j,k}} p_{i,j,k}\right)} = \Delta B_{\left(\sum_{i=1}^{N_{j,k}} p_{i,j,k}\right)}$, $B_{j,k} = \Delta B_{\left[\left(\sum_{i=1}^{N_{j,0}} p_{i,j,0}\right)+\left(\sum_{i=1}^{N_{j,k}} p_{i,j,k}\right)\right]}$, $B_{\left(\sum_{i=1}^{N_{j,k}} p_{i,j,k}\right)} \neq B_{j,k} - B_{j,0}$	$C_{\left(\sum_{i=1}^{N_{j,k}} p_{i,j,k}\right)} = \Delta C_{\left(\sum_{i=1}^{N_{j,k}} p_{i,j,k}\right)}$, $C_{j,k} = C_{j,0} + C_{\left(\sum_{i=1}^{N_{j,k}} p_{i,j,k}\right)}$	1. Pass if $\frac{B_{j,k}}{C_{j,k}} > 1$ 2. Repeat for each corridor j

be used as inputs for corridor-based project selection to achieve maximized overall benefits under budget constraints.

14.4.5 Warehousing delivery system mobility improvements

In addition to mobility for passenger travel, it is also important to maintain a desirable level of mobility for freight shipments and inventory management while minimizing costs. The following example illustrates evaluation of alternative warehouse delivery scenarios aimed at minimizing transportation and inventory costs, which are highly influenced by the shipment time.

EXAMPLE 14.3

An auto parts factory located in City A has a shipping pattern to five Midwest cities, B, C, D, E, and F in the U.S. with major manufacturing plants. Evaluate four alternative warehousing delivery alternatives as follows:

I. Direct Less Than Truckload (LTL) shipment to and from each city
II. Direct shipment via Truckload (TL) truck to a warehouse/distribution center (DC) in Indianapolis with further distribution from the DC-LTL direct to each city
III. Direct shipment via TL truck to DC in City C with further TL distribution by a weekly peddling run going in turn to each city
IV. Direct shipment via railroad to DC in City C with further distribution using the most cost effective truck delivery to each city, either direct or peddling as determined in alternatives II or III above

The map is shown below

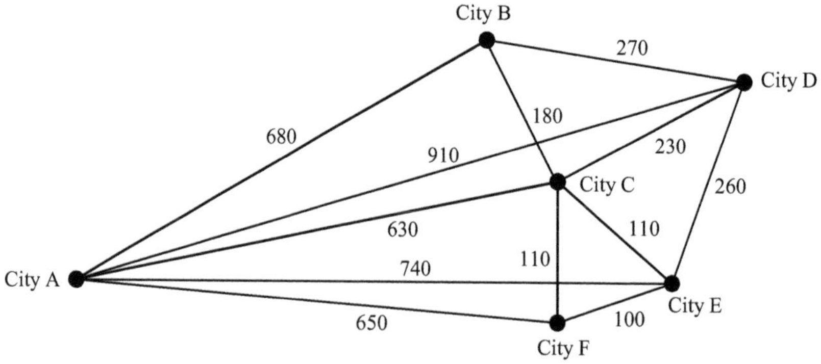

The principal cost elements are transportation cost and inventory cost, comprised of holding safe stock cost, in-transit inventory cost, and mean demand cost. The following terms are defined:

S = Purchase price, \$1500 per unit
W = Weight per shipment, in pounds for all units shipped and each unit weighing 300 lbs
D = Distance in miles
L = Order lead time, 7 days
C = Order cost, \$50 per order
G = Backorder cost, \$150 per unit

Package 2 units fit one standard pallet (40″ × 48″ × 53″);

TR = Truck rate, $= 0.0001(2.88D \cdot W^{-0.15} + 10120W^{-0.33})$ (in \$ per pound) for both LTL and TL trucks
DT = Delivery truck time, one day if <300 miles, two days if >300 miles and <600 miles, and 3 days if >600 miles for TL; and 2 days for LTL for all shipping distances
RR = Railroad rate, 5.7 cents per ton-mile
RT = Rail transit time, including local transportation of drayage, five days
H = Holding cost, \$6 per unit per week as a fraction of purchase price
IC_{IT} = In-transit inventory cost, 25% of holding cost for the fraction of time used for shipping
IC_{MD} = Mean demand inventory cost, where $IC_{MD} = H \cdot (Q_0/2)$
Q_0 = Economic order quantity (EOQ), units per order, where $Q_0 = \sqrt{2CR/H}$
R = Quantity demanded per week, units per week
T = Average order interval Q_0/R, no. of week(s) per order
S_{saf} = Safety stock, units per week, Poisson distribution $= S_{tot} - R$, where $P(S_{tot}) = (H \cdot Q_0)/(G \cdot R \cdot \sqrt{L/7}) = (R^{S_{tot}} \cdot e^{R})/(S_{tot}!)$
H_{saf} = Cost of holding the safety stock, in dollars per week or per year

Notes:

1. Total cost of ordering quantity Q is $TC(Q) = P \cdot R + C \cdot (R/Q) + H \cdot (Q/2)$
   Solve for min $TC(Q)$, we get $Q_0 = \sqrt{2CR/H}$
2. Assume each order is shipped separately, that is, no. of orders equals to no. of shipments
3. Use 7 days per week and 50 weeks per year (i.e., 350 days per year)

Solution

1. *Alternative I*: Direct delivery from City A (DC) to each city
   Assuming that EOQ applies, develop an annual transportation cost based on truck distance weight formula, and then calculate inventory cost in three quantities: holding safety stock, in-transit, and mean demand

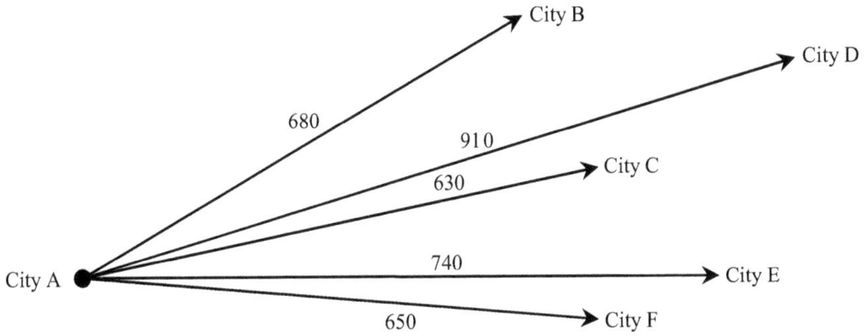

a. Annual transportation cost

City A to	R (uts/wk)	Q_0	D	W	$/lb	$/ship	Shipments/ year	Transportation cost
City B	30	23	680	6900	0.1068	736.61	65.22	48039.96
City C	15	16	630	4800	0.1126	540.44	46.88	25333.23
City D	20	19	910	5700	0.1299	740.62	52.63	38979.86
City E	10	13	740	3900	0.1277	498.20	38.46	19161.52
City F	10	13	650	3900	0.1202	468.96	38.46	18036.74
							Total	149551.31

Taking the first row as an example:

$$Q_0 = \sqrt{\frac{2CR}{H}} = \sqrt{\frac{2\times 50\times 30}{6}} = 22.36, \text{ round it to 23 unit/order}$$

$$W = 300\cdot Q_0 = 300\times 23 = 6900\text{lbs}$$

$$\begin{aligned} TR &= 0.0001\cdot(2.88\cdot D\cdot W^{-0.15} + 10120\cdot W^{-0.33}) \\ &= 0.0001\times(2.88\times 680\times 6900^{-0.15} + 10120\times 6900^{-0.33}) \\ &= \$0.1068/\text{lb} \end{aligned}$$

$$\text{Unit shipment cost} = TR\cdot W = 0.1068\times 6900 = \$736.61/\text{ship}$$

$$\text{Per year shipment} = \frac{R}{Q_0}\times 50 = \frac{30}{23}\times 50 = 65.22 \text{ shipments/year}$$

(Not rounded, considering multi-year operations)

$$\begin{aligned} \text{Transportation cost} &= \text{Unit shipment cost}\times\text{Per year shipment} \\ &= 736.61\times 65.22 = \$48039.96/\text{year} \end{aligned}$$

Repeat the same calculations for all rows.

b. Annual inventory cost

City A to	R (uts/wk)	Q_0	$P(S_{tot})$	S_{tot}	S_{saf}	Inventory cost: Safe stock	In transit	Mean demand
City B	30	23	0.0307	37	7	2100	41925.47	4500
City C	15	16	0.0427	20	5	1500	15066.96	2250
City D	20	19	0.038	26	6	1800	22556.39	3000
City E	10	13	0.052	14	4	1200	8241.76	1500
City F	10	13	0.052	14	4	1200	8241.76	1500
				Subtotal		7800	96032.34	12750
				Total			116582.34	

Taking the first row as an example:
Given lead time $L=7$ days,

$$P(S_{tot}) = \frac{H\cdot Q_0}{G\cdot R\sqrt{L/7}} = \frac{6\times 23}{150\times 30\times\sqrt{7/7}} = 0.0307$$

$$\text{Solve } P(S_{tot}) = \frac{R^{S_{tot}}}{S_{tot}!}\cdot e^{-R} \text{ for } S_{tot}$$

$$\text{Solve } 0.0307 = \frac{30^{S_{tot}}}{S_{tot}!}\cdot e^{-30} \text{ for } S_{tot}, S_{tot} = 37 \text{ unit/week}$$

$S_{saf} = S_{tot} - R = 37 - 30 = 7$ unit/week

Safe stock $= H_{saf} \times 50 = S_{saf} \cdot H \times 50 = 7 \times 6 \times 50 = \$2100$/year

$DT = 2$ days/shipment for LTL

$$FTS \text{ (Fraction of time for shipping)} = \frac{DT \times \text{Per year shipment}}{350 \text{ days/year}} = \frac{2 \times 65.22}{350} = 0.373$$

In transit $= Q_0 \cdot$ Per year shipment $\cdot H \times 50 \times FTS \times 25\% = 23 \times 65.22 \times 6 \times 50 \times 0.373 \times 0.25 = \$41964.2$/year (This value is \$41925.47 when rounding error is eliminated)

$$\text{Mean demand} = \frac{H \cdot Q_0}{2} = \frac{\$6 \times 23}{2} \text{ per order} = \frac{\$6 \times 23}{2} \times 65.22 = \$4500/\text{year}$$

Repeat the same calculations for all rows.

2. *Alternative II*: Use the Central Distribution Center in City C
   - Examine the Cost from City A to the DC in City C. A single EOQ will be developed for City C as the DC. Since we are going to ship by full truck load, each truck will carry a maximum of 44 pallets at two high and two abreast. Using weekly demand of 85 for TL and the truck formula figure out transportation cost for the year
   - Using the LTL trucks, calculate the cost to truck from DC to the cities served
   - Using $R = 85$ and the actual TL order quantity, calculate safety stock needed in City C
   - Look at the stock inventory delivery plan from a delivery pattern that never lets the stock go below the safety stock for City C. Calculate the average inventory
   - Determine the safety stock in each city noting that the lead time is down to two days from 7 days

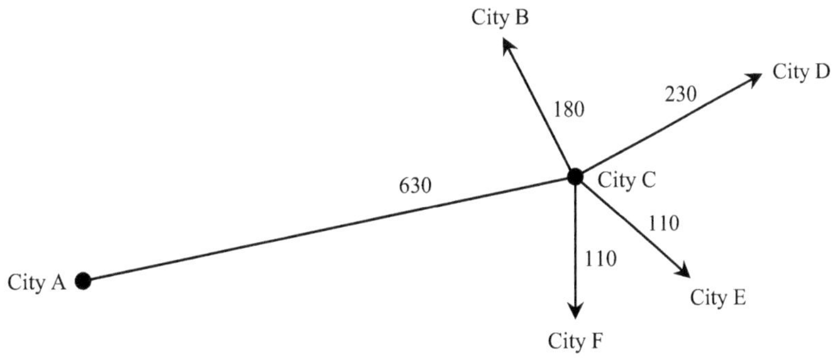

a. Annual transportation cost

City A to	R (uts/wk)	Q_0	D	W	\$/lb	\$/ship	Shipments/year	Transportation cost
City C	85	38	630	11400	0.0911	1038.27	111.84	116122.23
							Total	116122.23

$R = 30+15+20+10+10 = 85$ unit/week

$$Q_0 = \sqrt{\frac{2CR}{H}} = \sqrt{\frac{2\times 50\times 85}{6}} = 37.63\text{, round it to 38 unit/order}$$

$W = 300\times 38 = 11400\text{lbs}$

$$\begin{aligned}TR &= 0.0001\cdot(2.88\cdot D\cdot W^{-0.15}+10120\cdot W^{-0.33})\\ &= 0.0001\times(2.88\times 630\times 11400^{-0.15}+10120\times 11400^{-0.33})\\ &= \$0.0911/\text{lb}\end{aligned}$$

Unit shipment cost $= TR\cdot W = 0.0911\times 11400 = \$1038.27/\text{ship}$

$$\text{Per year shipment} = \frac{R}{Q_0}\times 50 = \frac{85}{38}\times 50 = 111.84\text{ shipments/year}$$

$$\begin{aligned}\text{Transportation cost} &= \text{Unit shipment cost}\times\text{Per year shipment}\\ &= 1038.27\times 111.84 = \$116122.23/\text{year}\end{aligned}$$

Similarly, the transportation cost from DC to individual cities can be calculated as summarized below:

City C to	*R (uts/wk)*	*Q_0*	*D*	*W*	*\$/lb*	*\$/ship*	*Shipments/year*	*Transportation cost*
City B	30	23	180	6900	0.0685	472.75	65.217	30831.28
City D	20	19	230	5700	0.0764	435.55	52.632	22923.71
City E	10	13	110	3900	0.0753	293.49	38.462	11288.08
City F	10	13	110	3900	0.0753	293.49	38.462	11288.08
							Total	76331.14

b. Annual inventory cost

						Inventory cost		
City A to	*R (uts/wk)*	*Q_0*	*$P(S_{tot})$*	*S_{tot}*	*S_{saf}*	*Safe stock*	*In transit*	*Mean demand*
City C	85	38	0.0335	91	6	1800	305568.61	12750
						Total	320118.61	

Given lead time $L=2$ days,

$$P(S_{tot}) = \frac{H\cdot Q_0}{G\cdot R\sqrt{L/7}} = \frac{6\times 38}{150\times 85\times\sqrt{2/7}} = 0.0335$$

$$\text{Solve } P(S_{tot}) = \frac{R^{S_{tot}}}{S_{tot}!}\cdot e^{-R}\text{ for }S_{tot}$$

$$\text{Solve } 0.0335 = \frac{85^{S_{tot}}}{S_{tot}!}\cdot e^{-85}\text{ for }S_{tot},\ S_{tot} = 91\text{ unit/week}$$

$S_{saf} = S_{tot} - R = 91 - 85 = 6$ unit/week

Safe stock $= H_{saf} \times 50 = S_{saf} \cdot H \times 50 = 6 \times 6 \times 50 = \$1800/\text{year}$

$DT = 3$ days/shipment for TL with 630 miles of shipping distance

$$FTS(\text{Fraction of time for shipping}) = \frac{DT \times \text{Per year shipment}}{350\,\text{days/year}} = \frac{3 \times 111.84}{350} = 0.959$$

In transit $= Q_0 \cdot$ Per year shipment $\cdot H \times 50 \times FTS \times 25\% = 38 \times 111.84 \times 6 \times 50 \times 0.959 \times 0.25 = \$305675/\text{year}$ (This value is \$305568.61 when rounding error is eliminated)

$$\text{Mean demand} = \frac{H \cdot Q_0}{2} = \frac{\$6 \times 38}{2} \text{per order} = \frac{\$6 \times 38}{2} \times 111.84 = \$12750/\text{year}$$

The inventory cost from DC to individual cities can be calculated, $DT = 2$ days/shipment for LTL. The results are summarized below:

						Inventory cost		
DC to	*R (uts/wk)*	Q_0	$P(S_{tot})$	S_{tot}	S_{saf}	*Safe stock*	*In transit*	*Mean demand*
City B	30	23	0.0574	33	3	900	41925.47	4500
City D	20	19	0.0711	23	3	900	22556.39	3000
City E	10	13	0.0973	12	2	600	8241.76	1500
City F	10	13	0.0973	12	2	600	8241.76	1500
				Subtotal		3000	80965.37	10500
				Total			94465.37	

3. *Alternative III*: The same transition costs from City A to City C as the distribution center apply for the peddling run as suggested below:

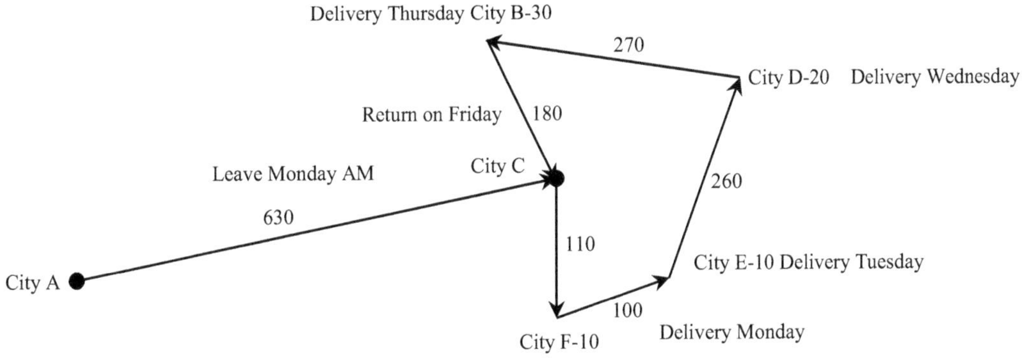

Van cost: \$5000 three years 7%
Tractor cost: \$75000 six year 7%
Driver: \$45000 per year
Hotel expanse: \$60.00 per day on the road
Per diem: \$6.0 per quarter day
Assume that there is a backhaul from City B to City C on Friday each week for \$500
Note that the lead time for calculating safety stock will depend on the day of delivery

a. Annual transportation cost

City A to	R (uts/wk)	Q_0	D	W	\$/lb	\$/ship	Shipments/ year	Transportation cost
City C	85	38	630	11400	0.0911	1038.27	111.84	116122.23
							Total	116122.23

City C to	R (uts/wk)	Q_0	D	W	L (day)	\$/lb	\$/ship	Shipments/ year	Transportation cost
City F	10	13	110	3900	1	0.0753	293.49	50	14674.50
City E	10	13	210	3900	2	0.0836	325.98	50	16299.18
City D	20	19	470	5700	3	0.0953	543.22	50	27161.05
City B	30	25	740	7500	4	0.1092	818.66	50	40933.16
								Total	99067.90

The truck pedals every week therefore the transportation cost is slightly different.
$D_F = 110$; $D_E = 110 + 100 = 210$; $D_D = 110 + 100 + 260 = 470$...
$Q_{0F} = 13$; $Q_{0E} = 13$; $Q_{0D} = 19$; $Q_{0B} = 10 + 10 + 20 + 30 - 13 - 13 - 19 = 25$
$W_F = 300 \times Q_{0F} = 300 \times 13 = 3900$;...

Per year shipment = 1 shipment/week × 50 weeks/year = 50 shipments/year

b. Annual inventory cost

City A to	R (uts/wk)	Q_0	$P(S_{tot})$	S_{tot}	S_{saf}	Inventory cost: Safe stock	Inventory cost: In transit	Inventory cost: Mean demand
City C	85	38	0.0335	91	6	1800	305568.61	12750
						Total	320118.61	

City C to	R (uts/wk)	Q_0	$P(S_{tot})$	S_{tot}	S_{saf}	Inventory cost: Safe stock	Inventory cost: In transit	Inventory cost: Mean demand
City F	10	13	0.1376	10	0	0	6964.29	1500
City E	10	13	0.0973	12	2	600	13928.57	1500
City D	20	19	0.058	24	4	1200	30535.71	3000
City B	30	25	0.0411	35	5	1500	53571.43	4500
				Subtotal		3300	105000	10500
					Total		118800	

Taking City F as an example the lead time $L = 1$ day

$$P(S_{tot}) = \frac{H \cdot Q_0}{G \cdot R\sqrt{L/7}} = \frac{6 \times 13}{150 \times 10 \times \sqrt{1/7}} = 0.1376$$

$$\text{Solve } P(S_{tot}) = \frac{R^{S_{tot}}}{S_{tot}!} \cdot e^{-R} \text{ for } S_{tot}$$

Solve $0.1376 = (10^{S_{tot}}/S_{tot}!) \cdot e^{-10}$ for S_{tot}. There is no feasible solution for S_{tot} meaning that since City F is the first city being served by the fully loaded truck for the pedaling delivery it does not need any safe stock. Therefore $S_{saf} = 0$ and $S_{tot} = R = 10$

Safe stock = 0
DT = L = 1 days/shipment

$$\text{FTS (Fraction of time for shipping)} = \frac{\text{DT} \times \text{Per year shipment}}{350 \text{ days/year}} = \frac{1 \times 50}{350} = 0.143$$

In transit = $Q_0 \cdot$ Per year shipment $\cdot H \times 50 \times FTS \times 25\% = 13 \times 50 \times 6 \times 50 \times 0.143 \times 0.25 = \$6971.25$/year (This value is \$6964.29 when rounding error is eliminated.)

$$\text{Mean demand} = \frac{\text{H} \cdot \text{R}}{2} = \frac{\$6 \times 10}{2} \text{ per week} = \frac{\$6 \times 10}{2} \times 50 = \$1500/\text{year}$$

c. Other costs

Item	*Cost*	*Equivalent annual cost (\$/year)*
Van	\$5000 for 3 years 7%	$5000 \times \left[\frac{0.07(1+0.07)^3}{(1+0.07)^3 - 1}\right] = 1905.26$
Tractor	\$75000 for 6 years 7%	$75000 \times \left[\frac{0.07(1+0.07)^6}{(1+0.07)^6 - 1}\right] = 15734.68$
Driver	\$45000/year	45000
Hotel	\$60/per day on the road	60×4 nights/week $\times 50 = 12000$
Per Diem	\$6.0 per quarter day	$6 \times 4 \times 5 \times 50 = 6000$
Backhaul	\$500 per week	$500 \times 50 = 25000$
Total		55639.94

4. *Alternative IV*: Now look at the train to provide the line haul from City A to the DC in City C. Send a boxcar when needed. Assume one boxcar will hold 46.308 tons of product (enough for supplying two weeks of demand of 170 units and additional weekly safety stock amounting to one week's demand in City C to offset rail reliability concerns) namely shipping 340 units in one boxcar good for two weeks' demand of 170 units per week. Rail reliability is such that safety stock should amount to about one week's worth in City C as the distribution center

   Drayage and classification in City A: three days
   Line haul City C: one day
   Switching to set car: one day

   In total a 5-day rail transit time is needed. From the DC in City C to each city choose the lower total transportation and inventory costs between looking at direct delivery as in Alternative II and the peddling run as in Alternative III.

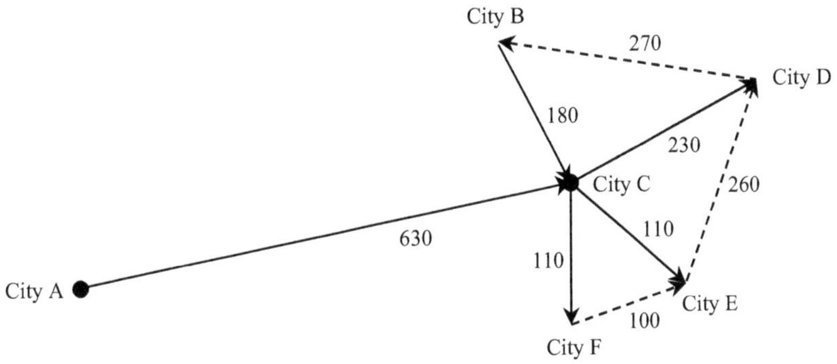

a. Annual transportation cost

City A to	R (uts/wk)	Q_0	D	W	$/ship	Shipments/year	Transportation cost
City C	85	170	630	51000	831.46	25	20786.5
						Total	20786.5

Rail rate $= \$0.057 \times 46.308 \times 630 = \$831.46$/shipment

b. Annual inventory cost

							Inventory cost	
City A to	R (uts/wk)	Q_0	$P(S_{tot})$	S_{tot}	S_{saf}	Safe stock	In transit	Mean demand
City C	85	170	–	170	85	25500	113839.29	12750
						Total	152089.29	

Safe stock $= H_{saf} \times 50 = S_{saf} \cdot \mathrm{H} \times 50 = 85 \times 6 \times 50 = \$25500$/year

$RT = 5$ days/shipment

$$\text{FTS(Fraction of time for shipping)} = \frac{DT \times \text{Per year shipment}}{350\,\text{days/year}} = \frac{5 \times 25}{350} = 0.357$$

In transit $= Q_0 \cdot$ Per year shipment $\cdot H \times 50 \times FTS \times 25\% = 170 \times 25 \times 6 \times 50 \times 0.357 \times 0.25 = \$113794$/year (This value is \$113839.29 when rounding error is eliminated)

$$\text{Mean demand} = \frac{\mathrm{H} \cdot \mathrm{Q}_0}{2} = \frac{\$6 \times 170}{2}\text{ per order} = \frac{\$6 \times 170}{2} \times 25$$
$$= \$12750/\text{year}$$

From the distribution center to other cities:

Cost of direct distribution (from Scenario II): 76331.14 + 94465.37 = \$170797/year

Cost of pedaling (from Scenario III): 99067.90 + 118800 + 55639.94 = \$273 508/year

Therefore direct distribution from DC in City C to all other cities is the choice.

5. Conclusion
   The annual total costs for the four alternatives are as below:
   Alternative I: 149551.31 + 116582.34 = \$266133.65/year
   Alternative II: 116122.23 + 76331.14 + 320118.61 + 94465.37 = \$607037.35/year
   Alternative III: 116122.23 + 99067.9 + 320118.61 + 118800 = \$654108.74/year
   Alternative IV: 20786.5 + 152089.29 + 170797 = \$343672.79/year
   Therefore Alternative I is the best auto parts delivery choice.

PROBLEMS

14.1 Explain the causes of recurrent nonrecurrent and emergency event-triggered congestion and the focuses of mitigating each type of congestion.

14.2 Elaborate on differences between people mobility and traffic mobility for passenger transportation; and between goods and traffic mobility for freight transportation.

14.3 List primary performance measures used for passenger and freight mobility management.

14.4 What is the importance of considering multimodal integration as a means of transportation system capacity expansion?

14.5 One 4-leg signalized intersection is always oversaturated during rush hours. This intersection is located around the edge of an urban area. Since it connects two major arterials some countermeasures have to be found to ease the congestion problem. Help the local agency find some solutions from the following perspectives respectively.

a. Demand management
b. Capacity expansion
c. Efficient system utilization

14.6 Contraflow is a practical way to "increase" roadway capacity and potentially reduce congestion and delays. Explain what the difficulties of this method are and how to decide the start/end points. What are the corresponding facilities required?

14.7 In order to alleviate an Interstate highway congestion issue during peak hours an agency has come up with a low-cost solution: opening a new BRT line running on that highway during peak hours. Specifically these buses can use both regular vehicular travel lanes and shoulders to by-pass slow traffic therefore reducing customers' travel time. This is called bus-on-shoulder (BOS) service. This new transit service has a 12-year life-span with 2% travel growth. Table P14.1 summarizes the expected cost items as well as the time.

Table P14.1 Alternative cost profile

Items	*Equipment purchasing*	*Facility preparation*	*Bus purchasing*	*Fare collection*	*Bus driver training*	*System operating*
Amount (dollars)	1300000	840000	2560000	680000	83000	1420000
Spent year	0	0	0	1–12	0	1–12

Compute the total net present worth of the above agency costs. Assume a discount rate of 5%.

14.8 For the above BOS problem continue the following calculations:

a. The traffic demand forecasting model gives the following predictions along with costs. These predictions are made for year 1.

Without BOS the predicted travel demand is 87600000 VMT with the average cost of 147.2/kVMT. With BOS implemented the BOS will attract 25460000 VMT from the current highway users and the equivalent average travel cost of using BOS is approximately 0.5/PMT with average occupancy 1.25 passengers per vehicle. Compute the user benefits for the travelers shifted from auto to transit.

b. Further study shows that the newly built BOS can induce extra travel demands. As travel demand forecasting models predicted the induced travel demand of highway auto travel is 12380000 VMT at an average cost of 136.8/kVMT. What are the user benefits of these travelers?

c. The transit can obtain agency benefits from advertisement business from this line estimated to be 200000 per year. No salvage value is left at year 12. Update the present worth of agency cost and compute the net present worth and B/C ratio values of this project. Comment the viability of this project based on your calculation.

d. If the traffic demand is increased by 5% per year how will the result change? Assume the fare collected and highway VMT is increased by 5% per year. State other assumptions made to justify your result.

Chapter 15

Economic analysis of highway safety improvements

The analysis of highway safety improvements includes identifying high crash sites, proposing countermeasure treatments, quantifying the effects of safety improvements achieved from the proposed treatments in terms of crash reductions, and estimating the economic benefits associated with the crash reductions.

15.1 METHODS FOR IDENTIFYING HIGH VEHICLE CRASH SITES

For a highway network that can be partitioned into highway segments and intersections, safety analysis begins with identifying high crash segment or intersection sites. Intuitively, a high crash site has a higher risk of vehicle crashes compared to other sites with similar characteristics. Different approaches can be used in high crash site identification.

15.1.1 Vehicle crash rate method

Higle and Witkowski (1988) suggested that black spots are usually identified by comparing the vehicle crash rate with the regional mean. If the vehicle crash rate is higher than the regional mean plus the SD, the area is usually referred to as a black spot. This is the simplest technique to identify black spots.

$$k > \mu + \sigma \tag{15.1}$$

where

k = Expected vehicle crash rate

μ = Average crash rate of a region

σ = Standard deviation of vehicle crash rate

15.1.2 Rate quality control method

The rate quality control method is a statistical method for identifying black spots. This method consists of calculating three different parameters for each highway segment. The three parameters are crash rate, crash frequency, and severity index (Norden et al., 1956). Each of these values is compared with a critical value. The crash rate is compared with one critical value, the crash frequency with another critical value, and the severity value with a third critical value. If a certain highway segment shows higher values than the critical ones for all these three parameters, the segment is considered to be a black spot.

15.1.3 Autocorrelation techniques

Recently, autocorrelation techniques have been applied to vehicle crash analysis. Autocorrelation techniques are applied to find out the clustering of black zones. It is obvious that a cluster of minor black spots represents a more severe problem than just one spot with a high crash rate. Autocorrelation techniques are useful in characterizing such black spots. Various autocorrelation methods are in use in crash analysis. Popular methods include clustering and Moran's I analyses that allow grouping of concentrated zones of crashes.

K-mean clustering analysis: *K*-mean clustering analysis is widely used for crime analysis and disease outbreak analysis. By predefining *K* clusters that categorize different levels of crashes such as low, medium, and high, each observation in the data set is assigned to a cluster such that the observation has the closest value to the mean value of all observations in the cluster. The best result is achieved by maximizing the distance of observations in different clusters and minimizing the distance of observations within the same cluster (MacQueen, 1967). A general *k*-mean clustering analysis is represented by minimizing the following:

$$J = \sum_{k=1}^{K} \sum_{i=1}^{nk} (X_i - \mu_k)^2 \tag{15.2}$$

where

X_i = Number of vehicle crash counts at site i in the data set
μ_k = Mean value of vehicle crashes in cluster k
k = Cluster k, $k = 1, 2, \ldots, K$
n_k = Number of observations in cluster k
n = Total number of observations, $n = \sum_{k=1}^{K} n_k$

Moran's I analysis: Moran's *I* analysis is a weighted correlation coefficient used to detect departures from spatial randomness that indicates spatial patterns such as clusters, and other kinds of pattern such as geographic trend (Moran, 1950). The analysis can be conducted for global spatial autocorrelation in group-level data as well as local spatial autocorrelation. When values of observations in nearby areas are similar, Moran's *I* value will be large and positive. Whereas when the values of observations in nearby areas are dissimilar, Moran's *I* value will be negative. Moran's *I* value lies between −1 and 1. Negative one represents a high dissimilarity between the observations, whereas positive one indicates a clustering of very similar values. Moran's *I* value is computed by

$$I = \frac{n \cdot \left[\sum_{i=1}^{n} \sum_{j=1}^{n, j \neq i} (w_{ij} \cdot z_i \cdot z_j) \right]}{\left[\sum_{i=1}^{n} \sum_{j=1}^{n, j \neq i} w_{ij} \right] \cdot \left[\sum_{i=1}^{n} \left(z_i^2 \right) \right]} \tag{15.3}$$

where

w_{ij} = Weighting scale between sites i and j
z_i = Deviation of value of vehicle crashes at site i from mean value of cluster k to which the site belongs, $z_i = \mu_k - X_{i,n_k}$
z_j = Deviation of value of vehicle crashes at site j from mean value of cluster k to which the site belongs, $z_j = \mu_k - X_{j,n_k}$

μ_k = Mean value of vehicle crashes in cluster k, $k = 1, 2, \ldots, K$
n_k = Number of observations in cluster k
X_{i,n_k}, X_{j,n_k} = Number of vehicle crash counts at sites i and j in cluster k
n = Total number of observations, $n = \sum_{k=1}^{K} n_k$

Local Moran's I analysis: This method is a special case of Moran's I that is decomposed to consider the contribution of each observation. It was first used to study the spatial pattern of conflict on the African continent by identifying "hot spots" at the local level (Anselin, 1995). When the neighboring observations are similar, the local Moran's I value will have a positive value. Dissimilar neighboring values of observations means the local Moran's I value is negative. A higher local Moran's I magnitude indicates a greater similarity or dissimilarity depending on the sign. The local Moran's I is defined as

$$I_i = z_i \cdot \sum_{j=1}^{n} (w_{ij} \cdot z_j) \tag{15.4}$$

where
I_i = Local Moran's I value at location i
w_{ij} = Weighting scale between sites i and j
z_i = Deviation of number of vehicle crashes at site i from mean value
z_i = Deviation of number of vehicle crashes at site j from mean value where j is the neighboring site i

15.2 TYPICAL HIGHWAY SAFETY IMPROVEMENT COUNTERMEASURES

15.2.1 Highway segment safety improvement countermeasures

To evaluate safety of highway segments, all potential sites need to be screened to identify the most critical sites and consequently address corresponding issues. Some of the proven strategies to improve highway segment safety are discussed below.

Vertical alignments: There are more crashes on gradients compared to level segments. In order to address safety concerns of high crashes on gradients, which are exacerbated by higher travel speeds, the vertical curvature of roadways is an important design criterion for consideration. Intuitively, crash frequency would decrease by reducing gradients. Reducing the length of the approaching segment to crest curves could make drivers reduce their driving speed, which in turn enhances safety.

Cross-section improvements: Depending on the roadway functional classification, traffic volume, speed, and other important variables such as proximity to commercial business districts, road designers should come up with appropriate lane width, especially on undivided roadways. The safest configuration is for lanes to be neither too narrow nor too wide. Rather, lane widths should correlate with traffic demand.

Lane widening: Where there are more trucks and other heavy vehicles, widening at least one lane can increase overall safety. In these circumstances, there should be additional informative devices letting drivers of heavy vehicles know that they are not allowed to take other lanes.

Shared lanes: Some roadway segments, preferably not along high-speed roads, could have lanes where vehicles and bikes share the road. This makes drivers more cautious and use lower driving speeds.

Paved shoulders: Up to standard wide shoulders would assist with drivers who unintentionally veer out of the travel lane. It is essential to provide wide enough shoulders at least on horizontal curves to ensure motorists' safety.

Rumble strips: This is a cost-effective solution to warn drivers of lane configurations and also lane departure. Rumble strips have demonstrated reduced run-off-road and other crash types. This is particularly useful to mark shoulders.

Safety edges: This is a road paving technique that would provide a small degree slope toward the road's edge. It reduces height difference between the pavement surface and materials off the road. This is an extremely important and low-cost solution for reducing potential severe crashes.

Sidewalks: Ensuring there is a sidewalk adjacent to the roadway can improve pedestrians' safety. Additionally, when drivers see a sidewalk along the road, they understand they are in a more populated pedestrian area and need to drive slower.

Roundabouts: At locations where installation of a traffic signal is not warranted, it might be a good solution to build a roundabout to reduce driving speed as well as making it easier for side-streets to get onto the mainline.

Speed bumps: There are certain locations, such as school zones, where in addition to roadside traffic signs additional safety control strategies are required. Speed bumps could be installed on the roadway to make drivers reduce speed.

Signs: According to roadway geometric design, appropriate traffic signs should be designed and installed along the roadways to provide drivers with traffic advisory information. Location of traffic signs needs to be established based on crash history and traffic volume. Stop and yield signs must be installed on side street phases. Additionally, the responsible agencies need to check visibility of roadside signs to ensure, for example, they are not blocked by trees. This is especially important at accident-prone sites.

Variable message signs: To address real-time circumstances such as inclement weather and special events, variable message signs (VMSs) may be required to reflect the suggested speed limit according to the actual circumstances. This is because normal warning signs may not be clearly visible in bad weather or drivers may not pay attention to them when they are in the middle of a special event. For example, in contrast to normal speed limit signs, VMSs have the potential to describe and justify the actual traffic or weather conditions when drivers would comply with the suggested (i.e., requested) speed limit.

Roadside barriers: Run-off-road can cause severe crashes. Roadside barriers have the potential to reduce severe crashes. Depending on terrain conditions, different types of barriers could be installed at sharp horizontal curves, roads with higher operational speed, and historically unsafe segments.

Median barriers: To separate opposing traffic movements on divided roadways, median barriers can reduce cross-median crashes. Cable median barriers are an example of this strategy.

Retroreflective materials: Materials such as reflectors are especially useful at night to guide drivers through a curve. Providing this kind of delineation on concrete and metal barriers would also help drivers to safely get through work zones. Additionally, reflective materials could be installed on roadside objects, which would prevent potential crashes.

Flashing beacons: To increase drivers' awareness in school zones or similar locations such as in proximity to an especial event, flashing beacons along with a warning sign (e.g., speed limit) would smooth traffic movement and provide sufficient room, for example, for pedestrians to use the roadway.

Speed cameras: In order to change speed limit and/or take additional actions to improve safety along roadways, speed cameras can be used to detect and record driving speeds. This data would further be used for forensic investigation of crashes.

Road safety audits: These are important to make sure roads are designed or planned to be designed where safety concerns are addressed. Road safety professionals review the existing and planned-for-future infrastructure to identify potential problems before they happen. Existing roads need to be under a systematic 6-month and 12-month maintenance program when each segment is investigated in detail and inspectors generate work orders to repair any observed issues.

Maintenance: Repairing road defects such as potholes and cracking would certainly smooth travel lanes' surfaces and reduce vehicles' abrupt hiccups. This would prevent sudden improper maneuvering, which has been demonstrated as a main cause of roadway crashes. Especially in winter, spreading salt and sand is a suggested procedure to ensure roadway roughness is up to standard and vehicles do not slide and crash. Some road segments may have to be completely shut down due to excessive snow and ice issues and traffic should be re-directed to ensure that safety is achieved.

15.2.2 Safety improvement countermeasures for intersections

Intersections could be considered more critical locations compared to highway segments. This is due to the existence of several conflicting movements in one limited area where vehicles need to be discharged one after another or concurrently depending on phasing design of the intersection. There are two general categories of intersections: signalized and non-signalized. As the name implies, signalized intersections are those equipped with traffic signals to control different movements. At unsignalized intersections, traffic signs control each movement. The following strategies are some of the main treatments to improve users' safety at intersections.

Sight distance: Drivers, especially those making left and right turns, need to see far ahead of their moving direction. This is to ensure there is not any conflict with other movements. To this end, bushes and trees in medians and roadsides should be trimmed to create clear sight distance for motorists.

Intersection channelization: Using hard islands, gore areas, and clear striping, 3-legged and 4-legged intersections should be channelized to make drivers aware of the permitted and restricted areas. This would help avoid conflicts of right and left turns with through traffic movements well before they get close to the intersection.

Intersection striping: Crosswalks, stop bars, and other markings required to split different movements should be clearly visible for motorists and pedestrians to avoid potential conflicts. Roadway striping needs to show desired reflectivity, especially at night, when drivers should to be able to see road limits and shoulders. This needs to be regularly inspected throughout the year, especially in winter, when snow and ice interrupt traffic movement.

Pedestrian refuge islands: This strategy can improve pedestrians' safety, especially in downtown areas and where the road is very wide (e.g., more than two lanes in each direction) and pedestrians may not be able to complete crossing the road at once.

Red-light running cameras: Depending on crash history and existing traffic volume at an intersection, red-light running enforcement cameras may need to be installed for desired phases. If that is the case and the corresponding agency decides to install the camera, a traffic sign needs to go with it way before approaching the intersection to let motorists know that the intersection is equipped with a red-light running camera. This way, drivers would be more cautious in terms of following traffic rules.

Retroreflective boards on signal backplates: Red-light running has been found to be a serious issue causing traffic accidents. Retroreflective boards on backplates of traffic signals would make traffic signals more visible to motorists, particularly at night. It would consequently reduce red-light running and improve safety.

Signs: Placing stop and yield signs on minor roads and ramps would minimize possible conflicts between side street traffic with mainlines. Additionally, it is suggested to install a "do not block intersection" sign on the mast arm letting drivers know that they are not supposed to block the intersection if there is no room for them to cross it, even if they have a green light. Other signs such as those showing status of left turns, either permitted or protected, are necessary to improve intersections' safety. In case of flashing yellow arrow (FYA), the appropriate sign according to the Manual on Uniform Traffic Control Devices (MUTCD) should be installed on the mast arm.

Lighting: It is suggested to provide enough lighting at intersection areas where motorists can see other vehicles and make a safe decision to either continue their movement or stop/yield for other drivers. This has noticeable effects at signalized intersections where left turns are not protected (i.e., they are either protected-permissive or permissive only) and at non-signalized intersections.

Signal head alignment: All signal heads should be aligned appropriately facing the desired approach. For example, left turn signal heads need to be installed such that motorists will not get confused whether it is for left turners or through traffic. Signals are preferred to be installed on the near side of the intersection instead of the far side to prevent cars from stopping in the crosswalk, since they cannot see the signal from there.

Supplementary signal heads: In case the traffic signal is not visible to approaching motorists, for example, on curves, a supplementary signal could be installed either on a mast arm or pole where it would be clearly visible to oncoming drivers.

Placement and number of signal heads: Following state and federal standards, each approach lane may need to have its signal head installed on a mast arm. This would overcome concerns such as line-of-sight blockage and invisibility of different displays as issues associated with pole-mounted signals.

Flashing yellow arrow: A major possible safety problem at intersections, especially at large ones and those in the downtown area, is due to conflicts between left turns and opposing through traffic. Recent studies have shown that installation of FYA improves safety of intersections.

Minimum green times: Each phase at the intersection should receive appropriate "minimum green time" (sometimes referred to as minimum initial or min green) such that drivers could safely clear the intersections. Short "min greens" such as 5 seconds may not give drivers enough time to complete their movement. This would, in turn, create red-light running and consequently deteriorated safety.

Yellow time: It is important to consider drivers' reaction time in signal timing design. When getting closer to end of the green time, drivers would continue their movements and try to discharge through the intersection. To ensure there is not conflict between different movements, each phase should receive a good amount of yellow time. This can be calculated for each approach separately.

All-red time: As a part of clearance time at each intersection, there is a certain amount of time during which none of the phases can move. This time is called all-red time and depending on various parameters is usually designed as a couple of seconds.

Protected left turn: In cases when there is not enough sight distance for left turners, it is suggested to make this as a protected left turn instead of protected-permissive or permissive left turns. In these circumstances, the left turn should have its own timing schedule in terms of min green, yellow time, and so on.

Right turn signal: Depending on intersection geometric design, traffic demand, and the phasing diagram of the intersections, right turns may have to be protected. In other words, there might be situations where right turns are not allowed during red signal indication. This is to avoid conflicts of right turns with other active phases. The use of right on red is highly discouraged or should be eliminated as it encourages drivers to look left as they glide through a crosswalk, which leads to striking pedestrians crossing from the right.

Pedestrian crossing time: It is extremely important to make sure pedestrians have enough time to cross the road. Pedestrian time consists of two parts: walk and flashing do not walk times. Following the standard equations, and with collected data from the site, pedestrian times should be calculated.

Countdown timers for pedestrians: In addition to "walking person" (symbolizing walk) and "upraised hand" (symbolizing don't walk), it would be more beneficial to embed a countdown timer into the pedestrian signal head letting pedestrians know the remaining time to complete their walk.

Audible pedestrian signals: To assist pedestrians who are visually impaired, audible pedestrian signals advise when they have right-of-way to cross a road.

Upraised pavement markings: Depending on geometric design of the intersection, especially where there is a merging lane, agencies may prefer to install upraised marking. It could be either reflective or non-reflective. Upraised marking would guide traffic flow toward the travel lane.

15.3 EFFECTIVENESS OF SAFETY IMPROVEMENTS

15.3.1 Methods for estimating effectiveness of safety improvements

Highway safety analysis is the passive pursuit of observing the safety consequences of some safety improvement measures implemented for certain purposes. These are observational studies, not statistical experiments where researchers deliberately design experiments to answer hypothetical questions. The natural domain of an observational before–after study is the circumstance when the sites that are changed by the treatment retain much of their original attributes. In contrast, the natural domain of observational cross-section studies is the circumstance when the treatment substantially alters the sites. Often, before–after studies attract more attention to investigators as changes of sites are generally made gradually.

Two basic tasks need to be accomplished in an observational before–after study: (i) the task of predicting what would have been the safety of the site in the "after" treatment period had a treatment not been implemented; and (ii) the task of estimating what the safety of the treated site in the "after" treatment period was (Hauer, 2002). Obviously, the difference between the two is the true effect of such a treatment. The observational before–after analysis consists of four steps:

- Predict the expected number of vehicle crashes of a specific site (λ) in an "after" period had it not been treated; and estimate the expected number of target vehicle crashes of the site (μ) in the "after" treatment period
- Estimate the variances of λ and μ, $V(\lambda)$ and $V(\mu)$
- Estimate reduction in expected number of vehicle crashes $\delta = \lambda - \mu$ and treatment effectiveness index $\theta = \mu/\lambda$
- Estimate the variances of δ and θ, $V(\delta)$ and $V(\theta)$. Specifically, the true reduction in expected number of vehicle crashes and effectiveness of a treatment are then $\delta \pm \sqrt{V(\delta)}$ and $\theta \pm \sqrt{V(\theta)}$, respectively

Table 15.1 Vehicle crash counts and expected crashes for treatment group before and after treatment

		Treatment group	
Period	*Data*	*Crash counts*	*Expected crashes*
Before	Site *i*	K_i	k_i
	All sites	$K=\sum_i K_i$	$k=\sum_i k_i$
After	Site *i*	M_i	μ_i λ_i (if not treated)
	All sites	$M=\sum_i M_i$	$\mu=\sum_i \mu_i=\sum_i M_i, V[\mu]=\sum_i V[\mu_i]$ $\lambda=\sum_i \lambda_i(\text{if not treated}), V[\lambda]=\sum_i V[\lambda_i]$

Conventional methods to accomplish Steps 1–4 include (i) Naïve before–after study; (ii) comparison-group before–after study; (iii) empirical Bayesian (EB) approach; (iv) EB naïve before–after study; and (v) EB comparison-group before–after study.

15.3.2 Naïve before: After study

Suppose that some treatment has been implemented on sites 1, 2, ... , i, ... , n. Vehicle crash counts and expected crashes for all sites "before" and "after" treatment periods are shown in Table 15.1.

In practice, the duration of data collection and traffic volumes on each site in the "before" and "after" periods may change and remaining factors stay the same in both periods. This requires some modification in the estimation.

Denote

$d_{a,i}$ = Duration of data collection in "after" period for site i
$d_{b,i}$ = Duration of data collection in "before" period for site i
$r_{d,i}$ = Ratio of durations "after" and "before" periods for site i, $r_{d,i}=d_{a,i}/d_{b,i}$
$w_{a,i}$ = Average traffic volume in "after" period for site i
$V[w_{a,i}]$ = Variance of traffic volumes in "after" period for site i
$w_{b,i}$ = Average traffic volume in "before" period for site i
$V[w_{b,i}]$ = Variance of traffic volumes in "before" period for site i
$r_{w,i}$ = Ratio of traffic volumes "after" and "before" periods for site i, $r_{w,i}=w_{a,i}/w_{b,i}$
$V[r_{w,i}]$ = Variance of traffic volume ratios in "after" and "before" periods for site i,

$$V\left[r_{w,i}\right]=r_{w,i}^2\left[\frac{V\left[w_{a,i}\right]}{w_{a,i}^2}+\frac{V\left[w_{b,i}\right]}{w_{b,i}^2}\right]$$

The four-step analysis of naïve before–after study is summarized in Table 15.2.

15.3.3 Comparison-group before–after study

To better predict the true effect of vehicle crashes' reduction as a result of a treatment, it is necessary to utilize additional information for analysis. One way to do this is using

Table 15.2 Method of naïve before–after analysis

Procedure		*Estimate*
Step 1.	Expected vehicle crashes had not treatment been applied λ and vehicle crashes after treatment μ in "after" treatment period	$\hat{\lambda}=\sum_{i=1}^{n}\hat{\lambda}_i=\sum_{i=1}^{n} r_{d,i}\cdot r_{w,i}\cdot K_i$ $\hat{\mu}=\sum_{i=1}^{n}\hat{\mu}_i=\sum_{i=1}^{n} M_i$
Step 2.	Variances of λ and μ	$V[\hat{\lambda}]=\sum_{i=1}^{n}V[\hat{\lambda}_i]=\sum_{i=1}^{n}\left[\hat{\lambda}_i^2\left(\frac{1}{K_i}+\frac{\hat{V}[\hat{r}_{w,i}]}{\hat{r}_{w,i}^2}\right)\right]$ $V[\hat{\mu}]=\sum_{i=1}^{n}V[\hat{\mu}_i]=\sum_{i=1}^{n} M_i$
Step 3.	Reduction of vehicle crashes δ and effectiveness index θ	$\hat{\delta}=\hat{\lambda}-\hat{\mu}$ $\hat{\theta}=\frac{\hat{\mu}}{\hat{\lambda}}\cdot\frac{1}{[1+V[\hat{\lambda}]/\hat{\lambda}_i^2]}$
Step 4.	Variances of δ and θ	$V[\hat{\delta}]=V[\hat{\lambda}]+V[\hat{\mu}]$ $V[\hat{\theta}]=\frac{\hat{\theta}^2\left[\frac{V[\hat{\lambda}]}{\hat{\lambda}^2}+\frac{V[\hat{\mu}]}{\hat{\mu}^2}\right]}{\left[1+\frac{V[\hat{\lambda}]}{\hat{\lambda}^2}\right]^2}$
Result:	True reduction in vehicle crashes due to treatment: $\hat{\delta}\pm\sqrt{V[\hat{\delta}]}$ Effectiveness of treatment: $\hat{\theta}\pm\sqrt{V[\hat{\theta}]}$	Note: Estimates of λ, μ, δ, and θ are designated by carets (^).

a comparison group with sites that remain untreated and that are similar to the treated sites. The treated sites form the "treatment" group, while the untreated sites belong to the "comparison" group. The hope is that the change from "before" and "after" in the safety of the comparison group is indicative of how the number of vehicle crashes on the treatment group would have changed. This is based on two assumptions: (i) factors affecting occurrence of vehicle crashes have changed from the "before" to "after" period in the same manner in both treatment and comparison groups; and (ii) changes of such factors influencing occurrence of vehicle crashes in the same way on both treatment and comparison groups. In the analysis process, the comparison group needs to have the same data collection durations for the "before" and "after" periods as compared to those for the treatment group.

Let crash counts and expected crashes for all sites in the treatment and comparison groups in the "before" and "after" treatment periods be as shown in Table 15.3.

Denote

$D_{a,i}$ = Duration of data collection in "after" period for site i

$D_{b,i}$ = Duration of data collection in "before" period for site i

$r_{d,i}$ = Ratio of durations "after" and "before" periods for site i, $r_{d,i}=D_{a,i}/D_{b,i}$

$W_{Ta,i}$, $V[W_{Ta,i}]$ = Mean and variance of traffic volumes in "after" period for site i in treatment group

$W_{Tb,i}$, $V[W_{Tb,i}]$ = Mean and variance of traffic volumes in "before" period for site i in treatment group

Table 15.3 Vehicle crash counts and expected crashes for treatment and comparison groups before and after treatment

		Treatment group		Comparison group without treatment	
Period	Data	Crash counts	Expected crashes	Crash counts	Expected crashes
Before	Site i	K_i	k_i	A_i	α_i
	All sites	$K=\sum_i K_i$	$k=\sum_i k_i$	$A=\sum_i A_i$	$\alpha=\sum_i \alpha_i$
After	Site i	μ_i	μ_i λ_i (if not related)	B_i	β_i
	All sites	$M=\sum_i M_i$	$\mu=\sum_i \mu_i=\sum_i M_i, V[\mu]=\sum_i V[\mu_i]$ $\lambda=\sum_i \lambda_i$(if not treated)$, V[\lambda]=\sum_i V[\lambda_i]$	$B=\sum_i B_i$	$\beta=\sum_i \beta_i$

Note: $V[\cdot]$ stands for variance.

$r_{Tw,i}$ = Mean ratios of traffic volumes in the "after" and "before" periods for site i in treatment group, $r_{Tw,i}$

$V[r_{Tw,i}]$ = Variance of traffic volume ratios in the "after" and "before" periods for site i in treatment group,

$V[r_{Tw,i}]$ $= r_{Tw,i}^2\left[\frac{V[W_{Ta,i}]}{W_{Ta,i}^2}+\frac{V[W_{Tb,i}]}{W_{Tb,i}^2}\right]$

$W_{Ca,i}$, $V[W_{Ca,i}]$ = Mean and variance of traffic volumes in "after" period for the site in comparison group corresponding to site i in treatment group

$W_{Cb,i}$, $V[W_{Cb,i}]$ = Mean and variance of traffic volumes in "before" period for site in comparison group corresponding to site i in treatment group

$r_{Cw,i}$ = Mean ratio of traffic volumes "after" and "before" periods for site in comparison group corresponding to site i in treatment group, $r_{Cw,i}$

$V[r_{Cw,i}]$ = Variance of $r_{Cw,i}$, $V\left[r_{Cw,i}\right]=r_{Cw,i}^2\left[\frac{V[W_{Ca,i}]}{W_{Ca,i}^2}+\frac{V[W_{Cb,i}]}{W_{Cb,i}^2}\right]$

$r_{T,i}$ = Ratio of expected crash counts for site i in treatment group should there be no treatment applied in "after" period and "before" period, $r_{T,i}=\lambda_i/(r_{d,i}\cdot r_{Tw,i}\cdot k_i)$

$r_{C,i}$ = Ratio of expected crashes "after" and "before" periods for site in comparison group corresponding to site i in treatment group, $r_{C,i}=\beta_i/(r_{d,i}\cdot r_{Cw,i}\cdot \alpha_i)$

ω_i = Odds ratio for site i in treatment group, $\omega_i=r_{C,i}/r_{t,i}$ $V[\omega_i]$

$V[\omega_i]$ = Variance of ω_i, $V[\omega_i]=s^2\left[\frac{K_iB_i}{M_i(r_{d,i}\cdot r_{d,i}\cdot A_i)(1+(1/M_i)+(1/(r_{d,i}\cdot r_{Cw,i}\cdot A_i)))}\right]$

$$\times\left(1+\frac{1}{M_i}+\frac{1}{r_{d,i}\cdot r_{Cw,i}\cdot A_i}+\frac{1}{B_i}\right)$$

if >0; and 0 otherwise, where $S^2[\cdot]$ represents sample variance of multi-time period crashes.

The four steps of the comparison-group before–after study are presented in Table 15.4.

Table 15.4 Method of comparison-group before–after study

Procedure	*Estimate*
Step 1. Expected crashes had not treatment been applied λ and crashes after treatment μ in "after" treatment period	$\hat{\lambda}=\sum_{i=1}^{n}\hat{\lambda}_i=\sum_{i=1}^{n} r_{d,i}\cdot r_{Tw,i}\cdot\frac{B_i}{1+r_{d,i}\cdot r_{Cw,i}\cdot A_i}\cdot K_i$ $\hat{\mu}=\sum_{i=1}^{n}\hat{\mu}_i=\sum_{i=1}^{n}M_i$
Step 2. Variances of λ and μ	$V[\hat{\lambda}]=\sum_{i=1}^{n}V[\hat{\lambda}_i]=\sum_{i=1}^{n}\left[\hat{\lambda}_i^2\left(\frac{1}{K_i}+\frac{1}{r_{d,i}\cdot r_{Cw,i}\cdot A_i}+\frac{1}{B_i}+V[\omega_i]\right)\right]$ $V[\hat{\mu}]=\sum_{i=1}^{n}V[\hat{\mu}_i]=\sum_{i=1}^{n}M_i$
Step 3. Reduction in crashes δ and effectiveness index θ	$\hat{\delta}=\hat{\lambda}-\hat{\mu}$ $\hat{\theta}=\frac{\hat{\mu}}{\hat{\lambda}}\cdot\frac{1}{\left[1+V[\hat{\lambda}]/\hat{\lambda}_i^2\right]}$
Step 4. Variances of δ and θ	$V[\hat{\delta}]=V[\hat{\lambda}]+V[\hat{\mu}]$ $V[\hat{\theta}]=\frac{\hat{\theta}^2\left[\frac{V[\hat{\lambda}]}{\hat{\lambda}^2}+\frac{V[\hat{\mu}]}{\hat{\mu}^2}\right]}{\left[1+\frac{V[\hat{\lambda}]}{\hat{\lambda}^2}\right]^2}$
Result: True reduction in vehicle crashes due to treatment: $\hat{\delta}\pm\sqrt{V[\hat{\delta}]}$ Effectiveness of treatment: $\hat{\theta}\pm\sqrt{V[\hat{\theta}]}$	Note: Estimates of λ, μ, δ, and θ are designated by carets (^).

15.3.4 EB approach

In both the naïve and comparison-group before–after study approaches, one of the two key assumptions is that, for any treated site, crash counts in the "before" treatment period are an unbiased estimate of expected vehicle crashes. In observational studies, there is likely to be a link between the decision to make a treatment to a site and recorded historical crash counts. For instance, a site with high crash history may receive higher priority for treatment. This link causes a so-called "selection" bias. It makes crash counts a biased estimate of expected vehicle crashes. In addition, the real task of analysis is to estimate the expected vehicle crashes in the "before" treatment period for as many years into the past as possible and to use the estimated values to predict vehicle crashes in the "after" treatment period had the treatment not been made. The question becomes how to determine a reasonable time window for "before" period estimation. To remedy selection bias and "before" period time window problems, the EB approach is advocated. The approach helps to deal with the above problems in that two separate pieces of information are used to estimate site vehicle crashes: (i) crash counts of the site that reflect its safety; and (ii) safety of other sites with similar characteristics of the site for which we have crash data (Box and Tiao, 1992; Makridakis et al., 1997).

Denote

K = Vehicle crash counts in a period of interest for a site
k = Expected number of vehicle crashes for the site in a given period
$E[k]$, $V[k]$ = Mean and variance of a reference population of sites having similar characteristics of the site under investigation and expected k crashes
$E[k|K]$, $V[k|K]$ = Mean and variance of subpopulation of sites from the reference population all recorded K crashes

Since the site under investigation has similar characteristics and the same crash record as all other sites of the subpopulation, its expected number of k crashes can be any of the k's in this subpopulation of sites that all recorded K vehicle crashes with equal probability. Hence, the best estimate of the expected number of k crashes for this site is $E[k|K]$ and the variance of this estimate is $V[k|K]$. The two parameters can be estimated from $E[k]$, $V[k]$, and K as below

$$E[k \mid K] = \alpha \cdot E[k] + (1 - \alpha) \cdot K \tag{15.5}$$

$$V[k \mid K] = (1 - \alpha) \cdot E[k \mid K] \tag{15.6}$$

with

$$\alpha = \frac{1}{1 + [(r_d \cdot V[k])/E(k)]}$$

where

r_d = Ratio of duration of K crash counts and duration of expected k vehicle crashes.
r_d = 1 if K and k pertain to the same time duration

The analysis then concentrates on estimating $E[k]$ and $V[k]$, where $E[k] = E[K]$ and $V[k] = V[K] - E[k]$. Common methods for estimating $E[k]$ and include the simple method of sample moments and more complex multivariate regression methods.

Method of sample moments: Consider a reference subpopulation of n sites, of which $n(K)$ sites have recorded $K = 0, 1, 2, \ldots$ vehicle crashes during a specific period. The sample mean and sample variance are

$$\bar{K} = \frac{\sum_{i=1}^{n} [K \cdot n(K)]}{n} \tag{15.7}$$

$$S^2 = \frac{\sum_{i=1}^{n} [(K - \bar{K})^2]}{n - 1} \tag{15.8}$$

As number of sites n increases, the sample mean $\bar{K}$ and sample variance S^2 approach true mean $E[K]$and variance $V[K]$. This give estimates of $E[K]$ and $V[K]$ below (Hauer, 2002)

$$E[k] = E[K] = \frac{\sum_{i=1}^{n} [K \cdot n(K)]}{n} \tag{15.9}$$

Table 15.5 Multivariate regression method for safety analysis

Procedure		*Estimate*
Step 1.	Calibrate a multivariate statistical model for the estimation of $E[k]$	Vehicle crash counts and measured characteristics (covariate values) for n sites serve as data. Fit a multivariate regression model that yields estimates $\hat{E}[k]$ of $E[k]$ as a function of the specific values that the covariates assume, that is, $\hat{E}[k] = f(\text{covariates})$
Step 2.	Compute the residual for each site	The residual is the difference between the crash count K and the fitted value $\hat{E}[k]$. The difference between squared residual $[K - \hat{E}[k]]^2$ and fitted value $\hat{E}[k]$ is the estimate of variance $V[k]$ for this site, that is, $\hat{V}[k] = [K - \hat{E}[k]]^2 - \hat{E}[k]$. One $\hat{V}[k]$ value is computed for each site
Step 3.	Calibrate multivariate statistical model for the estimation of $V[k]$, that is, $\hat{V}[k] = g(\text{covariates})$	Since $\hat{E}[k] = f(\text{covariates})$ is constructed, a simple function between $\hat{V}[k]$ and $\hat{E}[k]$ can be developed. The function can take the form $\hat{V}[k] = \frac{[\hat{E}[k]]^2}{b}$

$$V[k] = V[\mathrm{K}] - E[k] = \frac{\sum_{i=1}^{n}[(K - \bar{K})^2]}{n-1} - \frac{\sum_{i=1}^{n}[K \cdot n(K)]}{n} \tag{15.10}$$

Multivariate regression method: It is often necessary to estimate the mean and variance of sites having k vehicle crashes for which a sizeable reference population is not available. In this case, it is common to estimate the vehicle crashes of sites with specific characteristics by the multivariate statistical model. Table 15.5 presents the main steps.

15.3.5 EB naïve before–after study

In the EB naïve before–after analysis, the estimates of expected vehicle crashes $\hat{\lambda}_i$ and variance $V[\hat{\lambda}_i]$ for site i in "after" period had no treatment been applied, are enhanced. Instead of directly using the vehicle crashes counts K_i for site i in "before" period, the estimates of expected accounts for site i in "after" period had no treatment been applied conditioned on K_i, $E[k_i|K_i]$, and its variance $V[E[k_i|K_i]]$ are used, see Table 15.6. The remaining steps of naïve before–after study are kept unchanged in the EB naïve before–after study.

15.3.6 EB comparison-group before–after study

As illustrated in Figure 15.1, the circumstance here is that a treatment has been implemented to n sites that may have the same "before" and "after" periods and similar characteristics. Each site has a separate comparison group.

The EB comparison-group before–after analysis improves the estimation of expected vehicle crashes $\hat{\lambda}_i$ for site i in "after" period had no treatment been made and its variance $V[\hat{\lambda}_i]$. As in Table 15.7, this is done by replacing K_i by $E[k_i|K_i]$ and $1/K_i$ by $V[E[k_i|K_i]]/E[k_i|K_i]^2$ in the equations.

Table 15.8 shows details of applying the EB comparison-group before–after analysis (Roshandeh et al., 2013, 2016).

Table 15.6 Means and variances of naïve and EB naïve before–after studies

Parameter	*Naive before–after study*	*EB naive before–after study*
Expected crashes	$\hat{\lambda}=\sum_{i=1}^{n}\hat{\lambda}_i=\sum_{i=1}^{n} r_{d,i}\cdot r_{w,i}\cdot K_i$	$\hat{\lambda}=\sum_{i=1}^{n}\hat{\lambda}_i=\sum_{i=1}^{n} r_{d,i}\cdot r_{w,i}\cdot E[k_i \mid K_i]$
Variance	$V[\hat{\lambda}]=\sum_{i=1}^{n}V[\hat{\lambda}_i]$	$V\left[\hat{\lambda}\right]=\sum_{i=1}^{n}V\left[\hat{\lambda}_i\right]$
	$=\sum_{i=1}^{n}\left[\hat{\lambda}_i^2\left(\frac{1}{K_i}+\frac{\hat{V}[\hat{r}_{w,i}]}{\hat{r}_{w,i}}\right)\right]$	$=\sum_{i=1}^{n}\left[\hat{\lambda}_i^2\left(\frac{\hat{V}[\hat{r}_{w,i}]}{\hat{r}_{w,i}^2}+\frac{V[E[k_i \mid K_i]]}{(E\left[k_i \mid K_i\right])^2}\right)\right]$

where

$$E[k_i|K_i]=\hat{\alpha}_i\cdot\hat{E}[k_i]+(1-\hat{\alpha}_i)\cdot K_i$$

$$V[E[k_i|K_i]]=(1-\hat{\alpha}_i)\cdot E[k_i|K_i]$$

$$\hat{\alpha}_i=\frac{1}{1+[(r_{d,i}\cdot\hat{V}[k_i])/\hat{E}[k_i]]}$$

$$\hat{E}[k_i]=\frac{\sum_{i=1}^{n}K_i}{n}$$

$$\hat{V}[k_i]=\frac{\sum_{i=1}^{n}[(K_i-\hat{E}[k_i])^2]}{n-1}-\hat{E}[k_i]$$

$r_{d,i}$ = Ratio of duration of K_i crash counts and duration of expected k_i vehicle crashes for site i. $r_{d,i}$ = 1 if K_i and k_i pertain to the same time duration

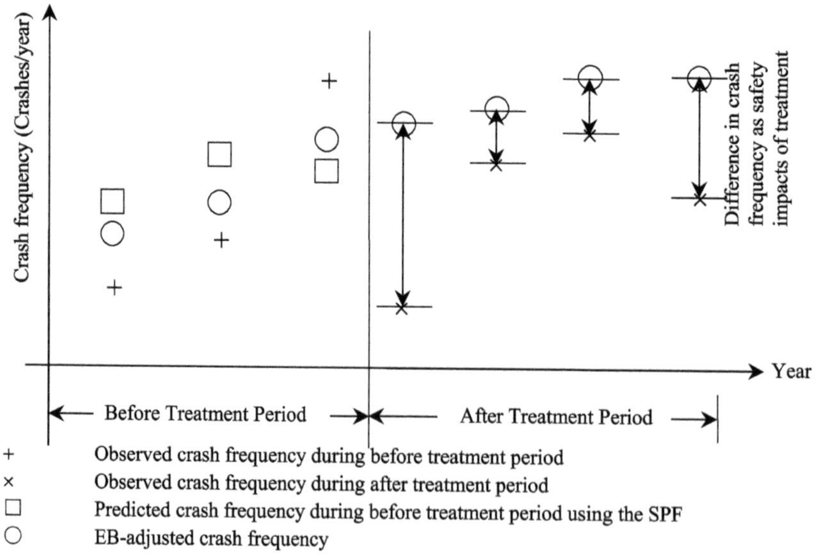

Figure 15.1 Illustration of EB comparison group before-after safety analysis. (Adapted from Bamzai, R. et al. 2009. *Safety Impacts of Highway Shoulder Attributes in Illinois*. Department of Civil, Architectural, and Environmental Engineering, Illinois Institute of Technology, Chicago, IL.)

Table 15.7 Means and variances of comparison-group and EB comparison-group before–after studies

Parameter	*Comparison-group before–after study*	*EB comparison-group before–after study*
Expected Crashes	$\hat{\lambda} = \sum_{i=1}^{n} \hat{\lambda}_i = \sum_{i=1}^{n} \left[r_{d,i} \cdot r_{Tw,i} \left(\frac{B_i}{1 + r_{d,i} \cdot r_{Cw,i} \cdot A_i} \right) \cdot K_i \right]$	$\hat{\lambda} = \sum_{i=1}^{n} \hat{\lambda}_i = \sum_{i=1}^{n} \left[r_{d,i} \cdot r_{Tw,i} \left(\frac{B_i}{1 + r_{d,i} \cdot r_{Cw,i} \cdot A_i} \right) \cdot E[k_i \mid K_i] \right]$
Variance	$V[\hat{\lambda}] = \sum_{i=1}^{n} V[\hat{\lambda}_i] = \sum_{i=1}^{n} \left[\hat{\lambda}_i^2 \left(\frac{1}{K_i} + \frac{1}{r_{d,i} \cdot r_{Cw,i} \cdot A_i} + \frac{1}{B_i} + V(\omega_i) \right) \right]$	$V[\hat{\lambda}] = \sum_{i=1}^{n} V[\hat{\lambda}_i] = \sum_{i=1}^{n} \left[\hat{\lambda}_i^2 \left(\frac{V[E[k_i \mid K_i]]}{(E[k_i \mid K_i])^2} + \frac{1}{r_{d,i} \cdot r_{Cw,i} \cdot A_i} + \frac{1}{B_i} + V(\omega_i) \right) \right]$

where

$$E[k_i|K_i] = \hat{\alpha}_i \cdot \hat{E}[k_i] + (1 - \hat{\alpha}_i) \cdot K_i$$

$$V[E[k_i|K_i]] = (1 - \hat{\alpha}_i) \cdot E[k_i|K_i]$$

$$\hat{\alpha}_i = \frac{1}{1 + [(r_{d,i} \cdot \hat{V}[k_i])/\hat{E}[k_i]]}$$

$$\hat{E}[k_i] = \frac{\sum_{i=1}^{n} K_i}{n}$$

$$\hat{V}[k_i] = \frac{\sum_{i=1}^{n} \left[\left(K_i - \hat{E}[k_i] \right)^2 \right]}{n - 1} - \hat{E}[k_i]$$

$r_{d,i}$ = Ratio of duration of K_i crash counts and duration of expected k_i vehicle crashes for site *i*. $r_{d,i} = 1$ if K_i and k_i pertain to the same time duration

Table 15.8 Computational procedure for *EB* comparison-group before–after analysis

Procedure		*Estimate*
Step 1.	Predict multi-year crash frequency for each treated site in the before treatment period	Apply the appropriate SPF for the computation
Step 2.	Calculate the expected multi-year crash frequency for each treated site in the before treatment period	$EB_{seg,i,B} = w_{seg,i} \times CF_{seg,i,P,B} + (1 - w_{seg,i}) \times CF_{seg,i,O,B}$ $EB_{int,i,B} = w_{int,i} \times CF_{int,i,P,B} + (1 - w_{int,i}) \times CF_{int,i,O,B}$ $w_{seg,i} = \frac{1}{1 + \left[k_{seg} \times (T \times L_{seg,i,B}) \times \sum_{t=1}^{T} CF_{seg,i,P,B,t}\right]}$ $w_{int,i} = \frac{1}{1 + \left[k_{int} \times T \times \sum_{t=1}^{T} CF_{int,i,P,B,t}\right]}$
Step 3.	Calculate crash adjustment factor for each treated site in the after treatment period	$r_{seg,i,A} = \left(\frac{\sum_{a=1}^{A} AADT_{i,a}}{\sum_{b=1}^{B} AADT_{i,b}}\right) \times \left(\frac{L_{seg,i,A}}{L_{seg,i,B}}\right)$ $r_{int,i,A} = \left[\frac{\sum_{a=1}^{A} (AADT_{i,major,a} + AADT_{i,minor,a})}{\sum_{b=1}^{B} (AADT_{i,major,b} + AADT_{i,minor,b})}\right]$
Step 4.	Calculate the expected crash frequency for each site over the entire after period in the absence of treatment	$EB_{seg,i,A} = EB_{seg,i,B} \times r_{seg,i,A}$ $EB_{int,i,A} = EB_{int,i,B} \times r_{int,i,A}$
Step 5.	Calculate the overall odds ratio for the treated sites	$OR_{seg} = \left[\frac{\sum_{i=1}^{N} CF_{seg,i,O,A}}{\sum_{i=1}^{N} EB_{seg,i,A}}\right] \times \frac{1}{1 + \left[\frac{Var(EB_{seg,A})}{\left(\sum_{i=1}^{N} EB_{seg,i,A}\right)^2}\right]}$ $OR_{int} = \left[\frac{\sum_{i=1}^{N} CF_{int,i,O,A}}{\sum_{i=1}^{N} EB_{int,i,A}}\right] \times \frac{1}{1 + \left[\frac{Var(EB_{int,A})}{\left(\sum_{i=1}^{N} EB_{int,i,A}\right)^2}\right]}$
Step 6.	Calculate the average safety improvement level for the treated sites	$Eff_{seg} = 100 \times (1 - OR_{seg})$ $Eff_{int} = 100 \times (1 - OR_{int})$

(Continued)

Table 15.8 (Continued) Computational procedure for *EB* comparison-group before–after analysis

Procedure	*Estimate*
Step 7. Calculate the variance of the safety improvement level for the treated sites	$Var(Eff_{seg}) = \left[\frac{\sum_{i=1}^{N} CF_{seg,i,O,A}}{\sum_{i=1}^{N} EB_{seg,i,A}}\right]^2 \times \left[\frac{1}{\left(\sum_{i=1}^{N} CF_{seg,i,O,A}\right)^2} + \frac{Var(EB_{seg,A})}{\left(\sum_{i=1}^{N} EB_{seg,i,A}\right)^2}\right] \times \frac{1}{\left[1 + \frac{Var(EB_{seg,A})}{\left(\sum_{i=1}^{N} EB_{seg,i,A}\right)^2}\right]}$ $Var(Eff_{int}) = \left[\frac{\sum_{i=1}^{N} CF_{int,i,O,A}}{\sum_{i=1}^{N} EB_{int,i,A}}\right]^2 \times \left[\frac{1}{\left(\sum_{i=1}^{N} CF_{int,i,O,A}\right)^2} + \frac{Var(EB_{int,A})}{\left(\sum_{i=1}^{N} EB_{int,i,A}\right)^2}\right] \times \frac{1}{\left[1 + \frac{Var(EB_{int,A})}{\left(\sum_{i=1}^{N} EB_{int,i,A}\right)^2}\right]}$
Step 8. Test for statistical significance of the safety improvement level for all treated sites	$Test_{seg} = Eff_{seg} / \left(100 \times \sqrt{Var(Eff_{seg})}\right)$ $Test_{int} = Eff_{int} / \left(100 \times \sqrt{Var(Eff_{int})}\right)$
Result: If $\lvert Test_{int}\rvert$ or $\lvert Test_{seg}\rvert < 1.7$	Treatment effect is not significant at 90% confidence level
If $\lvert Test_{int}\rvert$ or $\lvert Test_{seg}\rvert \geq 1.7$	Treatment effect is significant at 90% confidence level
If $\lvert Test_{int}\rvert$ or $\lvert Test_{seg}\rvert \geq 2.0$	Treatment effect is significant at 95% confidence level

Denote

$EB_{seg,i,B}$, $EB_{int,i,B}$ = EB-adjusted multi-year crash frequency before treatment for segment or intersection i

$EB_{seg,i,A}$, $EB_{int,i,A}$ = EB-adjusted multi-year crash frequency after treatment for segment or intersection site i

$w_{seg,i}$, $w_{int,i}$ = Weighting factor between SPF predicted and field observed multi-year crash frequencies for segment or intersection site i

k_{seg}, k_{int} = Overdispersion parameter of the crash frequency for urban street segment per mile per year or for urban intersection per year

$r_{seg,i,A}$, $r_{int,i,A}$ = Adjustment factor for segment or intersection i after treatment

OR_{seg}, OR_{int} = The odds ratio as the effect of safety improvements to all segments or intersections

$CF_{seg,i,P,B,t}$, $CF_{int,i,P,B,t}$ = Predicted crash frequency before treatment for segment or intersection site i in year t

$CF_{seg,i,P,B}$, $CF_{int,i,P,B}$ = SPF predicted multi-year crash frequency before treatment with further adjustments according to the crash modification factors for segment or intersection site i

$CF_{seg,i,O,B}$, $CF_{int,i,O,B}$ = Field observed multi-year crash frequency before treatment for segment or intersection site i

$CF_{seg,i,O,A}$, $CF_{int,i,O,A}$ = Observed or predicted multi-year crash frequency after treatment at segment or intersection site i

$AADT_{i,b}$, $AADT_{i,a}$ = $AADT$ on urban street segment i before and after treatment

$AADT_{i,major,b}$ and $AADT_{i,minor,b}$ = $AADT$ on major and minor approaches of urban intersection i before and after treatment

$L_{seg,i,B}$ = Length of the urban street segment i before treatment

$L_{seg,i,A}$ = Length of the urban street segment i after treatment

Eff_{seg}, Eff_{int} = Average safety improvement level for the treated segment or intersection sites

$Var(Eff_{seg})$, $Var(Eff_{int})$ = Variance of the safety improvement level for the treated segment or intersection sites

$Var(EB_{seg,A})$, $Var(EB_{int,A})$ = Variance of EB-adjusted multi-year crash frequency after treatment for all urban street segments or intersections

$Test_{seg}$, $Test_{int}$ = Test for statistical significance of the safety improvement level for all treated segment or intersection sites

i = $1, 2, \ldots, N$

t = $1, 2, \ldots, T$

Case Study: Safety impacts of intersection signal timing optimization for an urban street network

This case study focuses on evaluating the safety impacts of intersection signal timing optimization for a densely populated urban street network in the U.S. It begins with collecting multi-year field observed data on vehicle-to-vehicle and vehicle-to-pedestrian crashes at individual urban intersections and vehicle-to-vehicle crashes at urban street segments, as well as vehicle volumes at intersections and on street segments, all for the period of before signal timing optimization (i.e., the before treatment period).

Then, it identifies appropriate SPFs to predict vehicle-to-vehicle and vehicle-to-pedestrian crashes at urban intersections and street segments. The regional travel demand forecasting model is executed to obtain the traffic volume on each roadway segment or intersection approach after intersection signal timing optimization (i.e., the after treatment period) for the urban street network. The signal timing optimization is achieved by adjusting the green splits of signal timing plans simultaneously for all intersections in the network hour-by-hour during a typical weekday to minimize vehicle and pedestrian delays according to simulated vehicle volumes utilizing individual intersections.

Next, the field-observed (before treatment period) and simulated (after treatment period) traffic volumes along with the appropriate SPFs are used to estimate crash frequencies at each urban intersection and on each street segment over the multi-year period before and after treatments. The safety impacts in terms of changes in vehicle-to-vehicle and vehicle-to-pedestrian crashes are assessed to confirm the statistical significance of the impacts.

a. *SPFs for predicting crashes at urban intersections and on street segments*: As a key step of applying the proposed EB method for safety impacts analysis, SPFs need to be utilized to predict vehicle-to-vehicle and vehicle-to-pedestrian crashes at intersections and vehicle-to-vehicle crashes at street segments before and after treatment. Historically, Poisson and negative binomial modeling techniques have been used for SPF calibration. The Poisson regression model assumes that the variance of crash frequencies for a given time period is equal to the mean. Conversely, this assumption might not always be supported by the dataset. To overcome this limitation, the negative binomial modeling technique is typically used by adding a quadratic term to the variance in the negative binomial distribution to capture the extra Poisson variation due to variables that are not included in the model (Jovanis and Chang, 1985). Further, the Poisson or negative binomial model may exhibit null crash occurrence. Zero-inflated Poisson, zero-inflated negative binomial models, and zero-state Markov switching count-data models have been developed to account for the zero-crash cases (Long, 1997; Lord et al., 2005, 2007; Malyshkina and Mannering, 2010). The following briefly describes SPFs for predicting fatal, injury, and PDO crashes at an urban street network documented in the 2010 HSM (AASHTO, 2010a,b):

For predicting urban intersection vehicle-to-vehicle crashes, the SPF is of the following specification

$$CF_{v\text{-}to\text{-}v,int} = e^{\alpha_0 + \alpha_1 \cdot \ln(AADT_{major}) + \alpha_2 \cdot \ln(AADT_{minor})} \tag{15.11}$$

where

$CF_{v\text{-}to\text{-}v,int}$ = The vehicle-to-vehicle crash frequency at an urban intersection, crashes/int/year
$AADT_{major}$ = Annual average daily traffic (AADT) on urban intersection major street, veh/day
$AADT_{minor}$ = AADT on the urban intersection minor street, veh/day
α_0, α_1, α_2 = Model coefficients (see Table 15.9)

For predicting urban intersection vehicle-to-pedestrian crashes, the SPF is as follows:

$$CF_{v\text{-}to\text{-}p,int} = e^{\alpha_0 + \alpha_1 \cdot \ln(AADT_{total}) + \alpha_2 \cdot \ln(AADT_{minor}/AADT_{major}) + \alpha_3 \cdot \ln(N_{ped}) + \alpha_4 \cdot C} \tag{15.12}$$

where

$CF_{v\text{-}to\text{-}p,int}$ = The vehicle-to-pedestrian crash frequency for an urban signalized intersection, crashes/int/year
$AADT_{major}$ = AADT on the urban intersection major street, veh/day
$AADT_{minor}$ = AADT on the urban intersection minor street, veh/day
$AADT_{total}$ = The daily total of vehicular traffic entering to the urban intersection, veh/day
N_{ped} = The daily total number of pedestrians crossing all urban intersection approaches, pedestrians/day
C = a constant value taken as 700 for a signalized intersection experiencing a middle level of pedestrian traffic and 1500 for a medium to high level of pedestrian traffic
α_0, α_1, α_2, α_3, α_4 = Model coefficients (see Table 15.9)

For predicting urban street segment vehicle crashes, the SPF is of the following general form

$$CF_{v\text{-}to\text{-}p,seg} = e^{\alpha_0 + \alpha_1 \cdot \ln(AADT) + \alpha_2 \cdot \ln(L)} \tag{15.13}$$

where

$CF_{v\text{-}to\text{-}p,seg}$ = The vehicle-to-vehicle crash frequency for an urban street segment, crashes/seg/year

AADT = AADT on the urban street segment, veh/day

$\alpha_0, \alpha_1, \alpha_2$ = Model coefficients (see Table 15.9)

Table 15.9 summarizes coefficients of SPFs based on the 2010 HSM (AASHTO, 2010a,b) that are employed for the present study. As can be seen in the summary table, the negative binomial approach is calibrated for all models. For multi-vehicle crash predictions, the model coefficients of AADT on intersection major street approaches and street segments are greater than one. For single-vehicle crash and vehicle-to-pedestrian crash predictions, the model coefficients of all other predictors are smaller than or equal to one. This indicates that AADT on the intersection major approaches and on street segments are the most influential toward the vehicle crash potential. Except for the SPF predicting multi-vehicle PDO crashes on urban street segments, the overdispersion factors for all other SPFs are lower than one, ranging from 0.24 to 0.99.

When applying the SPF for predicting the crash frequency at a specific urban intersection or street segment, a crash modification factor (CMF) may need to be employed to modify the SPF predicted crash frequency to account for the impact of any geometric design characteristic or traffic control feature of the study site that differs from the base condition assumed by the SPF. The value of CMF might be greater than, equal to or lower than 1.0 if the aforementioned impacts are associated with a higher, equivalent or lower level of crash frequency compared to the base condition, respectively.

For a typical urban four-leg signalized intersection, the frequency of vehicle-to-vehicle crashes predicted by SPF needs to be adjusted using the CMFs accounting for the number of approaches with left-turn and right-turn lanes, protected phases for left-turn movements, right-turn prohibition, lighting installation, and red-light running photo enforcement. Furthermore, the frequency of vehicle-to-pedestrian crashes at an intersection may also need to be adjusted using CMFs pertinent to the existence of bus stops, schools, and liquor stores adjacent to the intersections that could potentially increase the crashes. Similarly, the factors that could affect single vehicle and multiple vehicle crashes at the urban street segment include on-street parking, median width, and roadside fixed objects for motorized and non-motorized guidance. Table 15.10 presents the CMFs used in this study.

b. *Target crash types affected by signal timing optimization*: While signal timing optimization could potentially influence fatal, injury, and PDO crashes at urban intersections and street segments, it may not necessarily affect all types of crashes. For urban intersections, there are four types of crashes that are more likely to be affected by signal timing optimization: (i) angle; (ii) rear-end; (iii) sideswipe with one or more vehicles in the same or opposite directions; and (iv) head-on crashes. Since urban street segments interconnect intersection approaches, the aforementioned types of crashes on street segments are also likely to be influenced by the treatment. In addition, the single-vehicle fixed-object crash type on urban street segments might be correlated with adjustments of intersection signal timing plans. Thus the above five types of crashes (angle, rear-end, sideswipe, head-on, and single vehicle fixed-object) are treated as target crash types for safety impacts analysis in the present study. The proportions of target crash types might vary for urban intersections and street segments in general, change by intersection or street segment, and also fluctuate over different years at the same intersection or street segment. Thus,

Table 15.9 Coefficients of SPFs for urban intersections and street segments

Urban facility type	Crash type		Vehicle-to-vehicle crashes								Vehicle-to-pedestrian crashes					
			Single vehicle				Multiple vehicles									
	Crash severity	Model type	α_0	α_1	α_2	k	α_0	α_1	α_2	k	α_0	α_1	α_2	α_3	α_4	k
4-leg signalized intersection	Fatal and injury	Negative binomial					−13.14	1.18	0.22	0.33						
	PDO	Negative binomial					−11.02	1.02	0.24	0.44						
	Total	Negative binomial					−10.99	1.07	0.23	0.39	−9.53	0.4	0.26	0.45	0.04	0.24
Street segment	Fatal and injury	Negative binomial	−7.37	0.61	1.0	0.54	−12.08	1.25	1.0	0.99						
	PDO	Negative binomial	−8.50	0.84	1.0	0.97	−12.53	1.38	1.0	1.08						

Table 15.10 Crash modification factors for adjusting crash predictions

Urban facility type		*Design and traffic control features*	*Crash modification factors*	*Sources*
4-Leg signalized intersection	Vehicle-to-vehicle crashes	Approaches with left-turn lanes	0.81	AASHTO (2010a,b)
		Approaches with right-turn lanes	0.92	
		Approaches with right-turn prohibitions	0.96	
		Protected left-turn phasing	0.94	
		Lighting at intersection	0.91	
		Red-light running photo enforcement	0.86–0.98	Lee et al. (2014)
	Vehicle-to-pedestrian crashes	1–2 bus stops within 1,000-ft	2.78	AASHTO (2010a,b)
		Any school within 1,000-ft	1.35	
		1–8 alcohol sales within 1,000-ft	1.12	
Street segment	Vehicle-to-vehicle crashes	Median width	1.01	Harkey et al. (2008)
		On street parking	$1 + p_{pk} \times (f_{pk} - 1.0)$	Bonneson et al. (2005)
		Street-side fixed objects	$f_{offset} \cdot D_{fo} \cdot p_{fo} + 1 - p_{fo}$	Zegeer and Cynecki (1984)

Note:

p_{pk} = Proportion of curb length with on-street parking = $0.5L_{pk}/L$
L_{pk} = Sum of curb length with on-street parking for both sides of the streets combined
L = Length of street segment
f_{pk} = A factor depending upon type of parking (parallel or angle) and land use (commercial or institutional)
f_{offset} = Fixed-object offset factor, 0.0044–0.232
D_{fo} = Fixed object density, objects/mi
p_{fo} = Proportion of fixed-object crashes out of total crashes

the safety impacts of the treatment within an urban street network can be assessed in terms of changes in fatal, injury, and PDO crashes for the target crash types.

c. *Data collection and processing*: Detailed data were collected on vehicle crashes and traffic volumes associated with intersections and street segments of the study area over the period of 2004–2010. Table 15.11 presents the temporal distribution of vehicle crashes by crash severity level and type. For intersections, the total number of crashes fluctuated from 2004 to 2010 with the highest number of crashes recorded in 2004 and the lowest in 2008. Specifically, about 2% are fatal and injury type A, 26% injury type B and C, and 72% PDO crashes. More than 50% of crashes at intersections are angle and rear-end and these two types of crashes roughly take the equal share, approximately 13% are sideswipe, over 1% head-on, and the remaining 34% are other types of crashes. For street segments, the total number of crashes also varied for the period 2004–2010 with the highest number of crashes recorded in 2010 and the lowest in 2008. For single- and multi-vehicle crashes on street segments classified by crash severity level, about 1% are fatal and injury type A, 6% injury type B and C, and 93% PDO crashes. For vehicle crashes on street segments classified by type, more than 3% are angle, 13% rear-end, approximately 8% sideswipe, less than one percent head-on, 5% fixed-object, and close to 71% other

Table 15.11 Distribution of vehicle crashes by severity level and type

Facility type	*Crash distribution*		*2004*	*2005*	*2006*	*2007*	*2008*	*2009*	*2010*	*Average*	
Intersections	Severity level	Fatal	9	9	4	8	2	2	0	5	0%
		Injury A	345	289	263	210	197	245	220	253	2%
		Injury B, C	5082	4508	4315	3379	2902	2669	2390	3606	26%
		PDO	11739	10431	9596	8406	8778	11229	11088	10181	72%
		Total	17175	15237	14178	12003	11879	14145	13698	14045	
	Type	Angle	5070	4436	4187	3290	3038	2883	3024	3704	26%
		Head-on	192	176	150	105	141	185	186	162	1%
		Rear-end	3990	3397	3127	2591	3236	4334	4486	3594	26%
		Sideswipe	2141	1814	1817	1505	1547	1909	2127	1837	13%
		Other	5782	5414	4897	4512	3917	4834	3875	4747	34%
		Total	17175	15237	14178	12003	11879	14145	13698	14045	
Street segments	Severity level	Fatal	0	3	1	1	0	4	4	2	0%
		Injury A	32	40	114	25	38	31	34	45	1%
		Injury B, C	253	251	230	295	306	263	327	275	6%
		PDO	3947	3964	3853	4186	3802	3877	4165	3971	93%
		Total	4232	4258	4198	4507	4146	4175	4530	4292	
	Type	Angle	212	179	167	51	135	123	71	134	3%
		Head-on	27	18	12	12	25	23	11	18	0%
		Rear-end	634	455	465	771	481	458	717	569	13%
		Sideswipe	265	305	323	430	269	280	447	331	8%
		Fixed object	120	149	133	301	212	219	336	210	5%
		Other	2974	3152	3098	2942	3024	3072	2948	3030	71%
		Total	4232	4258	4198	4507	4146	4175	4530	4292	

types of crashes. In general, most of the intersections are 4-leg and each approach maintains two through movement lanes in each direction. The AADT ranges from 5149 to 73938 vehicles daily with an average of 13,880 vehicles per day.

d. *Safety impacts analysis results*

   i. *Safety impacts on urban intersections after signal timing optimization*: Table 15.12 summarizes the average level, SD, and statistical significance of safety impacts in terms of reductions in crashes at urban intersections. The positive value obtained for the average level of safety impacts indicates that a crash reduction is reached after the treatment. The estimated results reveal that, for all weighting scenarios used for calculating vehicle and pedestrian delays in signal timing optimization, vehicle-to-vehicle and vehicle-to-pedestrian crashes at intersections have reduced for all crash severity levels and target crash types. The crash reductions remain fairly stable for different weighting scenarios. For vehicle-to-vehicle crashes, a high extent of crash reductions is achieved for PDO crashes as compared to those of fatal and injury crashes. For fatal and injury crashes combined, reductions are more significant for angle and rear-end crashes at over 12% for each target crash type, followed by sideswipe crashes at slightly over 10% and head-on crashes at nearly 10%. For PDO crashes, a similar reduction trend is discovered. Specifically, crash reductions are more significant for angle and rear-end crashes, at approximately 50%–60% for each target crash type, followed by sideswipe crashes at over 35%, and head-on crashes at about 25%. For vehicle-to-pedestrian crashes, the reduction in fatal and injury crashes is around 18%.

Table 15.12 Safety impacts of signal timing optimization on intersections

	Vehicle-to-vehicle crashes								*Vehicle-to-pedestrian crashes*
	Fatal and injury				*PDO*				*Fatal and injury*
Relative weights of vehicle vs. pedestrian delays (w)	*Angle*	*Head-on*	*Rear-end*	*Side swipe*	*Angle*	*Head-on*	*Rear-end*	*Side swipe*	
	Average level of safety impacts (Eff_{int}, % reductions in crashes)								
100	8.38	3.86	7.89	4.80	57.14	19.32	45.61	30.81	11.99
90	12.84	9.77	12.68	10.07	59.41	25.07	48.84	35.16	17.80
80	12.65	9.89	12.48	9.81	59.38	25.37	48.84	35.11	17.98
70	12.66	9.74	12.54	10.07	59.36	25.15	48.82	35.23	17.90
60	12.66	9.74	12.54	10.08	59.36	25.14	48.82	35.24	17.89
50	12.58	9.65	12.44	9.97	59.32	25.07	48.76	35.16	17.80
40	12.68	9.77	12.55	10.08	59.37	25.18	48.83	35.24	17.93
30	12.54	9.63	12.42	9.95	59.31	25.05	48.75	35.15	17.78
20	12.49	9.58	12.36	9.89	59.29	25.02	48.72	35.11	17.74
10	12.16	9.24	12.00	9.46	59.14	24.76	48.52	34.81	17.51
(w)	*Standard error of the safety impacts ($100 \times \sqrt{Var(Eff_{int})}$, % reductions in crashes)*								
100	4.17	5.01	3.16	3.04	1.94	4.29	1.85	2.21	2.60
90	3.96	4.72	2.99	2.87	1.84	4.01	1.75	2.07	2.44
80	3.97	4.71	3.00	2.88	1.84	4.00	1.75	2.08	2.43
70	3.97	4.71	3.00	2.87	1.84	4.00	1.75	2.07	2.44
60	3.97	4.71	2.99	2.86	1.84	4.00	1.75	2.07	2.44
50	3.97	4.71	3.00	2.87	1.84	4.01	1.75	2.07	2.44
40	3.96	4.72	3.00	2.87	1.84	4.00	1.75	2.07	2.43
30	3.97	4.72	3.00	2.87	1.84	4.01	1.75	2.07	2.44
20	3.98	4.72	3.00	2.88	1.84	4.01	1.75	2.08	2.44
10	3.99	4.74	3.02	2.88	1.85	4.03	1.76	2.08	2.45
(w)	*Statistical significance of safety impacts (Abs[$Test_{int}$])*								
100	2.01	0.77	2.50	1.58	29.42	4.50	24.60	13.97	4.61
90	3.24	2.07	4.24	3.51	32.30	6.25	27.95	16.95	7.30
80	3.19	2.10	4.16	3.41	32.27	6.35	27.96	16.92	7.39
70	3.19	2.07	4.18	3.51	32.24	6.28	27.93	17.01	7.35
60	3.19	2.07	4.19	3.52	32.24	6.28	27.93	17.01	7.34
50	3.17	2.05	4.15	3.47	32.19	6.25	27.86	16.95	7.30
40	3.20	2.07	4.19	3.51	32.26	6.29	27.94	17.01	7.37
30	3.16	2.04	4.14	3.47	32.17	6.25	27.86	16.95	7.29
20	3.14	2.03	4.12	3.44	32.14	6.24	27.82	16.92	7.27
10	3.05	1.95	3.98	3.28	31.94	6.15	27.60	16.70	7.16

The standard errors of safety impacts of all crash severity levels and target crash types are between 2% and 5%. Except for vehicle-to-vehicle crashes corresponding to fatal and injury severity levels and sideswipe type for the scenario of assigning 100% weight to vehicle delays as the basis of intersection signal timing optimization, the test statistics show that intersection safety improvements for all weighting scenarios are statistically significant at the 95% confidence level.

ii. *Safety impacts on urban street segments after signal timing optimization*: Table 15.13 lists the results of safety impacts on street segments. Apart from single vehicle crashes and fixed object PDO crashes for the scenario of assigning 100% weight to vehicle delays, crash reductions are reached for all crash severity levels and target crash types. Similarly, reductions in crashes on street segments have not varied significantly corresponding to different

Table 15.13 Safety impacts of signal timing optimization on street segments

Relative weights of vehicle vs. pedestrian delays (w)	*Single-vehicle crashes*		*Multiple vehicle crashes*							
	Fetal and injury	*PDO*	*Fetal and injury*				*PDO*			
	Fixed objects	*Fixed objects*	*Angle*	*Head-on*	*Rear-end*	*Side swipe*	*Angle*	*Head-on*	*Rear-end*	*Side swipe*
	Average level of safety impacts (Eff_{seg}, % reductions in crashes)									
100	11.60	−1.10	7.20	5.10	61.70	1.60	3.40	0.40	37.90	2.60
90	12.30	3.80	7.40	5.20	63.40	1.60	3.80	0.40	42.80	3.00
80	12.10	3.80	7.30	5.20	63.20	1.60	3.90	0.40	42.90	3.00
70	12.20	3.90	7.30	5.20	63.30	1.60	3.90	0.40	42.90	3.00
60	12.20	3.80	7.30	5.20	63.30	1.60	3.90	0.40	42.90	3.00
50	12.20	3.80	7.30	5.20	63.30	1.60	3.90	0.40	42.80	3.00
40	12.20	3.90	7.30	5.20	63.30	1.60	3.90	0.40	42.90	3.00
30	12.20	3.80	7.30	5.20	63.30	1.60	3.90	0.40	42.80	3.00
20	12.10	3.80	7.30	5.20	63.20	1.60	3.80	0.40	42.80	3.00
10	12.10	3.60	7.30	5.20	63.10	1.60	3.80	0.40	42.60	3.00
(w)	*Standard error of the safety impacts ($100 \times \sqrt{Var(Eff_{seg})}$, % reductions in crashes)*									
100	82.86	55.00	3.85	2.73	32.99	0.86	2.41	0.28	26.88	1.84
90	82.00	54.29	3.52	2.48	30.19	0.76	2.12	0.22	23.91	1.68
80	80.67	54.29	3.48	2.48	30.10	0.76	2.17	0.22	23.83	1.67
70	81.33	55.71	3.48	2.48	30.14	0.76	2.17	0.22	23.83	1.67
60	81.33	54.29	3.48	2.48	30.14	0.76	2.17	0.22	23.83	1.67
50	81.33	54.29	3.49	2.49	30.29	0.77	2.17	0.22	23.78	1.67
40	81.33	55.71	3.48	2.48	30.14	0.76	2.17	0.22	23.83	1.67
30	81.33	54.29	3.49	2.49	30.29	0.77	2.17	0.22	23.78	1.67
20	80.67	54.29	3.49	2.49	30.24	0.77	2.12	0.22	23.91	1.68
10	86.43	51.43	3.53	2.51	30.48	0.77	2.13	0.22	23.93	1.69
(w)	*Statistical significance of safety impacts ($Abs[test_{seg}]$)*									
100	0.14	0.02	1.87	1.87	1.87	1.87	1.41	1.41	1.41	1.41
90	0.15	0.07	2.10	2.10	2.10	2.10	1.79	1.79	1.79	1.79
80	0.15	0.07	2.10	2.10	2.10	2.10	1.80	1.80	1.80	1.80
70	0.15	0.07	2.10	2.10	2.10	2.10	1.80	1.80	1.80	1.80
60	0.15	0.07	2.10	2.10	2.10	2.10	1.80	1.80	1.80	1.80
50	0.15	0.07	2.09	2.09	2.09	2.09	1.80	1.80	1.80	1.80
40	0.15	0.07	2.10	2.10	2.10	2.10	1.80	1.80	1.80	1.80
30	0.15	0.07	2.09	2.09	2.09	2.09	1.80	1.80	1.80	1.80
20	0.15	0.07	2.09	2.09	2.09	2.09	1.79	1.79	1.79	1.79
10	0.14	0.07	2.07	2.07	2.07	2.07	1.78	1.78	1.78	1.78

weighting scenarios utilized for computing vehicle and pedestrian delays in intersection signal timing optimization. For single-vehicle fixed object crashes, reductions are about 12% for fatal and injury crashes and 4% for PDO crashes. For multi-vehicle crashes, reductions are generally higher for fatal and injury crashes than PDO ones. For fatal and injury crashes combined, crash reductions are at about 63% for rear-end crashes, followed by angle crashes at slightly over 7%, head-on crashes at about 5%, and sideswipe crashes at nearly 2%. For PDO, the reduction is most significant for rear-end crashes at about 43%. The crash reductions for remaining target crash types including angle, sideswipe, and head-on are much lower, ranging from 0.4% to nearly 4%.

The standard errors of single-vehicle fixed object crash reductions are at 83% for fatal and injury crashes and at 55% for PDO crashes, respectively. The test statistics indicate that single-vehicle fixed object crashes for all crash severity levels are statistically insignificant for different weighting scenarios. For multi-vehicle crashes, reductions in fatal and injury for all target crash types are found to be statistically significant for all weighting scenarios. However, reductions in PDO for all target crash types are statistically insignificant for the weighting scenario of assigning a weight of 100% to vehicle delays in intersection signal timing optimization.

Source: Adapted from Roshandeh, A.M. et al. 2016. *Elsevier Journal of Traffic and Transportation Engineering* 3(1), 16–27.

PROBLEMS

15.1 Explain what "regression-to-mean" biases are in highway safety impacts analysis. Explain why the EB comparison group before-after analysis method could address such biases.

15.2 Design a detailed analytical procedure for applying the EB comparison group before and after analysis method to evaluate safety impacts of intersection red light running photo enforcement where two types of treatments are classified:

a. Type I treatment sites with intersection signal modernization and red light running photo enforcement;

b. Type II treatment sites with intersection signal modernization only.

The difference in safety impacts of the above two types of treatments is the safety impacts of intersection red light running photo enforcement.

15.3 Design a detailed analytical procedure for applying the EB comparison group before and after analysis method to evaluate safety impacts of shoulder paving where two types of treatments are classified:

a. Type I treatment sites with pavement resurfacing and shoulder paving;

b. Type II treatment sites with pavement resurfacing only.

The difference in safety impacts of the above two types of treatments is the safety impacts of shoulder paving.

15.4 The shoulder paving can be further split into three cases: (a) re-paving the same width of the paved shoulder; (b) widening the paved width of the paved shoulder; and (c) adding a new paved shoulder. Design a refined analytical procedure for applying the EB comparison group before and after analysis method to evaluate safety impacts of shoulder paving as the difference in safety impacts of the two types of treatments.

Chapter 16

Transportation environmental impacts analysis

With the enactment of legislation such as Intermodal Surface Transportation Efficiency Act (ISTEA) of 1991 and Transportation Equity Act for the 21st Century (TEA-21) of 1998, transportation decision-making in the U.S. needs to be preceded by an investigation of the environmental impacts of the proposed intervention. The same is true for other transportation modes in the U.S. and transportation projects in polluted cities in certain developing countries, particularly when multilateral agencies such as the World Bank fund such projects.

An agency may generally seek to evaluate either the existing air quality situation at a given time (with no intent of any transportation intervention) or the estimated or actual air quality impact after a transportation intervention. In the case of an intervention, air quality impacts are measured in terms of the resulting concentrations of selected air pollutants after the implementation of the project or policy vis-à-vis projected base case concentrations (assuming the intervention does not take place).

16.1 AIR QUALITY IMPACTS

16.1.1 Pollutant types, sources, and effects

Air pollutants can be categorized into primary and secondary pollutants. Primary air pollutants are emitted directly into the atmosphere from sources. They can have effects both directly and as precursors of secondary air pollutants. Secondary air pollutants are produced in the air by the interaction of two or more primary pollutants or by reaction with normal atmospheric constituents, with or without photoactivation. Examples of secondary air pollutants are ozone, formaldehyde, and acid mist.

Natural sources of air pollution include forest fires and volcanoes; anthropogenic sources include power generation, fuel use, slash-and-burn agricultural practices, and transportation. The main air pollutants from transportation sources are listed as follows:

Carbon Monoxide (CO): CO is a colorless, odorless toxic gas formed by incomplete combustion of fossil fuels. The greatest sources of CO to outdoor air are cars, trucks, and other vehicles or machinery that burn fossil fuels. Breathing air with a high concentration of CO reduces the amount of oxygen that can be transported in the blood stream to critical organs like the heart and brain. These organs are especially vulnerable to the effects of CO when exercising or under increased stress. In these situations, short-term exposure to elevated CO may result in reduced oxygen to the heart accompanied by chest pain also known as angina.

Carbon Dioxide (CO_2): CO_2 enters the atmosphere through burning fossil fuels (coal, natural gas, and oil), solid waste, trees, and wood products, as well as a result of certain chemical reactions. The major environmental impact of CO_2 is the greenhouse effect, which traps more heat and causes the planet to become warmer that it would be naturally.

Ozone (O_3): Ozone is a gas composed of three atoms of oxygen (O_3). O_3 occurs both in the Earth's upper atmosphere and at ground level. O_3 can be good or bad, depending on where it is found. Good O_3, also called stratospheric O_3, occurs naturally in the upper atmosphere, where it forms a protective layer that shields us from the sun's harmful ultraviolet rays. Bad O_3, or ground level O_3, is not emitted directly into the air but is created by chemical reactions between oxides of nitrogen (NO_X) and volatile organic compounds (VOC). This happens when pollutants emitted by cars, power plants, industrial boilers, refineries, chemical plants, and other sources chemically react in the presence of sunlight. Breathing O_3 can trigger a variety of health problems including chest pain, coughing, throat irritation, and airway inflammation. It also can reduce lung function and harm lung tissue. O_3 affects sensitive vegetation and ecosystems, including forests, parks, wildlife refuges, and wilderness areas. Particularly, O_3 harms sensitive vegetation during the growing season.

Chlorofluorocarbons (CFCs): CFCs are nontoxic, nonflammable chemicals containing atoms of carbon, chlorine, and fluorine. CFCs have no natural source but were entirely synthesized for such diverse uses as refrigerants, aerosol propellants, and cleaning solvents. They can destroy the stratospheric O_3 layer, thus reducing the protection it offers the Earth from the sun's harmful ultraviolet (UV) rays. CFCs also contribute to global warming as greenhouse gases, although the amounts emitted are relatively small.

Sulfur Oxides (SO_2): SO_2 is a colorless water-soluble pungent and irritating gas. All combustion processes of products containing sulfur yield SO_2 emissions. The transportation share is essentially contributed by engines running on diesel and fossil fuels. Short-term exposures to SO_2 can harm the human respiratory system and make breathing difficult. At high concentrations, gaseous SO_X can harm trees and plants by damaging foliage and decreasing growth. It is also the major factor responsible for acid rain.

Nitrogen Oxides (NO_X): Nitrogen oxides are a group of gases that is composed of nitrogen and oxygen. Two of the most common nitrogen oxides are nitric oxide (N_2O) and nitrogen dioxide (NO_2). NO_X are released into the air from motor vehicle exhaust or burning diesel fuel. Longer exposures to NO_X in smog can trigger serious respiratory problems, including damage to lung tissue and reduction in lung function. It can cause collapse, rapid burning, and swelling of tissues in the throat and upper respiratory tract, difficult breathing, throat spasms, and fluid build-up in the lungs. NO_2 and other NO_X interact with water, oxygen, and other chemicals in the atmosphere to form acid rain. Acid rain harms sensitive ecosystems such as lakes and forests.

Particle Matters ($PM_{2.5}/PM_{10}$): Particle matter (PM) is a mixture of solids and liquid droplets floating in the air. Some particles are released directly from a specific source, while others form in complicated chemical reactions in the atmosphere. Particles come in a wide range of sizes. $PM_{2.5}$ are 2.5 μm in diameter or smaller, and can only be seen with an electron microscope. $PM_{2.5}$ are produced from all types of combustion, including motor vehicles, diesel engines, and other sources. PM_{10} are 2.5–10 μm in diameter. Sources include crushing or grinding operations, dust stirred up by vehicles on roads, and toxic dust that accumulates on roadways from degrading of tires and especially brake pads. These particles less than or equal to 10 μm in diameter are so small that they can get into the lungs, potentially causing serious health problems. Numerous scientific studies connect particle pollution exposure to a variety of health issues, including irritation of the eyes, nose, and throat; coughing; chest tightness and shortness of breath; reduced lung function; and so forth.

Lead: Sources of lead emissions vary from one area to another. In the transportation industry, the major sources of lead emitted into the air are vehicle and aircraft engines operating on leaded fuel. Other sources are waste lead-acid batteries used in vehicles. Lead can adversely affect the nervous system, kidney function, immune system, reproductive and developmental systems, and the cardiovascular system. Lead exposure also affects the oxygen-carrying capacity of the blood. The lead effects most commonly encountered in current populations are neurological effects in children and cardiovascular effects. Elevated lead in the environment can result in decreased growth and reproductive rates in plants and animals, and neurological effects in vertebrates.

16.1.2 Classification of emission models

Emission estimation models used by transportation sectors can be classified as shown in Figure 16.1. This classification standard is based on the level of detailed factors considered as the input of the model inputs (Elkafoury et al., 2014).

Static (top–down) models consider aggregated parameters such as the average speed of vehicles and average emission factor as inputs, regardless of detailed activities and data of emission source. Static models can be further categorized into average speed models and aggregated emission factor models. Average speed models assume that average emissions over a trip vary against the overall average speed of the trip. It cannot estimate detailed emission variation as the average speed of the trip fails to describe the instantaneous speed of a certain investigated vehicle over the trip. Moreover, the average speed models fail to differentiate the cases that have different driving behaviors and vehicle dynamics but have the same average speed. The aggregate emission models use a single emission factor to represent a variety of vehicles and a general driving condition, without considering the power source and specifications of vehicles. Aggregated emission factor models generate emission rates for a large spatial area based on vehicle kilometers of traveled (VKT) or VMT or amount of fuel consumed. Generally, static models are usually used in modeling emissions of large network level transportation planning projects. Static models have advantages including simplicity and ease of application, but their major problem is the lack of accuracy.

Dynamic (bottom–up) models include more microscopic variables, such as the traffic situation and instantaneous speeds of vehicles. These models are created according to more detailed and micro-level data, including the instantaneous status of vehicle dynamics and fuel combustion rate. Therefore, this dynamic modeling has better accuracy performance in evaluating emission dynamically. Similarly, dynamic models can be further classified into traffic situation models and instantaneous models. Traffic situation models consider aggregated traffic flow dynamics on different types of roadway segment in the emission

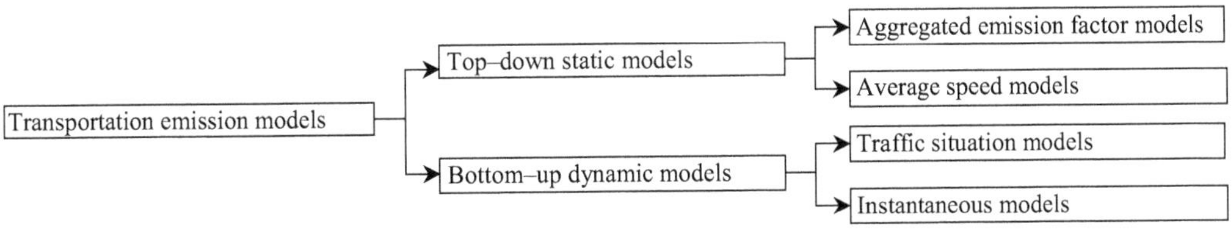

Figure 16.1 Classification of transportation emission models.

Table 16.1 Classification of transportation emission models

Examples	*Mobile (2000)*	*Moves (2010)*	*CMEM (2006)*	*COPERT 4 (2012)*	*IVE (2008)*
Type	Static and average speed	Static and modal	Dynamic and instantaneous	Static and modal	Static and modal
Emission factor unit	g/mile	g/s, g/mile, g/km	g/s	g/s, g/mile, g/km	g/s, g/mile, g/km
Inputs					
Speed-time profile			√		
Modal variables		√		√	√
Cruise speed		√		√	√
Road type	√	√		√	√
Link flow		√		√	√
Number of lanes		√		√	√
Link length		√		√	√
Average speed	√	√		√	√
Climatic conditions		√		√	√
Inspection and maintenance		√		√	√
Detailed traffic composition	√	√	√	√	√
Link VKT/VMT	√	√	√	√	√
Road grade		√		√	√

estimation models. Traffic situation involves road types, traffic flow conditions, vehicle specifications, and vehicle speed changes. Instantaneous models are proposed to model vehicle emissions over a very short time period related to vehicle operation at that moment. In these models, power-based approach uses the engine power to estimate the emission rate, which is considered as the result of the instantaneous speed and acceleration rate of a vehicle. Dynamic models are usually incorporated into large-scale traffic simulation models in order to obtain network-wide emissions. Dynamic models, compared with static models, are more accurate concerning detailed inputs. Nevertheless, calibrated dynamic models could be only applied in a specific area. It is unlikely that multiple regions could share one calibrated dynamic model.

Table 16.1 summarizes a comparison of existing popular emission models in terms of type, emission factor unit, and inputs (Szeto et al., 2012). The static models include Motor Vehicle Emission Simulator (MOVES), Computer Programme to calculate Emissions from Road Transport (COPERT), and International Vehicle Emissions (IVE) models. The dynamic emission model only has Comprehensive Modal Emission Model (CMEM). Selection of a model should rely on the study objective, data availability, and the desired level of accuracy.

16.1.3 Modeling monetary costs of air emissions

There are at least three ways to infer the costs of air emissions: (i) direct estimation of damage, which traces the links between air emissions and adverse consequences, and attempts to place economic values on those consequences; (ii) willingness-to-pay method, which builds up relations between the price that the public are willing to pay and the resulting air quality improvement; and (iii) estimation of the cost of recovering the air quality as desired, where pollution costs are inferred from the costs of meeting pollution regulations.

16.1.3.1 Direct damage estimation

In the method of direct damage estimation, several links in the causal chain must be separately measured (Hall and Winer, 1992). A pollutant released into the air changes the ambient concentrations of that pollutant spatially and temporally, probably changing the concentrations of other chemicals. The spatial and temporal distribution of a pollutant is determined by a variety of factors including atmospheric characteristics, geographical features, and the ambient concentration of other chemicals. The consequential ambient concentrations then affect human beings, animals, plants, and infrastructure depending on the pollutant type and concentrations. As described in the previous section, each kind of pollutant could interact with human beings, animals, and infrastructures in multiple ways. However, most studies focus on the impact of human health deterioration. The results may be physical and/or psychological effects: coughing, erosion of stone, retarded plant growth, injury to young, loss of pleasurable views, and so forth. Finally, these effects could be quantified into economic values, which are considered as the emission costs.

16.1.3.2 Willingness-to-pay method

It is widely acknowledged that the social cost of varying economic outcomes can be practically measured by adding together individuals' willingness to pay (WTP) for that variation within given economic circumstances. In this case, the cost of emission is estimated by assessing the extent to which people's living quality is affected and how much individuals are willing to pay to avoid this problem. The underlying assumption is that the public is consistently aware of a valuations system and how the price they pay will change the adverse impact of air pollution. For example, a research objective could be to determine how much people would be willing to pay to reduce the risk and severity of lung diseases caused by air pollution, assuming full information and clear choices are provided. There are many approaches in which we can observe the quantity of WTP, mostly in the labor market, where there are jobs with changing risks of exposure to pollutants and compensating wage differentials.

16.1.3.3 Cost of air quality recovery

Cleaning up air pollution involves the installation of air scrubbers. The costs of cleaning the air are estimated based on the total cost of planning, purchasing, installing, operating, and maintaining air scrubbers. In the context of transportation, air scrubbers are installed along the polluting line source. Many factors would affect the air scrubber installation intervals, including the traffic characteristics (e.g., traffic volume, heavy vehicle percentage, average speed, etc.), the environment traffic characteristics (e.g., temperature, humidity, wind speed, wind direction, etc.), and the scrubber specifications (e.g., capacity, etc.). Air pollution costs estimated in this manner can be rather more excessive than in other approaches.

16.1.4 Air emission cost values

16.1.4.1 Earlier efforts

Several studies have studied the effects of conventional air emissions on human health and the associated costs (Matthews and Lave, 2000). These studies determine the quantitative health effects on humans by using the medical and lost work cost caused by air emissions. Table 16.2 summarizes several estimates of the social cost of sulfur dioxide emissions from several earlier researches using this approach (Matthews et al., 2001). Further, the costs are converted to 1992 constant dollars.

Table 16.2 Estimates of external costs of SO_2 emissions (dollars)

Study	*Valuation/ton* SO_2
Zuckerman and Ackerman (1995)	4700
CEC (1993)	3700
BPA (1987)	2350
EPRI (1990)	2080
NPSC (1993)	1800
Elkins (1985)	1760
CEC (1989)	1300
Cifuentes and Lave (1993)	770
Rowe et al. (1995)	940
Repetto (1990)	850
[Median of Studies]	[1800]

Note: CEC, California Energy Commission; BPA, Bonneville Power Administration; EPRI, Electric Power Research Institute; NPSC, Public Service Commission of Nevada

Matthews and Lave (2000) have compiled similar estimate ranges for the other conventional pollutants and has adjusted the figures to 1992 dollars as shown in Table 16.3.

Note that vehicles could produce other greenhouse gases besides CO_2, such as methane (CH_4) and nitrous oxide (N_2O) from the tailpipe. Moreover, some hydro fluorocarbon (HFC) will leak from air conditioners. These gases have higher global warming potential (GWP) than CO_2 although the amount of these gases is smaller. In order to describe the global warming effects of greenhouse gases, one standardized metric system should be designed, which is CO_2 equivalent as shown in Table 16.3. Another important factor mentioned is GWP, which is multiplied by the amount of the gas to obtain the corresponding CO_2 equivalent. For vehicle-related greenhouse gases, these GWP are summarized in Table 16.4. Unlike CO_2, emissions of CH_4 and N_2O depend on vehicle miles of travel rather than per mile fuel consumption. The amount of leaked HFC from vehicular air conditioners is even more unpredictable. On average, in terms of GWPs, CO_2 emissions account for 95%–99% of the total greenhouse gas emissions from a passenger vehicle and CH_4, N_2O, and HFC emissions account for the rest.

16.1.4.2 FHWA HERS-ST valuation

The Highway Economic Requirements System-State Version (HERS-ST) is a software package developed by FHWA that facilitates decision-making. HERS-ST can predict the investment needed if certain highway system performance levels are to be achieved. On the one

Table 16.3 Unit social damage estimates for some air emissions (1992 Dollars)

		External costs ($/ton of air emissions)			
Species	*Number of studies*	*Minimum*	*Median*	*Mean*	*Maximum*
CO	2	1	520	520	1050
GWP (in CO_2 equivalents)	4	2	14	13	23
NO_X	9	220	1060	2800	9500
SO_2	10	770	1800	2000	4700
PM	12	950	2800	4300	16200
VOCs	5	160	1400	1600	4400

Table 16.4 Global warming potentials for vehicle-related greenhouse gases

Greenhouse gas	*Abbreviation*	*GWP*
Methane	CH_4	25
Carbon dioxide	CO_2	1
Nitrous oxide	N_2O	298
Hydro fluorocarbon	HFC-134a	1430

hand, HERS-ST can assess the highway system performance resulting from a certain investment level. HERS-ST is also capable of calculating the monetary cost of damages caused by air pollution produced by motor vehicles.

Damage costs for individual air pollutants: As implied by the name, the estimated costs are based on the damage to human health and property per unit of each pollutant. The source of these cost values is a widely cited study (McCubbin and Delucchi, 1996). The cost value of a certain pollutant is derived by dividing the estimated total annual damage costs resulting from highway vehicles in the atmosphere by the total amount of that pollutant released annually by highway vehicles. Therefore, these values are nationwide average damage costs. As shown in Table 16.5, these values are used to simply represent rough estimates of the marginal cost in total health and property as the corresponding pollutant varied by one ton (Lee and Burris, 2005).

Adjustments to per-ton damage cost: Intuitively, an area with a different population density should have a different damage cost even if the concentration of the pollutants is the same. However, the damage costs presented in Table 16.5 give average levels, which fail to reflect the spatial variation of the cost values. Therefore, an adjustment factor is introduced to reflect local variation in damage costs caused by the variation in population and property exposed to air pollutions. Whether the study area is in an urban or rural area is a simple and effective measure to differentiate the population and property densities. Moreover, securing this information is quite easy, which makes it more practical when other more detailed information is not presented.

Instinctively, higher population and property density should lead to a higher damage cost value than average due to the higher number of people and properties that can be hurt by air pollutants. This is not necessarily true. Note that three pollutants (CO, PM, and road dust) tend to remain national average values for urban areas and be scaled downward for rural areas. In contrast, the other three more widely dispersed pollutants (O_3, NO_X, and SO_2) tend to be scaled upward from their national averages for urban areas and remain the national averages for rural areas.

Table 16.5 Air pollutant damage costs and adjustment factors used in HERS (2000 dollars)

		Adjustment factor	
Pollutant	*Damage costs ($/kg)*	*Urban*	*Rural*
Carbon monoxide	0.11	1	0.5
Nitrogen oxides	4.00	1.5	1
Sulfur dioxide	9.26	1.5	1
Fine particulate matter ($PM_{2.5}$)	5.32	1	0.5
Volatile organic compounds	3.03	1.5	1
Road dust	5.32	1	0.5

Air pollution costs per vehicle-mile: These damage costs per ton (dollars/ton) of each pollutant were multiplied by the emission rate expressed in the unit of ton per mile (ton/mile) by vehicle class and speed for each HERS section, with the per-ton damage cost for each pollutant adjusted using its urban or rural location. The resulting values for each pollutant were then summed to determine total air pollution damage costs per VMT for each highway functional classification (Interstate, principal arterial, minor arterial, major collector, other freeway/expressway, and collector) and land area (urban and rural), vehicle type (four-tire vehicle, single-unit truck, and combination truck), and average speed at 5-mph intervals. Interpolation could be used to find any damage costs for other speeds. Tables 16.6 and 16.7 give examples of such cost values for different highway functional classifications (Tables 16.8 and 16.9).

Table 16.6 Emission damage costs associated with rural Interstate highways

	Emission damage cost (2000 $ per vehicle-mi)		
Speed (mph)	*Four-tire vehicles*	*Single-unit trucks*	*Combination trucks*
5	0.03283	0.05347	0.14932
10	0.02140	0.04081	0.13186
15	0.01740	0.03529	0.12004
20	0.01613	0.03204	0.11209
25	0.01459	0.03031	0.10701
30	0.01537	0.02941	0.10419
35	0.01506	0.02907	0.10333
40	0.01505	0.02915	0.10432
45	0.01508	0.02959	0.10727
50	0.01512	0.03041	0.11254
55	0.01519	0.03163	0.12072
60	0.01529	0.03336	0.13287
65	0.01540	0.03581	0.15068
70	0.01550	0.03820	0.16855

Table 16.7 Emission damage costs associated with rural other principal arterials

	Emission damage cost (2000 $ per vehicle-mi)		
Speed (mph)	*Four-tire vehicles*	*Single-unit trucks*	*Combination trucks*
5	0.03584	0.05605	0.12750
10	0.02338	0.04224	0.10851
15	0.01935	0.03617	0.09577
20	0.01713	0.03257	0.08717
25	0.01602	0.03066	0.08169
30	0.01533	0.02964	0.07865
35	0.01490	0.02927	0.07771
40	0.01489	0.02936	0.08181
45	0.01492	0.02986	0.08196
50	0.01497	0.03077	0.08762
55	0.01505	0.03215	0.09643
60	0.01515	0.03413	0.10952
65	0.01520	0.03693	0.12868
70	0.01524	0.03967	0.14825

Table 16.8 Emission damage costs associated with urban Interstate highways

	Emission damage cost (2000 $ per vehicle-mi)		
Speed (mph)	*Four-tire vehicles*	*Single-unit trucks*	*Combination trucks*
5	0.02562	0.04511	0.14771
10	0.01646	0.03578	0.13023
15	0.01353	0.03124	0.11871
20	0.01291	0.02861	0.11083
25	0.01256	0.02714	0.10579
30	0.01231	0.02641	0.10301
35	0.01209	0.02619	0.10215
40	0.01209	0.02637	0.10314
45	0.01216	0.02689	0.10607
50	0.01222	0.02775	0.11128
55	0.01229	0.02901	0.11940
60	0.01239	0.03077	0.13144
65	0.01249	0.03322	0.14908
70	0.01258	0.03563	0.16677

The purpose of the legislation is to protect public health from different types of air pollution caused by various pollution sources including transportation. In the U.S., the major regulations regarding air quality are based on the Clean Air Act (CAA), which was enacted in 1970 and was revised in 1977 and 1990. Depending on the law, the U.S. Environmental Protection Agency (EPA) has nationally established air quality standards. In terms of the transportation planning in an urban system, states and planners are required to follow the standards. Furthermore, the federal government has significant authority and responsibility for monitoring and controlling air pollutants.

Table 16.9 Emission damage costs associated with urban other freeways/expressways

	Emission damage cost (2000 $ per vehicle-mi)		
Speed (mph)	*Four-tire vehicles*	*Single-unit trucks*	*Combination trucks*
5	0.02550	0.04503	0.14537
10	0.01637	0.03571	0.12802
15	0.01346	0.03118	0.11660
20	0.01284	0.02856	0.10880
25	0.01249	0.02710	0.10383
30	0.01224	0.02637	0.10107
35	0.01202	0.02616	0.10022
40	0.01203	0.02633	0.10119
45	0.01209	0.02685	0.10407
50	0.01215	0.02771	0.10920
55	0.01223	0.02896	0.11719
60	0.01232	0.03071	0.12904
65	0.01242	0.03314	0.14639
70	0.01251	0.03553	0.16380

Table 16.10 Some major U.S. legislative and regulatory acts for air quality control

Year	*Act*	*Description*
1955	The Air Pollution Control Act	This act is the first federal legislation regarding air pollution in the U.S.
1963	The Clean Air Act	This act is a federal program related to air pollution control for Public Health Service and authorized research into techniques for monitoring and controlling air pollution
1967	The Air Quality Act	This act was established in order to expand federal government activities for monitoring and controlling studies and stationary source inspections Enforcement proceedings were initiated in the area involving Interstate air pollution
1970	Clean Air	This act authorized the development of comprehensive federal and state regulations to limit emissions from both industrial sources and mobile sources. In order to implement the requirements of this act, the U.S. EPA was established
1977	Clean Air Act Amendments	This act additionally contained provisions for the Prevention of Significant Deterioration (PSD) of air quality
1990	Clean Air Act Amendments	This act increased the authority and responsibility of the federal government for regulation and control of air pollutants and authorized new regulatory programs regarding control of acid deposition (acid rain) and O_3 protection
1991	Intermodal Surface Transportation Efficiency Act	This act requires that transportation plans conform to air quality enhancement initiatives and provided state and local governments with the funding and flexibility to improve air quality through the development of a balanced, environmentally sound, intermodal transportation program
2005	SAFETEA-LU Act	This act improves the air quality conformity process in the frequency of conformity determinations and conformity horizons

16.1.5 Regulations and legislations

The Air Pollution Control Act of 1955 was the first act in the history of federal legislation with respect to the air quality impacts of transportation. In 1963, the CAA was passed (subsequently amended in 1965 and several times later) to update and improve emission standards. The Air Quality Control Act of 1967 initiated the establishment of air quality criteria. The CAA amendments of 1970 made the federal government involved in controls in individual states for regulating and reducing motor vehicle and aircraft emissions. In a 1977 amendment to the CAA, penalties were explicitly established for areas that failed to meet air quality standards. The 1990 CAA strengthened conformity requirements. The ISTEA in 1991 reinforced the CAA90 requirement. In the Safe, Accountable, Flexible, Efficient Transportation Equity Act: A Legacy for Users (SAFETEA-LU) of 2005, the air quality conformity process was further improved. Table 16.10 gives an overview of these acts.

16.1.6 U.S. air quality standards

The CAA is a U.S. federal law designed to control air pollution on a national level. This act, after the latest update in 1990, leads the U.S. EPA to design National Ambient Air Quality Standards (NAAQS) for each pollutant that is considered deteriorating to human health and

Table 16.11 National ambient air quality standards

Pollutant		*Primary/ secondary*	*Average time*	*Level*	*Form*
Carbon monoxide (CO)		Primary	8 hours	9 ppm	Not to be exceeded more than once per year
			1 hour	35 ppm	
Lead		Primary and secondary	Rolling 3 month average	0.15 μg/m$^{3(1)}$	Not to be exceeded
Nitrogen dioxide (NO_2)		Primary	1 hour	100 ppb	98th percentile of 1-hour daily maximum concentrations averaged over 3 years
		Primary and secondary	1 year	53 ppb$^{(2)}$	Annual mean
Ozone (O_3)		Primary and secondary	8 hours	0.070 ppm$^{(3)}$	Annual fourth-highest daily maximum 8-hour concentration averaged over 3 years
Particle pollution (PM)	$PM_{2.5}$	Primary	1 year	12.0 μg/m^3	Annual mean, averaged over 3 years
		Secondary	1 year	15.0 μg/m^3	Annual mean averaged over 3 years
		Primary and secondary	24 hours	35 μg/m^3	98th percentile averaged over 3 years
	PM_{10}	Primary and secondary	24 hours	150 μg/m^3	Not to be exceeded more than once per year on average over 3 years
Sulfur dioxide (SO_2)		Primary	1 hour	75 ppb$^{(4)}$	99th percentile of 1-hour daily maximum concentrations averaged over 3 years
		Secondary	3 hours	0.5 ppm	Not to be exceeded more than once per year

the environment. Two types of NAAQS with different objectives are identified in the CAA, where primary standards aim to protect public health, especially in protecting populations who are sensitive to air quality, such as persons with disabilities, pregnant women, children, and the elderly. Secondary standards aim to protect public welfare, protecting animals, crops, vegetation, and buildings.

The NAAQS are established for six major conventional pollutants, also known as "criteria" air pollutants. Moreover, the standards are updated periodically. The latest NAAQS are listed in Table 16.11 (EPA, 2016). Units involved are parts per million (ppm) by volume, parts per billion (ppb) by volume, and micrograms per cubic meter of air (μg/m^3).

EXAMPLE 16.1

A state transportation agency plans to update the signal timing plan in a small area. The major objective is to improve the mobility performance within the study area. But the agency also wants to know the environmental benefit/cost after updating the signal timing. The volume data before changing the signals were collected and integrated on an hourly basis. The field data is then put into a traffic simulation system to further generate mobility and emission data to be analyzed. The simulated data is summarized in Table 16.12.

Table 16.12 Data used in emission benefit/cost analysis

	Unit cost	*Before*	*After*
VMT	–	43560	49837
CO	$520/ton	5137 g	4133 g
NO_2	$1060/ton	8256 g	6952 g
CO_2	$2/ton	18426 kg	20964 kg
PM10	$2800/ton	203 g	215 g

Taking CO as an example, the total costs of CO before and after implementing the new signal timing are computed as follows:

$$Cost(\text{before}) = 520 \times 5137 \times 10^{-6} = \$2.67$$

$$Cost(\text{after}) = 520 \times 4133 \times 10^{-6} = \$2.15$$

The per VMT costs before and after signal timing update are then computed.

$$UCost(\text{before}) = \frac{\$2.67}{43560} = \frac{\$0.000061295}{VMT}$$

$$UCost(\text{after}) = \frac{\$2.15}{49837} = \frac{\$0.000043141}{VMT}$$

The benefit comes from the reduction of emission cost in CO. The difference of consumer surplus can be used as the benefit of CO emission reduction as illustrated in Figure 16.2. The supply curve shifts because of the signal timing update. The equilibrium state changes

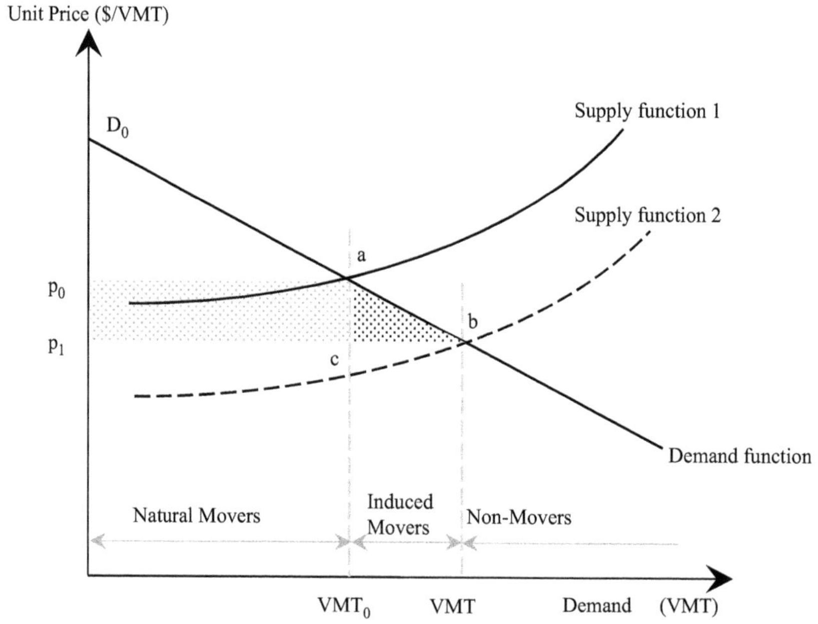

Figure 16.2 Illustration of consumer surplus and user benefits.

from points a to b with the unit price changing from p_0 to p_1. The area of the shaded trapezoid is the user benefit to be computed, where the triangle abc is the user benefit of induced movers and the remaining area represents the user benefit of existing movers.

$$Benefit(\mathrm{CO}) = \left[\frac{\$0.000061295 - \$0.000043141}{VMT}\right] \times (43560 + 49837) \times 0.5$$

$$Benefit(\mathrm{CO}) = \$0.88$$

Similarly, the benefit of other emissions reduction can be computed using the same approach.

$$Benefit(\mathrm{NO_2}) = \frac{1060}{10^6} \times \left(\frac{8256}{43560} - \frac{6952}{49837}\right) \times 0.5 \times (43560 + 49837) = \$2.48$$

$$Benefit(\mathrm{CO_2}) = \frac{2}{10^3} \times \left(\frac{18426}{43560} - \frac{20964}{49837}\right) \times 0.5 \times (43560 + 49837) = \$0.22$$

$$Benefit(\mathrm{PM_{10}}) = \frac{2800}{10^6} \times \left(\frac{203}{43560} - \frac{215}{49837}\right) \times 0.5 \times (43560 + 49837) = \$0.05$$

If more detailed data is provided, one could use this approach to compute the potential benefit/loss for each segment and intersection within each short time interval within a day, ending by adding together all values to obtain the total benefit or loss. Note that, for intersections, the demand is expressed in terms of the number of entering vehicles rather than VMT. The unit price also needs to change to per entering vehicle cost.

16.2 NOISE IMPACTS

Today, noise pollution is considered a serious problem that hurts human health physically and mentally. The annoyance caused by noise has been understood for a long time. However, it is quite recent that the transportation decision-making process has environmental factors involved. In fact, the transportation industry is a major source of noise in our daily lives. Specifically, the noise pollution problem is worsened as the number of vehicles driving on roads continues to grow.

16.2.1 Sources of highway traffic noise

Compared to other adverse environmental externalities of transportation operations, noise affects the most people. It has been estimated that over 18 million people are affected every year in the U.S., and 100 million worldwide. Excessive noise associated with transportation systems leads to a reduction in real-estate values, constitutes a nuisance, and in extreme cases, can result in health problems. It is important to understand the nature of noise impacts as well as the required analytical techniques to assess the level of expected noise due to transportation system operations. Another unique feature of noise pollution is its spontaneity and lack of duration (Cohen and McVoy, 1982). That is, noise generated at a particular time is not affected by prior activity, nor does it affect future activities, and unlike other pollutants, noise leaves no residual effects that are evidential of its unpleasantness. For this reason, there is a tendency to overlook or to underestimate the problem of noise pollution. The sources of noise generated from a moving vehicle are as follows:

Vehicle/air interaction: When a vehicle is in motion, friction between the body of the vehicle and the surrounding air induces a gradient in the air pressure field and thereby generates noise. The more aerodynamic the vehicle, the less the drag and consequently the lower the level of noise generated in such manner.

Tire/pavement interaction: Tire/pavement generation is a direct result of the friction and small impacts that occur as the tire rolls along the pavement surface. Such noise is more pronounced for concrete pavements and less for asphalt concrete pavements. Also, pavement surface grooving may increase or decrease noise generation depending on the direction of grooving. The condition of the tire and pavement also affect traffic noise that is generated between the vehicle tire and pavement.

Vehicle engines: As vehicles travel along highways, the sound generated by the operation of their engines is also a major source of highway noise. Such noise levels are particularly significant in areas where vehicles tend to accelerate or decelerate. Also, higher speeds also translate to higher noise levels.

Furthermore, the operation of larger vehicles such as trucks and higher speeds are generally associated with higher noise levels.

Vehicle exhaust systems: Faulty vehicle exhaust systems lead to higher noise levels, especially in cases of malfunctioning mufflers. The incidence of noise from this source is often closely related to that from vehicle engines: high speeds, speed changes, and large vehicles are associated with higher levels of exhaust noise.

Brakes: Braking also constitutes a significant noise source, particularly for large trucks. Noise generated from all highway traffic propagates over distances and has an acoustic spectrum that typically ranges from 250 to 4000 Hz. As this frequency range is discernible by the human ear, highway noise causes great discomfort to humans, and the mitigation of such impacts is therefore sought through the use of regulations (such as decreased speed limits), construction practice (such as blacktopping existing concrete pavements), and direct intervention (such as constructing barriers to shield adjacent properties from highway traffic noise). As discussed in the next section, there are certain aspects of the weather that impede or enhance the propagation of sound from the highway to neighboring properties.

16.2.2 Mitigation of transportation noise

It has been proven that the noise problem has three basic elements: a sound source, a transmission path, and a receiver. These three elements lead the solution in three different directions. The most desirable way to minimize exposure to noise is to manage the source of noise if applicable. Promoting the use of quiet tires, electronic vehicles, and appropriate pavement material are noise-source-oriented approaches. But in most cases, the most corrective countermeasure is to control the noise path. Noise wall installation is a typical example of this noise-path-oriented approach. The receiver-oriented approaches tend to solve this problem at the policy level, where a maximum value is assigned to each land use activity category, such as residences, schools, churches, and so on.

In general, noise results from the friction of tires at high speed and is also emitted from the engine during acceleration for heavy trucks. Therefore, pavement-type selection can reduce tire and pavement interaction and lessen the noise impacts. Also, insulation in the engine compartment is used and mufflers are used for exhaust noise to reduce engine noise. For example, the use of contra-noise can reduce noise by generating a wave form of equal amplitude but in opposite phase to the cancelled engine or exhaust noise. The techniques utilize speakers used in engine compartments to cancel engine noise or electronic mufflers for exhaust noise. The tire/pavement generation is a direct result of the friction and small impacts that occur as the tire rolls along the surface. Therefore, a smoother pavement generates less noise. Also, open-graded asphalt is utilized for tire noise control. The techniques reduce the generation of noise by diffusing normally reflected waves and transmitting less pressure fluctuation. It is possible to adopt certain traffic management techniques for speed reduction and banning truck operations in certain streets to achieve noise mitigation.

Altering the path of the noise can also reduce noise. For instance, some European paved light rail train (LRT) tracks are converted to green (grass or sedum) tracks primarily for sound attenuation. Noise levels can decrease as the distance between the noise source and the receiver increase because of geometric spreading effects. Such an abatement measure is enhanced by enforcing sufficient right-of-way distances. Establishing greenbelt buffer zones also contributes to the reduction of noise levels at the receiver locations. A more cost-effective method for noise control can be achieved by changes in the vertical or horizontal alignment so that the transportation facility avoids noise sensitive areas. Also, the technique to diffract sound waves by using a noise barrier can be utilized to alleviate noise, which is especially efficient in the area near noise sources.

16.2.3 Federal regulations

Table 16.13 summarizes regulations and guidelines that provide the legal authority and guidance for noise analysis procedures.

The 1969 National Environmental Policy Act (NEPA) established the environment-related decision-making framework for actions at the federal level. The key step in the decision-making process is the assessment of potential adverse environmental effects caused by the actions to be studied. Nevertheless, the criteria for evaluating the noise impacts are not specified by NEPA.

The FHWA has the obligation to plan and design a highway project, and simultaneously protect public health and welfare. The FHWA is responsible for administering the federal-aid highway program according to federal laws and regulations. The FHWA developed the noise standards and abatement requirements for highway traffic noise, as required by the Federal-Aid Highway Act of 1970 (Public Law 91-605, 84 Stat. 1713). These standards, as stated in Part 772 of Title 23 of the Code of Federal Regulations (23 CFR 772) Procedures for Abatement of Highway Traffic Noise and Construction Noise, applies to highway projects where federal funding is successfully obtained by the state transportation agencies in

Table 16.13 Regulation and guidelines addressing transportation-related noise

Year	*Regulation/guidelines*
1969	National Environmental Policy Act
1970	Federal-Aid Highway Act
1972	Noise Control Act
2010	Federal Highway Administration (FHWA) Noise Standards—23 CFR Part 772 "Procedures for Abatement of Highway Traffic Noise and Construction Noise"
2010 (revised in 2011)	FHWA Policy and Guidance—"Highway Traffic Noise: Analysis and Abatement Guidance"

the project. The regulation requires the highway agency to examine traffic noise impacts in areas next to federally funded highways projects including construction, reconstruction, maintenance, and expansion. Abatement must be considered in the first place if the impacts are identified. In addition, feasible and reasonable countermeasures should be a part of the concerns in the project design stage. As of now, the FHWA Guidance Manual, Highway Traffic Noise: Analysis and Abatement Guidance still helps state transportation agencies establish their localized policies.

The Noise Control Act of 1972 is a statute initiating a federal program (noise control program) of regulating emissions from virtually all sources, including commercial products, aircraft, railroads, and motor vehicles. The financial support of federal noise control programs was terminated in 1981. In 1982, the state and local governments took over the primary responsibility for addressing noise pollution. The U.S. EPA retains the authority to take the lead in noise impact assessment approaches, conducting research and publishing information.

Seven distinct activity categories based on land use are created and used by FHWA to assess potential noise impacts as defined by 23 CFR 772. Noise abatement criteria (NAC) are designed for five activity categories identified out of seven in which noise abatement needs to be assessed. Noise impact levels can be assessed from a variety of perspectives, but the most general approach, also used by NAC, is based on noise levels interfering with normal speech communication. Thus, the NAC strive for a balance of desirability and achievability.

Traffic noise impact takes place on a project in the following cases: (1) The predicted build noise level is greater than or equal to the minimum noise level specified in the NAC listed in Table 16.14; (2) The predicted noise level is substantially higher than the existing noise level.

16.2.4 Modeling monetary costs of noise emissions

To quantify the costs due to noise or the benefit due to noise reducing measures, a damage function approach (DFA) is applied. An overview of DFA applied to noise is given in Table 16.15.

The DFA for noise described in Table 16.15 is a general procedure. This framework is capable of considering many complicating factors, including nonlinear relationships in exposure-response functions (ERFs) and value functions, and different initial noise levels. Table 16.15 considers the impact of noise only in terms of annoyance, but this framework can be used to consider other forms of noise impacts' descriptions depending on ERFs.

16.2.4.1 Hedonic price

The hedonic price (HP) method can be used to monetize economic benefits or costs associated with environmental quality, including air and water pollution or environmental amenities, such as aesthetic views or proximity to recreational sites. The main advantage of the HP method is that it can reflect actual human behavior in the housing market. Specifically, the HP method can reflect how much an individual is willing to pay for a home with better environmental characteristics including noise conditions. The general limitation is that the method assumes that people have the chance to select the combination of features they prefer, given their income. However, the housing market may be affected by outside influences, like taxes, interest rates or other factors. Therefore, the price given by this method is implicit.

Table 16.14 FHWA noise abatement criteria on hourly weighted sound level

Activity category	*Leq(h)*	*Evaluation location*	*Description of activity category*
A	57	Exterior	Lands on which serenity and quiet are of extraordinary significance and serve an important public need and where the preservation of those qualities is essential if the area is to continue to serve its intended purpose
B	67	Exterior	Residential undeveloped lands permitted for this activity category are also included
C	67	Exterior	Active sport areas, amphitheaters, auditoriums, campgrounds, cemeteries, day care centers, hospitals, libraries, medical facilities, parks, picnic areas, places of worship, playgrounds, public meeting rooms, public or nonprofit institutional structures, radio studios, recording studios, recreation areas, sites under Section 4(f) of USDOT Act of 1966, schools, television studios, trails, and trail crossings. Undeveloped lands permitted for this activity category are also included
D	52	Interior	Auditoriums, day care centers, hospitals, libraries, medical facilities, places of worship, public meeting rooms, public or nonprofit institutional structures, radio studios, recording studios, schools, and television studios
E	72	Exterior	Hotels, motels, offices, restaurants/bars, and other developed lands, properties or activities not included in A–D or F. Undeveloped lands permitted for this activity category are also included
F	–	–	Agriculture, airports, bus yards, emergency services, industrial, logging, maintenance facilities, manufacturing, mining, rail yards, retail facilities, shipyards, utilities (water resources, water treatment, electrical), and warehousing
G	–	–	Undeveloped lands that are not permitted

Note: Leq(h)—Hourly equivalent sound level

The implicitness of the estimated price could be a result of many factors. The price estimated could be significantly affected by the model specification, meaning that if other external effects of transportation are not considered in the model, the estimated impact of noise on property prices could include these impacts as well. The price estimated is also sensitive to estimation procedures and functional form, including the level of information about the

Table 16.15 Overview of a damage function approach

Steps	*Description*
1	Identify noise mitigating measures, described in terms of change in time, location, frequency, level, and sources of noise
2	Select proper noise dispersion models
3	By using noise dispersion models, estimate the changed exposures to noise at different geographical locations; measured in dB(A) and noise indicators
4	Build exposure-response functions (ERFs), establishing relationships between decibel levels and levels of annoyance, ischemic heart disease, subjective sleep quality, and other impacts of noise. ERF can be expressed by percentage of exposed persons per year that are highly annoyed (HA)
5	Apply ERFs to obtain the overall noise impact change (e.g., expressed in a number of HA persons). In this process, more information might be required such as number of dwellings, household size, and so on
6	Economic valuation techniques are used to set an economic value for a unit of noise impact change (e.g., expressed in dollars per number of HA persons per year)
7	Economic benefits/costs are then computed by multiplying the unit economic value obtained in Step 6 by the quantity of the impact obtained in Step 5

response of the market to a certain noise level. Moreover, as implied by empirical studies, these estimated prices may vary from location to location depending on the market.

16.2.4.2 Avoidance costs

Cost avoidance is the calculated value of the difference between what we actually spend on utilities and what we would have spent had we maintained our old habits. For example, one would like to maintain the noise level in the house at a certain level. The noise level has been 45 dB(A), but a new railroad is about to be constructed in the neighborhood with a predicted noise level of 58 dB(A). Therefore, one has to implement some countermeasures such as installing noise reduction windows. If one spends 400 dollars on these countermeasures to keep the noise level at 45 dB(A), the 400 dollars is the avoidance cost (AC) in this case.

The major limitation of this method is that only in a few certain cases can the results be interpreted as a proxy for welfare changes from variations in noise level. Therefore, this method is not as popular as HP.

16.2.4.3 Contingent valuation

Contingent valuation (CV) is a method of estimating the value of a good that a person expects. The approach is quite straightforward: asking people to directly report their WTP to obtain this specified good or willingness to accept (WTA) to give up this good, rather than observing behaviors in the market. Because this method investigates the willingness by asking, where no actual transactions are made, CV is very practical for commodities whose transactions do not exist or are hard to be observed in regular markets under the desired conditions.

However, when this method is applied to noise, the CV studies could be very difficult in creating a good CV survey for valuing noise level reduction. Particularly, a scientific and understandable description of the noise reduction is hard to provide to the respondent. Besides, it is hard to make respondents accept WTP questions since they think it is unfair that they should pay to reduce noise created by others.

16.2.4.4 Noise emission cost values

Two main units are widely used in the noise valuation studied: (1) An economic value per decibel per year, measured by the noise depreciation sensitivity index (NDSI), defined as the average percentage change in property prices per decibel; (2) An economic value per year per person or household annoyed by noise. It could be measured in value per person "highly annoyed" (HA) or in value per person "annoyed."

The first unit is commonly used in HP studies, which is highly related to the housing market. Most of these studies report results in terms of the NDSI in terms of the average percentage change in property prices per decibel. To interpret this capitalized value into an annual value, several assumptions about time horizon and discount rate have to be made. Using monthly rent as the dependent variable in the HP study could avoid making these assumptions. However, the rental market might be controlled, therefore, the rental prices may not be able to fully reflect the difference in noise level.

The second unit is commonly used in CV and choice experiments studies, since this unit could best reflect the response of the people exposed to noise.

Table 16.16 gives some examples of noise valuation from past studies. Note that these values cannot be used elsewhere because of the high localization of the method (Navrud, 2002).

Table 16.16 Studies estimate external costs of noise

Study	*Location*	*Noise type*	*Results*
Palmquist (1980, 1982)	Kingsgate North King County Spokane	Traffic noise	0.48 NDSI 0.30 NDSI 0.08 NDSI
Nelson (1982)	Review of existing studies	Traffic noise	0.40 NDSI
Kanafani (1983)	Review of existing studies	Traffic noise	0.06%–0.12% of GDP
Levesque (1994)	Winnipeg	Aircraft noise	1.3 NDSI

EXAMPLE 16.2

The table below is part of a noise analysis for a roadway project which included horizontal and vertical alterations. Suppose the state noise policy specifies the following criteria to define the occurrence of traffic noise impact: (1) Design year traffic noise levels are predicted to approach, meet or exceed the NAC; or (2) Design year traffic noise levels are predicted to substantially increase (greater than 14 dB(A)) over existing noise levels.

Identify the receptors where a traffic noise impact occurs and the external cost caused by the noise brought about by the project.

Receptor	*Description of activity category*	*Activity category*	*Evaluation location*	*Existing noise level (dB[A])*	*Build noise level (dB[A])*	*Property price (dollars)*	*Noise value*
R_1	Undeveloped lands	G	–	40	55	23000k	0.26
R_2	Library	D	Interior	51	66	13400k	0.66
R_3	Residential	B	Exterior	60	71	12845k	0.46
R_4	Office	E	Exterior	57	71	13230k	0.76

The computation is summarized in the following table.

Receptor	*Increase from the existing to build scenario*	*Allowable noise level*	*Meet criterion 1*	*Meet criterion 2*	*Impact*
R_1	$55-40=15$	–	–	$15>14$, Yes	Yes
R_2	$66-51=15$	52	$66>52$, Yes	$15>14$, Yes	Yes
R_3	$71-60=11$	67	$71>67$, Yes	$11<14$, No	Yes
R_4	$71-57=13$	72	$71<72$, No	$13<14$, No	No

Therefore, noise abatement must be considered in the R_1, R_2, and R_3.

To compute the external cost caused by the noise brought about by the project. For R_1, R_2, and R_3, the evaluation requires more information about the cost of noise abatement measures and corresponding noise level after implementing those measures. Take R_4 as an example. The external cost is computed as

$$\$13230000 \times 0.76\% \times 13 = \$1307124$$

Suppose the life cycle of this project is 20 years and the interest rate is 5%. The external cost could be annualized as follows:

$$\$1307124 \times \left[\frac{i(1+i)^n}{(1+i)^n - 1}\right](i = 5\%, n = 20) = \$104887/\text{year}$$

16.3 IMPACTS OF CHANGES IN CLIMATE AND EXTREME WEATHER CONDITIONS ON THE TRANSPORTATION SYSTEM

16.3.1 Impact of climate change on transportation

Significant global climate change phenomena include global warming, rising sea levels, and El Niño. Five climate changes have been identified that can significantly affect the transportation system: (i) increases in very hot days and heat waves; (ii) increases in arctic temperatures; (iii) rising sea levels; (iv) increases in intense precipitation events; and (v) increases in hurricane intensity.

Traditional climate change impact methods may no longer be valid to guide transportation project investment because of the presence of new types of climate change and varying corresponding human activities.

Typically, transportation infrastructure is designed for local weather patterns with reasonable assumptions and predictions. Climate change will increase the possibility of extreme weather, which challenges transportation infrastructure and management systems. Specifically, extreme weather could speed up the deterioration rate of infrastructure. Weather events such as blizzards and thunderstorms can lead to network-wide congestion and even gridlock. In the meantime, traffic safety is also challenged because of the friction features of the infrastructure, visual conditions, and other environmental characteristics that are changed.

For different transportation modes and locations, the impact varies from case to case. Therefore, emergency strategies must be prepared beforehand in the planning stage. Table 16.17 summarizes the potential impacts on transportation of five typical climate change conditions (NRC, 2008).

16.3.2 Definition of extreme weather

Extreme weather is unpredictable and unexpected severe weather that does not usually occur. In recent years, extreme weather events have increased due to human-induced factors

Table 16.17 Impacts of climatic changes on transportation

Climate change	*Examples of impacts on infrastructure*	*Examples of impacts on operations*
Increasing in very hot days and heat waves	Thermal expansion on bridge expansion joints and paved surfaces; concerns regarding pavement integrity (e.g., softening), traffic-related rutting, migration of liquid asphalt; rail-track deformities	Shorter periods of construction activity due to health and safety concerns
Increases in arctic temperatures	Thawing of permafrost, causing subsidence of roads, rail beds, bridge supports (cave-in), pipelines, and runway foundations	Longer ocean shipping season and more ice-free ports in northern regions
Rising sea levels	Inundation of roads, rail lines, and airport runways in coastal areas; more frequent or severe flooding of underground tunnels and low-lying infrastructure	More severe storm surges, requiring evacuation or changes in development patterns
Increases in intense precipitation events	Increases in road washout, damages to rail-bed support structures, and landslides and mudslides that damage roadways and tracks; increases in scouring of pipeline roadbeds and damage to pipelines	Increases in weather-related delays and traffic disruptions
Increases in hurricane intensity	Greater probability of infrastructure failures; increased threat to stability of bridge decks; impacts on harbor infrastructure from wave damage and storm surges	More frequent interruptions in air service More frequent and potentially more extensive emergency evacuations

Table 16.18 Damage by disaster type, 1980–2011 (2011 Dollars)

Type of disaster events	*Number of disaster events*	*Adjusted damage ($ billions)*	*Percent damage (%)*	*Percent frequency (%)*
Tropical cyclones	31	417.9	47.4	23.3
Droughts/heatwaves	16	210.1	23.8	12.0
Severe local storms	43	94.6	10.7	32.3
Nontropical floods	16	85.1	9.7	12.0
Winter storms	10	29.3	3.3	7.5
Wildfires	11	22.2	2.5	8.3
Freezes	6	20.5	2.3	4.5

such as global warming. Studies indicate that the threat of extreme weather will keep increasing in the future.

Extreme weather has multiple definitions. For instance, the climate modeling community often defines extremes according to a threshold (e.g., the temperature of a day exceeding 95°F), and/or distribution or we can say, the probability to happen (1 out of 100 chance). The Draft of the National Climate Assessment notes, "Researchers use different definitions depending on which characteristics of extreme they are choosing to explore at any one time." The hydrology community defines extreme events according to average annual exceedance probabilities or average recurrence intervals. Intuitively, extreme weather events will cause massive welfare loss. Table 16.18 presents damage information of extreme weather events by disaster type (Smith and Katz, 2013).

16.3.3 Weather and climate considerations

Even though accurate prediction of extreme weather is still a challenge, understanding the magnitude, extent, and timing of impending extreme weather can help us better predict extreme weather and manage for emergencies. More information and detail are useful to prepare better operational plans and response strategies. Emergency participants should identify the following specific needs:

- Framework of extreme weather recognition and forecasting system
- Real-time data of actual extreme weather conditions during the entire event
- A reliable centralized place to gather and disseminate information to public
- Develop emergency traffic management strategies system, including traffic supply, demand strategies, and long-term strategies
- Enough budget for extreme weather management, responding and recovery if applicable
- Regular and table-top exercise and drill
- Consideration for people with special needs

16.3.4 Costs of extreme weather

Extreme weather events may cause severe types of cost to transportation facilities and transportation dependent populations, especially when people are traveling by road. Although it is literally impossible to erase the negative effects caused by extreme weather, modern technology seeks to mitigate the damage and therefore the cost as much as possible.

Much money is required to immediately repair or protect transportation facilities when experiencing an extreme weather event, resulting in a significant amount of costs to transportation agencies. Specifically, extreme weather events jeopardize transportation operations, cause deterioration of infrastructure, and shorten their life spans. Transportation agencies need to manage the ever-increasing costs of extreme weather and the requirement of fast recovery from these extreme weather events. Agencies have to bear all these costs. Moreover, if the extreme weather event disrupts traffic flow, causing extra delays or detours, the user also bears some costs.

Benefit-cost analysis, net present value, and life-cycle cost analysis are major methods to compare one project alternative with others. Since the value of money today will not be the same tomorrow, the time value of money should be considered when using these methods. An essential parameter we need to know when deciding the net present value is the timing of the investment; however, most of the time it is difficult to know the exact timing for investment for extreme weather protection or preparation. Therefore, when using analytical methods, the use of proper prediction and probabilistic analysis are significant factors that need to be considered carefully (Meyer et al., 2013).

16.3.5 Extreme weather management

In several instances, recurring pressures on state transportation officials to prepare for, manage, and recover from extreme weather events have caused organizational change, including new management activities (e.g., staff assigned emergency management responsibilities), modified standard operating procedures, and expanded staff training in managing and administering recovery efforts.

16.3.5.1 Response

Coordination among agencies ensures the success of extreme weather response. It is common that multiple jurisdictions and agencies will be involved during extreme weather events. Coordination and cooperation are highly needed to protect road users and transportation facilities. Poor coordination among agencies will cause the response to be inefficient and available resources cannot be utilized efficiently. Therefore, agency cooperation is the premise of successful extreme weather response.

If an extreme weather event is unavoidable, traffic management strategies are required to make road use more efficient and help users on the road to escape from the impacted area. From the traffic supply side, the typical methods include the following:

16.3.5.2 Emergency traffic signal timing plan

Traffic signal timing is a technique to determine the optimal ROW plans by selecting appropriate values for traffic signal timing parameters. Essential parameters include but are not limited to cycle length, number of phases, green splits, and offsets between adjacent intersections. Optimized and effective signal timing plans ensure the flexibility of traffic flow and are consistent with traffic needs for all approaches and minimization of total delays, fuel consumption, and intersection-related vehicle stop-and-go conditions. In the case of an extreme weather event, time is of essence. Reducing time is meant to save more lives.

16.3.5.3 Conflict point minimization

Conflict points refer to the points that roadway users may cross, diverge from or merge with other roadway users at an intersection. Drivers are likely to make mistakes due to misperception, and sometimes vehicle crashes may be triggered by movement conflicts. For a typical four-leg unsignalized intersection, there are 32 conflict points, including 16 crossing points, 8 margining points, and 8 diverging points. If bicycles and pedestrians are considered, the number will rise to 48. During extreme weather events, only normal signalized control is not enough. Additional traffic control measures are required to minimize the conflict points.

16.3.5.4 Contraflow, road shoulder utilization, and work zone removal

During an extreme weather event, the traffic demand pattern reflects driver behavior that can be quite different. Temporarily using the road shoulder and removing the work zone could make full use of available capacity to accommodate varying traffic demand. Contraflow lanes could be used to deal with the problem with significantly imbalanced inbound and outbound traffic demand.

PROBLEMS

16.1 List major types of vehicle air pollutants and explain their impacts on human bodies and the feasibility of transportation projects.

16.2 What are the differences between static and dynamic emission models? What are the corresponding pros and cons? What kinds of data are required for each model to estimate the emission impacts?

16.3 Assume a four-tire vehicle drive at a stable speed of 55 mph in the rural area. For each trip, the pollutants of CO, NO_X, and SO_2 are 0.5 kg for each. Estimate the total emission costs for each trip.

16.4 Using the method introduced in Example 16.1 and the following information, compute the benefits.

Table P16.1 Emission and VMT variations

	Unit cost	*Before*	*After*
VMT	–	30000	35000
CO	$520/ton	2512 g	2048 g
NO_2	$1060/ton	4365 g	3846 g

16.5 You are asked to mitigate impacts of vehicle air and noise pollution. What measures would be implemented?

16.6 Increases in fuel price are one way to reduce running vehicles, if we only consider light-duty vehicles. 10% fuel price increasing will lead to 5% reduction of running vehicles on an Interstate road and the corresponding speed will increase from 40 mph to 45 mph. Estimate the benefits.

16.7 What are the climate change impacts on transportation? How could you mitigate the impacts?

16.8 What are the effective response measurements in extreme weather events?

Chapter 17

Transportation decision models

17.1 GENERAL CONCEPTS

17.1.1 Model and decision analysis

According to Weber (1947), models should be "ideal types" of reality. However, a transportation system is much too complex to be simply copied in miniature into a model. It would be utopian to look for an all-purpose model that is simultaneously suitable for testing hypotheses, forecasting, clarifying transportation engineering, and making decisions. The reality is that the model needs to be formulated on no more than a few aspects.

Decision models considered in this book are mainly conceptual schedules and algorithms intended to make rational decision making easy or at least feasible. In other words, they are derived by formulating the "ideal types" of the concept of decision and of tangent concepts, such as rationality, way of behavior, consequence of action, utility, state, and so on. Nevertheless, the method is not purely inductive but a complicated combination of inductive and deductive elements.

Owing to the ideal-typical character of the decision concept, "concrete choices" are not classifiable into rational decisions and nonrational decisions. Factual decisions are rather combinations of "truly rational" choices, on the one hand, and impulsive–instinctive, strictly traditional, stochastic, and inconsistent ways of behavior on the other. Such difficulty is an essential obstacle to the suitability of decision models in explaining and forecasting a transportation system, but it hardly interferes with their normative, prescriptive, and operational application (Menges, 1974).

17.1.2 Specification, sensitivity, and abstraction degree

Generally, the decision model must be operationally utilizable and is an abstract setup of relationships. Both properties, however, raise difficulties with respect to the establishment of decision models.

Let us denote the decision model by M and assume it is a set of possible empirical situations $s_1, s_2, \ldots$

$$M = \{s_1, s_2, \ldots\}$$

Let the actual empirical situation be represented by S. Now, the question is whether S is included in M. Due to the ideal-typical character of M, this question would be given a negative answer in principle. Nevertheless, a model that contains an s_i which comes to S is expected. If the model M has this property, it is called *well specified*; otherwise, it is called *mis-specified*. As illustrated in Figure 17.1, the decision model M_1 is better specified than the

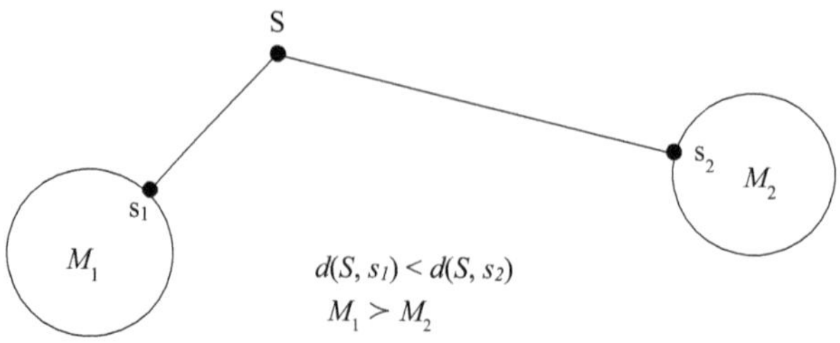

Figure 17.1 Illustration of well-specified and mis-specified decision models.

decision model M_2 with respect to the empirical situation S, because the distance d between S and s_1 (the nearest point of M_1) is shorter than between S and s_2 (the nearest point of M_2). The postulate "Since M_1 is better specified than M_2, choose M_1" has been designed as accommodation principle (Menges, 1974).

Should one follow the accommodation principle in solving a given decision problem? This choice is also a decision problem and can possibly be surrendered to formal calculi. However, the accommodation principle is not automatically rational. As in the above example, it is possible that the model M_2 is less sensitive than M_1 regarding changes in the assumptions. Thus, in M_2, contrary to M_1, the utility function in the whole remains unchanged when certain hypotheses on the model are modified. Is then M_2 to be preferred? What is rational? The answer to this question again relies on the preference system of the decision maker and cannot be given in general. The conflict between proper specification and insensitivity in setting up a decision model originates in two components of the model: norm and explanation. These two elements often complement each other, but when they are antagonistic, a difficult pre-decision must be made.

In addition, the unavoidable fact that each decision model is abstract and idealized implies another difficulty, namely, the determination of the degree of abstraction. If a model is too abstract, it may be an empty form without any reference to reality and with a low degree of operability. Conversely, if the model is highly particular, it will suit one specific situation but not others. Therefore, there are at least some objective criteria that should be followed. For example, making the model as concrete as possible based on objective knowledge of the decision situation, widening the above basis by adding further information if it is too narrow to allow concretization up to ensuring operability of the model, deciding against the introduction of another a priori assumption which cannot be examined, leaving the model as abstract as it is in the case of doubt, and so on.

17.1.3 Definitions of certainty, risk, and uncertainty

The decisions may involve input factors for analysis under certainty, risk, and uncertainty. For cases under certainty, all factors are with single values, which are deterministic. The outcome of analysis is therefore deterministic. Engineering systems, however, are almost always designed, constructed, and operated under unavoidable conditions of risk and uncertainty.

For the case under risk, an input factor may contain multiple possible values and a probability distribution can be assigned to the possible values. In this case, the probabilistic risk assessment can be performed to ultimately establish its mathematical expectation (or expected value). The expected value of the input factor can then be used for analysis.

For the case under uncertainty, an input factor may contain multiple possible values. However, a reliable probability distribution may not be determined for the possible values or the full range of possible values may not even be known. Hence, the mathematical expectation cannot be established. In this case, risk-based analysis cannot be directly conducted (Klir, 2006; Kurowicka and Cooke, 2006; Luenberger, 2006).

As a practical matter, the decision analysis typically involves multiple input factors for analysis. Some may be under certainty, some may be under risk, and the remaining may be under uncertainty. Thus, the decision analysis deals with mixed cases of certainty, risk, and uncertainly, which need to be treated on a case-by-case basis. The following sections introduce models and methods dealing with decision making under certainty, risk, and uncertainty, respectively.

17.2 DECISION MAKING UNDER CERTAINTY

17.2.1 Linear programming

The type of decision situations to which linear programming (LP) is applicable can be characterized as follows:

$$\text{Min} \quad V = cx^T \tag{17.1}$$

Subject to

$$Ax^T \leq b \tag{17.2}$$

$$x \geq 0 \tag{17.3}$$

where the objective quantity V is linearly dependent upon certain variables, c is a n-dimensional row vector, A is an $m \times n$ matrix, b is a m-dimensional column vector, and x is an n-dimensional row vector of variables or unknowns. The objective function (17.1) is known to the decision maker, and it makes no difference at all whether the problem is to minimize or to maximize the objective quantity. As in constraints (17.2), the variables are subject to a system of linear constraints. Constraints (17.3) characterize "states of the world" by variables that cannot be negative values.

The simplex algorithm, Dantzig–Wolfe decomposition algorithm, and the Karmarkar algorithm are often employed to solve the LP decision problem (Luenberger, 1973; Hillier and Lieberman, 2006; Bazaraa et al., 2009).

17.2.1.1 Simplex algorithm

$$\begin{array}{rcl} \text{Min } c_{1\times n}x_{n\times 1} & & \text{Max } -cx \\ s.t.\ A_{m\times n}x_{n\times 1} = b_{m\times 1} & \Rightarrow & s.t.\ Ax = b \\ x \geq 0 & & x \geq 0 \end{array}$$

1. Initialization step
   Choose a starting basic feasible solution (BFS) with basis B
2. Main step
   i. Solve the system $BX_B = b$

$$\begin{cases} X_B = B^{-1}b = \bar{b} \\ X_N = 0 \\ z = C_B X_B \end{cases}$$

ii. Solve the system $\omega_B = C_B$, $\omega = C_B B^{-1}$ (ω is referred to as the vector of simplex multipliers). Calculate $z_j - c_j = \omega a_j - c_j$ for all non-basis variables.

Let $z_k - c_k = \max_{j \in R}\{z_j - c_j\}$ (R is the current set of indices associated with the non-basic variables).

If $z_k - c_k \le 0$, then stop. The current BFS is an optimal solution. Otherwise, go to step 3 with x_k as the entering variable.

iii. Solve the system $By_k = a_k$, $y_k = B^{-1}a_k$

If $y_k \le 0$, then stop and the optimal solution is unbounded.

$$\left\{\begin{pmatrix} \bar{b} \\ 0 \end{pmatrix} + x_k \begin{pmatrix} -y_k \\ e_k \end{pmatrix}, \quad x_k \ge 0\right\}$$

where e_k is an $n - m$ vector of zeros except for a 1 at the kth position.

iv. If $y_k \nleq 0$, go to step 4

Let x_k enter the basis B, the index r of the following variable x_{B_r} which leaves the basis B is determined by the following minimum ratio test:

$$\frac{\bar{b}_r}{y_{rk}} = \min_{1 \le i \le m}\left\{\frac{\bar{b}_i}{y_{ik}} : y_{ik} > 0\right\}$$

Update the basis B where a_k replaces a_{B_r}, update the index set R, and repeat step 1.

$$\min z$$

$$① \; z - C_B X_B - C_N X_N = 0$$

$$② \; BX_B + NX_N = b$$

$$X_B, X_N \ge 0$$

$$③ = B^{-1} ②$$

$$X_B + B^{-1}NX_N = B^{-1}b$$

$$④ = C_B \times ③$$

$$C_B X_B + C_B B^{-1} N X_N = C_B B^{-1} b$$

$$⑤ = ① + ④$$

$$z - C_B X_B - C_N X_N + C_B X_B + C_B B^{-1} N X_N = C_B B^{-1} b$$

$$z - 0X_B + (C_B B^{-1} N - C_N) X_N = C_B B^{-1} b$$

	z	X_B	X_N	Right hand side (RHS)
z	1	0	$C_B B^{-1} N - C_N$	$C_B B^{-1} b$ Row 0
X_B	0	I	$B^{-1}N$	$B^{-1}b$ Rows 1-m

	Z	x_{B_1}	…	x_{B_r}	…	x_{B_m}	…	x_j	…	x_k	…	RHS
z	1	0	…	0	…	0	…	$z_j - c_j$	…	$z_k - c_k$	…	$C_B\bar{b}$
x_{B_1}	0	1	…	0	…	0	…	y_{1j}	…	y_{1k}	…	$\bar{b}_1$
…	…	…	…	…	…	…	…	…	…	…	…	…
x_{B_r}	0	0	…	1	…	0	…	y_{rj}	…	y_{rk}	…	$\bar{b}_r$
…	…	…	…	…	…	…	…	…	…	…	…	…
x_{B_m}	0	0	…	0	…	1	…	y_{mj}	…	y_{mk}	…	$\bar{b}_m$

Let $z_k - c_k = \max\{z_j - c_j, j \in R\}$.
If $z_k - c_k \leq 0$, then stop and the current solution is optimal.
Otherwise, examine y_k.

a. If $y_k \leq 0$, then stop and the optimal solution is unbounded.

$$\left\{\begin{pmatrix} B^{-1}\bar{b} \\ 0 \end{pmatrix} + x_k \begin{pmatrix} -y_k \\ e_k \end{pmatrix}, x_k \geq 0\right\}$$

where e_k is a vector of zeros except for a 1 at the kth position.

b. If $y_k \nleq 0$, determine the index r as follows:

$$\frac{\bar{b}_r}{y_{rk}} = \min_{1\leq i\leq m}\left\{\frac{\bar{b}_i}{y_{ik}} : y_{ik} > 0\right\} = \max_{1\leq i\leq m}\left\{\frac{y_{ik}}{\bar{b}_i} : y_{ik} \geq 0\right\} = \max_{1\leq i\leq m}\left\{-\frac{\bar{b}_i}{y_{ik}} : y_{ik} > 0\right\}$$

	z	x_{B_1}	… x_{B_r}	… x_{B_m}	…	x_j	…	x_k	…	*Right-hand side (RHS)*
z	1	0	… 0	… 0	…	$(z_j - c_j) - \left(\frac{y_{rj}}{y_{rk}}\right)\cdot(z_k - c_k)$	…	$-\frac{(z_k - c_k)}{y_{rk}}$	…	$C_B\bar{b} - \left(\frac{\bar{b}_r}{y_{rk}}\right)\cdot(z_k - c_k)$
x_{B_1}	0	1	… 0	… 0	…	$y_{1j} - \left(\frac{y_{rj}}{y_{rk}}\right)\cdot y_{1k}$	…	$-\frac{y_{1k}}{y_{rk}}$	…	$\bar{b}_1 - \left(\frac{\bar{b}_r}{y_{rk}}\right)\cdot y_{1k}$
…	…	…	… …	… …	…	…	…	…	…	…
x_{B_r}	0	0	… 1	… 0	…	$\frac{y_{rj}}{y_{rk}}$	…	$\frac{1}{y_{rk}}$	…	$\frac{\bar{b}_r}{y_{rk}}$
…	…	…	… …	… …	…	…	…	…	…	…
x_{B_m}	0	0	… 0	… 1	…	$y_{mj} - \left(\frac{y_{rj}}{y_{rk}}\right)\cdot y_{mk}$	…	$-\frac{y_{mk}}{y_{rk}}$	…	$\bar{b}_m - \left(\frac{\bar{b}_r}{y_{rk}}\right)\cdot y_{mk}$

EXAMPLE 17.1

Solve the following LP problem using the simplex algorithm.

$$\begin{aligned} &\text{Min} \quad x_1 + x_2 - 4x_3 \\ &s.t \quad x_1 + x_2 + 2x_3 \leq 9 \\ &x_1 + x_2 - x_3 \leq 2 \\ &-x_1 + x_2 + x_3 \leq 4 \\ &x_1, x_2, x_3 \geq 0 \end{aligned}$$

$$\begin{aligned} &\min x_1 + x_2 - 4x_3 + 0x_4 + 0x_5 + 0x_6 \\ &s.t \quad x_1 + x_2 + 2x_3 + x_4 + 0 + 0 = 9 \\ &x_1 + x_2 - x_3 + 0 + x_5 + 0 = 2 \\ &-x_1 + x_2 + x_3 + 0 + 0 + x_6 = 4 \\ &x_1, x_2, x_3, x_4, x_5, x_6 \geq 0 \end{aligned}$$

$$C_B = [0,0,0], \quad C_N = [1,1,-4]$$

$$B^{-1} = \begin{bmatrix} 1 & 0 & 0 \\ 0 & 1 & 0 \\ 0 & 0 & 1 \end{bmatrix}_{3\times3}, \quad b = \begin{bmatrix} 9 \\ 2 \\ 4 \end{bmatrix}_{3\times1}, \quad y = \begin{bmatrix} 1 & 1 & 2 \\ 1 & 1 & -1 \\ -1 & 1 & 1 \end{bmatrix}_{3\times3}$$

$$\min c^T x$$
$$s.t\ \ Ax = b$$
$$x \geq 0$$

$$c^T = [1 \quad 1 \quad 4 \quad 0 \quad 0 \quad 0]$$

$$x = \begin{bmatrix} x_1 \\ x_2 \\ x_3 \\ x_4 \\ x_5 \\ x_6 \end{bmatrix} \quad b = \begin{bmatrix} 9 \\ 2 \\ 4 \end{bmatrix} \quad A = \begin{bmatrix} 1 & 1 & -4 & 1 & 0 & 0 \\ 1 & 1 & -1 & 0 & 1 & 0 \\ -1 & 1 & 1 & 0 & 0 & 1 \end{bmatrix} \quad x = \begin{bmatrix} 0 \\ 0 \\ 0 \\ 9 \\ 2 \\ 4 \end{bmatrix}$$

Initialization

I	*z*	x_1	x_2	x_3	x_4	x_5	x_6	*RHS*
z	1	1	1	−4	0	0	0	0
x_4	0	1	1	2	1	0	0	9
x_5	0	1	1	−1	0	1	0	2
x_6	0	−1	1	1	0	0	1	4

Iteration 1

1. Let $z_k - c_k = \max\{z_1 - c_1, z_2 - c_2, z_3 - c_3\} = 4 > 0$
   x_3 should replace one of x_4, x_5, and x_6.
2. Examine $y_{jk} = \{2, -1, 1\}$
3. For $y_{jk} \nleq 0$, that is, $y_{jk} = \{2, -1\}$ correspond to x_4, x_6, determine which one should be replaced by x_3

The index γ is determined as follows $\frac{\overline{b_i}}{y_{ik}} = \min\left\{\frac{b_4}{y_4}, \frac{b_6}{y_6}\right\} = \min\left\{\frac{9}{1}, \frac{4}{1}\right\} = 4$, x_6 should be replaced by x_3

I	*z*	x_1	x_2	x_3	x_4	x_5	x_3	*RHS*
z	1	$-1-\left(\frac{-1}{1}\right)\times 4 = 3$	$-1-\left(\frac{1}{1}\right)\times 4 = -5$	$-\left(\frac{4}{1}\right) = -4$	0	0	0	$0-\left(\frac{4}{1}\right)\times 4 = -16$
x_4	0	$1-\left(\frac{-1}{1}\right)\times 2 = 3$	$1-\left(\frac{1}{1}\right)\times 2 = -1$	$\left(\frac{-2}{1}\right) = -2$	1	0	0	$9-\left(\frac{4}{1}\right)\times 2 = 1$
x_5	0	$1-\left(\frac{-1}{1}\right)\times(-1) = 3$	$1-\left(\frac{1}{1}\right)\times(-1) = 2$	$-\left(\frac{-1}{1}\right) = -1$	0	1	0	$2-\left(\frac{4}{1}\right)\times(-1) = 6$
x_3	0	$\frac{-1}{1} = -1$	$\frac{1}{1} = 1$	$\frac{1}{1} = 1$	0	0	1	$\frac{4}{1} = 4$

Iteration 2

1. Let $z_k - c_k = \max\{3, -5, -4\} = 3 > 0$
   x_1 should replace one of x_4, x_5, and x_3
2. Examine $y_{jk} = \{3, 0, -1\}$
3. For $y_{jk} > 0$, x_4 should be replaced by x_1

I	z	x_1	x_2	x_3	x_1	x_5	x_3	*RHS*
z	1	$-\left(\frac{3}{3}\right)=-1$	$-5-\left(\frac{-1}{3}\right)\times 3=-4$	$-4-\left(\frac{-2}{3}\right)\times 3=-4$	0	0	0	$-16-\left(\frac{1}{3}\right)\times 3=-17$
x_1	0	$\frac{1}{3}$	$\frac{-1}{3}$	$\frac{-2}{3}$	1	0	0	$\frac{1}{3}$
x_5	0	$-\frac{0}{3}=0$	$2-\left(\frac{-1}{3}\right)\times 0=2$	$1-\left(\frac{-2}{3}\right)\times 0=1$	0	1	0	$6-\left(\frac{1}{3}\right)\times 0=6$
x_3	0	$-\left(\frac{-1}{3}\right)=\frac{1}{3}$	$1-\left(\frac{-1}{3}\right)\times(-1)=\frac{2}{3}$	$1-\left(\frac{-2}{3}\right)\times(-1)=\frac{1}{3}$	0	0	1	$4-\left(\frac{1}{3}\right)\times(-1)=\frac{13}{3}$

$z_k - c_k = \{-1, -4, -2\} < 0$, current solution of 13/3 is optimal.

17.2.2 Integer programming

If all variables in the previous LP decision problem are integers, then this problem is called a (linear) integer programming (IP) decision problem written as (Luenberger, 1973; Nemhauser and Wolsey, 1988; Wolsey, 1998; Winston and Venkataramanan, 2002)

$$\text{Min} \quad V = cx^T \tag{17.4}$$

Subject to

$Ax^T \leq b$

$x \geq 0$ and integer

If some but not all variables are integers, it is called a mixed IP decision problem; and if all variables are restricted to 0–1 values, it is called a 0–1 programming decision problem. Generally, branch-and-bound, column generation, branch-and-cut, cutting plane, as well as some heuristic algorithms can be used to solve such problems (Rayward-Smith et al., 1996).

17.2.2.1 Knapsack problem

The knapsack problem (KP), also called the capital budgeting problem (Lorie and Savage, 1955), could be categorized into four basic models: simple KP, multi-choice KP, multidimension KP, and multi-choice multidimension KP (Martello and Toth, 1990). The inputs of these four basic models are deterministic somehow, but some inputs may not be under certainty in the decision-making stage. Therefore, the stochastic KP should be introduced.

In the simple 0–1 KP, the consumer has a set of item candidates. The consumer needs to pick one or more items from the candidate pool. The goal is to maximize the total reward gained with the constraint that the knapsack cannot hold more than a certain volume.

$$\text{Max} \quad \sum_{i=1}^{n} B_i x_i \tag{17.5}$$

$$\text{Subject to} \quad \sum_{i=1}^{n} c_i x_i \leq C$$

$$x_i = 0/1$$

where

x_i = The decision variable indicating whether the item i
n = The number of items in the candidate pool
B_i = The reward gained from taking item i
c_i = The size of the item i
C = The size capacity of the knapsack

In the context of project selection, the item is a project. The size could be a constraint, usually budget or other resources. The reward could be the benefit in terms of profit or safety improvement and so forth.

In a more generalized form of the KP, multi-choice knapsack problem (MCKP), the consumer has a set of N classes, where each class contains a number of items. The consumer needs to pick exactly one item from each class. The consumer faces a "multi-choice" problem because he/she has a set of choices for each class. For example, the consumer needs to select one item from a number of items of class 1, one item from a number of items of class 2, and so on, to maximize the reward gained with the constraint that the cart cannot hold more than a certain volume. The general formulation of MCKP is as below:

$$\text{Max} \quad \sum_{j=1}^{N}\sum_{i=1}^{n_j} B_{ij}x_{ij} \tag{17.6}$$

$$\text{Subject to} \quad \sum_{i=1}^{n_j} x_{ij} = 1 \quad j = 1,2,\ldots,N$$

$$x_{ij} = 0/1$$

$$\sum_{j=1}^{N}\sum_{i=1}^{n_j} c_{ij}x_{ij} \le C$$

where

x_{ij} = The decision variable indicating whether the item i in class j is taken
n_j = The number of items in class j
N = The number of classes
B_{ij} = The reward gained from taking item x_{ij}
c_{ij} = The size of the item i in class j
C = The size capacity of the knapsack

In the context of project selection, each class represents a project in the larger pool of candidate projects. The choices for each facility include the several possible interventions or projects including the do-nothing alternative. The reward is measured in terms of multiple criteria such as cost, condition, and so on. The "size constraint" of the knapsack corresponds to the budget constraint for the program period.

Another variation of the KP is called multidimensional knapsack problem (MDKP). The consumer seeks to select from a set of distinct items subject to more than one size constraint, and each item has a known weight, volume, and width. For example, the shopping cart cannot hold more than a certain weight, a certain volume or a certain width, the shopper cannot spend beyond those limits, and so on. This gives the multidimensionality aspect to the problem. Mathematically, the MDKP can be stated as

$$\text{Max} \quad \sum_{i=1}^{n} B_i x_i \tag{17.7}$$

$$\text{Subject to} \quad \sum_{i=1}^{n} c_{id} x_i \le C_d$$

$$x_i = 0/1$$

where

x_i = The decision variable indicating whether the item i
n = The number of items in the candidate pool
B_i = The reward gained from taking item i
d = The dimension index, $d = 1,2, \dots, D$
c_{id} = The size measure in dimension d of the item i
C_d = The size capacity in dimension d of the knapsack

In the context of project selection, a scenario with multiple "size" constraints could be one having a budget constraint, network-wide condition constraint (minimum condition target), network-wide safety constraint, and so on. The goal is to choose one item from one class such that the profit is maximized without exceeding the capacities of the knapsack. If the number of size constraints is one and there is only one item in each class, then the problem reduces to a simple 0–1 KP.

This is a further generalization of the KP is multi-choice multidimensional knapsack problem (MCMDKP), which contains both the multi-choice (more than one item or activity in each class) and the multidimensional (more than one size constraint) aspects as explained above. The formulation of MCMDKP is generally as follows:

$$\text{Max} \quad \sum_{j=1}^{N} \sum_{i=1}^{n_j} B_{ij} x_{ij} \tag{17.8}$$

$$\text{Subject to} \quad \sum_{i=1}^{n_j} c_{ijd} x_{ij} \le C_d \quad d = 1,2,\dots,D$$

$$\sum_{i=1}^{n_j} x_{ij} = 1 \quad j = 1,2,\dots,N$$

$$x_{ij} = 0/1$$

where

x_{ij} = The decision variable indicating whether the item i in class j is taken
n_j = The number of items in the class j
B_{ij} = The reward gained from taking item i in class j
d = The dimension index, $d = 1,2, \dots, D$
c_{ijd} = The size measure in dimension d of the item i in class j
C_d = The size capacity in dimension d of the knapsack

17.2.3 Nonlinear programming

Soon after the first works on LP had been published, it was realized that the assumption of linear restrictions and linear objective functions often conflict with reality. If the objective function is not a linear function, or some of the constraints are not linear constraints, such a programming problem is called a nonlinear programming (NLP) problem (Bazaraa et al., 2006), which is written as

$$\text{Min} \quad f(x) \tag{17.9}$$

Subject to

$$g_i(x) \leq b_i, \quad j = 1, 2, \ldots, m$$

It is found that there exist no exact algorithms that are comparable to the simplex method in LP. All methods of NLP so far developed are iterative. The most important class of iterative solution algorithm for NLP is based on the Karush–Kuhn–Tucker (KKT) conditions, also known as the Kuhn–Tucker conditions, which are first-order necessary conditions for a solution in NLP to be optimal, provided that some regularity conditions are satisfied (Winston and Venkataramanan, 2002). The KKT conditions were originally named after Harold W. Kuhn and Albert W. Tucker, who first published the conditions in 1951. It was then found that these conditions for solving NLP had been stated by William Karush in his master's thesis in 1939 (Sundaram, 1996; Boyd and Vandenberghe, 2004; Nocedal and Wright, 2006).

For simplicity, if $x = (\bar{x}_1, \bar{x}_2, \ldots, \bar{x}_n)$ is an optimal solution to the above NLP problem, it must satisfy m constraints in Equation 17.5 and in the meanwhile, there must exist multipliers $\bar{\lambda}_1, \bar{\lambda}_2, \ldots, \bar{\lambda}_m$ satisfying

$$\frac{\partial f(\bar{x})}{\partial x_j} + \sum_{i=1}^{m} \bar{\lambda}_i \frac{\partial g_i(\bar{x})}{\partial x_j} = 0, \quad j = 1, 2, \ldots, n \tag{17.10}$$

$$\bar{\lambda}_i [b_i - g_i(\bar{x})] = 0, \quad i = 1, 2, \ldots, m \tag{17.11}$$

$$\bar{\lambda}_i \geq 0, \quad i = 1, 2, \ldots, m$$

Consider the NLP problem (standard form) in the following:

$$\begin{aligned} &\text{Optimize} && f(x) \\ &\text{Subject to} && g_i(x) \leq 0, \quad i = 1, 2, 3, \ldots m \\ & && h_j(x) = 0, \quad j = 1, 2, 3, \ldots l \end{aligned}$$

where $f(x)$ is the objective function, $g_i(x)$ are the inequality constraints, and $h_j(x)$ are equality functions. Suppose that the objective and constraints functions are continuously differentiable at x^* which is a local optimal solution and satisfy necessary conditions

1. Feasibility. $g_i(x^*) \leq 0$ and $h_j(x^*) = 0$.
2. Stationary
   For maximizing problem: $\nabla f(x^*) = \sum_{i=1}^{m} \mu_i \nabla g_i(x^*) + \sum_{j=1}^{l} \lambda_j \nabla h_j(x^*)$
   For minimizing problem: $-\nabla f(x^*) = \sum_{i=1}^{m} \mu_i \nabla g_i(x^*) + \sum_{j=1}^{l} \lambda_j \nabla h_j(x^*)$
3. Complementary slackness: $\mu_i g_i(x^*) = 0$, $i = 1, 2, 3, \ldots m$.
4. Positive Lagrange multiplier: $\mu_i \geq 0$, $i = 1, 2, 3, \ldots m$.

EXAMPLE 17.2

Solve the following NLP using the KKT conditions:

$$\begin{aligned}\text{Min} f &= x_1^2 + x_2^2 - 14x_1 - 6x_2\\ \text{subject to}\, g_1 &= x_1 + x_2 - 2 \le 0\\ g_2 &= x_1 + 2x_2 - 3 \le 0\end{aligned}$$

Solution

$$L(x_1, x_2, \mu_1, \mu_2) = x_1^2 + x_2^2 - 14x_1 - 6x_2 + \mu_1(x_1 + x_2 - 2) + \mu_2(x_1 + 2x_2 - 3)$$

$$\begin{aligned}L_{x_1} &= 2x_1 - 14 + \mu_1 + \mu_2 = 0\\ L_{x_2} &= 2x_2 - 6 + \mu_1 + 2\mu_2 = 0\\ L_{\mu_1} &= x_1 + x_2 - 2 = 0\\ L_{\mu_2} &= x_1 + 2x_2 - 3 = 0\end{aligned}$$

If g_1 and g_2 are active,

$$x_1 = 1, \quad x_2 = 1, \quad \mu_1 = 20, \quad \mu_2 = -8$$

$$f(1,1) = -18, \quad g_1(1,1) = 0, \quad g_2(1,1) = 0$$

λ cannot be negative

If none of g_1 and g_2 are active,

$$L_{x_1} = 2x_1 - 14 = 0$$

$$L_{x_2} = 2x_2 - 6 = 0$$

$$x_1 = 7, \quad x_2 = 3$$

$g_1(7, 3) = 8$, $g_2(7, 3) = 10$ do not satisfied the constricts

If only g_1 is active,

$$L_{x_1} = 2x_1 - 14 + \mu_1 = 0$$

$$L_{x2} = 2x_2 - 6 + \mu_1 = 0$$

$$L_{\mu_1} = x_1 + x_2 - 2 = 0$$

$$x_1 = 3, \quad x_2 = -1, \quad \mu_1 = 8$$

$$g_1(3,-1) = 0, \quad g_2(3,-1) = -2, \quad f(3,-1) = -26$$

If only g_2 is active,

$$L_{x_1} = 2x_1 - 14 + \mu_2 = 0$$

$$L_{x_2} = 2x_2 - 6 + 2\mu_2 = 0$$

$$L_{\mu_2} = x_1 + 2x_2 - 3 = 0$$

$$x_1 = 5, \quad x_2 = -1, \quad \mu_2 = 4$$

$$g_1(5,-1) = 2 \ge 0, \quad g_2(5,-1) = -2$$

Then, the optimal solution is $f(3, -1 = -26)$.

17.2.4 Parametric programming

The point is to ensure an optimal solution also in those cases where individual coefficients in the objective function or in the constraint are variables which depend on fixed parameters. This class of problems is called a parametric programming problem (Saaty and Gass, 1954; Jagannathan, 1966). For simplicity, if the ordinary objective function is

$$F = \alpha_0 + \alpha_1 X_1 + \alpha_2 X_2 + \cdots + \alpha_n X_n \tag{17.12}$$

then the parameter variant is that one or more coefficients rely on the time t or another factor expressed as parameter θ; the dependence may be linear of the first order,

$$\alpha_i(t) = \alpha_i + \delta_i t \tag{17.13}$$

or

$$\alpha_i(\theta) = \alpha_i + \gamma_i \theta \tag{17.14}$$

where δ_i and γ_i are constant for fixed $i(i = 1, 2, \ldots, n)$.

Thus, an additional set of coefficients, δ_i or γ_i, like the original coefficients α_i, needs to be optimized

$$F(X_1, \ldots, X_n; t \text{ or } \theta) \rightarrow \text{Max/Min} \tag{17.15}$$

subject to the constraints.

17.2.5 Dynamic programming

Let us denote a multistage decision process by $[a, f(a, x)]$, where a denotes a state, $f(a, x)$ represents a state transition function, and x is a decision vector. It is obvious that the state transition function depends on both state a and decision vector x. Suppose that there are n stages in the time horizon and let x_i be the decision vector at ith stage. The goal is to determine the decision vector $x = (x_1, x_2, \ldots, x_n)$ which drives the process from the initial stage a_0 to a_{n-1} considering all constraints so that it minimizes or maximizes the objective function $F(f_1(a_0, x_1), \ldots, f_n(a_{n-1}, x_n))$. The standard form of dynamic programming (DP) problem is

$$\text{Min/Max}\, F(f_1(a_0, x_1), \ldots, f_n(a_{n-1}, x_n)) \tag{17.16}$$

$$\begin{aligned} &\text{Subject to} \\ &a_i = f(a_{i-1}, x_i), \quad i = 1, 2, \ldots, n-1 \\ &a_i \in A_i, i = 0, \quad 1, \ldots, n-1 \\ &x_i = X_i(a_{i-1}), \quad i = 1, 2, \ldots, n \end{aligned}$$

where A_i denotes the state space, $X_i(a_{i-1})$ is the decision space.

The DP model is basically a recursive equation linking the different stages of the decision problem in a manner that guarantees each state's optimal solution is also optimal for the entire problem, on the basis of which both the forward and backward recursive can be used and yield the same solution (Bellman and Dreyfus, 2015).

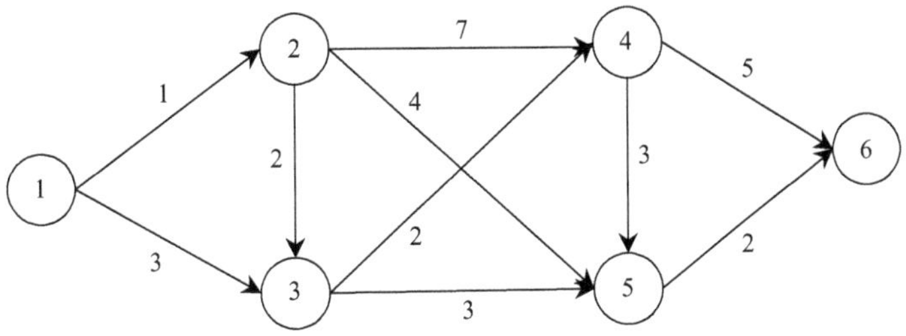

Figure 17.2 An example of the shortest path problem.

EXAMPLE 17.3

Solve the shortest path problem using backward and forward DP models (Figure 17.2).

Solution

1. Backward DP formulation for shortest path problem
   Denote: $f(i)$ = length of a shortest path from node $i (i = N - 1, N - 2, \ldots, 2,1)$ to node N
   a. For $i = N$, $f(N) = 0$, $f(6) = 0$
   b. For $i = N - 1$, $f(N - 1)$ = the shortest path from $N - 1$ to N,

$$f(N-1) = C_{N-1,N} + f(N) = C_{N-1,N}$$

$$f(5) = f(5 \to 6) = 2$$

   c. For $i = N-2$,

$$\begin{aligned} f(N-2) &= \min\{C_{N-2,N-1} + f(N-1), C_{N-2,N} + f(N)\} \\ &= \min_{j \in [N-1,N]} \{C_{N-2,j} + f(j)\} \\ f(4) &= \min_{j \in [5,6]} \{C_{4,5} + f(5), C_{4,6} + f(6)\} \end{aligned}$$

$$\begin{cases} f(i) = \min_{j \in [i+1,N]} \{C_{ij} + f(j)\}, i = N-1, N-2, \ldots, 1 \\ f(i) = 0 \end{cases}$$

Denote: $j^*(i)$ = node(s) immediately follow(s) node i on the optimal path(s) from node i to node N

$$i = 6, f(6) = 0, \quad j^*(6) = \varnothing$$

$$i = 5, f(5) = \min_{j \in [6]} \{C_{5,6} + f(6)\} = 2 + 0 = 2, j^*(5) = 6$$

$$i = 4, f(4) = \min_{j \in [5,6]} \{C_{4,5} + f(5), C_{4,6} + f(6)\} = \min\{3 + 2, 5 + 0\} = 5, j^*(4) = [5,6]$$

$$i = 3, f(3) = \min_{\substack{j \in [4,5] \\ j \neq 6, \text{ link } 3 \to 6 \text{ infeasible}}} \{C_{3,4} + f(4), C_{3,5} + f(5)\} = \min\{2 + 5, 3 + 2\} = 5, j^*(3) = 5$$

$$\begin{aligned} i = 2, f(2) &= \min_{\substack{j \in [3,4,5] \\ j \neq 6, \text{ link } 2 \to 6 \text{ infeasible}}} \{C_{2,3} + f(3), C_{2,4} + f(4), C_{2,5} + f(5)\} \\ &= \min\{2 + 5, 7 + 5, 4 + 2\} = 6, j^*(2) = 5 \end{aligned}$$

$$i = 1, f(1) = \min_{j \in [2,3]} \{C_{1,2} + f(2), C_{1,3} + f(3)\} = \min\{1 + 6, 3 + 5\} = 7, j^*(1) = 2$$

2. Forward DP formulation for shortest path problem
   Denote: $g(j)$ = length of a shortest path from node 1,2,3 ... to $j(j = 2, \ldots, N)$.

$$\begin{cases} g(j) = \min_{j\in[1,2,\ldots,j-1]} \{C_{ij} + g(i)\} \\ g(1) = 0 \end{cases}$$

Denote: $i^*(j)$ = node (s) immediately proceeds node j on the optimal path(s) from node 1 to node j. Namely, $i^*(j) \in [1, 2, \ldots, j-1]$ and $i^*(j)$ on the shortest path

$$j = 1,\ g(1) = 0,\ i^*(1) = \varnothing$$

$$j = 2,\ g(2) = \min_{i\in[1]}\{C_{1,2} + g(1)\} = 1 + 0 = 1,\ i^*(2) = 1$$

$$j = 3,\ g(3) = \min_{i\in[1,2]}\{C_{1,3} + g(1), C_{2,3} + g(2)\} = \min\{0 + 3, 1 + 2\} = 3,\ i^*(3) = [1,2]$$

$$j = 4,\ g(4) = \min_{i\in[2,3]}\{C_{2,4} + g(2), C_{3,4} + g(3)\} = \min\{1 + 7, 3 + 2\} = 5, i^*(4) = 3$$

$$j = 5,\ g(5) = \min_{i\in[2,3,4]}\{C_{2,5} + g(2), C_{3,5} + g(3), C_{4,5} + g(4)\}$$

$$= \min\{1 + 4, 3 + 3, 5 + 3\} = 5, i^*(5) = 2$$

$$j = 6,\ g(6) = \min_{i\in[4,5]}\left\{C_{4,6} + g(4), C_{5,6} + g(5)\right\} = \min\{5 + 5, 5 + 2\} = 7, i^*(6) = 5$$

17.3 DECISION MAKING UNDER RISK

17.3.1 Decision tree analysis

A decision tree is one way to display an algorithm by using a tree-like graph to illustrate consequences of events and corresponding possibilities. In addition, it is the most commonly used tool for risk decision making. The optimal alternative is selected in a systematic way and the relative desirability of each consequence is evaluated by utility value. Typically, a decision tree consists of three types of nodes.

Decision nodes: The branches that emanate from these nodes represent alternative courses of action about which the decision maker must make a choice, typically represented by squares.

Chance nodes: The branches leaving these nodes represent chance events, which represent outcomes of nature, typically represented by circles.

End nodes: End of an option or alternative, typically represented by triangles, but sometimes omitted (Figure 17.3).

Figure 17.4 illustrates an example of a decision tree with two alternatives. As defined, the decision tree starts from a decision node (the square node), which means a mutually exclusive choice should be made between alternative 1 and 2. Moreover, the sum of conditional probabilities at each chance node should be equal to 1, $P(\theta_1|A_1) + P(\theta_2|A_1) = 1$ for this example. Sometimes the corresponding possibilities are given or easy to estimate, sometimes it will be expressed as a continuous random variable, and the probability density function (PDF) is applied to calculate the possibilities of the branches. The desirability of each consequence is illustrated as a utility value, and more details about utility value will be provided in the following sections. In this example, each alternative has two possible consequences, illustrated as circle nodes in Figure 17.4. θ_1 and θ_2 are two corresponding outcomes. Utility values of four possible consequences are expressed as $u(A_i, \theta_j)$.

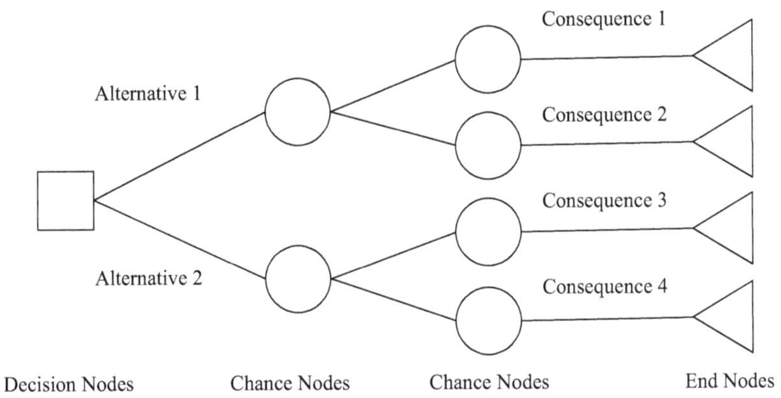

Figure 17.3 The decision tree framework.

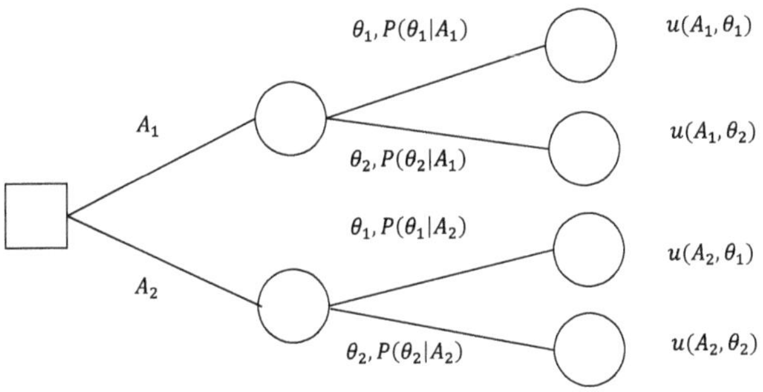

Figure 17.4 A decision tree example.

A_i = Alternative i
θ_j = Outcome j
$P(\theta_j|A_i)$ = Conditional probability
$u(A_i,\theta_j)$ = Corresponding utility value

17.3.1.1 Expected value of outcome

Expected value of outcome is typically denoted as $E(A_i)$. It is the significant measurement to select the preferred alternative. Intuitively, the maximum profit or the minimum cost will be selected. The expected value of the outcome is the sum of all possible outcomes with considering possibilities (Ang and Tang, 1984a,b).

$$E(A_i) = \sum_{j=1}^{n} P(\theta_j|A_i) \cdot u(A_i, \theta_j) \tag{17.17}$$

Take Figure 17.4 above as example,

$$E(A_1) = u(A_1, \theta_1) \cdot P(\theta_1|A_1) + u(A_1, \theta_2) \cdot P(\theta_2|A_1) \tag{17.18}$$

$$(A_2) = u(A_2,\theta_1)\cdot P(\theta_1|A_2) + u(A_2,\theta_2)\cdot P(\theta_2|A_2) \tag{17.19}$$

If the utility value represents cost here, we choose $\min(E(A_1), E(A_2))$.
If the utility value represents benefit here, we choose $\max(E(A_1), E(A_2))$.

17.3.1.2 Expected value of opportunity loss

Instead of maximizing the benefit, expected value of opportunity loss (EOL) is considered to minimize the lost opportunities when choosing alternatives. Suppose utility values represent benefits in the example, and θ_1 is the better situation and θ_2 is the worse situation, meaning it confers lower benefits.

$$M_j = \max\{u(A_i,\theta_j)\}, \quad i = 1,2,3\ldots n \tag{17.20}$$

$$\begin{aligned} EOL_j &= (M_1 - u(A_j,\theta_1))\cdot P(\theta_1|A_j) + (M_2 - u(A_j,\theta_2))\cdot P(\theta_2|A_j) + \cdots \\ &\quad + (M_n - u(A_n,\theta_2))\cdot P(\theta_n|A_j) \end{aligned} \tag{17.21}$$

In this example,

$$M_1 = \mathrm{Max}\{u(A_1,\theta_1), u(A_2,\theta_1)\}$$

$$M_2 = \mathrm{Max}\{u(A_1,\theta_2), u(A_2,\theta_2)\}$$

Determine the expected value of opportunity loss,

$$EOL_1 = (M_1 - u(A_1,\theta_1))\cdot P(\theta_1|A_1) + (M_2 - u(A_1,\theta_2))\cdot P(\theta_2|A_1) \tag{17.22}$$

$$EOL_2 = (M_1 - u(A_2,\theta_1))\cdot P(\theta_1|A_2) + (M_2 - u(A_2,\theta_2))\cdot P(\theta_2|A_2) \tag{17.23}$$

Since less EOL is preferred, the alternative with $\min(EOL_1, EOL_2)$ will be selected.

EXAMPLE 17.4

A new roadway is going to be constructed in a city. Two possible alternatives are considered. A larger capacity roadway will cost more than a smaller capacity roadway. However, demand will keep increasing in the following years. A smaller capacity roadway alternative has a higher probability of needing to be extended in 5 years than the larger capacity roadway alternative. The cost to construct and extend roadway and possibilities to extend are given below.

Construction cost: larger capacity roadway $1.5 million; smaller capacity roadway $1.0 million
Extension cost: larger capacity roadway $0.5 million; smaller capacity roadway $1.2 million
Possibilities to extend: larger capacity roadway 30%; smaller capacity roadway 80%
Determine which alternative is preferred?

Solution

Step 1

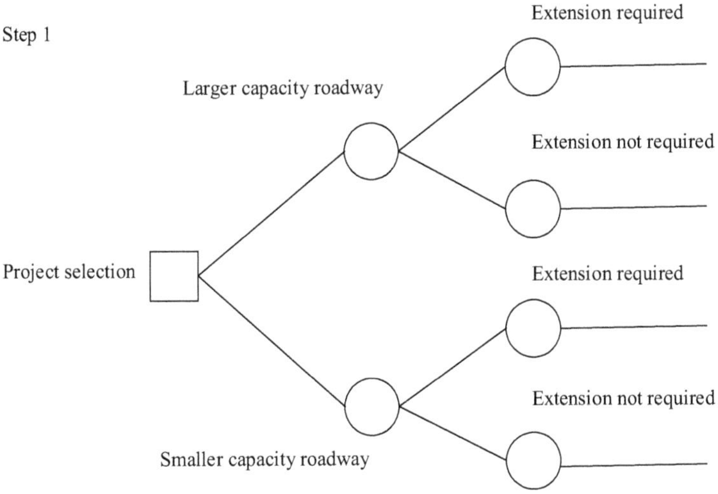

Step 2

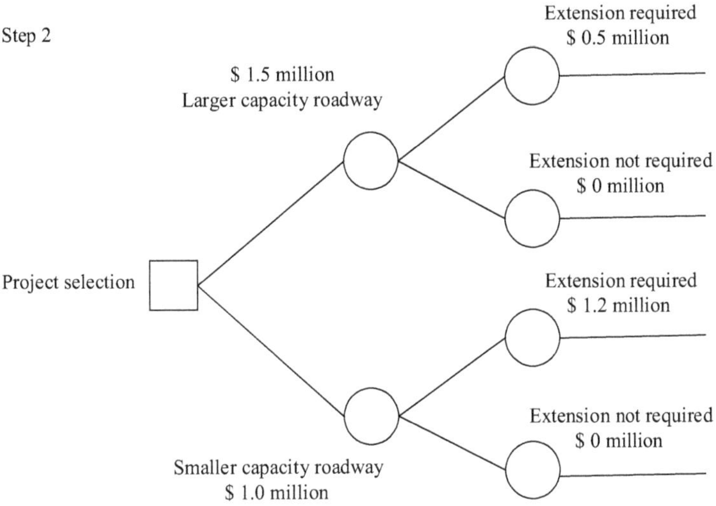

Step 3

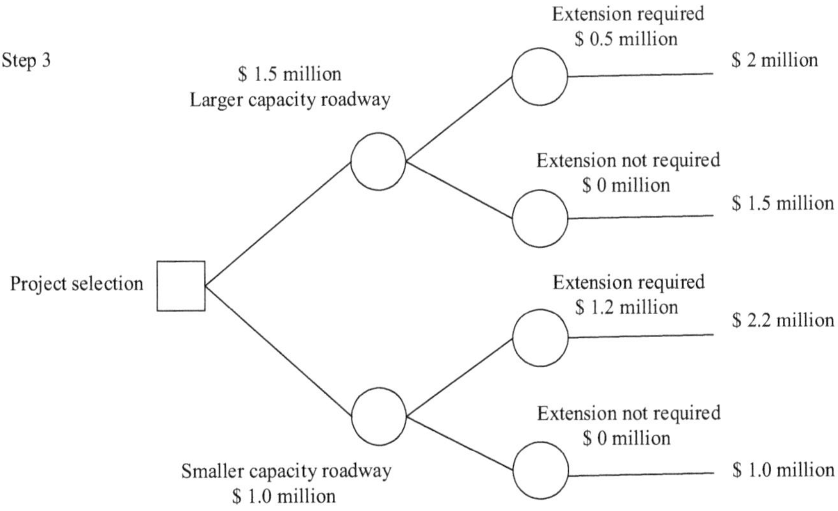

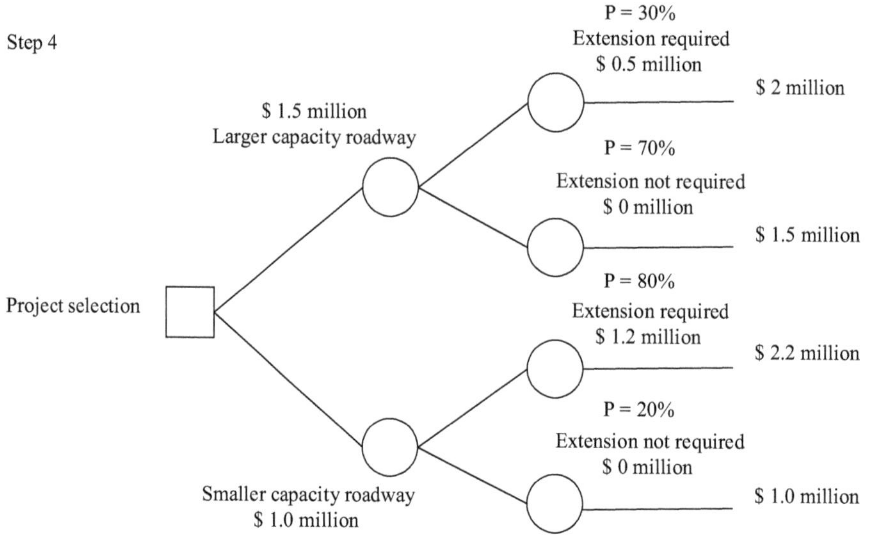

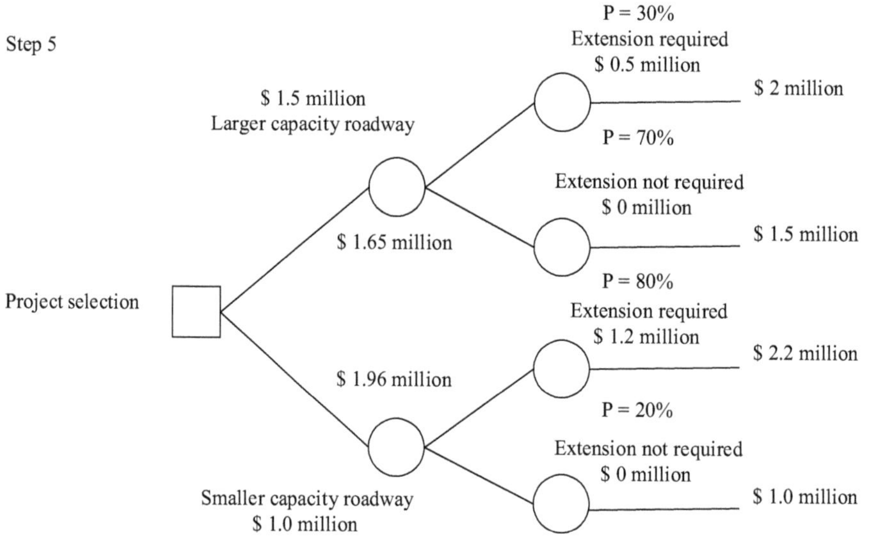

Therefore, the larger capacity road project is selected.

17.3.2 Bayesian models

17.3.2.1 Bayes' theorem for discrete probability models

Bayes' theorem provides a convenient way to determine conditional probabilities in some situations. For simplicity, suppose that the possible values of a parameter θ were assumed to be a set of discrete values $\theta_i(i = 1, 2, \ldots, n)$ with relative likelihoods $p_i = P(\Theta = \theta_i)$ (Θ is the random variable whose values represent possible values of the parameter θ). Then, if additional information becomes available (such as the results of a series of tests or surveys), the prior assumption on the parameter θ may be modified formally through Bayes' theorem as follows (Robbins, 1964; Box and Tiao, 1992; Paté-Cornell, 2002; Winkler, 2003; Grinstead and Snell, 2012).

Table 17.1 Some discrete distributions

Distribution	*Description*	*Parameter*	*Density function*	*Mean and variance*
Bernoulli	Take the value 1 with probability p and the value 0 with probability $q = 1 - p$	p $0 < p < 1$	$p^x(1-p)^{1-x}$	$E(X) = p$ $Var(X) = p(1-p)$
Binomial	The probability of exactly x occurrence among n trails in a Bernoulli sequence with the probability of occurrence of p in each trial	p $0 < p < 1$	$\binom{n}{x} p^x(1-p)^{n-x}$, $x = 0, 1, 2 \ldots n$	$E(X) = np$ $Var(X) = np(1-p)$
Geometric	In a Bernoulli sequence, the number of trails a specific event occurs for the first time	p $0 < p < 1$	$(1-p)^{x-1} p$, $x = 0,1,2 \ldots$	$E(X) = 1/p$ $Var(X) = (1-p)/p^2$
Hypergeometric	The probability of x successes in N draws without replacement	n, a, N	$\dfrac{\binom{a}{x}\binom{n-a}{N-x}}{\binom{n}{N}}$	$E(X) = \dfrac{na}{N}$
Negative binomial	The number of successes in a sequence of independent and identically distributed Bernoulli trials before a specified number of failures occurs	p $0 < p < 1$	$\binom{x-1}{n-1} p^n(1-p)^{x-n}$	$E(X) = \dfrac{n}{p}$
Poisson	The possible occurrence of events at point in time and/or space	v	$\dfrac{(\upsilon t)^x}{x!} e^{-\upsilon t}$, $x = 0, 1, 2 \ldots$	$E(X) = vt$ $Var(X) = vt$

Let ξ denote the observed outcome of the new information, the updated probability mass function (PMF) for Θ can then be obtained by Bayes' theorem

$$P(\Theta = \theta_i | \xi) = \frac{P(\xi | \Theta = \theta_i) \cdot P(\Theta = \theta_i)}{\sum_{i=1}^{n} P(\xi | \Theta = \theta_i) \cdot P(\Theta = \theta_i)}, \quad \forall i = 1, 2, \ldots, n \tag{17.24}$$

where $P(\xi|\Theta = \theta_i)$ is the likelihood of the experimental outcome ξ if $\Theta = \theta_i$, that is, the conditional probability of obtaining a particular experimental outcome assuming that the parameter is θ_i; $P(\Theta = \theta_i)$ represents the prior probability of $\Theta = \theta_i$, that is, prior to the availability of the new information ξ; $P(\Theta = \theta_i|\xi)$ denotes the posterior probability of $\Theta = \theta_i$, that is, the probability that has been revised in view of the new information outcome ξ.

The expected value of Θ is then commonly used as Bayesian estimator of the parameter, that is,

$$\hat{\theta} = E(\Theta|\xi) = \sum_{i=1}^{n} \theta_i P(\Theta = \theta_i|\xi) \tag{17.25}$$

Table 17.1 lists several commonly used discrete distributions.

EXAMPLE 17.5

A traffic engineer is interested in the average rate of vehicle crashes v at an improved road intersection. Suppose that from his previous experience with similar road and traffic

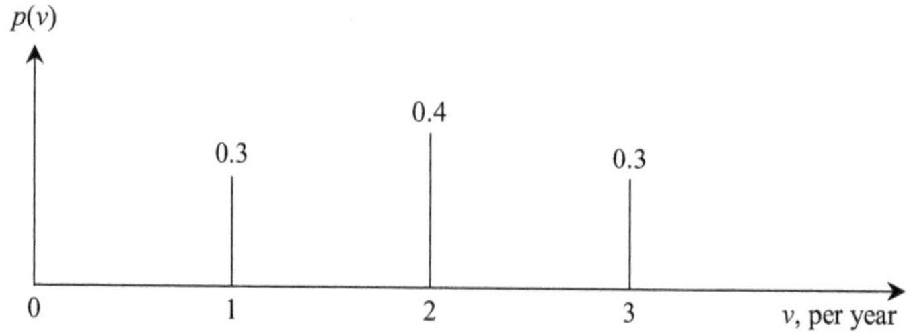

Figure 17.5 Prior distribution of v.

conditions, he deduced that the expected accident rate would between 1 and 3 per year, with an average of 2, and the prior PMF as shown in Figure 17.5. Occurrence of vehicle crashes is assumed to be a Poisson process. During the first month after completion of the intersection, one crash occurred. In the light of this observation, please revise the estimation for v (Table 17.1).

Solution

Let ξ be the event that one crash occurred in one month. The posterior probabilities then are

$$P(\nu = 1|\xi) = \frac{P(\xi|\nu = 1) \cdot P(\nu = 1)}{\sum_{i=1}^{3} P(\xi|\nu = 1) \cdot P(\nu = 1)}$$

$$= \frac{e^{-1/12}(1/12) \times (0.3)}{e^{-1/12}(1/12) \times (0.3) + e^{-1/6}(1/6) \times (0.4) + e^{-1/4}(1/4) \times (0.3)} = 0.166$$

Similarly,

$$P(\nu = 2|\xi) = 0.411$$

$$P(\nu = 3|\xi) = 0.423$$

Hence, the updated value of v is

$$\hat{v} = E(\nu|\xi) = 0.166 \times 1 + 0.411 \times 2 + 0.423 \times 3 = 2.26$$

17.3.2.2 Bayes' theorem for continuous probability models

In the previous section, the possible values of the parameter θ were limited to a discrete set of values, which was purposely assumed to simplify the presentation of the concepts underlying the Bayesian method of estimation (Robbins, 1964; Box and Tiao, 1992; Paté-Cornell, 2002; Winkler, 2003; Grinstead and Snell, 2012). In many situations, however, the value of a parameter might be in a continuum of possible values (e.g., cashes). Thus, it would be appropriate to assume the parameter to be a continuous random variable in the Bayesian estimation, which is discussed in the following:

Let Θ be the random variable for the parameter of a distribution with a prior density function $f(\theta)$. The prior probability that θ will be between θ_i and $(\theta_i + \Delta\theta)$ then is $f(\theta) \cdot \Delta\theta$. If ξ is a new observed outcome, the prior distribution $f(\theta)$ needs to be revised according to ξ using Bayes' theorem, getting the posterior probability that θ will be in $(\theta_i, \theta_i + \Delta\theta)$ as

$$Pf(\theta_i|\xi)\cdot\Delta\theta = \frac{P(\xi|\theta_i)f(\theta_i)\cdot\Delta\theta}{\sum_{i=1}^{n} P(\xi|\theta_i)f(\theta_i)\cdot\Delta\theta} \tag{17.26}$$

where, $P(\xi|\theta_i) = P(\xi|\theta_i < \theta \leq \theta_i + \Delta\theta)$. Without loss of generality, this yields

$$f(\theta|\xi) = \frac{f(\xi|\theta)f(\theta)}{\int_{-\infty}^{\infty} f(\xi|\theta)f(\theta)d\theta} \tag{17.27}$$

where $f(\xi|\theta)$ is the conditional probability or likelihood of observing the experimental outcome ξ assuming that the value of the parameter is θ.

Analogous to the discrete case, the expected value of Θ is commonly used as the point estimator of the parameter. The updated estimation of the parameter θ, in the light of new observed data ξ, is given by

$$\hat{\theta} = E(\Theta|\xi) = \int_{-\infty}^{\infty} \theta f(\xi|\theta)d\theta \tag{17.28}$$

Table 17.2 lists several commonly used discrete distributions.

EXAMPLE 17.6

Suppose that the vehicle travel speed in one segment follows a normal distribution. The prior parameters of observed speed distribution are mean $a = 60$ mph and standard deviation $b = 10$ mph. If additional $n = 10$ observations are gotten with a mean $m = 50$ mph and a standard deviation $\sigma = 5$ mph, please determine the posterior parameters.

Solution

Since the prior probability of the parameter is a normal distribution, the posterior probability will also be a normal distribution. Then, the posterior parameters can be determined by

$$\hat{a} = \frac{a\sigma^2 + mnb^2}{\sigma^2 + nb^2} = \frac{60\times25 + 50\times10\times100}{25+10\times100} = 50.24$$

$$\hat{b} = \sqrt{\frac{\sigma^2 b^2}{\sigma^2 + nb^2}} = \sqrt{\frac{25\times100}{25+10\times100}} = 1.56$$

Therefore, the expected travel speed is updated from 60 to 50.24 mph, while the standard deviation goes down from 10 to 2.44 mph. This provides a firmer distribution for travel speed as a performance measure for the computation (Table 17.2).

17.3.3 Stochastic process and Markov chain

17.3.3.1 Stochastic process

Many processes occurring in nature and those being studied in engineering and economics can be realistically described by time-dependent random variables. A set of random variables

Table 17.2 Some continuous distributions

Distribution	Description	Parameter	Density function	Mean and variance
Beta	A probability distribution appropriate for a random variable whose value is bounded in the beta distribution	a, b, q, r	$\frac{1}{B(q,r)} \frac{(x-a)^{q-1}(b-x)^{r-1}}{(b-a)^{q+r-1}}$, $a \le x \le b$	$E(X) = a + \frac{q}{q+r} \cdot (b-a)$ $Var(X) = \frac{qr}{(q+r)^2(q+r+1)} \cdot (b-a)^2$
Cauchy	Widely used as the canonical example of a "pathological" distribution since both expected value and variance are undefined		$\frac{1}{\pi(1+x^2)}$	$E(X) = 0$ $Var(X) = \infty$
Chi-square	A special case of the gamma distribution and widely used in inferential statistics	n	$\frac{1}{2^{n/2}\Gamma(n/2)}$	$E(X) = n$ $Var(X) = 2n$
Erlang	If $n = 1$, the Erlang distribution simplifies to the exponential distribution	n	$\frac{1}{\beta^n (n-1)!}$	$E(X) = n\beta$ $Var(X) = n\beta^2$ $Var(X) = n\beta_2$
Exponential	If occuring according to a Poisson distribution, then the time T_1 until the first occurrence of events has an exponential distribution	λ	$\lambda e^{-\lambda x}$, $x \ge 0$	$E(X) = 1/\lambda$ $Var(X) = 1/\lambda^2$
F	Frequently used in statistical test	n	$\frac{\Gamma\left(\frac{(n_1+n_2)}{2}\right) n_1^{\frac{n_1}{2}} n_2^{\frac{n_2}{2}} x^{\frac{n_1}{2}-1}}{\Gamma\left(\frac{n_1}{2}\right)\Gamma\left(\frac{n_2}{2}\right)(n_2+n_1 x)^{(n_1+n_2)/2}}$	$E(X) = \frac{n_2}{n_2 - 2}$
Gamma	Exponential distribution and chi-squared distribution are special cases of the gamma distribution	α, β	$\frac{1}{\beta^\alpha \Gamma(\alpha)} x^{\alpha-1} e^{-x/\beta}$	$E(X) = \alpha\beta$ $Var(X) = \alpha\beta^2$
Laplace	Two exponential distributions spliced together	β	$\frac{1}{2\beta} e^{-\lvert x \rvert/\beta}$	$E(X) = 0$
Lognormal	Lognormal distribution is a continuous probability distribution of a random variable whose logarithm is normally distributed	λ, ξ	$\frac{1}{\sqrt{2\pi}\xi x} e^{-\frac{1}{2}\left(\frac{\ln x - \lambda}{\xi}\right)^2}$, $x \ge 0$	$E(X) = e^{\lambda + \frac{1}{2}\xi^2}$ $Var(X) = E^2(X)(e^{\xi^2} - 1)$

(Continued)

Table 17.2 (Continued) Some continuous distributions

Distribution	*Description*	*Parameter*	*Density function*	*Mean and variance*
Normal (Gaussian)	The normal distribution is the probability distribution that plots all of its values in a symmetrical fashion, and most of the results are situated around the probability's mean	μ, σ	$\frac{1}{\sqrt{2\pi}\sigma} e^{-\frac{1}{2}\left(\frac{x-\mu}{\sigma}\right)^2}$, $-\infty \le x \le +\infty$	$E(X) = \mu$ $Var(X) = \sigma^2$
Rayleigh	The Rayleigh distribution is a special case of the Weibull distribution with a scale parameter of 2	α	$\frac{x}{\alpha^2} \cdot e^{-\frac{1}{2}\left(\frac{x}{\alpha}\right)^2}$, $x \ge 0$	$E(X) = \sqrt{\frac{\pi}{2}}\alpha$ $Var(X) = \left(2 - \frac{\pi}{2}\right)\alpha^2$
Student-*t*	Estimating the mean if a normal distribution with small sample size and the standard deviation is unknown	n	$\frac{\Gamma\left(\frac{n+1}{2}\right)\left(1+\frac{x^2}{n}\right)^{-(n+1)/2}}{\sqrt{n\pi}\Gamma\left(\frac{n}{2}\right)}$	$E(X) = 0$ $Var(X) = \frac{n}{n-2}$
Triangular	The triangular distribution is a continuous probability distribution with lower limit *a*, upper limit *b*, and mode *u*	a, b, u	$\frac{2}{h-a}\left(\frac{x-a}{u-a}\right), a \le x \le u$ $\frac{2}{b-a}\left(\frac{b-x}{b-u}\right), u \le x \le b$	$E(X) = (a + b + u)/3$ $(a^2 + b^2 + u^2 - ab - au - bu)$
Uniform	Uniform distribution refers to a probability distribution for which all of the values that a random variable can take on occur with equal probability	a, b	$\frac{1}{b-a}$, $a < x < b$	$E(X) = (a + b)/2$ $Var(X) = (b - a)^2/12$
Weibull	Weibull distribution interpolates between the exponential distribution and the Rayleigh distribution	α, β	$\beta\alpha x^{\alpha-1} e^{-\beta x^{\alpha}}$	$E(X) = \beta^{-1/\alpha}\Gamma(1 + 1/\alpha)$

depending on one parameter is called a stochastic process (Heyman and Sobel, 1982; Bronshtein and Semendyayev, 2013). The parameter, in general, can be considered as time t, so the random variable can be denoted by X_t and the stochastic process is given by the set

$$\{X_t | t \in T\}$$

where the set of parameter values is called the variable space T, and the set of values of the random variables is the state space Z.

If both the parameter space and the state space are discrete, that is, the state variable X_t and the parameter t only have finite and countably infinite different values, then the stochastic process is called a stochastic chain. In this case, the different states and different parameter values can be numbered as

$$Z = \{1, 2, \ldots, i, i+1, \ldots\} \tag{17.29}$$

$$T = \{t_0, t_1, \ldots, t_m, t_{m+1}, \ldots\} \text{ with } 0 \le t_0 < t_1 < \cdots < t_m < t_{m+1} < \cdots$$

where the times t_0, t_1, ... are not necessarily equally spaced.

17.3.3.2 Markov chain and transition probabilities

If the probability of the different values of X_{m+1} in a stochastic chain depends only on the state at t_m, then the stochastic chain is further called a Markov chain (Gilks et al., 1995). The Markov property is defined by the requirement that

$$P(X_{m+1} = j | X_1 = x_1, \ldots, X_m = x_m) = P(X_{m+1} = j | X_m = x_m) \tag{17.30}$$

where, $X_1 = x_1$, ... , $X_m = x_m$ represent all previous states of the system. Consider a Markov chain and times t_m and t_{m+1}. The conditional probability

$$p_{i,j}(m, m+1) = P(X_{m+1} = j | X_m = i) \tag{17.31}$$

is called the transition probability which determines the probability by which the system changes from the state $X_m = i$ at t_m to the state $X_{m+1} = j$ at t_{m+1}.

If the state space of a Markov chain is finite, that is, $Z = \{1, 2, \ldots, N\}$, then the transition probabilities $p_{i,j}(t_1, t_2)$ between the states at t_1 and t_2 can be represented by the so-called transition probability matrix

$$P(t_1, t_2) = \begin{bmatrix} p_{1,1}(t_1,t_2) & p_{1,2}(t_1,t_2) & \cdots & p_{1,N}(t_1,t_2) \\ p_{2,1}(t_1,t_2) & p_{2,2}(t_1,t_2) & \cdots & p_{2,N}(t_1,t_2) \\ \vdots & \vdots & \cdots & \vdots \\ p_{N,1}(t_1,t_2) & p_{N,2}(t_1,t_2) & \cdots & p_{N,N}(t_1,t_2) \end{bmatrix} \tag{17.32}$$

where times t_1 and t_2 are not necessarily consecutive.

17.3.3.3 Homogeneous Markov chains

If the transition probability of the Markov chain does not depend on the time, that is,

$$p_{i,j}(t_m, t_{m+1}) = p_{i,j}$$

then the Markov chain is called homogeneous or stationary. A homogeneous Markov chain with a finite state space $Z = \{1, 2, \ldots, N\}$ has the transition probability matrix

$$P = \begin{bmatrix} p_{1,1} & p_{1,2} & \cdots & p_{1,N} \\ p_{2,1} & p_{2,2} & \cdots & p_{2,N} \\ \vdots & \vdots & \cdots & \vdots \\ p_{N,1} & p_{N,2} & \cdots & p_{N,N} \end{bmatrix} \tag{17.33}$$

where, (1) $p_{i,j} \geq 0$ for all i, j and (2) $\sum_{j=1}^{N} p_{ij} = 1$, for all i.

Being independent of time, $p_{i,j}$ gives the transition probability from the state i to the state j during time unit. Probabilities in each row of the matrix P constitute a probability vector. Let $p(t)$ be the probability vector at time t, which denotes the distribution of the states of a homogeneous Markov chain at time t, then the distribution probability after k time intervals can be calculated by

$$\bar{p}(t+k) = \bar{p}(t) \cdot P^k \tag{17.34}$$

Particularly for $t = 0$, it can be written in the form

$$\bar{p}(k) = \bar{p}(0) \cdot P^k \tag{17.35}$$

That is, a homogeneous Markov chain is uniquely determined by the initial distribution $\bar{p}(0)$ and the transition probability matrix P. Therefore, this matrix can be used in predicting or projecting future conditions in the probability sense for the same environment.

EXAMPLE 17.7

It is well known that weather conditions have a crucial effect on traffic flow. Consider a city that never has two good weather days in a row. If it has a good weather day, it is just as likely to have snow as rain the next day. If it has snow or rain, it has an even chance of having the same the next day. If there is change from snow or rain, only half of the time is this a change to a good weather day. Let the initial probability vector u equal (1/3,1/3,1/3), which denotes the probability distribution for rain, good weather, and snow days today, please calculate the distribution of weather conditions after 3 days.

Solution

With the above information, the transition probability matrix of the homogeneous Markov chain can be obtained.

$$\begin{array}{ll} & \begin{array}{ccc} \text{rain} & \text{good weather} & \text{snow} \end{array} \\ P = \begin{array}{l} \text{rain} \\ \text{good weather} \\ \text{snow} \end{array} & \begin{bmatrix} 1/2 & 1/4 & 1/4 \\ 1/2 & 0 & 1/2 \\ 1/4 & 1/4 & 1/2 \end{bmatrix} \end{array}$$

Then, the distribution of the states after 3 days can be calculated

$$u^{(3)} = u \cdot P^3 = [1/3 \quad 1/3 \quad 1/3] \begin{bmatrix} 1/2 & 1/4 & 1/4 \\ 1/2 & 0 & 1/2 \\ 1/4 & 1/4 & 1/2 \end{bmatrix}^3 = [0.401 \quad 0.198 \quad 0.401]$$

That is, if the probability for rain, good weather, and snow today is all equal to 1/3, then it will be 0,401, 0.198, and 0.401 after 3 days, respectively.

17.3.3.4 Nonhomogeneous Markov models

As suggested by its name, the nonhomogeneous Markov chain model does not assume a homogeneous behavior of the stochastic process (Aalen and Johansen, 1978). In other words, the transition probability matrix P is not a constant but a function of time. An example appropriate for such a treatment is a scenario of special climate (such as a very warm winter that accelerates steel corrosion) in a certain year that significantly alters the mechanism of bridge element deterioration in the bridge management application (Fu and Devaraj, 2008).

Based on the homogeneous Markov chain concept, the transition probability matrix of a nonhomogeneous Markov chain over k time intervals is defined as the product of k one-time interval transition probability matrixes

$$\bar{p}(k) = \bar{p}(0).\prod_{i=1}^{k} P^{(i)} \tag{17.36}$$

where, $P^{(i)}$ denotes the transition probability matrix of the ith one-time interval.

For better understanding of this transition nature, a simple arithmetic method that uses the observed state change data over a period and thereby estimates the transition probabilities is introduced here as an illustration.

EXAMPLE 17.8

Let us consider an example for concrete deck (in square meters) for all bridges of a state-maintained highway network with an inspection interval of 2 years. The distribution for these bridges is given in Table 17.3.

It indicates that for these bridges at the beginning of the 2-year period, 24896 m^2 of concrete decks are in state 1, 34104 in state 2, 15800 in state 3, 10300 in state 4, and 5200 in state 5. Two years later, the same concrete decks would be expected to have 17399 in state 1, 34801 in state 2, 17200 in state 3, 13200 in state 4, and 7700 in state 5.

Solution

By comparison of $\{X^{(0)}\}$ and $\{Y^{(1)}\}$, it is seen that 17399 m^2 of the concrete decks remain in state 1 after 2 years of service. In other words, 7497 (=24896–17399) of the 24896 m^2 deteriorated to state 2. Out of 34104 of the concrete decks in state 2, 27304 [=34801–(24896–17399)] would stay in state 2 and 6800 (=34104–27304) would deteriorate to state 3. The analysis can be repeated until reaching the last state (Table 17.4).

Accordingly, the transition probability $p_{i,j}$ for the bare concrete deck of bridges from state i at t_{n-2} to state j at t_n can be estimated as follows:

Table 17.3 An example of the bridge deck condition state distributions

$\{X^{(0)}\}$	*x(0) 1*	*x(0) 2*	*x(0) 3*	*x(0) 4*	*x(0) 5*
m^2	24896	34104	15800	10300	5200
$\{Y^{(1)}\}$	y(1) 1	y(1) 2	y(1) 3	y(1) 4	y(1) 5
m^2	17399	34801	17200	13200	7700

Table 17.4 An example of estimating transition probabilities

$\{X^{(0)}\}$	$x(0)$ 1	$x(0)$ 2	$x(0)$ 3	$x(0)$ 4	$x(0)$ 5
m^2	24896	34104	15800	10300	5200
$\{Y^{(1)}\}$	$y(1)$ 1	$y(1)$ 2	$y(1)$ 3	$y(1)$ 4	$y(1)$ 5
m^2	17399	34801	17200	13200	7700
Quantity that stayed in the same state	17399	27304	10400	7800	5200

$$p = \begin{array}{c} \text{state} \\ \text{in} \\ 0 \\ \text{year} \end{array} \begin{array}{c} \\ \\ \end{array} \overset{\text{state in 2 years}}{\begin{bmatrix} p_{1,1} = \frac{17399}{24896} & p_{1,2} = \frac{24896 - 17399}{24896} & p_{1,3} = 0 & p_{1,4} = 0 & p_{1,5} = 0 \\ p_{2,1} = 0 & p_{2,2} = \frac{27304}{34104} & p_{2,3} = \frac{34104 - 27304}{34104} & p_{2,4} = 0 & p_{2,5} = 0 \\ p_{3,1} = 0 & p_{3,2} = 0 & p_{3,3} = \frac{10400}{15800} & p_{3,4} = \frac{15800 - 10400}{15800} & p_{3,5} = 0 \\ p_{4,1} = 0 & p_{4,2} = 0 & p_{4,3} = 0 & p_{4,4} = \frac{7800}{10300} & p_{4,5} = \frac{10300 - 7800}{10300} \\ p_{5,1} = 0 & p_{5,2} = 0 & p_{5,3} = 0 & p_{5,4} = 0 & p_{5,5} = \frac{5200}{5200} \end{bmatrix}}$$

The subsequent matrixes can be done in the same manner. It should be emphasized that this estimation approach is useful in understanding the concept of transition, particularly when used for small data sets. When used for a larger data set, however, it loses reliability due to lack of a statistical basis, and some other approaches should be on the stage.

17.3.4 Simulation models

Simulation is the process of replicating the real world based on a set of assumptions and conceived models of reality. For engineering purposes, simulation may be used to predict or study the performance and/or response of a system. With a prescribed set of values for the system parameters, the simulation process yields a specific measure of performance or response. Through repeated simulations, the sensitivity of the system performance to variation in the system parameters may be examined or assessed. By this procedure, simulation can be applied to appraise alternative designs or determine optimal designs.

17.3.4.1 Monte Carlo method

For problems involving random variables with known or assumed probability distributions, Monte Carlo simulation can be used. This involves repeating a simulation process, using in each simulation a particular set of values of the random variables generated in accordance with the corresponding probability distributions. By repeating the process, a sample of solutions, each corresponding to a different set of values of the random variables, is obtained. A sample from the Monte Carlo simulation is similar to a sample of experimental observations (Metropolis and Ulam, 1949).

The automatic generation of the requisite random numbers with specified distributions can be accomplished systematically for each variable by first generating a uniformly distributed random number between 0 and 1.0, and through appropriate transformations obtaining the corresponding random number with the specified probability distribution. The basis is as follows:

Suppose a random variable X with cumulative distribution function (CDF) $F_X(x)$. Then, at a given cumulative probability $F_X(x) = u$, the value of X is

$$x = F_X^{-1}(u) \tag{17.37}$$

Now suppose that u is a value of the standard uniform variate, U, with a uniform PDF between 0 and 1.0. Then,

$$F_U(u) = u \tag{17.38}$$

that is, the cumulative probability of $U \leq u$ is equal to u.

Therefore, if u is a value of U, the corresponding value of the variate X obtained by (·) will have a cumulative probability,

$$\begin{aligned} P(X \leq x) &= P[F_X^{-1}(U) \leq x] \\ &= P[U \leq F_X(x)] \\ &= F_U[F_X(x)] = F_X(x) \end{aligned} \tag{17.39}$$

which means that if $(u_1, u_2, \ldots, u_n)$ is a set of values from U, the corresponding set of values obtained through (·), that is,

$$x_i = F_X^{-1}(u_i), \quad i = 1, 2, \ldots, n \tag{17.40}$$

will have the desired CDF $F_X(x)$.

EXAMPLE 17.9

Consider the exponential distribution with the CDF

$$F_X(x) = 1 - e^{-\lambda x} \quad x > 0$$

Apply the Monte Carlo method.

Solution

The inverse function is

$$x = F_X^{-1}(u) = -\frac{1}{\lambda}\ln(1-u)$$

In this case, once the standard uniformly distributed random number u_i, $i = 1, 2, \dots, L$, are generated, the corresponding exponentially distributed random numbers can be obtained as

$$x_i = -\frac{1}{\lambda}\ln(1 - u_i)$$

Since $(1 - u_i)$ is also uniformly distributed, the required random numbers may be generated by

$$x_i = -\frac{1}{\lambda}\ln u_i, \quad i = 1, 2, \dots, L$$

17.3.4.2 Latin hypercube method

From the preceding section, it can found that the Monte Carlo method is entirely random in principle, that is, any given sample value may fall anywhere within the range of the input distribution. This implies that a great number of samples are typically required in traditional Monte Carlo to achieve good accuracy. Therefore, various techniques have been developed to improve Monte Carlo accuracy. Particularly, Latin hypercube method is a widely used technique to generate controlled random samples, the main idea of which is to make sampling point distribution close to the PDF.

Latin hypercube is the generalization of Latin square, which is a square grid containing sample points if (and only if) there is only one sample in each row and each column, to an arbitrary number of dimensions, whereby each sample is the only one in each axis-aligned hyperplane containing it. Obviously, this method can be used to generate "better" simulation results, with lower standard error levels, with fewer trials. For complex models with many random variables, this means one can generate results in less time (McKay et al., 1979).

EXAMPLE 17.10

Consider sampling a function of two variables, the range of each variable is divided into five equal probable intervals. Five sample points can then be selected randomly to satisfy the Latin square requirement, as seen in Figure 17.6. Note that this forces the number of divisions, 5, to be equal for each variable.

17.3.4.3 Variance reduction techniques

In Bayesian estimation, whether the data of an additional variable can reduce the uncertainty of a state variable is of interest, especially considering that the data from simulations (e.g., Monte Carlo method and Latin hypercube method) are often subject to sampling errors. It is desired to estimate the uncertainty reduction of the state variable before conducting Bayesian models to assure that the data collection is not waste of effort. To handle such a problem, variance reduction techniques are often used. The main ones include antithetic variates, correlated sampling, and control variates, and so on.

a. *Antithetic variates*: The method of antithetic variates attempts to reduce variance by introducing negative correlation between pairs of observation. Suppose Z' and Z'' are two unbiased estimators of Z, for example, from two separate samples. The two estimators may be combined to form another estimator

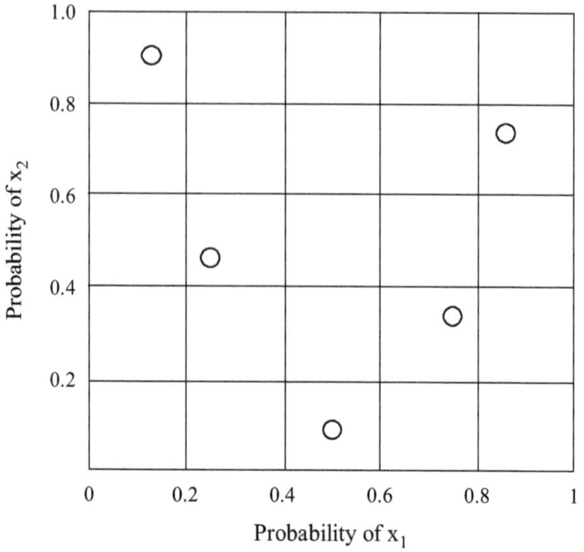

Figure 17.6 Illustration of the two-dimensional Latin hypercube method.

$$Z_A = \frac{Z' + Z''}{2} \tag{17.41}$$

The expected value of Z_A is

$$E(Z_A) = \frac{E(Z') + E(Z'')}{2} = \frac{Z + Z}{2} = Z \tag{17.42}$$

which means that Z_A is an unbiased estimator.
The corresponding variance is

$$Var(Z_A) = \frac{Var(Z') + Var(Z'') + 2Cov(Z', Z'')}{4} \tag{17.43}$$

Obviously, the accuracy of the estimator Z_A can be improved in two cases: (1) if the estimators Z' and Z'' are statistically independent, that is, $Cov(Z', Z'') = 0$; (2) if the estimators Z' and Z'' are negatively correlated estimators, that is, $Cov(Z', Z'') < 0$. A numerical sampling procedure to ensure negative correlation between Z' and Z'' is the following antithetic variates method developed by Hammersley and Morton (1956).

Suppose that a set of n uniformly distributed random numbers $u_1, u_2, \ldots, u_n$ is generated to obtain the estimator Z'. The related set of random numbers $1 - u_1, 1 - u_2, \ldots, 1 - u_n$ is subsequently generated to obtain Z''. The resulting Z' and Z'' will be negatively correlated.

EXAMPLE 17.11

Consider a transportation network of two connecting branches with respective travel times T_1 and T_2, such that the total travel time is $T = T_1 + T_2$. Suppose T_1 follows a standard uniform distribution. Whereas T_2 is exponentially distributed with a mean travel time of 1.0. It is desired to estimate the mean travel time $E(T)$ by Monte Carlo simulation.

Solution

Two sets of n random numbers u_{11}, u_{12}, … , u_{1n} and u_{21}, u_{22}, … , u_{2n} are first generated from the standard uniform distribution. Then, the values of T_1 and T_2 are

$$t''_{1i} = 1 - u_{1i}$$

$$t''_{2i} = -\ln(1 - u_{2i})$$

and

$$t''_1 1 - u_{1i} - \ln(1 - u_{2i}) \quad \forall i = 1,2,\ldots,n$$

The combined estimator is

$$Z_A = \frac{1}{2}(T' + T'') = \frac{1}{2n}\sum_i [(T'_{1i} + T'_{2i}) + (T''_{1i} + T''_{2i})]$$

whose variance is

$$Var(Z_A) = \frac{n}{4n^2} \cdot Var[(T'_{1i} + T'_{2i}) + (T''_{1i} + T''_{2i})]$$

But,

$$T'_{1i} + T''_{1i} = U_{1i} + (1 - U_{1i}) = 1$$

Therefore,

$$Var(Z_A) = \frac{1}{4n} \cdot Var(T'_{2i} + T''_{2i}) = \frac{1}{4n} \cdot [Var(T'_{2i}) + Var(T''_{2i}) + 2Cov(T'_{2i}, T''_{2i})]$$

Observe that

$$Var(T'_{2i}) = Var[-\ln U] = \int_0^1 (\ln u)^2\,du - \left[\int_0^1 -\ln u\,du\right]^2 = 2 - 1 = 1$$

$$Var(T''_{2i}) = Var[-\ln(1-U)] = \int_0^1 [\ln(1-u)]^2\,du - \left[\int_0^1 -\ln(1-u)\,du\right]^2 = 2 - 1 = 1$$

and

$$Cov(T'_{2i}, T''_{2i}) = E[\{T'_{2i} - E(T'_{2i})\}\{T''_{2i} - E(T''_{2i})\}] = E[(T'_{2i} - 1)(T''_{2i} - 1)]$$

$$= E[\{\ln U + 1\}\{\ln(1-U) + 1\}] = 1 + \int_0^1 \ln(1-u)\ln u\,du + \int_0^1 \ln(1-u)\,du + \int_0^1 \ln u\,du$$

$$= 1 + \left(2 - \frac{\pi^2}{6}\right) - 1 - 1 = 1 - \frac{\pi^2}{6}$$

Hence,

$$Var(Z_A) = \frac{1}{4n}\left[1+1+2-\frac{\pi^2}{6}\right] = \frac{0.178}{n}$$

As seen, the application of the antithetic variates technique significantly improves the accuracy of the estimated expected travel time $E(T)$.

b. *Correlated sampling*: The correlated sampling method is an effective variance reduction technique to compare the difference in performance between alternative system designs. Suppose the performance of a design A is

$$Z_A = g(A, X)$$

where $A = (a_1, a_2, \dots, a_m)$ is a set of design values for design A; $X = (x_1, x_2, \dots, x_n)$ is a set of random variables.

Similarly, the performance of another design B is

$$Z_B = g(B, X)$$

where $B = (b_1, b_2, \dots, b_m)$ is a set of design values for design B.
The difference in performance between design A and B, therefore, is

$$Z = Z_A - Z_B$$

Because Z_A and Z_B could be highly correlated, the method of correlated sampling may be used to estimate the statistics of Z, for example, its mean-value, $\bar{Z}$.
The variance of $\bar{Z}$ is

$$Var(\bar{Z}) = Var(\bar{Z}_A) + Var(\bar{Z}_B) - 2Cov(\bar{Z}_A, \bar{Z}_B) \tag{17.44}$$

Therefore, if Z_A and Z_B are positively correlated, that is, $Cov(Z_A, Z_B) > 0$,

$$Var(\bar{Z}) < Var(\bar{Z}_A) + Var(\bar{Z}_B) \tag{17.45}$$

The random numbers z_{Aj} and z_{Bj} will be positively correlated if they are generated as follows:

$$\begin{aligned} z_{A_j} &= g[A, F_{x1}^{-1}(u_1), F_{x2}^{-1}(u_2), \dots, F_{xn}^{-1}(u_n)] \\ z_{B_j} &= g[B, F_{x1}^{-1}(u_1), F_{x2}^{-1}(u_2), \dots, F_{xn}^{-1}(u_n)] \end{aligned} \tag{17.46}$$

where $(u_1, u_2, \dots, u_n)$ is an independent set of uniformly distributed random numbers.

EXAMPLE 17.12

Suppose the performance function of a system associated with design A is $Z_A = 3X_1 + 2X_2$ and the corresponding performance function associated with design B is $Z_B = 2X_1 + 4X_2$, that is, $A = \{3, 2\}$ and $B = \{2, 4\}$. Moreover, assume that X_1 and X_2 are uniformly distributed variates. Estimate the mean difference in performance between design A and B.

Solution

Two sets of n uniformly distributed random numbers $u_{11}, u_{12}, \ldots, u_{1n}$ and $u_{21}, u_{22}, \ldots, u_{2n}$ are first generated for X_1 and X_2, respectively. Then, according to the method of correlated sampling,

$$Z_{A_j} = 3u_{1j} + 2u_{2j}$$

and

$$Z_{B_j} = 2u_{1j} + 4u_{2j}$$

The estimator of the mean difference is

$$\bar{Z} = \bar{Z}_A - \bar{Z}_B = \frac{1}{n}\left[\sum_j (3U_{1j} + 2U_{2j}) - \sum_j (2U_{1j} + 4U_{2j})\right],$$

whose variance is

$$Var(\bar{Z}) = \frac{1}{n^2}\left\{Var\left[\sum_j (3U_{1j} + 2U_{2j})\right] + Var\left[\sum_j (2U_{1j} + 4U_{2j})\right]\right\}$$

$$-2Cov\left[\sum_j (3U_{1j} + 2U_{2j}), \sum_j (2U_{1j} + 4U_{2j})\right]$$

$$= \frac{1}{n}\{[9 \cdot Var(U_1) + 4 \cdot Var(U_2)] + [4 \cdot Var(U_1) + 16 \cdot Var(U_2)]\}$$

$$-2E[(3U_1 + 2U_2)(2U_1 + 4U_2)] + 2E(3U_1 + 2U_2)E(2U_1 + 4U_2)$$

$$= 0.417/n$$

c. *Control variates*: The control variates method is a variance reduction technique which exploits information about the errors of the estimations for known quantities to reduce the error of an estimate for an unknown quantity. Suppose $\bar{Z}$ is an estimator. Sometimes, estimation accuracy may be gained through an indirect estimator $\bar{Y}$. For example,

$$\bar{Y} = \bar{Z} - \beta(C - \mu_c) \tag{17.47}$$

where C is the random variable with known mean μ_c and is correlated with $\bar{Z}$; β is a coefficient. C is called a controlled variate for $\bar{Z}$. Observe that

$$E(\bar{Y}) = E(\bar{Z}) - \beta[E(C) - \mu_c] = E(\bar{Z}) \tag{17.48}$$

Hence, if $\bar{Z}$ is an unbiased estimator, then $\bar{Y}$ is also an unbiased estimator. Moreover, the variance of $\bar{Y}$ is given by

$$Var(\bar{Y}) = Var(\bar{Z}) - \beta^2 Var(C) - 2\beta Cov(\bar{Z}, C) \tag{17.49}$$

Therefore, if

$$\beta^2 Var(C) < 2\beta Cov(\bar{Z}, C)$$

Then,

$$Var(\bar{Y}) < Var(\bar{Z})$$

This implies that the indirect estimator $\bar{Y}$ is more accurate than the direct estimator $\bar{Z}$. In fact, the maximum reduction of variance may be achieved by selecting a value of β such that $Var(\bar{Y})$ is minimized. This is obtained by

$$\frac{\partial Var(\bar{Y})}{\partial \beta} = 2\beta Var(C) - 2Cov(\bar{Z},C) = 0 \tag{17.50}$$

then the optimal value of β is

$$\beta^* = \frac{Cov(\bar{Z},C)}{Var(C)} \tag{17.51}$$

and the corresponding optimum $Var(\bar{Y})$ is

$$Var(\bar{Y}) = Var(\bar{Z}) + \frac{Cov^2(\bar{Z},C)}{Var^2(C)} Var(C) - \frac{2Cov^2(\bar{Z},C)}{Var(C)} = Var(\bar{Z})[1 - \rho^2_{\bar{Z},C}] \tag{17.52}$$

where $\rho_{\bar{Z},C}$ is the correlation coefficient between $\bar{Z}$ and C.

It is found that the reduction in variance increases as $\rho_{\bar{Z},C}$ increases. Therefore, if C can be selected in order that it is greatly dependent on $\bar{Z}$, for example, through a good approximate model, considerable improvement in the accuracy of estimation may be achieved. All too often, to apply the control variates method, the value of β^* needs to be estimated earlier from simulations.

17.3.5 Stochastic programming

Stochastic programming (SP) includes the class of problems where the customary problem of mathematical programming in its essential structure is preserved, but where random instead of deterministic variables appear either in objective function and/or in some of constraints (Heyman and Sobel, 1984; Kall and Wallace, 1994; Birge and Louveaux, 1997). For simplicity, if a problem

$$F(X_0, Z) = \min_{x \in \{X\}} F(X, Z) \tag{17.53}$$

is to be interpreted like an ordinary mathematical programming problem, and particularly Z is a set of constraints.

This problem is no longer directly solvable, however, if X or Z represent random variables or vectors. In this case, the problem becomes choosing the random variable X_0 from the set $\{X\}$ of the decision variables X for which $\Phi[F(X,Z)]$ is minimized, that is,

$$\Phi[F(X_0, Z)] = \min_{x \in \{X\}} \Phi[F(X, Z)] \tag{17.54}$$

The natural idea of solving the above SP problem is to find a "constitute program" that is deterministic and nevertheless provides an optimal solution to the original stochastic program. Of course, this implies that the distribution probability or density function over the random variables $F(X,Z)$ or its relevant parameters are known. The two-stage programming and chance-constraint programming are often viewed as the typical examples of SP, and the reader is referred to Sakawa et al. (2011).

17.3.6 Loss functions

Loss function, also called cost function, is a function that maps an action or values of one or more variables onto a real number intuitively representing some "cost" associated with the action. In some situations, decisions need to be made on the basis of the value of the loss function. Although there exists an infinite number of loss functions, only a few of them have met with special interest in theory and practice (Menges, 1974).

17.3.6.1 Linear loss function

The linear loss function is of the following specification:

$$\nu = \alpha + \beta z \, (\alpha, \beta \in R) \tag{17.55}$$

is most universally applicable and simple in practical use. In sufficiently small regions of z, it can serve as an approximation of every other loss function.

17.3.6.2 Cramer and Bernoulli loss functions

The Cramer and Bernoulli loss functions are the two oldest types. The Cramer loss function is written as

$$\nu = \alpha - \beta\sqrt{z} \tag{17.56}$$

and the Bernoulli loss function reads

$$v = \alpha - \beta \log z \tag{17.57}$$

where v denotes the loss and z the values of the consequences of actions, with fixed parameters $\alpha(\alpha \geq 0)$ and $\beta(\beta \geq 0)$.

Both types of loss functions have the property that the marginal loss increases as z increases.

17.3.6.3 The quadratic loss function

The quadratic function can be written as

$$\nu = \alpha - \beta z + \lambda z^2, (\alpha \geq 0, \beta \geq 0, \lambda > 0) \tag{17.58}$$

Its characteristic property is that deviations, in both directions, of the assumed state from the true state are equally weighted, and that the weights grow larger, in quadratic progression, the greater the deviation.

17.3.6.4 Gaussian sum loss function

The Gaussian sum loss function is given as

$$\nu = \frac{1}{\sqrt{\alpha}} \int_{-\infty}^{z} e^{\frac{z^2}{\alpha}} dz, \;\; (\alpha > 0) \tag{17.59}$$

As z goes from $-\infty$ to ∞, v increases from 0 to 1.

17.3.7 Markowitz mean–variance models

The Markowitz portfolio theory provides a method to analyze how good a given portfolio is based on only the means and the variance of the returns of the assets contained in the portfolio (Markowitz, 1991; Luenberger, 1997; Bodie et al., 2006). For simplicity, suppose there are N risky assets whose rates of return are given by the random variables $R_1, R_2, \ldots, R_N$, that is,

$$R_n = \frac{S_n(1) - S_n(0)}{S_n(0)}, \quad n = 1, 2, \ldots, N \tag{17.60}$$

where, $S_n(1)$ and $S_n(0)$ is the final and initial value of the nth asset, respectively.

Let $w = (w_1, w_2, \ldots, w_N)^T$, w_n denotes the proportion of wealth invested in asset n, with $\sum_{n=1}^{N} w_n = 1$. The rate of return of the portfolio is

$$R_p = \sum_{n=1}^{N} w_n R_n \tag{17.61}$$

If μ_n and σ_n^2 denotes the mean and variance of R_n and σ_{ij} the variance between R_i and R_j, then

$$\mu_p = E(R_p) = \sum_{n=1}^{N} E(w_n R_n) = \sum_{n=1}^{N} w_n \mu_n \tag{17.62}$$

and

$$\sigma_p^2 = Var(R_p) = \sum_{i=1}^{N} \sum_{j=1}^{N} w_i w_j cov(R_i, R_j) = \sum_{i=1}^{N} \sum_{j=1}^{N} w_i \sigma_{ij} w_j \tag{17.63}$$

It can be found that for different choice of $w_1, w_2, \ldots, w_N$, different μ_p and σ_p^2 will be obtained. The set of all possible (μ_p, σ_p^2) combinations is called an attainable set. Those (μ_p, σ_p^2) with minimum σ_p^2 for a given μ_p or more and maximum μ_p for a given σ_p^2 or less are called efficient set (or efficient frontier). Since an investor wants a high profit and a small risk, he/she aims to maximize μ_p and minimize σ_p^2, therefore, he/she should choose a portfolio which gives a (μ_p, σ_p^2) combination in the effcient set, depending on how risk averse he/she is. The Markowitz mean–variance model laid the foundation for modern portfolio theory but it is not used in practice. The main reason for this is that it requires a huge amount of data. More useful models have, however, been developed from the Markowitz model by use of approximation.

17.4 DECISION MAKING USING UTILITY THEORY

17.4.1 Utility function

The conception of utility is widely used in the majority of economies. It is a measurement of preferences of some types of merchandise or service, and in transportation the service refers to the choice of transportation modes among auto, transit, and others. The desirability of an alternative transportation mode depends on several factors like waiting time, operating time, and out-of-pocket cost. Utility theory partially reflected the demand and satisfaction

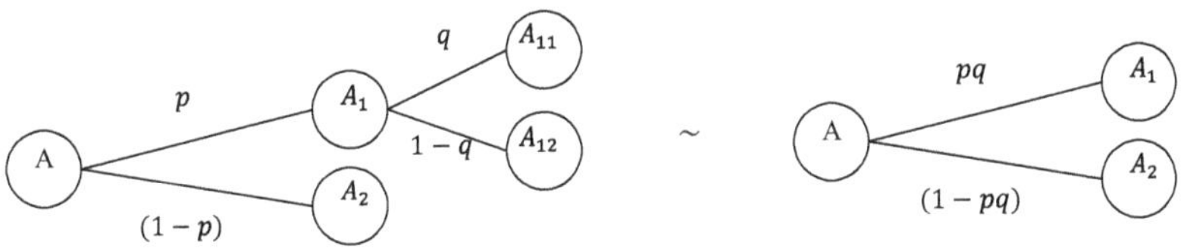

Figure 17.7 Illustration of utility model decomposability.

of specific transportation mode and quantifies the degree of preferences. Therefore, traffic engineers consider utility to be revealed in the willingness of transportation users to choose different transportation modes (Dyer et al., 1992; Keeney and Raiffa, 1993).

Typically, the following symbols define the degree of preferences between two options. Several axioms are given as well.

$A \succ B$, A is preferred to B
$A \sim B$, indifference between A and B
$A \succeq B$, A is preferred at least as much as B

Transitivity: If $A \succ B$ and $B \succ C$, then $A \succ C$.

Continuity: If $A \succ B \succ C$, there existed a value p, between 0 and 1, to make indifferent between choose B and choose A or C with probabilities p and $1 - p$, respectively.

Decomposability: A series of branches may be replaced by a single branch (Figure 17.7).

17.4.2 Utility function properties

As mentioned, the utility theory is a measurement of preferences; utility function is the way to quantify it. In this way, the preferences will be expressed numerically, which makes reading and understanding more intuitive. Typically, utility function is denoted as u. Important properties of utility functions are shown below.

If $A \succ B$, then $u(A) > u(B)$
If $A = B$, then $u(A) = u(B)$
If $A \prec B$, then $u(A) < u(B)$

If the possibilities to choose A_1 and A_2 are p and $(1 - p)$, then $u(A) = u(A_1) \cdot p + u(A_2) \cdot (1 - p)$

If $p_1 > p_2$ and $A_1 \succ A_2$ then $u(A_1) > u(A_2)$, $p_1 \cdot (u(A_1) - u(A_2)) > p_2 \cdot (u(A_1) - u(A_2))$, $p_1 \cdot (u(A_1) - u(A_2)) + u(A_2) > p_2 \cdot (u(A_1) - u(A_2)) + u(A_2)$, $u(A_1) \cdot p_1 + u(A_2) \cdot (1 - p_1) > p_2 + u(A_2)^*$ $(1 - p_2)$, therefore $u(A|p = p_1) > u(A|p = p_2)$.

If the utility function undergoes a linear transformation, the relative utility value will be retained or the preference decision will not be changed.

If $u(A) > u(B)$, then $a + b \cdot u(A) > a + b \cdot u(B)$, where a is a constant and b is a positive constant.

17.4.3 Utility function applications

A major application of utility functions in transportation is the discrete travel mode choice. In the four-step travel demand forecasting process, mode choice helps traffic engineers determine what travel mode will be selected, and what is the corresponding proportion. The value of the utility is used to compare different travel modes. The higher the utility value, more users will choose the travel mode.

A utility model in transportation is typically expressed as the linear weighted sum of the independent variables of their transformation, which can be written as

$$u = a_0 + a_1 x_1 + a_2 x_2 + \cdots + a_n x_n \tag{17.64}$$

where

u = Utility derived from choice
a_i = Model parameters
x_i = Attributes

If two modes, auto and transit, are considered, the formula to calculate the proportion of auto mode will be below. This approach is called logit model, which is widely used to predict the traffic demand for each transportation mode.

$$p_{Auto} = \frac{e^{u_{Auto}}}{e^{u_{Auto}} + e^{u_{Transit}}}$$

If more modes, like bike, carpool or even walking are considered, the idea will remain the same.

$$p_{Auto} = \frac{e^{u_{auto}}}{e^{u_{Auto}} + e^{u_{Transit}} + e^{u_{Bike}} + \cdots + e^{u_{Walking}}}$$

EXAMPLE 17.13

Utility functions for auto and transit modes are given. Determine the modal splits based on the binomial logit modeling principle (Table 17.5).

$$u = a_k - 0.35t_1 - 0.08t_2 - 0.005c$$

where

a_k = Mode specific variable
t_1 = Total travel time, min
t_2 = Waiting time, min
c = Out-of-pocket cost, cents

Solution

$$u_{auto} = -0.46 - 0.35 \times 20 - 0.08 \times 8 - 0.005 \times 320 = -9.7$$

$$u_{Transit} = -0.07 - 0.35 \times 30 - 0.08 \times 6 - 0.005 \times 100 = -11.55$$

$$p_{Auto} = \frac{e^{u_{Auto}}}{e^{u_{Auto}} + e^{u_{Transit}}} = \frac{e^{-9.7}}{e^{-9.7} + e^{-11.55}} = 0.86$$

$$p_{Transit} = \frac{e^{u_{Transit}}}{e^{u_{Auto}} + e^{u_{Transit}}} = \frac{e^{-11.55}}{e^{-9.7} + e^{-11.55}} = 0.14$$

Therefore, 86% of users will choose auto and 14% will choose transit.

Table 17.5 Utility function coefficients

Variable	*Auto*	*Transit*
a_k	−0.46	−0.07
t_1	20	30
t_2	8	6
c	320	100

17.5 DECISION MAKING UNDER UNCERTAINTY

17.5.1 Decision-making criteria (Menges, 1974)

17.5.1.1 Minimax criterion

To conduct analysis under uncertainty, typically the first step is to build a payoff table. Payoff is a quantitative measure of possible results of a particular state of natural for different alternatives. It can be the benefit (revenue) or the cost depending on specific scenarios. Denote a_i as alternative i, s_j as state j of natural, p_{ij} as the payoff of state j of alternative i, v_{ij} is the maximum payoff of each alternative. By using minimax criterion, alternatives with Min Max v_{ij} will be chosen.

Note that in this example, payoff is considered a cost. If payoff refers to the benefit, the idea of the method is still available, but modifications are required (Table 17.6).

In this example, the payoff (cost) is given in three states of natural for four different alternatives. For each state of natural, we pick out the maximum cost for each alternative. Then the alternative with minimum–maximum cost will be chosen. Therefore, alternative 4 with $v_{ij} = 70$ is selected.

17.5.1.2 Savage criterion

Savage criterion is also known as minimax regret criterion. It is widely used in decision theory. The purpose is to minimize the regret of the worst case. Across multiple alternatives, using the minimax regret criterion, the alternative with minimum maximum-regret will be chosen. Regret refers to opportunity loss or possible loss as well. It is the difference between the best payoff in a specific state of natural and the actual payoff one can get.

If payoff refers to benefit,

$$v_{ij} = \max_{i \in [1,m]} \{p_{ij}\} - p_{ij}$$

If payoff refers to cost,

$$v_{ij} = p_{ij} - \min_{i \in [1,m]} \{p_{ij}\}$$

Table 17.6 Payoff table

	s_1	s_2	…	s_n	*Max payoff*
a_1	p_{11}	p_{12}	…	p_{1n}	v_{1j}
a_2	p_{21}	p_{22}	…	p_{2n}	v_{2j}
…	…	…	p_{ij}	…	v_{ij}
a_m	p_{m1}	p_{m2}	…	p_{mn}	v_{mj}

Table 17.7 An example for applying the minimax criterion

Alternatives	*State 1*	*State 2*	*State 3*	*Max cost*
Alternative 1	80	70	40	80
Alternative 2	100	0	40	100
Alternative 3	10	90	50	90
Alternative 4	50	60	70	70

Table 17.8 Payoffs of the example for applying the savage criterion

Alternatives	*State 1*	*State 2*	*State 3*	*Maximum*
Alternative 1	70	70	0	70
Alternative 2	90	0	0	90
Alternative 3	0	90	10	90
Alternative 4	40	60	30	60

For the example shown in Table 17.7, the cost is given in three states of natural for four different alternatives. For each state of natural, we first compute v_{ij} for each alternative. The minimum cost of state 1 will be \$10, possible loss for each alternative will be \$70, \$90, \$0, \$40. The method is the same for the other states. The maximum possible loss is listed in the last column of the second table. By using savage criterion, the alternative with minimum regret will be chosen, in this case, it is alternative 4 (Table 17.8).

17.5.1.3 Hurwicz alpha criterion

The Hurwicz alpha criterion is a classical decision-making criterion under uncertainty. Alpha is the coefficient of optimism: the larger the alpha, the more the optimism. For the extreme cases, alpha is equal to 0 and 1, which represent extreme pessimism and optimism, respectively. Therefore, the value of alpha will change the selection of alternatives significantly. The expected value of payoff (payoff refers to cost here) is illustrated as

$$E_P = \alpha P_{\min} + (1-\alpha)P_{\max} \tag{17.65}$$

where

E_P = Expected value of payoff
$P_{\max}$ = Maximum payoff
$P_{\min}$ = Minimum payoff
α = Coefficient of the optimum

Alternative with maximum expected value of payoff will be chosen.

Using the same example of Table 17.7, maximum and minimum costs for alternative 1 are 80 and 40, respectively. For other alternatives, the corresponding cost is easy to get. If α is chosen as 80%, the expected value of cost will be 48, 20, 26, and 54. Then, alternative 2 is the optimal one with minimum cost (Table 17.9).

17.5.1.4 Hodges and Lehmann criterion

This criterion is an improvement of the minimax method with consideration for the experience factor. If the decision maker does not trust the priori distribution on the state set, the minimax is the criterion considered; if the decision maker trusts the priori distribution for

Table 17.9 An example of applying the Hurwicz alpha criterion

α = 80%	*State 1*	*State 2*	*State 3*	*Expected value of payoff*
Alternative 1	80	70	40	$80 \times 0.2 + 40 \times 0.8 = 48$
Alternative 2	100	0	40	$100 \times 0.2 + 0 \times 0.8 = 20$
Alternative 3	10	90	50	$90 \times 0.2 + 10 \times 0.8 = 26$
Alternative 4	50	60	70	$70 \times 0.2 + 50 \times 0.8 = 54$

Table 17.10 An example of applying the Hodges and Lehmann criterion

α = 80%, h = 60%	R_{ij}	H_{ij}
Alternative 1	50	$50 \times 0.6 + 80 \times 0.4 = 62$
Alternative 2	60	$60 \times 0.6 + 100 \times 0.4 = 76$
Alternative 3	40	$40 \times 0.6 + 90 \times 0.4 = 60$
Alternative 4	60	$60 \times 0.6 + 70 \times 0.4 = 64$

some reason, the criterion that should be considered is Bayes. The Hodges and Lehmann criterion combine the two criterions by giving a weight to each of them. The weight is the confidence parameter on each criterion, illustrated as h, and the value is between 0 and 1. The Hodges and Lehmann criterion is

$$H_{ij} = h \times R_{ij} + (1 - h) \times \max P_{ij} \tag{17.66}$$

where, R_{ij} represents the Bayesian risk expectation.

If the P_{ij} and R_{ij} refer to the expected cost (payoff), the optimal solution will be the alternative with minimum H_{ij}. In this example, alternative 3 is the optimal option (Table 17.10).

17.5.1.5 The adaption criterion

This method was proposed by Menges and Behara (1962) and Menges (1963) for unstable decision scenarios by applying various mixture of the minimax and Bayes criterions. The adaption criterion begins with the decomposition of the time-depend state space $\Omega(t)$ as

$$\Omega(t) = \Omega_1(t) \bigcup \Omega_2(t) \bigcup \cdots \bigcup \Omega_r(t)$$

and the corresponding risk function is r_t. The time density $\gamma(t)$ is considered. A priori distribution λ^* lies in Γ^*, Γ^* is a subclass Γ, which is the space of a priori distributions. Then, $\lambda^* \in \Gamma^* \subseteq \Gamma$. According to this criterion, alternatives with minimum costs will be chosen.

$$\int_T \gamma(t) \max_{\lambda \in \Gamma^*(t)} \sum_{i=1}^{r} \int_{\Omega_i(t)} r_t \, d\lambda \, dt \tag{17.67}$$

17.5.1.6 Schneeweiss criterion

This method is similar to the adaption criterion, with considering two decision criteria minimax and Bayes, but in a different way. The state spaces are decomposed into subspaces with a hierarchy. According to whether the probabilities are known or not, either Bayes or minimax is applied to individual state.

17.5.2 Game theory models

In many situations, two or more decision makers simultaneously choose an action, and the action chosen by each player affects the rewards earned by the other players. Game theory is useful for making decisions in these cases where two or more decision makers have conflict interests. Some of the main ideas of the context of two-person games will be introduced as follows.

17.5.2.1 Games and mathematical programming

In the context of two-person games, each player has several strategies. If the first player chooses strategy $a_i (i = 1, 2, \dots, m)$ while the second player chooses strategy $b_j (j = 1, 2, \dots, n)$, then player 1 gains v_{ij} while player 2 gains z_{ij}. This outcome is represented by (v_{ij}, z_{ij}). Two-person games where the player's interests are completely opposed are called zero-sum $(v_{ij} + z_{ij} = 0)$ or constant-zero $(v_{ij} + z_{ij} = c, c > 0)$ games, that is, one player's gain is the other player's loss. Games where the players' interests are not completely opposed are called variable-sum (also nonzero-sum) games. Some two-person games admit pure strategies whereas others require mixed strategies. A pure strategy is the one in which the players choose the same strategy every time they play the game. A mixed strategy is the one where players introduce a random element in their choice of a strategy, thus leaving the opponent guessing (Myerson, 1997; Karlin, 2003).

It was soon remarked that, after game theory was introduced, there existed very strong formal analogies between game theory and LP. To bring out the interdependence and correspondence in detail, let us start with the generalized two-person zero-sum game. Suppose the loss matrix is given for the decision maker (player 1), as in Table 17.11. Let player 1 chooses strategy a_i with the probability $p_i (i = 1, 2, \dots, m)$, and player 2 with strategy b_j and probability $q_j (j = 1, 2, \dots, n)$,

$$\sum_{i=1}^{m} p_i = 1, \quad 0 \le p_i \le 1$$

and

$$\sum_{j=1}^{m} q_i = 1, \quad 0 \le q_i \le 1$$

The vectors $p = (p_1, p_2, \dots, p_m)$ and $q = (q_1, q_2, \dots, q_n)$ are called mixed strategies of player 1 and player 2. Thus, the problem before player 1 is to choose a mixed strategy p so

Table 17.11 Loss matrix of player 1

		Actions of the opponent (player 2)			
Payoffs losses		b_1	b_2	...	b_n
Actions of the decision maker (player 1)	a_1	v_{11}	v_{12}	...	v_{1n}
	a_2	v_{21}	v_{22}	...	v_{2n}
	...	...	...	...	...
	a_m	v_{m1}	v_{m2}	...	v_{mn}

that his expected loss becomes as small as possible. This loss expectation, however, also depends on the choice of player 2, that is, if player 2 plays only pure strategy b_j, then

$$V(p,b_j) = \sum_{i=1}^{m} v_{ij}p_i \tag{17.68}$$

Let $V_a = \text{Max}\{V(p, b_j)\}$, then the following restrictions hold:

$$v_{11}p_1 + v_{21}p_2 + \cdots + v_{m1}p_m \leq V_a$$

$$v_{12}p_1 + v_{22}p_2 + \cdots + v_{m2}p_m \leq V_a$$

$$v_{1n}p_1 + v_{2n}p_2 + \cdots + v_{mn}p_m \leq V_a$$

The problem before player 1 is to choose his p so that V_a is minimized. To solve this problem, let us set

$$x_i = \frac{p_i}{V_a}$$

and obtain the restrictions as

$$v_{11}x_1 + v_{21}x_2 + \cdots + v_{m1}x_m \leq 1;$$
$$v_{12}x_1 + v_{22}x_2 + \cdots + v_{m2}x_m \leq 1;$$
$$\ldots$$
$$v_{1n}x_1 + v_{2n}x_2 + \cdots + v_{mn}x_m \leq 1;$$

and the objective function is

$$x_1 + x_2 + \cdots + x_m = \frac{1}{V_a} \tag{17.69}$$

V_a is to be minimized, that is, $1/V_a$ maximized, with respect to the $x_i \geq 0$ (linear maximum program).

The problem before player 2 is analogous. The payoff expectation for player 2, if player 1 plays only pure strategy a_i, is

$$V(a_i,q) = \sum_{j=1}^{n} v_{ij}q_j \tag{17.70}$$

Let $V_b = \text{Min}\{V(a_i, q)\}$, then the following restrictions hold:

$$v_{11}q_1 + v_{12}q_2 + \cdots + v_{1n}q_n \geq V_b;$$
$$v_{21}q_1 + v_{22}q_2 + \cdots + v_{2n}q_n \geq V_b;$$
$$\ldots$$
$$v_{m1}q_1 + v_{m2}q_2 + \cdots + v_{mn}q_n \geq V_b.$$

By setting $y_j = q_i/V_b$, the following linear minimum program can be obtained, determine y_j ($j = 1, 2, \dots, n$), subject to the constraints,

$$v_{11}y_1 + v_{12}y_2 + \cdots + v_{1n}y_n \geq 1;$$
$$v_{21}y_1 + v_{22}y_2 + \cdots + v_{2n}y_n \geq 1;$$
$$\dots$$
$$v_{m1}y_1 + v_{m2}y_2 + \cdots + v_{mn}y_n \geq 1$$

so that the objective function

$$y_1 + y_2 + \cdots + y_n = \frac{1}{V_b} \tag{17.71}$$

is minimized.

17.5.2.2 Two-person zero-sum games

As stated above, a two-person zero-sum game is a situation in which each player's gain or loss of utility is exactly balanced by the loss or gain of the utility of another player (Nash, 1953). If the total gains of the players are added up and the total losses are subtracted, they will sum to zero. Suppose the losses of the first player's utility is estimated as in Table 17.12. Let us see how the players play the two-person zero-sum game using the pure strategies.

For player 1's viewpoint, he would like player 2 to choose strategy b_4 because, with a_1, this would yield a favorable outcome, +5, for it. If, however, player 1 has decided for one of the strategies a_1 or a_4, he will suffer considerable losses in case player 2 chooses b_3 or b_1. Player 2 faces the same challenges. If he chooses b_3 in the hope of obtaining +40, player *a* might choose a_2, and the gain is only +2. Rationally, player 2 cannot choose the conciliatory strategy b_4, as strategy b_1 yields better outcomes no matter which strategy is adopted by player 1 (strategy b_1 dominates b_4).

Employment of the minimax criterion requires that player 1 chooses the strategy which minimizes the opponent's maximal gains. Player 2, by this rule, must choose the strategy that maximizes his minimal gains. Thus, player 1's minimax strategy coincides with player 2's maximin strategy, and the ensuring loss of player 1, the value of the game, corresponds to +12. Strategies with this property are called equilibrium strategies. For player 2, b_1 is the best response to a_3, and for player 1, a_3 is also the best response to b_1. The minimax solution is (a_3, b_1).

However, not every conflict situation possesses a minimax equilibrium if only pure strategies are admitted. If the above loss matrix is modified so that the outcome of (a_3, b_3) is +19, then +12 is no longer the maximal value in the third row, and a_3, with +15, becomes the minimax

Table 17.12 An example of loss matrix of player 1

		Strategies of player 2				
Losses		b_1	b_2	b_3	b_4	*Max* b_j
Strategies of player 1	a_1	+20	+25	+40	−5	+40
	a_2	+15	+14	+2	+4	+15
	a_3	+12	+8	+10	+11	+12
	a_4	+35	+10	+5	0	+35
Min a_i		+12	+8	+2	−5	

Table 17.13 Advised loss matrix

		Strategies of play 2	
Losses		b_1	b_3
Strategies of player 1	a_2	+15	+2
	a_3	+12	+19

strategy of player 1, while b_1, with +12, remains player 2's maximin strategy. The pure strategies (a_2, b_1), hence, are no longer equilibrium strategies. Considering the new loss matrix, it is found that b_4 is dominated by b_1 and, thus, can be omitted. a_1 is equally dominated by a_2 and a_3 and, thus is out of the consideration. Furthermore, it can be found that b_2 is dominated by b_1 and, hence, a_4 by a_2. What remains is the following 2 × 2 loss matrix (Table 17.13).

Now, let us admit the mixed strategies. For player 2's mixed strategies $(qb_1, (1-q)b_3)$ against player 1's pure strategies a_2 and a_3, the following LP model can be obtained:

$$\text{Max}\, V_b$$

Subject to

$$V_b \le 15q + 2(1-q)$$
$$V_b \le 12q + 19(1-q)$$
$$0 \le q \le 1$$

Then, $q = 0.85$ is obtained, which means that the strategy $(0.85b_1, 0.15b_3)$ is to be the maximin strategy for player 2. By this strategy, his expected gain $V_b = 13.05$.

Similarly, for player 1's mixed strategy $(pa_2, (1-p)a_3)$ against player 2's pure strategies b_1 and b_3, the following LP model is obtained:

$$\text{Min}\, V_a$$

Subject to

$$V_a \ge 15p + 2(1-p)$$
$$V_a \le 2p + 19(1-p)$$
$$0 \le p \le 1$$

Thus, $p = 0.35$ is obtained. This means that the strategy $(0.35a_2, 0.65a_3)$ is the minimax strategy for player 1. By this strategy, his expected loss V_a is also equal to 13.05.

Therefore, the mixed strategies $(0.35a_2, 0.65a_3)$ and $(0.85b_1, 0.15b_3)$ are equilibrium strategies. As seen, the problems before the players, in the context of two-person zero-sum games, are LP problems. Upon certain assumptions of the gambling behavior of the opponents, they are even dual programming problems, and possess one and only one equilibrium point (Table 17.14).

17.5.2.3 Two-person nonzero-sum games

As suggested by the name, a two-person nonzero-sum game is a situation in which two interacting players' aggregate gains and losses can be less than or more than zero. That is, the utility scales of two players may differ so that the zero-sum restrictions cannot be satisfied. Thus, to every pair of strategies, there exist two individual utility values for two players, just as seen in Table 17.14 (Scodel et al., 1959).

Table 17.14 An example of loss matrix of player 1

Utility of player 2 *Utility of player 1*		*Strategies of player 2*			
		b_1	b_2	b_3	b_4
Strategies of player 1	a_1	+5 −6	+7 −9	+13 −25	−14 +18
	a_2	0 −2	+1 −1	+4 +12	−7 +10
	a_3	−1 +1	+1 +4	−4 +3	+7 +2
	a_4	−2 +21	−2 +3	−8 +9	−12 +15

Rule out the strategies in which one player is dominated by others, viz., b_4 is dominated by b_2, a_1 is dominated by a_2 and a_3, b_1 are dominated by b_2.

A general answer to the question whether there exist equilibrium strategies for this case is found in the Nash theorem, which states that the mixed extension of any finite n-person game possesses at least one equilibrium point. A pair of strategies (a^*, b^*) is said to be the equilibrium point, if it holds that

$$V_b(a^*,b^*) \geq V_b(a^*,b_j),\ j = 2,3 \tag{17.72}$$

and

$$V_a(a^*,b^*) \geq V_a(a_i,b^*), \quad i = 2,3,4$$

Practically speaking, an equilibrium point is a rule of action where, if all players expect for one to follow the rule, the remaining one cannot do anything better than to also follow the rule. In our example, there are only two pairs of pure equilibrium strategies, (a_2, b_3) and (a_3, b_2). The equilibrium points are not equivalent for

$$V_b(a_2,b_3) = 4 \neq 1 = V_b(a_3,b_2)$$

and

$$V_a(a_2,b_3) = 12 \neq 4 = V_a(a_3,b_2)$$

Therefore, the problem is not solvable in the sense of Nash. Nash-solvability requires that all equilibrium points are equivalent. In practice, one will nevertheless come to a solution. Looking at the equilibrium pairs, it can be found that both players derive greater utility from (a_2, b_3) than from (a_3, b_2). They will therefore agree on (a_2, b_3). However, once player 1 is not sure that player 2 chooses b_3, he would hesitate to choose a_2. Such doubts are dropped if both players are willing to cooperate. On the additional assumption of cooperation, then (a_2, b_3) is a unique solution.

17.5.3 Shackle's model

Shackle's model overcomes the limitation of risk-based decision analysis for its inability to compute the mathematical expectation for an input factor for computing the impacts of a

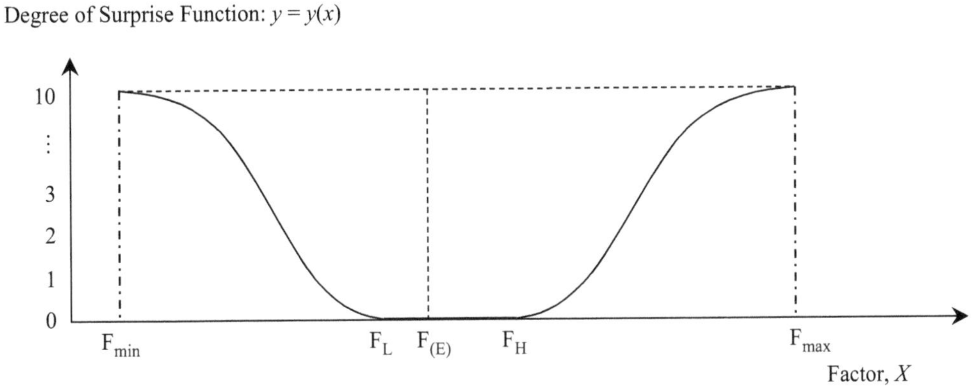

Figure 17.8 Illustration of a typical degree of surprise function. Note: F_{min}, F_{max}-minimum and maximum values of a factor for assessing the impacts; F_L, F_U-lower and upper extreme values of the factor with no degree of surprise; $F_{(E)}$-expected outcome of the factor; x-deviation of a possible outcome X from $F_{(E)}$, $x = X - F_{(E)}$. (Adapted from Li, Z. and P. Kaini. 2007. Optimal Investment Decision-Making for Highway Transportation Asset Management under Risk and Uncertainty. Report MRUTC 07-10. Department of Civil, Architectural, and Environmental Engineering, Illinois Institute of Technology, Chicago, IL; Li, Z. and S. Madanu. 2009. *ASCE Journal of Transportation Engineering* 135(8), 516–526; Li, Z. and K.C. Sinha. 2009. *ASCE Journal of Transportation Engineering* 135(3), 129–139.)

certain intervention (Shackle, 1949; Li and Sinha, 2004, 2009a, Li and Madanu, 2009). For instance, the issue of uncertainty associated with the input factor of predicted traffic in estimating transportation project benefits. The model is mainly based on three pillars. The first element is the use of degree of surprise instead of probability to measure uncertainty. That is, degree of surprise measures the individual's degree of uncertainty concerning the hypothesized outcomes (gains and losses) brought about by a particular behavior. The second element is the priority weight, which is the measure by which the decision maker evaluates the degree of surprise corresponding to the gains and losses. The third is that it identifies and standardizes the focus gain and focus loss values relative to an expected outcome from maximum priority weights.

Degree of surprise: The degree of surprise reflects the decision-maker's reaction to degree of uncertainty regarding possible outcomes of a factor for assessing the impacts resulting from any investment option, with gains and losses from the expected outcome being considered separately (Figure 17.8). A degree of surprise function for a factor for computing a specific item of project impacts can be established using the following steps:

- Assume a range of s possible outcomes of a factor X for assessing the impacts from an investment option ($X = F_1, F_2, \ldots, F_S$ ranging from F_{min} to F_{max})
- Denote $F_{(E)}$ as the expected outcome for the factor for assessing the impacts
- Let the deviation of an outcome of the factor X relative to the expected outcome $F_{(E)}$ to be x, $x = X - F_{(E)}$
- Assign a value as degree of surprise y, ranging from 0 for no surprise to 10 for highest surprise, to reflect the decision-maker's degree of belief for a given outcome X as captured by the deviation x
- Establish a degree of surprise function $y = f(x)$

Priority function: The priority function indicates the weight assigned to the deviation of any outcome of the factor for assessing the impacts from the expected outcome and degree

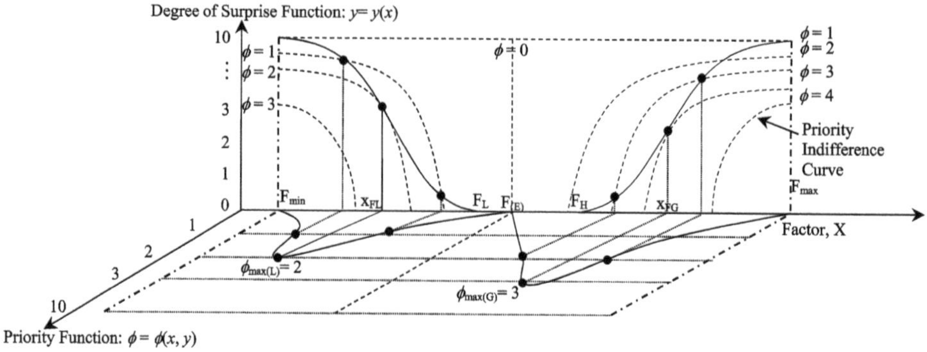

Figure 17.9 Illustration of a typical priority function. Note: F_{min}, F_{max}-minimum and maximum values of a factor for assessing the impacts; F_L, F_U-lower and upper extreme values of the factor with no degree of surprise; $F_{(E)}$-expected outcome of the factor; X-deviation of a possible outcome X from $F_{(E)}$, $x = X - F_{(E)}$, *xFG, xFL*-focus gain and focus loss. (Adapted from Li, Z. and P. Kaini. 2007. Optimal Investment Decision-Making for Highway Transportation Asset Management under Risk and Uncertainty. Report MRUTC 07–10. Department of Civil, Architectural, and Environmental Engineering, Illinois Institute of Technology, Chicago, IL; Li, Z. and S. Madanu. 2009. *ASCE Journal of Transportation Engineering* 135(8), 516–526; Li, Z. and K.C. Sinha. 2009. *ASCE Journal of Transportation Engineering* 135(3), 129–139.)

of surprise pair (x, y) or in Shackle's terminology, the power of any pair to attract the attention of the decision maker (Figure 17.9). A priority function for a number of possible outcomes related to the factor can be developed as follows:

- Determine a priority weight index ϕ by jointly considering the deviation of each outcome of the factor and degree of surprise pair (x, y) using an index of 0 for lowest priority weight and indices of 2, 3, 4, 5, ... for higher priorities
- Denote the decision-maker's priority function by $\phi = \phi(x, y)$ and the function possesses the following properties: $(\partial \varnothing / \partial x) > 0$; $(\partial \varnothing / \partial y) < 0$. A priority function can be defined in the following function forms $\phi = \alpha \cdot x^{0.5} - \beta \cdot y^2$, $\phi = \alpha \cdot x - \beta \cdot y^2$ or $\phi = \alpha \cdot x^{0.5} - \beta \cdot y$, where α and β are coefficients with respect to the deviation of the factor from the expected outcome and degree of surprise
- Priority function ϕ is a saddle-shaped curve that maintains a maximum priority weight on the gain side from expected outcome and a maximum priority weight on the loss side from expected outcome. The deviations of the two outcomes corresponding to the two maximum priority weights are called focus gain (x_{FG}) and focus loss (x_{FL}) values

Standardized focus gain and loss values: The focus gain and loss values involve uncertainty because they have nonzero degrees of surprise. It is, therefore, necessary to filter out such uncertainty to establish the standardized focus gain and loss values with zero degree of surprise. The standardization process can be accomplished by using the priority indifference curves at both the gain and loss side from the expected outcome that retains the maximum priority weights consistent with those of the focus gain and focus loss values.

Notations:

x = Deviation of a possible outcome of a factor X from the expected outcome $F_{(E)}$

$y(x)$ = Degree of surprise function, set $y(x) = c \cdot x^2$

$\varnothing_1(x, y)$ = Priority indifference curve, set $\varnothing_1(x, y) = \alpha_1 \cdot x^{0.5} - \beta_1 \cdot y^2 = k (k > 0)$

$\varnothing_2(x, y)$ = Maximum priority indifference curve on the gain side, $\varnothing_2(x, y) = \alpha_2 \cdot x^{0.5} - \beta_2 \cdot y^2 = \varnothing_{max(G)}$

x_{SG}	=	Standardized gain value on indifference curve $\varnothing_1(x, y)$ with no surprise
x_{FG}	=	Focus gain value on maximum priority indifference curve $\varnothing_2(x, y)$
x_{SFG}	=	Standardized focus gain value on maximum priority indifference curve $\varnothing_2(x, y)$ with no surprise

A, B, C are points on $\varnothing_1(x, y)$, and O is a point on $\varnothing_2(x, y)$.

The purpose is to find the standardized focus gain x_{SFG} from the underlying focus gain x_{SFG} on the maximum priority indifference curve $\varnothing_2(x, y)$. As $\varnothing_2(x, y)$ only intersects with the degree of surprise function $y(x)$ at point O, it would be impractical to further progress the standardization process. This is because it is impossible to simultaneously calibrate two parameters α_2 and β_2 for $\varnothing_2(x, y)$ solely on one point on the curve. To overcome this restriction, the indifference curve $\varnothing_1(x, y)$ closest to $\varnothing_2(x, y)$ that intersects with the degree of surprise function $y(x)$ twice at points A and B can be utilized. As shown in Figure 17.10, when the priority indifference curve $\varnothing_1(x, y)$ approaches $\varnothing_2(x, y)$ (i.e., $\varnothing_1(x, y) = k \to \varnothing_{\max(G)}$), the standardized gain value x_{SG} for $\varnothing_1(x, y)$ will overlap with the standardized focus gain x_{SFG}. Hence, the process reduces to establishing a mathematical expression for the standardized gain value x_{SG}.

For points A and B on priority indifference curve $\varnothing_1(x, y)$, we have

$$\alpha_1 \cdot x_A^{0.5} - \beta_1 \cdot y_A^2 = k \tag{17.73}$$

$$\alpha_1 \cdot x_B^{0.5} - \beta_1 \cdot y_B^2 = k \tag{17.74}$$

Substituting $y_A = c \cdot x_A^2$ and $y_B = c \cdot x_B^2$ into Equations 17.73 and 17.74, we obtain

$$\alpha_1 = \frac{k(x_B^4 - x_A^4)}{(x_B^4 x_A^{0.5} - x_B^{0.5} x_A^4)} \tag{17.75}$$

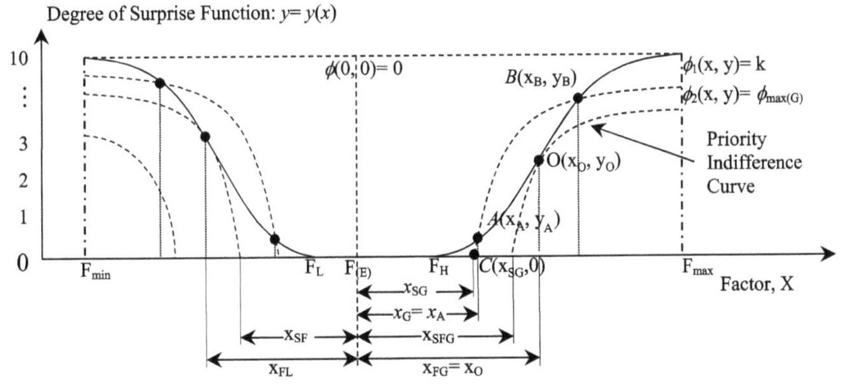

Figure 17.10 Illustration of standardized focus gain and loss values. Note: F_{min}, F_{max}-minimum and maximum values of a factor for assessing the impacts; F_L, F_U-lower and upper extreme values of the factor with no degree of surprise; $F_{(E)}$-expected outcome of the factor; x-deviation of a possible outcome X from $F_{(E)}$, $x = X - F_{(E)}$; x_{FG}, x_{SFG}-focus gain and standardized focus gain; x_{FL}, x_{SFL}-focus loss and standardized focus loss. (Adapted from Li, Z. and P. Kaini. 2007. Optimal Investment Decision-Making for Highway Transportation Asset Management under Risk and Uncertainty. Report MRUTC 07-10. Department of Civil, Architectural, and Environmental Engineering, Illinois Institute of Technology, Chicago, IL; Li, Z. and S. Madanu. 2009. *ASCE Journal of Transportation Engineering* 135(8), 516–526; Li, Z. and Sinha, K.C. 2009. *ASCE Journal of Transportation Engineering* 135(3), 129–139.)

For point $C(x_{SG}, 0)$ on $\varnothing_1(x, y)$, we get $\varnothing_1(x,y) = \alpha_1 \cdot x_{SG}^{0.5} - \beta_1 \cdot 0^2 = \alpha_1 \cdot x_{SG}^{0.5}$. Thus,

$$x_{SG} = \left(\frac{\varnothing_1(x_{SG},0)}{\alpha_1}\right)^2 \quad \text{and} \quad x_{SFG} = \left(\frac{\varnothing_{\max(G)}}{\alpha_1}\right)^2 = \left(\frac{\varnothing_{\max(G)} \cdot \left(x_B^4 x_A^{0.5} - x_B^{0.5} x_A^4\right)}{k(x_B^4 - x_A^4)}\right)^2$$

Following this procedure, the standardized focus gain and loss values for a factor for assessing the impacts, x_{SFG} and x_{SFL}, corresponding to the maximum priority indices, $\varnothing_{\max(G)}$ and $\varnothing_{\max(L)}$, on the gain side and loss side from the expected outcome can be determined.

17.5.4 Leontief input–output model

The original Leontief input–output (I/O) model is a structure for studying the equilibrium phenomenon of economic behaviors. The system to be studied is composed of a number of subsystems or separate economic elements. This section gives a brief skeleton of the Leontief input–output model to trace the change of resources and commodities in one economy (Leontief, 1986).

The economic system is supposed to contain a group of interacting sectors or industries, where industry represents the source of one commodity. To produce commodities, an industry needs inputs from the outside, including commodities from interacting industries. Each industry must produce sufficient commodities constrained by two demands, where interacting demands mean demands from industries within the group and external demands mean demands from industries across the group. An equilibrium-competitive status of economy, with fixed coefficients for a given time slot, is assumed.

Define the following notations:

x_j = The overall output of commodity j, $j = 1,2, \ldots, n$
r_i = The overall input of resource i, $i = 1,2, \ldots, m$
x_{kj} = The amount of the commodity k used in the production of the commodity j
r_{ij} = The amount of the resource i used in producing the commodity j

Leontief I/O model assumes that the inputs of both commodities and resources required for producing a commodity are proportional to the output of that commodity. The linear relationships can be formulated as follows:

$$x_{kj} = a_{kj}x_j, \quad j,k = 1,2, \ldots, n \tag{17.76}$$

$$r_{ij} = b_{ij}x_j, \quad i = 1,2,\ldots,m; \quad j = 1,2, \ldots, n \tag{17.77}$$

Furthermore, the commodity produced is used either as input for producing other commodities or as final outputs, c_k. Therefore, the overall output of one commodity can be formulated as Equation 17.78

$$x_k = \sum_{j=1\,(j\neq k)}^{n} x_{kj} + c_k, \quad k = 1,2, \ldots, n \tag{17.78}$$

The Leontief Equation 17.79 is obtained by combining the balance Equation 17.76 and the overall output Equation 17.78

$$x_k = \sum_{j=1\,(j\neq k)}^{n} \alpha_{kj}x_j + c_k, \quad k = 1,2, \ldots, n \tag{17.79}$$

Thus, the Leontief equation for resources is shown as follows:

$$r_i = \sum_{j=1}^{n} r_{ij} = \sum_{j=1}^{n} b_{ij} x_j, \quad i = 1,2,\ldots,m \tag{17.80}$$

Since the demand for the resource i cannot exceed its supply, then

$$\sum_{j=1}^{n} b_{ij} x_j \leq r_i, \quad r_i \geq 0; \quad i = 1,2,\ldots,m \tag{17.81}$$

The above basic model of the economy was adapted to develop the Leontief-based input–output infrastructure model, although there is a major difference in the interpretation of the variables.

17.5.5 Network flow model

In project selection problems, an agency will often encounter a situation in which two or more projects are to be undertaken at the same time. Traditional analyses presume the projects undertaken simultaneously are independent, and the overall benefits and costs resulting from implementing these projects thus can be considered as the sum of benefits and costs brought about by each individual project. The truth is, transportation projects change the transportation network supplies in a variety of ways, resulting in network-wide traffic redistribution and further changing the user benefits and costs.

The agency costs of construction, rehabilitation, and maintenance, as well as user costs of vehicle operation, travel time, crashes, and air emissions associated with a highway segment can be computed as a function of vehicle volumes, compositions, and speeds for a particular time interval of the day such as on an hourly basis that could be eventually extended to the pavement or bridge facility service life cycle. Uncertainty associated with projected traffic, project costs, and the discount rate that may vary during the course of the facility service life cycle are considered in the computation. In general, project benefit estimation mainly involves establishing the traffic details for all highway segments in the network directly and indirectly affected by the project before and after project implementation, calculating project life-cycle agency and user costs and overall benefits, and incorporating uncertainty considerations into the analysis.

This section mainly introduces the network flow model including user equilibrium (UE) and system optimum (SO). These two methods are generally used for solving traffic assignment problems (Dafermos and Sparrow, 1969; Manheim, 1979; Sheffi, 1985; Nagurney, 2000; Bobzin, 2006; Larson and Odoni, 2007).

The network flow model is mainly based on three pillars. The first pillar is the performance function, which describes how the level of service (LOS) (such as travel time, path distance, fare) deteriorates with increasing traffic volume. The second pillar is the demand function, which describes how the traffic volume increases with the improved LOS. Note that the LOS here differs from the LOS defined in HCM (TRB, 2010).

UE: The UE is the same as Nash equilibrium, and occurs when the current traffic volume could equalize the performance function and demand function. Therefore, the question of interests here is how to distribute the current travel demand among all possible paths so that the UE could be achieved.

The UE conditions proposed by Wardrop (1952) form the basis of the mechanism of UE as:

- For each O–D pair, the journey times used on all the routes actually used are equal, and less than those which would be experienced by a single vehicle on any unused route (Wardrop's First Principle)

- Assume each vehicle travels on the path with the minimum travel time from origin to destination among all path options
- No vehicle can experience a lower travel time by unilaterally changing paths when UE is reached

Beckmann et al. (1956) formulated an optimization problem for deterministic UE represented as follows:

$$\text{Min} \quad \sum_{a \in A} \int_0^{x_a} t_a(\omega) d\omega \tag{17.82}$$

$$\text{Subject to} \quad \sum_{k \in P_{rs}} f_k^{rs} = q_{rs} \quad r \in R, s \in J$$

$$x_a = \sum_{r \in R} \sum_{s \in J} \sum_{k \in P_{rs}} f_k^{rs} \delta_{ak}^{rs} < C_a \quad a \in A$$

$$f_k^{rs} \geq 0, k \in P_{rs}, r \in R, s \in J$$

where
A = The set of links in the network
a = The link index, $a \in A$
R = The set of origins in the network
r = The origin index, $r \in R$
J = The set of destinations in the network
s = The destination index, $s \in J$
P_{rs} = The set of paths collecting O–D pair rs
k = The path/route index, $k \in P_{rs}$
f_k^{rs} = The flow on path k between O–D pair rs
q_{rs} = The travel demand for O–D pair rs
δ_{ak}^{rs} = The 0–1 value indicating the link usage. $\delta_{ak}^{rs} = 1$, if path k using link a; otherwise, $\delta_{ak}^{rs} = 0$
x_a = The flow on link a
C_a = The capacity of link a

The first constraints are the fixed travel demand conservation constraints. The second constraints define the link flow using all path flows and limit the link flow using link capacity. The third constraints are nonnegativity constraints.

EXAMPLE 17.14

Use UE pattern to find the flow on each link of the network. Assuming the travel demand from this origin–destination is 140 (Figure 17.11).

$$\text{Min} \quad \int_0^{x_1} \omega d\omega + \int_0^{x_2} \omega d\omega + \int_0^{x_3} (\omega + 1) d\omega = 0.5(x_1^2 + x_2^2 + x_3^2) + x_3$$

$$\text{Subject to} \quad f_1 + f_2 + f_3 = 140$$

$$x_1 = f_1 \leq 80$$

$$x_2 = f_2 \leq 100$$

$$x_3 = f_3 \leq 120$$

$$x_1, x_2, x_3, f_1, f_2, f_3 \geq 0$$

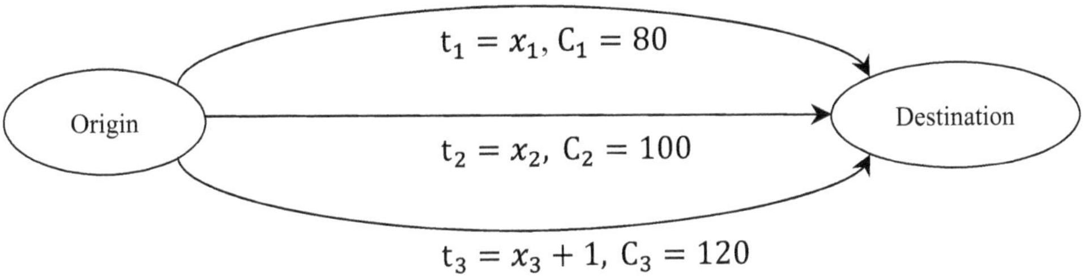

Figure 17.11 An example of the network flow problem.

The solution is $(x_1, x_2, x_3) = (47, 47, 46)$.

SO: The UE minimization program expressed in (5–10) lacks an intuitive interpretation. It is merely an efficient method for solving the UE equations. But the objective function of SO is the total travel time spent in the network. The flow pattern that solves this program minimizes this objective function while satisfying the same constraints of UE. The objective function is expressed as follows:

$$\text{Min} \sum_{a \in A} x_a \cdot t_a(x_a) \tag{17.83}$$

The assignment pattern of SO that minimizes the total travel time does not generally represent an equilibrium situation. In other words, at the SO flow assignment, motorists may be able to reduce their travel time by unilaterally altering the path. Such a situation is unlikely to be stable and thus should not be used as a model of actual behavior and equilibrium.

EXAMPLE 17.15

Use the SO assignment to find the flow pattern in the case illustrated in UE session.

$$\text{Min} \quad (x_1^2 + x_2^2) + x_3(x_3 + 1)$$

$$\begin{aligned} \text{Subject to} \quad & f_1 + f_2 + f_3 = 140 \\ & x_1 = f_1 \leq 80 \\ & x_2 = f_2 \leq 100 \\ & x_3 = f_3 \leq 120 \\ & x_1, x_2, x_3, f_1, f_2, f_3 \geq 0 \end{aligned}$$

The solution is $(x_1, x_2, x_3) = (46.83, 46.83, 46.33)$. As we can see, the solution is slightly different from the UE assignment even in this simple network.

In sum, UE is solving a problem with the set of marginal travel times, while SO is solving the same problem with the travel time itself. Based on this understanding, we could further compare the similarity and dissimilarity of the UE and SO flow patterns.

In cases where the travel demand over the network is low, the links are uncongested. The marginal travel time on each link is very small. That is, adding one more vehicle will not make the travel time performance significantly worse. Taking the example of the BPR function, by Bureau of Public Roads (BPR), which was the predecessor of FHWA, the fourth power function of the vehicle increases slowly in the beginning and rapidly in the end. In these uncongested cases, the UE and SO flow patterns are close to each other since

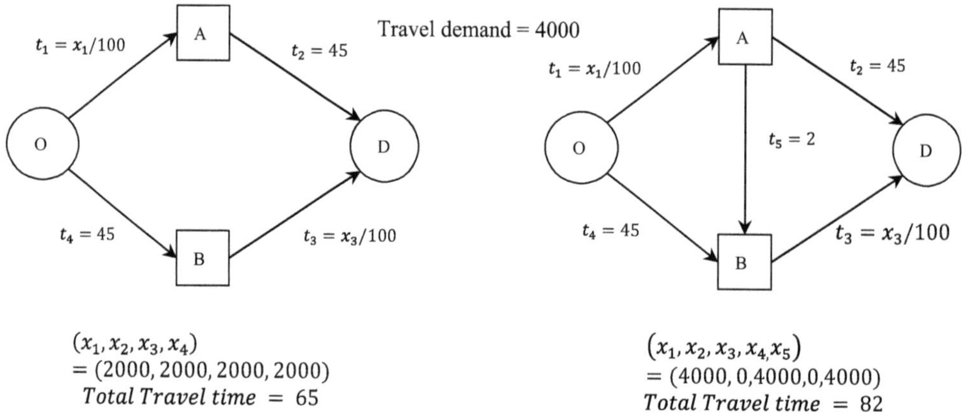

Figure 17.12 An example of the Braess' paradox.

the marginal travel times of the additional travel demand are low. However, as the travel demand in the network increases, the dissimilarity of UE and SO flow patterns becomes significant. The reason is that some of the links run out of capacity, where the marginal travel time becomes very large. Therefore, the difference between solving marginal travel times (UE) and travel times (SO) is increasing.

A failure to realize the fundamental difference between the normative SO flow pattern and the descriptive UE flow pattern can lead to pseudo-paradoxical scenarios. The most famous of these is known as Braess' paradox (Sheffi, 1985; Nagurney, 2000).

Braess' paradox: Braess' paradox, proposed by German mathematician Dietrich Braess, argues that adding additional capacity to a network may reduce overall performance when each participant chooses the shortest path selfishly. The reason is that the UE of such a system is not necessarily optimal. That is, UE is not SO. The paradox states: "For each point of a road network, let there be given the number of cars starting from it, and the destination of the cars. Under these conditions one wishes to estimate the distribution of traffic flow. Whether one street is preferable to another depends not only on the quality of the road, but also on the density of the flow. If every driver takes the path that looks most favorable to him, the resultant running times need not be minimal. Furthermore, it is indicated by an example that an extension of the road network may cause a redistribution of the traffic that results in longer individual running times." In the case of Braess' paradox, drivers will always choose their shortest path until reaching UE regardless of the reduction in overall travel time performance. Figure 17.12 illustrates an example of Braess' paradox.

17.6 MULTICRITERIA DECISION-MAKING MODELS AND METHODS

17.6.1 Introduction

Nearly all kinds of decision-making problems require one to simultaneously consider several different objectives that might be mutually conflicting. This also happens in the transportation industry. For example, a transit authority is planning to purchase several new buses to improve service quality. The decision maker may be concerned with the price of the bus, its safety features, the attractiveness of appearance, and fuel-economic efficiency. Clearly, the good-looking and energy-saving bus with low costs and great safety performance would be favored by the agency. Nonetheless, these characteristics are evidently in conflict since great safety features must lead to a higher price of the bus. Therefore, the agency has to be willing to make trade-offs among the various criteria in order to find the most optimal solution. In detail, the agency must know how much it could and will sacrifice in the price to gain more safety features.

Multicriteria problems are frequently met in a lot of engineering practice. Simple problems with a few objectives and criteria can usually be solved sufficiently using intuition or by various processes of choice. Nevertheless, the complex problems with a large number of objectives and alternatives remarkably enlarge the need for more analytical techniques. In addition, some decisions made by agencies need to be justified to the public, which also calls for analytical techniques. Complex decisions with many objectives and alternatives make it difficult for decision makers to articulate trade-off information and keep the consistency of measures. The analytical techniques assist decision makers in ranking and structuring their preferences (Steuer, 1989; Mollaghasemi, 1997).

17.6.1.1 Multicriteria decision-making process

Most decision scenarios contain the following elements: (i) criteria, goals or objectives to be achieved; (ii) needs to be satisfied; (iii) constraints and requirements affected by the decision; (iv) decision options or alternatives; (v) the background environment for the decision; and (vi) decision-maker(s)-related characteristics. Various factors are also involved in decision scenarios. The decision maker is the individual or group authorized to finalize the decision. Stakeholders are the individuals or groups that can affect or be affected by the decisions. An analyst is a group or individual that combines the subjective and objective inputs gathered from the decision makers and stakeholders and converts them into meaningful outputs to support the final decision making.

The multicriteria decision-making (MCDM) process refers to the process of problem solving. The purpose is to assist the decision maker to enhance the understanding of complex decision problems and to make proper decisions in a holistic way. Two concepts that need to be distinguished are a good decision and a good outcome from a decision. A good decision is one that is rationally made according to a comprehensive understanding of the problem and a thorough study of the related aspects. Outcomes, to some extent, may be good or bad. A good decision could also lead to a bad outcome as long as the adverse consequence could be anticipated. The decision-making process must be able to report the possibility of the bad outcome, and both the uncertainty and the pros and cons should be balanced by the decision maker to achieve an informed decision.

Figure 17.13 illustrates the steps and related actions of MCDM. The first step is to define criteria, goals or objectives as more concrete statements of goals. This step aims to assess the current situation and find general needs. The second step is to identify attributes (qualitative) or performance measures (quantitative) that relate to each criterion or objective proposed in the previous step. Note that an attribute (or a performance measure) is a qualitative judgment (or quantitative measurement) that can reveal the level of achievement for a certain criterion or objective. For example, if the objective is to alleviate the congestion issue on a highway segment, the associated performance could be quantitatively measured in average travel time in minutes. The criteria or objectives could be subjectively assessable (e.g., the level of satisfaction, the comfort of the service) or objectively measurable (e.g., length, price). The third step is to identify a set of candidate alternatives. The last step is to evaluate each alternative based on a predetermined set of decision rules. Based on the evaluation, a rank of alternatives should be reported. The top alternative is viewed as the optimal solution of the problem (Steuer, 1989; Mollaghasemi, 1997).

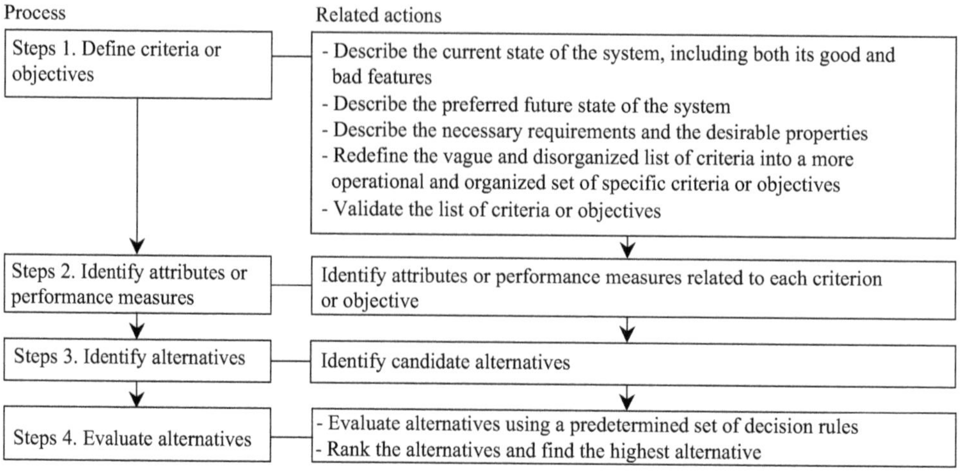

Figure 17.13 Multicriteria decision-making process.

17.6.1.2 Multiattribute decision-making and multi-objective optimization problems

Multicriteria decision problems are broadly classified into two types: (i) multiple attribute problems denote the problems with a relatively small number of alternatives, where the alternatives are expressed using qualitative attributes; and (ii) multiple-objective optimization problems denote the problems with a very large number of feasible alternatives, where the objectives and the constraints are expressed by quantitative performance measures in the form of decision variables.

Multiattribute decision problems generally have a limited set of n alternatives with a fairly large set of p attributes. The goal of such problems can help the decision-maker (i) establish the list of candidate alternatives; (ii) filter out good alternatives among the studied alternatives; and (iii) rank alternatives in terms of preference.

To solve multiattribute problems, appropriate attributes must be selected. This process will serve as the fundamental basis for alternatives' evaluation. Then, the importance of the criteria should be determined, where corresponding weights are assigned. Subsequently, alternatives are ranked in terms of the criteria.

There is no ideal number of attributes to be considered. Too many attributes may contain too many details in the problem, while too few attributes may result in missing important ones. A good set up of attributes should have the following characteristics: (i) an attribute should be sufficient to indicate the level of achieving the objective and be measurable; (ii) the set of attributes should be able to represent all related aspects of the decision problem; (iii) the set of attributes could be used meaningfully in the further analyses; (iv) the set of attributes should simplify the decision process by decomposing the problem into handy components; (v) no part of the decision problem should be considered twice; and (vi) another smaller set of attributes that could describe the same decision problem does not exist. After the attributes are determined, structuring them into a hierarchy is highly recommended, especially when the attribute's size is large (Steuer, 1989; Mollaghasemi, 1997).

Multi-objective optimization problems generally involve maximizing multiple-objective functions defined over a set of feasible decisions. The mathematical description of such problem is shown as below

$$\text{Maxmize} \quad f(x) = [f_1(x), f_2(x), \ldots, f_p(x)] \tag{17.84}$$

$$\text{Subject to} \quad g_j(x) \leq 0 \quad j = 1, 2, \ldots, m$$

where

x = An n-dimensional vector of decision variables
$f_i(x)$, $i = 1,2, \ldots, p$ = An n-dimensional vector of decision variables
$g_j(x)$, $j = 1,2, \ldots, m$ = An n-dimensional vector of decision variables

Because of the conflict among the objectives given in the above, the decision variables, x, may not be able to maximize all of the objectives simultaneously in most cases. Instead, some solutions are good in improving some objectives only, whereas other solutions are better for a different subset of the objectives. If solutions appear to be worse in at least one objective and the same in the remaining objectives, then these solutions could be eliminated. In choosing a solution, the decision maker must be willing to sacrifice the performance of some objectives to trade off the increase of another objective. This is also known as making trade-offs among the objectives.

17.6.1.3 Definitions, terminology, and notation

Before discussing the classes of solution methodologies, this section will introduce some general notation and definitions that are commonly used in MCDM.

Pareto-optimal (non-dominated or efficient) solution: Let X denote the set of all feasible solutions. An efficient (non-dominated) solution is a feasible solution, $x^* \in X$, for which there does not exist any other feasible solution, $x \in X$, that provides the same or better value in each of the objectives. Using the mathematical expression, there does not exist $x \in X$, the following statement holds: $f_i(x) \geq f_i(x^*)$ for all $i = 1,2, \ldots, p$, and there exists at least one i that $f_i(x) > f_i(x^*)$, assuming a larger $f_i(x)$ is preferred. Generally, an efficient solution is not unique.

Superior solution: This refers to a solution that maximizes all objectives at the same time, assuming this is a maximization problem. Using mathematical expression, a superior solution, $x^0 \in X$, should have such properties: $f_i(x^0) \geq f_i(x)$ for all $x \in X$ and all $i = 1,2, \ldots, p$. Although the superior solution is the ideal solution, a superior solution hardly exists because of the conflict among objectives.

Most preferred (best-compromise) solution: A most preferred solution is an efficient solution selected according to the decision-maker's overall preference function. The solution varies from decision maker to decision maker since it is oriented to the decision-maker's preference.

Trade-off or marginal rate of substitution: The trade-off ratio, also known as the marginal rate of substitution, between objectives f_i and f_j at a certain point is the ratio $(\partial U/\partial f_i)/(\partial U/\partial f_j)$, where U stands for the utility defined in terms of f_i and f_j by the decision maker to express the preference. This value is used as follows: when the ratio is r at a given point, a decrement of r units in f_i with one-unit increment in f_j makes no difference to the existing situation from the perspective of decision maker.

Ordinal ranking versus cardinal ranking: Ordinal ranking creates only a sequential ranking of alternatives without noting the magnitude of difference. Cardinal ranking not only gives a sequential ranking of alternatives but also provides the magnitude of difference.

Aspiration levels, goals: An aspiration level is a value describing the desirable level of an objective. A goal is the result of using an objective associated with an aspiration value. For example, if an agency states the desire to control the maintenance cost of a highway segment below $4 million per year, then $4 million is the aspiration level and the statement they made is the goal.

17.6.1.4 Classes of solution methods

All methods used to solve multicriteria decision problems generally contain two processes: (a) articulation of the decision-maker's preference structure among the set of criteria; and (b) preference structure optimization. Therefore, based on the sequence of executing these two processes, the methods could be classified into three classes: (i) prior articulation of preferences; (ii) progressive articulation of preferences; and (iii) posterior articulation of preferences (Evans, 1984).

In the first class, the decision-maker's preference information is obtained via detailed interviews before the optimization process. In the second class, acquiring decision-maker's preference and optimization processes are mutually interspersed. That is, the decision maker provides personal opinions on the multidimensional outcome space, which enables the formulation of a single-criterion problem, whose solution forms a new multidimensional outcome space to be presented. This repetitive process stops when the best-compromise solution is converged or the decision maker terminates the process. The third class seeks all efficient solutions and presents them to the decision makers for selection. Table 17.15 summarizes the features of three classes of methods for solving MCDM problems.

17.6.2 Methods based on the prior articulation of preferences

This section will introduce MCDM approaches that are based on the prior articulation of preferences. Some approaches are appropriate for solving only multiattribute problems, such

Table 17.15 MCDM method classifications

Class	*Prior articulation of preferences*	*Progressive articulation of preference*	*Posterior articulation of preference*
Description	The preference structure is obtained prior to the optimization	The preference structure extraction is interspersed with optimization	The preference structure extraction is after the generation of efficient solutions
Examples	Goal programming, method using a MAV or utility function	STEP method, Tchebycheff method	DEA
Advantages	It is relatively simple to derive solution	The preference information is acquired gradually throughout the process, which makes the process more natural	The process is fairly easy to understand
Disadvantages	It's hard for the decision maker to provide preference structure required	The process is more complicated and less transparent to users	The algorithms are usually complex and difficult to understand; large size MCDM is hard to solve in reality; possibly, a large size of efficient solutions could be generated, resulting in the difficulty in choosing one from them

as scoring methods, the analytic hierarchy process (AHP), and outranking method. Some approaches are appropriate only for multiple-objective programming problems, such as goal programming. Others can be used for both multiattribute and multi-objective programming problems, such as utility-based methods.

17.6.2.1 Scoring methods

Scoring methods are one of the most intuitive and commonly used tools for solving multiattribute decision problems. These methods assign weights to the criteria and then rate the alternatives against each criterion. Generally, weight-and-rate methods include two subcategories depending on how these weights are determined. The first approach is to assign the weights arbitrarily or empirically. The second approach is to assign weights based on analytical procedures supported by justified theories. There are a variety of methods belonging to each of these two categories (Dujmovic, 1977).

The general procedure of scoring method is as follows:

Step 1: Weighting—The decision maker should assign weights, $w_i (i = 1, 2, \ldots, m)$, to each of the m attributes. The weight could be represented on a scale of 1–100 to indicate the importance of each attribute quantitatively, where the highest weight means the highest level of importance. Note that a weight should never be zero. Zero weight means the attribute is not important at all, so then this attribute should be removed from the consideration.

Step 2: Scaling—The decision maker should assess the performance of each of the n alternatives associated with each of the m attributes. The output of this step is an $m \times n$ matrix A, where the element $a_{ij} (i = 1, 2, \ldots, m; j = 1, 2, \ldots, n)$ is a value indicating the how good the alternative j performs on attribute i. Similarly, the rating process also needs a scale to unify all the attributes with difference measures and units. A scale of 0–100 could be a good option for this step, where 0 means zero achievement while 100 means achieving the best performance. Note that zero exists in this scale because it might happen that one alternative does not contain a certain attribute.

Step 3: Amalgamation—The decision maker needs to combine all the information obtained above by taking the weighted sum of each alternative, getting an overall rate of performance or worth v_j. It could be presented as in Equation 17.85. The alternative with the highest worth is selected as the best option for this problem.

$$v_j = \sum_{i=1}^{m} w_i a_{ij} \tag{17.85}$$

EXAMPLE 17.16

A transit authority plans to buy some new buses to improve service quality. The decision maker takes the following features into account: attractiveness of appearance, fuel economies, price, and reliability. This authority has identified four buses from different manufacturers as candidate alternatives.

Solution

Step 1: Weighting—The decision maker ranks the importance of these attributes: Price > Fuel economies > Reliability > Attractiveness. The corresponding attribute weights assigned are 10, 8, 6, 4, respectively.

Table 17.16 Transit performance ratings

Criteria	*Alternatives*			
	Bus 1	*Bus 2*	*Bus 3*	*Bus 4*
Attractiveness	96	64	52	44
Fuel economy	52	72	64	86
Price	61	44	86	36
Reliability	49	86	55	95

Step 2: Scaling—Among all attributes concerned, attractiveness and reliability are rated subjectively on a 0–100 scale, and fuel economies and price are objectively measured by gas mileage (in miles per gallon) and dollars spent, respectively. In order to keep the consistency, fuel economies and price need to be converted into a score on a 0–100 scale. The rating result is given as follows (Table 17.16).

Step 3: Amalgamation—Compute the weighted sum of each alternative.

$$v_1 = [96, 52, 61, 49] \times [10, 8, 6, 4]^T = 96 \times 10 + 52 \times 8 + 61 \times 6 + 49 \times 4 = 1938$$

$$v_2 = [64, 72, 44, 86] \times [10, 8, 6, 4]^T = 1824$$

$$v_3 = [52, 64, 86, 55] \times [10, 8, 6, 4]^T = 1768$$

$$v_4 = [44, 86, 36, 95] \times [10, 8, 6, 4]^T = 1724$$

As shown in the computational result, the first alternative has the highest worth, meaning it is the best choice for this problem.

17.6.2.2 Multiattribute value functions

Preference-based methods are used to solve MCDM problem where the rank of alternatives is desirable. These methods require the decision-maker's preference over the set of attributes to be mathematically converted into real-valued functions. Such function is called a value or utility function. Based on the value or utility function, a cardinal ranking of alternatives could be developed (Dyer and Sarin, 1979; Chankong and Haimes, 1983; Li and Sinha, 2009b; Sinha et al., 2009).

Preference functions can be subcategorized into utility functions and value functions based on utility theory. Specifically, a preference function under certainty is called a value function, while a preference function under uncertainty is called a utility function.

A multiattribute value (MAV) function is used when the decision problem is not related to uncertainty. In this approach, the development of a value function is the core to the ranking development. The value function v mathematically denotes the decision-maker's preferences to each alternative x in set of candidates X, associated with m attributes or objectives. This value function generates a number to each alternative, measuring an alternative's worth from the perspective of decision-maker's preferences. The alternatives are ranked based on the outcome generated from value function. Then, the alternative with the highest functional value is selected as the best option.

Suppose the decision maker is dealing with two different alternatives $x' = (x_1', x_2', \ldots, x_m')$ and $x'' = (x_1'', x_2'', \ldots, x_m'')$. The decision maker is indifferent between these two alternatives if and only if $v(x') = v(x'')$ and x' is considered to be more preferred than x'' (noted as $x' \succ x''$) if and only if $v(x') > v(x'')$.

Directly assessing the value function $v(x) = v(x_1, x_2, \ldots, x_m)$ is very difficult since m attributes are involved in the assessment. The direct assessment requires the decision maker to consider all the attributes comprehensively, and then assign a value representing how much the alternative meets the preference. The complexity of this problem increases rapidly as the increment of the attribute size. Therefore, the direct assessment of an MAV function, $v(x_1, x_2, \ldots, x_m)$, is decomposed into m single-attribute value functions to simplify the problem, $v(x_1), v(x_2), \ldots, v(x_m)$. Although the single-attribute value function is also difficult to get, the decomposition indeed reduces the complexity of the problem.

After the decomposition, we need to find a way to relate the MAV function, $v(x_1, x_2, \ldots, x_m)$, to these single-attribute value functions. One simple way to do so is using the following function:

$$v(x_1, x_2, \ldots, x_m) = a_1 \cdot v_1(x_1) + a_2 \cdot v_2(x_2) + \cdots + a_m \cdot v_m(x_m) \tag{17.86}$$

where

a_i = the measure of relative importance of attribute i, and $\sum_{i=1}^{m} a_i = 1$

x_i = a measure of performance of this alternative on attribute i

$v_i(x_i)$ = the single-attribute value function associated with attribute i

Note that before using this additive function, one requirement on decision-maker's preference must be strictly satisfied. This requirement is known as corresponding trade-offs condition. Two attributes x_1 and x_2 satisfy the corresponding trade-offs condition if an increment in x_1 by y units is worth z units in x_2, irrespective of the value of x_1, x_2, y, z. To use the additive form of value function to deal with multiattribute decision problem with two attributes, these two attributes should satisfy the corresponding trade-offs condition.

In the case of three or more attributes, a condition called mutual preferential independence must be satisfied to use the additive form. Consider three attributes x_1, x_2, and x_3. The pair of x_1, x_2 is defined to be preferentially independent of x_3, if preferences for different values of attributes x_1 and x_2 do not depend on the level of x_3. In the case of four or more attributes, any three of the attributes set must be preferentially independent.

It is highly possible that the mutual preferential independence cannot hold, and the additive form for the value function cannot be used. In this case, other forms of value function may be used instead. For example, the multiplicative decomposition could be used as substitution. The form is

$$c \cdot v(x_1, x_2, \ldots, x_m) + 1 = [c \cdot a_1 \cdot v_1(x_1) + 1][c \cdot a_2 \cdot v_2(x_2) + 1] \cdots [c \cdot a_m \cdot v_m(x_m) + 1] \tag{17.87}$$

In the form, $\sum_{i=1}^{m} a_i \neq 1$ because the mutual preferential independence condition does not hold. Moreover, the new scaling constant c should be assessed.

There are some other forms of the MAV function used to deal with the case where the mutual preferential independence condition does not hold, such as quasi-additive and multilinear forms. MAV is a mathematical translation of a decision-maker's preference. Therefore, the process of assessing the MAV requires interaction between decision maker and analyst. The decision maker needs to assess the magnitude of preference of each attribute. In general, the single-attribute value function can be calibrated using data collected by two approaches: (i) direct questioning approach; and (ii) midpoint splitting approach (Chankong and Haimes, 1983; Li and Sinha, 2009b; Sinha et al., 2009).

Direct questioning approach: The direct questioning approach aims to derive the value outcomes corresponding to fixed attribute values. The general steps are as follows: first, ask the decision maker to give an interval to a given attribute i, denoting $[x_i^L, x_i^U]$, where the value function assigned to the lower bound x_i^L and upper bound x_i^U is 0 and 1, respectively, denoting $v_i(x_i^L)=0$, $v_i(x_i^U)=1$. Then, ask the decision maker to determine a preferred value for $v_i[0.5(x_i^U-x_i^L)]$ for the attribute value $0.5(x_i^U-x_i^L)$. Next, the decision maker is asked to assigned value outcomes $v_i[0.25(x_i^U-x_i^L)]$ for the attribute value $0.25(x_i^U-x_i^L)$ and for $v_i[0.75(x_i^U-x_i^L)]$ for the attribute value $0.75(x_i^U-x_i^L)$, respectively. Repeat this process in addition to the five pairs of attribute values and value function outcomes as needed. Use this approach to determine the single-attribute value function for each attribute.

Midpoint splitting approach: An approach that is widely used for developing a single-attribute value function is called the mid-value splitting method. The general steps are as follows: firstly, ask the decision maker to give an interval to a given attribute i, denoting $[x_i^L, x_i^U]$, the value function assigned to the lower bound x_i^L and upper bound x_i^U is 0 and 1, respectively, denoting $v_i(x_i^L)=0$, $v_i(x_i^U)=1$. Then ask the decision maker to give a value of the attribute i, $x_i^{0.5}$, where $v_i(x_i^{0.5})=0.5$. The following step is likewise, asking $x_i^{0.25}$ and $x_i^{0.75}$ where $v_i(x_i^{0.25})=0.25$ and $v_i(x_i^{0.75})=0.75$. Repeat this process until there are enough points to fit a nice curve. Use this approach to determine the single-attribute value function for each attribute.

After all single-attribute value functions are determined, the weights, a_i, need to be determined. The weights are obtained by asking the decision maker to identify two points whose overall value is the equivalent. Taking the case with three attributes as an example, two points selected are x' and x'', with identical overall value, that is, $a_1v_1(x_1')+a_2v_2(x_2')+a_3v_3(x_3')=a_1v_1(x_1'')+a_2v_2(x_2'')+a_3v_3(x_3'')$. We can find the value of a_1, a_2, a_3 by solving the two linear independent equations in the above form, along with the equation $\sum_{i=1}^{m} a_i=1$. For a general case with m attributes, the decision maker needs to provide $m-1$ pairs of points with the same overall value.

Up until now, the single-attribute value functions then could compose into a MAV function.

In the multiattribute problem with limited size of alternatives, each alternative is taken into the MAV function to compute the corresponding overall value. The one with the highest overall value is selected as the best option.

However, in the multi-objective problems with a limited size of alternatives, the optimal solution is found by using mathematical programming. In this case, alternatives are various decision variables, x, and the m attributes are expressed by m objective functions, $f_1(x), \ldots,$ $f_m(x)$. The problem can be expressed in the mathematic programming form, that is,

$$\text{Max } v[f_1(x), f_2(x), \ldots, f_m(x)] \tag{17.88}$$

$$\text{Subject to} \quad x \in X$$

17.6.2.3 Multiattribute utility functions

A utility function is a special case of a value function that incorporates the decision-maker's preference of choice. Multiattribute utility theory (MAUT) is a methodology that facilitates the decision maker's search for the best option from the set of feasible alternatives under uncertainty. Since it is an uncertainty-related problem, the likelihood of decision outcome is expressed in terms of PDFs over the attribute space. The general idea of using MAUT is the development of a multiattribute utility function, $u(x_1, x_2, \ldots, x_m)$, representing the decision-maker's preference over the probability distributions defined on the attribute space. And the expected value of this function, called "expected utility," is used as a measure of value of each attribute (Dyer et al., 1992).

Before using the utility theory, the following axioms must be satisfied.

Ordering of outcomes and transitivity: For any two alternatives, the decision maker can state one is preferred to the other or they are equivalent. This kind of preference order is transitive. For example, option A is preferred to option B. Option B is preferred to option C. Then, we are safe to state option A is preferred to option C.

Reduction of compound uncertain events: A decision maker is indifferent between a compound uncertain event and a simple uncertain event as determined by reduction by using standard probability manipulation.

Continuity: For any three preference consequences, $A_1 > A > A_2$, the decision maker can set a preference probability p of receiving A_1 versus A_2 that he or she feels indifferent to receiving A for sure.

Substitutability: The decision maker is indifferent to any original uncertain event that includes outcome A and one formed by substituting A for an uncertain event perceived to be equivalent.

Monotonicity: If the decision maker needs to choose between two lotteries with the same outcomes, the lottery with the higher probability will be chosen for better consequence.

Invariance: All that are needed to determine decision-maker's preferences among uncertain events are the payoffs and the associated probabilities.

Finiteness: No outcomes are considered infinitely good or bad.

Suppose that the decision maker is dealing with two different alternatives $x' = (x_1', x_2', \ldots x_m')$ and $x'' = (x_1'', x_2'', \ldots x_m'')$. The decision maker is indifferent to these two alternatives if and only if $E[u(x')] = E[u(x'')]$ and x' is considered to be more preferred than x'' (noted as $x' \succ x''$) if and only if $E[u(x')] > E[u(x'')]$, where $E[u(x')]$ denotes the expected utility of alternative x'.

The assessment of a multiattribute utility function is much more difficult than that of MAV function because of the uncertainty. Nonetheless, a major advantage of MAUT function over MAV function is that MAUT could consider the uncertainty appearing in the decision outcomes.

Similar to MAV function determination, the assessment of a utility function is usually accomplished by decomposing the multiattribute utility function into m single-attribute functions.

The assessment process used for utility functions requires the decision maker to establish a certain equivalent (CE) to a lottery. A CE to a given lottery is the certain value where the decision maker is indifferent between CE and the lottery. For example, one investor has two options to invest his five million. The first one is that he could gain 80 million for sure. The second option is uncertainty related. That is, he may gain 150 million with a probability of 0.75 or gain nothing with a probability of 0.25. If the investor is indifferent between these two options, we could say his CE is 80 million in this case. Note that the expected gain is 112.5, which is greater than CE. Therefore, this investor is risk-averse. If his CE is greater than the expected value, this investor is a risk-taker. If his CE is equal to the expected value, then the investor is risk-neutral.

Before embarking on MAUT development, one of the fundamental concepts in MAUT should be discussed: utility independence. An attribute, x_1, is utility independent of attribute x_2 if preferences for lottery that are affected by the value of attribute x_1 are independent of the value of x_2. Various utility-independence conditions imply specific forms of utility functions. However, additive (Equation 17.89) and multiplicative forms (Equation 17.90) are generally used in practice.

$$u(x_1, x_2, \ldots, x_m) = k_1 \cdot u_1(x_1) + k_2 \cdot u_2(x_2) + \cdots + k_m \cdot u_m(x_m) \tag{17.89}$$

where

$u(x_1, x_2, \ldots, x_m)$ = Overall utility on a scale from 0 to 1

$u_i(x_i)$ = The single-attribute utility function associated with attribute i, on a scale from 0 to 1

k_i = The scaling positive constant, $\sum_{i=1}^{m} k_i = 1$

$$c \cdot u(x_1, x_2, \ldots, x_m) + 1 = [c \cdot k_1 \cdot u_1(x_1) + 1][c \cdot k_2 \cdot u_2(x_2) + 1] \cdots [c \cdot k_m \cdot u_m(x_m) + 1] \qquad (17.90)$$

where

k_i = The scaling constant could be greater or less than 1

c = The constant determined by solving Equation 17.91

$$c + 1 = \prod_{i=1}^{m} (1 + c \cdot k_i) \qquad (17.91)$$

The assessment of the overall utility function requires the assessment of m single-attribute utility functions. The curve of the single-attribute utility function is determined by the risk attitude of the decision maker. Figure 17.14 presents an example of the shape of utility. The shape of risk-prone utility is convex, while the shape of risk-neutral is a straight line and the shape of risk-averse is concave.

In order to obtain the utility curve, various points on the curve should be determined by the decision maker. The general procedure is as follows. For a particular attribute x_i, to determine how $u_i(x_i)$ changes against x_i, the first step is to identify two values x_i^L and x_i^U, which correspond to the worst and best value of x_i, respectively. Then, assign $u_i(x_i^L) = 0$ and $u_i(x_i^U) = 1$. The next step is to ask the decision maker the following question: you have a probability of p to achieve x_i^U and a probability of $1 - p$ to achieve x_i^L, what least certain amount of x would be acceptable instead of taking a risk? Once the answer is obtained, we could get $u_i(x) = p$. Repeat asking this question with p changed all the time until we have enough points to fit the utility curve nicely.

After the single-attribute utility function has been assessed, the scale constants k_i, which indicates the relative importance of each attribute, should be assessed using the following steps. The decision maker first will be asked to choose an attribute as the base attribute, denoted as attribute i. Then, select another attribute j. The following step is to create two imaginary alternatives. One alternative has the attributes $x_i = x_i^U$ and $x_j = x_j^L$ and another alternative has $x_i = x_i^L$ and $x_j = x_j^L$. Then increase the value of x_j from x_j^L in the second alterative gradually until the decision maker is indifferent to the two alternatives. Note down the value of x_j as x_j^P and form Equation 17.92 with respect to k_i and k_j. Repeat this process

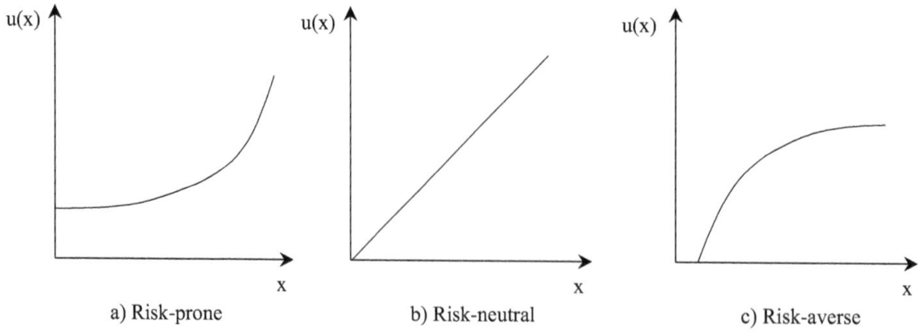

Figure 17.14 Illustration of utility functions.

by iterate j. In the end, we could form $m-1$ equations, which could be used to solve for all scaling constant k_i, along with equation $\sum_{i=1}^{m} k_i = 1$.

Once the multiattribute utility function has been assessed, the alternative can be ranked in terms of the expected overall utility. It is computed by the following:

$$E[u(x)] = \int_{-\infty}^{+\infty} u(x)f(x)dx \tag{17.92}$$

where

$f(x)$ = The set of conditional PDF for alternative x, $f(x) = [f_1(x_1), f_2(x_2), \ldots, f_m(x_m)]$

The alternative with the highest expected utility is considered the best alternative for this problem.

17.6.2.4 Analytic hierarchy process

The analytic hierarchy process (AHP) concerns both subjective and objective aspects in searching for the best alternative. This method is used to achieve a ratio-scale cardinal ranking of alternatives for multiattribute decision problems (Cambron and Evans, 1991).

Three principles that form the foundation of AHP are decomposition, comparative judgment, and synthesis of priorities. The decomposition principle means the problem should be decomposed into a hierarchy to unveil the important elements of the problem. The comparative judgments principle requires pairwise comparisons on a degree of relative importance of the elements within a given level. The judgments are organized into a matrix form and this matrix could be used to derive the ratio-scale local priorities at the given level. The synthesis of priorities principle collects each derived ratio-scale local priorities from all levels in the hierarchy and constructs a global set of priorities for the alternative (Saaty, 1990; Saaty and Vargas, 2000, 2006).

EXAMPLE 17.17

Let us take an example to explain the process of AHP and how these three principles work collaboratively.

A city is in negotiation with an auto glass manufacturer to build a recycle facility. Now they have entered into the phase of location selection. Three attributes are considered as top priorities: Costs, land lost, and jobs created. This city has identified 4 locations as candidate alternatives.

Step 1: Decomposition—This step is going to decompose the problem into a hierarchy. Figure 17.15 is the hierarchy decomposed.

Step 2: Comparative judgments—Table 17.17 presents pairwise judgments between attributes. The city considers jobs created as 7 times more important than costs and 5 times more important than land lost. The cost is 3 times more important than land lost.

Similarly, Table 17.18 gives a pairwise comparison of alternatives related to costs. As we can see, alternative I is 3 times preferred to alternative II, 7 times preferred to alternative III, and 5 times preferred to alternative IV with respect to costs. Tables 17.19 and 17.20 provides information on pairwise comparison of alternatives related to land lost and jobs created, respectively.

Once we have the comparative matrix, we could derive the local priority. Table 17.21 gives the process to perform the computation.

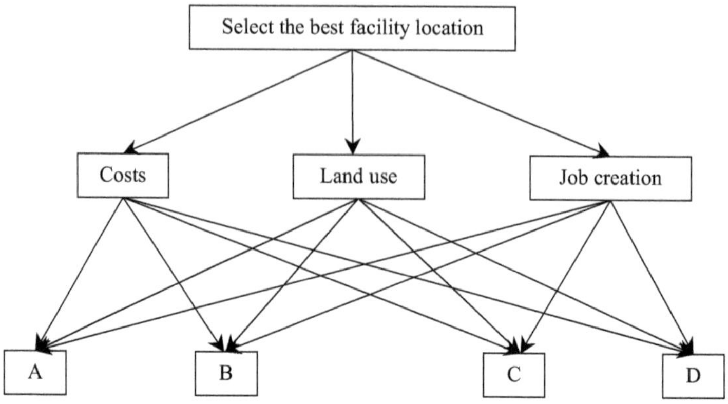

Figure 17.15 Complete hierarchy for facility location selection.

Table 17.17 Pairwise comparison of attributes

Attributes	*Costs*	*Land use*	*Jobs creation*
Costs	1	3	1/7
Land use	1/3	1	1/5
Jobs creation	7	5	1

Table 17.18 Pairwise comparison of alternatives related to costs

Costs	*I*	*II*	*III*	*IV*
I	1	3	7	5
II	0.33	1	1.5	1.25
III	0.14	0.67	1	0.83
IV	0.20	0.80	1.20	1

Table 17.19 Pairwise comparison of alternatives related to land use

Land use	*I*	*II*	*III*	*IV*
I	1	3	9	5
II	0.33	1	4	1.20
III	0.11	0.25	1	0.30
IV	0.20	0.83	3.33	1

Table 17.20 Pairwise comparison of alternatives related to jobs created

Jobs creation	*I*	*II*	*III*	*IV*
I	1	0.08	0.04	0.07
II	12.5	1	0.5	0.83
III	25	2	1	1.67
IV	15	1.20	0.60	1

Table 17.21 Pairwise comparison of alternatives related to costs

Attributes	*Costs*	*Land use*	*Job creation*	*Column entry divided by Corresponding column sum*			*Row sum of last 3 columns*	*Normalized weights*
Costs	1	3	1/7	3/25	1/3	5/47	0.560	0.187
Land use	1/3	1	1/5	1/25	1/9	7/47	0.300	0.100
Job creation	7	5	1	21/25	5/9	35/47	2.140	0.713
Sum	25/3	9	47/35					

Table 17.22 Summary of local priorities in each level

Attributes	*Costs*	*Land use*	*Job creation*
Priority	0.187	0.100	0.713
I	0.60	0.60	0.02
II	0.17	0.20	0.23
III	0.10	0.05	0.47
IV	0.13	0.15	0.28

Perform the same computation to get the local priorities of alternatives I, II, III, and IV. Table 17.22 summarizes the local priority or weight in each level.

Step 3: Synthesis of priorities—This step is to compute the global priority by using all local priority. They are computed as follows:

$$\text{I: } 0.60\times0.187+0.60\times0.100+0.02\times0.713=0.186$$
$$\text{II: } 0.17\times0.187+0.20\times0.100+0.23\times0.713=0.216$$
$$\text{III: } 0.10\times0.187+0.05\times0.100+0.47\times0.713=0.359$$
$$\text{IV: } 0.13\times0.187+0.15\times0.100+0.28\times0.713=0.239$$

Thus, alternative III is the best choice.

I: One important feature of AHP is its capability of measuring the consistency of the decision-maker's judgment. This feature uniquely exists in AHP among all commonly-used approaches. If the decision-maker states attribute 1 is more important that attribute 2, and attribute 2 is more important than attribute 3, then the statement that attribute 1 is more important than attribute 3 should be expected because of transitivity. However, if the decision-maker states attribute 3 is more important than attribute 1, then the decision-maker is considered inconsistent. AHP uses the consistency ratio (CR) to measure the degree of inconsistency, as Equation 17.93. Generally, it is acceptable if CR is less than 0.1.

$$\text{CR} = \frac{\lambda_{\text{max}} - N}{(N-1)(\text{R.I.})} \tag{17.93}$$

where

λ_{max} = The dominate eigenvalue of attribute comparative matrix
N = The dimension of the comparison matrix
R.I. = Random indices, the value is determined by the comparison matrix sizes N, as summarized in Table 17.23.

The CR for the previous example is computed as follows:
For costs:
Pairwise comparison matrix of alternatives related to costs × Local priority vector =

Table 17.23 Random indices for computing the consistency ratio

N	1	2	3	4	5	6	...
R.I.	0.00	0.00	0.058	0.90	1.12	1.24	...

$$\begin{bmatrix} 1 & 3 & 7 & 5 \\ 1/3 & 1 & 2 & 1 \\ 1/7 & 0.5 & 1 & 1 \\ 1/5 & 1 & 1 & 1 \end{bmatrix} \times \begin{bmatrix} 0.60 \\ 0.17 \\ 0.10 \\ 0.13 \end{bmatrix} = \begin{bmatrix} 2.46 \\ 0.70 \\ 0.40 \\ 0.52 \end{bmatrix}$$

The resulting vector is then divided by its local priority vector, that is

$$\begin{bmatrix} 2.46/0.60 \\ 0.70/0.17 \\ 0.40/0.10 \\ 0.52/0.13 \end{bmatrix} = \begin{bmatrix} 4.10 \\ 4.12 \\ 4.01 \\ 4.00 \end{bmatrix}$$

The dominant eigenvalue λ_{max} is computed by averaging the elements in the resulting vector.

$$\lambda_{max}(\text{Costs}) = \frac{4.10 + 4.12 + 4.01 + 4.00}{4} = 4.06$$

$$\text{R.I.}(\text{Costs}) = \text{R.I.}(N = 4) = 0.9$$

$$\text{C.R.}(\text{Costs}) = \frac{4.06 - 4}{(4-1)(0.9)} = 0.02 < 0.1$$

Repeat the same process in computing the remaining consistency ratios.

$$\text{C.R.}(\text{Land use}) = 0.06 < 0.1$$

$$\text{C.R.}(\text{Jobs creation}) = 0.003 < 0.1$$

Therefore, the decision-maker was highly consistent when performing the pairwise comparison.

Even with the prevalent use of AHP, this approach also has some limitations. One of the main criticisms is called "rank reversal." This term refers to the reversal of the preference ranking of the alternatives when new alternatives are added into the candidate list. For example, an analysis using AHP may indicate alternative I is preferred to alternative II when alternative III is not an option. However, when alternative III is introduced to be an option, it may indicate alternative II is preferred to alternative I.

17.6.2.5 Outranking methods

Outranking methods can also solve multiattribute decision problems. These methods generally need less information from the decision maker. Outranking methods are a class of MCDM techniques that generate an ordinal ranking of the alternatives. Sometimes, the ranking might be a partial ordering. In this section, we mainly introduce the ELimination

Et Choix Traduisant la REalité (ELimination and Choice Expressing Reality, ELECTRE) I method first proposed by Bernard Roy in 1960s (Roy, 1968). This method has evolved over time into ELECTRE II and ELECTRE III (Roy, 1991).

ELECTRE I method allows the decision maker to select the preferred alternatives in favor of most criteria without causing an unacceptable level for any one criterion. This is a process particularly suitable for multiattribute problems (Roy, 1968). It assesses the non-dominated alternatives and filters out a subset of the non-dominated solutions where the dissension and discordance caused are acceptable by the decision maker. The construction of this subset is based on defining an outranking relationship that catches the decision-maker's preference. Preference relationships between two alternatives are developed for each criterion. These preference relations are then synthesized for each alternative to generate the outranking relationship, based on which a graph is plotted where each node represents a non-dominated alternative. This graph is further used to determine which of the alternatives is preferred.

The following notations are defined:

I = The set of m criteria
$w(k)$ = The weight on criterion k, representing the decision-maker's preference, $\sum w(k) = 1$
$B(i, j)$ = The set that contains the indices of the criteria for which alternative i is preferred or equally preferred to alternative j
$c(i, j)$ = Concordance between alternative i and alternative j
C = The matrix form of $c(i, j)$
a_{ik} = The original value on criterion k of alternative i, converted to quantitative if the element is qualitative
x_{ik} = The standard value on criterion k of alternative i
y_{ik} = The weighted standard value on criterion k of alternative i
$d(i, j)$ = Discordance index between alternative i and alternative j
D = The matrix form of $d(i, j)$
p = Minimal acceptable concordance
q = Maximal discordance allowed

The three fundamental concepts used in ELECTRE I are concordance, discordance, and threshold value. The concordance between any two alternatives (i, j) is a weighted measure of the number of criteria for which alternative i is preferred or equally preferred to alternative j. The concordance is computed by using Equation 17.94

$$c(i,j) = \sum_{k \in A(i,j)} w(k) \tag{17.94}$$

The discordance index is a measure of the dissatisfaction of choosing alternative i over alternative j, defined for each pair of alternatives. An interval scale is commonly defined for each criterion to obtain the discordance index. The purpose is to assess the level of discomfort resulting from improving one criterion by a given interval while worsening another criterion by a certain level. For a qualitative criterion where an ordinal scale (best, medium, worst) is used, numerical measures are assigned in describing this criterion. The discordance index between the alternative and that defined in Equation 17.95 is computed

$$d(i,j) = \frac{\max\limits_{k \in B(j,i)} \{|y_{ik} - y_{jk}|\}}{\max\limits_{i,j,k} \{|y_{ik} - y_{jk}|\}} \tag{17.95}$$

The discordance index is then synthesized with the concordance index to define the outranking relation. At this stage, a graph indicating mutual relationships among the alternatives is plotted where an arrow going from i to j indicates that alternative i is preferred to alternative j if and only if $d(i, j) \leq q$ and $c(i, j) \geq p$. These two values, q and p, are the threshold values.

The general step of the ELECTRE I approach is introduced as follows:

Step 1: Form the original matrix A of elements a_{ik}. This step will convert qualitative measure to quantitative measure. It is not necessary that a higher value of a_{ik} indicats a higher performance in criterion.

Step 2: Form the standard decision matrix X of elements x_{ik}. The computation of x_{ik} is based on Equation 17.96.

$$x_{ik} = \frac{a_{ik}}{\sqrt{\sum_i a_{ik}^2}} \tag{17.96}$$

Step 3: Form a weighted standard decision matrix Y of elements y_{ik}. The computation of y_{ik} is based on Equation 17.97.

$$y_{ik} = w(k) \times x_{ik} \tag{17.97}$$

Step 4: Find all concordance indices and discordance indices based on Equations 17.96 and 17.97. Then form matrices C and D.

Step 5: Set the threshold values p and q. These could be obtained by asking the decision maker or computed by taking the average value of matrix C and D, respectively.

Step 6: Check all the values in C using threshold p and check D using threshold q. Then filter the alternative pairs that satisfy the threshold conditions.

Step 7: Create the outranging relation diagram. Using the remaining alternative pairs (i, j).

EXAMPLE 17.18

A state transportation agency plans to build an expressway connecting two counties, and three attributes will be taken into account: construction cost, duration, and user costs caused by construction. Currently, four project alternatives with different alignments are identified. Table 17.24 summarizes the rating of alternatives with respect to criteria, along with the weights assigned by the decision maker. Apply the ELECTRE I method for project selection decision making.

Table 17.24 Summary of rating of alternatives and criteria weights

Criteria (Index)	*Construction cost (1)*	*Construction duration (2)*	*User cost (3)*
Weights	0.5	0.35	0.15
Alternative 1	350M	24 months	Low
Alternative 2	460M	19 months	Medium
Alternative 3	420M	20 months	High
Alternative 4	395M	22 months	Medium

Solution

Step 1: Form the matrix A. The user cost is measured qualitatively. Then, a value function could be used to translate it into a quantitative value, say, low, medium, and high will become 2, 5, 8, respectively.

$$A = \begin{bmatrix} 350 & 24 & 2 \\ 460 & 19 & 5 \\ 420 & 20 & 8 \\ 395 & 22 & 5 \end{bmatrix}$$

Step 2: Form the matrix X

$$X = \begin{bmatrix} 0.43 & 0.56 & 0.18 \\ 0.56 & 0.45 & 0.46 \\ 0.51 & 0.47 & 0.73 \\ 0.48 & 0.52 & 0.46 \end{bmatrix}$$

Step 3: Use the weights to form matrix Y

$$Y = \begin{bmatrix} 0.43\times 0.5 & 0.56\times 0.35 & 0.18\times 0.15 \\ 0.56\times 0.5 & 0.45\times 0.35 & 0.46\times 0.15 \\ 0.51\times 0.5 & 0.47\times 0.35 & 0.73\times 0.15 \\ 0.48\times 0.5 & 0.52\times 0.35 & 0.46\times 0.15 \end{bmatrix} = \begin{bmatrix} 0.21 & 0.20 & 0.03 \\ 0.28 & 0.16 & 0.07 \\ 0.25 & 0.16 & 0.11 \\ 0.24 & 0.18 & 0.07 \end{bmatrix}$$

Step 4: Find all concordance indices and discordance indices

$$B(1,2) = \{1,3\}$$

$$c(1,2) = \sum_{k\in B(1,2)} w(k) = 0.5 + 0.15 = 0.65$$

Similarly,

$$B(1,3) = \{1,3\}$$
$$c(1,3) = \sum_{k\in B(1,3)} w(k) = 0.65$$

...

$$C = \begin{bmatrix} - & 0.65 & 0.65 & 0.65 \\ 0.35 & - & 0.35 & 0.5 \\ 0.35 & 0.5 & - & 0.35 \\ 0.35 & 0.65 & 0.65 & - \end{bmatrix}$$

$$\max_{i,j,k}\{|\, y_{ik} - y_{jk} \,|\} = 0.08$$

$$d(2,1) = \frac{\text{Max}\{0.28 - 0.21, 0.07 - 0.03\}}{0.8} = \frac{0.07}{0.8} = 0.875$$
$$d(3,1) = \frac{\text{Max}\{0.25 - 0.21, 0.11 - 0.03\}}{0.8} = \frac{0.8}{0.8} = 1$$

...

$$D=\begin{bmatrix} - & 0.5 & 0.5 & 0.25 \\ 0.875 & - & 0.375 & 0.5 \\ 1 & 0.5 & - & 0.5 \\ 0.5 & 0.25 & 0.25 & - \end{bmatrix}$$

Reorganize the concordance and discordance in Table 17.25.

Step 5: Find the two threshold values by taking the averages

$$p=\frac{\sum_{i\neq j} c_{ij}}{4(4-1)}=\frac{6.3}{12}=0.525$$

$$q=\frac{\sum_{i\neq j} d_{ij}}{4(4-1)}=\frac{6}{12}=0.5$$

Step 6: Filter out the pairs (i, j), where $c_{ij} > p$ and $d_{ij} <_q$

In this case, there are five alternative pairs that satisfy the threshold conditions. These are (1,2), (2,3), (1,4), (4,2), (4,3).

Step 7: Create a diagram, drawing an arrowed arc from node i to node j for all alternative pairs listed above, as shown in Figure 17.4. We can see alternative 2 dominates (is preferred to) alternative 3. Moreover, alternative 1 dominates any other alternatives, and alternative 4 dominates alternatives 2 and 3. Therefore, this gives a complete ordinal ranking. Alternative 1 is the best option.

This ordinal ranking depends mostly on the threshold value set by the decision maker. Figure 17.16 also illustrates another two cases. Suppose that the decision maker sets a minimum concordance (p) index of 0.60 and a maximum discordance

Table 17.25 Concordance (C) and discordance (D) matrix

	Alternative 1		*Alternative 2*		*Alternative 3*		*Alternative 4*	
	C	*D*	*C*	*D*	*C*	*D*	*C*	*D*
Alternative 1	–	–	0.65	0.5	0.65	0.5	0.65	0.25
Alternative 2	0.35	0.875	–	–	0.35	0	0.5	0.5
Alternative 3	0.35	1	0.5	0.375	–	–	0.35	0.5
Alternative 4	0.35	0.5	0.65	0.25	0.65	0.125	–	–

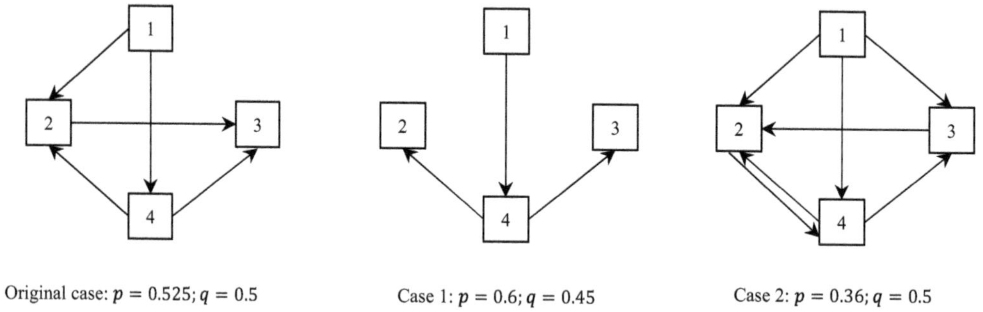

Figure 17.16 Preference ordering of alternatives.

index (q) of 0.45. Obviously, the following alternative pairs satisfy both conditions: (4, 3), (1, 4), (4, 2). Suppose that the decision maker sets a minimum concordance (p) index of 0.36 and a maximum discordance index (q) of 0.5. Then, the following alternative pairs satisfy both conditions: (1,2), (3,2), (4,2), (1,3), (4,3), (1,4), (2,4). Create the diagram indicating these arcs for both cases. In case 1, we can say alternative 1 is preferred to alternative 4 and alternative 4 is preferred to alternatives 2 and 3. But we cannot judge whether alternative 2 is preferred to alternative 3. In case 2, however, we find a cycle between alternative 2 and alternative 4. This cycle means these two alternatives are equally preferred, which differs from the conclusion drawn above. In the two extra cases, we see the preference ordering is partially given.

Because of the potential of ELECTRE I to produce a partial preference ordering, other more advanced methods have been developed by modifying the ELECTRE I method, such as ELECTRE II, ELECTRE III, ELECTRE IV, ELECTRE IS, and TRI (ELECTRE Tree). Furthermore, other great outranking methods have been developed from other perspectives, such as Qualitative Flexible Multiple Criteria Method (QUALIFLEX) (Paelinck, 1976, 1977, 1978), Preference Ranking Organization Method for Enrichment Evaluations (PROMETHEE) (Brans, 1982), and Regime method (Hinlopen et al., 1983).

17.6.2.6 Goal programming

Goal programming is a method to solve MCDM problems. This approach facilitates solving multi-objective problems by assigning each a priority. This approach is used when the relationship between the objectives and the decision variables can be expressed mathematically.

In the goal programming approach, the decision maker is asked to specify the goal and the priorities for the achievement of each goal. The idea is to find a single point with all constraints satisfied that achieve all goals "as closely as possible." A single point that achieves all goals probably does not exist in practice (Tamiz, 2012).

A goal refers to a criterion and a numerical target level, which the decision maker desires to achieve on that criterion. Goal programming models are very similar to LP models, but linear programs have one objective and goal programs can have several objectives.

EXAMPLE 17.19

A logistics company has two distribution centers in a city, noted as centers A and B. There are 100 labor hours at most to devote to the two distribution centers together. For each distribution center, one labor hour can prepare one container for shipping. The quota requires a minimum of 60 containers at distribution center A and 80 containers at distribution center B. The packing of container at distribution center A has twice the marginal profit as that at distribution center B. The goals of the labor hour assignment are set as follows: (i) try to utilize all labor hours; (ii) try to achieve the maximum overall profit; and (iii) try to minimize overtime operation.

Solution

Denote the labor hours assigned at distribution centers A and B as x_1 and x_2, respectively. The goals stated above could be mathematically expressed as

$$x_1 + x_2 \leq 100$$

$$x_1 \geq 60$$

$$x_2 \geq 80$$

The first expression represents the labor hours constraint. The second and third expressions are the basic demand constraint for both distribution centers. These three expressions can also be expressed using deviational variables, d^- and d^+, where d^- is the negative deviation and d^+ is the positive deviation.

$$x_1 + x_2 + d_1^- - d_1^+ = 100$$

$$x_1 + d_2^- - d_2^+ = 60$$

$$x_2 + d_2^- - d_2^+ = 80$$

where

d_1^+ = The number of labor hours over the available 100 labor hours
d_1^- = The number of unused labor hours
d_2^-, d_3^- = The shortage of containers prepared in meeting the demand at distribution centers *A* and *B*, respectively
d_2^+, d_3^+ = The extra containers prepared at distribution centers *A* and *B*

The next step is to assign priorities to each goal. Suppose ranking of importance among these three goals are goal 1, goal 2, and goal 3. Then the priority factors p_1, p_2, and p_3 are assigned by the decision maker to three goals. The complete goal programming can be expressed as

$$\text{Min} \quad p_1 d_1^- + 2p_2 d_2^- + p_2 d_3^- + p_3 d_1^+ \tag{17.98}$$

$$\begin{aligned} \text{Subject to} \quad & x_1 + x_2 + d_1^- - d_1^+ = 100 \\ & x_1 + d_2^- - d_2^+ = 60 \\ & x_2 + d_2^- - d_2^+ = 80 \\ & x_1, x_2, d_1^-, d_1^+, d_2^-, d_2^+, d_3^-, d_3^+ \geq 0 \end{aligned}$$

The first term in the objective function, $p_1 d_1^-$, represents the weighted "closeness" of goal 1, which is to use as many labor hours as possible. The second term, $2p_2 d_2^- + p_2 d_3^-$, represents the weighted "closeness" of goal 2, which is to maximize the overall profit. The third term, $p_3 d_1^+$, represents the weighted "closeness" of goal 3, which is to most utilize the available labor hours. As illustrated in this example, the objective is to minimize the weighted sum of all unwanted goal deviations. Now, the goal programming becomes an LP problem. This could be solved by using the simplex method.

The example above uses the weighted goal programming approach (also known as Archimedean goal programming or non-preemptive goal programming), which is designed for problems where all goals are quite important, with only modest differences in importance measured by assigned weights to the goals. Another most commonly used type is preemptive goal programming (or lexicography goal programming), where major differences exist in the importance of goals. In preemptive goal programming, the goals are listed in the order of importance. It begins by dealing only with the most important goal; then it deals with the second goal, adding a constraint that the first goal is also reached. Similarly, it then deals with the third goal without hurting the previous more important goals. It repeats these steps until all goals are dealt with.

Chebyshev goal programming (or minimax goal programming or fuzzy goal programming) is another popular type of goal programming, which is often used when the prime concern is a balance between the achievements of the goals (Zimmermann, 2001). This approach is quite practical in multi-stakeholder problems. The basic idea is to minimize the maximum unwanted weighted, normalized deviation.

Goal programming is not perfect. The main criticism claims that the solution to a goal programming might be dominated. It is possible that a better solution in favor of some or all objectives than the resulting solution could be obtained through goal programming.

17.6.3 Methods based on the progressive articulation of preferences

17.6.3.1 Introduction

The techniques based on progressive articulation of preference follow a common pattern. A subset of the non-dominated alternatives is presented to the decision maker. Then the decision maker is asked to comment on these alternatives with his or her local preference. These comments enable the formulation of a single-criterion sub-problem, which is then solved for a new non-dominated solution. The new non-dominated solution and the outcome are then offered to the decision maker for further selection. This iterative process terminates either when a best-compromise solution is converged or the decision maker is satisfied with the current solution. The ultimate objective is to find a satisfactory solution within a satisfactory amount of computation time.

Most of interactive approaches generate a subset of non-dominated solutions (Zionts and Wallenius, 1976). There are two major approaches for generating such solutions. They are (1) weighting-based methods; and (2) constraint-based methods.

The general formulation of a multi-objective programming problem is given as follows:

$$\text{Max} \quad f(x) = [f_1(x), f_2(x), \ldots, f_p(x)] \tag{17.99}$$

$$\text{Subject to} \quad g_j(x) \leq 0 \quad j = 1, 2, 3, \ldots, m$$

where

x = An n-dimensional vector of decision variables

$f_i(x)$ = The ith objective function

$g_j(x)$ = The jth constraint function

The main idea of the weighting-based method is to convert the set of multiple objectives into a single objective with weights assigned. The weighting-based method transforms a multi-objective problem into a single-objective problem expressed in Equation 17.15.

$$\text{Max} \quad f(x) = w_1 f_1(x) + w_2 f_2(x) + \cdots + w_p f_p(x) \tag{17.100}$$

$$\begin{aligned} \text{Subject to} \quad & g_j(x) \leq 0 \quad j = 1, 2, 3, \ldots, m \\ & w_1 + w_2 + \cdots + w_p = 1 \\ & w_i \geq 0 \quad i = 1, 2, 3, \ldots, p \end{aligned}$$

This approach could at least generate some possible non-dominated solutions if the weights are varied. All of the non-dominated solutions could be generated through exploring all possible weights if the problem is convex.

The main idea of the constraint-based method is to transfer the set of multiple objectives into a single objective by treating all but one of the objectives as inequality constraints. Therefore, the problem expressed by Equation 17.14 could be transferred into the problems expressed by Equation 17.16.

$$\text{Max} \quad f_k(x) \tag{17.101}$$

$$\text{Subject to} \quad g_j(x) \le 0 \quad j = 1,2,3,\dots,m$$
$$f_i(x) \le \varepsilon_i \quad i = 1,2,3,\dots,p, i \ne k$$

where

$f_x(x)$ = The primary objective function selected from the set of multiple objectives
$f_i(x)$ = The remaining objective functions in the set of multiple objectives
ε_i = The value selected for objective functions so that the problem has feasible solutions

Unlike the weighting approach, it turns out that all of the non-dominated solutions can be enumerated if all possible feasible values of ε_i are explored, even for problems that are not convex.

17.6.3.2 Surrogate worth trade-off method

The surrogate worth trade-off (SWT) method is a constraint-based method that aids the decision makers in choosing the most preferred solution after several solutions have been a generated by the constraint approach. It recognizes that given any current set of objective levels attained, it is much easier for the decision maker to assess the relative value of the trade-off of marginal increases and decreases between any two objectives than their absolute values (Sakawa, 1980).

The method consists of two phases: (1) identification and generation of non-dominated solutions which form the trade-off functions in the objective surface and (2) the search for a preferred solution in the non-dominated solutions. The preferred decision is located by interaction with the decision maker to assess the indifference band by the use of the newly introduced surrogate worth function.

The trade-off function can be found from the values of the dual variables associated with the constraints in a reformulated problem. The problem is reformulated as follows:

$$\text{Max} \quad f_k(x)$$

$$\text{Subject to} \quad g_j(x) \le 0 \quad j = 1,2,3,\dots,m$$
$$f_i(x) \le \varepsilon_i \quad i = 1,2,3,\dots,p, i \ne k$$

where

$$\varepsilon_i = \bar{f}_i - \bar{\varepsilon}_i \quad i = 1,2,3,\dots,p, i \ne k$$
$$\bar{\varepsilon}_i > 0 \quad i = 1,2,3,\dots,p, i \ne k$$

$\bar{f}_i$ are the feasible ideal solutions of the following single-objective problems

$$\text{Max} \quad f_i(x)$$

$$\text{Subject to} \quad g_j(x) \le 0 \quad j = 1,2,3,\dots,m$$

$\bar{\varepsilon}_i$ are the deviations from the ideal values and will be varied parametrically in the process of constructing the trade-off functions. The generalized Lagrangian, L, to the problem above is

$$L = f_k(\boldsymbol{x}) + \sum_{j=1}^{m} \mu_j g_j(\boldsymbol{x}) + \sum_{i=1, i \neq k}^{p} \lambda_{ki}(f_i(\boldsymbol{x}) - \varepsilon_i) \tag{17.102}$$

where

u_j, λ_{ki} = The generalized Lagrange multipliers. Specifically, λ_{ki} is the Lagrange multiplier associated with the kth objective function and the ith objective-constraint; this is also known as the trade-off rate.

Another important concept is the surrogate worth function, w_{ki}, which provides the interface between the decision maker and the model. The value w_{ki} is the decision-maker's assessment of how much on an ordinal scale (where a scale from −10 to +10 is commonly used with zero meaning indifferent attitude) the decision maker prefers trading λ_{ki} marginal units of the objective f_k for one marginal unit of the objective f_i, without hurting the attainment of other objectives. In other words, w_{ki} is the decision-maker's preference level of the trade-off rates, decision variables, and objectives. It could be defined as follows: (1) $w_{ki} > 0$ when λ_{ki} marginal units of $f_k(\boldsymbol{x})$ are preferred over one marginal unit of $f_i(\boldsymbol{x})$ given each of the remaining objectives $f_j(\boldsymbol{x})$ attains the satisfaction level of ε_j; (2) $w_{ki} = 0$ when λ_{ki} marginal units of $f_k(\boldsymbol{x})$ are equally preferred to one marginal unit of $f_i(\boldsymbol{x})$ given each of the remaining objectives $f_j(\boldsymbol{x})$ attains the satisfaction level of ε_j; (3) $w_{ki} < 0$ when λ_{ki} marginal units of $f_k(\boldsymbol{x})$ are less preferred to one marginal unit of $f_i(\boldsymbol{x})$ given each of the remaining objectives $f_j(\boldsymbol{x})$ attains the satisfaction level of ε_j.

In general, the surrogate worth function assigns a scalar value to any given non-dominated solution in objective function space. Once the indifference band λ_{ki}^* (where $w_{ki} = 0$) has been determined, the next step is to determine an $\boldsymbol{x}^*$ that for all λ_{ki}^* ($k = 1, 2, \ldots, p; i = 1, 2, \ldots, p; i \neq k$). The corresponding $f_i^*(\boldsymbol{x})$ is the value of $f_i(\boldsymbol{x})$ so that $\lambda_{ki}^*(f_i^*(\boldsymbol{x}) - \varepsilon_i) = 0$. The optimal decision vector $\boldsymbol{x}$ is then obtained by solving Equation 17.59.

$$\text{Max} \quad f_k(\boldsymbol{x})$$

$$\begin{aligned} \text{Subject to} \quad & g_j(x) \le 0 & & j = 1, 2, 3, \ldots, m \\ & f_i(x) \le f_i^*(x) & & i = 1, 2, 3, \ldots, p, i \neq k \end{aligned}$$

EXAMPLE 17.20

Apply the SWT method to solve the problem formulated in Equation 17.61.

$$\text{Max} \quad f_1 = x_1 x_2$$

$$\text{Min} \quad f_2 = (x_1 - 4)^2 + x_2^2$$

$$\begin{aligned} \text{Subject to} \quad & g_1(x) = x_1 + x_2 \le 30 \\ & x_1, x_2 \ge 0 \end{aligned}$$

Solution

Step 1: Find the ideal solutions. Solve two single-objective optimization problems. The results are $\overline{f_1} = 156.25$ at $x_1 = x_2 = 12.5$ and $\overline{f_2} = 0$ at $x_1 = 4, x_2 = 0$.

Step 2: Find non-dominated solutions. Standardize the formulation.

$$\text{Max} \quad f_1 = x_1 x_2$$

$$\text{Subject to} \quad (x_1 - 4)^2 + x_2^2 \le \varepsilon$$
$$x_1 + x_2 \le 30$$
$$x_1, x_2 \ge 0$$

The Lagrangian of this problem is

$$L = x_1x_2 + \mu_1(x_1 + x_2 - 30) + \lambda_{12}[(x_1 - 4)^2 + x_2^2 - \varepsilon]$$

The corresponding KKT conditions are

$$\frac{\partial L}{\partial x_1} = x_2 + \mu_1 - 2\lambda_{12}(x_1 - 4) = 0$$

$$\frac{\partial L}{\partial x_2} = x_1 + \mu_1 - 2\lambda_{12}x_2 = 0$$

$$\frac{\partial L}{\partial \lambda_{12}} = (x_1 - 4)^2 + x_2^2 - \varepsilon \le 0$$

$$\frac{\partial L}{\partial \mu_1} = x_1 + x_2 - 30 \le 0$$

$$\mu_1(x_1 + x_2 - 30) = 0$$

$$\lambda_{12}[(x_1 - 4)^2 + x_2^2 - \varepsilon] = 0$$

After solving the expressions above, it can be implied that

$$\lambda_{12} = \frac{x_2}{2(x_1 - 4)} = \frac{x_1}{2x_2}$$

$$x_2 = \sqrt{x_1(x_1 - 4)}$$

$$(x_1 - 4)^2 + x_2^2 = \varepsilon$$

These three expressions jointly define the non-dominated solution space, as listed in Table 17.26.

Step 3: Find the surrogate function. In this step, Table 17.26 is presented to the decision maker. Then, the decision maker is asked to fill the last row of this table. Based on the information presented, the decision maker is asked to give the subjective

Table 17.26 Non-dominated solution space

x_1	4	5	6	7	8	9	10	11
x_2	0	2.24	3.46	4.58	5.66	6.71	7.75	8.78
λ_{12}	1.118	0.866	0.764	0.707	0.670	0.645	0.626	0.612
f_1	0	11.18	20.78	32.08	45.26	60.37	77.46	96.53
f_2	0	6	16	30	48	70	96	126
W_{12}	–	+10	+7	+4	+1	−2	−5	−8

surrogate worth for each solution listed. The results are given in Table 17.26 as well.

Step 4: Find the indifference solution(s). According to the trend inferred from Table 17.26, one could find the solution with $w_{12} = 0$. In the example, this solution is between $x_1 = 8$ and $x_1 = 9$. Using linear interpolation, one could estimate the preferred solution is $x_1 = 8.33$ and $x_2 = 6.01$. If there is more than one solution that satisfies $w_{12} = 0$, all of these solutions are taken as preferred solutions.

17.6.3.3 STEP method

The STEP method (STEM) is an interactive method that can be used to identify the best-compromise solution for multi-objective LP problems (Benayoun et al., 1971). The mathematical formulation of this problem is

$$\text{Max}\left[\sum_{i=1}^{n} c_{1i}x_i, \sum_{i=1}^{n} c_{2i}x_i, \ldots, \sum_{i=1}^{n} c_{mi}x_i\right] \tag{17.103}$$

$$\text{Subject to} \quad \sum_{i=1}^{n} a_{ji}x_i \le b_j \quad j = 1,2,3,\ldots,m$$

$$x_i \ge 0 \qquad i = 1,2,3,\ldots,n$$

The general process follows these steps.

Step 1: Construction of a payoff table—A payoff table (Table 17.27) is constructed before the first interaction cycle. Denote f_j^*, $j = 1,2, \ldots, p$, to be the feasible ideal solutions of the following single-objective LP problems expressed in (17.23). The corresponding decision variable $x = x_j^*$.

$$\text{Max } f_j(x) = \sum_{i=1}^{n} c_{ji}x_i \tag{17.104}$$

$$\text{Subject to} \quad \sum_{i=1}^{n} a_{ji}x_i \le b_j \quad j = 1,2,3,\ldots,m$$

$$x_i \ge 0 \qquad i = 1,2,3,\ldots,n$$

In Table 17.27, the element $z_{ji} = f_j(x_i^*)$. Note that the diagonal element $z_{jj} = f_j(x_j^*) = f_j^*$.

Table 17.27 A payoff table of STEM

	f_1	f_2	…	f_j	…	f_m
f_1				z_{1j}		
f_2				z_{2j}		
…				…		
f_j	z_{j1}	z_{j2}	…	f_j^*	…	z_{jm}
…				…		
f_m				z_{mj}		

Step 2: Calculation step—The kth feasible solution to Equation 17.24 is found, which is considered as the "nearest" to the ideal solution f_j^*.

$$\text{Min} \quad d$$

$$\text{Subject to} \quad \pi_j \left[f_j^* - f_j(x) \right] \leq d \quad j = 1,2,3,\ldots,m$$
$$x \in X^k$$
$$d \geq 0$$

where

X^k = The feasible region that includes $\sum_{i=1}^{n} a_{ji} x_i \leq b_j, j = 1,2,3,\ldots,m$ and $x \geq 0$, plus all constraints added in the previous $(k-1)$ cycles

π_j = The factor indicating the relative importance of the distances to the optima $f_j^* - f_j(x)$. This factor is computed by Equation 17.105

$$\pi_j = \frac{\alpha_j}{\sum_{i=1}^{m} \alpha_i} \tag{17.105}$$

where

$$\alpha_j = \frac{f_j^* - f_j^{\min}}{f_j^*} \left(\sum_{i=1}^{n} c_{ji}^2 \right)^{-0.5} \quad \text{if } f_j^* > 0$$

$$\alpha_j = \frac{f_j^{\min} - f_j^*}{f_j^{\min}} \left(\sum_{i=1}^{n} c_{ji}^2 \right)^{-0.5} \quad \text{if } f_j^* \leq 0$$

Step 3: Decision step—The compromise-solution x^k is presented to the decision maker. The decision maker compares the objective vector with the ideal objective and identifies the satisfactory and unsatisfactory objectives. The decision maker will relax a satisfactory objective to improve the attainment level of unsatisfactory objectives. The key in this step is to modify the feasible region, X^k, and set it to be the new feasible region, X^{k+1}, in the next iteration.

$$X^{k+1} = \begin{cases} X^k \\ f_j(x) \geq f_j(x^k) - \Delta f_j \\ f_i(x) \geq f_i(x^k) \qquad i \neq j, i = 1,2,\ldots,m \end{cases} \tag{17.106}$$

where

$f_i(x)$ = The satisfactory objective to be relaxed

Δf_j = The amount to be relaxed associated with objective $f_j(x)$

The weight α_j is set to zero in the next iteration.

The STEM iterative computation procedure is given below (Figure 17.17).

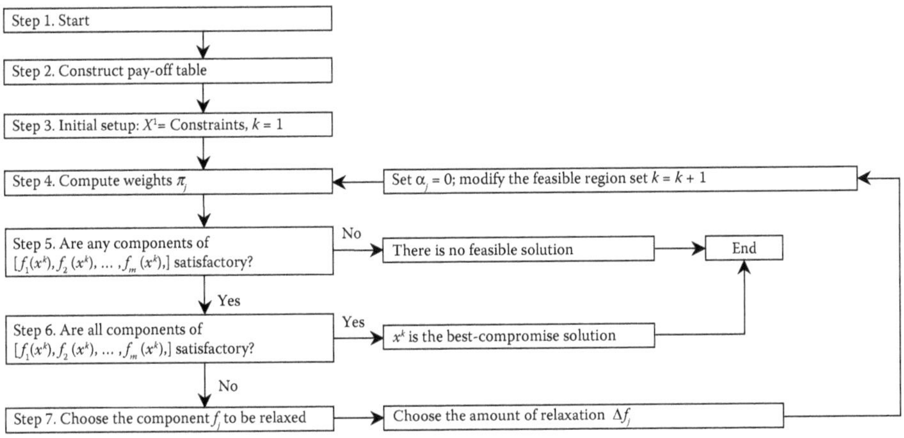

Figure 17.17 Iterative computation process of the STEM method.

EXAMPLE 17.21

Apply the STEM for solving the following problem:

$$\text{Max} \quad f_1 = 4x_1 + 3x_2$$

$$\text{Max} \quad f_2 = x_1$$

$$\begin{aligned} \text{Subject to} \quad & g_1(x) = x_1 + x_2 \leq 40 \\ & g_2(x) = 2x_1 + x_2 \leq 50 \\ & x_1, x_2 \geq 0 \end{aligned}$$

Solution

Step 1: Construction of a payoff table

$$f_1(x_1^*) = f_1(10,30) = 130; f_2(x_1^*) = f_2(10,30) = 10$$

$$f_2(x_2^*) = f_2(25,0) = 25; \quad f_1(x_2^*) = f_1(25,0) = 100$$

The payoff table is

	f_1	f_2	x_1	x_2
f_1	130	10	10	30
f_2	100	25	25	0

Step 2 (Iteration 1): Calculation step

Compute weights

$$\alpha_1 = \frac{f_1^* - f_1^{\min}}{f_1^*}\left(\sum_{i=1}^{2} c_{1i}^2\right)^{-0.5} = \frac{130 - 100}{130}(4^2 + 3^2)^{-0.5} = \frac{3}{65}$$

$$\alpha_2 = \frac{f_2^* - f_2^{\min}}{f_2^*}\left(\sum_{i=1}^{2} c_{2i}^2\right)^{-0.5} = \frac{250-100}{250}(1^2+0^2)^{-0.5} = \frac{3}{5}$$

$$\pi_1 = \frac{\alpha_1}{\alpha_1+\alpha_2} = 0.0714$$

$$\pi_2 = \frac{\alpha_2}{\alpha_1+\alpha_2} = 0.9286$$

Solve the LP problem

Min d

$$\text{Subject to} \quad \begin{aligned} & 0.0714\times[130-(4x_1+3x_2)] \le d \\ & 0.9286\times[25-x_1] \le d \\ & x \in X^1 \\ & d \ge 0 \end{aligned}$$

where

$$X^1 = \{x \mid g_1(x) \le 40; g_2(x) \le 50\}$$

The result of this problem is

$$x^1 = (x_1^1, x_2^1) = (23,4)$$

$$f^1 = (f_1^1, f_2^1) = (104,23)$$

Step 3: Decision step. The decision maker is present the solution $f^1 = (f_1^1, f_2^1) = (104,23)$ to be compared with $f^* = (f_1^*, f_2^*) = (130,25)$. Suppose that the decision maker if satisfied with f_2^1 only. Then, the amount of relaxation of f_2^1, Δf_2, is asked. Suppose that the decision maker gives $\Delta f_2 = 4$. Then, the feasible region is updated.

$$X^2 = \begin{cases} X^1 \\ f_2(x) \ge f_2(x^1) - \Delta f_2 \\ f_i(x) \ge f_i(x^k) \quad i \ne j, i = 1,2,\ldots,m \end{cases} = \begin{cases} X^1 \\ x_1 \ge 23-4 = 19 \\ 4x_1+3x_2 \ge 104 \end{cases}$$

Step 4 (Iteration 2): Calculation step
Compute weights $\pi_1 = 1$ and $\pi_2 = 0$
Solve the LP problem

Min d

$$\text{Subject to} \quad \begin{aligned} & 1\times[130-(4x_1+3x_2)] \le d \\ & 0\times[25-x_1] \le d \\ & x \in X^2 \\ & d \ge 0 \end{aligned}$$

The result of this problem is

$$x^2 = (x_1^2, x_2^2) = (19,12)$$
$$f^2 = (f_1^2, f_2^2) = (112,19)$$

Step 5: Decision step. The decision maker is presented the solution $f^2 = (f_1^2, f_2^2) = (112,19)$ to be compared with $f^* = (f_1^*, f_2^*) = (130,25)$. Suppose that the decision maker is satisfied with both f_1^2 and f_2^2. Then the best-compromise solution is found to be $(x_1, x_2) = (19,12)$.

17.6.3.4 Geoffrion–Dyer–Feinberg method

The Geoffrion–Dyer–Feinberg (GDF) interactive algorithm is a search procedure that moves iteratively toward better solutions via interactions with the decision maker at each iteration. In this method, it is not necessary to know the overall preferences of the decision-maker explicitly. Throughout the interviews, the decision maker provides local preference information to be used for determining the direction toward improved solutions. Once this direction is found, the decision maker is asked to provide information to determine the step length to move forward. This process will be terminated when the decision maker is satisfied with the current solution or when the improvement between two steps is less than a given value (Geoffrion et al., 1972).

The GDF algorithm is based on the well-known Frank–Wolfe algorithm.

Consider the following multi-objective programming problem:

$$\text{Max} \quad U(f_1(x), f_2(x), \ldots, f_p(x)) \tag{17.107}$$

$$\text{Subject to} \quad x \in X$$

where

$f_i(x) i = 1,2, \ldots, p$ are functionally independent criterion functions, which are assumed concave and continuously differentiable

U = The decision-maker's overall preference function defined over the objective space, which is also assumed to be concave and continuously differentiable. Note that this preference function may not be known explicitly at the outset

In order to solve the direction-finding problem using the Frank–Wolfe algorithm, the following problem must be solved:

$$\text{Max} \quad \nabla_x U(f_1(x^k), f_2(x^k), \ldots, f_p(x^k)) \cdot z \tag{17.108}$$

$$\text{Subject to} \quad z \in X$$

where x^k is an initial feasible solution and the solution z^k results in the determination of the optimal direction, d^k, by using $d^k = z^k - x^k$.

Unfortunately, because the preference function U is not known, the direction-finding problem cannot rely on the gradient of U. The following approach is suggested to solve this problem:

$$\text{Max} \quad \sum_{i=1}^{p} w_i^k \nabla f_i(x^k) \cdot z \tag{17.109}$$

$$\text{Subject to} \quad z \in X$$

where

$$w_i^k = \left(\frac{\partial U}{\partial f_i}\right)^k / \left(\frac{\partial U}{\partial f_1}\right)^k$$

w_i^k is the trade-off ratio of f_i and against criterion f_1 at point x^k. Once these trade-offs are obtained from the decision maker, the direction-finding problem can be solved. This direction is again obtained by letting $d^k = z^k - x^k$, where x^k is an initial feasible solution chosen by the user and z^k is a solution to the problem.

The next step is to find a proper step length to move forward. This can be achieved by presenting the decision maker with the vector $[f_1(x^k + td^k), f_2(x^k + td^k), \ldots, f_p(x^k + td^k)]$ for various values of t between 0 and 1. Once the decision maker selects a preferred solution, the step length could consequently be found. At this point, the decision maker has the option to decide whether to terminate the search or return to the direction-finding problem.

17.6.3.5 Zionts and Wallenius method

The Zionts and Wallenius method assumed that all the objective functions are concave and the constraints form a convex set; nonlinear functions are linearly standardized. The utility function is also assumed to be unknown to the decision maker but is tacitly a linear function of the objective functions. This method uses this function on an interactive basis (Zionts and Wallenius, 1976).

The first step of this method is to arbitrarily choose a set of positive multipliers and create a composite objective function or utility function using these multipliers. The composite objective function with constraints is then solved for a non-dominated solution. A subset of efficient variables is then selected from the set of nonbasic variables. Note that an efficient variable is one which, when introduced into the basis, cannot increase one objective without hurting one or more remaining objectives. A set of trade-offs is defined for each efficient variable by which some objectives are increased and others reduced. Such trade-offs are then presented to the decision maker for judging whether the trade-offs are desirable, undesirable or indifferent. His/her answers will help the construction of a new set of multipliers and the corresponding non-dominated solution. A new set of trade-offs is obtained at the new solution after repeating this process. This method guarantees a convergence to an overall optimum associated with the decision-maker's implicit utility function.

17.6.3.6 Interactive and visual goal programming

In this approach, the decision maker begins with specifying the type of goal, which can be maximization, minimization or fixing a value. The next step is to specify the aspiration level for each of the goals. The goals are then grouped based on priority levels, which give the importance ranking of objective groups. Then, within each group, the decision maker provides penalty weights for those priority levels with more than one goal. After the initial setup, the goal program is solved using the preemptive goal programming approach. Next, the solution is presented to the decision maker. At this point, the decision maker has options to make any modifications in the goal type, aspiration level, priority or weights if needed. The solution to the new problem is then found and presented to the decision maker for further modifications. This interactive process is repeated until the decision maker is content with the solution.

Visual interactive goal programming (VIG) is a visual search procedure extended from interactive goal programming. One advantage of VIG over the traditional interactive goal programming is the capability of visualizing the effect of changing the aspiration levels at each stage. VIG enables the analyst to search the efficient (Pareto-optimal) frontier of a multi-objective LP problem, by which the system presents only efficient solutions to the decision maker (Korhonen and Laakso, 1986).

Thanks to modern computer science, the VIG system could present the values of the objectives as bar graphs on the computer screen. The length of the bars changes dynamically as the solution moves on the efficiency frontier. By allowing the user to change the right-hand side of the equations one at a time, the VIG system can visualize how the modification would affect other objectives.

17.6.4 Methods based on the posterior articulation of preferences

The techniques based on the posterior articulation of preferences are the least popular among the MCDM approaches. Most of the methods in this category are applied to multi-objective problems. These methods first try to enumerate all non-dominated solutions to the problem. The decision maker is provided with the non-dominated solutions and solicited to select the preferred one. As for multi-objective problems, there usually exists an infinite number of elements in the set of non-dominated solutions. Therefore, selecting a single preferred solution could be remarkably troublesome, although there exist techniques that can reduce the size of the set of non-dominated solutions.

Similar to progressive methods, posterior methods generally conduct the generation of non-dominated solutions through either the weighting or the constraint approach. The decision maker is then solicited to provide some information regarding the non-dominated solutions. Most of the time, obtaining the information in a posterior method is more difficult than that in the interactive methods due to the size of the set of non-dominated solutions. Because of the similarity between posterior methods and progressive methods and the low popularity of posterior methods, one method will be briefly described: data envelopment analysis (DEA) method.

17.6.4.1 Data envelopment analysis

The decision-making units (DMUs), inputs, and outputs jointly form the fundamental elements of a DEA. A DMU is any entity where measurable inputs and outputs are assigned. For example, DMUs may be departments, agencies, governments or stakeholders. The input can be the amount of resources, travel time costs, and other types of assets. The outputs may be economic benefit, throughput, and other performance indicators. The viewpoint is that a growth in an input is expected to lead to a growth in an output. The viewpoint also believes that it is desirable to minimize inputs because the inputs collection is resources driven and thus costly. In the context of the MCDM, the DMUs can be viewed as the alternatives and the output can be viewed as the criteria. Sometimes, the inputs could also be viewed as attributes to be minimized or maximized (Charnes et al., 1985).

DEA usually infers several useful results by constructing the so-called envelopment surface (also known as efficient frontier). This could facilitate the determination of efficient DMUs among all DMUs. The envelopment surface indicates the best level of output that can be achieved. Moreover, it shows the minimum amount of inputs required to achieve a given output level.

DEA also creates an inefficiency metric. The inefficiency measure is based on the distance between the current point and the envelopment surface, which further could identify the sources and degrees of the inefficiencies. It can also map how to move an inefficient solution to the efficient frontier.

17.6.5 Conclusion

This section provides brief comparisons of MCDM problems and solution methods. It provides readers with a capsule looking into several existing methods, as well as their characteristics and applications. Table 17.28 summarizes the primary MCDM methods categorized by various properties.

Table 17.28 Comparisons of primary MCDM methods

	Type of problem			*Nature of information*	
Technique	*Multi-objective*	*Multiattribute*	*Uncertainty*	*Prior/progressive/posterior*	*Information required from the decision maker*
Scoring method		√		Prior	Score on attributes and objectives
MAV function	√	√		Prior	Weights and value functions on attributes
Multiattribute utility function	√	√	√	Prior	Weights and utility functions on attributes using lotteries
Analytic hierarchy process		√		Prior	Pairwise comparisons of attributes
ELECTRE I, II, III		√		Prior	Rating scale for criteria weights and discordance index
Goal programming	√			Prior	Priorities and aspiration levels on goals
DEA		√		Posterior	No decision-maker's preferences are used
SWT method	√			Progressive	Limits on objectives and worth values on trade-offs
STEP method	√			Progressive	Degree of satisfaction of objectives
GDF method	√			Progressive	Trade-off between objectives and solution preferences
Zionts and Wallenius	√			Progressive	Desirability of trade-offs
Interactive goal programming	√			Progressive	Priorities on goals, penalty weights

PROBLEMS

17.1 Given the following linear programming problem:

$$\text{Max} \quad x_1 + 2x_2$$

$$\begin{aligned} \text{Subject to} \quad -x_1 + 3x_2 &\leq 10 \\ x_1 + x_2 &\leq 6 \\ x_1 - x_2 &\leq 2 \\ x_1 + 3x_2 &\leq 6 \\ x_1, x_2 &\geq 0 \end{aligned}$$

Solve by using the Simplex method.

17.2 List the purposes of traffic assignment.

17.3 Two networks are shown below, use UE assignment to find the flow pattern and the travel path cost of each case. Compare the results. (This problem is known as Braess' paradox.)

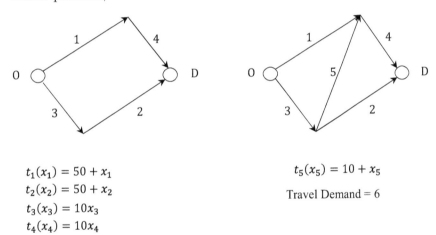

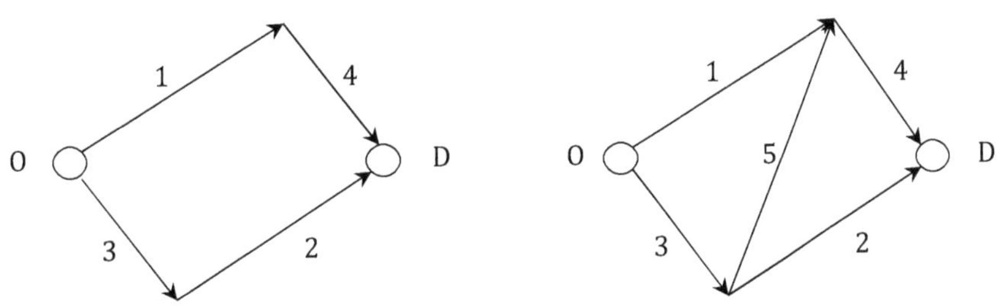

$t_1(x_1) = 50 + x_1$
$t_2(x_2) = 50 + x_2$
$t_3(x_3) = 10x_3$
$t_4(x_4) = 10x_4$

$t_5(x_5) = 10 + x_5$

Travel Demand = 6

17.4 Explain the advantages and disadvantages of each of the decision principles Minimax, Savage, Hurwicz Alpha, Hodges and Lehmann Criterion, and the Adaption Criterion.

17.5 Given the information on a transportation project decision matrix as shown in Table P17.1, determine the recommended alternative under the principles of Minimax and Hurwicz Alpha principles. Note: A = Alternative; and S = State of nature.

Table P17.1 Transportation project decision matrix

	S1	S2	S3	S4
A1	−90	70	80	190
A2	−70	100	90	220
A3	100	−50	100	200
A4	150	50	70	150

17.6 A roadway is going to be constructed on a partially leveled hill. A wall is required to prevent the rainwater from damaging the road. The probability of a rain greater than a given amount and the associated damage and construction costs are shown in Table P17.2. Determine the lowest annual cost plan. The service life is 25 years with a 4% discount rate.

Table P17.2 Rainfall damage and construction costs

Rainfall (inch)	*Possibility of Greater rainfall*	*Wall construction cost*	*Annual damage of Greater rainfall*
1.0	0.6	15000	2000
2.0	0.3	18000	3000
3.0	0.1	22000	5000
4.0	0.005	30000	8000

17.7 A transportation investment company has 5 million to invest. Putting the money in banks is assured to get an interest rate of 6.35% per year. If investing in a transportation project, two scenarios may happen with a half-half possibility. Corresponding cash flows are shown in Table P17.3. Determine if the money should be invested in the transportation project.

Table P17.3 Cashflow of the project

Year	*Scenario I*	*Scenario I*
0	−5.0 million	−5.0 million
1	0.3 million	0.5 million
2	0.7 million	0.8 million
3	1.3 million	1.3 million

17.8 Find examples in practice that could be solved by the knapsack, two-stage knapsack, stochastic knapsack, and hypergraph knapsack models, respectively.

17.9 A simplified road network is shown below. Corresponding travel costs are marked on the line. Determine the most economic route from the origin of node 1 to the destination of node 7.

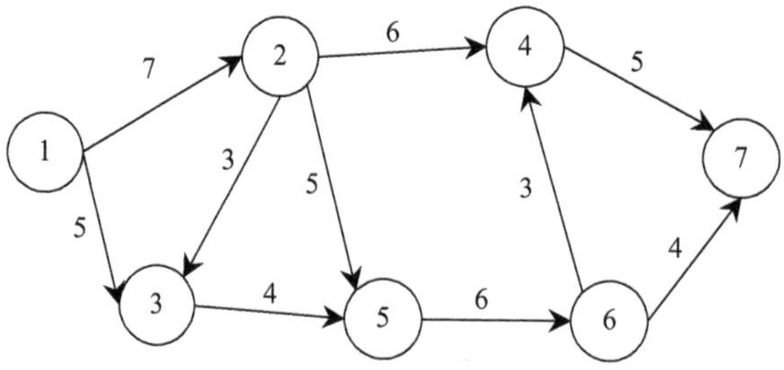

17.10 Solve for the user-equilibrium flow pattern in the network depicted in the following figure, where:

$$t_1 = 2 + x_1^2$$

$$t_2 = 3 + 7x_2$$

$$t_3 = 1 + 3x_3^2$$

$$t_4 = 3 + 4x_4^2$$

$$t_5 = 12 + 0.5x_5$$

$$t_6 = 4 + 2x_6 + x_6^2$$

$$q = 36$$

a. Requirements:
b. Find the objective function
c. Find all three types of constraints (i. Non-negative; ii. Path-link; iii. Demand)
d. Solve for the link flow (round your final answers to the nearest integers)
e. Solve for the path cost for all paths using your rounded link flows
f. Which path will the next traveler (from node 1 to node 4) choose in UE assignment?

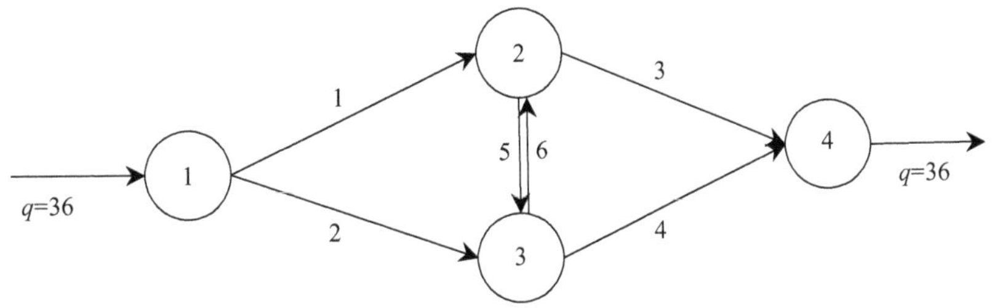

17.11 Compare the three typical types of multicriteria decision-making methods by listing the pros and cons for each of them.

Chapter 18

Transportation project selection

18.1 OVERVIEW OF PROJECT SELECTION METHODS

Transportation project selection aims to identify a subset of economically feasible projects to yield the maximized overall benefits subject to budget and other constraints. Ranking, prioritization, and optimization are the typical methods used for project selection (FHWA, 1991; Haas et al., 1994).

18.1.1 Ranking

Ranking is the simplest method of single-year priority setting for transportation project selection. It mainly includes two steps. First, it determines project items for facility preservation or improvement. For each set of candidate projects, the best alternative for each candidate project is identified. The associated cost is estimated. The next step involves prioritizing the candidate projects according to one or more criteria, such as facility condition, initial cost, least present cost and timing, life-cycle cost, benefit–cost ratio, and cost-effectiveness (Zimmerman, 1995). It generates a ranked list of candidate projects proposed for implementation, the associated project cost, and a cut-off line established based on the budget level. As the project timing is not considered in the ranking process, the long-term impacts of varying timing of project implementation from year to year cannot be readily evaluated.

18.1.2 Prioritization

Compared with the ranking method, multiyear prioritization is a more complex method for project selection that could potentially derive an optimal solution for highway resource allocation. This method uses condition deterioration models or remaining service life models for analysis. It also requires defining threshold values for needs assessment during the analysis period. Common prioritization methods include marginal cost-effectiveness, incremental benefit-to-cost ratio, and remaining service life analysis. Multiyear prioritization differs from the ranking procedure in that it considers alternatives, performance models, and implementation timing, as well as varying funding levels for analysis (FHWA, 1991). This helps derive more rigorous decision results.

18.1.3 Optimization

18.1.3.1 Optimization models

The optimization formulation for transportation project selection is known in the literature as the capital budgeting problem (Lorie and Savage, 1955; Weingartner, 1963). In a broader context, this problem falls into the category of the MCMDKP (Sinha and Zoltners, 1979), including its special classes of the MCKP, which concerns multiple budget sources and a single-year analysis period and the MDKP. Specifically, the budget is available and constrained from different sources and the analysis is conducted for a multiyear decision-making period. Notable deterministic optimization models developed in the last several decades include integer programming (IP) (Isa Al-Subhi et al., 1989; Weissmann et al., 1990; Zimmerman, 1995; Neumann, 1997), goal/compromise programming (Geoffroy and Shufon, 1992; Ravirala and Grivas, 1995), and multi-objective optimization (Teng and Tzeng, 1996; Li and Sinha, 2004). Over time, stochasticity of input factors of the above model formulations has been studied gradually. For instance, Friesz and Fernandez (1979) introduced a stochastic model to handle the demand uncertainty. Studies also addressed the impact of uncertainty induced by data and predictability of transportation facility performance models using simulation(de la Garza et al., 1998; Salem et al., 2003) and Markovian chain/dynamic programming approaches (Feighan et al., 1988; Harper et al., 1990; Ben-Akiva et al., 1991; Cesare et al., 1994).

18.1.3.2 Solution algorithms

The optimization formulation for project selection as the MCMDKP problem is nondeterministic polynomial hard (NP-hard) in the sense that no nondeterministic polynomial time algorithm exists in which the time required for the optimal solution grows exponentially with the problem size (Martello and Toth, 1990). Developing an efficient solution algorithm for solving the MCMDKP formulation is crucial to the success of model implementation. Solution algorithms can be classified into two groups: exact algorithms and heuristic algorithms.

The exact algorithms are mainly based on branch-and-bound, dynamic programming, and a hybrid of the two techniques. Sinha and Zoltners (1979) presented a branch-and-bound algorithm for the MCKP problem that resided with a quick solution of linear programming (LP) relaxation and its efficient, subsequent re-optimization from branching. Armstrong et al. (1983) conducted a computational study based on the branch-and-bound algorithm developed by Sinha and Zoltners wherein data structures, sorting techniques, and fathoming criteria were investigated. Dyer et al. (1995) developed a hybrid algorithm that combined dynamic programming and the branch-and-bound algorithm to solve the MCKP problem. In this algorithm, Lagrangian duality was used in a computationally efficient manner to compute tight bounds on every active node in the search tree. Klamroth and Wieck (2001) also proposed a dynamic programming approach to find all non-inferior solutions to the MCMDKP problem. Osorio and Glover (2001) presented a method of logic cuts from dual surrogate constraint analysis before solving the MDKP problem with branch-and-bound.

Heuristic algorithms may solve the problem close to optimal in polynomial time but do not guarantee optimality. Toyoda (1975) introduced a heuristic algorithm based on Lagrangian relaxation for an approximate solution to the MCMDKP problem. Magazine and Oguz (1984) presented a polynomial time-generalized Lagrangian multiplier approach based on Toyoda's algorithm. Volgenant and Zoon (1990) further extended the algorithm by establishing an upper bound to the optimal solution through simultaneous computation of two Lagrangian multipliers. Aggarwal et al. (1992) proposed a two-stage algorithm that leads to an optimal solution in a relatively low depth of search. Freville and Plateau (1994) introduced a subgradient heuristic algorithm for the MDKP problem. Teng and Tzeng (1996) suggested an effective distance heuristic optimization algorithm for the MDKP problem involving project interdependence relationships. Chu and Beasley (1998) presented an algorithm that incorporated problem-specific knowledge into the genetic algorithm for the

MDKP problem. Akbar et al. (2001) developed two heuristic algorithms for solving the MCMDKP problem by sorting the project benefits in nondecreasing order.

18.2 ISSUES OF PROJECT SELECTION

18.2.1 Budget constraints

Transportation project selection depends on several factors. One is the available budget for the multiyear project selection and programming period. Transportation agencies generally maintain a number of management programs to handle different systems management issues, such as pavement and bridge preservation, safety improvements, roadside improvements, system expansion, intelligent transportation systems (ITSs), multimodal facilities, and maintenance. A certain level of budget is designated to each program per year and the budget for each program is not to be transferred to different programs for use. Candidate projects are proposed to compete for funding within the eligible programs. Hence, project selection is constrained by the yearly budget or cumulative budget designated for each program in a multi-year period.

18.2.2 Budget uncertainty

In addition to the issue budget constraints by program and by year, the annual budget may be involved with uncertainty. This is because investment decisions are usually based on an estimated budget many years ahead of project implementation. With the passage in time, the estimated budget will be updated and revised based on additional information made available. Project selection needs to incorporate budget uncertainty consideration to ensure decision outcomes are realistic.

18.2.3 Project implementation scenarios

To mitigate traffic disruption at the construction stage, multiple projects within one transportation corridor or transit route might be bundled together for implementation. In addition, some large-scale projects might have to be postponed for a few years due to reasons such as ROW acquisition delays, design changes, and significant environmental impacts. As such, project selection could be carried out using highway segment-based, transportation corridor-based or deferment-based project implementation scenarios.

18.2.4 Risk tolerance of overall project benefits

During project selection, the decision maker aims to choose a portfolio of projects that minimizes the total risk of all projects subject to total budget constraint and lower bound constraints on the expected benefits of individual projects.

18.2.5 Multi-project benefit interdependency

When multiple projects are implemented in adjacent locations or in an overlapping time sequence, they may trigger traffic redistribution across the transportation network that would eventually change agency and user costs of all projects directly and indirectly affecting network links and nodes. Therefore, the overall benefits of interdependent projects may be greater than, equal to or smaller than the direct addition of individually estimated benefits of related projects.

18.3 KNAPSACK MODELS FOR PROJECT SELECTION

The knapsack problem often arises in resource allocation where there are financial constraints. Since it was first studied in 1896 (Mathews, 1896), it has been widely applied in fields such as civil engineering, computer science, applied mathematics, cryptography, and complexity theory. The knapsack model formulation is particularly suited for transportation project selection.

18.3.1 Basic model

As mentioned in the above, optimization models for project selection can be formulated as the zero/one integer doubly constrained KP. The doubly constrained multidimensions refer to doubly constrained budgets for different management programs and the budget for each program is constrained in each year of the multi-year analysis period (Martello and Toth, 1990). The objective is to select a subset from all economically feasible candidate projects to achieve maximized total benefits under various constraints. The zero/one value of a decision variable implies rejection or selection of a proposed project.

Denote

x_i = Decision variable for project i, $i = 1, 2, \dots, N$
a_i = Benefits of project i, $i = 1, 2, \dots, N$
c_{ikt} = Costs of project i using budgets from management program k in year t
X = Decision vector for all decision variables, $X = [x_1, x_2, \dots, x_N]^T$
A = Vector of benefits of N projects, $A = [a_1, a_2, \dots, a_N]^T$
C_{kt} = Vector of costs of N projects using budget from management program k in year t
C_{it} = $[c_{1kt}, c_{2kt}, \dots, c_{Nkt}]^T$
i = $1, 2, \dots, N$
k = $1, 2, \dots, K$
t = $1, 2, \dots, M$

Superscript T stands for transpose of a vector

A basic deterministic model as a doubly constrained KP formulation under the yearly constrained budget scenario is formulated as

$$\text{Max} \quad A^T \cdot X \tag{18.1}$$

$$\text{Subject to} \quad C_{kt}^T \cdot X \le B_{kt} \tag{18.2}$$

where X is a decision vector with 0/1 integer decision variables.

As expression 18.1, the objective function helps select a subset of all candidate projects to achieve maximized total benefits. Expression 18.2 lists budget constraints by management program and by year. The zero/one integrality constraints for the decision variables are used for rejection or selection of individual projects. For the cumulative budget scenario, budget constraints by analysis year are reduced to a single-period constraint. Only the budget constraints by management program are retained. The notation B_{kt} is replaced by $\sum_{t=1}^{M} B_{kt}$, accordingly.

18.3.2 A stochastic model addressing budget uncertainty

18.3.2.1 Treatments of budget uncertainty

As in Figure 18.1, consider a multiyear project selection period of t_Ω years. The transportation agency makes a first round of investment decisions many years ahead of the project

Implementation Year	1 to t_1	t_1+1 to t_2	···	t_{L-2}+1 to t_{L-1}	t_{L-1}+1 to t_L	t_L+1 to t_{L+1}	···	$t_{\Omega-1}$+1 to t_Ω
Budget	1 possibility	s_2 possibilities	···	s_{L-1} possibilities	s_L possibilities	s_{L+1} possibilities	···	s_Ω possibilities
Stage 1:	Deterministic (initially estimated budget)							
Stage 2:	Deterministic				Stochastic			
...				...				
Stage L-1:		Deterministic			Stochastic			
Stage L:			Deterministic			Stochastic		
Stage L+1:				Deterministic			Stochastic	
...				...				
Stage Ω:				Deterministic				Stochastic

Figure 18.1 Budget attributes in an Ω-stage recourse project selection process. (Adapted from Li, Z. and P. Kaini. 2007. Optimal investment decision-making for highway transportation asset management under risk and uncertainty. Report MRUTC 07-10. Department of Civil, Architectural, and Environmental Engineering, Illinois Institute of Technology, Chicago, IL; Zhou, B. et al. 2014. *ASCE Journal of Transportation Engineering* 140(5), 04014007; Li, Z. et al. 2013. *ASCE Journal of Transportation Engineering* 139(7), 686–696; Li, Z. et al. 2012. *TRB Journal of Transportation Research Record* 2285, 36–46; Li, Z. et al. 2010. *Wiley Journal of Computer-Aided Civil and Infrastructure Engineering* 25(6), 427–439; Li, Z. 2009. *ASCE Journal of Transportation Engineering* 135(6), 371–379.)

implementation period using estimated budgets for all years. With time passage, updated budget information on the first few years would become available, motivating the agency to refine its investment decisions. In each refined decision-making process, the annual budget for each management program for the first few years that could be accurately determined is treated as a deterministic value, while the budgets for the remaining years without accurate information are still processed as stochastic budgets.

Assume that the multiyear budgets are refined Ω times and each time an increasing number of years with accurate budget information from the first project implementation year onward is obtained. Hence, Ω-decision stages are involved. Without loss of generality, we assume a discrete probability distribution of budget possibilities for each year where no accurate budget estimates are available. For first stage decisions, the multiprogram, multiyear budget matrix is made up of the expected budgets for all years that can be best estimated at the time of decision making. For second-stage decisions, accurate information on budgets for years 0 to t_1 is known and the budgets are treated as deterministic, and there are ($p_2 = s_2 \cdot s_3 \ldots s_{L-1} \cdot s_L \cdot s_{L+1} \ldots s_\Omega$) possible budget combinations for the remaining years from $t_1 + 1$ to t_Ω. For the generic stage L decisions, budgets up to year t_{L-1} are deterministic and there are ($p_L = s_{L+1} \ldots s_\Omega$) possible combinations for years $t_{L-1} + 1$ to t_Ω. The final stage has deterministic budgets up to year $t_{\Omega-1}$ and $p_\Omega = s_\Omega$ budget possibilities from year $t_{\Omega-1} + 1$ to t_Ω.

18.3.2.2 A stochastic model using budget recourse functions

The stochastic model with multi-stage budget recourses is formulated as a deterministic equivalent program that combines first stage decisions based on the initially estimated budgets with expected values of recourse functions for the remaining stages (Birge and Louveaux, 1997).

Denote

x_i = Decision variable for project i, $i = 1, 2, \ldots, N$

a_i = Benefits of project i, $i = 1, 2, \ldots, N$

c_{ikt} = Costs of project i using budgets from management program k in year t

ξ_L = Randomness associated with budgets in stage L and decision space

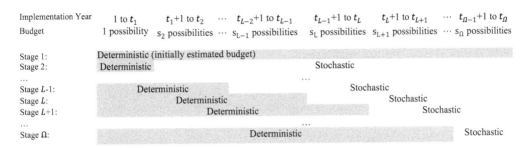

$X_L(p)$ = Decision vector using budget $B^L_{kt}(p)$ in stage L, $X_L(p) = [x_1, x_2, \dots, x_N]^T$
A = Vector of benefits of N projects, $A = [a_1, a_2, \dots, a_N]^T$
C_{kt} = Vector of costs of N projects using budget from management program k in year t
C_{it} = $[c_{1kt}, c_{2kt}, \dots, c_{Nkt}]^T$
$Q(X_L(p), \xi_L)$ = Recourse function in stage L
$E_{\xi_L}[Q(X_L(p), \xi_L)]$ = Mathematical expectation of the recourse function in stage L
$B^L_{kt}(p)$ = The pth possibility of budget for management program k in year t in stage L
$p\,(B^L_{kt}(p))$ = Probability of having budget scenario $B^L_{kt}(p)$ occur in stage L
$E\,(B^L_{kt})$ = Expected budget in stage L, where
p = $1, 2, \dots, p_L$, where $p_L = s_L \cdot s_{L+1} \cdots s_\Omega$
L = $1, 2, \dots, \Omega$
i = $1, 2, \dots, N$
k = $1, 2, \dots, K$
t = $1, 2, \dots, M$

Superscript T stands for transpose of a vector

The stochastic model with Ω-stage budget recourses under yearly constrained budgets is as

$$\text{Maximize} \quad A^T \cdot X_1 + \sum_{\omega=2}^{\Omega} E_{\xi\omega}[Q_\omega(X_\omega(p), \xi_\omega)] \tag{18.3}$$

Stage 1

$$\text{Subject to} \quad C^T_{kt} \cdot X_1 \le E\left(B^1_{kt}\right) \tag{18.4}$$

X_1 is a decision vector with 0/1 integer elements.

Stage 2

$$E_{\xi 2}[Q_2(X_2(p), \xi_2)] = \max\left\{A^T \cdot x_2(p) \mid B^2_{kt}(p) = E\left(B^2_{kt}\right)\right\} \tag{18.5}$$

$$C^T_{kt} \cdot X_2(p) \le B^2_{kt}(p) \tag{18.6}$$

$$X_1 + X_2(p) \le 1 \tag{18.7}$$

X_1 and $X_2(p)$ are decision vectors with 0/1 integer elements.

...

Stage L

$$E_{\xi L}[Q_L(X_L(p), \xi_L)] = \max\left\{A^T \cdot x_L(p) \mid B^L_{kt}(p) = E\left(B^L_{kt}\right)\right\} \tag{18.8}$$

$$\text{Subject to} \quad C^T_{kt} \cdot X_L(p) \le B^L_{kt}(p) \tag{18.9}$$

$$X_1 + X_2(p) + \dots + X_L(p) \le 1 \tag{18.10}$$

$X_1, X_2(p), \ldots, X_L(p)$ are decision vectors with 0/1 integer elements.

...

Stage Ω

$$E_{\xi\Omega}[Q_\Omega(X_\Omega(p), \xi_\Omega)] = \max\left\{A^T \cdot x_\Omega(p) \mid B_{kt}^\Omega(p) = E\left(B_{kt}^\Omega\right)\right\} \tag{18.11}$$

$$\text{Subject to} \quad C_{kt}^T \cdot X_\Omega(p) \leq B_{kt}^\Omega(p) \tag{18.12}$$

$$X_1 + X_2(p) + \cdots + X_L(p) + \cdots + X_\Omega(p) \leq 1 \tag{18.13}$$

$X_1, X_2(p), \ldots, X_L(p), \ldots, X_\Omega(p)$ are decision vectors with 0/1 integer elements.

In the objective function as Equation 18.3, the first term is for total project benefits in the first stage decisions using initial budgets, and the second term is for the expected value of total project benefits for the remaining (Ω-1)-stage recourse decisions. Equations 18.4, 18.6, 18.9, and 18.12 are employed to satisfy budget constraints by management program and year for investment decisions at each stage. Equations 18.5, 18.8, and 18.11 compute the expected values of optimal project benefits that use one possible budget closest to the updated budget. Equations 18.7, 18.10, and 18.13 ensure that one project is selected at most once in the multistage decision process.

For the cumulative budget constraint scenario, budget constraints by management program are maintained. The notations $B_{kt}^L(p)$, $p\left(B_{kt}^L(p)\right)$, and $E\left(B_{kt}^L\right)$are replaced by $\sum_{t=1}^M B_{kt}^L(p)$, $p\left(\sum_{t=1}^M B_{kt}^L(p)\right)$, and $E\left(\sum_{t=1}^M B_{kt}^L\right)$ where

$$E\left(\sum_{t=1}^{M} B_{kt}^L\right) = \sum_{p=1}^{p_L}\left[p\left(\sum_{t=1}^{M} B_{kt}^L(p)\right)\cdot \sum_{t=1}^{M} B_{kt}^L(p)\right] \; (L = 1,2,\ldots,\Omega), \text{ respectively.}$$

18.3.2.3 Enhanced stochastic model considering alternative project implementation scenarios

The above stochastic model can be further enhanced by considering alternative project implementation scenarios, which include (i) bundled implementation of candidate projects by highway or transit segment; (ii) bundled implementation of candidate projects by freeway corridor, major urban arterial corridor, or transit corridor; and (iii) deferred implementation of some large-scale projects.

Segment-based project implementation approach: Multiple projects within one or more highway segments or in transit segments between two or more stops/stations might be tied together for actual implementation to reduce traffic disruption at the construction stage. The first step for applying this approach is to identify the list of highway or transit segments in the highway or transit network to be considered for segment-based project implementation. Next, all projects within one or more highway or transit segments are tied together to form one "project group" and they are either all rejected or selected for implementation. For example, if three projects $(i+1)$, $(i+2)$, and $(i+3)$, belong to one "project group" S_g, the respective zero/one decision variables $x_{(i+1)}$, $x_{(i+2)}$, and $x_{(i+3)}$ are replaced by a zero/one decision variable x_{Sg}. For those isolated projects that do not belong to any of the identified "project groups," they are still treated as stand-alone projects that are designated with unique zero/one decision variables.

Suppose that g "project groups" are identified from N candidate projects as

$1, 2, \ldots, i$ (i isolated projects)
$i+1, i+2, \ldots, i+n_1$ (n_1 projects in "project group" S_1)

$i + n_1 + 1, i + n_1 + 2, \ldots, i + n_1 + n_2$ (n_2 projects in "project group" S_2)
$i + n_1 + n_2 + 1, i + n_1 + n_2 + 2, \ldots, i + n_1 + n_2 + n_3$ (n_3 projects in "project group" S_3)
...
$i + n_1 + n_2 + \cdots + n_{g-2} + 1, i + n_1 + n_2 + \cdots + n_{g-2} + 2, \ldots, i + n_1 + n_2 + \cdots + n_{g-2} + n_{g-1}$ (n_{g-1} projects in "project group" S_{g-1}), $i + n_1 + n_2 + \cdots + n_{g-2} + n_{g-1} + 1, i + n_1 + n_2 + \cdots + n_{g-2} + n_{g-1} + 2, \ldots, N$ ($N-i-n_{g-1}$ projects in "project group" S_g)

The decision vector in stage L decisions $X_L(p) = [x_1, x_2, \ldots, x_i, \ldots, x_N]^T$ in the stochastic model is thus replaced by $X_L(p) = [x_1, x_2, \ldots, x_i, x_{S1}, x_{S2}, x_{S3}, \ldots, x_{S(g-1)}, x_{Sg}]^T$ ($L = 1, 2, \ldots, \Omega$). This implies that the basic stochastic model could still be used, but the size of the decision vector $X_L(p)$ is reduced from having N decision variables to $(i + g)$ decision variables. Each decision variable still takes a zero/one integer value representing the rejection or selection of an isolated project or multiple projects in a segment-based "project group." The benefits of all constituent projects of a segment-based "project group" are added together as the overall benefits of the "project group."

Corridor-based project implementation approach: As an extension of segment-based project implementation approach, the tie-ins of multiple projects within one or more highway or transit segments could be further expanded to a freeway corridor, a major urban arterial corridor or a transit corridor. First, the list of corridors in the network to be considered for corridor-based project implementation is identified. Then, all candidate projects in the same corridor that are grouped by segment are further bundled into one corridor-based "grand project group." All constituent projects in the same "grand project group" are either all rejected or selected for implementation. For those isolated projects that do not belong to any of the identified segment-based "project groups" or corridor-based "grand project groups," they are still handled as stand-alone projects associated with unique decision variables.

Suppose that N candidate projects are classified as $1, 2, \ldots i$ isolated projects and $S_1, S_2, S_3, S_4, \ldots, S_{g-2}, S_{g-1}, S_g$ segment-based "project groups." The decision vector in stage L decisions is $X_L(p) = [x_1, x_2, \ldots, x_i, x_{S1}, x_{S2}, x_{S3}, x_{S4}, \ldots, x_{S(g-2)}, x_{S(g-1)}, x_{Sg}]^T$ ($L = 1, 2, \ldots, \Omega$). Further assume that all projects in "project groups" S_2 and S_3 are in one freeway corridor and all projects in "project groups" S_{g-1} and S_g are in one urban arterial corridor. This creates two corridor-based "grand project groups" for possible implementation: "grand project group" C_1 that combines "project groups" S_2 and S_3; and "grand project group" C_2 that joins "project groups" S_{2g-1} and S_g. Hence, the decision vector in stage L decisions $X_L(p) = [x_1, x_2, \ldots, x_i, x_{S1}, x_{S2}, x_{S3}, x_{S4}, \ldots, x_{S(g-2)}, x_{S(g-1)}, x_{Sg}]^T$ in the stochastic model that uses segment-based project implementation approach for project selection is further reduced to $X_L(p) = [x_1, x_2, \ldots, x_i, x_{S1}, x_{C1}, x_{S4}, \ldots, x_{S(g-2)}, x_{C2}]^T$ ($L = 1, 2, \ldots, \Omega$).

This implies that the enhanced stochastic model incorporating the segment-based project implementation approach could still be used for the stochastic model utilizing the corridor-based project implementation approach. However, the size of the decision vector $X_L(p)$ is reduced from having $(i + g)$ decision variables to $(i + g-2)$ decision variables. Each decision variable still takes a zero/one integer value representing the rejection or selection of an isolated project, multiple projects in a segment-based "project group," or multiple projects in a corridor-based "grand project group." The benefits of all constituent projects of a corridor-based "grand project group" are combined to obtain its overall benefits.

Deferment-based project implementation approach: Practically, some large-scale projects may have a high chance of being deferred for a few years due to various reasons such as land acquisition, resettlement, and funding availability. In the application of the deferment-based project implementation approach, the basic stochastic model remains unchanged and the decision vector in stage L decisions $X_L(p) = [x_1, x_2, \dots, x_N]^T$ in the stochastic model is kept the same. For projects involving deferred implementation, the project benefits and costs are adjusted according to number of years deferred. The deferred projects would compete for funding with other unaffected projects in the newly designated implementation years using the adjusted project benefits and costs.

18.3.2.4 Solution algorithm

This section first presents a theorem of Lagrange multipliers and briefly discusses the essential part of the proposed heuristic algorithm extended from the heuristic of Volgenant and Zoon (1990), which uses two Lagrange multipliers, on how (suboptimal) values for multiple Lagrange multipliers can be determined. It then discusses the improvement of the upper bound for the optimum of the proposed model (Li, 2009; Li et al., 2010b).

Theorem of the Lagrange multipliers: The stage L optimization can be reformulated in the following:

$$\text{Max} \quad z(X_L) = A^T \cdot X_L \tag{18.14}$$

$$\text{Subject to} \quad {C_{kt}}^T \cdot X_L \leqq {B_{kt}}^L \tag{18.15}$$

where X_L is stage L decision vector with zero/one integer elements for rejecting or selecting individual projects.

Given nonnegative, real Lagrange multipliers λ_{kt}, the Lagrange relaxation of Equation 18.14, $z_{LR}(\lambda_{kt})$, can be written as

$$\begin{aligned} \text{Objective} \quad z_{LR}(\lambda_{kt}) &= \text{Max}\, A^T \cdot X_L + \sum_{k=1}^{K}\sum_{t=1}^{M}\left[\lambda_{kt} \cdot \left(B_{kt}^L - C_{kt}^T \cdot X_L\right)\right] \\ &= \text{Max}\left(A^T - \sum_{k=1}^{K}\sum_{t=1}^{M}\left(\lambda_{kt} \cdot C_{kt}^T\right)\right)X_L + \sum_{k=1}^{K}\sum_{t=1}^{M}\left(\lambda_{kt} \cdot B_{kt}^L\right) \end{aligned} \tag{18.16}$$

subject to X_L with zero/one integer elements.

Because $\sum_{k=1}^{K}\sum_{t=1}^{M}\left(\lambda_{kt} \cdot B_{kt}^L\right)$ in Equation 18.16 is a constant, optimization can just be concentrated on the first term, namely, maximizing

$$\left(A^T - \sum_{k=1}^{K}\sum_{t=1}^{M}\left(\lambda_{kt} \cdot C_{kt}^T\right)\right)X_L \tag{18.17}$$

The solution to Equation 18.17 is X_L^*, where

$$X_L^* = \begin{cases} 1, & \text{if}\left(A^T - \sum_{k=1}^{K}\sum_{t=1}^{M}\left(\lambda_{kt} \cdot C_{kt}^L\right)\right) > 0 \\ 0, & \text{otherwise} \end{cases} \tag{18.18}$$

Then X_L^* maximizes $z(Y_L) = A^T \cdot X_L$, subject to X_L with zero/one integer elements.

In order to obtain an optimal solution by maximizing $z(X_L) = A^T \cdot X_L$, only subject to X_L with zero/one integer elements, the following condition needs to be satisfied:

$$\sum_{k=1}^{K}\sum_{t=1}^{M}\left[\lambda_{kt}\cdot\left(B_{kt}^{L} - C_{kt}^{T}\cdot X_L\right)\right] = 0 \tag{18.19}$$

In this regard, stage L optimization operations need to focus on determining Lagrange multipliers λ_{kt} such that (i) X_L^* obtained in Equation 18.18 is feasible to the original model, that is, $C_{kt}^T \cdot X_L \leqq B_{kt}^L$ is valid; and (ii) condition (18.19) is satisfied to maintain optimality to the original model as Equations 18.14 and 18.15.

The heuristic algorithm: At the recourse decision stage L, the heuristic initializes the Lagrange multiplier values to zero and all variables to the value one so that Equation 18.18 is satisfied. In general, this solution is not feasible, because constraints of the proposed model as Equation 18.15 are violated. In each of the iterations, the constraint that has the largest ratio of the remaining total benefits and costs is first determined. Then, the corresponding multiplier value is increased as much as necessary to violate Equation 18.18 for just one variable, and the variable will be reset to zero. This step is repeated until the solution becomes feasible. An improvement step "polishes" the solution obtained.

Denote X_L^* is the optimal decision vector at stage L, $s\left(X'_{(L-1)}\right)$ is the set of projects selected at stage $L-1$, $s\left(X'_L\right)$ is the set of projects selected at stage L, and $S\left(X'_L\right)$ is the set of projects selected at stage $L-1$ so that each of these projects at least uses the budget from year 1 to $t_{(L-1)}$, where the budget at stage L remains the same as that at stage $L-1$ for the period from year 1 to $t_{(L-1)}$, which means that project $i \in s\left(X'_{(L-1)}\right)$ and $c_{ikt} > 0$ for any k and at least one t $(t = 1, 2 \ldots\ t_{(L-1)})$ and $S\left(X'_L\right) \subseteq s\left(X'_{(L-1)}\right)$, and $\bar{S}\left(X''_L\right)$ is the set of projects not selected at stage $L-1$ or selected projects that do not use budget between year 1 and year $t_{(L-1)}$ (complement of $S\left(X'_L\right)$). In full, the heuristic has the following steps:

Step 0: Initialization and normalization

- For stage 1, set $X_0^* = \{0, 0, \ldots\ 0\}$ (no project selected at stage 0). Hence, $s(X'_0) = S(X'_1) = \varphi$
- For stage L, use budget $B_{kt}^L(p)$ for computation such that $\Delta B^L(p) = \min\left\{\sum_{k=1}^{K}\sum_{t=1}^{M}[B_{kt}^L(p) - E(B_{kt}^L)]^2\right\}$ and perform the following calculations for project $i \in S(X'_L)$: (i) sort the projects by benefits (A_i) in descending order, set $\lambda_{kt} = 0$ for all k, t and $x_i = 1$; (ii) normalize cost and budget matrices by setting $c'_{ikt} = c_{ikt} / \left(B_{kt}^L(p)\right)$ for all k, t and $B_{kt}^L(p) = 1$ for all k, t; and (iii) compare sum of normalized costs with normalized budgets $C_{kt} = \sum_{i=1}^{N} c'_{ikt}$. If $C_{kt} \leq 1$ for all k, t, go to Step 4. Otherwise, go to Step 1.

Step 1: Determine the most violated constraint k, t

Set $C'_{kt} = \text{maximum}\ \{C_{kt}\}$ for all k, t

Step 2: Compute the increase of Lagrange multiplier value λ_{kt}

$$\theta_i = \begin{cases} \dfrac{A_i - \sum_{k=1}^{K}\sum_{t=1}^{M}(\lambda_{kt}\cdot c'_{ikt})}{\sum_{k=1}^{K}\sum_{t=1}^{M}\left(c'_{ikt}\cdot \dfrac{C_{kt}}{C'_{kt}}\right)}, c_{ikt} > 0 & \text{for all project } i \in S(X'_L) \\ \infty, \ \text{otherwise} & \end{cases}$$

Select project $i \in S(X'_L)$ that has the lowest θ_i and let $\theta'_i = \min\{\theta_1, \theta_2, \ldots, \theta_i, \ldots\}$

Step 3: Enlarge λ_{kt} by $\theta'_i \cdot C_{kt}/C'_{kt}$ and reset x_i to the value zero

$$\text{Let } \lambda_{kt} = \lambda_{kt} + \left(\theta'_i \cdot \frac{C_{kt}}{C'_{kt}}\right) \quad \text{and} \quad C_{kt} = C_{kt} - c'_{ikt} \quad \text{for all } k, t$$

Reset $x_i = 0$ for project $i \in S(X'_L)$ and shift project i from $S(X'_L)$ to $S(X''_L)$. If $C_{kt} \le 1$ for all k, t, go to Step 4. Otherwise, go to Step 1.

Step 4: Improve current solution

For the solution generated in Step 3, check whether the projects with zero-variable values can have the value 1 without violating the constraints $C_{kt} \le 1$. If so, choose the project with the highest benefits and add it to the selected project list. Repeat this step until no project can be added. Update the set of projects selected at stage L, $s(X'_L) = \{i|$ for all $x_i = 1\}$, and this establishes an improved solution.

Step 5: Further improved solution with budget carryover

In each year of the multiyear project selection period, a small fraction of budget might be left after project selection. Such an amount could be carried over to the immediately following year, one year at a time, to repeat Steps 1–4 to further improve the solution. Update the set of projects selected at stage L, $s(X'_L) = \{i|$ for all $x_i = 1\}$, and this finds an improved solution with budget carryover.

If $L = \Omega$, stop. X^*_L is final. Otherwise, repeat Steps 1–5.

The upper bound improvement: Let X^s_L be the solution obtained in Step 3 in the above, this solution could be substituted into Equation 18.16. This gives an upper bound for the objective function z^U as

$$Z^U = A^T \cdot X^s_L + \sum_{k=1}^{K}\sum_{t=1}^{M}\left[\lambda_{kt} \cdot \left(B_{kt} - C^T_{kt} \cdot X^s_L\right)\right] \tag{18.20}$$

The upper bound depends on the non-violated budget constraints with positive Lagrange multipliers. At the beginning of an iteration, suppose that more than one non-violated constraints have positive Lagrange multipliers. Denote I^s the index of the constraint with the largest value of $\lambda_{kt}\left(B_{kt} - C^T_{kt}X^s_L\right)$. The question is then whether the value of Lagrange multiplier $\lambda_{kt(I^S)}$ can be chosen smaller so that the influence of constraint I^s in the computation of the upper bound for the objective function is reduced. Obviously, there is no influence if the multiplier value is set to zero. However, if a smaller value of $\lambda_{kt(I^S)}$ is used, some other Lagrange multiplier value must be increased in order to satisfy the condition in Equation 18.18. In the proposed algorithm, we have heuristically chosen the multiplier $\lambda_{kt(i')}$ that is associated with the selected project maintaining the least extent of loss in the "benefit-to-cost" ratio if removed, where the index *i'* is determined by $\theta'_i = \min\{\theta_1,\theta_2,\ldots,\theta_i,\ldots\}$ in Step 2.

In the execution of the proposed algorithm, only the decision variable x_i's with the value one are set to zero, that is, a project selected previously is removed in the current iteration. For the two non-violated constraints with positive multipliers $\lambda_{kt(I^S)}$ and $\lambda_{kt(i')}$, the trade-offs of decreasing $\lambda_{kt(I^S)}$ and increasing $\lambda_{kt(i')}$ satisfy the following conditions:

$$b_i - \sum_{k=1}^{K}\sum_{t=1}^{M}(\lambda_{kt} \cdot c_{ikt}) + \alpha_1 \cdot c_{ikt(I^s)} \cdot \left(\frac{C_{kt}}{C'_{kt}}\right) - \alpha_2 \cdot c_{ikt(i')} \cdot \left(\frac{C_{kt}}{C'_{kt}}\right) \le 0, \quad \text{for all } x_i = 0 \tag{18.21}$$

$$b_i - \sum_{k=1}^{K}\sum_{t=1}^{M}(\lambda_{kt} \cdot c_{ikt}) + \alpha_1 \cdot c_{ikt(I^s)} \cdot \left(\frac{C_{kt}}{C'_{kt}}\right) - \alpha_2 \cdot c_{ikt(i')} \cdot \left(\frac{C_{kt}}{C'_{kt}}\right) \ge 0, \quad \text{for all } x_i = 1 \tag{18.22}$$

where α_1 and α_2 are respective changes in the values of $\lambda_{kt(I^S)}$ and $\lambda_{kt(i')}$.

For a specific project i with $x_i = 1$, the decision variable x_i will be changed from one to zero only when Equation 18.22 holds with equality. For the purpose of determining (α_1, α_2) pair, two conditions must be satisfied: (i) α_1 is maximal; and (ii) Equation 18.22 holds with equality. Having obtained the values of α_1 and α_2, a project i with $x_i = 1$ that satisfies the equality condition is removed by setting its decision variable x_i to zero. The values of α_1 and α_2 can be established by the following procedure.

The inequalities in Equations 18.21 and 18.22 define the lower and upper boundaries of the feasible region for (α_1, α_2) pair. The lower bound function $f_L(\alpha_1)$ and upper bound function $f_U(\alpha_1)$ for α_1, can be defined as

$$f_L(\alpha_1) = \max\left\{\left[b_i - \sum_{k=1}^{K}\sum_{t=1}^{M}(\lambda_{kt}c_{ikt}) + \alpha_1 c_{ikt(I^s)}\left(\frac{C_{kt}}{C'_{kt}}\right)\right] / \sum_{k=1}^{K}\sum_{t=1}^{M}\left[c_{ikt}\left(\frac{C_{kt}}{C'_{kt}}\right)\right]\right\}, \quad \text{for all } x_i = 0 \tag{18.23}$$

$$f_U(\alpha_1) = \min\left\{\left[b_i - \sum_{k=1}^{K}\sum_{t=1}^{M}(\lambda_{kt}c_{ikt}) + \alpha_1 c_{ikt(I^s)}\left(\frac{C_{kt}}{C'_{kt}}\right)\right] / \sum_{k=1}^{K}\sum_{t=1}^{M}\left[c_{ikt}\left(\frac{C_{kt}}{C'_{kt}}\right)\right]\right\}, \quad \text{for all } x_i = 1 \tag{18.24}$$

This is identical to determine the α_1 value that will have the function $g(\alpha_1) = f_U(\alpha_1) - f_L(\alpha_1)$ reach zero value. The function is continuous and piecewise linear with a computational complexity of $O(N)$, where N is the total number of projects. A numerical method for computing zero of the function $g(\alpha_1)$ can be found in Bus and Dekker (1975).

Case Study 1: Multiyear statewide highway programming under budget uncertainty

Project data: Eleven-year data on 7380 projects proposed for a statewide highway investment decisions during 1996–2006 has been collected to implement the knapsack model for project selection. 23% of the projects are for bridge preservation, 19% for pavement preservation, 46% for safety and roadside improvements, 6% for major/new construction, 1% for ITS installations, and 4% for system maintenance. As seen in Table 18.1, benefits of the projects are estimated as the total of reductions in agency costs of construction, repair, and maintenance, as well as decreases in the costs of vehicle operation, travel time, crashes, and air emissions. On average, the present worth amount of project-level life-cycle benefits for the 7380 projects is 6.64 million dollars per project and the average benefit-to-cost ratio is 5.16.

Budget data: The annual average budgets are approximately 700 million dollars with 4% increment per year. Of the total budgets, 47% are designated for new construction, 28% for pavement preservation, 9% for bridge preservation, 9% for maintenance, and the remaining 7% for safety improvements, roadside improvements, ITS installations, and miscellaneous programs. The initially estimated budgets for the project implementation period are found to have been updated three times by the state transportation agency. This provided 4-stage budget recourses in the application of the proposed heuristic approach for project section. The budget adjustments are mainly made on pavement preservation, bridge preservation, new construction, and maintenance programs, with changes varying from −32% to +60%.

Project selection results: Table 18.2 presents project selection results. The total benefits for all system goals combined are between 36.7 and 38.2 billion dollars. Of the total benefits, approximately 26% are for agency cost, 13% for vehicle operating cost, 9% for mobility, 41% for safety, and 11% for the environment. Regardless of project implementation options, the total benefits for the use of stochastic budgets are higher than those of deterministic budgets. For stochastic budgets, there are virtually increases in benefits for all system goals, totaling 0.34 billion dollars. The majority of the benefit increases are for safety and agency cost goals.

Table 18.1 Facility life-cycle benefits of some pavement and bridge projects

Project no.	Lanes	Length (miles)	AADT	Work type	Project cost	Project benefit items (%)					Total benefits
						Agency costs	Vehicle operation	Mobility	Safety	Environment	
1	4	0.11	69200	Bridge widening	2291000	4	22	1	55	19	11703264
2	4	0.50	32630	Pavement resurfacing	4620000	2	33	1	37	27	6365844
3	2	2.06	3170	Pavement resurfacing	3000000	3	27	1	46	23	15545501
4	2	3.70	16770	Added travel lanes	750000	2	30	7	34	26	4806134
5	2	13.63	4190	Pavement resurfacing	11573000				100		63943225
6	4	2.53	11150	Pavement rehabilitation	151000	10	32	1	31	27	1505738
7	4	0.78	2664	Pavement replacement	196000	52	20	18	5	5	736046
8	2	9.46	1100	Pavement rehabilitation	131000				100		353545
9	2	0.15	8291	Bridge widening	108000	13	28	4	30	26	254516
10	2	1.10	13994	Pavement resurfacing	2757000		26	24	28	22	5702627
...	...	...	...	...	...						...

Table 18.2 Total project benefits considering budget uncertainty and project implementation options

Total benefits (in billion dollars)		*Project benefits by highway system goal*					
Budget	*Project implementation option*	*Agency costs*	*VOC*	*Mobility*	*Safety*	*Environment*	*Total*
Deterministic	Segment-based	9.78	4.78	3.46	15.46	4.23	37.7
	Corridor-based	9.34	4.3	3.35	15.14	4.19	36.8
	Deferment-based	9.19	4.99	3.27	15.51	4.38	37.3
Stochastic	Segment-based	9.87	4.86	3.52	15.66	4.30	38.2
	Corridor-based	9.44	4.74	3.34	15.23	4.20	37.0
	Deferment-based	9.27	5.03	3.29	15.59	4.42	37.6

Irrespective of budget uncertainty considerations, the total benefits of selected projects are lowest for using the corridor-based project implementation approach, and highest for using the segment-based project implementation approach. As compared to the total benefits of projects selected using the segment-based approach, total benefits of projects selected using the corridor-based approach reduced for all system goals, with a total reduction of 1.53 billion dollars. Also, total benefits of projects selected using the deferment-based approach increased for the goals of vehicle operating cost and the environment (0.84 billion dollars) and decreased for the remaining goals (−1.08 billion dollars), with a total reduction of 0.24 billion dollars. When comparing the total benefits of projects selected using the corridor-based approach, total benefits of projects selected using the deferment-based approach decreased for the goals of agency cost and the environment (−0.49 billion dollars) and increased for the remaining goals (1.77 billion dollars), with a net increase of 1.28 billion dollars.

Source: Adapted from Li, Z. 2009. *ASCE Journal of Transportation Engineering* 135(6), 371–379.

18.3.3 Two-step knapsack model addressing risk tolerance

Factors involved with estimation of project benefits, including project costs, traffic forecasts, and discount rate may vary in the service life cycle of any transportation facility, and this will result in changes in the project benefit estimates. For cost-efficiency considerations, a lower and upper bound can be set for the project benefits as being desirable for implementation. In the case that two projects are jointly implemented, the total benefits involve an even wider range of lower and upper bounds, meaning the total benefits are under risk. In this case, the covariance can be used to measure the extent of associated risk regarding total benefits, when jointly implementing two projects where each project's benefit follows a probability distribution. Furthermore, the sum of all covariance values associated with all possible combinations of jointly implementing any two projects captures the holistic risk of the overall benefits brought by the candidate projects.

The two-step knapsack model uses the Markowitz mean-variance model to determine the lower bound for the total risk of the expected project benefits under the budget constraint (Markowitz, 1991; Zhou et al., 2013). The second step uses an enhanced knapsack model to maximize the overall benefits under the budget constraint, the lowest bound of total risk associated with project benefits as a chance constraint, and integrality constraints. The addition of the constraint pertaining to the total risk to the basic knapsack model is the contribution of the study, which ensures that the sub-collection of economically feasible projects selected not only has maximum benefits but also has minimum possible risk associated with the total benefits.

18.3.3.1 Markowitz mean-variance model for step-one optimization

The Markowitz mean-variance analysis is a standard risk management tool used in random variables' comparison. A systematic trade-off between the expected return and the specified risk measure is evaluated. In the context of transportation, because of the randomness inherited with some factors such as traffic volumes, project costs, and discount rate, the project benefits are consequently random and are dependent upon the probability distribution of these factors. In many cases, the optimal project benefits cannot be obtained as are desired in theory. Particular to transportation project selection, the model aims to select a portfolio of projects that minimizes the total risk of the expected benefits of all projects subject to constraints of an available budget expressed in 100%, as indicated in Equation 18.26 and the minimum benefits to be expected from individual projects. The total risk of the overall project benefits is readily characterized by the summation of covariance values of the overall project benefits. The model is formulated as

$$\text{Min } R = \sum_{i=1}^{N}\sum_{j=1}^{N}\left[\left(\frac{\sum_{k=1}^{K}\sum_{t=1}^{T} C_{ikt}}{\sum_{k=1}^{K}\sum_{t=1}^{T} B_{kt}}\right)\cdot\left(\frac{\sum_{k=1}^{K}\sum_{t=1}^{T} C_{jkt}}{\sum_{k=1}^{K}\sum_{t=1}^{T} B_{kt}}\right)\cdot cov(a_i,a_j)\cdot x_i\cdot x_j\right] \tag{18.25}$$

$$\text{Subject to } \sum_{i=1}^{n}\left[\left(\frac{C_{ikt}}{B_{kt}}\right)\cdot x_i\right]\leq 100\% \tag{18.26}$$

$$E(\partial_i)\geq b_i \tag{18.27}$$

$$x_i,\ x_j = 0/1$$

where

R = Total risk of overall project benefits
x_i = 0/1 decision variables for rejection or selection of project i
C_{ikt} = Costs of project i using budget from highway management program k in year t
B_{kt} = Total available budget for management program k in year t
$E(\partial_i)$ = Expected benefits of project i
b_i = Preset threshold value of the benefits of project i
i = 1, 2, … , N
j = 1, 2, … , N
k = 1, 2, … , K
t = 1, 2, … , T

In Equation 18.25, the objective function is to seek the lowest level of total risk of jointly implementing any two projects. Equation 18.26 lists budget constraints by management program and by year. Equation 18.27 requires that the expected benefits of a project cannot be less than a preset threshold value before being considered for possible implementation. The zero/one integrality constraints for the decision variables are used to capture whether an individual project is selected. The proportion of the total budget by management program and by year B_{kt} to be used by project i is calculated as shown in Table 18.3.

In order to establish the covariance matrix representing the risk of the expected benefits of jointly implementing any two projects, the probability distribution of each individual project's benefit needs to be determined. With no loss of generality, we can pre-assume three levels of benefits for each project—low (a_L), medium (a_M), and high

Table 18.3 Proportion of annual total program-specific budget to be used by a project

Project	*Benefits*	*Costs*	*Proportion of budget use by project*
1	a_1	C_{1kt}	C_{1kt}/B_{kt}
2	a_2	C_{2kt}	C_{2kt}/B_{kt}
3	a_3	C_{3kt}	C_{3kt}/B_{kt}
...	...	...	...
N	a_N	C_{Nkt}	C_{Nkt}/B_{kt}
		$\sum_{i=1}^{N} C_{ikt} \geq B_{kt}$	$\frac{\sum_{i=1}^{N} C_{ikt}}{B_{kt}} \geq 1$

(a_H) with corresponding probabilities $P(a_L)$, $P(a_M)$, and $P(a_H)$. The three levels of low (l), medium (m), and high (h) are determined based on the probability of the expected individual benefit and a threshold value set according to the current policy and the decision makers' preference. The covariance value of each pair of projects to be jointly implemented is computed as

$$cov(\partial_i, \partial_j) = E(\partial_i, \partial_j) - E(\partial_i) \cdot (\partial_j)$$
$$= \sum_{S=1}^{3} \sum_{T=1}^{3} \left[\partial_{i,S} \cdot \partial_{j,T} \cdot P(\partial_{i,S}, \partial_{j,T}) \right] - \left[\sum_{S=1}^{3} \partial_{i,S} \cdot P(\partial_{i,S}) \right] \left[\sum_{T=1}^{3} \partial_{j,T} \cdot P(\partial_{j,T}) \right] \quad (18.28)$$

18.3.3.2 Enhanced knapsack model for step-two optimization

The result of the Markowitz mean-variance model can be further converted into an additional chance constraint to enhance the basic knapsack model. The enhanced model is formulated as follows:

$$\text{Max} \sum_{i=1}^{n} \partial_i x_i \quad (18.29)$$

$$\text{Subject to} \sum_{i=1}^{n} C_{ikt} \cdot x_i \leq B_{kt} \quad (18.30)$$

$$\left\{ \sum_{i=1}^{N} \sum_{j=1}^{N} \left[\left(\frac{\sum_{k=1}^{K} \sum_{t=1}^{T} C_{ikt}}{\sum_{k=1}^{K} \sum_{t=1}^{T} B_{kt}} \right) \cdot \left(\frac{\sum_{k=1}^{K} \sum_{t=1}^{T} C_{jkt}}{\sum_{k=1}^{K} \sum_{t=1}^{T} B_{kt}} \right) \cdot cov(\partial_i, \partial_j) \cdot x_i \cdot x_j \right] \leq R_{min} \right\}_{\text{chance } \alpha} \quad (18.31)$$

$$x_i = 0/1$$

where

x_i = 0/1 decision variables for rejection or selection of project i
a_i = Benefits of project i

C_{ikt} = Costs of project i using budget from highway management program k in year t
B_{kt} = Total available budget for highway management program k in year t
R_{min} = Lowest total risk of expected project benefits established in step-one optimization using the Markowitz mean-variance model
chance α = A chance constraint ensuring that the condition is satisfied for α percent of the time
i = $1, 2, \ldots, N$
j = $1, 2, \ldots, N$
k = $1, 2, \ldots, K$
t = $1, 2, \ldots, T$

Equations 18.29 and 18.30 of the enhanced knapsack model are the same as Equations 18.1 and 18.2 of the basic model. Equation 18.31 is the new added chance constraint that have the total risk of the expected project benefits controlled within an acceptable range at α percent probability.

18.3.3.3 Solution algorithms

Two solution algorithms are involved in solving the proposed two-step knapsack model. The first algorithm applied is a modified simplex algorithm. This algorithm is used to solve the Markowitz mean-variance model in the first step. The second is a heuristic algorithm used for solving the enhanced knapsack problem in the send step. These two algorithms are briefly introduced in the following sections.

Modified simplex algorithm: The Markowitz mean-variance model has a quadratic objective function. Conventionally, it is solved by specialized algorithms such as Wolfe's adaptation of the simplex method (Wolfe, 1959). The Markowitz mean-variance model is rewritten as

$$\text{Min } \frac{1}{2}\sum_{i=1}^{N}\sum_{j=1}^{N}\left[\left(\frac{\sum_{k=1}^{K}\sum_{t=1}^{T} C_{ikt}}{\sum_{k=1}^{K}\sum_{t=1}^{T} B_{kt}}\right)\cdot\left(\frac{\sum_{k=1}^{K}\sum_{t=1}^{T} C_{jkt}}{\sum_{k=1}^{K}\sum_{t=1}^{T} B_{kt}}\right)\cdot cov(\partial_i,\partial_j)\cdot x_i\cdot x_j\right]-\sum_{i=1}^{N}\partial_i\cdot x_i \tag{18.32}$$

$$\text{Subject to } \sum_{i=1}^{N}\left[\left(\frac{C_{ikt}}{B_{kt}}\right)\cdot x_i\right]\leq 100\% \tag{18.33}$$

$$E(a_i)\cdot x_i \geq b_i \tag{18.34}$$

$$x_i \geq 0 \tag{18.35}$$

where

x_i = 0/1 decision variables for rejection or selection of project i
C_{ikt} = Costs of project i using budget from highway management program k in year t
B_{kt} = Total available budget for highway management program k in year t
$cov(\partial_i, \partial_j)$ = Covariance value of overall benefits expected to be generated from two projects i and j

$E(a_i)$	= Expected benefits of project i; b_i = Preset threshold value of the benefits of project i
i	$= 1, 2, \dots, N$
j	$= 1, 2, \dots, N$
k	$= 1, 2, \dots, K$
t	$= 1, 2, \dots, T$

Because all variables $x_1, x_2, \dots, x_N$ are nonnegative, the Wolfe's modified simplex method is used to solve a LP problem derived from the Markowitz mean-variance model. The solution to the LP problem is therefore the optimal solution of the Markowitz mean-variance problem. The LP formulation is as follows:

$$\text{Min } Z = \sum_{i=1}^{N}\sum_{k=1}^{K}\sum_{t=1}^{T} z_{ikt} \tag{18.36}$$

$$\text{Subject to } \sum_{i=1}^{N}\left[\left[\left(\frac{\sum_{k=1}^{K}\sum_{t=1}^{T} C_{ikt}}{\sum_{k=1}^{K}\sum_{t=1}^{T} B_{kt}}\right)\left(\frac{\sum_{k=1}^{K}\sum_{t=1}^{T} C_{jkt}}{\sum_{k=1}^{K}\sum_{t=1}^{T} B_{kt}}\right) cov(\partial_i,\partial_j)x_i\right]\right] + C_{ikt}^{T}u_{kt} - y_{ikt} + z_{ikt} = \partial_i^T \tag{18.37}$$

$$\sum_{i=1}^{N} C_{ikt}\cdot x_i + v_{kt} = B_{kt} \tag{18.38}$$

$$E(a_i)\cdot x_i - e_i = b_i \tag{18.39}$$

$$\sum_{i=1}^{N} x_i^T \cdot y_{ikt} + u_{kt}\cdot v_{kt} = 0 \tag{18.40}$$

$$x_i \ge 0, y_{ikt} \ge 0, u_{kt} \ge 0, v_{kt} \ge 0, e_i \ge 0 \tag{18.41}$$

where

z_{ikt}	= Nonnegative artificial variables
y_{ikt}	= Dual variables; u_{kt}, v_{kt}
e_i	= Nonnegative slack or excessive variables
a_i	= Benefits of project i
$cov(\partial_i, \partial_j)$	= Covariance matrix of paired project benefit estimates
b_i	= Preset threshold value of the benefits of project i
i	$= 1, 2, \dots, N$
j	$= 1, 2, \dots, N$
k	$= 1, 2, \dots, K$
l	$= 1, 2, \dots, L$
t	$= 1, 2, \dots, T$

In Equation 18.36, the objective function minimizes the sum of all artificial variables, preferably to be zero. Equation 18.37 defines Karush–Kuhn–Tucker (KKT) conditions for the constrained optimization. Equations 18.38 and 18.39 are canonical forms of Equation 18.30 or 18.33 and Equation 34, respectively. Equation 18.40 is the complementary slackness constraint. Equation 18.41 is the nonnegativity constraint for all variables. The main steps of the solution algorithm to the LP formulation are as follows:

Step 1: Modify the constraints

Modify the constraints so that the right-handside of each constraint is nonnegative. This requires that each constraint with a negative right-hand side be multiplied by −1.

Step 2: Convert inequality constraints into the standard form

Identify each constraint that is now an "=" or "≥" constraint and convert each inequality constraint to the standard form. If constraint i is a "≤" constraint, add a slack variable u_{kt} and v_{kt}. If constraint i is a "≥" constraint, add an excessive variable e_i.

Step 3: Add artificial variables

For each "=" or "≥" constraint identified in Step 2, add an artificial variable z_{ikt}.

Step 4: Solve for the LP formulation

Solve for the LP formulation by satisfying complementary slackness requirements

$$\sum_{i=1}^{N}\left(x_i^T \cdot y_{ikt}\right) = 0 \quad \text{and} \quad u_{kt} \cdot v_{kt} = 0$$

If the optimal value $Z > 0$, the LP has no feasible solution. The solution $x = \{x_1, x_2, \ldots, x_N\}$ to which $Z = 0$ is the optimal solution to the original Markowitz mean-variance model. More details of the solution algorithm can be found in Hillier and Lieberman (2006) and Bazaraa et al. (2009).

Heuristic algorithm for solving the step-two enhanced knapsack model: This heuristic algorithm stems from the heuristic algorithm developed by Volgenant and Zoon (1990) and Markowitz (1991), using the Lagrangian relaxation technique. It establishes the near optimal solution by simultaneously computing multiple Lagrangian multipliers corresponding to the budget constraints for each management program category and by number of years of analysis, as well as total risk in project selection. This significantly improves the computational efficiency as compared to the algorithm developed by Volgenant and Zoon that uses two Lagrangian multipliers.

Denote X^* as the optimal decision vector, $s(X')$ as the set of projects selected, and $s(X'')$ as the set of projects not selected. The heuristic has the following steps:

Step 0: Initialization and normalization

Set $X^* = \{0, 0, \ldots, 0\}$ (no project selected in the beginning). Hence, $s(X') = \varphi$.

Use budget B_{kt} to perform the following calculations: (i) sort the projects by benefits a_i in descending order, set $\lambda_{kt} = 0$ and $\mu_{kt} = 0$ for all k, t, and $x_i = 1$; (ii) normalize cost and budget matrices by setting $c'_{ikt} = c_{ikt}/B_{kt}$ for all k, t and $B_{kt} = 1$ for all k, t; (iii) compare sum of normalized costs with normalized budgets $C_{kt} = \sum_{i=1}^{N} c'_{ikt}$; and (iv) compare sum of total project risk with the lowest total risk of expected project benefits. If $C_{kt} \leq 1$ is satisfied for all k, t and the chance constraint

$$\sum_{i=1}^{n}\sum_{j=1}^{N}\left[\left(\frac{\sum_{k=1}^{K}\sum_{t=1}^{T} C_{ikt}}{\sum_{k=1}^{K}\sum_{t=1}^{T} B_{kt}}\right)\cdot\left(\frac{\sum_{k=1}^{K}\sum_{t=1}^{T} C_{jkt}}{\sum_{k=1}^{K}\sum_{t=1}^{T} B_{kt}}\right)\cdot cov(\partial_i, \partial_j)\cdot x_i \cdot x_j\right] / R_{min} \leq 1$$

is satisfied for α percentage of the time, go to Step 4. Otherwise, go to Step 1.

Step 1: Determine the most violated constraint k, t

Set $C'_{kt} = maximum\ \{C_{kt}\}$ for all k, t

Step 2: Compute the increase of Lagrange multiplier value λ_{kt} and μ_{kt}

$$\theta = \begin{cases} \partial_i - \sum_{k=1}^{K}\sum_{t=1}^{M}(\lambda_{kt} \cdot c'_{ikt}) / \sum_{k=1}^{K}\sum_{t=1}^{M}\left(c'_{ikt} \cdot \frac{C_{kt}}{C'_{kt}}\right), c'_{ikt} > 0 & \text{for all project } i \in s(X') \\ \infty, \text{otherwise} \end{cases}$$

Select project $i \in s(X')$ that has the minimum θ_i and let $\theta'_i = \min\{\theta_1, \theta_2, \ldots, \theta_i, \ldots\}$

Step 3: Increase λ_{kt} and μ_{kt} by $\theta'_i \cdot (C_{kt}/C'_{kt})$ and reset x_i the value zero

Let $\lambda_{kt} = \lambda_{kt} + (\theta'_i \cdot (C_{kt}/C'_k))$, $\mu_{kt} = \mu_{kt} + (\theta'_i \cdot (C_{kt}/C'_{kt}))$ and $C_{kt} = C_{kt} - c'_{ikt}$ for all k, t.
Reset $x_i = 0$ for project $i \in s(X')$ and shift project i from $s(X')$ to $s(X'')$.

If $C_{kt} < 1$ is satisfied for all k, t, and the chance constraint

$$\sum_{i=1}^{n}\sum_{j=1}^{N}\left[\left(\frac{\sum_{k=1}^{K}\sum_{t=1}^{T} C_{ikt}}{\sum_{k=1}^{K}\sum_{t=1}^{T} B_{kt}}\right) \cdot \left(\frac{\sum_{k=1}^{K}\sum_{t=1}^{T} C_{jkt}}{\sum_{k=1}^{K}\sum_{t=1}^{T} B_{kt}}\right) \cdot cov(\partial_i, \partial_j) \cdot x_i \cdot x_j\right] / R_{min} \leq 1$$

is satisfied for α percentage of the time, go to Step 4. Otherwise, go to Step 1.

Step 4: Improve the solution

For the feasible solution obtained in Step 3, check whether the projects with decision variable values of zero can have the value 1 without violating the constraints $C_{kt} \leq 1$. When this is the case, choose the project with highest benefits and add it to the selected project list. Repeat this step until no project with zero-variable value can be found and stop. Update the set of projects selected, $s(X') = \{i| \text{ for all } x_i = 1\}$, and this establishes an improved solution.

Step 5: Further improved solution with budget carryover

In each year of the multiyear project implementation period, a small amount of budget might be left after project selection. Such an amount can be carried over to the immediate following year, one year at a time, to repeat Steps 1–4 to further improve the solution. Update the set of projects selected, $s(X'_L) = \{i|\text{for all } x_i = 1\}$, and this provides an improved solution.

Case Study 2: Multiyear statewide highway programming addressing risk tolerance

Project benefits: Six-year data on 672 candidate projects proposed for a statewide Interstate highway programming during 2001–2006 has been collected to apply the two-step knapsack model for project selection addressing risk tolerance. As shown in Table 18.4, benefits of individual candidate projects are estimated as the total of reductions in agency costs of construction, repair, and maintenance, as well as decreases in the costs of vehicle operation, travel time, crashes, and air emissions. On average, the annualized medium-levelproject benefits are 6.6 million dollars per project with an average benefit-to-cost ratio of 5.2. The benefits are in the range of 7240 to 284578975. The low-level benefits are less than the medium-level benefits by approximately 25% and the high-levelbenefits are greater than the medium-level benefits by about 37%. Out of 672 projects, 639 are with expected benefits exceeding the threshold values, which are set at 70% of the medium-level benefits.

Covariance matrix: To execute the Markowitz mean-variance model, the covariance of the expected benefits for each pair of project $cov(a_i, a_j)$ is computed to capture the risk of achieving the expected

Table 18.4 Facility life-cycle benefits of some candidate projects

Contract no.	Lanes	Length (miles)	Daily traffic	Work type	Estimated project benefits		
					Low	Medium	High
1	2	0.065	4960	Highway maintenance	2680748	3574331	4531131
2	4	1.75	67750	Add travel lanes	38048478	50731305	63519764
3	4	19.10	11890	Wedge and level	871846	1162462	1508925
4	4	0.10	22760	Bridge deck overlay	3443397	4591197	5801616
5	6	0.10	89350	Bridge replacement	7600911	10134549	12734579
6	4	6.1	32910	Pavement resurfacing	1852358	2469811	3148909
7	4	12.00	53490	Pavement repair	5482952	7310603	9206704
8	6	1.55	112570	Pavement repair	3505980	4674640	5898343
9	2	0.06	13100	Bridge deck widening	6094593	8126125	10227752
10	2	0.15	12000	Bridge construction	1743938	2325251	2969840
...	...	...	...	...	...	...	...

benefits of the two projects i and j. For each project, the probabilities of low-, medium-, and high-level benefits $P(a_{i,L})$, $P(a_{i,M})$, and $P(a_{i,H})$ are set as 0.16, 0.68, and 0.16 according to information extracted from the historical data. This helps compute the joint probabilities of the expected benefits for two projects i and j as $P(a_{i,L}, a_{j,L}) = P(a_{i,L}, a_{j,H}) = P(a_{i,H}, a_{j,L}) = P(a_{i,H}, a_{j,H}) = 0.05$, $P(a_{i,M}, a_{j,L}) = P(a_{i,M}, a_{j,H}) = P(a_{i,L}, a_{j,M}) = P(a_{i,H}, a_{j,M}) = 0.06$, and $P(a_{i,M}, a_{j,M}) = 0.56$. Subsequently, the covariance of each pair of projects is computed using Equation 18.28. For all pairs of 672 projects, a 672×672 covariance matrix is established. The computed covariance values range from 443.22 to 402478620 with an average value of 19389271 and standard deviation of 25192730.

Project selection results: The basic knapsack model and the enhanced two-step knapsack model that incorporates the Markowitz mean-variance model are separately executed using the same dataset for project selection. First, the basic knapsack model is applied to identify the best sub-collection of candidate projects that can achieve maximized total benefits under budget constraints for each year of the 6-year analysis period. On average, 248.43 million dollars of budgets are designated in each year. Of those, 629 out of the 672 projects are selected for possible implementation. The annualized overall project benefits generated by the basic model are approximately 1.497 billion dollars and the total risk associated with the overall project benefits is 13.79 million square dollars.

The Markowitz mean-variance model for step-one optimization is applied to choose the portfolio of projects in order to minimize the total associated risk. The estimated low-, medium-, and high-level project benefits, covariance values, annual budget, and benefit threshold values are utilized as inputs of model applications. For the same level of budgets, the model solution has shown that 633 out of the 672 projects can be selected for possible implementation. The annualized overall benefits of the selected projects are about 1.441 billion dollars and the corresponding risk is 11.50 million. The above minimum total risk is then added as an additional constraint to enhance the basic knapsack model for the step-two optimization. Specifically, this constraint is treated as a chance constraint with an α value set as 0.9, indicating that the total risk of benefits expected to be achieved from the selected projects will be controlled within the lower bound risk level for 90% of the time. The results of step-two optimization indicate that 459 out of 672 projects can be selected for possible implementation. The annualized overall project benefits amount to 1.448 billion dollars, with the total risk reaching 11.06 million.

Compared with the results of overall project benefits obtained in the step-one optimization, the overall benefits achieved in the step-two optimization have increased by $(1.448 - 1.441)/1.441 = 1.0\%$ and the total risk has slightly reduced from the addition of the chance constraint imposed to the lower bound risk established from step-one optimization by (11.06 – 11.50)/11.496 = −4%. This reveals that the step-two optimization has produced a slightly better decision outcome with a marginally higher level of project benefits while maintaining the total risk of the expected benefits virtually at the same level.

Cross comparisons of results obtained from the basic and the enhanced knapsack models further reveal that the enhanced model can reduce the total risk of expected project benefits by (11.06 – 13.79)/13.79 = −20% and the reduction in the overall project benefits is about (1.448 – 1.497)/1.497 = −3%.

Source: Adapted from Zhou, B., et al. 2014. *ASCE Journal of Transportation Engineering* 140(5), 04014007.

18.3.4 Hypergraph knapsack model dealing with multi-project benefit interdependency

18.3.4.1 Multi-project benefit interdependency issue

The overall benefits of a highway project are estimated as net reductions in project-level facility life-cycle agency costs of construction, rehabilitation, and maintenance and in project user costs of vehicle operations, travel time, crashes, and vehicle emissions by comparing the modified transportation network and the original network. Accordingly, the network-wide impacts and interdependency relationship of two or more projects can be estimated as the difference in network-wide benefits of simultaneously implementing two or more projects and the sum of network-wide benefits obtained by separately implementing individual projects one at a time. Therefore, the level of project interdependence in multi-project benefits can be computed as (Li et al., 2012, 2013)

$$
\begin{aligned}
a_{ij} &= [a_i + a_j] + \Delta a_{ij} \\
a_{ijk} &= \left[a_i + a_j + a_k\right] - [\Delta a_{ij} + \Delta a_{ik} + \Delta a_{jk}] + \Delta a_{ijk} \\
&\dots \\
a_{1,2,\dots,N} &= \left[a_1 + a_2 + \dots + a_N\right] \\
&\quad - [\Delta a_{12} + \Delta a_{13} + \dots + \Delta a_{(N-1),N}] \\
&\quad + [\Delta a_{123} + \Delta a_{124} + \dots + \Delta a_{(N-2),(N-1),N}] \\
&\quad \dots \\
&\quad + (-1)^{N+1} \Delta a_{12\dots\dots N}
\end{aligned}
\tag{18.42}
$$

where

$a_{1,2,\dots,N}$ = Network-wide benefits of simultaneously implementing projects $1, 2, \dots, N$

$a_1, a_2, \dots,$ and a_N = Network-level benefits of separately implementing project $1, 2, \dots, N$ one at a time

$\Delta a_{12}, \Delta a_{123}, \dots, \Delta a_{1,2,\dots,N}$ = Differences between the overall benefits of simultaneously implementing multiple projects and the sum of benefits of separately implementing individual projects one at a time

18.3.4.2 The model formulation

The zero/one integer hypergraph knapsack model with project interdependency considerations is formulated as

$$\text{Max} \quad \sum_{i=1}^{N} a_i \cdot x_i - \left[\sum_{(i,j)=1}^{C_2^N} \Delta a_{ij} \cdot x_{ij} + \sum_{(i,j,k)=1}^{C_3^N} \Delta a_{ijk} \cdot x_{ijk} + \cdots + \sum_{(1,2,\ldots,N)=1}^{C_N^N} \Delta a_{12\ldots N} \cdot x_{12\ldots N} \right] \quad (18.43)$$

Subject to

$$\sum_{i=1}^{N} c_{ikt} \cdot x_i \leq B_{kt} \quad (18.44)$$

$$x_{ij} + x_{ijk} + \cdots + x_{ij,\ldots,N} \leq 1 (i = 1, 2, \ldots, N) \quad (18.45)$$

$$(x_i + x_j) - 2 \cdot x_{ij} \geq 0 \quad (18.46)$$

$$(x_1 + x_2 + \cdots + x_{N-1}) - (N-1) \cdot x_{1,2,\ldots,(N-1)} \geq 0 \quad (18.47)$$

$$(x_1 + x_2 + \cdots + x_{N-1}) - N \cdot x_{1,2,\ldots,N} \geq 0 \quad (18.48)$$

$$x_1, x_2, \ldots, x_N, x_{1,2,\ldots,N} = 0/1$$

where

$a_1, a_2, \ldots$, and a_N = Network-level benefits of separately implementing project 1, 2, …, N one at a time

$\Delta a_{12}, \Delta a_{123}, \ldots, \Delta a_{1,2,\ldots,N}$ = Differences between the overall benefits of simultaneously implementing multiple projects and the sum of benefits of separately implementing individual projects one at a time

$x_1, x_2, \ldots, x_N, x_{ij}, x_{ijk}$, and $x_{1,2,\ldots,N}$ = 0/1

c_{ikt} = Costs of project i using budget from management program k in year t

i = $1, 2, \ldots, N$

k = $1, 2, \ldots, K$

t = $1, 2, \ldots, M$

The objective function is an expression (18.43) that maximizes the overall benefits of the subset of candidate projects selected for implementation. The budget constraints are given by expression (18.44). Expression (18.45) requires that at most one possible combined project implementation scenario could be ultimately selected. The constraints as expressions (18.46) through (18.48) ensure that $x_{1,2\ldots,N} = 1$ if and only if $x_1 = x_2, \ldots, = x_N = 1$, indicating that the interdependencies in overall benefits of multiple projects will only be considered when all constituent projects are selected for implementation.

Case Study 3: Toll highway system capital investment program dealing with multi-project benefit interdependency

The board of directors of a state toll highway authority has identified six capital investment projects for the tollway system that accommodate a significant portion of travel in a metropolitan area. A high fidelity regional demand forecasting model is available for traffic assignment analysis that could help determine the traffic redistribution effects from project implementation. That is, the regional demand forecasting model can be executed using the regional daily travel demand to obtain link-based traffic volumes, vehicle compositions, and speeds for the entire regional network before and after project implementation.

Project implementation combinations: As a practical matter, the six tollway projects could be implemented separately one at a time or jointly by implementing two, three, four, and five projects at a time, to all six tollway projects together. As such, the number of modified regional transportation networks relevant to all project implementation scenarios could become $C_1^6 + C_2^6 + C_3^6 + C_4^6 + C_5^6 + C_6^6 = 63$. For each of the 63 project implementation scenarios, the geometric designs, capacities, and speed limits of highway segments within the project physical range were modified accordingly. In addition, a buffer zone extending significantly beyond the project physical range can be created for each of the 63 project implementation scenarios to include all tollway and non-tollway segments expected to be indirectly affected by project implementation. Then, the regional demand forecasting model can be applied by assigning the regional daily travel demand to each of the 63 modified regional transportation networks to establish the redistributed link-specific traffic volumes, vehicle compositions, and speeds for the modified highway network after project implementation. Table 18.5 presents a data sample associated with one candidate tollway project.

Estimated project benefits: The benefits of each project include agency-itemized user benefits such as reductions in vehicle operating costs, travel time, crashes, and air emissions. For each project, the agency and user benefits are separately estimated using the life-cycle cost analysis approach with risk and uncertainty considered for tollway facility construction and treatment costs, annual traffic growth rate, user costs, and a discount rate based on historical records. Tables 18.6 presents annualized project benefits and interdependencies for all implementation scenarios. For instance, the project interdependency level of jointly implementing projects 1, 2, and 3 with risk and uncertainty considerations is quantified as $\Delta a_{123}/(a_1 + a_2 + a_3) = 79506619/(162168188 + 89408673 + 12566679) = 30\%$. Project interdependencies are found to exist for all joint project implementation scenarios and could be either positive or negative. Consequently, overall benefits of simultaneously implementing multiple projects could be greater or smaller than total project benefits from separately implementing them one at a time.

Project selection results: The project benefits computed for 63 project implementation scenarios a_i, a_{ij}, a_{ijk}, a_{ijkl}, a_{ijklm}, and a_{123456}; and interdependencies of jointly implementing multiple projects Δa_i, Δa_{ij}, Δa_{ijk}, Δa_{ijkl}, Δa_{ijklm}, and Δa_{123456} $(i, j, k, l, m = 1, 2, 3, 4, 5, 6)$ using deterministic and risk/uncertainty-based life-cycle cost analysis approaches are separately applied to the proposed hypergraph knapsack model to

Table 18.5 Directional daily traffic, truck traffic, and vehicle speeds of some transportation network links affected by one candidate tollway project (2010 values)

Link		*Daily traffic*		*Truck traffic*		*Vehicle speed (kmph)*	
Category	*ID*	*Before*	*After*	*Before*	*After*	*Before*	*After*
Directly affected (61)	1	61689	78642	9503	19809	88.49	88.19
	2	49727	61424	7660	15580	88.49	88.47
	3	58028	75386	8939	14138	88.28	88.10
	4	58033	80612	8940	19078	88.13	88.13
	...	...	...	...	...	...	...
	61	49718	58477	7659	13791	88.49	88.47
Indirectly affected (59)	1	33632	47049	4907	7585	88.49	88.41
	2	35300	48385	5150	7842	88.70	88.31
	3	40522	51744	5912	9065	88.49	87.98
	...	...	...	...	...	...	...
	59	28164	35805	4109	6332	88.85	88.49

Note: The number in the bracket of the first column refers to the total number of segments.

Table 18.6 Project benefits and interdependencies for different implementation scenarios

Implementation scenario	*Annualized benefits (dollars)*					
	Total		*Interdependencies*			
1	181908958	162168188				
2	92179614	89408673				
3	16432599	12566679				
4	350234692	324963005				
5	216721103	166477912				
6	524278048	332668720				
7	257035710	252337880	257035710	252337880	17052862	−761019
8	257561907	249012454	257561907	249012454	−59220350	−74277587
9	544511508	532276002	544511508	532276002	−12367858	−45144809
10	406361930	361276828	406361930	361276828	−7731869	−32630728
11	829939570	516218684	829939570	516218684	−123752564	−21381776
12	14314329	5957791	14314329	5957791	94297884	96017561
13	408796067	369982955	408796067	369982955	33618239	44388723
14	95620765	67279632	95620765	67279632	213279952	188606953
15	499476231	373753661	499476231	373753661	116981431	48323732
16	422850674	243111563	422850674	243111563	−56183383	94418121
17	139210594	68160311	139210594	68160311	93943108	110884280
18	712239672	411076077	712239672	411076077	−171529025	−65840678
19	388463853	311324896	388463853	311324896	178491942	180116021
20	803746986	626017519	803746986	626017519	70765754	31614206
21	940298028	572249099	940298028	572249099	−199298877	−73102467
22	195498287	184636921	333413659	322671204	95022884	79506619
23	645483427	625751563	564859858	528845274	−21160163	−49211697
24	264021626	242452932	494996779	438441408	226788049	175601841
25	864106524	579834659	722344987	562475566	−65739904	4410922
26	633058863	473407214	591865226	550992805	−84482614	26290658
27	428588680	314317264	374545751	364132329	−13526020	26895515
28	756220539	609612650	1043520610	566694565	−33600934	−102209063
29	661333651	564474439	678003640	640403287	87531102	89134666
30	1093212540	898856553	1084985524	775655652	−36790842	−79056640
31	1022214955	782065215	1154384573	667679396	−99306846	−120750395
32	545322551	382613498	300638519	236438811	−86475646	44324859
33	156357527	53232398	92788161	88165336	168975789	215220866
34	400743072	327585241	825287160	463202288	232147189	107058831
35	554618406	452679742	338262279	295907741	104517003	128169848
36	988409834	850187215	723609450	519566920	−21717480	−103146817
37	600496859	450568954	934898165	562713438	232681906	137986351
38	274245074	88502386	676280047	534094384	309143320	415505210
39	1011739407	628075303	927097925	652129856	−120794068	42123101
40	847172851	486310941	944575443	565174546	−89741101	25402370
41	1134996907	948846067	997511960	560745447	−43763064	−124736430
42	793135139	603585768	678842206	589855217	−152379276	−14479223

(Continued)

Table 18.6 (Continued) Project benefits and interdependencies for different implementation scenarios

Implementation scenario	*Annualized benefits (dollars)*					
	Total		*Interdependencies*			
43	365732418	106919308	491371533	316304689	141509856	323702144
44	599555091	524375698	953554088	631062774	215244128	72436562
45	919244861	527242974	894577584	536362491	−78200494	215774804
46	1143293051	954614890	895586798	670571601	5308261	−45406304
47	628944709	578810492	906836979	667004516	386143014	171913001
48	748386976	345942301	910121174	670403052	16910376	320233483
49	1992892909	721624594	2069511877	689385173	−920038612	110741998
50	1298002650	674822398	1529417326	660509781	−358661942	−940899
51	2212362616	1117418738	2213926438	842539492	−939219815	−131140913
52	390434221	227494619	329146945	316283743	285133787	365921650
53	1937891397	809595884	1853100492	651034193	−954766444	−49988807
54	494522998	421524656	890912308	602303693	355088366	179597328
55	1302556882	1173730612	1160436806	792056396	−119143425	−260212302
56	706909713	686489044	845565281	766693812	400756729	150187272
57	989115766	151436213	804017094	738400712	−131638800	604148244
58	1869675488	886910290	2213769709	746070313	−704641577	34864975
59	536151565	433539104	1467244116	719618690	495368757	329751068
60	1331520567	836543299	2220237470	1102274224	33801848	239143199
61	2355540366	648534213	2185415752	994018263	−1065964966	350310291
62	1145072098	821840823	1598157824	877305751	54773958	104244166
63	2380470484	349534564	3329846507	1060946795	−998715470	738718613

Note: The numbers in the first column—project IDs; D—deterministic; and RU—risk and uncertainty.

find the best sub-collection of projects for implementation at various budget levels. Table 18.7 presents project selection results for budget levels varying from 10% to 100% of $17.37 billion equivalent to the total costs of all six candidate tollway projects.

The overall benefits of projects selected for implementation at various budget levels with risk and uncertainty considerations are about 27% lower than the benefits estimated without such considerations. The overall project benefits generally follow an incremental trend along with budget increases. However, no additional network-wide benefits could be produced after the budget level reaches 80% of $17.37 billion (i.e., $13.9 billion), suggesting that the total amount of tollway capital investments should be controlled within $13.9 billion. This differs significantly from the basic knapsack model for project selection that neglects the effect of multi-project interdependencies that may be either positive or negative, leading to underestimating or overestimating the overall project benefits.

Source: Adapted from Li, Z. et al. 2013. *ASCE Journal of Transportation Engineering* 139(7), 686–696; Li, Z. et al. 2012. *TRB Journal of Transportation Research Record* 2285, 36–46.

18.3.5 Two-stage hypergraph knapsack model

The two-stage hypergraph knapsack model comprises a Markowitz mean-variance model to establish the minimum overall risks of projects that could be achieved. At the second stage, the hypergraph knapsack model is enhanced by adding the output of stage-one optimization as an additional constraint.

Table 18.7 Overall project benefits, benefit-to-cost ratios, and the best sub-collection of projects selected for implementation for a given budget level

Budget level ($17.37 billion)	*Benefit estimation approach*	*Overall annual benefits*	*Benefit-to-cost ratio*	*Selected projects*
10%	Deterministic	16432599	9.59	3
	Risk/uncertainty	12566679	6.98	3
20%	Deterministic	540710647	17.85	3 + 6
	Risk/uncertainty	345235399	10.85	3 + 6
30%	Deterministic	940298028	22.98	56
	Risk/uncertainty	572249099	15.40	56
40%	Deterministic	956730627	22.44	3 + 56
	Risk/uncertainty	584815778	15.07	3 + 56
50%	Deterministic	1062474364	9.11	4 + 36
	Risk/uncertainty	661657772	15.53	2 + 56
60%	Deterministic	1290532720	10.14	4 + 56
	Risk/uncertainty	897212104	7.50	4 + 56
70%	Deterministic	1363148702	10.59	34 + 56
	Risk/uncertainty	815360662	7.76	34 + 56
80%	Deterministic	1484809536	9.81	14 + 56
	Risk/uncertainty	1104525101	7.28	14 + 56
90%	Deterministic	1484809536	9.81	14 + 56
	Risk/uncertainty	1104525101	7.28	14 + 56
100%	Deterministic	1484809536	9.81	14 + 56
	Risk/uncertainty	1104525101	7.28	14 + 56

Note: The numbers in the fifth column are project IDs. For instance, 14 + 56 means jointly implementing projects 1 and 4 and jointly implementing projects 5 and 6.

Markowitz mean-variance model as Equations 18.25 through 18.27 in Section 18.3.3.1 is utilized for stage-one optimization to determine the lowest level of overall risk of all candidate projects R_{min}, which is further added as one more constraint for the hypergraph knapsack model in step-two calculation as Equations 18.43 through 18.48 in Section 18.3.4.2.

18.4 TRADE-OFF ANALYSIS IN TRANSPORTATION PROJECT SELECTION

The transportation decision-making process involves multiple stakeholders that often possess conflicting preferences. Effective decision outcomes could only be reached by considering such conflicts in a holistic manner. A truly optimal solution to all stakeholders rarely exists and finding the most compromised solution becomes essential. To this end, it often involves trade-off analysis of candidate projects that assigns different levels of priorities for various transportation performance goals, different travel modes, different types of facilities, or facilities versus system operations with fixed and varying budgets to determine which portion of the transportation system would benefit, which portion would be adversely affected, and what is the best combination of investments across modes, facilities, and system operation aspects that would produce the highest overall benefits (Bai et al., 2015).

18.4.1 Surrogate worth trade-off method

The surrogate worth tradeoff (SWT) method considers the decision-maker's perspectives of the transportation system as a whole and the management requirements for specific facility categories using a tangible measure in terms of a percentage of the available budget being allocated to a facility category or a prioritized list of projects be considered for investment (Haimes and Hall, 1974; Cohon and Mark, 1975; Roshandeh et al., 2015).

The multi-objective optimization of transportation project selection with N performance criteria, M constraints, and K nonnegative real decision variables via the SWT method is formulated as

$$\text{Max} \quad \{F_1(x), F_2(x), \dots, F_N(x)\} \tag{18.49}$$

$$\text{Subject to} \quad g_m(x) \geq 0, x \in R^2 \quad \text{and} \quad x \geq 0 \tag{18.50}$$

where

F_i = Objective function i; $m = 1, 2, \dots, M$
g_m = Constraint m, $m = 1, 2, \dots, M$
x = $\{x_1, x_2, \dots, x_K\}$ with K nonnegative real decision variables ($K \leq N$); $i = 1, 2, \dots, N$

18.4.1.1 SWT method execution procedure

A solution to the multi-objective function given in Equation 18.49 requires a decision maker who selects a preferred solution from the non-inferior solutions (Haimes and Hall, 1974; Cohon and Mark, 1975). The solution procedure for applying the SWT method generally covers the following computational steps.

ε-Constraint form: The first step in applying the SWT method is converting Equations 18.49 and 18.50 into the ε-constraint form presented by Equation 18.51

$$\begin{aligned} &\text{Maximize} \quad F_i(x) \\ &\text{Subject to} \quad F_j(x) \geq \varepsilon_j \\ &g_m(x) \geq 0, x \in R^2 \quad \text{and} \quad x \geq 0 \end{aligned} \tag{18.51}$$

where

ε_j = The lowest variability level of objective function j
$i, j = 1, 2, \dots, N$

The levels of satisfactory ε_j can be varied parametrically to evaluate the impact on the single objective function $F_i(x)$. The ith objective function $F_i(x)$ can be replaced by the jth objective function $F_j(x)$, and the solution procedure gets repeated. The iterative procedure facilitates generating non-inferior solutions for the original multi-objective optimization problem.

Trade-off function: The generalized Lagrange relaxation $L(x, \lambda_{ij}, u_m)$ to the ε-constraint problem can be written as

$$L(x, \lambda_{ij}, u_m) = F_i(x) + \sum_{j=1, j \neq i}^{N} \lambda_{ij}(F_j(x) - \varepsilon_j) - \sum_{m=1}^{M} u_m g_m(x) \tag{18.52}$$

where

λ_{ij} = Nonnegative real Lagrange multipliers corresponding to the dual variables associated with the jth constraint $F_j(x) \geq \varepsilon_j$, in the ε-constraint form of the original optimization problem

u_m = Nonnegative real Lagrange multipliers relevant to the dual variables associated with the mth constraint $g_m(x) \geq 0$, of the original multi-objective optimization problem

$i = 1, 2, \ldots, N$

$m = 1, 2, \ldots, M$

The optimal solution must satisfy the Kuhn–Tucker necessary conditions

$$\lambda_{ij}[F_j(x) - \varepsilon_j] = 0; \lambda_{ij} \geq 0, j \neq i; \quad i, j = 1, 2, \ldots, N \tag{18.53}$$

$$u_m g_m = 0; u_m \geq 0, g_m \geq 0; \quad m = 1, 2, \ldots, M \tag{18.54}$$

Equation 18.52 is solved using the Newton–Raphson method for R values of ε_j ($j \neq i$; $i, j = 1, 2, \ldots, N$) while other ε_k ($k \neq j \neq i$; $i, j, k = 1, 2, \ldots, N$) are fixed at some levels, where the initial value of ε_j is taken such that $\varepsilon_j \in \{\min F_j(x), \max F_j(x)\}$. Only those values of $\lambda_{ij}^r > 0$ and active constraints $F_j^r(x) = \varepsilon_j^r$ ($r = 1, 2, \ldots, R$) are considered. The set of positive Lagrange multipliers, that is, $\lambda_{ij}^r > 0$, is associated with the non-inferior set of solutions and represents the set of trade-off ratios between the principal objective function and each of the constraining objective functions (Haimes, 2015). Regression analysis or interpolation can be performed to yield the tradeoff functions $\lambda_{ij}[F_j(x)]$ or $\lambda_{ij}(x)$ for the decision variable x_k in $F_j(x)$.

Surrogate worth function: In the multi-objective analysis, it is assumed implicitly that the decision maker maximizes the utility, which in transportation project selection is a monotonic increasing function of the objective functions for benefit maximization. The surrogate worth function assigns a scalar value to any given non-inferior solution. There is a close relationship between the surrogate worth function and the partial derivatives of the utility functions as explained below.

Given a decision variable x and the associated consequences $F(x)$, the utility is given by

$$U = U[F_1(x), F_2(x), \ldots, F_N(x)] \tag{18.55}$$

For a small change in F_i, a linear expression can be established for Equation 18.55 as

$$\Delta U = \frac{\partial U}{\partial F_1}\Delta F_1 + \frac{\partial U}{\partial F_2}\Delta F_2 + \cdots + \frac{\partial U}{\partial F_N}\Delta F_N \tag{18.56}$$

However, for non-inferior points, we obtain

$$\Delta F_i = \sum\nolimits_{j=1, j\neq i}^{N} \frac{\partial F_i}{\partial F_j}\Delta F_i = -\sum\nolimits_{j=1, j\neq i}^{N} \lambda_{ij}\Delta F_i \tag{18.57}$$

Eliminating ΔF_i in Equation 18.57 from Equation 18.56 yields

$$\Delta U = \sum\nolimits_{j=1, j\neq i}^{N} \left[\left(\frac{\partial U}{\partial F_j} + \frac{\partial U}{\partial F_i}\lambda_{ij}\right)\Delta F_i\right] = \sum\nolimits_{j=1, j\neq i}^{N} (\Delta U_{ij} \cdot \Delta F_i), \quad \text{where } \Delta U_{ij} = \frac{\partial U}{\partial F_j} + \frac{\partial U}{\partial F_i}\lambda_{ij} \tag{18.58}$$

$$\text{Let} \quad a_j = \frac{\partial U}{\partial F_j}, \quad \text{then} \quad \Delta U_{ij} = a_j + a_i \lambda_{ij} \tag{18.59}$$

The surrogate worth function W_{ij} is a monotonic increasing function of ΔU_{ij} with property of $W_{ij} = 0 \leftrightarrow \Delta U_{ij} = 0$, and can be written as

$$W_{ij} = h_j \cdot \Delta U_{ij} = h_j \cdot (a_j + a_i \lambda_{ij}) \tag{18.60}$$

where

h_j = A monotonic function with a range of −10 to +10 and the property that $h_j(0) = 0$ and $j \neq i$

$i, j = 1, 2, \ldots, N$

If $a_i = (\partial U/\partial F_i)$ and $a_j = (\partial U/\partial F_j)$ in Equation 18.60 are constants or vary slightly from F_i and F_j, it can be assumed that W_{ij} depends only on λ_{ij}. The surrogate trade-off functions $W_{ij}(\lambda_{ij})$ can be derived by regression analysis or interpolation. Setting $W_{ij}(\lambda_{ij}) = 0$, the values of λ_{ij}^* are derived to be used subsequently for finding the optimal set of decision vectors.

Optimal set of decision vectors: The surrogate worth functions $W_{ij}(\lambda_{ij})$ establish the values of λ_{ij}^* corresponding to non-inferior solutions $x^* = \{x_1^*, x_2^*, \ldots, x_K^*, \ldots\}$. Substituting the values of λ_{ij}^* into the tradeoff functions $\lambda_{ij}(x)$, the optimal set of decision vector x^* are generated. Alternatively, the optimal set of decision vector x^* can be found by solving the following problem:

$$\text{Max} \quad F_i(x) + \sum_{j=1, j\neq i}^{N} \left[\lambda_{ij}^* \cdot F_j(x)\right] \tag{18.61}$$

$$\text{Subject to} \quad g_m(x) \geq 0, x \in R^2 \quad \text{and} \quad x \geq 0$$

where

F_i = Objective function i

g_m = Constraint m

$x = \{x_1, x_2, \ldots, x_K, \ldots\}$, decision vector with K nonnegative real decision variables

$i = 1, 2, \ldots, N$

$m = 1, 2, \ldots, M$

18.4.2 The e-STEP tradeoff analysis method

The multi-objective optimization of transportation project selection among N projects associated with M transportation facilities is generally a maximization of benefits of K different performance criteria that can be formulated as

$$\text{Maximize } \{F_1(x), \ldots, F_l(x), \ldots, F_K(x)\} \tag{18.62}$$

$$\text{Subject to} \quad \sum_{i=1}^{M} \left(c_i^j \cdot x_i^j\right) \leq b_j, \quad i = 1, 2, \ldots, M \tag{18.63}$$

$$\sum_{i=1}^{M} \sum_{j=1}^{N} \left(c_i^j \cdot x_i^j\right) \leq B, \quad j = 1, 2, \ldots, N \tag{18.64}$$

$$x_i^j = 0/1$$

where

$F_l(x)(l = 1, 2, \ldots, K)$ = K performance criteria considered for the tradeoff analysis and investment decisions; without loss of generality, all objective functions are maximized

x = $\{x_1^1 x_2^1, \ldots, x_i^j, \ldots, x_M^N\}(i = 1,2, \ldots, M, j = 1,2, \ldots, N)$ with the decision variable $x_i^j = 0/1$ indicating without or with the use of budget from transportation program j by project i

c_i^j = Cost of investment project i using budget from program j

b_j = Available budget for program j

B = The total budget

18.4.2.1 e-STEP method execution procedure

Figure 18.2 shows the iterative computation procedure for executing the e-STEP method (Truong et al., 2017).

Step 1: Establish the payoff matrix for multiple objective functions

The multiple objective functions that characterize the performance criteria can be utilized to assess changes in the values of individual performance criteria as impacts caused by potential investments. The extent of quantified impacts is used as the basis of investment decisions. As the initiation of the e-STEP model execution, a payoff matrix is created to show how the values of the remaining objective functions are changed by optimizing one objective function in each time under the same set of constraints. Given the objective functions of the kth and lth performance criteria $F_k(x)$ and $F_l(x)(k, l = 1, 2, \ldots, K$, and $k \neq l)$, z_l^k can be defined as the value of $F_l(x)(l = 1, 2, \ldots, K))$ when the optimality is reached for $F_k(x)$, denoted as $F_k^*(k = 1, 2, \ldots, K$, and $k \neq l)$. In this way, two ideal solution vectors can be defined, which are X^* corresponding to all $z_l^k(k, l = 1, 2, \ldots, K$, and $k \neq l)$ and Z^* concerning $F_k^*(k = 1, 2, \ldots, K)$ that is exhibited on the main diagonal of the payoff matrix.

Step 2: Calculate the relative importance of performance criteria

The relative importance of performance criterion l, λ_l, can be established by synthesizing the prior relative importance with the informational importance, which is of the following specification:

$$\lambda_l = \frac{w_l \times \tilde{\lambda}_l}{\sum_{l=1}^{K}(w_l \times \tilde{\lambda}_l)} \tag{18.65}$$

where

w_l = The priori normalized relative importance of performance criterion l that reflects the decision maker's preference, $0 \leq w_l \leq 1$. In an extreme case with no priori preference given to any performance criterion, w_l can be set at equal among all K performance criteria, namely, $w_l = 1/K$

$\tilde{\lambda}_l$ = Context-dependent of normalized informational importance for performance criterion l, $0 \leq \tilde{\lambda}_l \leq 1$

For calculating the informational importance of $\tilde{\lambda}_l$, the concept of entropy measure is employed. The computation involves the following: (1) degree of closeness to ideal point d_l^k; (2) normalized entropy measure $e(d_l)$; and (3) normalized informational importance $\tilde{\lambda}_l$. The degree of closeness d_l^k for performance criterion l is computed by

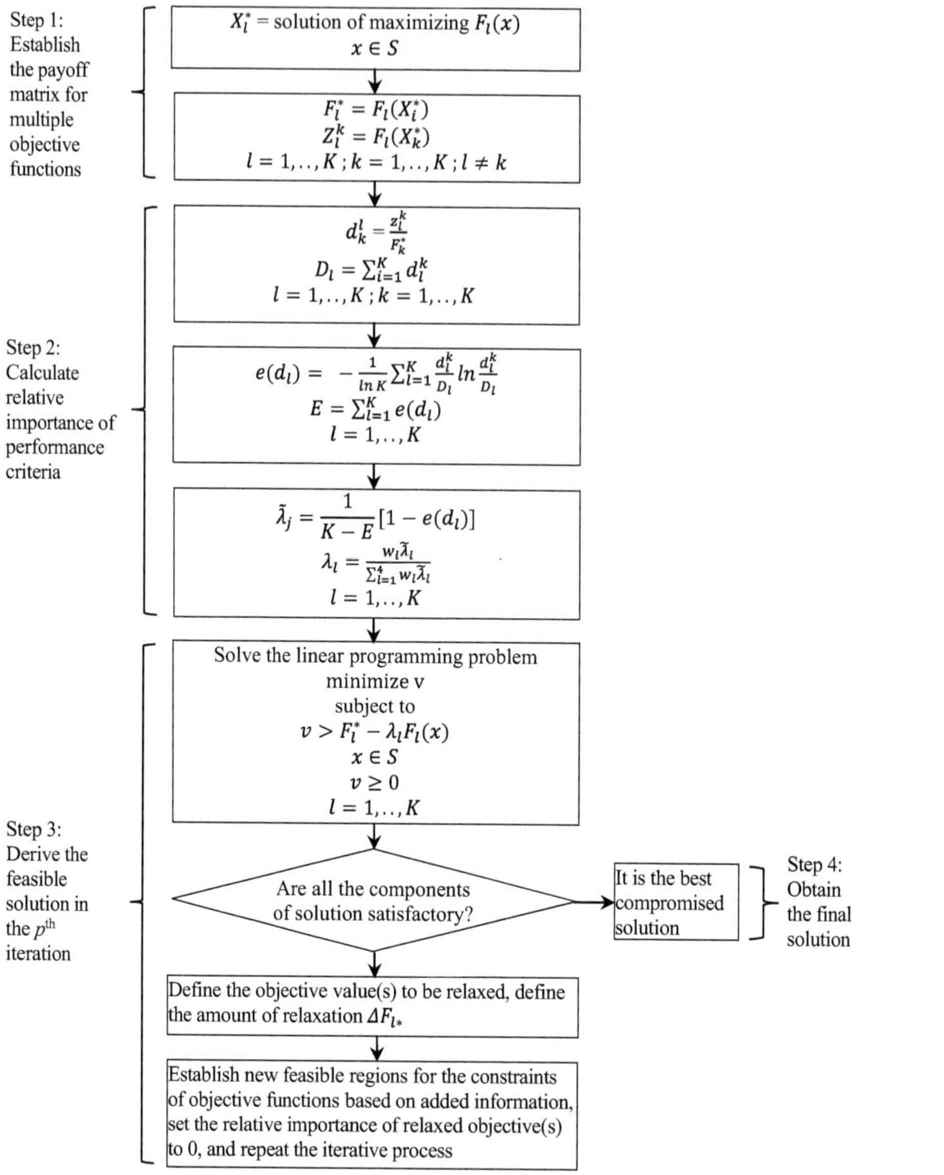
Step 1:
Establish the payoff matrix for multiple objective functions
X_l^* = solution of maximizing $F_l(x)$
$x \in S$
$F_l^* = F_l(X_l^*)$
$Z_l^k = F_l(X_k^*)$
$l = 1,..,K ; k = 1,..,K ; l \neq k$
$d_k^l = \frac{z_l^k}{F_k^*}$
$D_l = \sum_{l=1}^{K} d_l^k$
$l = 1,..,K ; k = 1,..,K$
Step 2:
Calculate relative importance of performance criteria
$e(d_l) = -\frac{1}{ln\,K}\sum_{l=1}^{K}\frac{d_l^k}{D_l}ln\frac{d_l^k}{D_l}$
$E = \sum_{l=1}^{K} e(d_l)$
$l = 1,..,K$
$\tilde{\lambda}_j = \frac{1}{K-E}[1 - e(d_l)]$
$\lambda_l = \frac{w_l\tilde{\lambda}_l}{\sum_{l=1}^{4} w_l\tilde{\lambda}_l}$
$l = 1,..,K$
Solve the linear programming problem
minimize v
subject to
$v > F_l^* - \lambda_l F_l(x)$
$x \in S$
$v \geq 0$
$l = 1,..,K$
Step 3:
Derive the feasible solution in the pth iteration
Are all the components of solution satisfactory?
It is the best compromised solution
Step 4:
Obtain the final solution
Define the objective value(s) to be relaxed, define the amount of relaxation ΔF_{l*}
Establish new feasible regions for the constraints of objective functions based on added information, set the relative importance of relaxed objective(s) to 0, and repeat the iterative process

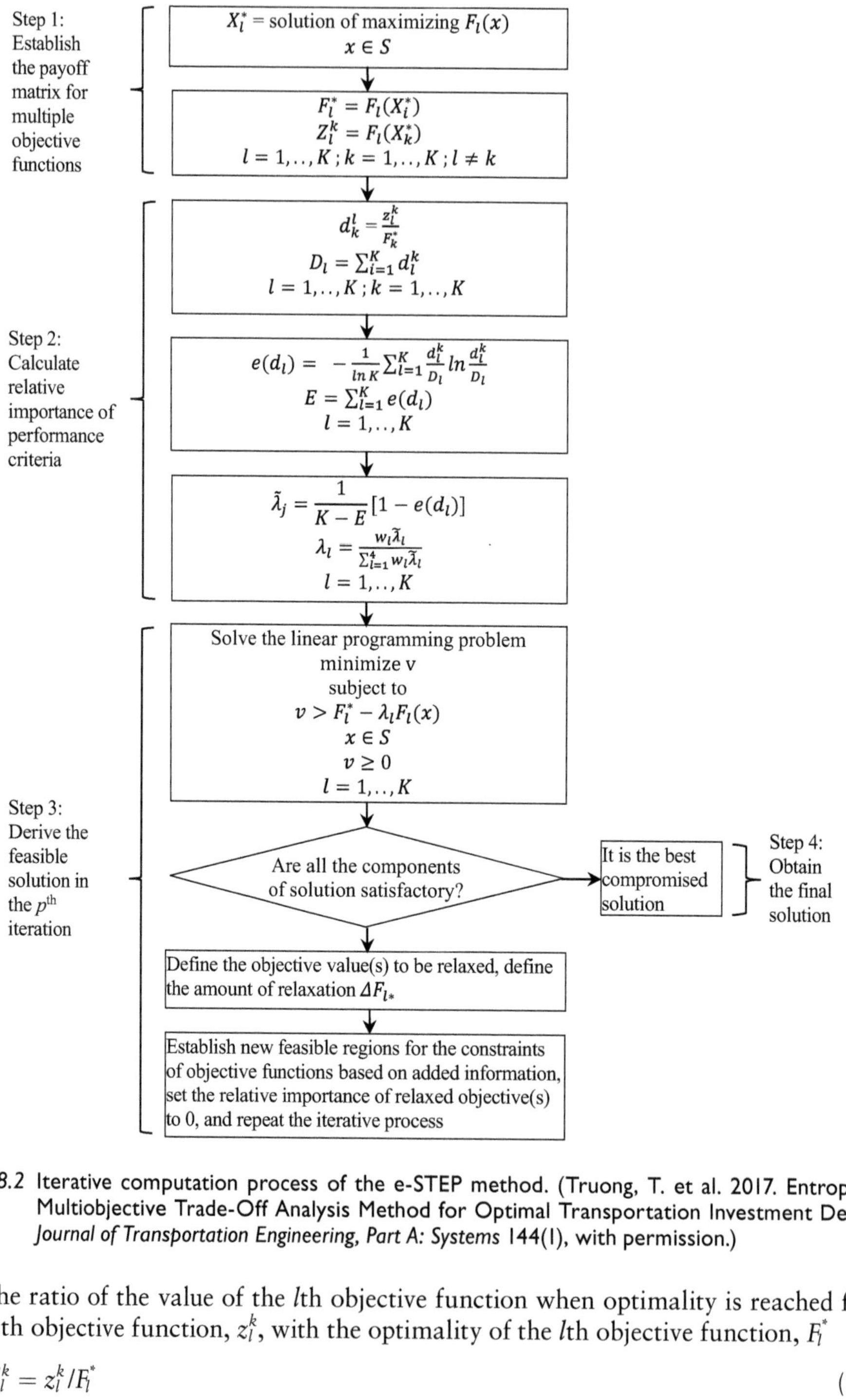

Figure 18.2 Iterative computation process of the e-STEP method. (Truong, T. et al. 2017. Entropy-STEP Multiobjective Trade-Off Analysis Method for Optimal Transportation Investment Decisions. *Journal of Transportation Engineering, Part A: Systems* 144(1), with permission.)

the ratio of the value of the lth objective function when optimality is reached for the kth objective function, z_l^k, with the optimality of the lth objective function, F_l^*

$$d_l^k = z_l^k / F_l^* \tag{18.66}$$

The normalized entropy measure $e(d_l)$ of performance criterion l is calculated by

$$e(d_l) = -\frac{1}{\ln K}\sum_{l=1}^{4}\frac{d_l^k}{D_l}\ln\frac{d_l^k}{D_l} \tag{18.67}$$

with

$$0 \le d_l^k \le 1$$

$$0 \le e(d_l) \le 1$$

$$D_l = \Sigma_{l=1}^K d_l^k$$

$$k, l = 1, 2, \ldots, K$$

Naturally, the larger the normalized entropy measure $e(d_l)$ is, the less information is conveyed. Therefore, the reverse value $1 - e(d_l)$ should be used to calculate informational importance. The normalized informational importance of performance criterion l is computed by

$$\tilde{\lambda}_l = \frac{1}{K - E}[1 - e(d_l)] \quad (18.68)$$

where

$E = \Sigma_{l=1}^K e(d_k)$

$l = 1, 2, \ldots, K$

For performance criterion l, the prior importance w_l available and informational importance $\tilde{\lambda}_l$ computed by Equation 18.7 can be synthesized using Equation 18.4 to arrive at its normalized relative importance. The normalized relative importance values for all performance criteria are used for the iterative calculations in the subsequent step to obtain the feasible solution.

Step 3: Derive the feasible solution in the *p*th iteration

In the *p*th iteration, the feasible solution X^p of the decision vector is determined as the "nearest" to the ideal solution X^* in the minimax sense. As such, solution to the original multi-objective optimization model is equivalent to deriving a solution for the following LP formulation:

$$\text{Minimize} \quad v \quad (18.69)$$

$$\text{Subject to} \quad v_l \ge F_l^* - \lambda_l \cdot F_l(x) \quad (18.70)$$

$$X^p \in S^p \quad (18.71)$$

$$v_l \ge 0 \quad (18.72)$$

where

v = $\{v_1, v_2, \ldots, v_k\}$, as the set of gaps of the weighed objective function values with the respective ideal values for all performance criteria

v_l = Gap of the weighed objective function value with its ideal value for the lth performance criterion

X^p = Feasible solution in the pth iteration

S^p = Feasible region of the solution in the pth iteration; in the first iteration, $S^1 = S$, and is defined by constraints as shown in Equations 18.63 and 18.64

Step 4: Obtain the final solution

The solution X^p derived by solving the above LP formulation is the "best compromise" solution given the relative importance determined in the previous step. In the case where the optimality of some of the objective functions $F_l(X^p)$ of Z^p are satisfactory and others are not, the decision maker can carry out trade-offs by relaxing ΔF_l^* amount of a satisfactory objective function F_l^* to allow for an improvement of the unsatisfactory objective(s) in the next iteration.

In the subsequent iteration for solving the above linear model, the feasible region is adjusted by adding more constraints to the original constraints, which is of the following specification:

$$S^{p+1} = \begin{cases} S^i \\ F_l^*(X) \geq F_l^*(X^p) - \Delta F_l^* \\ F_l(X) \geq F_l(X^p), l \neq l^* \end{cases} \quad (18.73)$$

When solving the revised model, the relative importance of the satisfactory objective function F_l^* is assigned to zero since the optimality of the objective function has already been reached and becomes the least important. The best compromised solution for other objective functions is calculated until no further improvement can be made in the consecutive iterations.

Case Study 4: Tradeoff analysis in project selection using e-STEP and SWT methods

Project benefits and costs: As an extension of Case Study 3 that deals with tollway capital investment decision making by the state toll highway authority, we conduct a trade-off analysis of tollway project selection using the e-STEP method, in conjunction with the SWT method using the estimated annualized project-level benefits and costs for the 63 project implementation scenarios as shown in Table 18.8.

Multi-objective optimization formulations: Having estimated project agency benefits, itemized user benefits, and costs for different project implementation scenarios without and with considering risk and uncertainty of input factors for benefit and cost estimation, as well as budget constraints, the multi-objective optimization models can be formulated as

$$\text{Max} \begin{cases} \text{Agency benefits: } F_1(x) = \sum_{i=1}^{63} AB_i^C \cdot x_i, \\ \text{Vehicle operating cost reductions: } F_2(x) = \sum_{i=1}^{63} VOC_i^C \cdot x_i, \\ \text{Travel time savings: } F_3(x) = \sum_{i=1}^{63} TT_i^C \cdot x_i, \\ \text{Vehicle crash decreases: } F_4(x) = \sum_{i=1}^{63} VC_i^C \cdot x_i \end{cases} \quad (18.74)$$

or

$$\text{Max} \begin{cases} \text{Agency benefits: } F_1(x) = \sum_{i=1}^{63} AB_i^{CRU} \cdot x_i, \\ \text{Vehicle operating cost reductions: } F_2(x) = \sum_{i=1}^{63} VOC_i^{CRU} \cdot x_i, \end{cases}$$

Table 18.8 Annualized project benefits and costs for different project implementation scenarios

Project ID	Agency benefits		Veh opt. cost reductions		Travel time savings		Vehicle crash decreases		Project costs	
	Certainty	With RU	Certainty	With RU	Certainty	With RU	Certainty	With RU	Certainty	With RU
1	100267345	81530582	49206716	48613935	2678383827	2663518670	2895867689	2880247153	196603120	197003940
2	2181544	3769044	28787762	25774864	3629632060	3292270489	2169325968	2324816588	111537833	111840529
3	7322038	5206667	14357	11383	253938536	215181058	256298311	214023528	14376185	14376185
4	329445124	317176389	49358973	48613688	6268656576	6193694824	4339241723	4303027131	254204859	254723113
5	53045817	44047738	439451697	819892772	3551643699	3195260127	2794742921	2259513474	176483076	176842876
6	244752721	174042524	320497691	316502095	10231530984	10090261505	5712858088	5668179943	102763666	102973173
7	100402700	81530582	56739380	56208911	3943789489	3935010754	3999132832	3992255359	375384973	376150280
8	107724739	86737249	52668759	52148448	3871797026	3862130503	4082280277	4071879287	227432220	227895892
9	423268897	392373047	364414866	359610107	9120346471	9051610326	7195328454	7168056790	450807979	451727053
10	153448517	125578320	197183449	316463645	6141575768	5388473824	6314840103	5488347654	373086196	373846816
11	497673342	364027891	365246424	355570941	14029073738	13712354487	11209255928	11040599532	299366786	299977113
12	7322038	5206667	422175	421914	221864770	220672535	220957078	219914384	209610953	210038292
13	10477010	212977742	55709994	54073121	7512225938	7311396003	5350033656	4497884504	432986712	433869453
14	53045817	40079824	141005833	209439559	1586257079	1459218686	1227811389	1069013505	355264929	355989216
15	244752721	174042524	294012925	285278656	9657340618	9397559099	5528977960	5416155638	281545520	282119513
16	17799048	216448853	50014167	48395229	7660723895	7438876627	5642259757	4739030893	285033959	285615065
17	60367855	49254405	106554286	205649545	2129113941	2062150469	2088721237	1875172718	207312176	207734828
18	404592680	287703977	332231503	322255420	12707682227	12362298356	8962046172	8776961499	133592766	133865125
19	63522827	361224127	175222236	283091892	6854070765	6146780704	5175813152	4567659183	430687935	431565989
20	714574224	591039904	320606607	311778571	14532750692	14246115317	9646957787	9518779755	356968526	357696286
21	450316458	326545047	868329667	823069691	16486314077	12649087379	11801346310	9484249272	279246742	279816049
22	107724739	86737249	27986964	22998953	6153086451	5003099447	6217110830	5007283376	406214073	407042232
23	426504395	395488053	312426529	276684623	11733259419	9468206186	10094636806	7218030521	629589832	630873393
24	153448517	125578320	202362755	321352315	7583464622	6683112574	7671555944	6702273421	551868049	552993156
25	497673342	364027891	555573460	404879738	17720156440	12646980525	13626780482	9404320135	478148639	479123453
26	437169863	403913638	120533848	104767207	12537230713	10022138228	10900173587	7695639177	481637079	482619005

(Continued)

Table 18.8 (Continued) Annualized project benefits and costs for different project implementation scenarios

Project ID	*Agency benefits*		*Veh opt. cost reductions*		*Travel time savings*		*Vehicle crash decreases*		*Project costs*	
	Certainty	*With RU*	*Certainty*	*With RU*	*Certainty*	*With RU*	*Certainty*	*With RU*	*Certainty*	*With RU*
27	160770555	130784987	196196427	314755172	7606995992	6293863246	7699615302	5830533052	403915295	404738768
28	504995380	369234559	579723029	423475104	17901264740	12653671156	13869231296	9482194568	330195886	330869065
29	482893641	442754709	250712921	364700055	13400030554	11249119141	11622031085	8760954933	627291055	628569929
30	814976924	672570486	586583472	435813588	23867094053	18031760175	17949952127	12575372608	553571646	554700226
31	550719158	408075629	748185710	730651777	19513001008	14575692404	15535855147	11355423427	475849862	476819989
32	336767162	322383056	54257716	48379929	5147957583	4380553928	3876606614	2816825338	463815812	464761405
33	60367855	49254405	111703123	209177473	1524845021	1535715720	1219361664	1117534862	386094029	386881168
34	404592680	287703977	405092656	282075052	14860303127	10889267069	10312592508	7199061420	312374620	313011465
35	382490941	361224127	184837030	284154635	5749725415	5210688422	4274276737	3346595570	609469788	610712329
36	714574224	591039904	404853444	292502495	16246595602	12657531177	10879754075	7763035825	535750379	536842627
37	450316458	326545047	559312995	575949451	16462730148	12793786555	11963133178	9042878964	458028595	458962390
38	389812979	366430794	193025971	315300535	11402859320	10056344356	9295900854	7221075641	461517035	462457941
39	721896262	596246572	441318068	321078189	16613096200	12838070121	11338297211	8010624971	387797626	388588238
40	457638496	331751715	553406287	571982429	16481499029	12695926030	12246831296	9164878849	310075842	310708001
41	767620041	635087642	538905630	554050780	17638912351	14305821090	12382164747	10188545682	533451601	534539163
42	290085919	267857909	104398157	102640553	11374650423	10932801468	9102879143	7932297174	660418932	661765345
43	160770555	130784987	182688094	296684575	7648048810	6701863486	7502408310	6494820473	582697149	583885108
44	504995380	369234559	553174139	402502785	19525239305	14130110590	15342644190	10612563195	508977739	510015405
45	482893641	442754709	264012579	374877683	14702008792	12792294736	12709583603	10854851235	806072908	807716269
46	814976924	672570486	453795251	331710708	19446527767	15234530706	14498141840	11351992425	732353499	733846566
47	550719158	408075629	814124352	770286776	18052897300	14083492888	14482632700	11594109558	654631715	655966329
48	490215679	447961376	244450337	358872300	14860107627	12827372947	13053176211	11041313889	658120155	659461881
49	822298963	677777154	744697098	558530390	35534021095	27671770105	28120803205	21848172752	584400745	585592178
50	558041197	413282297	771111056	761540204	25552382612	19749984609	21327389288	17085912069	506678962	507711941
51	868022741	716618224	861145827	850127316	37564206348	30106358149	30482297292	24739327851	730054721	731543102

(Continued)

Table 18.8 (Continued) Annualized project benefits and costs for different project implementation scenarios

Project	*Agency benefits*		*Veh opt. cost reductions*		*Travel time savings*		*Vehicle crash decreases*		*Project costs*	
ID	*Certainty*	*With RU*	*Certainty*	*With RU*	*Certainty*	*With RU*	*Certainty*	*With RU*	*Certainty*	*With RU*
52	247315643	234067948	137801802	271934624	5727517453	5923331819	4229157328	3810044745	640298888	641604281
53	721896262	596246572	527857148	380852118	32851174934	25917404200	24305895893	19193644273	566579479	567734579
54	457638496	331751715	505635471	522268944	14183393073	11138466290	10398675499	7791510814	488857695	489854342
55	479177236	398943651	405151476	557949539	15231138954	15358640544	10287872026	9482127590	712233454	713685503
56	484058595	402414763	401295250	554084213	15265669397	15376290640	10454231432	9636535279	564280701	565431115
57	311304245	285144065	207058252	368644595	13590171941	13616615455	11235132894	10157263374	836902008	838608221
58	822298963	677777154	697061131	519857011	37885824527	29795590920	30379940186	23858488621	763182599	764738518
59	558041197	413282297	825808280	815324843	30000865258	23532042575	28044287335	22559772827	685460815	686858281
60	994554643	590086322	1274581854	1161416872	37342052163	37238487660	34021605908	31742647571	908836574	910689442
61	875344780	721824892	824866856	865333259	36972514893	29711738165	30198829059	24516074593	760883821	762435054
62	774942079	640294310	653327760	672209845	32132942207	26207421723	24616206922	20409522290	743062554	744577455
63	1294308848	721824892	1017705215	935002110	48230275759	38127250964	44118754180	32198865662	939665674	941581394

Note: Benefits of vehicle air emission decreases are negligible and are omitted from the list of itemized user benefits.

$$\text{Travel time savings: } F_3(x) = \sum_{i=1}^{63} TT_i^{CRU} \cdot x_i,$$

$$\left.\text{Vehicle crash decreases: } F_4(x) = \sum_{i=1}^{63} VC_i^{CRU} \cdot x_i \right\} \quad (18.75)$$

$$721237841 \le \sum_{i=1}^{63} c_i \cdot x_i \le 21983022693 \quad (18.76)$$

$$\sum_{i=1}^{63} c_i \cdot x_i = 1 \quad (18.77)$$

$$x_i = 0/1, i = 1,2,\ldots,63$$

where

$F_k(x)$ = $\{F_1(x), F_2(x), F_3(x), F_4(x)\}$, which are concerned with agency benefits and user benefits of vehicle operating cost reductions, travel time savings, and vehicle crash decreases

x = $\{x_1, x_2, \ldots, x_{63}\}$ contains decision variables $x_i = 0/1$ indicative of rejection or selection of the ith project implementation scenario

$AB_i^C, VOC_i^C, TT_i^C, VC_i^C$; $AB_i^{CRU}, VOC_i^{CRU}, TT_i^{CRU}, VC_i^{CRU}$ = agency benefits, user benefits of vehicle operating cost reductions, travel time savings, and vehicle crash decreases achieved without and with risk and uncertainty considerations for computation from the ith project implementation scenario

c_i = Cost associated with i^{th} project implementation scenario

Source: Adapted from Truong, T. et al. 2017. *ASCE Journal of Transportation Engineering* 144(1), DOI: 10.1061/JTEPBS.0000100.

e-STEP method execution: For applying the e-STEP computational procedure to each of the above models, each objective function needs to be optimized one at a time under the same set of constraints to derive the payoff values for the remaining objective functions that will help establish the payoff matrix. This shows to what extent the values of other objective functions would change when one objective function is optimized. For the current computational study, the objective of maximizing agency benefits is initiated, followed by maximizing user benefits of vehicle operating cost reductions, travel time savings, and vehicle crash decreases. Next, the prior importance of performance criteria concerning agency costs and user costs of vehicle operation, travel time, and crashes derived by Li and Sinha (2009b; Sinha et al., 2009) is adopted. The normalized values are $w_1 = 0.2688$, $w_2 = 0.2424$, $w_3 = 0.2546$, $w_4 = 0.2342$. Then, Equations 18.66 through 18.68 and 18.65 are executed to derive the degree of closeness d_l^k, normalized entropy measures of importance $e(d_l)$, informational importance $\tilde{\lambda}_l$, and relative importance λ_l, as summarized in Table 18.9.

After relative weights of performance criteria are refined, they could be used to help convert the multi-objective optimization formulation without dealing with the risk and uncertainty issue as expressions (18.74) and (18.76) through (18.78) and the multi-objective optimization formulation considering the risk and uncertainty issue as expressions (18.75) and

Table 18.9 Prior importance, entropy measures of importance, and relative importance for multiple performance criteria

Case	*Performance criterion*	*Priori importance* (w_j)	*Entropy measure of importance* ($e(d_j)$)	*Relative importance* (λ_j)
Certainty	Agency costs	0.2688	0.9957	0.28404
	Vehicle operating costs	0.2424	0.9963	0.22026
	Travel time	0.2546	0.9960	0.25452
	Vehicle crashes	0.2342	0.9958	0.24118
Incorporating risk and uncertainty	Agency costs	0.2688	0.9954	0.29221
	Vehicle operating costs	0.2424	0.9964	0.20817
	Travel time	0.2546	0.9958	0.25251
	Vehicle crashes	0.2342	0.9956	0.24711

(18.76) through (18.78) to LP models as described in Step 3 of the e-STEP method, which are of the following specifications:

$$\text{Min } v \tag{18.78}$$

Subject to:

$$\text{Agency benefits: } v \geq 1294308848 - 0.2840 \cdot \sum_{i=1}^{63} AB_i^C \cdot x_i \tag{18.79}$$

$$\text{Vehicle operating cost reductions: } v \geq 1274581854 - 0.2203 \cdot \sum_{i=1}^{63} VOC_i^C \cdot x_i \tag{18.80}$$

$$\text{Travel time savings: } v \geq 48230275759 - 0.2545 \cdot \sum_{i=1}^{63} TT_i^C \cdot x_i \tag{18.81}$$

$$\text{Vehicle crash decreases: } v \geq 44118754180 - 0.2412 \cdot \sum_{i=1}^{63} VC_i^C \cdot x_i \tag{18.82}$$

$$721237841 \leq \sum_{i=1}^{63} c_i \cdot x_i \leq 21983022693 \tag{18.83}$$

$$\sum_{i=1}^{63} c_i \cdot x_i = 1 \tag{18.84}$$

$$x_i = 0/1, \quad i = 1,2,\ldots,63$$

and
Subject to:

$$\text{Agency benefits: } v \geq 721824892 - 0.2922 \cdot \sum_{i=1}^{63} AB_i^{CRU} \cdot x_i \tag{18.85}$$

$$\text{Vehicle operating cost reductions: } v \geq 1161416872 - 0.2082 \cdot \sum_{i=1}^{63} VOC_i^{CRU} \cdot x_i \quad (18.86)$$

$$\text{Travel time savings: } v \geq 38127250964 - 0.2525 \cdot \sum_{i=1}^{63} TT_i^{CRU} \cdot x_i \quad (18.87)$$

$$\text{Vehicle crash decreases: } v \geq 32198865662 - 0.2471 \cdot \sum_{i=1}^{63} VC_i^{CRU} \cdot x_i \quad (18.88)$$

$$721237841 \leq \sum_{i=1}^{63} c_i \cdot x_i \leq 21983022693 \quad (18.89)$$

$$\sum_{i=1}^{63} c_i \cdot x_i = 1 \quad (18.90)$$

$$x_i = 0/1, \; i = 1,2,\ldots,63$$

Trade-off analysis results: For the first model without considering risk and uncertainty of input factors involving benefit and cost calculations, the optimal decision vector is $x^* = \left(x_1^* = x_2^* = \cdots = x_{62}^* = 0, x_{63}^* = 1\right)$. This shows that the maximized total benefits can be achieved by jointly implementing all six major investment projects. The annualized total benefits are \$94661044001. For the second model considering risk and uncertainty of input factors, the optimal decision vector is still $x^* = \left(x_1^* = x_2^* = \cdots = x_{62}^* = 0, x_{63}^* = 1\right)$. This indicates that the total benefits can be maximized by jointly implementing all six major investment projects, which are \$71982943628.

Comparisons of e-STEP and SWT method results: To apply the same dataset to the SWT method, the following efforts are performed. First, any one of the four objective functions could be chosen as a primary objective function. Without loss of generality, the objective function concerning agency benefits, $F_1(x)$, is treated as a primary objective function. Secondly, non-inferior solutions are obtained by varying the relative importance λ_1 and λ_j $(j = 2,3,4)$ with a step of 0.01 while λ_k $(k \neq j,\ k \neq \text{i})$ of the remaining objective functions are arbitrarily fixed at 0.01 by solving corresponding LP models. It is noted that the relative importance is varied and fixed in the way that their sum remains equal to 1.00. For instance, to study the SWT function $S_{12}(\lambda_2)$ of primary objective function $F_1(x)$ to the objective function $F_2(x)$, λ_3 and λ_4 are fixed at 0.01 while λ_1 and λ_2 are varied from 0.98 to 0.00 and 0.00 to 0.98, respectively. Next, for each case scenario of relative importance distribution, the degree of closeness values, μ_1 and μ_j, and their ratios $S_{1j} = \mu_1/\mu_j$, are calculated. An SWT function $S_{1j}(\lambda_j)$ between the ratio of the degree of closeness of primary objective function F_1 with respect to F_j, and λ_j is derived by performing regression analysis. Finally, the optimal relative importance $\lambda_j = \lambda_j^*$ of objective function F_j is evaluated by finding the roots of equation $S_{1j}(\lambda_j) = \mu_D$, where μ_D is the satisfaction level of the SWT function. Details of implementing the SWT method execution can be found in Roshandeh et al. (2015). Table 18.10 summarizes the decision outcomes generated by the two methods.

Table 18.10 Comparisons of decision outcomes using different methods

Decision case			*e-STEP method*	*SWT method*
Certainty	Relative importance for finding optimal solution		$\lambda_1 = 0.28404$, $\lambda_2 = 0.22026$, $\lambda_3 = 0.25452$, $\lambda_4 = 0.24118$	–
	SWT functions		–	$S_{12} = 1.230 - 6.783\lambda_{12} + 8.247\lambda_{12}^2$ $S_{13} = 0.993 + 0.101\lambda_{13} - 3.993\lambda_{13}^2$ $S_{14} = 1.003 - 0.037\lambda_{14} + 0.41\lambda_{14}^2$
	Optimal decision vector		$x_{63}^* = 1$ $x_1^* = x_2^* = \cdots = x_{62}^* = 0$	$x_{60}^* = 1$ $x_1^* = x_2^* = \cdots = x_{59}^*$ $= x_{61}^* = x_{62}^* = x_{63}^* = 0$
	Maximized annual benefits	Agency costs	1294308848	994554643
		Veh operation	1017705215	1274581854
		Travel time	48230275759	37342052163
		Crashes	44118754180	34021605908
		Total	94661044002	73632794568
	Difference		–	28.6%
Incorporating risk and uncertainty	Relative importance for finding optimal solution		$\lambda_1 = 0.29221$, $\lambda_2 = 0.20817$, $\lambda_3 = 0.25251$, $\lambda_4 = 0.24711$	
	STW functions		–	$S_{12} = -0.101 + 0.523\lambda_{12} + 1.562\lambda_{12}^2$ $S_{13} = 1.873 + 36.691\lambda_{13} - 3.873 + \lambda_{13}^2$ $S_{14} = 0.806 - 4.481\lambda_{14} + 5.602\lambda_{14}^2$
	Optimal decision vector		$x_{63}^* = 1$ $x_1^* = x_2^* = \cdots = x_{62}^* = 0$	$x_{60}^* = 1$ $x_1^* = x_2^* = \cdots = x_{59}^*$ $= x_{61}^* = x_{62}^* = x_{63}^* = 0$
	Maximized annual benefits	Agency costs	721824892	590086322
		Veh operation	935002110	1161416872
		Travel time	38127250964	37238487660
		Crashes	32198865662	31742647571
		Total	71982943628	70732638426
	Difference		–	1.8%

PROBLEMS

18.1 A local transportation agency is considering investments into 7 projects: A, B, C, D, E, F and G. As shown in Table P18.1, each project has a capital cost and an expected net present worth through its life-cycle as well as the prerequisite project(s). For example, project D can be selected only if project A is selected at the same time. The agency wishes to achieve the maximum net present worth by making a full use of the 4500 million available budget. Formulate the problem as an integer programming problem and solve it in Microsoft (MS)Excel Solver, LINDO or CPLEX.

Table P18.1 Project candidate details

	Projects						
	A	B	C	D	E	F	G
Initial cost (million)	700	800	1030	960	1260	1350	2150
NPW (million)	130	150	200	310	400	420	750
Prerequisite	–	–	–	A	B	C	C

18.2 Orders of 15, 20, and 10 steel components have been received from three bridge construction sites (A, B, and C). These orders are to be filled by steel mills (1), (2), and (3) which have respective availabilities of 35, 30, and 10 components. The shipping distances in miles from the various mills to each of the construction sites are given in Table P18.2:

Table P18.2 Shipping distance between sites and mills

		Sites		
		A	B	C
Mills	(1)	700	800	400
	(2)	100	200	600
	(3)	300	400	700

Assuming an equal shipping cost per mile for all components, how should the sites be supplied so that the total shipping cost is minimized? Formulate the problem as a linear programming problem.

18.3 A transit agency would like to select one or more service improvement projects among six project candidates over the next four years. Table P18.3 shows the detailed cash flow of all six alternatives (e.g., project A is a one-year project and only cost $4 million in year 1 with a net present worth of benefits of $2.5 million.) The objective is to select a subset of projects among these candidates in order to maximize the net present worth. Also, budget constraints over the four-year horizon need to be satisfied at the same time. Formulate the project selection program as an integer programming technique and solve for the optimization formulation using MS Excel Solver, LINDO or CPLEX.

Table P18.3 Cashflow of each project candidate

Projects	*Year 1*	*Year 2*	*Year 3*	*Year 4*	NPW
Budget	7.5	8.5	6	3	–
A	4	0	0	0	2.5
B	0	2	1	0	3.8
C	5	0	2	1	4.6
D	2	1	1	0	3.2
E	7	5	2	1	8.2
F	0	4	4	0	5.1

Chapter 19

Innovative financing and investment decisions

19.1 HISTORICAL OVERVIEW

In 1920, Arthur C. Pigou propounded the theory of taxation on industry causing negative externalities accounted in social costs but not covered in the private cost of the business activity. The theory of taxing an activity that is causing negative externalities was initially suggested for use by the RAND Corporation for freight transportation in 1949. This concept was formulated, developed, and championed in the U.S. and the United Kingdom during the 1950s. The prominent researchers involved in this process were James M. Buchanan, Alan Walters, and William Vickrey. They expanded on an idea by Pigou and proposed a flexible transit fare structure reflective of the ridership over time of a day for the New York subway system in 1952.

In later years, the idea was extended to the highway system. For instance, Roth and Thomson (1963) developed the concept of road pricing that could potentially be used as a travel demand management strategy to cure congestion. Allais and Roth (1968) extended the concept by working out details of charges to road users to achieve economic efficiency. About the same period, Smeed et al. (1964) recommended congestion pricing for busy roads in the United Kingdom, and made recommendations to consolidate various road costs, congestion costs, and social costs in a road pricing scheme in which the users pay not only for creating congestion on the road, but also for using and causing various social impacts. Their economic analysis indicated that the largest economic benefits of road pricing were the generated revenue, and demand management was a by-product. This translates to economic feasibility of the various infrastructure costs in today's times when budget is a major constraint to maintain and operate a highway system smoothly and efficiently. A major move toward pricing for highway facility damages were the highway cost allocation studies conducted in countries like Australia, Canada, and the U.S. (FHWA, 1998a). Small et al. (1989) developed a framework to estimate the user marginal cost associated with road pricing by considering pavement wear and durability, and congestion, and discussed its impacts on highway financing, investment policy, and decision making.

The road pricing principle was implemented to address congestion problems in Singapore in 1976 by using electronic devices to charge the users (FHWA, 2010). This was followed in Norway by implementing a cordon-based user charging system in the central business district (CBD) of Oslo in 1990. The system in Norway was to tax the users to generate revenues for roadway maintenance and was not aimed at congestion mitigation. The concept of road pricing conceptualized in the United Kingdom eventually got implemented in Durham and London as congestion mitigation strategies. However, there were cases where the implementation of road pricing was discontinued. An example is Hong Kong's electronic road pricing system, which had a successful run from 1983 to 1985, but was abandoned owing to public opposition (Litman, 2003, 2007, 2011, 2013).

More recently, the concept of road pricing was extended by taking negative externalities caused by transportation users/nonusers (or freight) into consideration. For example, Austria taxes heavy vehicles on its roadways, as they cause excessive pavement damages from higher axle loads. The heavy vehicle pricing system was implemented in 2004, dividing all types of vehicles into two groups. Those vehicles having axle loads higher than 3.5 tons are taxed and the others are exempt. Finland is currently in the process of implementing a road tax that reflects holistic taxation based on the concepts of road pricing, and is in the process of trial runs to address the issues of social equity, privacy, and user acceptance. Many other countries such as Australia, Germany, China, Italy, India, Sweden, United Arab Emirates, and the U.S. have used this strategy of taxation as a means of recovering facility damages and mitigating traffic congestion. Expanded the system of user charges to cover a broader range of external costs of transportation, such as safety and environmental costs, has also been proposed. The justification is that the users pay for their own marginal social costs (Maddison et al., 1996; Ramjerdi, 1996; Button, 2004; Ison, 2004).

Based on recent studies, road pricing was found to reduce cross-subsidies among users, increase the horizontal equity, and potentially solve the funding gap of total highway costs (FHWA, 1998c). Nevertheless, user acceptance is a major barrier. Charging vulnerable groups such as lower-income, elderly, and disabled users is a concern, but it depends on the percentage of lower-income users on the highway, the ride quality, and revenue utilization (Giuliano, 1994; Rajé, 2003; Parry, 2008; Burris et al., 2013). Kain (1994) indicated that road pricing could reduce the subsidies that users pay and increase their travel choices as well. Research studies also revealed that the issue of equity could be addressed by using a portion of the revenue generated from the road pricing to mitigate negative impacts created by the system (Levine and Garb, 2000; King et al., 2007). The resistance from all the other factions, such as the freight industry and government, are due to political motivations, distrust due to past mistakes, and inefficient usage of revenues (Samuel, 2000; Vassallo and Sánchez-Soliño, 2007; Bain, 2009).

19.2 MOTIVATION

Operating, preserving, and expanding the transportation system in an efficient, effective, and equitable manner to provide the highest level of user service is the utmost concern of transportation agencies today. In recent decades, technology and intelligence transfers among other engineering fields have made it possible to reduce transportation costs. However, the ever-growing travel demand, coupled with a much slower pace of growth in the budget caused by economic recession and revenue decrease over time, puts transportation agencies under enormous pressure to effectively restore the conditions of deteriorated transportation facilities, slow down the rate of performance degradation in system operations, and build the much-needed new capacity under the extremely tight budget.

The traditional sources of transportation funding are mainly based on fuel tax, vehicle registration fees, motor carrier registration fees and usage taxes, bonds, tolls, and parking charges. In the long run, fuel tax as the predominant portion of revenue sources is expected to decrease along with the expanded use of advanced technology vehicles to the transportation system. The funding shrinkage in the future further amplifies pressure facing the transportation agencies for facility preservation, system usage performance maintenance, and capacity expansion. Meanwhile, alternative funding channels have been sought to potentially reduce the funding gap. Particularly, weight-distance taxes, congestion pricing, and parking charge schemes have been adopted to recover costs of damages to highway facilities such as pavements and bridges by heavy vehicles, mitigate traffic congestion on major highways, and reduce traffic congestion (and air emissions) in densely populated urban areas.

Transportation management goals are multiple in nature. Apart from addressing goals of facility preservation and traffic mobility improvements, additional goals of agency and user costs, safety, and environmental impacts, as well as quality of life and the economic competitiveness that comes with it need to be addressed in a holistic manner. Thus, the current pricing schemes could potentially reduce the rate of deterioration in facility conditions and traffic congestion intensities, but they will still lead to a deteriorating trend of facility conditions and system usage performance. Also, the current funding sources and pricing schemes lack rigorous handling of impacts of new technologies on physical facilities, vehicles, and users/nonusers (or freight) of a transportation system. On the other hand, funding resource allocation decisions fail to embrace data-driven approaches that heavily rely on fine-grained data associated with facilities and system usage to capture the time-dependent or dynamic (i.e., time and space dependent) impacts of vehicles (and users/nonusers or freight) on facilities in time and space domains. Consequently, a truly performance-based management of the transportation system that connects facility supply, system usage, and impacts assessment with mitigation strategies in sustaining overall system performance at the best level cannot be achieved. Hence, pricing based financing schemes simultaneously addressing all system management goals that extend beyond the current road pricing practices need to be developed. In parallel, innovative investment decisions need to be made to support the best possible performance level for the entire transportation system using pricing-based revenue generated as the available budget.

Without loss of generality, the pricing scheme introduced hereinafter focuses on the highway transportation system that is typically predominant in a multimodal transportation system and also faces the largest funding gap. A similar pricing scheme can be adopted for other modes of transportation. The current funding scheme is set up to build and maintain a system that is inherently so inefficient that we can no longer afford to maintain it, and that when successful only generates the congestion that is the biggest barrier to economic growth and prosperity in cities. What if instead of looking for better ways to fund and manage our highways and parking lots, we were designing the kind of compact and convenient cities that people want to live in, the kind of cities that are inherently efficient to build and operate, where the attractiveness of congestion-immune transportation and walkable neighborhoods generates more than enough economic growth to fund the infrastructure projects that make it happen? What if instead of trying to manage a budget under hopelessly constrained circumstances we found a way to grow the budget? What if the established system of driving and parking wherever we go is so economically inefficient that we are leaving lots of money on the table—more than enough to build the kind of city we want to live in, the kind of city we can afford to manage in the long run, the kind of city that can compete for jobs and talent and prosperity with any city in the world?

19.3 PRICING BASED INNOVATIVE FINANCING FRAMEWORK

The concept behind pricing for transportation facility delivery and protection is to impose charges on various types of vehicles (or users/nonusers or freight) to recover costs of restoring functional and structural losses of physical facilities in their service life cycles. Meanwhile, the costs of vehicle operations, travel time, crashes, and environmental impacts incurred from system usage need to be included in the pricing scheme. To support sustainable transportation development, the total amount of user charges should be equivalent to the total

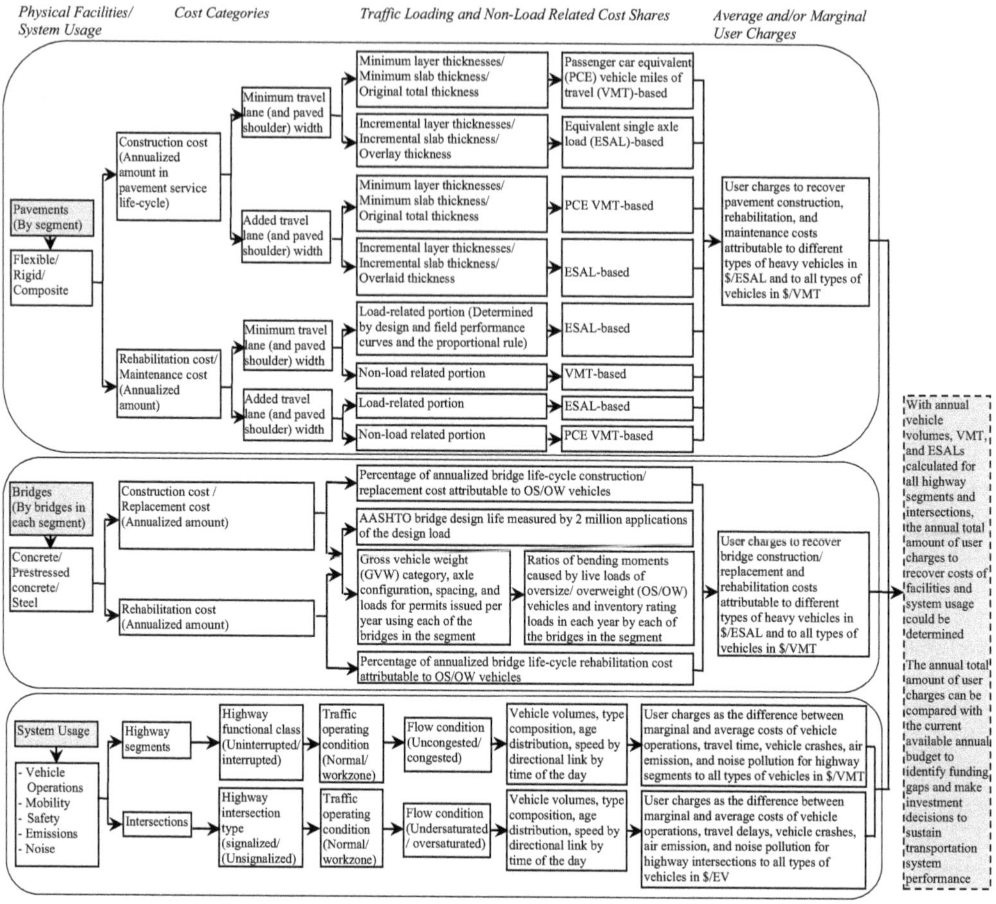

Figure 19.1 Framework of pricing-based innovative financing and resource allocation for transportation asset management.

cost to recover facility conditions and improve system usage performance in an efficient, effective, and equitable manner. Figure 19.1 illustrates the proposed pricing-based innovative financing framework and details follow in subsequent sections.

19.4 PRICING FOR SUSTAINMENT OF PHYSICAL FACILITY PERFORMANCE

19.4.1 Basic concept of facility-related user charges

The concept behind pricing for sustainment of transportation facility performance is to impose charges to various types of vehicles (or users/nonusers or freight) fairly to recover the cost of facility construction and subsequent costs of repair and maintenance treatments needed to restore deterioration of its functional and structural conditions in the service life cycle.

Physical facility performance is a manifestation of combined effects of cumulative traffic loading and non-load factors such as design standard, material type, construction quality, environmental condition, and climatic features. Taking the pavement facility as an example, according to the AASHTO pavement structure design procedure, pavement damages caused by the passage of each vehicle is measured by equivalent single axle load (ESAL) value or load equivalency factor (LEF), which takes into consideration the gross vehicle weight (GVW), axle configuration, spacing, and loading (AASHTO, 1998). Practically, a minimum thickness is required in pavement design to accommodate the non-load-related impacts and incremental thicknesses are considered to withstand excessive load-related damages. In addition, a wider travel lane (and paved shoulder) width is desirable to safely accommodate larger-sized vehicles (TRB, 1990; Lin et al., 1996; Huang, 2004).

With the total thickness provided from construction and timely implementation of maintenance and repair treatments in the subsequent service period, the facility design service life span could potentially be achieved. Thus, the cost of construction, repair, and maintenance could generally be split into load and non-load-related portions, which are correlated to cumulative damages measured by ESALs and cumulative usage in terms of total VMT. Therefore, the essence of user charges for sustaining physical facility performance is to establish marginal and/or average cost per ESAL of damages and per VMT of usage for facility construction, as well as repair and maintenance treatments in the service life cycle. The respective marginal and/or average cost per ESAL or per VMT could be adopted as user charge rates.

For a specific pavement segment or a bridge, the cumulative ESALs and VMTs could be estimated based on traffic volume, truck traffic volume and composition, segment length, and annual traffic and traffic loading growth rates for the given facility service life cycles and rehabilitation or replacement intervals. The cumulative ESALs can be multiplied by the user charge rates to obtain the load-related costs of construction, rehabilitation or replacement, and maintenance. Similarly, the total VMTs can be multiplied by non-load-related user charge rates to establish the non-load-related costs of construction, rehabilitation or replacement, and maintenance (Li et al., 2001, 2002a,b). The two portions of costs could be added together to arrive at the total amount of user charges to recover costs of damages caused by traffic loading and additional condition deterioration of the pavement segment or the bridge due to repetitive use. The same procedure could be applied to estimate the total amount of user charges to recover costs of all pavement segments and bridges in a highway network. The total amount can be converted to an equivalent uniform annualized amount to reflect the yearly amount of user charges that should be imposed to adequately manage pavement and bridge facilities.

19.4.2 User charges to recover pavement facility costs

19.4.2.1 Pavement construction cost

The pavement structural thickness in terms of layer thicknesses or structure number for a flexible or composite pavement and slab thickness for a rigid pavement is designed to accommodate cumulative damages caused by vehicles in the traffic stream for a given number of years of design life. As seen in Figure 19.2, the annualized pavement life-cycle construction

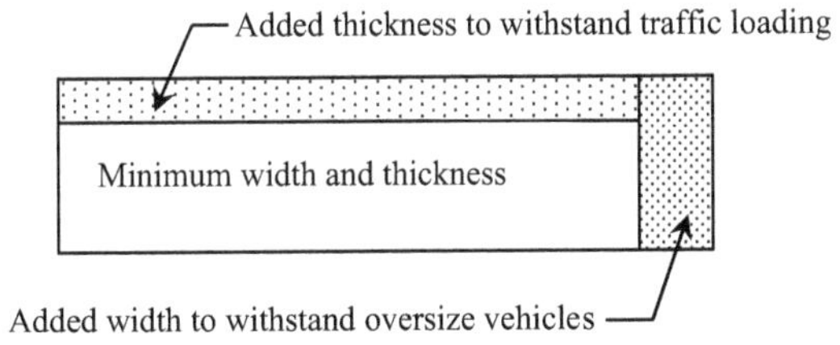

Figure 19.2 Illustration of different pavement construction cost components.

cost can be grouped into four categories, which are related to (i) minimum width and minimum thickness to accommodate all types of vehicles; (ii) added width and minimum thickness to serve for larger-sized vehicles; (iii) minimum width and incremental thickness in excess of the minimum thickness to handle extra damages caused by vehicles; and (iv) added width and incremental thickness to deal with larger sized and heavier weight vehicles. For the first type of construction costs, user charges should be applied to all vehicles based on miles of travel. The second type deals with added width to the base system to accommodate oversize vehicles, so the extra charges should be accountable to oversize vehicles, measured by cost unit per ESAL. Similarly, the extra pavement life-cycle cost of minimum width and incremental thickness should be charged to heavy vehicles with respect to their ESALs. Finally, the extra cost of the last type of construction cost should be assigned to both oversize and overweight vehicles participating in the network as unit user charge per ESALs. In this respect, the pavement construction cost for a highway segment is mainly affected by pavement material type, structural thickness, travel lane and paved shoulder width, and segment length (Winston et al., 1989; TRB, 1990; Lin et al., 1996; AASHTO, 1998; Huang, 2004).

In general, the total pavement construction cost can be calculated as

$$C_P^{const}(W,D) = C_0 + C_1 \cdot W + C_2 \cdot W \cdot D \tag{19.1}$$

where

$C_P^{const}(W,D)$	= Total pavement construction cost
C_0	= Capital construction cost
C_1	= Construction cost in terms of width with constant thickness
C_2	= Construction cost in terms of added thickness
W	= Pavement width, unit of 12 ft (or one lane)
D	= Pavement thickness, inches

Proportionality of load and non-load shares: The estimation of construction cost is mainly based on the loading that trucks and oversize/overweight vehicles apply to pavement surface. However, this cost does not reach 100% solely caused by traffic loading, but by other non-loading factors as well concerning pavement type, age, climatic features, subgrade material characteristics, drainage conditions, construction quality, and rehabilitation and maintenance treatments, which are deteriorated by the time of operations. The cost of the non-load portion could be estimated with respect to passenger car equivalency (PCE)–VMT of all vehicles.

The share of responsibility for load and non-load factors can be calculated using the linear proportional method proposed by Fwa and Sinha (1987), and can be illustrated as in Figure 19.3, where a = pure load-related responsibility portion (%); b = pure non-load-related responsibility portion (%); and c, d = shares of load and non-load-related in interaction, respectively.

Denote

$C_{P,ESAL}^{const}$	= Load-related portion of pavement construction cost
$C_{P,PCE-VMT}^{const}$	= Non-load-related portion of pavement construction cost
$AC_{P,ESAL}^{const}$	= Load-related average pavement construction cost per ESAL
$AC_{P,PCE-VMT}^{const}$	= Non-load-related average pavement construction cost per PCE–VMT
VMT_0	= Annual vehicle miles of travel in base year 0
$VMT(t)$	= Total vehicle miles of travel for the pavement in service for t years from the base year 0

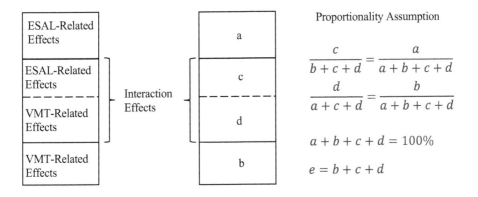

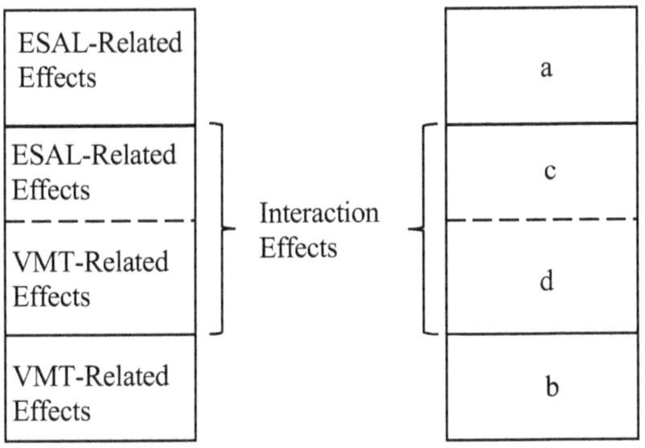

Proportionality Assumption

$$\frac{c}{b+c+d} = \frac{a}{a+b+c+d}$$

$$\frac{d}{a+c+d} = \frac{b}{a+b+c+d}$$

$$a+b+c+d = 100\%$$

$$e = b+c+d$$

Figure 19.3 Load and non-load shares of pavement construction cost. (Adapted from Fwa, T.F. and K.C. Sinha. 1985. A Performance-Based Approach for Determining Cost Responsibilities of Load-Related and Non-Load Related Factors in Highway Pavement Rehabilitation and Maintenance Cost Allocation. *TRB Journal for Transportation Research Record* 1028, 1–6.)

$VMT(N)$ = Total vehicle miles of travel in N-year pavement service life-cycle
VMT = Average annual VMT in N-year pavement service life-cycle
Q_0 = Annual ESAL value in base year 0
Q' = Small constant increment of annual ESAL value
$Q(t)$ = Cumulative ESAL value for the pavement in service for t years from the base year 0
$Q(N)$ = Cumulative ESAL value in N-year pavement service life-cycle
Q = Average annual ESAL value in N-year pavement service life-cycle
g_V = Annual traffic volume growth rate
g_L = Annual ESAL growth rate
N = Pavement design service life
i = Discount rate

The pavement construction cost to recover damages caused by heavy vehicles and to restore condition deterioration due to non-load factors are then estimated as

$$C_{P,ESAL}^{const} = (a+c) \cdot C_P^{const}(W,D) \tag{19.2}$$

$$C_{P,PCE-VMT}^{const} = (b+d) \cdot C_P^{const}(W,D) \tag{19.3}$$

Cumulative and annual travel with growth in N-year pavement service life-cycle: With the base year VMT_0 determined, we could consider a baseline annual growth rate of g_V that leads to a growth path of $VMT_0 \cdot e^{(g_V \cdot t)}$. Hence, the cumulative VMT in t years can be calculated as

$$\begin{aligned} VMT(t) &= \int_0^t [VMT_0 \cdot e^{(g_V \cdot \tau)}] d\tau \\ &= VMT_0 \cdot \left[\frac{e^{(g_V \cdot t)} - 1}{g_V} \right] \end{aligned} \tag{19.4}$$

Hence, the cumulative *VMT* in *N*-year pavement service life-cycle and average annual *VMT* become

$$VMT(N) = VMT_0 \cdot \left[\frac{e^{(g_V \cdot N)} - 1}{g_V}\right] \tag{19.5}$$

$$VMT = VMT_0 \cdot \left[\frac{e^{(g_V \cdot N)} - 1}{g_V \cdot N}\right] \tag{19.6}$$

Cumulative and annual traffic loading with growth in N-year pavement service life-cycle: For the calculation of annual Q, growth in annual traffic loading in the *N*-year pavement service life-cycle generally occurs due to increases in truck traffic volumes and axle loads. With the base year traffic loading Q_0 determined, we could consider a baseline annual growth rate of g_L that leads to a growth path of $Q_0 \cdot e^{(g_L \cdot t)}$ and a small constant increment Q' per year from that baseline path. Hence, the cumulative traffic loading in t years can be calculated as

$$Q(t) = \int_0^t \left[Q_0 \cdot e^{(g_L \cdot \tau)} + Q'\right] d\tau = Q_0 \cdot \left[\frac{e^{(g_L \cdot t)} - 1}{g_L}\right] + Q' \cdot t \tag{19.7}$$

Hence, the cumulative traffic loading in *N*-year pavement service life-cycle and average annual traffic loading become

$$Q(N) = Q_0 \cdot \left[\frac{e^{(g_L \cdot N)} - 1}{g_L}\right] + Q' \cdot N \tag{19.8}$$

$$Q = Q_0 \cdot \left[\frac{e^{(g_L \cdot N)} - 1}{g_L \cdot N}\right] + Q' \tag{19.9}$$

Average load and non-load-related costs of pavement construction: The average load-related cost of pavement construction for the passage of every ESAL and average non-load-related cost of pavement construction for the usage of every PCE–VMT are

$$AC_{P,ESAL}^{const} = \frac{i \cdot C_{P,ESAL}^{const}}{Q} = i \cdot (a + c) \cdot \frac{C_P^{const}(W, D)}{Q_0 \cdot \left[\frac{e^{(g_L \cdot N)} - 1}{g_L \cdot N}\right] + Q'} \tag{19.10}$$

$$AC_{P,PCE-VMT}^{const} = \frac{i \cdot C_{P,PCE-VMT}^{const}}{VMT} = i \cdot (b + d) \cdot \frac{C_P^{const}(W, D)}{\left[VMT_0 \cdot \left(\frac{e^{(g_V \cdot N)} - 1}{g_V \cdot N}\right)\right]} \tag{19.11}$$

EXAMPLE 19.1

Calculate the total cost of construction of a 5-mile, 2-lane flexible pavement, given that the cost of a construction portion to withstand trucks and overweight loading is $120000/lane-mile, and the percent of pure loading impacts to the pavement is 70%.

Solution

By applying the linear proportionality assumption, we have $a = 70\%$

$$\frac{c}{b+c+d} = \frac{a}{a+b+c+d} \Rightarrow \frac{c}{30\%} = \frac{70\%}{100\%} \Rightarrow c = (70\% \times 30\%))/100\% = 21\%$$

The total percentage of load-related damages to the pavement is $a + c = 70\% + 21\% = 91\%$

The total pavement construction cost is

$$C_P^{const}(W,D) = \frac{\$120000}{0.91} \times 2 \text{ lanes} \times 5 \text{ miles} = \$1318681$$

19.4.2.2 Pavement rehabilitation cost

Proportionality of load and non-load shares: Pavement performance in its service life cycle is the manifestation of combined effects of traffic loading and non-load factors. Conventionally, the pavement surface condition for a highway segment can be measured by the present serviceability index (PSI) or pavement condition index (PCI), which is established on the basis of roughness, cracking, and rutting measurements. The pavement performance trend can be assessed using PSI or PCI values against the cumulative ESALs in its service life cycle. The PSI or PCI values can be measured according to the pavement design curve and field performance curve, respectively.

As depicted in Figure 19.4, the concept of linear proportionality as present in the calculation of load and non-load shares of pavement construction costs can be adopted here. The area between the horizontal, zero-deterioration line and the design curve in terms of PSI-ESAL or PCI-ESAL losses represents the portion of pavement condition deterioration caused by pure traffic loading (portion *a*). The area between the design curve and field performance curve as PSI-ESAL or PCI-ESAL losses represents the portion of pavement condition deterioration

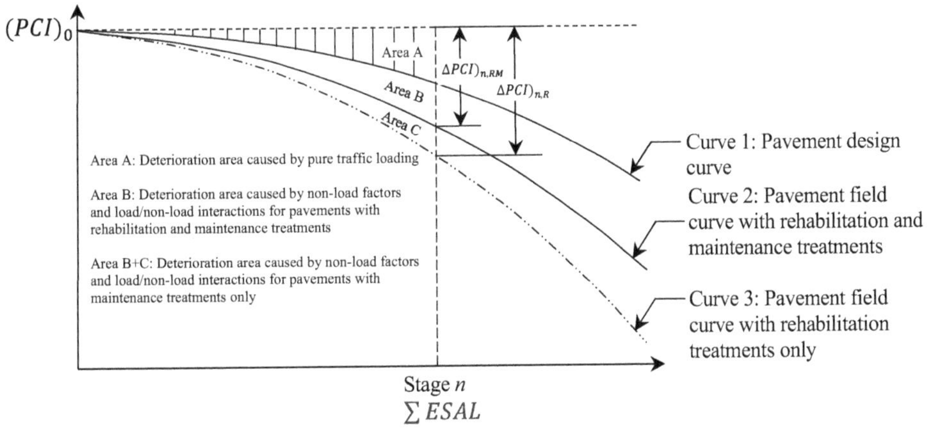

Figure 19.4 Pavement design and field performance curves. (Adapted from Fwa, T.F. and K.C. Sinha. 1985. A Performance-Based Approach for Determining Cost Responsibilities of Load-Related and Non-Load Related Factors in Highway Pavement Rehabilitation and Maintenance Cost Allocation. *TRB Journal for Transportation Research Record* 1028, 1–6.)

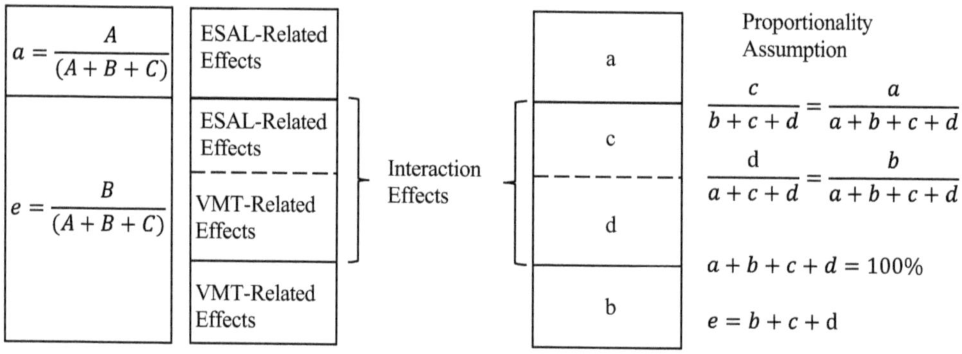

Figure 19.5 Load and non-load shares of pavement rehabilitation cost. (Adapted from Fwa, T.F. and K.C. Sinha. 1985. A Performance-Based Approach for Determining Cost Responsibilities of Load-Related and Non-Load Related Factors in Highway Pavement Rehabilitation and Maintenance Cost Allocation. *TRB Journal for Transportation Research Record* 1028, 1–6.)

caused by pure non-load factors (portion d), as well as the load and non-load interactions that can be further split into load-related interaction (portion c) and non-load-related interaction (portion d). Thus, the overall load portion of pavement condition deterioration is the sum of pure traffic loading and load-related interaction portions (portion a + portion c).

Based on the method described in the above, the design curve (Curve 1) and field performance curve involving rehabilitation treatments only (Curve 3) can be utilized to establish the load and non-load portions of pavement rehabilitation cost. Figure 19.5 illustrates the computation procedure.

Pavement condition deterioration over time and with traffic loading: According to Paterson (1987) and Newbery (1988), pavement condition deteriorates exponentially over time and linearly with cumulative ESALs. This can be expressed in the following:

$$\frac{\pi_0 - \pi_t}{\pi_0 - \pi_T} = \left(\frac{T}{T_0}\right) \cdot e^{mt} \quad (19.12)$$

$$\pi_t = [\pi_0 + Q(t)/Q(T)] \cdot e^{mt} \quad (19.13)$$

with

$$T_0 = \frac{Q(T)}{(\lambda \cdot Q)}$$

where

π_0 = Initial pavement condition, typically using PCI or international roughness index (IRI)
π_T = Pavement condition that requires rehabilitation treatment after T years of service from the base year 0
π_t = Pavement condition in year t
Q_0 = Annual ESAL value in base year 0
Q = Average annual ESALs over the rehabilitation interval

$Q(t)$ = Cumulative ESAL value for the pavement in service for t years from the base year 0
$Q(T)$ = Cumulative ESAL value in pavement rehabilitation interval
T_0 = Pavement rehabilitation interval without accounting for aging effects
T = Pavement rehabilitation interval considering aging effects
λ = The fraction of one-directional axel passages traveled
m = Pavement aging coefficient
t = Year t

Pavement rehabilitation interval considering aging effects: From Equation 19.8, $\pi_t = \pi_T$ when the pavement is in service for T-year rehabilitation interval, the pavement condition deterioration ratio reaches unity. This reduces the equation to the following:

$$T = T_0 \cdot e^{-mT} = \left[\frac{Q(T)}{(\lambda \cdot Q)}\right] \cdot e^{-mT} \tag{19.14}$$

Cumulative and annual travel and traffic loading with growth in T-year rehabilitation interval: Similar to the way in which growth in travel and traffic loading are handled in N-year pavement service life cycle, such growth also needs to be considered for the T-year rehabilitation interval. Using the base year travel VMT_0, annual traffic volume growth rate g_V, traffic loading Q_0, annual traffic volume growth rate g_L that leads to a growth path of $Q_0 \cdot e^{(g_L \cdot t)}$, as well as a small constant increment Q' per year from that baseline path, the cumulative and annual travel and traffic loading with growth can be computed as

$$VMT(T) = VMT_0 \cdot \left[\frac{e^{(g_V \cdot T)} - 1}{g_V}\right] \tag{19.15}$$

With the recurrence of n rounds of $VMT(T)$ that maintains the same annual growth rate g_V, the total travel in $(n \cdot T)$ years can be computed as follows:

$$\begin{aligned} VMT(n \cdot T) &= VMT_0 \cdot \left[\frac{e^{(g_V \cdot T)} - 1}{g_V}\right] \cdot \left[1 + e^{(g_V \cdot T)} + e^{(2 \cdot g_V \cdot T)} + \cdots + e^{((n-1) \cdot g_V \cdot T)}\right] \\ &= VMT_0 \cdot \left[\frac{e^{(g_V \cdot T)} - 1}{g_V}\right] \cdot \left[\frac{e^{(n \cdot g_V \cdot T)} - 1}{e^{(g_V \cdot T)} - 1}\right] \end{aligned} \tag{19.16}$$

$$VMT = VMT_0 \cdot \left[\frac{e^{(g_V \cdot T)} - 1}{g_V \cdot T}\right] \cdot \left[\frac{e^{(n \cdot g_V \cdot T)} - 1}{e^{(g_V \cdot T)} - 1}\right] \cdot \left[\frac{1}{n \cdot T}\right] \tag{19.17}$$

$$Q(T) = Q_0 \cdot \left[\frac{e^{(g_L \cdot T)} - 1}{g_L}\right] + Q' \cdot T \tag{19.18}$$

$$Q = Q_0 \cdot \left[\frac{e^{(g_L \cdot T)} - 1}{g_L \cdot T}\right] + Q' \tag{19.19}$$

Average and marginal load and non-load-related costs of pavement rehabilitation: The average and marginal load and non-load-related costs of pavement rehabilitation are calculated using the following procedure.

Denote

C_P^{rehab} = Pavement rehabilitation cost, in dollars
$PW_P^{rehab}(Q,W,D)$ = Present worth of pavement rehabilitation cost
$PW_{P,ESAL}^{rehab}$ = Load portion of present worth of pavement rehabilitation cost
$PW_{P,PCE-VMT}^{rehab}$ = Non-load portion of present worth of pavement rehabilitation cost
$C_{P,ESAL}^{rehab}$ = Load portion of annualized pavement rehabilitation cost
$C_{P,PCE-VMT}^{rehab}$ = Non-load portion of annualized pavement rehabilitation cost
$MC_{P,ESAL}^{rehab}$ = Load-related marginal pavement rehabilitation cost per ESAL
$AC_{P,PCE-VMT}^{rehab}$ = Non-load-related average pavement rehabilitation cost per PCE–VMT
T = Pavement rehabilitation interval, in years
VMT = Average annual VMT in the rehabilitation interval
Q = Average annual ESAL value in the rehabilitation interval
i = Discount rate

Since rehabilitation interval T is shorter than N-year pavement service life-cycle, multiple rounds of rehabilitation treatments are typically made in the pavement service life-cycle, with the first rehabilitation applied in year T from the base year 0. The present worth of total pavement rehabilitation cost during its service life cycle can then be estimated by converting the rehabilitation cost C_P^{rehab} every period of T years during the N-year pavement service life cycle to present value with a discount rate of i

$$PW_P^{rehab}(Q,W,D) = \frac{C_P^{rehab}}{(e^{i \cdot T} - 1)} \tag{19.20}$$

The present worth values of total pavement rehabilitation cost attributable to traffic loading and non-load factors are given as

$$PW_{P,ESAL}^{rehab} = (a+c) \cdot PW_P^{rehab} = (a+c) \cdot \frac{C_P^{rehab}}{(e^{i \cdot T} - 1)} \tag{19.21}$$

$$PW_{P,PCE-VMT}^{rehab} = (b+d) \cdot PW_P^{rehab} = (b+d) \cdot \frac{C_P^{rehab}}{(e^{i \cdot T} - 1)} \tag{19.22}$$

The annualized pavement rehabilitation cost attributable to traffic loading and non-load factors are

$$C_{P,ESAL}^{rehab} = i \cdot PW_{ESAL}^{rehab} = i \cdot (a+c) \cdot \frac{C_P^{rehab}}{(e^{i \cdot T} - 1)} \tag{19.23}$$

$$C_{P,PCE-VMT}^{rehab} = i \cdot PW_{P,PCE-VMT}^{rehab} = i \cdot (b+d) \cdot \frac{C_P^{rehab}}{(e^{r \cdot T} - 1)} \tag{19.24}$$

Marginal and average load and non-load-related costs of pavement rehabilitation: The load-related marginal pavement rehabilitation cost for the passage of every ESAL and non-load-related marginal pavement rehabilitation cost for the usage of every PCE–VMT are calculated as

$$MC_{P,ESAL}^{rehab} = \frac{\partial(i \cdot PW_{P,ESAL}^{rehab})}{\partial Q} = i \cdot \frac{\partial\left(PW_{P,ESAL}^{rehab}\right)}{\partial T} \cdot \left(\frac{dT}{dQ'}\right)_{Q'=0}$$

$$= i \cdot (a+c) \cdot \left[-\frac{\left(i^2 \cdot e^{(i \cdot T)} \cdot C_P^{rehab}\right)}{(e^{(i \cdot T)} - 1)^2}\right] \cdot \left[-\frac{(\lambda \cdot T^2)}{Q(T)}\right]$$

$$\cdot \left[\frac{e^{(m \cdot T)}}{(m \cdot T + (Q_0 \cdot e^{(g_L \cdot T)})/((Q_0 \cdot (e^{(g_L \cdot T)} - 1))/(g_L \cdot T)))}\right]$$

$$= (a+c) \cdot \left[\frac{((i \cdot T)^2 \cdot e^{(i \cdot T)})}{(e^{(i \cdot T)} - 1)^2}\right] \cdot \left[\frac{C_P^{rehab}}{(Q(T)/\lambda)}\right] \cdot \left[\frac{e^{(m \cdot T)}}{(m \cdot T + (g_L \cdot T \cdot e^{(g_L \cdot T)})/(e^{(g_L \cdot T)} - 1))}\right] \tag{19.25}$$

$$AC_{P,PCE-VMT}^{rehab} = \frac{i \cdot PW_{P,PCE-VMT}^{rehab}}{VMT} = \frac{i \cdot (b+d)}{(e^{i \cdot T} - 1)} \cdot \frac{C_P^{rehab}}{\left[VMT_0 \cdot \left[\frac{e^{(g_V \cdot T)} - 1}{g_V \cdot T}\right] \cdot \left[\frac{e^{(n \cdot g_V \cdot T)} - 1}{e^{(g_V \cdot T)} - 1}\right] \cdot \left[\frac{1}{n \cdot T}\right]\right]} \tag{19.26}$$

As an example, Table 19.1 summarizes load shares of pavement 3R (resurfacing, rehabilitation, and restoration) costs established by highway functional classification and pavement type from the 1997 Federal Highway Cost Allocation Study (USDOT, 1998).

19.4.2.3 Annual pavement routine maintenance cost

Proportionality of load and non-load shares: As seen in Figure 19.6, the design curve (Curve 1) and field performance curve involving rehabilitation and maintenance treatments (Curve 2) can be utilized to establish the load and non-load portions of pavement rehabilitation and maintenance costs based on the proportionality method described in the previous section.

Table 19.1 Load shares of pavement repair costs

Highway functional classification		*Flexible pavement (%)*	*Rigid pavement (%)*
Rural	Interstate	11.0	9.3
	Other principal arterials	12.1	15.7
	Minor arterials	12.2	13.7
	Major collectors	14.7	14.5
	Minor collectors	14.7	14.5
	Local	14.7	15.4
Urban	Interstate	10.1	7.9
	Other freeways/expressways	10.6	11.0
	Major arterials	11.5	12.8
	Minor arterials	12.7	16.3
	Collectors	13.9	20.5
	Local	13.9	20.5

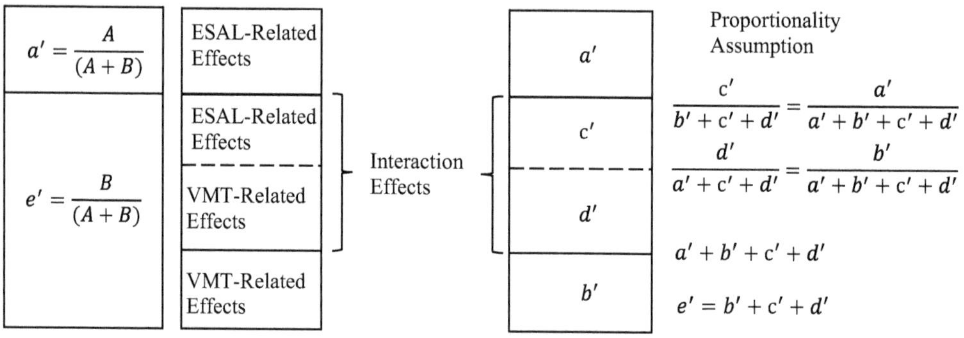

Figure 19.6 Load and non-load shares of pavement rehabilitation and maintenance costs. (Adapted from Fwa, T.F. and K.C. Sinha. 1985. A Performance-Based Approach for Determining Cost Responsibilities of Load-Related and Non-Load Related Factors in Highway Pavement Rehabilitation and Maintenance Cost Allocation. *TRB Journal for Transportation Research Record* 1028, 1–6.)

Using the proportionality rule, $(a + c)$ and $(b + d)$ are treated as the load and non-load shares of rehabilitation cost. Similarly, $(a' + c')$ and $(b' + d')$ are treated as the load and non-load shares of rehabilitation and maintenance costs. With both the rehabilitation cost and maintenance cost calculated, the load and non-load shares of maintenance cost can be derived.

Average load and non-load-related costs of pavement maintenance: The following notations are used for the calculation.

Denote

C_P^{maint} = Annual pavement routine maintenance cost, in dollars

$PW_P^{maint}(Q,W,D)$ = Present worth of annual pavement routine maintenance cost

$PW_{P,ESAL}^{maint}$ = Load portion of present worth of annual pavement routine maintenance cost

$PW_{P,PCE-VMT}^{maint}$ = Non-load portion of present worth of annual pavement routine maintenance cost

$C_{P,ESAL}^{maint}$ = Load portion of annual pavement routine maintenance cost

$C_{P,PCE-VMT}^{maint}$ = Non-load portion of annual pavement routine maintenance cost

$AC_{P,ESAL}^{maint}$ = Load-related average pavement routine maintenance cost per ESAL

$AC_{P,PCE-VMT}^{maint}$ = Non-load-related average pavement routine maintenance cost per VMT

VMT = Average annual VMT in the pavement service life-cycle

Q = Average annual ESAL value in the pavement service life-cycle

i = Discount rate

Similar to pavement rehabilitation, the present worth of annual pavement routine maintenance cost during its service life cycle can be estimated by converting the annual routine maintenance cost C^{maint} during the service life cycle to present value with a discount rate of r as follows:

$$PW_P^{maint}(Q,W,D) = \frac{C_P^{maint}}{(e^i - 1)} \quad (19.27)$$

The present worth values of total routine maintenance cost attributable to traffic loading and non-load factors are given as

$$
\begin{aligned}
PW_{P,ELSA}^{maint} &= (a' + c')\left[PW_P^{rehab} + PW_P^{maint}\right] - (a + c)PW_P^{rehab} \\
&= \left[(a' - a) + (c' - c)\right]\frac{C_P^{rehab}}{(e^{i \cdot T} - 1)} + (a' + c')\frac{C_P^{maint}}{(e^{i} - 1)}
\end{aligned}
\tag{19.28}
$$

$$
\begin{aligned}
PW_{P,PCE-VMT}^{maint} &= (b' + d')\left[PW_P^{rehab} + PW_P^{maint}\right] - (b + d)PW_P^{rehab} \\
&= \left[(b' - b) + (d' - d)\right]\frac{C_P^{rehab}}{(e^{i \cdot T} - 1)} + (b' + d')\frac{C_P^{maint}}{(e^{i} - 1)}
\end{aligned}
\tag{19.29}
$$

The annual pavement routine maintenance cost attributable to traffic loading and non-load factors are

$$
C_{P,ESAL}^{maint} = i \cdot PW_{ESAL}^{maint} = \left[(a' - a) + (c' - c)\right] \cdot \frac{i \cdot C_P^{rehab}}{(e^{i \cdot T} - 1)} + (a' + c') \cdot \frac{i \cdot C_P^{maint}}{(e^{i} - 1)} \tag{19.30}
$$

$$
C_{P,PCE-VMT}^{maint} = i \cdot PW_{PCE-VMT}^{maint} = \left[(b' - b) + (d' - d)\right]\frac{i \cdot C_P^{rehab}}{(e^{i \cdot T} - 1)} + (b' + d')\frac{i \cdot C_P^{maint}}{(e^{i} - 1)} \tag{19.31}
$$

The load-related average cost of pavement annual routine maintenance for the passage of every ESAL and non-load-related average cost of pavement annual routine maintenance for the usage of every PCE–VMT are given by

$$
\begin{aligned}
AC_{P,ESAL}^{maint} &= \frac{i \cdot PW_{P,ESAL}^{maint}}{Q} \\
&= \frac{i \cdot [(a' - a) + (c' - c)]}{(e^{i \cdot T} - 1)} \cdot \frac{C_P^{rehab}}{Q_0 \cdot \left[\dfrac{e^{(g_L \cdot N)} - 1}{g_L \cdot N}\right] + Q'} + \frac{i \cdot (a' + c')}{(e^{i} - 1)} \cdot \frac{C_P^{maint}}{Q_0 \cdot \left[\dfrac{e^{(g_L \cdot N)} - 1}{g_L \cdot N}\right] + Q'}
\end{aligned}
\tag{19.32}
$$

$$
\begin{aligned}
AC_{P,PCE-VMT}^{maint} &= \frac{i \cdot PW_{P,PCE-VMT}^{maint}}{VMT} \\
&= \frac{i \cdot [(b' - b) + (d' - d)]}{(e^{i \cdot T} - 1)} \cdot \frac{C_P^{rehab}}{\left[VMT_0 \cdot \left(\dfrac{e^{(g_V \cdot N)} - 1}{g_V \cdot N}\right)\right]} + \frac{i \cdot (b' + d')}{(e^{i} - 1)} \cdot \frac{C_P^{maint}}{\left[VMT_0 \cdot \left(\dfrac{e^{(g_V \cdot N)} - 1}{g_V \cdot N}\right)\right]}
\end{aligned}
\tag{19.33}
$$

19.4.3 User charges to recover bridge facility costs

19.4.3.1 Bridge construction cost

The construction cost of a bridge is determined by various factors such as dimensions (length, height, width), structure types (beam, cable-stayed, suspensions, etc.), number and length of spans, material types (concrete, steel), and the location. In general, the bridge construction cost can be presented as

$$C_B^{const} = \sum_j c_j N_j \tag{19.34}$$

where
C_B^{const} = Total bridge construction cost
c_j = Unit cost of each cost element j
N_j = Quantity of the j element

Similar to pavement construction cost, bridge construction cost is affected by both load and non-load-related factors. However, the effects of these factors to the bridge structure are different from those to the pavement. The linear proportionality method for determining the load and non-load shares of pavement construction cost can be employed to establish the load and non-load shares of bridge construction cost.

Denote
e = Pure load-related portion of bridge damages
f = Pure non-load-related portion of bridge damages
g, h = Load and non-load-related shares in the interaction portion, respectively
$C_{B,ESAL}^{const}$ = Load-related portion of bridge construction cost
$C_{B,PCE-VMT}^{const}$ = Non-load-related portion of bridge construction cost
$AC_{B,ESAL}^{const}$ = Load-related average bridge construction cost per ESAL
$AC_{B,PCE-VMT}^{const}$ = Non-load-related average bridge construction cost per PCE–VMT
VMT_0 = Annual vehicle miles of travel in base year 0
VMT = Average annual VMT in N_B-year bridge service life-cycle
Q_0 = Annual ESAL value in base year 0
Q' = Small constant increment of annual ESAL value
Q = Average annual ESAL value in N_B-year bridge service life-cycle
g_V = Annual traffic volume growth rate
g_L = Annual ESAL growth rate
N_B = Bridge service life-cycle
i = Discount rate

The bridge construction cost to withstand damage caused by traffic loading and restore deterioration resulting from non-load factors can be estimated as

$$C_{B,ESAL}^{const} = (e+g)\cdot C_B^{const} \tag{19.35}$$

$$C_{B,PCE-VMT}^{const} = (f+h)\cdot C_B^{const} \tag{19.36}$$

Average load and non-load-related costs of bridge construction: The load-related average cost of bridge construction for the passage of every ESAL and non-load-related average cost of bridge construction for the usage of every passenger car equivalent VMT can be computed by

$$AC_{B,ESAL}^{const} = \frac{i\cdot C_{B,ESAL}^{const}}{Q} = i\cdot(e+g)\cdot\frac{C_B^{const}}{Q_0\cdot\left[\dfrac{e^{(g_L\cdot N_B)}-1}{g_L\cdot N_B}\right]+Q'} \tag{19.37}$$

$$AC_{B,PCE-VMT}^{const} = \frac{i \cdot C_{B,PCE-VMT}^{const}}{VMT} = r \cdot (f+h) \cdot \frac{C_B^{const}}{\left[VMT_0 \cdot \left(\frac{e^{(g_V \cdot N_B)} - 1}{g_V \cdot N_B}\right)\right]} \tag{19.38}$$

19.4.3.2 Bridge component repair and routine maintenance costs

Marginal and average costs of bridge component repair and maintenance: Given that rehabilitation or replacement of a bridge deck, superstructure or substructure component in the amount of $C_B^{reh/rep}$ is made in every T years of the bridge service life cycle and routine maintenance is implemented annually in the amount of C_B^{maint}, the load-related marginal cost and non-load-related average cost of bridge component rehabilitation or replacement, as well as load-related and non-load-related average costs of annual routine maintenance can be determined using the following procedure.

Denote

$C_B^{reh/rep}$	= Bridge component rehabilitation or replacement cost, in dollars
C_B^{maint}	= Annual bridge routine maintenance cost, in dollars
$PW_B^{reh/rep}(Q,W,D)$	= Present worth of bridge component rehabilitation or replacement cost
$PW_B^{maint}(Q,W,D)$	= Present worth of annual bridge routine maintenance cost
$PW_{B,ESAL}^{reh/rep}$	= Load portion of present worth of bridge component rehabilitation or replacement cost
$PW_{B,PCE-VMT}^{reh/rep}$	= Non-load portion of present worth of bridge component rehabilitation or replacement cost
$PW_{B,ESAL}^{maint}$	= Load portion of present worth of annual bridge routine maintenance cost
$PW_{B,PCE-VMT}^{maint}$	= Non-load portion of present worth of annual bridge routine maintenance cost
$C_{B,ESAL}^{reh/rep}$	= Load portion of annualized bridge component rehabilitation or replacement cost
$C_{B,PCE-VMT}^{reh/rep}$	= Non-load portion of annualized bridge component rehabilitation or replacement cost
$C_{B,ESAL}^{maint}$	= Load portion of annual bridge routine maintenance cost
$C_{B,PCE-VMT}^{maint}$	= Non-load portion of annual bridge routine maintenance cost
$MC_{B,ESAL}^{reh/rep}$	= Load-related marginal bridge component rehabilitation or replacement cost per $ESAL$
$AC_{B,PCE-VMT}^{reh/rep}$	= Non-load-related average bridge component rehabilitation or replacement cost per PCE–VMT
$AC_{B,ESAL}^{maint}$	= Load-related average bridge routine maintenance cost per $ESAL$
$AC_{B,PCE-VMT}^{maint}$	= Non-load-related average bridge routine maintenance cost per VMT
T	= Bridge component rehabilitation or replacement interval, in years
VMT	= Average annual VMT in the bridge component rehabilitation or replacement interval
Q	= Average annual ESAL value in the bridge component rehabilitation or replacement interval
N_B	= Bridge service life-cycle
I	= Discount rate

$$MC_{B,ESAL}^{reh/rep} = \frac{\partial\left(i \cdot PW_{B,ESAL}^{reh/rep}\right)}{\partial Q} = i \cdot \frac{\partial\left(PW_{B,ESAL}^{reh/rep}\right)}{\partial T} \cdot \left(\frac{dT}{dQ'}\right)_{Q'=0}$$

$$= i \cdot (e+g) \cdot \left[-\frac{\left(i^2 \cdot e^{(i \cdot T)} \cdot C_B^{reh/rep}\right)}{(e^{(i \cdot T)}-1)^2}\right] \cdot \left[-\frac{(\lambda \cdot T^2)}{Q(T)}\right]$$

$$\cdot \left[\frac{e^{(m \cdot T)}}{(m \cdot T + (Q_0 \cdot e^{(g_L \cdot T)})/((Q_0 \cdot (e^{(g_L \cdot T)}-1))/(g_L \cdot T)))}\right]$$

$$= (e+g) \cdot \left[\frac{((i \cdot T)^2 \cdot e^{(i \cdot T)})}{(e^{(i \cdot T)}-1)^2}\right] \cdot \left[\frac{C_B^{reh/rep}}{(Q(T)/\lambda)}\right] \cdot \left[\frac{e^{(m \cdot T)}}{(m \cdot T + (g_L \cdot T \cdot e^{(g_L \cdot T)})/(e^{(g_L \cdot T)}-1))}\right] \tag{19.39}$$

$$AC_{B,PCE-VMT}^{reh/rep} = \frac{i \cdot PW_{B,PCE-VMT}^{reh/rep}}{VMT}$$

$$= \frac{i \cdot (f+h)}{(e^{i \cdot T}-1)} \cdot \frac{C_B^{reh/rep}}{\left[VMT_0 \cdot \left[\frac{e^{(g_V \cdot T)}-1}{g_V \cdot T}\right] \cdot \left[\frac{e^{(n \cdot g_V \cdot T)}-1}{e^{(g_V \cdot T)}-1}\right] \cdot \left[\frac{1}{n \cdot T}\right]\right]} \tag{19.40}$$

$$AC_{B,ESAL}^{maint} = \frac{i \cdot PW_{B,ESAL}^{maint}}{Q} = \frac{i \cdot (e+g)}{(e^i-1)} \cdot \frac{C_B^{maint}}{Q_0 \cdot \left[\frac{e^{(g_L \cdot N_B)}-1}{g_L \cdot N_B}\right] + Q'} \tag{19.41}$$

$$AC_{B,PCE-VMT}^{maint} = \frac{i \cdot PW_{B,PCE-VMT}^{maint}}{VMT} = \frac{i \cdot (f+h)}{(e^i-1)} \cdot \frac{C_B^{maint}}{\left[VMT_0 \cdot \left(\frac{e^{(g_V \cdot N_B)}-1}{g_V \cdot N_B}\right)\right]} \tag{19.42}$$

19.5 PRICING FOR SUSTAINMENT OF SYSTEM USAGE PERFORMANCE

19.5.1 Basic concept of system usage-related user charges

For estimating individual system usage cost components concerning vehicle operations, travel time, crashes, air emissions, and noise pollution, a highway network is partitioned into segments and intersections. Highway segments can be classified by functional classification and intersections may be signalized or unsignalized. Traffic operations may be under normal conditions or in the presence of work zones with added system usage costs. The flow condition may be uncongested or congested for highway segments and undersaturated or oversaturated for intersections.

Highway vehicles can be grouped into three categories: (i) passenger cars; (ii) trucks; and (iii) oversized/overweight (OS/OW) vehicles. Each category can have subcategories based on axle load and configuration. One possible way to determine the vehicle impacts on user costs is to employ the concept of PCE for each category of vehicles in determining its unit user cost for different user cost components per vehicle mile of travel on a highway segment or per vehicle entering an intersection. In the 2010 Highway Capacity Manual, the PCE

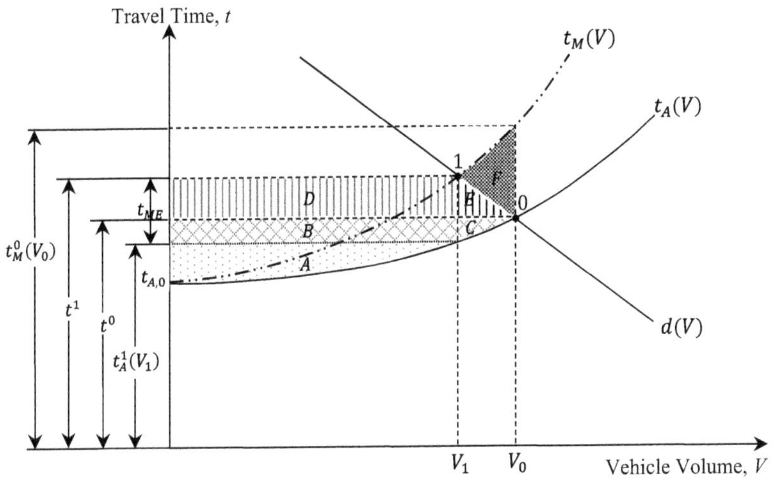

Figure 19.7 Short-run average, marginal social, and marginal external travel time.

factor is defined as the number of passenger cars displayed by a single heavy vehicle of a particular type under specified roadway, traffic, and control conditions (TRB, 2010).

Given the capacity of a highway segment, short-run vehicle operating speed could remain constant up to a certain vehicle volume level and decreases nonlinearly at a relatively low rate prior to reaching capacity. However, when the vehicle volume exceeds capacity in the presence of traffic congestion, the vehicle operating speed decreases drastically. As seen in Figure 19.7, the short-run marginal travel time exceeds short-run average travel time along with traffic volume increases. This is because the marginal social travel time t_M not only includes the average travel time incurred by the traveler, t_A, but also the additional travel time that the traveler causes to all other travelers in the traffic stream, which is called marginal external travel time, t_{ME}. The efficient utilization of highway capacity is achieved when the benefit of each trip equals to marginal social travel time t_M represented by the demand and supply equilibrium point "1" based on the marginal travel time function. If the equilibrium point is at point "0" based on the average travel time function, user charge at a rate of $\alpha \cdot t_{ME}$ (α is the value of travel time) needs to be imposed to shift the equilibrium from point "0" to point "1."

For the unpriced equilibrium point "0," the average travel time experienced by V_0 travelers is t^0. After shifting the equilibrium to point "1" by imposing the user charge, the average travel time experienced by V_1 travelers is reduced to $t_A^1(V_1)$. The generalized travel time for unpriced and priced equilibria are t^0 and $t^1 = t_A^1(V_1) + t_{ME}$, respectively. The quantity $(t^0 - t_{A,0}) \cdot V_0$ highlighted by shaded areas $A + B + C$ can be interpreted as the extra total travel time experienced by travelers in the unpriced equilibrium compared with the travel using the base free flow speed. Specifically, area A represents efficient extra travel time and areas $B + C$ reflect inefficient extra travel time, which is eliminated by imposing the user charge. Further, the gain in social surplus is depicted by shaded areas $C + E + F$. The change in consumer surplus as user benefits is represented by shaded areas $D + E$, which can be computed by $(t^1 - t^0 \cdot [(V_0 + V_1)]/2)$.

With user charge imposed to reduce inefficient travel time for sustainment of traffic mobility, the optimal flow is obtained as V_1 at the new equilibrium compared with the flow of V_0 at unpriced equilibrium. The average vehicle operating speed for travelers using the highway segment will increase from L/t^0 to $L/t_A^1(V_1)$ correspondingly.

Along with the increase in the efficiency of traffic mobility, this will lead to changes in the performance of vehicle operations, crashes, air emissions, and noise pollution. Likewise, user charges to these costs need to be applied to improve their efficiency levels as well. The volumes and average operating speeds corresponding to the unpriced and priced equilibria derived from pricing based efficient travel time analysis can be used as inputs to compute the marginal social cost, average cost, and marginal external cost. The computed marginal external costs corresponding to vehicle operations, crashes, air emissions, and noise pollution are the user charges needed to sustain their performance efficiency. Adopting user charges for all primary aspects of system usage concerning vehicle operations, travel time, crashes, air emissions, and noise pollution will support the overall system usage efficiency.

19.5.2 User charge to maintain efficient vehicle operating costs

Unit vehicle operating costs: Vehicle operating costs include costs of fuel consumption, oil consumption, tire wear, maintenance and repair, and depreciable value. Each of the five components can be categorized into three sub-operating costs: constant-speed operating cost, excess operating cost due to speed changes, and excess operating cost due to curves. The estimation of constant-speed operating cost relies on the average effective speed acquired from the speed model, average grades, and pavement conditions from the pavement condition deterioration model. For each vehicle type, constant-speed operating cost is estimated as the sum of five components representing costs for fuel, oil, tires, maintenance and repair, and vehicle depreciation per mile traveled shown in function. The general form of a unit vehicle operating cost function follows:

$$UVOC_{il} = f\left(C_{il}^{g}, C_{il}^{o}, C_{il}^{t}, C_{il}^{mr}, C_{il}^{I}, C_{il}^{pt}, C_{il}^{d}\right) \tag{19.43}$$

where

$UVOC_{il}$ = Unit vehicle operating cost of vehicle type i on highway segment l, dollars/VMT
C_{il}^{g} = Gas cost of vehicle type i on highway segment l, dollars/VMT
C_{il}^{o} = Oil cost of vehicle type i on highway segment l, dollars/VMT
C_{il}^{t} = Tire cost of vehicle type i on highway segment l, dollars/VMT
C_{il}^{mr} = Maintenance and repair cost of vehicle type i on highway segment l, dollars/VMT
C_{il}^{I} = Insurance cost of vehicle type i on highway segment l, dollars/yr
C_{il}^{pt} = Parking fees and tolls of vehicle type i on highway segment l, dollars/mi
C_{il}^{d} = Depreciation cost of vehicle type i on highway segment l

Cumulative and annual travel with growth in N-year facility service life-cycle: With the base year $VMT_{il,0}$ for vehicle type i on highway segment l determined, we could consider a baseline annual growth rate of g_V that leads to a growth path of $VMT_{il,0} \cdot e^{(g_V \cdot t)}$. The cumulative VMT in N-year facility service life cycle and average annual VMT become

$$VMT_{il}(N) = \int_{0}^{N} \left[VMT_{il,0} \cdot e^{(g_V \cdot \tau)}\right] d\tau = VMT_{il,0} \cdot \left[\frac{e^{(g_V \cdot N)} - 1}{g_V}\right] \tag{19.44}$$

$$VMT_{il} = VMT_{il,0} \cdot \left[\frac{e^{(g_V \cdot N)} - 1}{g_V \cdot N} \right] \tag{19.45}$$

Marginal, average, and marginal external costs and user charge: The present value of annual total vehicle operating costs for all vehicles on a highway segment l can be determined as

$$VOC_l = \sum_i (\phi_i \cdot UVOC_{il} \cdot VMT_{il}) \tag{19.46}$$

where
ϕ_i = PCE index of vehicle type i

It is noted that since depreciation is caused by wear and tear on the vehicle over time and by the change in demand and taste of users, its cost is assumed to be a function of vehicle's mileage and age. Meanwhile, vehicle maintenance, fuel, oil, and tire-wear costs, parking fees, and tolls depend mainly on the distance traveled.

At equilibrium, marginal, average, and marginal external vehicle operating costs are estimated in terms of distance traveled as follows:

$$VOC_{M,l} = \frac{\partial (VOC_l)}{\partial \sum_i (\phi_i \cdot VMT_{il})} \tag{19.47}$$

$$VOC_{A,l} = \frac{VOC_l}{\sum_i (\phi_i \cdot VMT_{il})} \tag{19.48}$$

$$VOC_{ME,l} = (VOC_{M,l} - VOC_{A,l})_{V=V_1} \tag{19.49}$$

The marginal external vehicle operating cost is the user charge that needs to be imposed, namely, $\tau_{VOC,l} = VOC_{ME,l}$.

19.5.3 User charges to maintain efficient travel time costs

Travel time can be estimated using static and dynamic models with added complexity. The following section proceeds from the simplest to the more complex models for estimating the travel time cost and determining user charges for maintaining efficient travel time cost. All of them use the same basic principle by explicitly addressing (i) a single highway segment with a single analysis time period; (ii) a single highway segment with multiple analysis time periods; (iii) a highway network with a single analysis time period; (iv) a travel path with schedule-varying dynamics; and (v) inclusion of intersection delay cost.

19.5.3.1 Static travel time costs

A single highway segment with a single time period: Without loss of generality and for an easier way to underline the principle of user charge for travel time cost, we assume an uninterrupted static traffic stream with homogeneous travelers on a single highway segment considering a single time period for analysis.

According to the widely used BPR function, travel time cost on the highway segment is estimated as

$$TC_l(v_l) = \alpha \cdot T_{f,l} \cdot \left[1 + a \cdot \left(\frac{v_l}{c \cdot c_l}\right)^b\right] \tag{19.50}$$

where

$TC_l(v_l)$ = Average travel time cost on link l, in dollars
$T_{f,l}$ = Travel time of traversing link l at a free flow speed, in hr
v_l = Traffic volume on link l
c_l = Capacity of link l
α = Value of travel time, dollars/hr
a, b, c = Model coefficients, the most commonly used values are $a = 0.15$, $b = 4.0$, and $c = 1$

At equilibrium, marginal, average, and marginal external travel time costs against the traffic volume are determined as follows:

$$TC_{M,l} = \frac{\partial(TC_l(v_l) \cdot v_l)}{\partial(v_l)} = \alpha \cdot T_{f,l} + \alpha \cdot T_{f,l} \cdot a \cdot (b+1) \cdot \left[\left(\frac{v_l}{c \cdot c_l}\right)^b\right] \tag{19.51}$$

$$TC_{A,l} = TC_l(v_l) = \alpha \cdot T_{f,l} + \alpha \cdot T_{f,l} \cdot a \cdot \left[\left(\frac{v_l}{c \cdot c_l}\right)^b\right] \tag{19.52}$$

$$TC_{ME,l} = TC_{M,l} - TC_{M,l} = \alpha \cdot T_{f,l} \cdot a \cdot b \cdot \left[\left(\frac{v_l}{c \cdot c_l}\right)^b\right] \tag{19.53}$$

The marginal external travel time cost is the user charge that needs to be imposed, that is, $\tau_{TC,l} = TC_{ME,l}$.

EXAMPLE 19.2

Given the demand function $v = 4500 - 100t$ and supply function $t = 10[1 + 0.15(v/2400^4)]$, calculate the marginal travel time cost and the optimal toll charge for a 20-mile long highway segment for congestion mitigation, with unit value of travel time at \$20 per hour.

Solution

Demand function $v = 4500 - 100t$ and supply function $t = 10[1 + 0.15(v/2400)^4]$

Total travel time of all vehicles becomes: $T = t \cdot v = 10[1 + 0.15(v/2400)^4] \cdot v = 10[v + (0.15v^5/2400^4)]$

Marginal time $TC_M = dT/dv = 10[1 + 0.75v^4/2400^4]$

The equilibrium of demand and supply functions yields: $v = 3088.6$ vph and $t = 14.1$ min

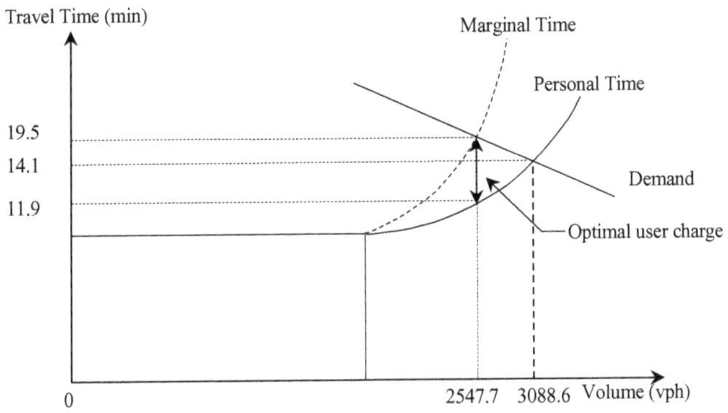

Figure 19.8 Determining the optimal user charges for efficient travel time cost.

The equilibrium of marginal time and supply functions yields $v = 2547.7$ vph and $t = 19.5$ min

For $v = 2547.7$ vph, the average cost $t = 10[1 + 0.15(2547.7/2400)^4] = 11.9$ min

Therefore, the user charge $\tau_{TC} = TC_{ME} = [(19.5 - 11.9)/60]\$20/\text{hr} = \$2.53$ for 20 miles, equivalent to 12.67 cent/mile, as seen in Figure 19.8.

As an example, Table 19.2 shows the marginal external travel time costs for different types of vehicle types using different classes of highways in the 1997 Federal Highway Cost Allocation Study (USDOT, 1998).

Similarly, Small's duration-dependent travel time model can be employed to account for impacts of the duration when traffic volume exceeds capacity on travel time estimation (Small, 1983). The model is as follows:

$$TC_l(v_l) = \begin{cases} \alpha \cdot T_{f,l}, & \text{if } v_l \le c_l \\ \alpha \cdot T_{f,l} + 0.5 \cdot \alpha \cdot p \cdot (v_l/c_l - 1), & \text{otherwise} \end{cases} \quad (19.54)$$

where

$TC_l(v_l)$ = Average travel time cost on link l, in dollars

$T_{f,l}$ = Travel time of traversing link l at a free flow speed, in hr

α = Value of travel time, dollars/hr

p = Peak period of fixed duration

v_l = Traffic volume on link l

c_l = Capacity of link l

Table 19.2 Marginal external travel time costs (cents per mile)

	Rural highways			*Urban highways*			*All highways*		
Vehicle type	*High*	*Middle*	*Low*	*High*	*Middle*	*Low*	*High*	*Middle*	*Low*
Automobiles	3.76	1.28	0.34	18.27	6.21	1.64	13.17	4.48	1.19
Pickups and vans	3.80	1.29	0.34	17.78	6.04	1.60	11.75	4.00	1.06
Buses	6.96	2.37	0.63	37.59	12.78	3.38	24.79	8.43	2.23
Single unit trucks	7.43	2.53	0.67	42.65	14.50	3.84	26.81	9.11	2.41
Combination trucks	10.87	3.70	0.98	49.34	16.78	4.44	25.81	8.78	2.32
All vehicles	4.40	1.50	0.40	19.72	6.71	1.78	13.81	4.70	1.24

At equilibrium, marginal, average, and marginal external travel time costs against the traffic volume are determined as follows:

$$TC_{M,l} = \frac{\partial(TC_l(v_l)\cdot v_l)}{\partial(v_l)} = \alpha\cdot T_{f,l} + \alpha\cdot p\cdot(v_l/c_l - 0.5) \tag{19.55}$$

$$TC_{A,l} = TC_l(v_l) = \alpha\cdot T_{f,l} + 0.5\cdot\alpha\cdot p\cdot(v_l/c_l - 1) \tag{19.56}$$

$$TC_{ME,l} = TC_{M,l} - TC_{M,l} = 0.5\cdot\alpha\cdot p\cdot(v_l/c_l) \tag{19.57}$$

The marginal external travel time cost is the user charge that needs to be imposed, that is, $\tau_{TC,l} = TC_{ME,l}$.

A single highway segment with multiple time periods: Inefficient travel time or traffic congestion could happen in multiple periods of the day, especially during peak hours. Consider the highway segment is serving vehicles in H different time periods where each of duration of time p_h has corresponding flow $v_{l,h}$ under static traffic conditions. The total vehicle volume and associated short-run total variable travel time cost during period h are $p_h \cdot v_{l,h}$ and $TC_{l,h}^V \cdot (p_h \cdot v_{l,h})$. The short-run marginal cost of adding another traveler to the traffic stream is

$$TC_{M,l,h} = \frac{\partial(TC_{l,h}^V \cdot (p_h \cdot v_{l,h}))}{\partial(p_h \cdot v_{l,h})} = \frac{\partial\left(TC_{l,h}^V \cdot v_{l,h}\right)}{\partial(v_{l,h})} = TC_{l,h}^V + v_{l,h}\cdot\frac{\partial\left(TC_{l,h}^V\right)}{\partial(v_{l,h})} \tag{19.58}$$

The user charge $\tau_{TC,l,h}$ that is equal to the marginal external travel time cost becomes

$$\tau_{TC,l,h} = TC_{ME,l,h} = TC_{M,l,h} - TC_{l,h}^V = mc_h(V_h) - c_h(V_h) = v_{l,h}\cdot\frac{\partial\left(TC_{l,h}^V\right)}{\partial(v_{l,h})} \tag{19.59}$$

Network optimum with a single time period: When dealing with travel time costs for a highway network with combination of j links, r and ρ routes, and m O–D pairs, the generalized travel costs with link-based user charges equivalent to the marginal external travel time cost $TC_{ME,l}$ under the user equilibrium principle can be presented as

$$\forall\delta_{rm} = 1: \begin{cases} \sum_{l=1}^{L}[\delta_{lr}\cdot(TC_{A,l}(v_l) + TC_{ME,l})] - d_m(v_m) \geq 0 \\ v_r \geq 0 \\ v_r\cdot\left\{\sum_{l=1}^{L}[\delta_{lr}\cdot(TC_{A,l}(v_l) + TC_{ME,l})] - d_m(v_m)\right\} = 0 \end{cases} \tag{19.60}$$

where

$TC_{A,l}(v_l)$ = Average travel time cost of link l carrying a vehicle volume of v_l
δ_m, δ_{pm} = 1 for any route r or route ρ that serves O–D pair m
$\delta_{lr}, \delta_{l,\rho}$ = 1 for any link l that is part of route r or route ρ
v_l = $\sum_{r=1}^{R}\delta_{lr}.v_r$, total vehicle volume on link l
v_m = $\sum_{r=1}^{R}\delta_{rm}.v_r$, total vehicle volume from O–D pair m
$d_m(v_m)$ = Marginal benefits of vehicle volume of v_m from O–D pair m

The user equilibrium in Equation 19.58 means that the generalized travel costs $TC_{A,l}(v_l) + TC_{ME,l}$ should be equal for all used routes of an O–D pair and equal to marginal benefits $d_m(v_m)$. Moreover, for the unused routes (if any), the generalized prices should not be lower than the used routes.

On the other hand, the optimal flow pattern under the system optimal principle can be found by maximizing social surplus W as below

$$\text{Max}_{r=1\ldots R} W = \sum_m \int_0^{v_m} d_m(v)dv - \left[\sum_l (v_l \cdot TC_{A,l}(v_l; v_{K,l})) - \sum_l (\rho \cdot K_l(v_{K,l}))\right] \quad (19.61)$$

Subject to $v_r \geq 0$ for all r

where

δ_{rm} = 1 for any route r that serves O–D pair m
δ_{lr} = 1 for any link l that is part of route r
v_l = $\sum_{r=1}^{R} \delta_{lr} \cdot v_r$, total vehicle volume on link l
v_m = $\sum_{r=1}^{R} \delta_{rm} \cdot v_r$, total vehicle volume from O–D pair m
$v_{K,l}$ = Added vehicle volume on link l after the capital expenditure K
ρ = Annualized factor of the capital expenditure K
$K_l(v_{k,l})$ = Capital expenditure

Taking the derivatives of W against route flow $v_r \geq 0$ for all r, the Kuhn–Tucker first-order conditions are obtained

$$\forall \delta_{rm} = 1: \begin{cases} \sum_{l=1}^{L}\left[\delta_{lr} \cdot \left(TC_{A,l}(v_l) + v_l \cdot \frac{\delta TC_{A,l}(v_l)}{\delta v_l}\right)\right] - d_m(v_m) \geq 0 \\ v_r \geq 0 \\ v_r \cdot \left\{\sum_{l=1}^{L}\left[\delta_{lr} \cdot \left(TC_{A,l}(v_l) + v_l \cdot \frac{\delta TC_{A,l}(v_l)}{\delta v_l}\right)\right] - d_m(v_m)\right\} = 0 \end{cases} \quad (19.62)$$

Comparing Expressions (19.59) and (19.57), it shows that the optimal flow pattern in the user equilibrium is identical to the flow pattern in the system optimum where the user charge as the marginal external travel time cost $TC_{ME,l}$ is replaced by $v_l \cdot (\partial TC_{A,l}(v_l))/\partial v_l$. In the system optimum condition, all used routes for an O–D pair have identical marginal travel time costs equivalent to marginal benefits for that O–D pair. There are no unused routes with marginal costs lower than this.

Therefore, the user charge for link l of a highway network $\tau_{TC,l}$ can be set as the marginal external travel time cost to maintain travel time efficiency.

$$\tau_{TC,l} = TC_{ME,l} = v_l \cdot \frac{\partial TC_{A,l}(v_l)}{\partial v_l} \text{ for all links } \quad l, l = 1, 2, \ldots, L \quad (19.63)$$

19.5.3.2 Dynamic travel time costs

Best travel time cost for a travel path in tolled equilibrium incorporating schedule-varying dynamics: Based on the queueing model dealing with a single pure bottleneck involving schedule-varying dynamics, the best travel time cost can be derived from unpriced equilibrium. The following briefly describes the procedure.

Denote

$TC^{\prime,0}(t)$ = Weighted total travel time cost incurred when departing at t incorporating trade-offs between actual and desirable arrival times

$TC^{\prime,0}(t')$ = Weighted total travel time cost incurred when exiting queue at t' incorporating trade-offs between actual and desirable arrival times

$T_D^0(\mathrm{t})$ = Travel delay from queueing at a bottleneck during peak period when departing at t

$TC_D^0(t')$ = Cost of travel delay from queueing at a bottleneck during peak period when exiting queue at t'

$T_S^0(t)$ = Schedule-varying trade-off time when departing at t

$TC_S^0(t')$ = Cost of schedule-varying trade-off time when exiting the queue at t'

t = Departure time

t_d = Desired arrival time

t' = Time at which a traveler exits from the queue at the bottleneck

t_m^{exit} = Queue exit time for the traveler experiencing the longest travel delay $T_{D,m}$

c_k = Capacity of a bottleneck location k

v_d = Constant departure rate during the peak interval $[t_p, t_{p'}]$

α = Value of travel time

β = Shadow price of early arrival, where $\beta < \alpha$

γ = Shadow price of late arrival

t_p = Peak interval beginning time

$t_{p'}$ = Peak interval ending time

p = Peak period fixed duration, $p = t_{p'} - t_p$

t_q = Queue beginning time

$t_{q'}$ = Queue dissipation time

When the peak duration interval $p = t_{p'} - t_p$ approaches zero, the interval beginning time t_p and ending time $t_{p'}$ will be identical to the time of the first traveler entering the queue t_q and to the last traveler exiting the queue $t_{q'}$, respectively. For the total number of travelers departing from the interval, $v_d \cdot p$, the peak starting and ending times are

$$t_q = t_m^{exit} - [\gamma/(\beta+\gamma)]\cdot(v_d/c_k)\cdot p \tag{19.64}$$

$$t_{q'} = t_m^{exit} + [\beta/(\beta+\gamma)]\cdot(v_d/c_k)\cdot p \tag{19.65}$$

For a traveler exiting the queue before the desired arrival time t_d, the unpriced equilibrium requires that the queue entry time minimizes the weighted total travel time cost $TC^{\prime,0}(t) = \alpha\cdot T_D^0(t) + \beta\cdot[(t_d - t) - T_D^0(t)]$. Setting $T_D^0(t)\cdot(\alpha-\beta) - \beta\cdot t = 0$, the rate of change in $T_D^0(t)$ against departure time t is $T_D^0(t)/t = [\beta/(\alpha-\beta)]$. Likewise, for a traveler exiting the queue after the desired arrival time t_d, the unpriced equilibrium requires that the queue entry time minimizes the weighted total travel time $TC^{\prime,0}(t) = \alpha\cdot T_D^0(t) + \gamma\cdot[T(t) - (t_d - t)]$. Setting $T_D^0(t)\cdot(\alpha+\gamma) + \gamma\cdot t = 0$, the rate of change in $T_D^0(t)$ against departure time t is $T_D^0(t)/t = [-\gamma/(\alpha+\gamma)]$. Setting $\left(v_a^{early}/c_k - 1\right) = \beta/(\alpha-\beta)$ and $\left(v_a^{late}/c_k - 1\right) = -\gamma/(\alpha+\gamma)$, we obtain

$$v_a^{early} = [\alpha/(\alpha-\beta)]\cdot c_k \tag{19.66}$$

$$v_a^{late} = [\alpha/(\alpha+\gamma)]\cdot c_k \tag{19.67}$$

Travel delay cost $TC_D^0(t')$, schedule-varying trade-off time cost $TC_S^0(t')$, and weighted total travel time cost $TC^{',0}(t')$ in unpriced equilibrium as functions of queue exit time t' can be computed as

$$TC_D^0(t') = \begin{cases} \beta \cdot (t' - t_q), & \text{if } t_q \le t' \le t_m^{exit} \\ \gamma \cdot \left[\dfrac{\beta}{(\beta+\gamma)} \cdot \left(\dfrac{v_d}{c_k} \right) \cdot p - \left(t' - t_m^{exit} \right) \right], & \text{if } t_m^{exit} \le t' \le t_{q'} \end{cases} \tag{19.68}$$

$$TC_S^0(t') = \begin{cases} \beta \cdot \left[\dfrac{\gamma}{(\beta+\gamma)} \cdot \left(\dfrac{v_d}{c_k} \right) \cdot p - (t' - t_q) \right], & \text{if } t_q \le t' \le t_m^{exit} \\ \gamma \cdot \left[\left(t' - t_m^{exit} \right) \right], & \text{if } t_m^{exit} \le t' \le t_{q'} \end{cases} \tag{19.69}$$

$$TC^{',0}(t') = \left[\dfrac{\beta \cdot \gamma}{(\beta+\gamma)} \right] \cdot \left(\dfrac{v_d}{c_k} \right) \cdot p, \; t_q \le t' \le t_{q'} \tag{19.70}$$

From the above, it shows that in unpriced equilibrium the queue entry rates v_a^{early} and v_a^{late} for travelers arriving before and after the desired arrival time t_d will ensure that the weighted total travel time cost $TC^{',0}(t')$ remains constant over all queue entry times from t_q to $t_{q'}$. Also, both the average travel delay cost $\overline{TC}_D^0$ and the average schedule-varying trade-off time cost $\overline{TC}_S^0$ are equal to one-half of the average weighted total travel time cost $\overline{TC}^{',0}$, which yield the following:

$$\overline{TC}_D^0 = \overline{TC}_S^0 = 0.5 \cdot [(\beta \cdot \gamma)/(\beta+\gamma)] \cdot (v_d/c_k) \cdot p \tag{19.71}$$

$$\overline{TC}^{',0} = [(\beta \cdot \gamma)/(\beta+\gamma)] \cdot (v_d/c_k) \cdot p \tag{19.72}$$

The total unpriced equilibrium travel time cost and marginal travel time cost equal to

$$\overline{TC}_{total}^{',0} = [(\beta \cdot \gamma)/(\beta+\gamma)] \cdot [(v_d \cdot p)^2/c_k] \tag{19.73}$$

$$\overline{TC}_M^{',0} = 2 \cdot [(\beta \cdot \gamma)/(\beta+\gamma)] \cdot (v_d/c_k) \cdot p = 2 \cdot \overline{TC}^{',0} \tag{19.74}$$

The optimal travel pattern for the best travel time cost should satisfy the following three criteria: (i) if exits occur, the exit rate should not be lower than the capacity c_k, or else the period of exits could be shortened and hence total of schedule-varying cost could be reduced; (ii) the queue entry and exit rates should both be equal to capacity c_k throughout the peak interval; and (iii) exits should occur between the same instants from t_q to $t_{q'}$ as in the unpriced equilibrium.

Therefore, some adjustments are needed for the patterns of queue entries and exits established from the unpriced equilibrium. This is achieved by introducing the user charge $\tau_{TC}(t')$ that is set as identical to the travel delay cost $TC_D^0(t')$ in unpriced equilibrium

$$\tau_{TC}(t') = \begin{cases} \beta \cdot (t' - t_q), & \text{if } t_q \le t' \le t_m^{exit} \\ \gamma \cdot \left[\dfrac{\beta}{(\beta+\gamma)} \cdot \left(\dfrac{v_d}{c_k} \right) \cdot p - \left(t' - t_m^{exit} \right) \right], & \text{if } t_m^{exit} \le t' \le t_{q'} \end{cases} \tag{19.75}$$

In tolled equilibrium, the pattern of queue exits remains unchanged, but the pattern of queue entries is altered to $v_a = c_k$. Since the user charge fully replaces the travel delay cost, it reduces travel delay cost to zero and still keeps the schedule-varying trade-off time cost unchanged. As a result, the average weighted total travel time cost is decreased by one-half. The user charge leads to the best travel time cost in tolled equilibrium with the benefits achieved being equal to the value of travel time savings.

19.5.3.3 Intersection delay time costs

In the above analysis of the user charge to maintain efficient travel time for a single highway segment with single and multiple analysis time periods, a highway network with a single analysis time period, and travel path with a single bottleneck incorporating schedule-varying dynamics, the focus has been on the calculation of travel time cost for a single highway segment, multiple highway segments or all highway segments for an entire highway network. In this respect, costs of vehicle delays at intersections typically, in the presence of a travel path, are not explicitly addressed. If the travel path consists of multiple intersections such as a travel path using a dense urban street network, the portion of intersection delay cost out of the total travel time could be significant, and therefore needs to be included into the estimation of total travel time cost.

With the vehicles entering intersection i in time interval $t \in [0, t]$ defined as $v_i(t)$ and the capacity being $c_{k,i}$, the queueing model for a single pure bottleneck analysis as Equations 5.46 and 5.47 can be employed to calculate the total entering vehicles (EV), total vehicle delay cost, marginal delay cost, average delay cost, and the marginal external delay cost as the use charge for every vehicle entering the intersection. The details are presented below:

$$EV_i(t) = \int_0^t v_i(\tau) \cdot d\tau \tag{19.76}$$

$$VDC_i^{int}(t) = \begin{cases} 0, & \text{if } v_i(t) \le c_{k,i} \\ \alpha \cdot \int_0^t v_i(\tau) \cdot [v_i(\tau)/c_{k,i} - 1] \cdot d\tau, & \text{otherwise} \end{cases} \tag{19.77}$$

$$VDC_{M,i}^{int} = \frac{\partial\left(VDC_i^{int}(t)\right)}{\partial(EV_i(t))} \tag{19.78}$$

$$VDC_{A,i}^{int} = \begin{cases} 0, & \text{if } v_i(t) \le c_{k,i} \\ \alpha \cdot \left\{ \left[\int_0^t (v_i(\tau)/c_{k,i} - 1) \cdot d\tau \right] / t \right\} \end{cases} \tag{19.79}$$

$$VDC_{ME,i}^{int} = VDC_{M,i}^{int} - VDC_{A,i}^{int} \tag{19.80}$$

where

$EV_i(t)$ = Number of vehicles entering intersection i in time interval t
$VDC_i^{int}(t)$ = Cost of vehicle delays at intersection i in time interval t
$VDC_{M,i}^{int}$ = Marginal cost of vehicle delays at intersection i
$VDC_{A,i}^{int}$ = Average cost of vehicle delays at intersection i
$VDC_{ME,i}^{int}$ = Marginal external cost of vehicle delays at intersection i

The marginal external vehicle delay cost at the intersection is the user charge that needs to be imposed, namely, $\tau_{VDC,i}^{int} = VDC_{ME,i}^{int}$ for intersection i.

19.5.4 User charges to maintain efficient vehicle crash costs

Highway vehicle crashes are analyzed separately for highway segments and intersections. Vehicle crash frequencies by fatal, injury, and PDO severity levels can be predicted for individual segments and intersections using safety performance functions and crash modification factors. A safety performance function could predict the fatal, injury or PDO crash frequency as a function of traffic exposure. The predicted crash frequency by severity level is further adjusted using crash modification factors to reflect the localized conditions of geometric design, consistency of geometric design, traffic control and safety hardware conditions, and roadside features.

Unit vehicle crash costs: With fatal, injury, and PDO crash frequencies established for a specific highway segment in terms of crashes per highway segment per year or crashes per intersection per year, the crash frequencies can be converted to crash rates in fatal, injury, and PDO crashes per million VMT for the highway segment or in fatal, injury, and PDO crashes per million EV for the intersection.

Using the monetary value per fatal, injury or PDO crash, the respective fatal, injury, and PDO crashes per million VMT for the highway segment or fatal, injury, and PDO crashes per million vehicles entering the intersection can be determined. These monetary values are unit vehicle crash costs, which could be separately created for different functional classes of highway segments and different types of intersections. The general form of a unit vehicle operating cost function follows:

$$UVCC_{f,l}^{s} = f\left(VCR_{f,l}^{s}, UR^{s}\right) \quad \text{or} \quad UVCC_{f,i}^{int,s} = f\left(VCR_{f,i}^{int,s}, UR^{s}\right) \tag{19.81}$$

where

$UVCC_{f,l}^{s}$ = Unit vehicle crash cost for highway segment l by facility classification f and crash severity level s, dollars/VMT

$UVCC_{f,i}^{int,s}$ = Unit vehicle crash cost for highway intersection i by facility classification f and crash severity level s, dollars/EV

$VCR_{f,l}^{s}$ = Vehicle crash rate for highway segment l by facility classification f and crash severity level s, crashes/million VMT

$VCR_{f,i}^{int,s}$ = Vehicle crash rate for intersection i by facility classification f and crash severity level s, crashes/million EVs

UR^{s} = Unit rate of crashes by crash severity level s, dollars/crash

f = Facility classification where highway segments are classified into rural/urban interstate, multilane (undivided and divided median), and two-lane highways highway intersections are classified into rural/urban signalized and unsignalized intersections

s = Crash severity level, including fatal, injury, and PDO levels

Cumulative and annual travel with growth in N-year facility service life cycle: For estimating the annual vehicle crash costs associated with a highway segment or an intersection, the annual VMT or annual EVs need to be computed. As shown in Equations 19.44 and 19.45, the base year $VMT_{f,l,0}$ for highway segment l within facility classification f, $EV_{f,i,0}$ for intersection i that belongs to facility classification f, and a baseline annual traffic growth rate of g_V can be utilized to compute the cumulative VMT and cumulative EV in N-year facility service life cycle. The average annual *VMT* and average annual EVs can then be calculated as $VMT_{f,l} = VMT_{f,l,0} \cdot [(e^{(g_V \cdot N)} - 1)/(g_V \cdot N)]$ and $EV_{f,i} = EV_{f,i,0} \cdot [(e^{(g_V \cdot N)} - 1)/(g_V \cdot N)]$.

Marginal, average, and marginal external costs and user charge: The present value of annualized total vehicle operating costs for all vehicles utilizing a highway segment (or an intersection) l can be determined as

$$VCC_{f,l} = \sum_s (UVCC^s_{f,l} \cdot VMT_{f,l}) \quad \text{or} \quad VCC^{int}_{f,i} = \sum_s (UVCC^{int,s}_{f,i} \cdot EV_{f,i}) \tag{19.82}$$

where

$VCC_{f,l}, VCC^{int}_{f,i}$ = Annualized vehicle crash cost for highway segment l (or intersection i) for facility classification f

At equilibrium, marginal, average, and marginal external vehicle crash costs are estimated as follows:

$$VCC_{M,f,l} = \frac{\partial(VCC_{f,l})}{\partial(VMT_{f,l})} \quad \text{or} \quad VCC^{int}_{M,f,i} = \frac{\partial\left(VCC^{int}_{f,i}\right)}{\partial(EV_{f,i})} \tag{19.83}$$

$$VCC_{A,f,l} = \frac{VCC_{f,l}}{VMT_{f,l}} \quad \text{or} \quad VCC^{int}_{A,f,i} = \frac{VCC^{int}_{f,i}}{EV_{f,i}} \tag{19.84}$$

$$VCC_{ME,f,l} = (VCC_{M,f,l} - VCC_{A,f,l})_{V=V_1} \quad \text{or} \quad VCC^{int}_{ME,f,i} = \left(VCC^{int}_{M,f,i} - VCC^{int}_{A,f,i}\right)_{V=V_1} \tag{19.85}$$

The marginal external vehicle crash cost is the user charge that needs to be imposed, namely, $\tau_{VCC,f,l} = VCC_{ME,f,l}$ for highway segment l by facility classification f or $\tau^{int}_{VCC,f,i} = VCC^{int}_{ME,f,i}$ for intersection i by facility classification f.

As an example, Table 19.3 summarizes the marginal external vehicle crash costs established by highway functional classification and vehicle type from the 1997 Federal Highway Cost Allocation Study (USDOT, 1998).

19.5.5 User charges to maintain efficient vehicle air emission costs

Vehicle air emissions are related to changes in ambient gas percentages and particulates resulting from highway vehicle usage. Highway vehicle emissions account for a substantial portion of all air emissions. The typical pollutants of vehicle air emissions include NMHC,

Table 19.3 Marginal vehicle crash costs (cents per *VMT*)

	Rural highways			*Urban highways*			*All highways*		
Vehicle type	*High*	*Middle*	*Low*	*High*	*Middle*	*Low*	*High*	*Middle*	*Low*
Autos	9.68	3.15	1.76	4.03	1.28	0.78	6.02	1.94	1.13
Pickups and vans	10.21	3.31	1.75	4.05	1.27	0.74	6.70	2.15	1.17
Buses	14.15	4.40	2.36	6.25	1.89	1.08	9.55	2.94	1.62
Single unit trucks	5.97	2.00	0.97	2.21	0.71	0.40	3.90	1.29	0.65
Combination trucks	6.90	2.20	1.02	3.67	1.16	0.56	5.65	1.79	0.84
All vehicles	9.52	3.09	1.68	3.98	1.26	0.76	6.12	1.97	1.11

CO, CO_2, NO_X, SO_2, and TSP or PM. The emission rate of each pollutant for vehicles using a highway segment can be estimated by traffic volumes, vehicle age and composition, operating speeds, and speed change cycles (Frey et al., 2006, 2008; Zhai et al., 2008).

Unit vehicle air emission costs: With vehicle air emission rates established for different types of pollutants including NMHC, CO, CO_2, NO_X, SO_2, and TSP or PM in terms of grams of pollutant per VMT for a specific highway segment and grams of pollutant per EV for an intersection, the monetary value per gram of pollutant can be employed to establish the unit vehicle air emission cost in dollars per VMT for the highway segment or dollars per EV for the intersection. These monetary values could be separately created for different functional classes of highway segments and different types of intersections. The general form of a unit vehicle air emission cost function follows:

$$UVEC^{p}_{f,k,l} = f\left(E^{p}_{f,k,l}, VMT_{f,k,l}, UR^{p}\right) \quad \text{or} \quad UVEC^{int,p}_{f,i} = f\left(E^{int,p}_{f,i}, EV_{f,i}, UR^{p}\right) \tag{19.86}$$

where

$UVEC^{p}_{f,k,l}$ = Unit vehicle emission cost of pollutant type p for vehicles of type k using highway segment l within facility classification f, dollars/VMT

$UVEC^{int,p}_{f,i}$ = Unit vehicle emission cost of pollutant type p for vehicles using intersection i within facility classification f, dollars/EV

$E^{p}_{f,k,l}$ = Estimated emission quantity of pollutant p for vehicles of type k using highway segment l within facility classification f, grams

$E^{int,p}_{f,i}$ = Estimated emission quantity of pollutant p for vehicles using intersection i within facility classification f, grams

$VMT_{f,k,l}$ = Vehicle miles of travel by vehicles of type k using highway segment l within facility classification f

$EV_{f,k,i}$ = Vehicles of type k entering intersection i within facility classification f

UR^{p} = Unit rate of vehicle emissions of pollutant type p, in dollars/gram

f = Facility classification where highway segments are classified into rural/urban interstate, multilane (undivided and divided median), and two-lane highways; highway intersections are classified into rural/urban signalized and unsignalized intersections

p = Pollutant type p, including NMHC, CO, CO_2, NO_X, SO_2, and TSPs

Cumulative and annual travel with growth in N-year facility service life cycle: As shown in Equations 19.44 and 19.45, the base year $VMT_{l,0}$ and $VMT_{k,l,0}$ by vehicle type k for highway segment l; $EV_{i,0}$ for intersection i; and a baseline annual traffic growth rate of g_V can be utilized to compute the cumulative VMT (cumulative EV) in N-year facility service life cycle. The average annual VMT or average annual EVs can then be calculated as

$$VMT_l = VMT_{l,0} \cdot \left[\frac{e^{(g_V \cdot N)} - 1}{g_V \cdot N}\right]; VMT_{k,l} = VMT_{k,l,0} \cdot \left[\frac{e^{(g_V \cdot N)} - 1}{g_V \cdot N}\right]; \text{and } EV_i = EV_{i,0} \cdot \left[\frac{e^{(g_V \cdot N)} - 1}{g_V \cdot N}\right]$$

Marginal, average, and marginal external costs and user charge: The present value of annualized total vehicle emission costs for vehicles of type k using a highway segment l within facility classification f or for vehicles entering an intersection i within facility classification f can be calculated as

$$VEC_{f,k,l} = \sum_{p}\left(UVEC^{p}_{f,k,l} \cdot VMT_{f,k,l}\right) \quad \text{or} \quad VEC^{int}_{f,i} = \sum_{p}\left(UVEC^{int,p}_{f,i} \cdot EV_{f,i}\right) \tag{19.87}$$

where

$VEC_{f,k,l}$, $VEC_{f,i}^{int}$ = Annualized vehicle emission costs for highway segment l or intersection i

At equilibrium, marginal, average, and marginal external vehicle emission costs are estimated as follows:

$$VEC_{M,f,k,l} = \frac{\partial(VEC_{f,k,l})}{\partial(VMT_{f,k,l})} \quad \text{or} \quad VEC_{M,f,i}^{int} = \frac{\partial\left(VEC_{f,i}^{int}\right)}{\partial(EV_{f,i})} \tag{19.88}$$

$$VEC_{A,f,k,l} = \frac{VEC_{f,k,l}}{VMT_{f,k,l}} \quad \text{or} \quad VEC_{A,f,i}^{int} = \frac{VEC_{f,i}^{int}}{EV_{f,i}} \tag{19.89}$$

$$VEC_{ME,f,k,l} = (VEC_{M,f,k,l} - VEC_{A,f,k,l})_{V=V_1} \quad \text{or} \quad VEC_{ME,f,i}^{int} = \left(VEC_{M,f,i}^{int} - VEC_{A,f,i}^{int}\right)_{V=V_1} \tag{19.90}$$

The marginal external vehicle emission costs are the user charges that need to be imposed on vehicles using highway segments and intersections, that is, $\tau_{VEC,f,k,l} = VEC_{ME,f,k,l}$ and $\tau_{VEC,f,i}^{int} = VEC_{ME,f,i}^{int}$.

19.5.6 Estimation of annual vehicle noise pollution costs

Hourly vehicle noise costs: The FHWA's traffic noise model could help determine the hourly equivalent sound level $L_{eq,t}$ caused by vehicles using a highway segment in hour t of a day. With noise depreciation sensitivity index (NDSI) introduced by Nelson (1982) that calculates the percentage reduction in the house value caused by the net increase in the equivalent noise level from the maximum acceptable noise level at 50 dB(A), the social costs of noise can be estimated. The hourly vehicle noise cost owing to house value depreciation is defined as

$$VNC_t = N_{HH} \cdot \left(\frac{i \cdot W_{avg}}{365 \times 24}\right) \cdot D \cdot (L_{eq,t} - L_{max}) \tag{19.91}$$

with

$$N_{HH} = [(RD) \cdot D \cdot (2 \times L)] / 5280$$

where

VNC_t = Vehicle noise cost in hour t, dollars/hr
$L_{eq,t}$ = Equivalent sound level, dB(A)
D = Distance to the highway, ft
N_h = Number of houses affected per mi²
RD = Average residential density around a highway, houses/mi²
L = Length of the relevant highway segment, mi
L_{max} = Maximum acceptable noise level, L_{max} = 50 dB(A)
d = Percentage discount in value per dB(A) increase in the ambient noise level, d = 0.4%
W_{avg} = Average house value, dollars/house
i = Discount rate

Annual vehicle noise costs: The annual noise cost function along highway segment l that belongs to highway functional classification f can be written as

$$VNC_{f,l} = 365\int_{1}^{24}\int_{50}^{L_{eq,max}}\left[\left(\frac{(RD)\cdot D\cdot(2\times L)}{5{,}280}\right)\cdot\left(\frac{i\cdot W_{avg}}{365\times 24}\right)\cdot D\cdot(L_{eq,f,l,t}-50)\right]dL_{eq,f,l,t} \quad (19.92)$$

where

$VNC_{f,l}$ = Annualized vehicle noise costs for highway segment l within highway functional classification f

$L_{eq,f,l,t}$ = Equivalent noise level for highway segment l within highway functional classification f in hour t, dB(A)

$L_{eq,max}$ = Maximum equivalent noise level, dB(A)

Cumulative and annual travel with growth in N-year facility service life-cycle: As shown in Equations 19.44 and 19.45, the base year $VMT_{f,l,0}$ for highway segment l that belongs to highway functional classification f and a baseline annual traffic growth rate of g_V can be utilized to compute the cumulative VMT in N-year facility service life-cycle. The average annual VMT can then be calculated as $VMT_{f,l} = VMT_{f,l,0}\cdot[(e^{(g_V\cdot N)}-1)/(g_V\cdot N)]$. Further, $VMT_{f,l,t}$ is defined as the hourly varying VMT corresponding to $VMT_{f,l}$.

At equilibrium, marginal, average, and marginal external vehicle noise costs are estimated as follows:

$$VNC_{M,f,l} = \frac{\partial(VNC_{f,l})}{\partial(L_{eq,f,l,t})}\cdot\frac{d(L_{eq,f,l,t})}{d(VMT_{f,l,t})} \quad (19.93)$$

$$VNC_{A,f,l} = \frac{VNC_{f,l}}{VMT_{f,l}} \quad (19.94)$$

$$VNC_{ME,f,l} = (VNC_{M,f,k,l} - VNC_{A,f,k,l})_{V=V_1} \quad (19.95)$$

The marginal external vehicle noise costs are the user charge that needs to be imposed on vehicles using highways, that is, $\tau_{VNC,f,l} = VNC_{ME,f,l}$. The above user charge per VMT can be refined to user charges per PCE–VMT value by incorporating PCE values for different types of vehicles in calculating the marginal and average vehicle noise costs. Table 19.4 presents marginal external vehicle noise costs per VMT established by FHWA expressed in 2000 dollar values.

Table 19.4 Marginal external vehicle noise costs (cents per mile)

Vehicle noise costs per VMT	*Rural highways*			*Urban highways*			*All highways*		
	High	*Middle*	*Low*	*High*	*Middle*	*Low*	*High*	*Middle*	*Low*
Automobiles	0.03	0.01	0.00	0.30	0.11	0.03	0.20	0.06	0.02
Pickups and vans	0.03	0.01	0.00	0.27	0.10	0.03	0.17	0.06	0.02
Buses	0.35	0.13	0.04	4.55	1.72	0.48	2.79	1.06	0.30
Single unit trucks	0.27	0.10	0.03	3.14	1.19	0.33	1.85	0.70	0.20
Combination trucks	0.68	0.26	0.07	9.86	3.73	1.05	4.24	1.61	0.45
All vehicles	0.08	0.03	0.01	0.64	0.24	0.07	0.42	0.16	0.05

19.6 EMERGING ISSUES

19.6.1 Pricing-based transportation financing

The pricing-based financing aims to apply user charges to various types of vehicles or users/non-users (or freight) to recover total costs of building and preserving transportation facilities calculated based on damages of traffic loading and additional condition deterioration caused by repetitive use in their service life cycles. In addition, pricing-based financing imposes user charges to individual system user cost components, including costs of vehicle operations, travel time, crashes, air emissions, and noise pollution with user charge rates set as the marginal external user costs to eliminate inefficient system usage. The pricing-based financing methodology will help secure funding needed to sustain transportation facility conditions and system usage performance. To fully benefit from the proposed pricing-based methodology for transportation financing, the following issues need to be explicitly addressed.

Consideration of microscopic traffic dynamics: For a relatively short time period such as one month, the population and transportation system in a given region will likely remain stable. The demand for passenger travel and freight movements will also not change drastically for the same short period. However, the use of vehicles is doubly constrained by diverse behavior of users and varying needs of freight movements on one side and the available capacity that could be provided by transportation facilities on the other side. As a result, vehicular traffic exhibits spatial and temporal dynamics. The fairness of user charges determined by the pricing-based methodology is governed by the accuracy (the extent to which data measurements reflect the truth) and precision (the consistency of repeated measurements) of data on physical conditions of transportation facilities and system usage levels by vehicles in support of user travel and freight movements.

For transportation facilities and system usage, data used for determining user charges may be: (i) directly collected from the field; (ii) derived using field data via integration, interpolation, and extrapolation; and (iii) predicted using travel demand forecasting, facility performance, and system usage performance models. The field collected data along with data derived based on it is inherited with errors from diverse sources, including data sampling methods, measurement equipment, and data processing. The model predicted data needs to be used with caution unless it is generated from properly calibrated and validated models. In the analysis process, traffic condition assessed using field or predicted traffic data affects both facility and system usage performance and it varies dynamically in both spatial and temporal dimensions. Therefore, the use of fine-grained cross-sectional traffic data on vehicle volumes, composition, speed, and speed changes, as well as high-fidelity data on vehicle or traveler trajectories become essential. New vehicle technologies benefited from recent development of Internet of Things (IoT) have made it possible to collect traffic flow and vehicle or traveler trajectory data in real-time. This helps greatly with validating travel demand forecasting models; obtaining highly reliable predicted traffic flow and vehicle or traveler trajectory data; and proactively computing impacts of dynamically varying traffic on physical facility conditions and system usage levels concerning vehicle operations, travel time, crashes, air emissions, and noise pollution.

While ensuring data accuracy and precision, the resolutions of fine-grained cross-sectional data and vehicle and traveler trajectory data have been improving over time. This makes it possible to calculate traffic impacts on transportation facilities and system usage in a much-refined manner. For instance, the time intervals of traffic and travel data measurements could be reduced from daily, hourly, 15 min, 5 min, 1 min, 1 s, to a fraction of 1 second. That is, individual vehicles and travelers could be traced on a fraction of one-second basis, shifting the analysis from the macroscopic or mesoscopic level to microscopic level. Consequently, the use of spatially and temporally varying traffic data that capture traffic dynamics at the microscopic level for impacts analysis will help determine user charges that ensure truly efficient, effective, and equitable transportation provision.

Further, user charges will influence travel demand such as reductions in the total demand; changes in certain types of trip frequencies; and shifts of departure time, destination location and arrival time, travel mode, and trip path. Trade-offs not only exist in travel schedule changes aimed to minimize the weighted total travel time cost, but also costs of facility construction and preservation versus system usage in relation to travel time, vehicle crashes, air emissions, and noise pollution that are to be minimized holistically. Consequently, trade-offs need to be considered in establishing user charges to preserve transportation facilities and to ensure system usage efficiency. Traffic dynamics in (current and future) time and space domains need to be included in the multilayer trade-offs involved with impacts analysis with added complexity to ensure that the user charges are highly efficient in impacts mitigation that will sustain the overall transportation system performance.

Influences of intelligent infrastructure technologies: In recent decades, significant progress has been made in search of more durable and cost-effective materials for transportation facility construction and preservation and in the use of Smart Objects (SO), M2M communication, and IoT rooted from advanced sensor, wireless communication, and microprocessor technologies for facility condition monitoring in real-time. This development helps validate models calibrated for facility performance predictions, conduct proactive investment decision-making, and provide a feedback process to validate user charges to recover costs of transportation facility delivery and preservation.

Effects of advanced technology vehicles: Accompanied by further development of advanced vehicle technologies, the market penetration of advanced technology vehicles like battery-operated vehicles, biofuel vehicles, and connected and automated or autonomous vehicles (CAVs) will likely increase. This will lead to significant changes in traffic flow patterns and traffic stream characteristics. Traffic impacts on physical facilities and system usage will change accordingly. As a result, user charges need to be kept abreast of changes in advanced vehicle technologies to ensure efficiency in managing transportation systems.

Changes in user/nonuser or freight behavior: The essence of transportation provision is to ensure safe and efficient movements of people and goods. As mentioned earlier, vehicles are doubly constrained by users in need of travel and freight to be shipped, as well as availability of facility capacity to accommodate vehicles. Changes in performance of transportation facilities and characteristics of vehicles, as well as population, social, and economic conditions will collectively result in changes in the behavior of people and goods movements. The behavioral changes should be incorporated into the performance analysis and modeling, impacts assessment, and user charge estimation.

Integration of physical transportation facility, vehicle, and user/nonuser or freight components: Apart from the influence of advanced vehicle technologies, additional influences have been occurring from technological advancements in aspects of physical transportation facilities and system users/nonusers or freight. The facility, vehicle, and user/nonuser or freight components of a transportation system are becoming increasingly virtually integrated through data collection, processing, analysis, and exchange of pertinent information extracted from data. Meanwhile, the facility-based life-cycle costing approach for impacts analysis will be extended to cover life-cycle cost analysis of vehicles in their service lives and system users/nonusers or freight in their life spans. That is, life-cycle cost analysis will collectively consider physical facility, vehicle, and user/nonuser or freight components of an entire transportation system. The system integration offers opportunities to create a bottom-up iterative feedback process for the top-down asset management decisions with the determination of efficient user charges being a key analytical component. This will help achieve adaptive management of the transportation system. The user charges based on the pricing-based methodology should act as a catalyst to support system component integration, adaptive transportation system management, and sustainable transportation development.

19.6.2 Dynamic transportation asset management

As described in the above, the pricing-based methodology for transportation financing needs to be robust in addressing issues of traffic dynamics, changing facility, vehicle, user/nonuser or freight features from new technologies, and system integration. On the other hand, the financing resource allocation should support most cost-efficient, effective, and equitable investment decisions for sustaining the overall transportation system performance. This demands moving one step beyond the current practice of static transportation asset management by practicing dynamic transportation asset management to capture impacts of temporal and spatial dynamics of vehicles and users/non-users or freight movements on transportation facility and system usage performance.

19.6.3 Implementation

19.6.3.1 Phased adoption of the new financing and decision-making principles

The current sources of transportation financing in the U.S. are primarily from fuel tax, complemented by bonds, motor vehicle taxes, toll, parking charges, and so forth. The growth of advanced technology vehicles in the traffic stream, coupled with increases in fuel economy, translates into decreases in fuel tax revenues that will lead to shrinkage of funding resources in the future. In parallel, the ever-growing travel demand and truck traffic loading will exacerbate the rate of deterioration in transportation facility conditions that have been declining over time. Meanwhile, degradation of traffic operating conditions has been underway owing to a much slower pace of system capacity expansion compared with the rate of travel demand growth. This requires increasing transportation funding for investments to potentially slow down the deteriorating trend of facility conditions and system usage performance. Hence, a balance needs to be sought between the resistance of travelers in accepting user charges and the need to secure new funding sources to fill in the growing gap between the needed and available funding resources.

From a practical point of view, a phased adoption of pricing-based financing and dynamic transportation asset management decision principles is desirable to gradually resolve the funding shortage issue and ensure efficiency, effectiveness, and equity of investment decisions. For the highway network, for example, pricing-based financing and funding allocation decisions may be applied first to the freeway and major arterial subnetworks that handle a predominant portion of total travel demand, and then be expanded to the entire network. In the interest of equity, the funding needs for the lower classes of highways may be subsidized by revenues from user charges to the higher classes of highways, and the funding gap of transit may be partially subsidized by highway revenues from user charges to promote multimodal development. The improved transit system may attract travelers shifting from auto to transit, leading to reductions in inefficient highway travel and traffic congestion, with benefits to all.

19.6.3.2 Public–private partnerships

The costs of major transportation improvements are typically extensive. Although the pricing-based financing methodology could help recover the total costs needed for sustaining facility and system usage performance, upfront funding is needed to fill in the current funding gap and a lead period is required to actually implement pricing-based financing and investment decision principles, since it requires legislative, policy, and procedural changes. Therefore, immediate funding sources need to be sought as well. In a situation of insufficient funding from transportation agencies, attracting funding from the private sector appears to be a natural choice. However, the essence of transportation provision is to offer the best level of public service, which is not profit-driven. It is, therefore, not suited to have the transportation system owned by private entities. An increasingly popular alternative is public–private partnerships (PPPs), with public entities owning the transportation systems and private entities participating in designing, building, operating, maintaining, and financing them in a mutually acceptable manner.

Although PPPs have not been used extensively in the U.S., they have been used successfully in other countries and their long-term experience can be learned to assist in the implementation of pricing-based transportation financing while avoiding the pitfalls of such endeavors. For this purpose, FHWA and AASHTO have jointly sponsored studies through the National Cooperative Highway Research Program (NCHRP) on scanning PPP cases in various countries to solicit insights into the best practices and lessons learned (NCHRP, 2009). The three aspects for the success of a PPP venture are a well-analyzed and structured agreement, the user acceptance of the project, and the mindset and skill base of the public sector.

U.S. laws and legislatures for PPPs: The U.S. laws and legislatures currently support various innovative financing mechanisms including PPPs. For instance, the Transportation Infrastructure Finance and Innovation Act (TIFIA) allows for direct credit assistance to major transportation projects from the traditional grant projects that do not generate revenue to projects that do generate revenue without governmental credit assistance. The states also provide revolving funds to support surface transportation projects known as state infrastructure banks, which provide credit assistance to attract public and private investments to help operate, maintain, expand, and upgrade transportation networks. Another provision is for public-interest projects where the state must have benefits and legal title vested in the government until after the bonds are paid. Such activities are eligible to the 63-20 public benefit corporations, which are nonprofit corporation authorized to issue tax-exempt debt on behalf of private project developers. Another financing tool facilitated by the government is private activity bonds, which essentially are tax-exempt bonds issued by the state or local government to help benefit qualified projects. According to the current U.S. laws, the transportation sector is eligible for tax-exempt bonds for a value up to $15 billion. For pricing-based financing to mitigate traffic congestion, a PPP does not necessarily dictate rates of user charges. However, flat-fee structure and variable pricing initiatives are the main and direct source of revenues for the PPP venture. Alternatively, the shadow tolling method can be adopted to make regular payments to the private partner, which are equal to the forecasted toll revenues.

Typical PPP financing structures and contract types: The primary categorization of PPPs is based on PPP ventures to existing facilities and new facilities, respectively. Traditionally, contracts of transportation projects follow the design–bid–build format, wherein the transportation agency designs the facility project by means of in-house staff or an external contractor and then bid the project for construction. This traditional arrangement ensures that the design risk, along with the operation and maintenance, is retained by the agency while the construction is carried out by an external contractor. New contract forms have become increasingly popular, utilizing innovative financing structures for facility project delivery, operations, and maintenance, which include (i) design–build–finance–operate; (ii) build–operate–transfer; and (iii) long-term lease concessions (NCHRP, 2009).

In addition, innovative contract types have been developed by (i) using a nontraditional form of project delivery; and (ii) introducing some forms of private debt or equity investment to ensure the quality of construction, in the form of performance guarantees for a specific period after the construction is finished. The primary contract structures include (i) design and build; (ii) design–build–operate–maintain; (iii) cost-plus-time bidding ($A + B$) contracting; (iv) construction manager/general contractor; and (v) construction management at risk. The *design and build* contract requires the contractor to assume both design and construction and the associated risks, while the public agency pays for the venture by a pre-decided payment structure. The *design–build–operate–maintain* contract demands that the contractor not only design and build, but also operate and maintain the facility for a predetermined contractual time. Another variant of the contract structure is $A + B$, which have an objective of minimizing the delivery time for projects with priority for fast completion, as in cases of dense urban areas to mitigate traffic disruptions during construction. The *construction manager/general contractor* contract is most suited for speedy bridge construction, by means of hiring two separate contractors for design and construction, but with the control of design lying with the public agency. The *construction management at risk* contract requires the need of a contract manager and a design contractor, and then the transportation agency and construction manager negotiate a contract as the design process progresses, facilitating better understanding and team relations over the process.

PROBLEMS

19.1 List all possible reasons that could ignite the need for seeking innovative financing schemes.

19.2 What is performance-based highway pavement management?

19.3 What is performance-based highway bridge management?

19.4 What is performance-based transit management?

19.5 What are the advantages and disadvantages of pursuing PPP models for project delivery? What should be the recommendation for successful PPP implementation?

19.6 What are the essential elements that are used to determine user price for the preservation of physical facility condition?

19.7 What are the essential elements that are used to determine user price for maintaining the sustainment of transportation system usage performance?

19.8 What are the key emerging issues of transportation system management in the 21st century?

19.9 What is dynamic transportation asset management? How will it support sustainable transportation development?

Chapter 20

Institutional issues

20.1 PERFORMANCE-BASED AGENCY REORGANIZATION

20.1.1 Theory of institutions and issues

In general, the theory of institutions developed by Williamson (1994) categorizes them as (i) informal; (ii) formal; (iii) governance; and (iv) resource allocation/employment related. The value system, customs and practices, and norms form the *Informal* category, which is slow to change but highly sensitive to sudden events such as acts of terrorism, while constitution, legislatures, public laws, and so on, form the *Formal* category, which is stable over the long term but can change quickly owing to radical movements or requirements. The third category of governance is formed by rules such as regulations, administrative orders, policy directives, and so on, which aid in how the government functions. The dynamics of interactions of various government as well as nonprofit organizations and their interactive behavior and patterns form the fourth category of resource allocation.

The transportation system and institutional issues within this complex dynamic system are threefold: (i) long-term issues; (ii) mid-term issues; and (iii) short-term issues based on the temporal effect of the issue (Rietveld and Stough, 2005). The state or local transportation agency has traditionally compartmentalized its institutional framework (modal), where each department dealt with a specific activity. The current institutional framework and hierarchy in the transportation sector specifically is not conducive for interagency data sharing and optimal decision-making. This limits the potential for achieving a sustainable transportation system that performs optimally.

20.1.2 Current scenario

The governance of the transportation system in the U.S. is stratified in terms of jurisdiction as being under federal, state, and local governance. The demarcation of the transportation network between the federal and other departments is straightforward in terms of the federal government operating and maintaining the freeway system. The state and local transportation agencies have tasks that might have overlaps and inherent redundancy/duplication of efforts owing to an institutional structure that lacks collaborative transparency. The agencies were formed and evolved primarily to meet the growing demand with objectives of mobility and accessibility when the automobile industry was growing. As the system grew larger owing to the demand, safety became a system goal and priority for decision-making. By the turn of the 21st century, the growing concern for the environment and social equity resulted in their inclusion as system goals. Eventually, with the advancement in technology, automobile engineering, computers, electrical engineering, and mechanical and construction engineering, the various fields of sciences were applied to the transportation system as well and the agencies had to expand and adapt to serve the changing user demand. Today, information technology and computers have created an inseparable and important component to the transportation system, which is the intelligent transportation systems (ITS) or the evolving smart transportation sector. This has forced many agencies to collaborate but still there are areas in which there is duplication of efforts owing to the lack of an institutional framework that encourages systematic field data collection and processing, proactive data prediction, effective data sharing, robust data analysis, and holistic decision-making.

20.1.3 Problems to be addressed

Transportation demand is not only growing but also changing every day with technological advancements in the tangential fields. To address and adapt to these changes is a very difficult task for the agencies, as they are formed based on policy and legislature, which require time to be altered and refined. This institutional inertia is one of the major issues that must be addressed if we were to optimize the performance of the transportation system. The current transportation legislation mandates the use of performance-based management, which requires collaboration among all the agencies and departments involved in the transportation system's management. Performance-based system management mandates the use of network-level analysis versus the traditionally used project-level analysis, which requires collaboration among local, state, and federal agencies; as in an urban land use these three interact to form the network. Network-level analysis ensures that the impact of a project will be analyzed in terms of induced demand in the network rather than only using estimated traffic volumes at the local project zone level. Network-level analysis ensures optimal system performance, even though the performance indicators might not be encouraging in terms of project-level impacts.

Another important issue to be dealt with is policy formulation and alignment across the entire vertical hierarchy, and standardized performance measures to ensure uniformity in performance evaluation of the system performance. The laws and legislatures mandate the use of performance measures and standards to achieve the system goals, which are outlined in the policy formulated for the future vision of the transportation system of a nation. The laws and legislatures not only support the future vision for the entire nation's system, but also ensure that the decision-making process is refined to make the system standardized, effective, and swift. It is also important to remember that the future vision is based on the demand forecasted as well as availability of the resources and funding. One of the important resources of a transportation system, particularly highways, is land, which is difficult to acquire, especially in an urban environment. Acquiring the necessary land and building the highway network are the two most costly components in a highway project. In an urban setting, it is generally recognized that capacity cannot be augmented, and the outlook should focus on operating the existing system efficiently rather than expanding it as a congestion mitigation strategy. This outlook has to be integrated along with policy goals across all agencies involved. When the subject of policy formulation is broached, there will always be a dichotomy in the interest at local and national levels. It is important to balance the two and at least align them so as to integrate the local agencies' priorities within the policy and collaborate with the federal agencies. At times, the national interest will supersede the local interest, and this should not be a hindrance for the future collaborative efforts between the federal and local agencies. This is certainly true of the highway system, but when the intermodal transportation system is studied and efforts are made to have seamless and efficient intermodal transportation systems, the collaboration and institutional transparency is even more important and challenging. Adding to these complex dynamics is the threat of natural disasters, emergency situations, and terrorism, which demand the collaboration of not only transportation management agencies, but emergency medical, law enforcement, and other related services as well.

20.1.4 Sustainable transportation system and issues

The agencies are still trying to adapt to the ever-changing scenario of user demand in which most sciences are merging in the quest to achieve a sustainable livable community with transportation at its core. In order to optimize any system, it is imperative that each dollar spent yields the maximum benefits, which cannot be achieved until not only the vertical hierarchy but the horizontal hierarchy as well is revamped to achieve optimal collaboration in terms of intra-agency as well as inter-agency performance. The barriers that have to be addressed for achieving sustainable transportation system are (i) policy barriers; (ii) legal barriers; (iii) institutional barriers; (iv) social and cultural barriers; (v) resources barriers; and (vi) physical barriers.

20.2 INTRA- AND INTER-AGENCY COORDINATION

The importance of coordination and its effect in the process of decision-making has been outlined by various research work. Fuller et al. (2004) studied the effect of multilevel governance frameworks on economic development. Richards et al. (1999) observed policy-related problems when institutes are interacting, rather than within a single institute. Stoker and Mossberger (1994) suggest that the alignment of several common priorities is required to achieve effective change along with a not very complex institutional framework.

The coordination within an agency and across various agencies has different sets of issues to deal with in a transportation system today, depending on the objective of collaboration, including: policy; funding and resource allocation; data collection, storage, analysis, and distribution; operations and maintenance; work zone management; and inventory management. Ensuring that the various barriers are addressed, it is essential to ensure institutional integration in terms of (i) horizontal integration; (ii) vertical integration; and (iii) spatial integration. *Horizontal integration* requires the policies for sustainable development be covered by one agency at all levels, while *vertical integration* deals with the interaction among various levels of agencies in a nation by means of standardization and understanding. The relationship among federal, state, and local agencies must be good to ensure a successful collaboration. *Spatial integration* ensures the coverage of all responsibilities within an extended urban area (Marsden and May, 2006).

20.2.1 Intra-agency coordination

Coordination is required within an agency at various hierarchical levels for the purpose of horizontal integration. The primary issue to be dealt with is internal resistance change. In order to overcome this hurdle, it is necessary to align the agency goals at all levels of the hierarchy and create objectives to help achieve the same. Once the objectives are established, the various levels of the agency are given flexibility to follow and adapt to the change in accordance with their norm and structure, but the adaptation is facilitated by a coordinator, ensuring that the objectives are realized in a positive manner. An example of this is when a new federal law mandates that the methods utilized by the state and local transportation agencies be changed, intra-agency coordination is required. When the transportation legislation mandated the use of performance-based management by the local agency to be eligible for federal funding, it required collaboration at every level of state transportation agencies. It is a difficult task to switch over from traditional methods that have been used for a long time. It also takes time to understand and train the staff to use the new method, and to ensure that past projects and data are converted if a standardized comparison is desired, which takes time and effort, and might be viewed as a burden by the affected agency in the beginning. Resistance to a paradigm change is only natural, and facilitation along with flexibility and training in the early phase of change helps overcome the issue.

20.2.2 Inter-agency coordination

This issue must be addressed when attempting vertical integration for the purpose of interagency collaboration. When different agencies are interacting, issues of territorialism, institutional friction, the informal structure of the agencies, aligning agency goals, vested interests, and power distribution among stakeholders are a few of the issues to be dealt with. An example of such a dynamic situation is a mega project for a city: the highway system, transit agencies, subcontracting consultants, and state and local transportation agencies are all involved. Such a myriad composition will lead to a struggle for responsibilities and decision-making power. Successful collaboration will require sharing of data, duties, and responsibilities. The subject of data sharing in interagency collaboration is especially sensitive, as each agency is skeptical and wants primarily to protect its own interests. Here the formulation of the project and its intended benefits to each involved agency is of prime importance for successful data collaboration. An important point to remember in such collaborations is that each involved agency has to be treated as a separate entity, and the notion of collaboration should not assume seamless functioning of the agencies. Each agency will help achieve the collaborative task as individual agencies by sharing only as demanded by the task.

20.3 AGENCY STAFF DEVELOPMENT

A prominent problem facing transportation agencies is the development of their staffs to keep abreast with the demand that technological advancements impose in terms of skill base and current required technical knowledge to provide the necessary support for optimal system performance. In most projects today, many of the responsibilities are subcontracted/contracted to private consulting firms that provide the required technological and technical expertise and support to the agency. The cash-strapped agency has limited resources, but the cost is eventually borne by the agency for the manpower hired by consulting firms. As a private entity, the consulting firm is profit-driven, and in a public-sector project, where public interests are to be safeguarded, sometimes more money is spent than if the agency had been able to finish the project by in-house staff—which would call for a necessary change in the agency staff, and it is always struggling to keep up with user demand in rapidly changing times and technology.

The agencies and industry have always provided feedback to academic institutions to shape the curriculum to meet the future requirements of the industry in terms of employment opportunities as well as required skill-set. However, technological advancements, the merging of various fields of engineering, and the increasingly integrated transportation system have made it imperative that experts in each relevant field be hired in-house by transportation agencies to curtail expenditures on consultants, and to better safeguard public interests.

Data collection and analysis efforts are a major part of the transportation agency's task of keeping the system operating efficiently. With the big data sets being collected, stored, and analyzed, a large number of consultants are hired, ranging from electrical, computers, hardware, data analysts, and many others. Not all consultants can possibly be eliminated, but the agencies surely need to develop their workforce toward the need for data processing and analysis. Most often consultants process data collected by public agencies and sell it back to the market for public use. This indicates a large gap of service that must be eliminated by the agency to ensure a more fiscally responsible management of the taxpayers' money for the data collection in the first place. Contractual laws can safeguard the data, but at times the processing makes it proprietary, and hence in-house data analysis staff can aid and eventually eliminate the need for consultants in this aspect, at least for system management.

This process of development cannot be done in a simple manner by hiring alone; the agencies need to work now with academic institutions to fashion a learning database from their varied experiences with the consultants and subcontractors. This will help the agency learn about the job expectations, and help their staff deal with the tasks and the technology being used, and their dependence on manpower and personnel versus technology to achieve the task. The current trend is that agency staff is attracted to the consulting firm and hired there eventually in managerial tasks, but the agencies need to learn the consultant's management practices in turn and hire in-house staff based on those experiences. Only then can they hope to meet the expectations for the future to be able to guard public interests and ensure optimal expenditures.

20.4 COMMUNICATIONS WITH STAKEHOLDERS

Conceptual development, the first step of a public project, requires dealing with various issues such as land acquisitions, alignments, environmental impacts, social acceptance, financing options and alternatives, public benefits, social equity, job creations, and so on. Once a project is conceptualized and choices pertaining to the above-listed factors are made to reach the next phase feasibility study of all possible alternatives, it is important to include the factor of communication with all involved stakeholders.

Stakeholders are all the parties involved in the initiated project who have an interest directly or indirectly either as users or nonusers and can decide the acceptance and success of the project. The essence of this process is transparency and information regarding each involved parameter and its impact on the stakeholder. The next important aspect of this communication is the information of benefits and dis-benefits relayed to stakeholders and methods for quantification of these impacts utilized for the feasibility study, to inform stakeholders of effects and impacts of the project, and their direct and indirect social impacts on them.

The goal of this communication is that all involved stakeholders are well informed about the agency's outlook, goals and objectives, parameters, performance measures that will be used to estimate the extent of goals to be achieved, tangible and intangible impacts of the project, their quantification process, funding sources and alternatives to the project, and strategies pertaining to the project and the reason for the initiative. Once stakeholders are all well informed, they can make a realistic decision based on facts rather than notions. The focus of this effort should be only on relaying the information, rather than convincing them. The information will eventually help them decide on their approval or disapproval of the project.

Once stakeholders form their decision, they can begin the next step of evaluating the reasons and causes of their decision. In case the decision is not favorable to the agency, this process of evaluation will help the agency understand the reasons and might help in bridging expectations of stakeholders and the agency's standing, to push the project to the next phase of design. In the worst case, if stakeholders cannot be convinced, the information will help the agency in the next project or the alternate venture for the same goal achievement. This process is not as simple as it seems, owing to heavy political influences at times, which are not based on facts, but rather lobbying by one faction of stakeholders who prioritize their profits over public benefits in general. The efforts on the part of the agency in this particular phase are to inform only, and tackle all opposition one stakeholder at a time. There are many methods and departments that deal with this phase effectively. These departments hold meetings on a regular basis with all various interest groups and stakeholders to align agency goals and objectives with stakeholder expectations and valid demands. The meetings may also include the local elected officials. These meetings also include surveys for interests and opinions for future projects and community development.

20.5 ACCEPTANCE BY USERS AND NONUSERS

The acceptance by users is key to the success of any initiative. The acceptance of the various factions of stakeholders is related to benefits they perceive from the venture. It is important to realize that to an individual group of stakeholders, the project is like any other project and the perceived profit will influence the decision to accept the project. Granovetter (1985) indicated that the social relations of the interacting entities dictate their economic decisions, rather than outside a social context and in accordance with or to the expected behavior scripted in the interaction by expectations. He also observed that interpersonal ties and not institutional structure decide the order and honesty in the relations. However, there is a problem that follows trust in interpersonal ties, which is dishonesty or malfeasance. This might not protect the public interest as intended at the start of the task. McGuire et al. (1993) recommended the use of historical case studies to observe how the interpersonal ties and organizational structure evolved to help make the interaction advantageous.

Keeping these hurdles and the objective of public interest in mind, the agency could try to communicate and eventually convince stakeholders to be onboard with the project. The users' acceptance is paramount to the success of the project. The myopic nature of the dynamics and analysis of the system sometimes render the task of balancing all stakeholders' interest to the same degree and to their satisfaction somewhat difficult. Once users accept the project, the feedback will help improve prospects for its success.

PROBLEMS

20.1 List various transportation stakeholders.

20.2 Formulate a definition for an institution, which is ideal in terms of function and achievements for a public agency.

20.3 What are the problems encountered during intra-agency integration?

20.4 What are the problems that are encountered during inter-agency integration?

20.5 What are the issues facing public agency in terms of workforce developments?

20.6 What is the purpose of communication with stakeholders?

20.7 What are the hurdles in the process of communication with stakeholders?

20.8 Describe an innovative approach according to you to improve interpersonal ties to initiate trust and transparency in the process of communicating with interest groups.

20.9 Why is acceptance of users important for a public agency?

References

AAA 1999. *Your Driving Costs.* American Automobile Association, Heathrow, FL.

Aalen, O.O. and Johansen, S. 1978. An empirical transition matrix for non-homogeneous Markov chains based on censored observations. *Scandinavian Journal of Statistics* 5(3), 141–150.

AASHTO 1978. *A Manual on User Benefit Analysis of Highway and Bus-Transit Improvements.* American Association of State Highway and Transportation Officials, Washington, D.C.

AASHTO 1990. *Guidelines for Pavement Management Systems.* American Association of State Highway and Transportation Officials, Washington, D.C.

AASHTO 1993. *AASHTO Guide for Design of Pavement Structures.* American Association of State Highway and Transportation Officials, Washington, D.C.

AASHTO 1997. *AASHTO Guide for Commonly-Recognized (CoRe) Structural Elements.* American Association of State Highway and Transportation Officials, Washington, D.C.

AASHTO 1998. *AASHTO Guide for Design of Pavement Structures*, 4th Edition. American Association of State Highway and Transportation Officials, Washington, D.C.

AASHTO 2001. *A Policy on Geometric Design of Highways and Street.* American Association of State Highway and Transportation Officials, Washington, D.C.

AASHTO 2002. *Transportation Asset Management Guide.* NCHRP Project 20-24(11). American Association of State Highway and Transportation Officials, Washington, D.C.

AASHTO 2003. *User Benefit Analysis for Highways.* American Association of State Highway and Transportation Officials, Washington, D.C.

AASHTO 2010a. *AASHTO Bridge Element Inspection Manual.* American Association of State Highway and Transportation Officials, Washington, D.C.

AASHTO 2010b. *Highway Safety Manual*, 1st Edition. American Association of State Highway and Transportation Officials, Washington, D.C.

AASHTO 2011. *A Policy on Geometric Design of Highways and Streets.* American Association of State Highway and Transportation Officials, Washington, D.C.

Adams, T.M., Koncz, N.A., and Vonderohe, A.P. 2001. *Guidelines for the Implementation of Multimodal Transportation Location Referencing Systems.* NCHRP Report 460. Transportation Research Board, National Academies Press, Washington, D.C.

Aggarwal, V., Deo, N., and Sarkar, D. 1992. The knapsack problem with disjoint multiple-choice constraints. *Naval Research Logistics* 39, 213–227.

Akbar, M.M., Manning, E.G., Shoja, G.C., and Khan, S. 2001. Heuristic Solutions for the Multiple-Choice Multi-Dimension Knapsack Problem. Working Paper. University of Victoria, Victoria, Canada.

Akcelik, R. 1991. Travel time functions for transport planning purposes: Davison's function, its time-dependent form and an alternative travel time function. *Australian Road Research* 21, 49–59.

Allais, M. and Roth, G. 1968. *The Economics of Road User Charges.* The World Bank, Washington, D.C.

Ang, A.H.S. and Tang, W.H. 1984a. *Probability Concepts in Engineering Planning and Design, Volume I: Basic Principles.* John Wiley and Sons, Hoboken, NJ.

Ang, A.H.S. and Tang, W.H. 1984b. *Probability Concepts in Engineering Planning and Design, Volume II: Decision Risk and Reliability.* John Wiley and Sons, Hoboken, NJ.

Anselin, L. 1995. Local indicators of spatial association-LISA. *Geographical Analysis* 27(2), 93–115.
Anwaar, A., Labi, S., Li, Z., and Shields, T. 2013. Aggregate and disaggregate statistical evaluation of the performance-based effectiveness of LTPP SPS-5 flexible pavement rehabilitation treatments. *Journal of Structure and Infrastructure Engineering* 9(2), 172–187.
APTA 1991. *Passenger Transport*. American Public Transportation Association, Washington, D.C.
Ardekani, S. and Herman, R. 1987. Urban network-wide traffic variables and their relations. *Transportation Science* 21, 1–16.
AREMA 2003. *Practical Guide to Railway Engineering*. American Railway Engineering Maintenance-of-Way Association, Lanham, MD.
Arminger, G., Clogg, C.C., and Sobel, M.E. 1995. *Handbook of Statistical Modeling for the Social and Behavioral Sciences*. Springer, New York, NY.
Armstrong, R.D., Kung, D.S., Sinha, P., and Zoltners, A.A. 1983. A computational study of a multiple-choice knapsack algorithm. *ACM Transactions on Mathematical Software* 9, 184–198.
Attoh-Okine, N.O. and Bowers, S. 2006. A Bayesian belief network model of bridge deterioration. *ICE Journal of Bridge Engineering* 159(2), 69–76.
Bahar, G., Masliah, M., Erwin, T., Tan, E., and Hauer, E. 2006. *Pavement Marking Materials and Markers: Real-World Relationship between Retroreflectivity and Safety over Time*. NCHRP Web-Only Document 92: Contractor's Final Report for NCHRP Project. Transportation Research Board, National Academies Press, Washington, D.C.
Bai, Q., Ahmed, A., Li, Z., and Labi, S. 2015. A hybrid Pareto frontier generation method for analyzing trade-offs among transportation performance measures. *Wiley Journal of Computer-Aided Civil and Infrastructure Engineering* 30(3), 163–180.
Bain, R. 2009. *Big Numbers Win Prizes: Twenty-One Ways to Inflate Toll Road Traffic and Revenue Forecasts*. Project Finance International, Leeds, United Kingdom.
Baker, W.T. and Blessing, W.E. 1974. *Highway Location Reference Methods*. NCHRP Synthesis of Highway Practice 21. Transportation Research Board, National Academies Press, Washington, D.C.
Bamzai, R., Machordomand, L., and Li, Z. 2009. *Safety Impacts of Highway Shoulder Attributes in Illinois*. Department of Civil, Architectural, and Environmental Engineering, Illinois Institute of Technology, Chicago, IL.
BaZant, Z.P. and Baweja, S. 1995. Creep and shrinkage prediction model for analysis and design of concrete structures-Model B3. *Materials and Structures* 28, 357–365, 415–430, 488–495.
Bazaraa, M.S., Jarvis, J.J., and Sherali, H.D. 2009. *Linear Programming and Network Flows*, 4th Edition. John Wiley and Sons, Hoboken, NJ.
Bazaraa, M.S., Sherali, H.D., and Shetty, C.M. 2006. *Nonlinear Programming: Theory and Algorithms*, 3rd Edition. Wiley-Interscience, Hoboken, NJ.
Beckmann, M., McGuire, C.B., and Winsten, C.B. 1956. *Studies in the Economics of Transportation*. Yale University Press, New Haven, CT.
Bein, P. 1997. *Monetization of Environmental Impacts of Roads*. B.C. Ministry of Transportation and Highways, Victoria, Canada.
Bellman, R.E. and Dreyfus, S.E. 2015. *Applied Dynamic Programming*. Princeton University Press, Princeton, NJ.
Ben-Akiva, M., Humplick, F., Madanat, S.M., and Ramaswamy, R. 1991. Latent performance approach to infrastructure management. *TRB Journal of Transportation Research Record* 1311, 188–195.
Ben-Akiva, M. and Lerman, S. 1985. *Discrete Choice Analysis: Theory and Application to Travel Demand*. The MIT Press, Cambridge, MA.
Benayoun, R., De Montgolfier, J., Tergny, J., and Laritchev, O. 1971. Linear programming with multiple objective functions: Step method (STEM). *Mathematical Programming* 1(1), 366–375.
Bentler, P.M. and Weeks, D.G. 1980. Linear structural equations with latent variables. *Psychometrika* 45, 289–308.
Berglund, B., Lindvall, T., and Schwela, D.H. 1999. *Guidelines for Community Noise*. World Health Organization, Geneva, Switzerland.
Bhatia, R. 2014. *Noise Pollution: Managing the Challenge of Urban Sounds*. Earth Journalism Network, Washington, D.C.

BHD 2010. *Maintenance Manual of Sidewalks and Bikeways*. Town of Bethlehem Highway Department, Bethlehem, NY.

Birge, J.R. and Louveaux, F. 1997. *Introduction to Stochastic Programming*. Springer-Verlag, New York, NY.

Black, A. 1995. *Urban Mass Transportation Planning*. McGraw-Hill, New York, NY.

Black, K.L., McGee, H.W., Hussain, S.F., and Rennilson, J.J. 1991. *Service Life of Retroreflective Traffic Signs*. Report No. FHWA-RD-90-10. Federal Highway Administration, U.S. Department of Transportation, Washington, D.C.

Boardman, A.E. and Lave, L.B. 1977. Highway congestion and congestion tolls. *Journal of Urban Economics* 4, 340–359.

Bobzin, H. 2006. *Principles of Network Economics*. Springer, Berlin, Germany.

Bodie, Z., Kane, A., and Marcus, A.J. 2006. *Investments*, 7th Edition. McGraw-Hill/Irwin, New York, NY.

Bonneson, J.A., Zimmerman, K., and Fitzpatrick, K. 2005. *Roadway Safety Design Synthesis*. No. FHWA/TX-05/0-4703-P1. Texas Transportation Institute, Texas A&M University System, College Station, TX.

Booz Allen and Hamilton Inc., Hagler Bailly Services Inc., and Brinckerhoff, P. 1999. *California Life-Cycle Benefit/Cost Model*. California Department of Transportation, Sacramento, CA.

Box, G.E.P. Hunter, J.S., and Hunter, W.G. 2005. *Statistics for Experimenters: Design, Innovation, and Discovery*, 2nd Edition. Wiley-Interscience, Hoboken, NJ.

Box, G.E.P. and Tiao, G.C. 1992. *Bayesian Inference in Statistical Analysis*. Wiley-Interscience, Hoboken, NJ.

Boyd, S. and Vandenberghe, L. 2004. *Convex Optimization*. Cambridge University Press, Cambridge, United Kingdom.

BPA 1987. *New Home Energy Conservation Programs Environmental Impact Statement*. Bonneville Power Administration, U.S. Department of Energy, Portland, OR.

BPR 1964. *Traffic Assignment Manual*. Bureau of Public Roads, U.S. Department of Commerce, Washington, D.C.

Brans, J.P. 1982. Lingenierie de la decision. Elaboration dinstruments daide a la decision. Methode PROMETHEE. Editors: Nadeau, R. and Landry, M. *Laide a la Decision: Nature, Instruments et Perspectives Davenir*. Presses de Universite Laval, Quebec, Canada, 183–214.

Breiman, L., Friedman, J.H., Olshen, R.A., and Stone, C.J. 1984. *Classification and Regression Trees*. Chapman and Hall, New York, NY.

Brockwell, P.J. and Davis, R.A. 1998. *Time Series: Theory and Methods*, 6th Printing Edition. Springer, Berlin, Germany.

Bronshtein, I.N. and Semendyayev, K.A.F. 2013. *Handbook of Mathematics*. Springer Science and Business Media, Berlin, Germany.

BTS 2014. *National Transportation Statistics*. Bureau of Transportation Statistics. U.S. Department of Transportation, Washington, D.C.

Burris, M., Lee, S., Geiselbrecht, T., Baker, R., and Weatherford, B. 2013. *Equity Evaluation of Sustainable Mileage-Based User Fee Scenarios*. Report 600451-00007-1. Texas Transportation Institute, Texas A&M University System, College Station, TX.

Burrow, M.P.N., Teixeira, P.F., Dahlberg, T., and Berggren, E.G. 2009. Track stiffness considerations for high speed railway lines. *Railway Transportation: Policies, Technology and Perspectives*, 1–55.

Bus, J.C.P. and Dekker, T.J. 1975. Two efficient algorithms with guaranteed convergence for finding a zero of a function. *ACM Transactions on Mathematical Software* 1(4), 330–345.

Bushell, M.A., Poole, B.W., Zegeer, C.V., and Rodriguez, D.A. 2013. *Costs for Pedestrian and Bicyclist Infrastructure Improvements*. Highway Safety Research Center, University of North Carolina, Chapel Hill, NC.

Button, K. 2004. *Road Pricing*. Center for ITS Implementation Research, Center for Transportation Policy, Operations and Logistics, School of Public Policy, George Mason University, Fairfax, VA.

Cafiso, S., La Cava, G., Montella, A., and Pappalardo, G. 2006. A procedure to improve safety inspections effectiveness and reliability on rural two-lane highways. *The Baltic Journal of Road and Bridge Engineering* 1(3), 143–150.

CALTRANS 2006. *Maintenance Manual, Volumes I and II*. California Department of Transportation, Sacramento, CA.

Cambridge Systematics, Inc. 2000. *A Guidebook for Performance-Based Transportation Planning*. NCHRP Report 446. Transportation Research Board, National Academies Press, Washington, D.C.

Cambridge Systematics, Inc., PB Consult, Inc., and Texas Transportation Institute. 2006. *Performance Measures and Targets for Transportation Asset Management*. NCHRP Report 551. Transportation Research Board, National Academies Press, Washington, D.C.

Cambron, K.E. and Evans, G.W. 1991. Layout design using the analytic hierarchy process. *Computers and Industrial Engineering* 20(2), 211–229.

Cascadia Center 2008. *Large Diameter Soft Ground Bored Tunnel Review: Review of Current Industry Soft Ground Bored Tunnel Practice*. Report REP/208085/S001. Cascadia Center of the Discovery Institute, Seattle, WA.

Catalina, A.J. 2016. Development of a Statistical Theory-Based Capital Cost Estimating Methodology for Light Rail Transit Corridor Evaluation Under Varying Alignment Characteristics. Ph.D. Dissertation. The University of Kentucky, Lexington, KY.

CEB 1993. CEB-FIP Model Code 1990. *CEB Bulletin d'Information* 2131214, 33–41. Comite Euro-International du Beton, Lausanne, Switzerland.

CEC 1989. Valuing Emission Reductions for ER90. Staff Issue Paper No. 3. California Energy Commission, Sacramento, CA.

CEC 1993. Externalities of the Fuel Cycle: External Project. Working Documents 1, 2, 5, and 9. Commission of the European Communities, Brussels, Belgium.

Cesare, M.A., Santamarina, C., Turkstra, C., and Vanmarcke, E.H. 1992. Modeling bridge deterioration with Markov chains. *ASCE Journal of Transportation Engineering* 118(6), 820–833.

Cesare, M.J., Santamaria, C., Turkstra, C.J., and Vanmarcke, E. 1994. *Risk-Based Bridge Management: Optimization and Inspection Scheduling*. Canadian Journal of Civil Engineering, Ottawa, Canada.

Chankong, V.V. and Haimes, Y.Y. 1983. *Multiobjective Decision Making: Theory and Methodology*. North-Holland, Amsterdam, Netherland.

Charnes, A., Cooper, W.W., Golany, B., Seiford, L., and Stutz, J. 1985. Foundations of data envelopment analysis for Pareto–Koopmans efficient empirical production functions. *Journal of Econometrics* 30(1/2), 91–107.

Chu, P.C. and Beasley, J.E. 1998. A genetic algorithm for the multidimensional knapsack problem. *Journal of Heuristic* 4, 63–86.

Cifuentes, L.A. and Lave, L.B. 1993. Economic valuation of air pollution abatement: Benefits from health effects. *Annual Review of Energy and the Environment* 18(1), 319–342.

Cohen, L.F. and McVoy, G.R. 1982. *Environmental Analysis of Transportation Systems*. John Willey and Sons, New York, NY.

Cohn, L., Wayson, R., and Harris, R. 1992. Environmental and energy considerations. In *Transportation Planning Handbook*. Editor: John D. Edwards. Institute of Transportation Engineers, Washington, D.C., 447–470.

Cohon, J.L. and Mark, D.H. 1975. A review and evaluation of multiobjective programming techniques. *Water Resource Research* 11(2), 208–220.

Collins, L. 1972. *Industrial Migration in Ontario; Forecasting Aspects of Industrial Activity through Markov chain analysis*. Statistics Canada, Ottawa, Canada, 147–154.

Cox, D.R. 1972. Regression models and life tables (with discussion). *Journal of the Royal Statistical Society*, B34, 87–220.

Dafermos, S.C. and Sparrow, F.T. 1969. The traffic assignment problem for a general network. *Journal of Research of the National Bureau of Standards* 73B(2), 91–118.

Daganzo, C.F. 1997. *Fundamentals of Transportation and Traffic Operations*, 1st Edition. Pergamon Press, Oxford, United Kingdom.

Debaillon, C., Carlson, P., He, Y., Schnell, T., and Aktan, F. 2007. *Updates to Research on Recommended Minimum Levels for Pavement Marking Retroreflectivity to Meet Driver Night Visibility Needs*. Report No. FHWA-HRT-07-059. Federal Highway Administration, U.S. Department of Transportation, Washington, D.C.

De La Garza, J.M., Drew, D.R., and Chasey, A.D. 1998. Simulating highway infrastructure management policies. *Journal of Engineering Management* 13(5), 64–72.
De Leur, P. and Sayed, T. 2002. Development of a road safety risk index. *TRB Journal of Transportation Research Record* 1784, 33–42.
Delucchi, M. and Hsu, S.L. 1998. The external damage cost of noise emitted from motor vehicles. *Journal of Transportation and Statistics* 1(3), 1–24.
DeStefano, P.D. and Grivas, D. 1998. A method for estimating transition probability in bridge deterioration models. *ASCE Journal of Infrastructure Systems* 4(2), 56–62.
Dujmovic, J.J. 1977. The preference scoring method for decision-making: Survey, classification, and annotated bibliography. *Informatica* 1(2), 26–34.
Dyer, J.S., Fishburn, P.C., Steuer, R.E., Wallenius, J., and Zionts, S. 1992. Multiple criteria decision making, multiattribute utility theory: The next ten years. *Management Science* 38(5), 645–654.
Dyer, M.E., Riha, W.O., and Walker, J. 1995. A hybrid dynamic programming/branch and-bound algorithm for the multiple-choice knapsack problem. *Journal of Computational and Applied Mathematics* 58(1), 43–54.
Dyer, J.S. and Sarin, R.K. 1979. Measurable multiattribute value functions. *Operations Research* 27(4), 810–822.
Ebers Öhn, W. and Conrad, J.R. 1998. Implementing a railway infrastructure maintenance system. In *Conference on Railway Engineering Proceedings: Engineering Innovation for a Competitive Edge*. CQU Publication Service, Rockhampton, Queensland, Australia, 395–402.
ECONorthwest and PBQD 2002. *Estimating the Benefits and Costs of Public Transit Projects: A Guidebook for Practitioners*. TCRP Report 78. Transit Cooperative Research Program, Transportation Research Board, National Academies Press, Washington, D.C.
ECORYS Transport and METTLE 2005. *Charging and Pricing in the Area of Inland Waterways: Practical Guideline for Realistic Transport Pricing*. Report for European Commission. Rotterdam, Netherland.
Elkafoury, A., Negm, A.M., Bady, M., and Aly, M.H.F. 2014. Review of transport emission modeling and monitoring in urban areas—Challenge for developing countries. In *Advanced Logistics and Transport (ICALT), 2014 International Conference IEEE*, Tunis, Tunisia, 23–28.
Elkins, C.L. 1985. Acid rain policy development. *Journal of the Air Pollution Control Association* 35(3), 202–204.
EPA 2011. *Development of Emission Rates for Light-Duty Vehicles in the Motor Vehicle Emissions Simulator (MOVES2010)*. U.S. Environmental Protection Agency, Washington, D.C.
EPA 2016. *National Ambient Air Quality Standards*. U.S. Environmental Protection Agency, Washington, D.C.
EPA 2017. *Light-Duty Vehicles and Light-Duty Trucks: Clean Fuel Fleet Exhaust Emission Standards*. Office of Transportation and Air Quality, U.S. Environmental Protection Agency, Washington, D.C.
EPRI 1990. *Efficient Electricity Use: Estimates of Maximum Energy Savings*. Report EPRI CU-6746s. Electric Power Research Institute, Charlotte, NC.
ESRI 2003. *Linear Referencing in ArcGIS: Practical Considerations for the Development of an Enterprisewide GIS*. Environmental Systems Research Institute, Inc., Redlands, CA.
Estes, A.C. and Frangopol, D.M. 1999. Optimum lifetime planning of bridge inspection/repair programs. *Journal of IABSE, Structural Engineering International* 9(3), 219–223.
Evans, G.W. 1984. An overview of techniques for solving multiobjective mathematical programs. *Management Science* 30(11), 1268–1282.
Ewing, R. and Cervero, R. 2001. Travel and the built environment: A synthesis. *TRB Journal of Transportation Research Record* 1780, 87–114.
Feighan, K.J., Shahin, M.Y., Sinha, K.C., and White, T.D. 1989. An application of dynamic programming and other mathematical techniques to pavement management systems. *TRB Journal of Transportation Research Record* 1215, 101–114.
Fekpe, E.S., Windholz, T., Beard, K., and Novak, K. 2003. *Quality and Accuracy of Positional Data in Transportation*. NCHRP Report 506. Transportation Research Board, National Academies Press, Washington, D.C.

FHWA 1985. *Traffic Control Systems Handbook.* Report No. FHWA-IP-85-11. Federal Highway Administration, U.S. Department of Transportation, Washington, D.C.

FHWA 1987. *Bridge Management Systems.* Demonstration Project No. 71. Federal Highway Administration, U.S. Department of Transportation, Washington, D.C.

FHWA 1991. *Pavement Management Systems.* Federal Highway Administration, U.S. Department of Transportation, Washington, D.C.

FHWA 1994. *Motor Vehicle Accident Costs.* Federal Highway Administration, U.S. Department of Transportation, Washington, D.C.

FHWA 1995. *Recording and Coding Guide for the Structure Inventory and Appraisal of the Nation's Bridges.* Report No. FHWA-PD-96-001. Federal Highway Administration, U.S. Department of Transportation, Washington, D.C.

FHWA 1998a. *FHWA Traffic Noise Model (FHWA TNM®) Technical Manual.* Federal Highway Administration, U.S. Department of Transportation, Washington, D.C.

FHWA 1998b. *Life-Cycle Cost Analysis in Pavement Design-In Search of Better Investment Decisions.* Federal Highway Administration, U.S. Department of Transportation, Washington, D.C.

FHWA 1998c. *1997 Federal Highway Cost Allocation Study Final Report.* Federal Highway Administration, U.S. Department of Transportation, Washington, D.C.

FHWA 1999. *Asset Management Primer.* Office of Asset Management, Federal Highway Administration, U.S. Department of Transportation, Washington, D.C.

FHWA 2000. *Highway Economic Requirements System.* Federal Highway Administration, U.S. Department of Transportation, Washington, D.C.

FHWA 2001a. *FHWA SMART Van.* Federal Highway Administration, U.S. Department of Transportation, Washington, D.C.

FHWA 2001b. *Data Integration Primer.* Federal Highway Administration, U.S. Department of Transportation, Washington, D.C.

FHWA 2003. *Freeway Management and Operations Handbook.* Report No. FHWA-OP-04-003. Federal Highway Administration, U.S. Department of Transportation, Washington, D.C.

FHWA 2004a. *Mitigating Traffic Congestion: The role of Demand-Side Strategies.* Federal Highway Administration, U.S. Department of Transportation, Washington, D.C.

FHWA 2004b. *Traffic Congestion and Reliability: Linking Solutions to Problems.* Federal Highway Administration, U.S. Department of Transportation, Washington, D.C.

FHWA 2005a. *Elements of a Comprehensive Signals Asset Management System.* Federal Highway Administration, U.S. Department of Transportation, Washington, D.C.

FHWA 2005b. *Roadway Safety Hardware Asset Management Systems Case Studies.* Report No. FHWA-HRT-05-073. Federal Highway Administration, U.S. Department of Transportation, Washington, D.C.

FHWA 2005c. *Traffic Congestion and Reliability: Trends and Advanced Strategies for Congestion Mitigation.* Federal Highway Administration, U.S. Department of Transportation, Washington, D.C.

FHWA 2007. *Maintaining Traffic Sign Retroreflectivity.* Report FHWA-SA-07-020. Federal Highway Administration, U.S. Department of Transportation, Washington, D.C.

FHWA 2009a. *Manual on Uniform Traffic Control Devices.* Federal Highway Administration, U.S. Department of Transportation, Washington, D.C.

FHWA 2009b. *Economics: Pricing, Demand, and Economic Efficiency: A Primer.* Office of Transportation Management, Federal Highway Administration, U.S. Department of Transportation, Washington, D.C.

FHWA 2010. *SafetyAnalystTM: Software Tools for Safety Management of Specific Highway Sites.* Report FHWA-HRT-10-063. Federal Highway Administration, U.S. Department of Transportation, Washington, D.C.

FHWA 2011. *Bridge Preservation Guide.* Report No. FHWA-HIF-11042. Federal Highway Administration, U.S. Department of Transportation, Washington, D.C.

FHWA 2013. *Practical Guide for Quality Management of Pavement Condition Data Collection.* Federal Highway Administration, U.S. Department of Transportation, Washington, D.C.

FHWA 2014. *Highway Performance Monitoring System Field Manual.* Federal Highway Administration, U.S. Department of Transportation, Washington, D.C.

FHWA 2016. *Highway Statistics.* Federal Highway Administration, U.S. Department of Transportation, Washington, D.C.

FHWA 2017. *Employment Impacts of Highway Infrastructure Investment.* Federal Highway Administration, U.S. Department of Transportation, Washington, D.C.

Fielding, G.J. 1987. *Managing Public Transit Strategically: A Comprehensive Approach to Strengthening Service and Monitoring Performance.* Jossey-Bass, San Francisco, CA.

Fielding, G.J. 1992. Transit performance evaluation in the U.S.A. *Transpiration Research Part A* 26(6), 483–491.

Ford, J.L. and Ghose, S. 1998. *Lottery Designs to Discriminate between Shackle's Theory, Expected Utility Theory and Non-Expected Utility Theories. Annals of Operations Research.* Kluwer Academic Publishers, Norwell, MA.

Fotheringham, A.S., Brunsdon, C., and Charlton, M. 2003. *Geographically Weighted Regression: The Analysis of Spatially Varying Relationships.* John Wiley and Sons, Hoboken, NJ.

Freville, A. and Plateau, G. 1994. An efficient preprocessing procedure for the multidimensional 0-1 knapsack problem. *Discrete Applied Mathematics* 49(1–3), 189–212.

Frey, H., Rouphail, N., and Zhai, H. 2006. Speed- and facility-specific emission estimates for on-road light-duty vehicles based on real-world speed profiles. *TRB Journal of Transportation Research Record* 1987, 128–137.

Frey, H., Rouphail, N., and Zhai, H. 2008. Link-based emission factors for heavy-duty diesel trucks based on real-world data. *TRB Journal of Transportation Research Record* 2058, 23–32.

Friesz, T. and Fernandez, E.J. 1979. A model of optimal transport maintenance with demand responsiveness. *Transportation Research Part B* 13(4), 317–339.

FSTINC 2014. *2014 Pedestrian Accessibility Study.* Report for Town of Lexington Department of Public Works, MA. Fay, Spofford and Thorndike, Inc., Burlington, MA.

FTA 1994. *Rail Modernization Study.* Federal Transit Administration, U.S. Department of Transportation, Washington, D.C.

FTA 2016. *Final Interim Policy Guidance Federal Transit Administration Capital Investment Grant Program.* Federal Transit Administration, U.S. Department of Transportation, Washington, D.C.

Fu, G. and Devaraj, D. 2008. *Methodology of Homogeneous and Non-Homogeneous Markov Chains for Modelling Bridge Element Deterioration.* Michigan Department of Transportation, Lansing, MI.

Fuller, C., Bennett, R., and Ramsden, M. 2004. Local government and the changing institutional landscape of economic development in England and Wales. *Environment and Planning C: Government and Policy* 22, 317–347.

Fwa, T.F. and Sinha, K.C. 1987. Estimation of environmental and traffic loading effects on highway pavements. *Australian Road Research* 17(4), 256–264.

Garber, N.J. and Hoel, L.A. 2014. *Traffic and Highway Engineering*, 5th Edition. Cengage Learning, Stamford, CT.

Gattuso, D. and Restuccia, A. 2014. A tool for railway transport cost evaluation. *Procedia—Social and Behavioral Sciences* 111, 549–558.

Geoffrion, A.M., Dyer, J.S., and Feinberg, A. 1972. An interactive approach for multi-criterion optimization, with an application to the operation of an academic department. *Management Science* 19(4), 357–368.

Geoffroy, D.N. 1996. Cost-effective Preventive Pavement Maintenance. *NCHRP Synthesis 223.* Transportation Research Board, National Academies Press, Washington, D.C.

Geoffroy, D.N. and Shufon, J.J. 1992. Network level pavement management in New York State: A goal-oriented approach. *TRB Journal of Transportation Research Record* 1344, 57–65.

George, K.P., Rajagopal, A.S., and Lim, L.K. 1989. Models for predicting pavement deterioration. *TRB Journal of Transportation Research Record* 1215, 1–7.

Gibbons, J. 1976. *Nonparametric Methods for Quantitative Analysis.* Holt, Rinehart and Winston Publisher, Austin, TX.

Gilks, W.R., Richardson, S., and Spiegelhalter, D. 1995. *Markov Chain Monte Carlo in Practice.* CRC Press, Boca Raton, FL.

Gion, L.C., Gough, J., Sinha, K.C., and Woods, R.E. 1993. *User's Manual for the Implementation of the Indiana Bridge Management System.* Purdue University, West Lafayette, IN.

Giuliano, G. 1994. *Equity and Fairness Considerations of Congestion Pricing. Curbing Gridlock,* Volume 2. Transportation Research Board, National Academies Press, Washington, D.C.

Goldberger, A.S. 1964. *Econometric Theory.* John Wiley and Sons, New York, NY.

Gopinath, D., Ben-Akiva, M., and Ramaswamy, M. 1996. Modeling performance of highway pavements. *TRB Journal of Transportation Research Record* 1449, 1–7.

Granovetter, M. 1985. Economic action and social structure: The problem of embeddedness. *American Journal of Sociology* 91, 481–510.

Greenberg, H. 1959. An analysis of traffic flow. *Operations Research* 7(1), 255–275.

Greene, W.H. 2017. *Econometric Analysis*, 8th Edition. Prentice-Hall, Upper Saddle River, NJ.

Greenshields, B.D. 1935. A study of highway capacity. *Proceedings of Highway Research Record* 14, 448–477.

Grinstead, C.M. and Snell, J.L. 2012. *Introduction to Probability.* American Mathematical Society, Providence, RI.

Guler, H. 2013. Decision support system for railway track maintenance and renewal management. *ASCE Journal of Computing in Civil Engineering* 27(3), 292–306.

Haas, R., Hudson, W.R., and Zaniewski, J.P. 1994. *Modern Pavement Management.* Krieger Publishing Company, Melbourne, FL.

Haimes, Y.Y. 2015. *Risk Modeling, Assessment, and Management*, 4th Edition. Wiley-Interscience, Hoboken, NJ.

Haimes, Y.Y. and Hall, W.A. 1974. Multiobjective in water resource systems analysis: The surrogate worth tradeoff method. *Water Resource Research* 10(4), 615–624.

Hall, J.V. and Winer, A.M. 1992. Valuing the health benefits of clean air. *Science* 255(5046), 812.

Hammersley, J. and Morton, K. 1956. A new Monte Carlo technique: Antithetic variates. *Mathematical Proceedings of the Cambridge Philosophical Society* 52(3), 449–475.

Harkey, D.L., Raghavan, S., Jongdea, B., Council, F.M., Eccles, K., Lefler, N., Gross, F., Persaud, B., Lyon, C., Hauer, E., and Bonneson, J. 2008. *Crash Reduction Factors for Traffic Engineering and ITS Improvement.* NCHRP Report No. 617. National Cooperative Highway Research Program, Transportation Research Board, National Academies Press, Washington, D.C.

Harper, W.V., Lam, J., Al-Salloum, A., Al-Sayyari, S., Al-Theneyan, S., Ilves, G., and Majidzadeh, K. 1990. Stochastic optimization subsystem of a network-level bridge management system. *TRB Journal of Transportation Research Record* 1268, 68–74.

Harvey, J.M. and Shaw, S.L. 2001. *Geographic Information Systems for Transportation: Principles and Applications.* Oxford University Press, New York, NY.

Harwood, D.H., Council, F.M., Hauer, E., Hughes, W.E., and Vogt, A. 2000. *Prediction of the Expected Safety Performances of the Rural Two-Lane Highways.* Publication FHWARD-99-207. Federal Highway Administration, U.S. Department of Transportation, Washington, D.C.

Hassan, Y., Easa, S.M., and Abd El Halim, A.O. 1996. Analytical model for sight distance analysis on three-dimensional highway alignments. *TRB Journal of Transportation Research Record* 1523, 1–10.

Hatry, H.P. 1980. Performance measurement principles and techniques: An overview for local government. *Public Productivity Review* 312–339.

Hauer, E. 1997. *Observational Before–After Studies in Road Safety: Estimating the Effect of Highway and Traffic Engineering Measures on Road Safety.* Pergamon Press, Elsevier Science, Oxford, United Kingdom.

Hauer, E. 2002. *SafetyAnalyst: Software Tools for Safety Management of Specific Highway Sites.* White Paper for Module 4-Evaluation. Federal Highway Administration, U.S. Department of Transportation, Washington, D.C.

Hawk, H. 2003. *Bridge Life Cycle Cost Analysis.* NCHRP Report 483. Transportation Research Board, National Academies Press, Washington, D.C.

Heyman, D.P. and Sobel, M.J. 1982. *Stochastic Models in Operations Research, Volume I: Stochastic Processes and Operating Characteristics*. McGraw-Hill Book Company, New York, NY.

Heyman, D.P. and Sobel, M.J. 1984. *Stochastic Models in Operations Research, Volume II: Stochastic Optimization*. McGraw-Hill Book Company, New York, NY.

Higle, J.L. and Witkowski, J.M. 1988. Bayesian Identification of Hazardous Locations (with discussion and closure). *TRB Journal of Transportation Research Record* 1185, 24–36.

Hillier, F.S. and Lieberman, G.J. 2006. *Introduction to Operations Research*, 8th Edition. McGraw-Hill, New York, NY.

Hinloopen, E., Nijkamp, P., and Rletveld, P. 1983. The regime method: A new multicriteria technique. Editor: Hansen, P. *Topics in Essays and Surveys on Multiple Criteria Decision Making*. Springer-Verlag, Berlin, Germany, 146–155.

Hoyle, R.H. 1995. *Structural Equation Modeling: Concepts, Issues, and Applications*. SAGE Publications, London, United Kingdom.

HRB 1961. *The AASHO Road Test*. Special Report. 61A: History and Description of Project. Highway Research Board, National Academies Press, Washington, D.C.

HRB 1962. *The AASHO Road Test*. Special Reports. 61B: Materials and Construction; 61C: Traffic Operations and Pavement Maintenance; 61D: Bridge Research; 61E: Pavement Research; 61F: Special Studies; and 61G—Final Summary. Highway Research Board, National Academies Press, Washington, D.C.

HRB 1972. *National Cooperative Highway Research Program Synthesis of Highway Practice 14: Skid Resistance*. Highway Research Board, National Academies Press, Washington, D.C.

Huang, Y. 2004. *Pavement Analysis and Design*, 2nd Edition. Pearson Prentice-Hall, Upper Saddle River, NJ.

INDOT 2002. *Indiana Department of Transportation Design Manual*. Indiana Department of Transportation, Indianapolis, IN.

Inman, R.P. 1978. A generalized congestion function for highway travel. *Journal of Urban Economics* 5, 21–34.

IPCC 2006. *2006 IPCC Guidelines for National Greenhouse Gas Inventories*. Intergovernmental Panel on Climate Change, Geneva, Switzerland.

Isa Al-Subhi, K.M., Johnston, D.W., and Farid, F. 1989. *Optimizing System-Level Bridge Maintenance, Rehabilitation, and Replacement Decisions*. North Carolina State University, Raleigh, NC.

Ison, S. 2004. *Road User Charging: Issues and Policies*. Ashgate Publishing, Farnham, Surrey, United Kingdom.

Jack Faucett Associates 1991. *Highway Economic Requirements System Technical Report*. Report for the Federal Highway Administration. U.S. Department of Transportation, Washington, D.C.

Jackson, N. and Mahoney, J.P. 1990. *Washington State Pavement Management System, Advanced Course in Pavement Management Notebook*. Federal Highway Administration, U.S. Department of Transportation, Washington, D.C.

Jagannathan, R. 1966. A simplex-type algorithm for linear and quadratic programming- A parametric procedure. *Econometric Society Journal of Econometrica* 34(2), 460–471.

Jiang, Y., Saito, M., and Sinha, K.C. 1988. *Bridge Performance Prediction Model Using the Markov Chain*. International Study of Highway Development and Management, Paris, France.

Jiang, Y. and Sinha, K.C. 1989. Bridge service life prediction model using the Markov chain. *TRB Journal of Transportation Research Record* 1223, 24–30.

Johnson, C. 1983. *Pavement Maintenance Management Systems*. American Public Works Association, Kansas City, MO.

Jovanis, P.P. and Chang, H. 1985. Modeling the relationship of crashes to miles traveled. *TRB Journal of Transportation Research Record* 1068, 42–51.

Kanafani, A. 1983. The Social Costs of Road Transport: A Review of Road Traffic Noise, Air Pollution and Accidents. Report *OECD DOC ENV/TE/84.3*. Paris, France.

Kain, J. 1994. Impacts of congestion pricing on transit and carpool demand and supply. In *Curbing Gridlock* 2. Editor: Nancy A. Ackerman. Transportation Research Board, Washington, D.C., 502–553.

Kall, P. and Wallace, S.W. 1994. *Stochastic Programming.* John Wiley and Sons, Chichester, West Sussex, United Kingdom.

Karlin, S. 2003. *Mathematics Methods and Theory in Games, Programming, and Economics.* Courier Corporation, North Chelmsford, MA.

Keeler, T.E. and Small, K.A. 1977. Optimal peak load pricing, investment, and service levels on urban expressways. *Journal of Political Economy* 85, 1–25.

Keeney, R.L. and Raiffa, H. 1993. *Decisions with Multiple Objectives: Preferences and Value Trade-Offs.* Cambridge University Press, Cambridge, United Kingdom.

King, A.D., Manville, M., and Shoup, D. 2007. The political calculus of congestion pricing. *Transport Policy* 14(2), 11–123.

Klamroth, K. and Wieck, M.M. 2001. Dynamic Programming Approaches to the Multiple Criteria Knapsack Problem. Working Paper. Clemson University, Clemson, SC.

Klir, G.J. and Parviz, B. 1992. Probability-possibility transformations: A comparison. *International Journal of General Systems* 21(3), 291–310.

Klir, G.J. 2006. *Uncertainty and Information: Foundations of Generalized Information Theory.* Wiley Interscience, Hoboken, NJ.

Korhonen, P. and Laakso, J. 1986. A visual interactive method for solving the multiple criteria problem. *European Journal of Operational Research* 24(2), 277–287.

Kurowicka, D. and Cooke, R. 2006. *Uncertainty Analysis with High Dimensional Dependence Modelling.* John Wiley and Sons, Hoboken, NJ.

Kutner, M.H., Nachtsheim, C.J., Neter, J., and Li, W. 2004. *Applied Linear Statistical Models,* 5th Edition. McGraw-Hill/Irwin, New York, NY.

Labi, S. 2001. Impact Evaluation of Highway Pavement Maintenance Activities. *PhD dissertation.* Purdue University, West Lafayette, IN.

Labi, S. Lamptey, G., and Li, Z. 2008. Decision support for optimal scheduling of highway pavement preventive maintenance within resurfacing cycle. *Journal of Decision Support Systems* 46(1), 376–387.

Labi, S. and Sinha, K.C. 2002. *Effect of Routine Maintenance on Capital Expenditures.* Purdue University, West Lafayette, IN.

Labi, S. and Sinha, K.C. 2005. Life cycle evaluation of highway pavement preventive maintenance. *ASCE Journal of Transportation Engineering* 131(10), 744–751.

Labi, S. and Sinha, K.C. 2008. *Updated Indiana Bridge Management System.* Purdue University, West Lafayette, IN.

Lamm, R., Beck, A., Ruscher, T., Mailaender, T., Cafiso, S., La Cava, G., and Matthews, W. 2006. *How to Make Two-Lane Rural Roads Safer. Scientific Background and Guide for Practical Application.* WIT Press, Southampton, United Kingdom.

Larson, R.C. and Odoni, A.R. 2007. *Urban Operations Research,* 2nd Edition. Dynamic Ideas, Charlestown, MA.

Lee, D.B. 2002. *Basic Theory of Highway Project Evaluation.* Appendix D of HERS Technique Report. Federal Highway Administration, U.S. Department of Transportation, Washington, D.C.

Lee, D.B. and Burris, M. 2005. *HERS-ST Highway Economic Requirements System-State Version: Technical Report.* Appendix F: Procedures for Estimating Air Pollution Costs. Federal Highway Administration, U.S. Department of Transportation, Washington, D.C.

Lee, Y., Li, Z., Zhang, S., Roshandeh, A.M., Patel, H., and Liu, Y. 2014. Safety impacts of red light running photo enforcement on signalized urban intersections. *Elsevier Journal of Traffic and Transportation Engineering* 1(5), 309–324.

Leontief, W. 1986. *Input–Output Economics.* Oxford University Press, Oxford, United Kingdom.

Levesque, T.J. 1994. Modelling the effects of airport noise on residential housing markets: A case study of Winnipeg International Airport. *Journal of Transport Economics and Policy,* 199–210.

Levine, J. and Garb, Y. 2000. *Evaluating the Promise and Hazards of Congestion Pricing Proposals: An Access Centered Approach.* Floersheimer Institute for Policy Studies, The Hebrew University, Jerusalem, Israel.

Li, J., Li, Z., and Zhou, B. 2010. A disaggregated approach for the computation of network level highway user costs. *International Journal of Pavement Research and Technology* 3(1), 24–33.

Li, Z. 2004. *Wisconsin Freeway Operations Assessment Guidelines*. Department of Civil and Environmental Engineering, University of Madison, Madison, WI.

Li, Z. 2009. Stochastic model and O(N^2) algorithm for highway investment decision-making under budget uncertainty. *ASCE Journal of Transportation Engineering* 135(6), 371–379.

Li, Z., Dao, H., Patel, H., Liu, Y., and Roshandeh, A.M. 2017. Incorporating traffic and safety hardware performance functions into risk-based highway segment safety evaluation. *PROMET Journal of Traffic and Transportation Engineering* 29(2), 143–153.

Li, Z. and Kaini, P. 2007. *Optimal investment decision-making for highway transportation asset management under risk and uncertainty*. Report MRUTC 07-10. Department of Civil, Architectural, and Environmental Engineering, Illinois Institute of Technology, Chicago, IL.

Li, Z., Kaul, H., Kapoor, S., Veliou, E., and Zhou, B. 2012. A new model for transportation investment decisions considering project interdependencies. *TRB Journal of Transportation Research Record* 2285, 36–46.

Li, Z., Labi, S., Karlaftis, M.G., Kepaptsoglou, K., Abbas, M., Zhou, B., and Madanu, S. 2010. A project-level life-cycle benefit/cost analysis approach for evaluating highway safety hardware improvements. *TRB Journal of Transportation Research Record* 2160, 1–11.

Li, Z., Labi, S., and Sinha, K.C. 2002a. *Highway Asset Management. Civil Engineering Handbook*, 2nd Edition. CRC Press LLC, Boca Raton, FL, 66-1–66-34.

Li, Z. and Madanu, S. 2009. Highway project-level life-cycle benefit/cost analysis under certainty, risk, and uncertainty: A methodology with case study. *ASCE Journal of Transportation Engineering* 135(8), 516–526.

Li, Z., Madanu, S., Wang, Y., Abbas, M., and Zhou, B. 2010. A heuristic approach for selecting highway investment alternatives. *Wiley Journal of Computer-Aided Civil and Infrastructure Engineering* 25(6), 427–439.

Li, Z., Roshandeh, A.M., Zhou, B., and Lee, S.H. 2013. Optimal decision-making of interdependent tollway capital investments incorporating risk and uncertainty. *ASCE Journal of Transportation Engineering* 139(7), 686–696.

Li, Z. and Sinha, K.C. 2004. Methodology for multicriteria decision making in highway asset management. *TRB Journal of Transportation Research Record* 1885, 79–87.

Li, Z., Shen, J., and Budiman, J. 2008. *Highway Tunnels: An Overview For Planners And Policymakers*. Report for Robert W. Galvin Mobility Initiative. Reason Foundation, Washington, D.C.

Li, Z. and Sinha, K.C. 2009a. Application of Shackle's model and system optimization for highway investment decision-making under uncertainty. *ASCE Journal of Transportation Engineering* 135(3), 129–139.

Li, Z. and Sinha, K.C. 2009b. A methodology for the determination of relative weighs of highway asset management system goals and performance measures. *ASCE Journal of Infrastructure Systems* 15(2), 95–105.

Li, Z., Sinha, K.C., and McCarthy, P.S. 2001. A methodology to determine load and non-load shares of highway pavement rehabilitation expenditures. *TRB Journal of Transportation Research Record* 1747, 79–88.

Li, Z., Sinha, K.C., and McCarthy, P.S. 2002b. A methodology to determine load and non-load shares of highway pavement routine maintenance expenditures. *Journal of Road and Transport Research* 11(2), 3–13.

Lighthill, M.J. and Whitham, G.B. 1955. On kinematic waves II: A theory of traffic flow on long crowded roads. *Proceedings of the Royal Society of London A: Mathematical, Physical and Engineering Sciences* 229(1178), 317–345.

Lin, P., Wu, Y., Huang, T., and Juang, C. 1996. Equivalent single-axle load factor for rigid pavements. *ASCE Journal of Transportation Engineering* 122(6), 462–467.

Lindly, J. and Wijesundera, R. 2003. *Evaluation of Profiled Pavement Markings*. Report No. 01465. University Transportation Center for Alabama, Tuscaloosa, AL.

Litman, T. 2002. Evaluating transportation equity. *World Transport Policy and Practice* 8(2), 50–65.

Litman, T. 2003. *London Congestion Pricing: Implications for Other Cities*. Victoria Transport Policy Institute, Victoria, Canada.

Litman, T. 2007. *Socially Optimal Transport Prices and Markets.* Victoria Transport Policy Institute, Victoria, Canada.

Litman, T. 2011. *Pricing for Traffic Safety: How Efficient Transport Pricing Can Reduce Roadway Crash Risk.* Victoria Transport Policy Institute, Victoria, Canada.

Litman, T. 2013. *Generated Traffic and Induced Travel Implications for Transportation Planning.* Victoria Transport Policy Institute, Victoria, Canada.

Litman, T. 2015. Evaluating household Chauffeuring burdens: Understanding direct and indirect costs of transporting non-drivers. *The 94th Annual TRB Meeting*, Washington, D.C.

Lomax, T., Turner, S., and Shunk, G. 1997. *Quantifying Congestion.* NCHRP Report 398. National Cooperative Highway Research Program, Transportation Research Board, National Academies Press, Washington, D.C.

Long, J.S. 1997. *Regression Models for Categorical and Limited Dependent Variables.* Sage Publications, Thousand Oaks, CA.

Long, S., Qin, R., Gosavi, A., and Ryan, T. 2011. *Life Expectancy Evaluation and Development of a Replacement Schedule for LED Traffic Signals.* Missouri University of Science and Technology, HDR Engineering, and Missouri Department of Transportation, Rolla, MO.

Lord, D., Washington, S.P., and Ivan, J.N. 2005. Poisson, Poisson-gamma and zero-inflated regression models of motor vehicle crashes: Balancing statistical fit and theory. *Accident Analysis and Prevention* 37(1), 35–46.

Lord, D., Washington, S.P., and Ivan, J.N. 2007. Further notes on the application of zero-inflated models in highway safety. *Accident Analysis and Prevention* 39(1), 53–57.

Lorie, J.H. and Savage, L.J. 1955. Three problems in rationing capital. *Journal of Business* 28(4), 229–239.

Luenberger, D.G. 1973. *Introduction to Linear and Nonlinear Programming.* Addison-Wesley, Reading, MA.

Luenberger, D.G. 1997. *Investment Science.* Oxford University Press, Oxford, United Kingdom.

Luenberger, D.G. 2006. *Information Science.* Princeton University Press, Princeton, NJ.

Lyons, W., Rasmussen, B., Daddio, D., Fijalkowski, J., and Simmons, E. 2014. *Nonmotorized Transportation Pilot Program: Continued Progress in Developing Walking and Bicycling Networks.* US Department of Transportation, Federal Highway Administration, U.S. Department of Transportation, Washington, D.C.

Lytton, R.L. 1988. *Concepts of Pavement Performance Prediction and Modeling.* Department of Civil Engineering, Texas A&M University, College Station, TX.

MacQueen, J. 1967. Some methods for classification and analysis of multivariate observations. *Proceedings of the Fifth Berkeley Symposium on Mathematical Statistics and Probability* 1(14), 281–297.

Madanat, S., Mishalani, R., and Ibrahim, W.H.W. 1995. Estimation of infrastructure transition probabilities from condition rating. *ASCE Journal of Infrastructure Systems* 1(2), 120–125.

Madanat, S.M., Karlaftis, M.G., and McCarthy, P.S. 1997. Probabilistic infrastructure deterioration models with panel data. *ASCE Journal of Infrastructure Systems* 3(1), 4–9.

Madanu, S. and Li, Z. 2007. *Case Study Report on Current Roadway Safety Hardware Asset Management Programs in Midwest States.* Illinois Institute of Technology, Chicago, IL.

Madanu, S., Li, Z., and Abbas, M. 2010. Life-cycle cost analysis of highway intersection safety hardware improvements. *ASCE Journal of Transportation Engineering* 136(2), 129–140.

Maddala, G.S. 1983. *Limited-Dependent and Qualitative Variables in Econometrics.* Cambridge University Press, Cambridge, United Kingdom.

Maddison, D., Pearce, D., Johansson, O., Calthorp, E., Litman, T., and Verhoef, E. 1996. *Blueprint 5: The True Cost of Transport.* Earthscan, London, United Kingdom.

Madejski, J. and Grabczyk, J. 1999. Track and rolling stock quality assurance related tools. In *Book: Structural Integrity and Passenger Safety.* Editor: Carlos A. Brebbia. Wessex Institute of Technology, Ashurst, United Kingdom, 85–114.

Madejski, J. and Grabczyk, J. 2002. Continuous geometry measurement for diagnostics of tracks and switches. *Proceedings of the International Conference on Switches.* Delft University of Technology, Delft, Netherland.

Magazine, M.J. and Oguz, O. 1984. A heuristic algorithm for the multidimensional zero-one knapsack problem. *European Journal of Operational Research* 16, 319–326.
Makridakis, S.G., Wheelwright, S.C., and Hyndman, R.J. 1997. *Forecasting: Methods and Applications*, 3rd Edition. John Wiley and Sons, Hoboken, NJ.
Malyshkina, N. and Mannering, F.L. 2010. Zero-state Markov switching count-data models: An empirical assessment. *Accident Analysis and Prevention* 42(1), 122–130.
Manheim, M.L. 1979. *Fundamentals of Transportation Systems Analysis, Volume 1: Basic Concepts.* The MIT Press, Cambridge, MA.
Markow, M.J. 2007. *Managing Selected Transportation Assets: Signals, Lighting, Signs, Pavement Markings, Culverts, and Sidewalks.* NCHRP Synthesis of Highway Practice 371. National Cooperative Highway Research Program, Transportation Research Board, National Academies Press, Washington, D.C.
Markowitz, H.M. 1991. *Portfolio Selection: Efficient Diversification of Investments*, 2nd Edition. Blackwell Publishing, Hoboken, NJ.
Marsden, G. and May, A. 2006. Do institutional arrangements make a difference to transport policy and implementation? Lessons from Britain. *Environment and Planning C: Government and Policy* 24(5), 771–789.
Martello, S. and Toth, P. 1990. *Knapsack Problems: Algorithms and Computer Implementations.* John Wiley and Sons, Chichester, West Sussex, United Kingdom.
Martin, T.C. 1994. *Pavement Behaviour Prediction for Life-Cycle Costing.* Research Report ARR 255. Australia Road Research Board, Melbourne, Australia.
Martincigh, L., Tonelli, C., and Fuller, T. 2010. The walking environment design: Indicators and measures. *Pedestrians' Quality Needs* 358, 277–304.
Mathews, G.B. 1896. On the partition of numbers. *Proceedings of the London Mathematical Society* 1(1), 486–490.
Matthews, H.S., Hendrickson, C., and Horvath, A. 2001. External costs of air emissions from transportation. *ASCE Journal of Infrastructure Systems* 7(1), 13–17.
Matthews, H.S. and Lave, L.B. 2000. Applications of environmental valuation for determining externality costs. *Environmental Science and Technology* 34(8), 1390–1395.
May, A. 1990. *Traffic Flow Fundamentals.* Prentice-Hall, Englewood Cliff, NJ.
McCarthy, P.S. 2001. *Transportation Economics: Theory and Practice: A Case Study Approach.* John Wiley and Sons, Hoboken, NJ.
McCubbin, D. and Delucchi, M. 1996. *The Social Cost of the Health Effects of Motor Vehicle Air Pollution.* Report 11 in the Series: The Annualized Social Cost of Motor Vehicle Use in the United States, based on 1990–1991 Data. Institute of Transportation Studies, University of California, Davis, CA.
McCullagh, P. and Nelder, J.A. 1989. *Generalized Linear Models.* Chapman and Hall, New York, NY.
McGuire, P., Granovetter, M., and Schwartz, M. 1993. Thomas Edison and the social construction of the early electricity industry in America. In *Explorations in Economic Sociology.* Editor: Richard Swedberg. Russell Sage Foundation, New York, NY.
McKay, M.D., Beckman, R.J., and Conover, W.J. 1979. Comparison of three methods for selecting values of input variables in the analysis of output from a computer code. *Technometrics* 21(2), 239–245.
McKelvey, R.D. and Zavoina, W. 1975. A statistical model for the analysis of ordinal level dependent variables. *Journal of Mathematical Sociology* 4(1), 103–120.
Menges, G. 1963. The adaptation of decision criteria and application patterns. In *Proceedings of the International Federation of Operational Research Societies Congress*, Oslo, Norway, 585–594.
Menges, G. 1974. *Economic Decision Making: Basic Concepts and Models.* Longman, London, United Kingdom.
Menges, G. and Behara, M. 1962. Das Bayes' sche Risiko bei sequentiellen Stichprobenentscheidungen. *SH* 3, 39–61.
Metropolis, N. and Ulam, S. 1949. The Monte Carlo method. *Journal of the American Statistical Association* 44(247), 335–341.

Meyer, M.D. and Miller, E.J. 2000. *Urban Transportation Planning*, 2nd Edition. McGraw-Hill, New York, NY.

Meyer, M.D., Rowan, E., Snow, C., and Choate, A. 2013. *Impacts of Extreme Weather on Transportation: National Symposium Summary*. American Association of State Highway and Transportation Officials, Washington, D.C.

Miaou, S.-P. 1996. *Measuring the Goodness-of-Fit of Accident Prediction Models*. Report No. FHWA-RD-96-040. Federal Highway Administration, U.S. Department of Transportation, Washington, D.C.

Miller, T.R. 1992. Benefit-cost analysis of lane marking. *TRB Journal of Transportation Research Record* 1334, 38–45.

Miwa, K. and Simon, H. 1993. Production system modeling to represent individual differences: Tradeoff between simplicity and accuracy in simulation of behavior. Prospects for Artificial Intelligence. *Proceedings of AISB93, the 9th Biennial Conference of the Society for the Study of Artificial Intelligence and the Simulation of Behavior*. Editors: Aaron Sloman, David Hogg, Glyn Humphreys, Allan Ramsay, and Derek Partridge. Birmingham, United Kingdom, 158–167.

Mohammed, D., Sinha, K.C., and McCarthy, P.S. 1997. Relationship between pavement performance and routine maintenance: Mixed logit approach. *TRB Journal of Transportation Research Record* 1597, 16–21.

Mohring, H. 1972. Optimization and scale economies in urban bus transportation. *American Economic Review* 62(4), 591–604.

Mollaghasemi, M. 1997. *Briefing: Making Multiple Objective Decisions*. IEEE Computer Society Press, Los Alamitos, CA.

Montella, A. 2005. Safety reviews of existing roads: Quantitative safety assessment methodology. *TRB Journal of Transportation Research Record* 1922, 62–72.

Montgomery, D. and Runger, G. 2006. *Applied Statistics and Probability for Engineers*, 4th Edition. John Wiley and Sons, New York, NY.

Moran, P.A. 1950. Notes on continuous stochastic phenomena. *Biometrika* 37(1/2), 17–23.

Morcous, G. and Lounis, Z. 2007. Probabilistic and mechanistic deterioration models for bridge management. In *Proceedings of the 2007 ASCE International Workshop on Computing in Civil Engineering*. Editors: Lucio Soibelman and Burcu Akinci. Pittsburgh, PA, 364–373.

Morcous, G., Rivard, H., and Hanna, A.M. 2002. Case-based reasoning system for modeling infrastructure deterioration. *ASCE Journal of Computing in Civil Engineering* 16(2), 104–114.

Mundrey, J.S. 2003. *Railway Track Engineering*, 4th Edition. Tata McGraw-Hill Publishing, New Delhi, India.

Myerson, R.B. 1997. *Game Theory: Analysis of Conflict*. Harvard University Press, Cambridge, MA.

Nagurney, A. 2000. *Sustainable Transportation Networks*. Edward Elgar, Northampton, MA.

Nash, J. 1953. Two-person cooperative games. *Econometric Society Journal of Econometrica* 21(1), 128–140.

Navrud, S. 2002. *The State-of-the-Art on Economic Valuation of Noise*. Final Report to European Commission DG Environment 14. Mannheim, Germany.

NCHRP 2005. *Crash Reduction Factors for Traffic Engineering and Intelligent Transportation System (ITS) Improvements: State-of-Knowledge Report*. NCHRP Research Results Digest 299. National Cooperative Highway Research Program, Transportation Research Board, Washington, D.C.

NCHRP 2008. Cost-effective performance measures for travel time delay, variation, and reliability. *National Cooperative Highway Research Program*. National Academies Press, Washington, D.C.

NCHRP 2009. *Public-Sector Decision Making for Public–Private Partnerships*. NCHRP Synthesis 391. National Cooperative Highway Research Program, Transportation Research Board, National Academies Press, Washington, D.C.

NEC 2008. *Employment Impacts of Increased Highway Infrastructure Investment*. The New England Council, Boston, MA.

Nelson, J.P. 1982. Highway noise and property values: A survey of recent evidence. *Journal of Transport Economics and Policy* 16(2), 117–138.

Nemhauser, G.L. 1966. *Introduction to Dynamic Programming.* John Wiley and Sons, Hoboken, NJ.
Nemhauser, G.L. and Wolsey, L.A. 1988. *Integer and Combinatorial Optimization.* Wiley-Interscience, Hoboken, NJ.
Neumann, L.A. 1997. *Methods for Capital Programming and Project Selection.* NCHRP Synthesis of Highway Practice 243. National Cooperative Highway Research Program, Transportation Research Board, National Academies Press, Washington, D.C.
Newbery, D.M. 1988. Road damage externalities and road user charges. *Econometrica* 56(2), 295–316.
Newell, G.F. 1971. *Applications of Queueing Theory.* Chapman and Hall, London, United Kingdom.
NHTSA 2000. *The Economic Impact of Motor Vehicle Crashes 2000.* National Highway Traffic Safety Administration, U.S. Department of Transportation, Washington, D.C.
Nocedal, J. and Wright, S.J. 2006. *Numerical Optimization.* Springer, New York, NY.
Norden, M., Orlansky, J., and Jacobs, H. 1956. Application of statistical quality-control techniques to analysis of highway-accident data. *Highway Research Board Bulletin* 117, 17–31.
NPSC 1993. *Adopted Regulation of the Public Service Commission of Nevada.* LCB File No. R085-93, Docket 89-651, Carson City, NV.
NRC 2008. *Potential Impacts of Climate Change on U.S. Transportation.* Report 290. Transportation Research Board, National Academies Press, Washington, D.C.
NSC 1995. *Estimating the Cost of Unintentional Injuries.* National Safety Council, Washington, D.C.
NYDOT 1992. *Comprehensive Report on Preventive Maintenance.* New York State Department of Transportation, Albany, NY.
NYDOT 1993. *Pavement Rehabilitation Manual- Volume II: Treatment Selection.* New York State Department of Transportation, Albany, NY.
NYDOT 1999. *Pavement Rehabilitation Manual- Volume III: Preventive Maintenance Treatments and Selection.* New York State Department of Transportation, Albany, NY.
O'Brien, L.G. 1989. *Evolution and Benefits of Preventive Maintenance Strategies.* NCHRP Synthesis of Highway Practice 153. Transportation Research Board, National Academies Press, Washington, D.C.
Ogden, K. 1996. *Safer Roads: A Guide to Road Safety Engineering.* Avebury Technical, Cambridge, MA.
O'Neill, W. and Harper, E. 1997. Location translation within a geographic information system. *TRB Journal of Transportation Research Record* 1593, 55–63.
Oppenheim, N. 1995. *Urban Travel Demand Modeling: From Individual Choices to General Equilibrium.* Wiley-Interscience, Hoboken, NJ.
Osorio, M.A. and Glover, F. 2001. *Logic Cuts Using Surrogate Constraint Analysis in the Multidimensional Knapsack Problem.* Working Paper. Imperial College of Science, Technology and Medicine, London; University of Colorado, Boulder, CO.
Paelinck, J.H.P. 1976. Qualitative multiple criteria analysis, environmental protection and multiregional development. *Papers of the Regional Science Association* 36, 59–74.
Paelinck, J.H.P. 1977. Qualitative multicriteria analysis: an application to airport location. *Environment and Planning* 9(8), 883–895.
Paelinck, J.H.P. 1978. Qualiflex: A flexible multiple-criteria method. *Economics Letters* 1(3), 193–197.
Palmquist, R.B. 1980. Impact of Highway Improvements on Property Values in Washington, WA-RD-37.1. National Technical Information Service. Springfield, VA.
Palmquist, R.B. 1982. Measuring environmental effects on property prices values with hedonic regressions. *Journal of Urban Economics* 11, 333–347.
Parry, I. 2008. *Pricing Urban Congestion.* RFF Discussion Paper 08-35. Resources for the Future, Washington, D.C.
Paté-Cornell, E. 2002. Risk and uncertainty analysis in government safety decisions. *Risk Analysis* 22(3), 633–646.
Paterson, W. 1987. *Road Deterioration and Maintenance Effects: Models for Planning and Management.* The Highway Design and Maintenance Standards Series. The Johns Hopkins University Press, Baltimore, MD.

Patidar, V., Labi, S., Sinha, K.C., Thompson, P.D., Shirole, A., and Hyman, W. 2007. *Multiple Objective Optimization for Bridge Management*. NCHRP Report 590. Transportation Research Board, National Academies Press, Washington, D.C.

Patten, R.S., Schneider, R.J., Toole, J.L., Hummer, J.E., and Rouphail, N.M. 2006. *Shared-Use Path Level of Service Calculator—A User's Guide*. Report No. FHWA-HRT-05-138. Federal Highway Administration, U.S. Department of Transportation, Washington, D.C.

Peshkin, D.G., Hoerner, T.E., and Zimmerman, K.A. 2005. *Optimal Timing of Pavement Preventive Maintenance Treatment Applications*. NCHRP Report 523. National Academies Press, Washington, D.C.

Pindyck, R.S. and Rubinfeld, D.L. 2000. *Econometric Models and Economic Forecasts*, 4th Edition. McGraw-Hill Publishing, New York, NY.

Pittou, M., Karlaftis, M.G., and Li, Z. 2009. Nonparametric binary recursive partitioning method for deterioration prediction of bridge deck elements. *Journal of Advances in Civil Engineering* 2009, Article ID 809767, 1–12.

Poister, T.H. 1997. *Performance Measurement in State Department of Transportation*. NCHRP Synthesis of Highway Practice 238. National Cooperative Highway Research Program, Transportation Research Broad, National Academies Press, Washington, D.C.

Poling, A., Lee, J., Gregerson, P., and Handly, P. 1994. Comparison of two sign inventory data collection techniques for geographic information systems. *TRB Journal of Transportation Research Record* 1429, 36–39.

Popović, Z., Lazarević, L., Brajović, L.J., and Gladović, P. 2014. Managing rail service life. *Metabk* 53(4), 721–724.

Proctor, S., Belcher, M., and Cook, P. 2001. *Practical Road Safety Auditing*. Thomas Telford, London, United Kingdom.

Profillidis, V.A. 2014. *Railway Management and Engineering*. Ashgate Publishing, Farnham, Surrey, United Kingdom.

Raimond, T. 1997. *Urban Freight Data Collection and Forecasting*. Australasian Transport Research (ATRF), Adelaide, Australia.

Rajé, F. 2003. *Impacts of Road User Charging/Workplace Parking Levy on Social Inclusion/Exclusion: Gender, Ethnicity and Lifecycle Issues*. Transport Studies Unit, University of Oxford, Oxford, United Kingdom.

Rajiv, B. 2014. *Noise Pollution: Managing the Challenge of Urban Sounds*. Earth Journalism Network. Online article: http://earthjournalism.net/resources/noise-pollution-managing-the-challenge-of-urban-sounds

Ramaswamy, R. and Ben-Akiva, M. 1997. Estimation of highway pavement deterioration from in-service pavement data. *TRB Journal of Transportation Research Record* 1570, 96–106.

Ramjerdi, F. 1996. *Road Pricing and Toll Financing, with Examples from Oslo and Stockholm*. Institute of Transport Economics, Oslo, Norway.

Ravirala, V. and Grivas, D.A. 1995. Goal-programming methodology for integrating pavement and bridge programs. *ASCE Journal of Transportation Engineering* 121(4), 345–351.

Rayward-Smith, V.J., Osman, I.H., Reeves, C.R., and Smith, G.D. 1996. *Modern Heuristic Search Methods*. John Wiley and Sons, Hoboken, NJ.

RDMTD 1999. *Signing and Marking of Substandard Horizontal Curves on Rural Roads*. Report 163. Road Directorate Ministry of Transport Denmark, Copenhagen, DK.

Reigle, J.A. 2000. Development of an Integrated Project-Level Pavement Management Model using Risk Analysis, Dissertation. West Virginia University, Morgantown, WV.

Repetto, R. 1990. Environmental productivity and why it is so important. *Challenge* 33(5), 33–38.

Richards, P.I. 1956. Shock waves on the highway. *Operations Research* 4(1), 42–51.

Richards, S., Barnes, M., Sullivan, H., Gaster, L., Leach, B., and Coulson, A. 1999. *Cross Cutting Issues in Public Policy and Public Service*. Report to the Department of Environment Transport and the Regions, London, United Kingdom.

Ries, T. 2000. Integrating traffic management data via an enterprise LRS. *GeoAnalytics Presentation at North American Travel Monitoring Exhibition and Conference*, Madison, WI.

Rietveld, P. and Stough, R. 2005. *Barriers to Sustainable Transport: Institutions, Regulations and Sustainability*. Spon Press, New York, NY.
Robbins, H. 1964. The empirical Bayes approach to statistical decision problems. *The Annals of Mathematical Statistics* 35(1), 1–20.
Rodriguez, M.M., Labi, S., and Li, Z. 2006. Enhanced bridge replacement cost models for Indiana's Bridge Management System. *TRB Journal of Transportation Research Record* 1958, 13–23.
Rose, E.R. and Carlson, P.J. 2005. Spacing chevrons on horizontal curves. *TRB Journal of Transportation Research Record* 1918, 84–91.
Roshandeh, A.M., Levinson, H.S., Li, Z., Patel, H., and Zhou, B. 2013. A new methodology for intersection signal timing optimization to simultaneously minimize vehicle and pedestrian delays. *ASCE Journal of Transportation Engineering* 140(5), 04014009.
Roshandeh, A.M., Li, Z., Neishapouri, M., Patel, H., and Liu, Y. 2015. A tradeoff analysis approach for multiobjective transportation investment decision-making. *ASCE Journal of Transportation Engineering* 141(3), 04014085.
Roshandeh, A.M., Li, Z., Zhang, S., Levinson, H.S., and Liu, Y. 2016. Vehicle and pedestrian safety impacts of signal timing optimization in a dense urban street network. *Elsevier Journal of Traffic and Transportation Engineering* 3(1), 16–27.
Roth, G.J. and Thomson, J.M. 1963. *Road Pricing, A Cure for Congestion?* University of Cambridge, Cambridge, United Kingdom.
Rowe, R.D., Lang, C.M., Chestnut, L.G., Latimer, D.A., Rae, D.A., Bernow, S.M., and White, D.E. 1995. *The New York Electricity Externality Study Volume I: Introduction and Methods*. Empire State Electric Energy Research Corporation, New York, NY.
Roy, B. 1968. Classement et choix en présence de points de vue multiples (la méthode ELECTRE)". *La Revue d'Informatique et de Recherche Opérationelle* (RIRO) 8, 57–75.
Roy, B. 1991. The outranking approach and the foundations of ELECTRE methods. *Theory and Decision* 31(1), 49–73.
Saaty, T.L. and Gass, S. 1954. Parametric objective function (Part 1). *Journal of the Operations Research Society of America* 2(3), 316–319.
Saaty, T.L. 1990. How to make a decision: The analytic hierarchy process. *European Journal of Operational Research* 48(1), 9–26.
Saaty, T.L. and Vargas, L.G. 2000. *Models, Methods, Concepts and Applications of the Analytic Hierarchy Process*. Springer, Berlin, Germany.
Saaty, T.L. and Vargas, L.G. 2006. *Decision Making with the Analytic Network Process: Economic, Political, Social and Technological Applications with Benefits, Opportunities, Costs and Risks*. Springer, Berlin, Germany.
Sadeghi, J. and Askarinejad, H. 2007. Influences of track structure, geometry, and traffic parameters on railway deterioration. *International Journal of Engineering* 20(3), 292–300.
Sadeghi, J. and Askarinejad, H. 2008. Development of improved railway track degradation models. *International Journal of Structure and infrastructure Engineering* 3(4), 675–688.
Sakawa, M. 1980. Multiobjective optimization for a standby system by the surrogate worth trade-off method. *Journal of the Operational Research Society* 31(2), 153–158.
Sakawa, M., Nishizaki, I., and Katagiri, H. 2011. *Fuzzy Stochastic Multiobjective Programming*. Springer Science and Business Media, Berlin, Germany.
Salem, O., AbouRizk, S., and Ariaratnam, S. 2003. Risk-based life-cycle costing of infrastructure rehabilitation and construction alternatives. *ASCE Journal of Infrastructure Systems* 9(1), 6–15.
Salford Systems 2003. CART6.0 ProEx Software, San Diego, CA.
Samuel, P. 2000. *Putting Customers in the Driver's Seat: The Case for Tolls*. Reason Public Policy Institute, Reason Foundation, Washington, D.C.
Savage, L.J. 1972. *The Foundations of Statistics*. Dover Publications, New York, NY.
Sayers, M.W. and Karamihas, S.M. 1996. *Interpretation of Road Roughness Profile Data*. Report No. FHWA-RD-96-101, Federal Highway Administration, U.S. Department of Transportation, Washington, D.C.
Schoech, W. 2007. Rolling contact fatigue mitigation by grinding. *Rail-Tech Europe: 6th International Exhibition and Seminars on Rail Technology*, Utrecht, Netherland.

Scodel, A., Minas, J.S., Ratoosh, P., and Lipetz, M. 1959. Some descriptive aspects of two-person non-zero-sum games. *Journal of Conflict Resolution* 3(2), 114–119.

Shackle, G.L.S. 1949. *Expectation in Economics*, 2nd Edition. Cambridge University Press, Cambridge, United Kingdom.

Shahin, M.Y., Darter, M.I., and Kohn, S.D. 1976. *Development of a Pavement Maintenance Management System, Volume I—Airfield Pavement Condition Rating.* U.S. Army Corps of Engineers Construction Engineering Research Laboratory, Champaign, IL.

Shaw, T. 2003. *Performance Measures of Operational Effectiveness for Highway Segments and Systems—A Synthesis of Highway Practice.* NCHRP Synthesis of Highway Practice 311. National Cooperative Highway Research Program, Transportation Research Board. National Academies Press, Washington, D.C.

Sheffi, Y. 1985. *Urban Transportation Networks: Equilibrium Analysis with Mathematical Programming Methods.* Prentice-Hall, Englewood Cliffs, NJ.

Shen, J., Rodriguez, A., and Gan, A. 2004. Development and Application of Crash Reduction Factors: State-of-the-Practice Review of State Departments of Transportation. In *TRB 83rd Annual Meeting Compendium of Papers, CD-ROM*, Transportation Research Board, National Research Council, Washington, D.C.

SHRP 1994. *Evaluation of the AASHTO Design Equations and Recommended Improvements.* Strategic Highway Research Program. National Academies Press, Washington, D.C.

Sianipar, P.R.M. and Adams, T.M. 1997. Fault-tree model of bridge element deterioration due to interaction. *ASCE Journal of Infrastructure Systems* 3, 103–110.

Siegel, S. and Castellan Jr., N.J. 1988. *Nonparametric Statistics for the Behavioral Sciences.* McGraw-Hill, Boston, MA.

Sinha, K.C. 1992. *Indiana Bridge Management System Technical Manual.* Joint Transportation Research Program, Purdue University, West Lafayette, IN.

Sinha, K.C. and Fwa, T.F. 1988. On the concept of total highway management. *TRB Journal of Transportation Research Record* 1229, 79–88.

Sinha, K.C. and Labi, S. 2011. *Transportation Decision Making: Principles of Project Evaluation and Programming.* John Wiley and Sons, New York, NY.

Sinha, K.C. Patidar, V., Li, Z., Labi, S., and Thompson, P. 2009. Establishing the weights of performance criteria—Case studies in transportation facility management. *ASCE Journal of Transportation Engineering* 135(9), 619–631.

Sinha, P., and Zoltners, A.A. 1979. The multiple-choice knapsack problem. *Operations Research* 27(3), 503–515.

Small, K.A. 1982. The scheduling of consumer activities: Work trips. *American Economic Review* 72, 467–479.

Small, K.A. 1983. Bus Priority and Congestion Pricing on Urban Expressways. *Research in Transportation Economics* 1, 27–74.

Small, K.A. and Verhoef, E.T. 2007. *The Economics of Urban Transportation*, 2nd Edition. Routledge, New York, NY.

Small, K., Winston, C., and Evans, C. 1989. *Road Work, A New Highway Pricing and Investment Policy.* Brookings Institution, Washington, D.C.

Smeed, R.J. 1968. Traffic studies and urban congestion. *Journal of Transportation Economics and Policy* 2, 33–70.

Smeed, R.J., Roth, G., Beesley, M.E., and Thompson, J.M. 1964. *Road Pricing: The Economic and Technical Possibilities.* Ministry of Transport, London, United Kingdom.

Smith, A.B. and Katz, R.W. 2013. U.S. billion-dollar weather and climate disasters: Data sources, trends, accuracy and biases. *Natural Hazards* 67(2), 387–410.

Song, B.-M., Han, B., Avram Bar-Cohen, A., Sharma, R., and Arik, M. 2010. Hierarchical life prediction model for actively cooled LED-based luminaire. *IEEE Transactions on Components and Packaging Technologies* 33(4), 728–737.

Steuer, R.E. 1989. *Multiple Criteria Optimization.* Robert E. Krieger Publishing Company, Malabar, FL.

Stoker, G. and Mossberger, K. 1994. Urban regime theory in comparative perspective. *Environment and Planning C: Government and Policy* 12(3), 195–212.

Sundaram, R.K. 1996. Inequality constraints and the theorem of Kuhn and Tucker. In *A First Course in Optimization Theory*. Cambridge University Press, Cambridge, United Kingdom, 145–171.

Szeto, W.Y., Jaber, X., and Wong, S.C. 2012. Road network equilibrium approaches to environmental sustainability. *Transport Reviews* 32(4), 491–518.

TAC 2004. *The Canadian Guide to In-Service Road Safety Reviews*. Transportation Association of Canada, Ottawa, Canada.

Tamiz, M. (Ed.). 2012. *Multi-Objective Programming and Goal Programming: Theories and Applications*. Springer Science and Business Media, Berlin, Germany.

Teng, J.Y. and Tzeng, G.H. 1996. A multiobjective programming approach for selecting non-independent transportation investment alternatives. *Transportation Research Part B* 30(4), 291–307.

Therneau, T.M. and Atkinson, E.J. 1997. *An Introduction to Recursive Partitioning Using the RPART Routines*. Mayo Foundation, Rochester, MN.

Thompson, P.D., Ford, K.M., Mohammad, A., Labi, S., Shirole, A., and Sinha, K.C. 2011. *Guide for Estimating Life Expectancies of Highway Assets*. NCHRP Project 08-71. National Cooperative Highway Research Program, National Academies Press, Washington, D.C.

Tobin, J. 1958. Estimation of relationships for limited dependent variables. *Econometrica* 26(1), 24–36.

Toyoda, Y. 1975 A simplified algorithm for obtaining approximate solutions to zero-one programming problems. *Management Science* 21, 1417–1427.

Transfund New Zealand 2003. *Safety Audits of Existing Roads: Developing a Less Subjective Assessment*. Transfund Report OG/0306/24S. Wellington, New Zealand.

TRB 1990. *Truck Weight Limits: Issues and Options*. TRB Special Report 225. Transportation Research Board, National Academies Press, Washington, D.C.

TRB 2010. *Highway Capacity Manual*, 5th Edition. Transportation Research Board, National Academies Press, Washington, D.C.

TRB 2015. *Funding and Managing the U.S. Inland Waterways System: What Policy Makers Need to Know*. TRB Special Report 315. Committee on Reinvesting in Inland Waterways: What Policy Makers Need to Know, Transportation Research Board, National Academies Press, Washington, D.C.

TRB 2016. *Highway Capacity Manual; A Guide for Multimodal Mobility Analysis*, 6th Edition. Transportation Research Board, National Academies Press, Washington, D.C.

Truong, T., Li, Z., and Kepaptsoglou, K. 2017. An entropy-STEM tradeoff analysis method for multiobjective transportation decision-making. *ASCE Journal of Transportation Engineering* 144(1), DOI: 10.1061/JTEPBS.0000100.

TTI 1990. *Technical Memorandum on Tasks 1 and 2*. NCHRP Project 7-12. Texas Transportation Institute, Texas A&M University, College Station, TX. National Academies Press, Washington, D.C.

Turner, S.M., Best, M.E., and Shrank, D.L. 1996. *Measures of Effectiveness for Major Investment Studies*. Report No. SWUTC/96/467106-1. Southwest Region University Transportation Center, Texas Transportation Institute, College Station, TX.

Turner, S.M., Eisele, W.L., Benz, R.J., and Holdener, D.J. 1998. *Travel Time Data Collection Handbook*. Report No. FHWA-PL-98-035. Federal Highway Administration, U.S. Department of Transportation, Washington, D.C.

UIC 1981. *Quantitative Evaluation of Geometric Track Parameters Determining Vehicle Behaviour*. Office of Research and Experiments, Union Internationale Des Chemins De Fer, or International Union of Railways, Paris, France.

USACE 2015. *Final Waterborne Commerce Statistics*. U.S. Waterborne Commerce Statistics Center, U.S. Army Corps of Engineers, New Orleans, LA.

USDOT 1992. *Characteristics of Urban Transportation Systems*. Report DOT-T-93-07. U.S. Department of Transportation, Washington, D.C.

USDOT 1998. *1997 Federal Highway Cost Allocation Study*. U.S. Department of Transportation, Washington, D.C.

Vassallo, J. and Sánchez-Soliño, A. 2007. A subordinated public participation loans for financing toll highway concessions in Spain. *TRB Journal of Transportation Research Record* 2297, 1–8.

Voelker, B. and Clark, S. 2008. *Comparison of North American Rail Asset Life.* Office of Rail Regulation Report No. ORR/RP/442/NATSLC 97074-00. Arup Texas, Inc., Houston, TX.

Vogt, A. 1999. *Crash Models for Rural Intersections: Four-Lane by Two-Lane Stop-Controlled and Two-Lane by Two-Lane Signalized.* Report FHWA-RD-99-128. Federal Highway Administration, U.S. Department of Transportation, Washington, D.C.

Volgenant, A. and Zoon, J.A. 1990. An improved heuristic for multidimensional 0–1 knapsack problems. *The Journal of the Operational Research Society* 41(10), 963–970.

Vonderohe, A.P., Chou, C.L., Sun, F., and Adams, T.M. 1997. *A Generic Data Model for Linear Referencing Systems.* NCHRP Research Results Digest 218. Transportation Research Board, National Academies Press, Washington, D.C.

Vonderohe, A.P., Travis, L., Smith, R.L., and Tsai, V. 1993. *Adaptation of Geographic Information Systems for Transportation.* NCHRP Report 359. Transportation Research Board, National Academies Press, Washington, D.C.

Vuchic, V.R. 2005. *Urban Transit: Operations, Planning, and Economics.* John Wiley and Sons, Hoboken, NJ.

Walker, J. 2008. Purpose-driven public transport: Creating a clear conversation about public transport goals. *Journal of Transport Geography* 16(6), 436–442.

Wardrop, J.G. 1952. Some theoretical aspects of road traffic research. *Proceedings of the Institution of Civil Engineers Part II* 1(2), 325–378.

Washington, S.P., Karlaftis, M.G., and Mannering, F.L. 2011. *Statistical and Econometric Methods for Transportation Data Analysis*, 2nd Edition. Chapman and Hall/CRC, Taylor and Francis Group, Boca Raton, FL.

WATIB 2017. *Sidewalk Program Criteria Guidelines.* Washington State Transportation Improvement Board, Olympia, WA.

Weber, M. 1947. *The Theory of Social and Economic Organization.* The Free Press, New York, NY.

Weingartner, H.M. 1963. *Mathematical Programming and the Analysis of Capital Budgeting Problems.* Prentice-Hall, Englewood Cliffs, NJ.

Weisbrod, G. and Reno, A. 2009. *Economic Impact of Public Transportation Investment.* American Public Transportation Association, Washington, D.C.

Weissmann, J., Harrison, R., Burns, N.H., and Hudson, W.R. 1990. *Selecting Rehabilitation and Replacement Bridge Projects, Extending the Life of Bridges.* ASTM STP, West Conshohocken, PA.

WERD 2003. Data Management for Road Administrations—A Best Practice Guide. *Road Data Subgroup, Conference of European Road Directors.* Western European Road Directors, Paris, France.

White, M.S. and Griffin, P. 1985. Piecewise linear rubber-sheet map transformations. *The American Cartographer* 12(2), 123–131.

Williamson, O. 1994. Transaction cost economics and organization theory. In *The Handbook of Economic Sociology.* Editors: Neil J. Smelser and Richard Swedberg. Russell Sage Foundation, New York, NY, 77–107.

Wilson, A.G. 1974. *Urban and Regional Models in Geography and Planning.* John Wiley and Sons, Hoboken, NJ.

Winkler, R.L. 2003. *An Introduction to Bayesian Inference and Decision*, 2nd Edition. Probabilistic Publishing, Gainesville, FL.

Winston, C., Small, K.A., and Evans, C.A. 1989. *Road Work: A New Highway Pricing and Investment Policy.* Brookings Institution Press, Washington, D.C.

Winston, W.L. and Venkataramanan, M. 2002. *Introduction to Mathematical Programming: Applications and Algorithms, Volume I*, 4th Edition. Thomson Learning, Stamford, CT.

Wolfe, P. 1959. The simplex method for quadratic programming. *Econometrica* 27(3), 382–398.

Wolsey, L.A. 1998. *Integer Programming.* Wiley-Interscience, Hoboken, NJ.

Wolshon, B. 2003. *Louisiana Traffic Sign Inventory and Management System.* Louisiana State University, Baton Rouge, LA.

Wolshon, B., Degeyter, R., and Swargam, J. 2002. Analysis and predictive modeling of road sign retroreflectivity Performance. *Presentation at the 16th Biennial Symposium on Visibility and Simulation.* University of Iowa, Iowa City, IA.

Wooldridge, J.M. 2015. *Introductory Econometrics: A Modern Approach*, 6th Edition. South-Western College Pub, Cincinnati, OH.
Wunsch, P. 1996. Cost and productivity of major urban transit systems in Europe: An exploratory analysis. *Journal of Transport Economics and Policy* 30(2), 171–186.
Yoder, E.J. 1985. *Principles of Pavement Design*. John Wiley and Sons, New York, NY.
Young, R.A. 2001. *Uncertainty and the Environment: Implications for Decision Making and Environmental Policy*. Edward Elgar, Cheltenham, Gloucestershire, United Kingdom.
Zaniewski, J.P., Butler, B.C., Cunningham, G.E., Paggi, M.S., and Machemehl, R. 1982. *Vehicle Operating Costs, Fuel Consumption, and Pavement Types and Condition Factors*. Final Report No. PB82-238676. Federal Highway Administration, U.S. Department of Transportation, Washington, D.C.
Zarembski, A.M. 2000. The implications of heavy axle load operations for track maintenance on short lines. In *AREMA 2000 Annual Conference Proceedings*, Lanham, MD, 1122–1151.
Zarembski, A.M. 2015. *Determination of the Impact of Heavy Axel Loads on Short Lines*. Delaware Center for Transportation, University of Delaware, Newark, DE.
Zegeer, C.V. and Cynecki, M.J. 1984. *Selection of Cost-Effective Countermeasures for Utility Pole Accidents: Users Manual*. Report No. FHWAIP-84-13. Federal Highway Administration, U.S. Department of Transportation, Washington, D.C.
Zhai, H., Frey, H., and Rouphail, N. 2008. A vehicle-specific power approach to speed- and facility-specific emissions estimates for diesel transit buses. *Environmental Science and Technology* 42(21), 7985–7991.
Zhang, Y.J., El-Sibaie, M., and Lee, S. 2004. FRA track quality indices and distribution characteristics. *2004 Annual American Railway Engineering Maintenance-of-Way Association (AREMA) Conference*, Nashville, TN.
Zhou, B., Li, Z., Patel, H., and Roshandeh, A.M. 2014. A risk-based two-step optimization model for optimal highway transportation investment decision-making. *ASCE Journal of Transportation Engineering*, 140(5), 04014007.
Zimmerman, H.J. 2001. *Fuzzy Set Theory and Its Applications*. Kluwer Academic Publishers, Boston, MA.
Zimmerman, K.A. 1995. *Pavement Management Methodologies to Select Projects and Recommend Preservation Treatments*. NCHRP Synthesis of Highway Practice 222. National Cooperative Highway Research Program, Transportation Research Board, National Academies Press, Washington, D.C.
Zionts, S. and Wallenius, J. 1976. An interactive programming method for solving the multiple criteria problem. *Management Science* 22(6), 652–663.
Zuckerman, B. and Ackerman, F. 1995. The 1994 Update of the Tellus Institute Packaging Study Impact Assessment Method. *SETAC Impact Assessment Working Group Conference*, Washington, D.C.

Index

Index